PENETRATION TESTING 1988

PROCEEDINGS OF THE FIRST INTERNATIONAL SYMPOSIUM ON PENETRATION TESTING / ISOPT-1 / ORLANDO / 20-24 MARCH 1988

PENETRATION TESTING 1988

Edited by

J.DE RUITER

Fugro Consultants International B.V.

VOLUME 1

International reference test procedures

Special lectures

Technical papers:

Standard penetration test

Dynamic probing / Weight sounding test

Dilatometer test / Pressuremeter test

A.A.BALKEMA / ROTTERDAM / BROOKFIELD / 1988

John H.Schmertmann, Chairman, Organizing Committee
John L.Davidson, General Secretary

The texts of the various papers in this volume were set individually by typists under the supervision of each of the authors concerned.

Published by
A.A.Balkema, P.O.Box 1675, 3000 BR Rotterdam, Netherlands
A.A.Balkema Publishers, Old Post Road, Brookfield, VT 05036, USA

For the complete set of two volumes ISBN 90 6191 801 4
For volume 1: ISBN 90 6191 802 2
For volume 2: ISBN 90 6191 803 0

Printed in the Netherlands

Penetration Testing 1988, ISOPT-1, De Ruiter (ed.)
© 1988 Balkema, Rotterdam, ISBN 90 6191 801 4

Contents

Technical papers: Standard penetration test

Technical papers: Dynamic probing / Weight sounding test

Technical papers: Dilatometer test / Pressuremeter test

Penetration Testing 1988, ISOPT-1, De Ruiter (ed.)
© 1988 Balkema, Rotterdam, ISBN 90 6191 801 4

Preface

After the Second European Symposium on Penetration Testing, ESOPT-II in 1982, the ISSMFE Technical Committee on Penetration Testing decided to continue these successful specialty conferences with a truly international event in 1988. Subsequently it accepted a proposal from the Department of Civil Engineering of the University of Florida to host the First International Symposium on Penetration Testing, ISOPT-1, at Disney World, Florida, from 20 to 24 March 1988. Under the auspices of the ISSMFE, this symposium is sponsored by the US National Society of ISSMFE, the ASCE Geotechnical Division, the ASCE Florida Section and the University of Florida Department of Civil Engineering, which assumed the financial responsibility.

The ISOPT-1 Organizing Committee feels honored that the ISSMFE Technical Committee on Penetration Testing has chosen to publish the latest versions of the proposed four international reference test procedures in the proceedings of this symposium. The previous drafts of these procedures were reviewed and revised by the Committee for verbal presentation at ISOPT-1 and preliminary publication herein. Thus, ISOPT-1 furnishes one more opportunity for exposure and professional comment and revision before the Committee submits them for publication by the ISSMFE in 1989.

The objective of ISOPT-1 is to focus on a wide range of aspects of penetration testing, with special emphasis on new developments since ESOPT-II, both in equipment and data interpretation. This scope is reflected in these proceedings, which include the papers associated with each of the nine special lectures delivered at the symposium plus the 96 technical papers contributed by 174 authors from many countries. All papers received a peer review before acceptance.

In view of the diverse nature of the contributions, the proceedings consist of three parts divided over two volumes: Part 1 – International reference test procedures; Part 2 – Special lectures; Part 3 – Technical papers.

A distinctly new feature of this symposium is the general interest in Marchetti dilatometer testing. This has led to a separate section in Part 3 on dilatometer and pressuremeter testing. The electric cone penetration test continues to attract great interest and the majority of the papers in Part 3 are devoted to this subject.

A subject index has been added at the end of the papers in Volume 2. It presents an attempt to summarize the essence of all contributions in a limited number of key words with a minimum of overlap. An author index concludes this second volume.

These proceedings are the result of hard and dedicated work of many individuals. Sincere thanks are due to the authors, the international team of peer reviewers, the publisher, the ISOPT-1 Organizing Committee, and last but not least, the secretarial staff of Fugro Consultants International. Without the support and cooperation of these and many others, the timely production of the proceedings would have been impossible.

The Editor

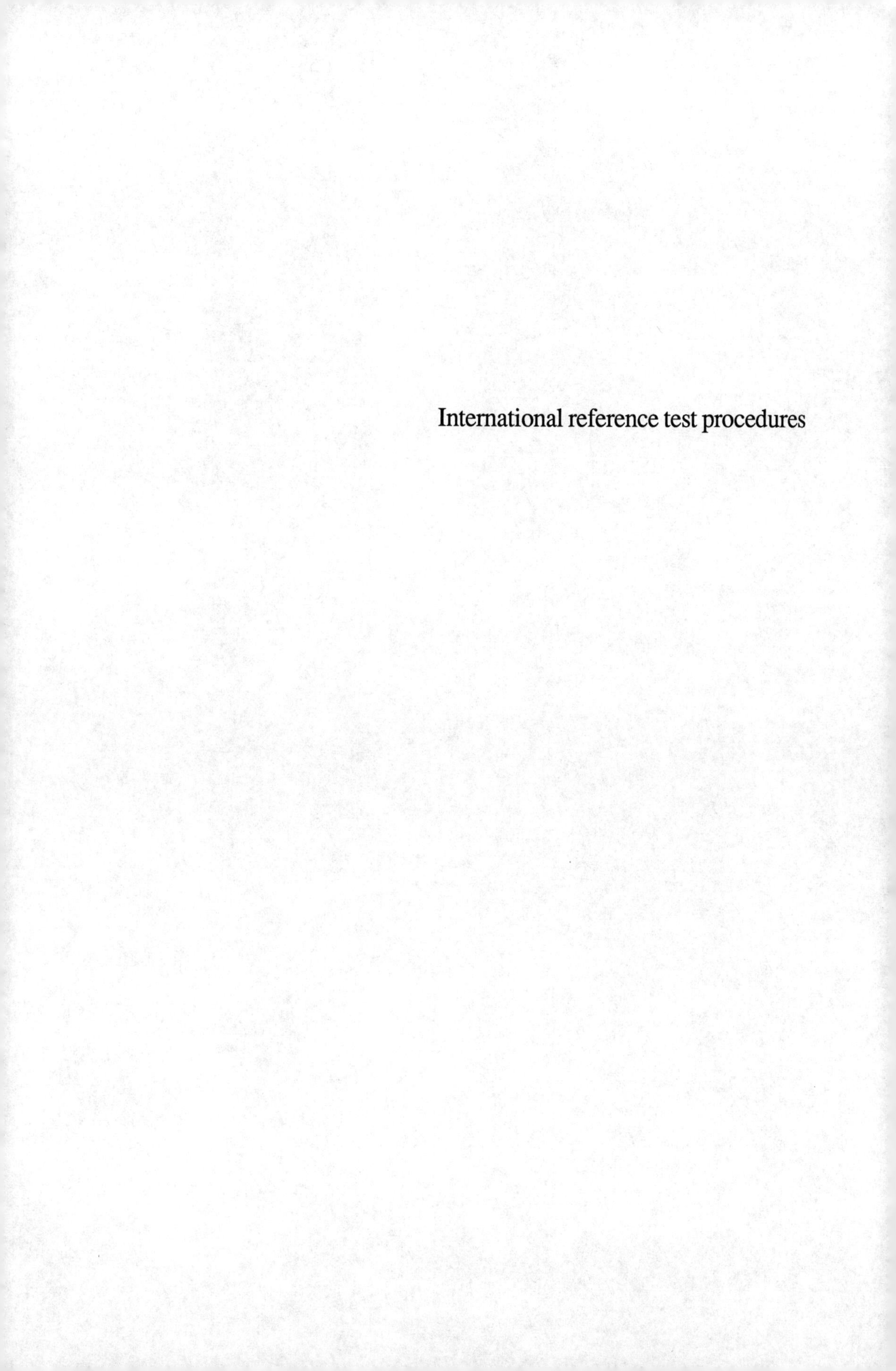

International reference test procedures

Penetration Testing 1988, ISOPT-1, De Ruiter (ed.)

Standard penetration test (SPT): International reference test procedure

ISSMFE Technical Committee on Penetration Testing
SPT Working Party:
L.Decourt (Brazil)
T.Muromachi (Japan)
I.K.Nixon (UK)
J.H.Schmertmann (USA)
S.Thorburn, Chairman (UK)
E.Zolkov (Israel)

This document presents recommendations of Working Party discussed at ISOPT-1, Orlando, USA, March 20-24 1988.
Document 1: Introduction and objectives; Document 2: Development and current use of the standard penetration test; Document 3: International reference test procedures; Document 4: Calibration methods.

DOCUMENT 1

INTRODUCTION AND OBJECTIVES

1 The standard penetration test (SPT) has maintained its popularity since its inception in 1927 and is used to a greater or lesser extent by at least sixteen of the ISSMFE member countries. The test is executed in a relatively simple manner by site personnel as part of a universal method of investigating ground by means of boreholes. The test equipment consists of few robust mechanical components, easily transported, handled, and maintained. Current opinion considers that the SPT will continue to be used as part of routine borehole investigations of sites since it permits an economic evaluation of ground conditions in both difficult and easy situations.

2 Reproducibility and accuracy of results are important requirements for any form of field test and variability of SPT results can result, inter alia, from poor boring methods and energy losses during impact.

In situations where comparison of results of tests is important, such as international research, it is essential that there be a reference test procedure specific enough to produce sufficient accuracy and consistency of results.

In 1977, the European Sub-Committee on Penetration Testing published and presented recommendations for a European Standard for the SPT to the IX ICSSMFE, Tokio, with the objective of maintaining the simplicity of the test but ensuring reproducibility of results and conformity with the ground conditions.

In 1982, the decision was taken at the Second European Symposium on Penetration Testing to form an ISSMFE Technical Committee on Penetration Testing because of the necessity for international co-operation in the preparation of reference test procedures.

The SPT Working party considered it important to present a list of the principles requiring compliance in order to permit comparability of test results from different member countries and within those countries, and leading to:

i) Better communication of information between the ISSMFE member countries.

ii) A better understanding of the differences in current test procedures.

iii) Better correlations between the work of site personnel, organisations, and countries.

3 The International Reference Test Procedures are intended to convey the simplest possible, complete description of all of the factors influencing the test results, recognising the need, as far as possible, for variations to suit national circumstances, including variations to suit particular ground conditions and environments. It is not a specification for obtaining results in a particular manner at a particular site – this is a standard.

The SPT Working Party consider that an International Standard could be inhibitive and unacceptable. Its adoption would mean a disruption to the successful application of individual methods of execution of the SPT combined with empirical knowledge at individual, local, and national levels over the past fifty years. Variations in test equipment and procedures may be used provided comparative tests demonstrate their suitability in comparison with the recommendations of the International Reference Test Procedures. The legal implications of the use of a standard also presents difficulties within the International Community and it is considered that the concept of an International Reference is more appropriate.

4 The overall description of the International Reference Test Procedures has been worded in a practical manner to help it gain universal understanding and acceptance notwithstanding language difficulties. The philosophy for individual sections was as follows:

a) *Scope*: The format most commonly used.

b) *Boring methods and test procedures*: It is intended

to represent the features essential for uniformity with all the safeguards gained from current experience to obviate disturbance in the ground to be tested.

c) *Sampler assembly, drive rods and hammer assembly*: Geometrical and material bounds to ensure comparable results according to the most common format being used. An energy input definition has not been proposed because of a lack of experience and of the possibility of its becoming part of a 'test' routine rather than the intention solely for 'equipment' calibration.

d) *Reporting of results*: The complete list of all the factors defining the method in use and the experimental observations/measurements essential for a full description of an individual test.

5 The SPT Working Party have produced a four part document comprising the following sections in order to convey the background to their decisions:

1) Introduction and Objectives;
2) Development and Current use of the SPT;
3) International Reference Test Procedures;
4) Calibration Methods.

In situations where comparisons of SPT results are important it is considered that energy input should be measured and methods of calibration have been presented for guidance.

It is not the intention of the SPT Working Party that these elaborate methods of calibration should be used during routine site investigations. Calibrations of drive rods and hammer assemblies may, however, require to be carried out for each rig and separately from the investigation work. They would be based upon the personnel and equipment involved with a particular project, and in the following situations:

i) when correlation of results between different rigs is required;

ii) when SPT results or equipment or techniques are involved in development work;

iii) when SPT results are to be quoted in research and technical papers.

DOCUMENT 2

DEVELOPMENT AND CURRENT USE OF THE STANDARD PENETRATION TEST

SYNOPSIS

To assist in the establishment of a sound basis for the formulation of the International Reference Test Procedures for the SPT, a study was made of its origin and how the method of the test has developed and has been applied in different countries. A commentary is presented on the use of the test results, concluding with applications within member countries.

1 ORIGIN OF THE SPT

Initially, the method arose from driving one inch nominal diameter pipe into the ground for sampling, as practised by Lt. Col. Charles R.Gow, whose company became a subsidiary of the Raymond Concrete Pile Company in 1922. According to a paper by G.E.A.Fletcher (1965), on the SPT, the two inch external diameter 'sampling spoon' was designed in 1927, based upon fieldwork by himself and research by H.A.Mohr, who was then District Manager of the Gow Company in New England, USA. He added that the method, including measurement of the driving resistance over a 12 inches, penetration using a 140 lb hammer dropping 30 inches, began appearing in specifications in 1930. Confirmation of when this method of the test had become established is contained in a Harvard University publication by Mohr (1937) on soil exploration and sampling, that showed a facsimile of a test boring report, dated February 1929, by the Gow Division of the Raymond Concrete Pile Company. The report displays a note at the base of the record sheet; 'Figures in the right hand column indicate the number of blows required to drive the sampling pipe one foot, using 140 lb. weight falling 30 inches.' There is no mention of a title for the test in this publication, only that it provides a 'rough idea of ground resistance and should always be done'. Mohr (1966), in his contribution to the discussion on Fletcher's paper (1965), showed a slide of typical washboring equipment of that period. The drive weight consisted of a square concrete block with a hardwood inset, all mounted on a guide spike.

Fletcher considered that the boring technique was the greatest obstacle to standardisation, and concluded that it was wrong to imply the test could be made reliable since experience had demonstrated that the data should be used with great restraint.

Neither Fletcher nor Mohr gave details of the design of the 2 inches diameter sampling spoon. The configuration and precise dimensions were given, however, as would be expected, by Hvorslev (1949) in his classic report on subsurface exploration and sampling. The diagram of the 'Raymond sampler' shows a split barrel 22 inches long or a solid one 34 inches long of seamless steel tubing, with a 'hardened tool steel' shoe and a vented (4 holes 5/8 inch diameter) sampler head, but without any ball check valve. The bore is marked 1 3/8 inches internal diameter, uniform throughout the barrel and shoe. There was no liner, although other samplers in the USA as far back as 1932 had included this facility. Hvorslev gave credit to H.A.Mohr for initiating the 'determination of the dynamic penetration resistance as a regular procedure........about 1927'. Nevertheless, it is recorded by Hvorslev that (in 1949) alongside the use of the Raymond sampler for the penetration test by 'other Divisions' of the Raymond Concrete Pile Company, the Gow Division continued using 2 inches casing filled with water and the sampler was simply a section of one inch extra heavy pipe, also used as a drill rod. The penetration resistance was measured from the depth to which the sampler sank under the weight of the drill rod. In the case of the other divisions 'the sampler was first forced about 6 inches into the soil in order to decrease the influence of the disturbed zone below the bottom of the borehole'.

The earliest reference to the method being called the 'Standard Penetration Test', to the authors' knowledge, is contained in Terzaghi's 1947 paper entitled 'Recent trends in subsoil exploration', presented to the 7th Texas

Conference on Soil Mechanics and Foundation Engineering. However when describing the well known basic features of the equipment it is of interest that the bore of the sampling spoon is given as 1 1/2 inches (38 mm) whereas in the first textbook reference to the test, namely the first edition (1948) of 'Soil Mechanics in Engineering Practice' by Terzaghi & Peck, the inside diameter is given the better known value of 1 3/8 inches (35.1 mm). In fact the configuration and dimensions of the sampling spoon in the book correspond with those given by Hvorslev for a split barrel sampler. Both the 1947 paper and the textbook refer also to the six inch initial 'seating drive'.

The standard penetration test is again referred to in 1953 by Peck, Hanson & Thornburn, in their book, 'Foundation Engineering'. The diagram of the 'split spoon sampler' is the same as given in 'Soil Mechanics in Engineering Practice' but no mention is made of a seating drive.

2 STANDARDISATION OF METHOD

2.1 *ASTM D1586*

The first ASTM description for the SPT was published in April 1958 as a 'Tentative method for penetration test and split barrel sampling of soils'. Casings or mud was mandatory in weak ground and the bore diameter was to be between 2 1/4 in.(57.2 mm) and 6 in.(152.4 mm). Although it laid down dimensions for a 2 in.(51 mm) external diameter sampler with a uniform bore of 1 3/8 in. (35 mm), other sizes could be used with the permission of the Engineer, and a note recorded 'split barrel may be 1 1/2 in.(38.1 mm) internal diameter and may contain a liner'.

The ASTM became a 'standard method' in 1967. Besides the well known precautions to obviate tests in disturbed ground, as included in previous editions of the ASTM, maintenance of the water level in the holes at or above ground water level was also added. To avoid 'whips' under hammer blows A size rods (1 5/8''OD)(41.5mm) or greater were advised, with stiffer than A for holes deeper than 50 ft. Another significant change was that the 'split barrel may be 1 1/2''(38.1 mm) internal diameter provided it contains a liner of 16 gauge wall thickness'. The current ASTM Standard carries no major changes since these earlier editions.

2.2 *Standards in different countries*

In order to illustrate how the SPT method has been adopted throughout the world, Table I has been compiled from a number of sources including, in the main, information supplied to the ESOPT I Conference. Complete accuracy is difficult in such matters but it is believed to project a fair indication of current usage. Eleven countries are represented, and it is clear that, while most National Standards follow closely the well established principles and contain the essential safeguards, such as having a clean hole, minimising disturbance and specifying hammer guide and an apparent free fall, there are a number of variations in the details, which are discussed below.

2.3 *Variations between standards and with proposed reference procedure*

To identify the differences between current Standards and, at the same time, to consider the adequacy of an international reference test procedure, Table I has been sub-divided to show what is believed to be the significant features of each part of the test method. Moreover, for convenience, the latter has been arranged under the same headings as in the draft procedure. Dealing with each of these in turn:

Scope: Variations in hammer weight and drop from the original 140 lb. (63.5 kg) and 30 in.(762 mm), have arisen in rounding up values when changing to suit metric units. In several cases where this has resulted in a heavier hammer the drop has been reduced, although not in the United Kingdom. Only one Standard, Czechoslovakia, was found that referred to energy input to be achieved if weight or drop is varied. The potential energy prescribed is 47.5 kgm with the deviation in the hammer weight not exceeding 25 per cent of the standard (63.5 kg).

Boring restrictions: Limits in hole diameter occur in about half the examples, in which case they are generally set between 57 mm (2 1/4 in) and 150 mm (6 in.) diameter. Otherwise there is general agreement on the need for a clean hole capability, only side discharge bits and casing or mud to support weak ground. The USA Standard also specifically bars jetting through the open sampler tube. Only two countries restrict the bailer diameter to reduce the continuous suction during withdrawal when the bailer is a close fit.

Sampler assembly: Uniformity in the open split tube sampler is good between all Standards. Several permit an enlarged bore but only for the insertion of a liner and not for use without it, notwithstanding the widespread omission of liner in the USA (Schmertmann 1979). Three countries include an option for using core retainers, but only Australia and the United Kingdom support the alternative fitting of a solid cone to replace the open shoe.

Drive rods: There is a surprisingly large amount of support for a minimum rod size equivalent to AW. For the deeper tests (below about 15 m) centering devices are specified in only two cases with the alternative of stiffer rods.

Hammer: Manual operation and visual judgement, including the use of rope and the cathead hoist, is still the most common permitted mode of operation, but virtually half the Standards also refer to means of ensuring the correct drop, e.g. using trip hammers; and while three countries advise this, it is mandatory in the United Kingdom. The drive head or anvil received few comments, with no accord between Standards.

Preparing borehole: There are still some deficiencies in the list of precautions to be taken, including the maintenance of the water level at or above ground water level. Raising the bailer slowly is least mentioned.

Executing test: Initial penetration (IP) below the base of the borehole under the dead weight of equipment prior to seating drive is recognised in three cases to be recorded. None state, however, the amounts when the seating drive or test drive should be omitted.

Only one country, Portugal, limits the number of

Table I. SPT standards in different countries.

Main items of specification	Australia	Canada	Czechoslovakia	Greece As earth manual 2nd ed. 1974	Japan	Mexico As ASTM 1586 dated 1967	Portugal	United Kingdom (BS 1377:1975)	USA (ASTM 1586-67) reapproved 1974	Poland	India
Scope											
Hammer wt (63.5) kg	●	●	●	●	●	64	64	65	●	65	65
Drop (760) mm	●	●	750	●	750	750	750	●	●	750	750
Boring restrictions			*separate standard for site exploration*		*do not disturb test zone*						
Clean hole capability	●	●	○	●	●		●	●	●		●
Side discharge bit only	●	●	○	●	○		○	●	●		●
Bailer dia. restricted	≤60%	○	○	○	○		○	≤90%	○		○
Casing/mud in weak soils	●	●	○	○	○	casing	casing	●	●		● prefer casing turned
Hole dia. restriction mm	○	57-152	○	75 min	65-150		○	○	57-152		55-150
Sampler assembly length mm	min	min		min							
Ext/int. dia. shoe & sampler	● | ● | 533	● | ● | 533	● | ● | 535	● | ● | 483	● | ● | 535	● | ● | –	*no details*	50 | ● | 533	● | ● | 533		50.8 | ● | 675
Drive shoe separate	●	●	●		●	○		●	●		●
Vent area mm^2/check valve	387 | ●	● | ●	● | ●		not specified | ●			390 | ●	● | ●		● | ●
Liner permitted	yes	yes*	no		no	yes		no	yes		?
Split or tube	either	split	split		split	tube		split	split		split
Hardened steel shoe	●	●	○		○			●	●		?
Solid cone	yes	no	no		no	○		yes optional	no		no
Other	optional thin core retainers	optional – core retainers	–		–			–	optional core retainers		–
Drive rods depth m						>15 m			stiffer over 15 m		
size, range dia. mm	AW/NW	AW*	○	AW/B	40.5 o.d. (just under A)	AW/NW	*no details*	AW min.	A	42/51 dia. mm	A min.
steadies, start/spacing (m)	15 1/4 6 (or stiffer rods)	no	no	○	no			15 3 (or stiffer rods)	no		10 | –
straight/tightly coupled	○ | ○	● | ●	○	○ | ●	○			○	○		
Drive weight											
Guide/freefall	● | ●	● | ●	● | ●	● | ●	● | ●	cathead	*no details*	● | ●	● | ●		● | ●
Trip mechanism	●*	○ no mention	● advised	● advised	● permitted	○		● mandatory	○		○
Anvil specification	●* domed	○	○	jar coupling	● 75 o.d.	○		● not defined	○	18 kg	○
Preparing borehole			*separate standard for site exploration*		*do not disturb test zone*						
clean hole	●	●	○	●	●		●	●	●		●
positive head	●	● sands & silts	○	○	○		○	●	●		●
Bailer raise slowly	●	○	○	○	○		○	○	●		○
casing at/above test level	●	●	○	○	○		●	○	●		●

Executing test											
initial gravity penetration (I.P.)	●	●	●	○	○		○	○	○		○
high IP/omit SD	○	○	○	○	○		○	○	○		○
excessive IP/omit N	○	○	○	○	○		○	○	○		○
SD max. blows (50)	○	○	○	○	○		60	○	○		○
SD procedure	●	●	light blows only	●	●	●	●	●	●		●
max. blows	60 incl. SD	100 incl. SD	○	50 incl. SD	50 during N test	50 per 100 mm	100 incl. SD	50 during N test	100 incl. SD or 50 for 50 mm or less		20 per 250 mm
Blow rate	○	○	○	○	○		○	○			○
Soil sample											
sample retention	●	●	●	●	●	●	●	●	●		●
sample labelling	●	●	●	●	●	●	●	●	●		●
Reporting											
boring method	●	●	○	○	○		○	●	●		●
rod size	○	○	○	○	○		○	○	○		○
hammer type	○	○	○	○	○		○	○	○		○
initial gravity penetration (L)	●	●	●	○	○		○	○	○		○
SD short	blows	–	○	○	○		blows & penet.	○	○		
SD – blows/L (mm)	every 152	every 152	no record	no record	no record		150	150	every 152		every 150
TD – blows/L (mm)			every 100 advised	300*	Penet./blow but if penet. < 20, report blows/100		300	every 75			
TD procedure (N)	●	●	●	●	●	●	●	●	●		●
short test drive	–	advance for final 20 blows	–	penet./50 blows	penet./50 blows			blows & penet.	blows & penet.		N = last 300 mm
casing data	casing size	casing used	○		○		depth base casing	casing size	casing used		

Legend: ● = as draft reference procedure ○ = not specified * = information in note

Table II. Usage and practice for SPT with some member countries.

Country	Usage		Materials range	Practice								
	Widely	Occasionally	S = sands & silts C = clays G = gravels WR = weak rocks	Own standards	Other basis	Main types of hammer	Solid cone	Wash boring	Percussion	Auger	Rotary bit	Casing/mud
Australia	●		S,C,G & WR Interbedded soil/rock. Suspect in gravels.	●		trip	●		●	●	●	●
Brazil	●		S,C & G	●		manual		●		●	●	●
Canada	●		Glacial till etc.	●		manual & trip						
Czechoslavakia		●	S & C	●		manual & trip					●	m
Greece		●	S & C		T & P48 EM74	manual					●	c
India	●		S & C	●		manual		●			●	m
Israel	●		S & WR Interbedded soil/rock. Kurkar		ASTM 67	trip		●		●		m
Italy		●	S,C & G		ASTM 67	trip						
Japan	●		S,C & G	●		manual & trip					●	●
Norway		●	S & G		T&P48						●	c
Poland		●		●								

Portugal	●		S,C & G Drilling mud in loose sands	●		manual & trip	●		●			c
South Africa	●		S,C,G & WR Spring retainer for sands		ASTM 67	manual & trip	●	●	●	●	●	m
Spain	●		S,C & G Solid cone for gravels		T&P48	manual	●		●	●	●	c
Turkey	●		S,C & G, also over-consolidated clays		ASTM 67	manual			●		●	
United Kingdom	●		S,C,G & WR Solid cone for G & WR	●		trip	●		●	●	●	c
U.S.A.	●		S,C & G	●		manual & trip		●		●	●	●
17. Totals:	**12**	**5**		**10**		**manual – 11** **trip – 10**	**5**	**5**	**6**	**7**	**12**	**–**

Notes: T&P 48 = Terzaghi, K. & R.B. Peck, 1948, Soil Mechanics in Engineering Practice
EM 74 = Earth Manual 1974
ASTM 67 = D 1586:1967, Standard method for Penetration Test and Split-barrel Sampling of soils

blows for the seating drive, but practically all have a limit for both seating and test drive, the highest being 100 and the lowest 50. No country appears to deal with the rate of application of blows of the hammer.

Soil sample: All countries specify sample retention and proper records of it.

Reporting: While all countries require basic information such as borehole and test depths, several do not call for the boring method to be recorded, and none ask for rod size or hammer type.

Some type of incremental blow count is specified usually and most commonly at 152 mm intervals of penetration, including both seating and test drives. The Japanese call for the penetration per blow during the test drive.

Where the maximum number of blows is reached before full penetration the usual record is the overall penetration achieved.

3 DEVELOPMENTS OF THE METHOD IN PRACTICE AND RESEARCH

3.1 *Practice*

The extent to which Standards reflect practice and the consistency achieved by their application are unknown. All standards refer, however, to the importance of reporting the significant features of the technique being used, and this requirement is essential in an International Reference Test Procedure that may be more broadly worded than a National Standard.

The national state-of-the-art reports submitted to the ESOPT I conference contained some references to the practice being followed within member countries and this has been used, therefore, together with other available data, to illustrate in Table II the usage and practice in 17 different countries. The test is reported to be widely used in 12 of these countries having their own Standards.

Techniques for driving the sampler assembly vary widely, from a direct pull and release of a guided weight using a rope over a simple pulley, to an automatic free fall hammer system (Riggs et al 1983) that is hydraulically powered to operate at any desired rate between 5 and 50 blows per minute. Table II shows, however, that many tests are still made using visual control of the drop and manual operation of a rope and cathead hoist with generally two or less nominal turns of the rope. Wide use is made in the USA of a visually operated purpose-built tubular 'safety hammer' with an enclosed anvil. It is mandatory to use the trip hammer in the UK and an automatic release system determines the amount of drop.

Five countries referred to the use of the solid cone, and generally preferred this for tests in gravel and weak rocks; or residual strata (South Africa and Australia).

The more common boring techniques used with the SPT were also featured in many of the state-of-the-art reports and it may be concluded from this information that there has been a definite trend away from the earlier use of simple hand-operated washboring equipment, with a tripod, towards rotary drill and auger rigs, probably often truck-mounted. In several countries, however, percussion boring equipment, with drop tools, is still preferred, as in the United Kingdom, where the test is often carried out in conjunction with 100 mm nominal diameter sampling. The use of continuous flight augers with a hollow stem has been another development in which the tests are made, but difficulties have arisen when using this technique in granular soils below the water table (Schmertmann 1975).

The Japanese Society (1983) carried out a special survey within Japan of the variations present in the equipment and procedure particularly with respect to the National J.I.S.Standard. 700 questionnaires were distributed to companies with 332 being returned. Amongst the more interesting features of the test it was revealed that almost all were using the standard rods but automatic trip hammers were not commonly employed. The borehole diameter was generally 66 mm with casing and mud maintained at the top of the drive pipe. The seating drive in hard ground was often left to the discretion of the operator. Penetration per blow was being recorded as specified for over 50% usage, otherwise it was per 100 mm of penetration.

A similar survey has been executed in the USA by Dr Paul Aldinger, CE Maguire Inc. Providence, Rhode Island.

Reference has already been made to the omission of the liner from the sampler in the USA leading to lower N values than might otherwise be expected.

3.2 *Research*

The effect of rod size has been the subject of research particularly in Japan (Japanese Society 1981) but also elsewhere including the USA. The principal conclusions were that the A type is normally adequate to lengths up to 20 metres.

The majority of research with respect to the test method in recent years however has concentrated upon the energy input due either to the different ways of using the rope and cathead hoist or to different types of trip hammer (Nixon, 1982, Kovacs et al 1983) and, whilst discussion on the results is outside the scope of this paper, sufficient evidence has been collected of significant variations occurring that it has led to the development of calibration methods which are the subject of Section No.4 of this report.

4 GENERAL COMMENTARY ON APPLICATION OF SPT

The standard penetration test (SPT) has been used extensively to provide information on the insitu condition of sand deposits. The resistance to penetration of a sampler driven by impulsive forces into sand, is considered to reflect the state of compaction (relative density) of the deposit, and empirical methods to permit interpretation of penetration resistance have been developed. Investigation of the accuracy of empirical relationships between penetration resistance and various soil characteristics is outwith the terms of reference of the Technical Committee and will be the subject of a separate study.

This commentary is, therefore, of a historical nature

and discusses the past and present uses of the test by the International Community.

The standard penetration test has become a popular and principal means of determining the permissible ground pressures to be used for the design of foundations in many member countries.

The following broad categories of types of foundations define the extent of use and application of the standard penetration test for design purposes, where surface deposits of sand are encountered:

i) Shallow Foundations;
ii) Deep Foundations;
iii) Raft Foundations.

The first two categories are concerned with foundations which generally impose relatively local and essentially individual stress fields in soils in contrast to raft foundations whose form of design and construction often involves interaction between the complete foundation system and a large mass of soil.

The first category may be considered to embrace pad and continuous footings founded on surface deposits of sand whose depth of embedment is equal to or less than the least width of the footings. Piers and piles form the second category and deep foundations in surface deposits of sand may be defined as elements whose depths of embedment are equal to or greater than ten times the least width or diameter of the piers and piles.

In practice, there is an intermediate category in which may be placed types of foundations whose depths of embedment in surface deposits of sand lie between these general criteria for shallow and deep foundations.

4.1 *Shallow foundations*

The performance of shallow foundations and the basis of empirical methods of analysis and design may be expected to relate to the characteristics of the first eight metres of sand forming surface deposits for typical dimensions of shallow footings.

For shallow foundations, therefore, most of the standard penetration tests used for the development of empirical methods may be expected to have been carried out within depths of eight metres below groundsurface in surface deposits of sand.

It is important to be aware of the various aspects of the standard penetration test which could have provided variable and unrepresentative resistance values.

4.1.1 *Preparation of borehole*

The formation of boreholes even in water-bearing fine sands to depths not exceeding about eight metres normally does not present too many difficulties for drillers, and, therefore, disturbance of sand due to the boring and cleaning operations at the elevations of the tests used for interpretative and research purposes should have been within reasonable limits in many situations.

Unacceptable disturbance during the initial stage of borehole preparation may be expected to have occurred in the following situations:

i) where the drillers failed to maintain sufficient hydrostatic head in a borehole and permitted sand to flow into the casing;
ii) where disturbance was caused by overboring;
iii) where overdriving of the casing occurred, and
iv) where the casing was withdrawn

Two conditions arise in situation (i), the first resulting from the drillers effectively cleaning the sand which entered the borehole down to the test elevation. The sand exposed by the cleaning operations is now in a loose condition due to disturbance and the penetration resistance values are seriously decreased and are atypical of the insitu state of the soil.

The second condition arises where the drillers failed to remove the sand which entered the casing and the sampler was seated on sand within the casing. The penetration resistance values can be greatly increased due to the condition of confinement within the casing.

The use of drilling mud as well as steel casing may be necessary to control effectively the stability of the borehole.

4.1.2 *Drive rods*

The length and weight of rods can affect the penetration resistances, due to the variations in the stiffness of the drive rod system. Comparative measurements made in Japan have indicated that there were no significant differences in blow count or energy transferred on impact for rods weighing between 4.33 and 10.03 kg/m.

The use of a large steel drive head can seriously increase the penetration resistance because of the decrease in energy transmitted to the rods.

4.1.3 *Execution of test*

The greatest cause of variability of penetration resistance at this stage of the operation is the variation in the amount of energy delivered to the drive rod system.

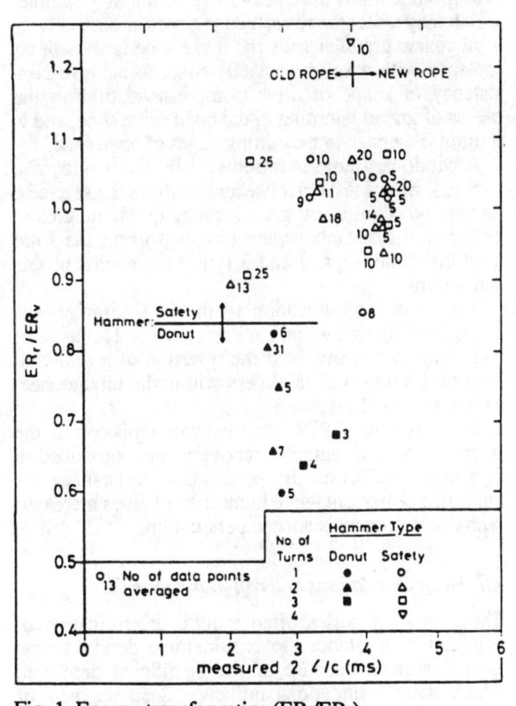

Fig. 1. Energy transfer ratios (ER_r/ER_v).

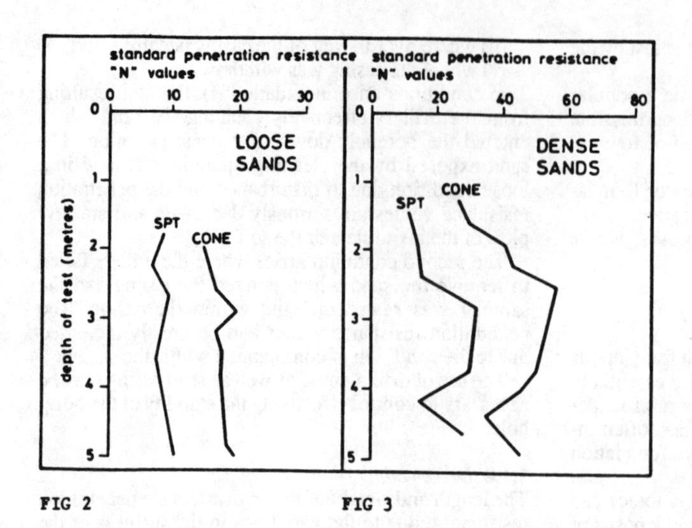

Figs 2 & 3. Comparison of dynamic penetration test results using standard sampler and 60° cone on standard sampler.

Fig.1 which was reproduced from Kovacs et al (1983), demonstrates that both the shape of the hammer and the number of turns on the cathead have a dominant influence on the amount of energy transferred to the sampler.

The energy transfer ratio represents the efficiency of transfer between the hammer energy immediately prior to impact and the energy delivered to the drill rod system.

4.1.4 *Tubular steel sampler*

Two modifications have been made to the SPT sampler which may affect the resistance to penetration.

In coarse granular soils the drive shoe is sometimes replaced with a solid steel 60° cone, to achieve consistency of shape of drive point instead of irregular pieces of gravel jamming in the open drive shoe, and to minimise damage to the cutting edges of open shoe.

A cautionary note is introduced by the results of a simple series of tests in loose and medium dense sands, having no significant gravel content, where greater penetration resistances were obtained using the solid steel 60° cone. Figs. 2 and 3 typify the results of this series of tests.

The second modification to the SPT sampler was made in an attempt to improve recovery and removal of soil samples and involved the insertion of a thin-wall liner of 1.5 mm wall thickness within the tubular steel sampler (ASTM D1586-67).

Schmertmann (1979) reported that omission of the liners improved sample recovery but produced a significant reduction in penetration resistances (N values). The percentage reduction in N values increased with decreasing resistance to penetration.

4.2 *Deep foundations and raft foundations*

These two categories often require determination of penetration resistances to considerable depths below ground surface because of the significant depths to which these foundations influence deep deposits of sand.

Similar problems with the execution of the test arise to those described for investigations for shallow foundations with the exception that the problems are greatly magnified because of the greater depths of drilling.

Good drilling and cleaning techniques and careful execution of the standard penetration test can result in typical values of penetration resistance being obtained when operating at depths of as much as 35 metres below groundsurface in deep sand deposits. Figure 4 compares the penetration resistances obtained from good quality ground investigation work (Curve 2) with poor quality work (Curve 1) at the same site. The latter curve reflects the serious disturbance caused by boring operations and completely misrepresents the state of compaction of the sand deposits.

4.3 *Stiff clays and weak rocks*

Considerable use has also been made of the standard penetration test as a means of assessing the strength and stiffness of rocks.

An extensive study has been made by Stroud (1974) concerning the standard penetration test (SPT) in insensitive stiff clays and soft rocks. The study indicated that the SPT could be used to estimate the properties of a variety of stiff clays and weak rocks.

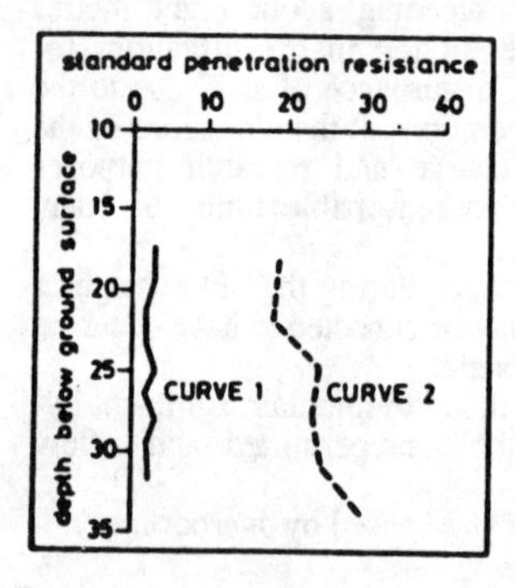

Fig. 4. Effects of borehole disturbance on SPT results.

Stroud correctly emphasised that for engineering design purposes the mass shear strength must be determined which takes into account the weakening effect of the system of discontinuities in stiff clays, and indicated that the mass shear strength of fissured London Clay may be only one quarter to one half of the shear strength of the intact material. To measure the mass shear strength of a fissured clay, a large enough volume of material must be tested to properly reflect the weakening effect of the system of discontinuities.

An empirical relationship has been presented by Stroud which, to a first approximation, gives the mass shear strength as a function of the standard penetration resistance (N value) i.e.

$$c = f_1 \times N$$

Figure 5 indicates the possible variation of f_1 with changes in the value of plasticity index.

It is important to note that for the interpretation of the SPT in weak rocks to be of any value on a universal basis, standardisation is essential. In the very large number of tests to which reference is made by Stroud in the development of his empirical relationships, it would appear that the standard splitspoon sampler was used. In hard materials where the full penetration of 450 mm was not achieved, the tests were stopped at 100 blows and the actual penetrations noted. The number of blows for 300 mm penetration was then obtained by extrapolation.

The solid steel 60° cone has been substituted by some investigators for the open drive shoe but this action may invalidate the empirical relationships developed by Stroud.

5 APPLICATION OF SPT WITHIN SOME MEMBER COUNTRIES

Advantage has again been taken in the main of the wide-ranging contributions to ESOPT I to review how the use of SPT results internationally. Out of 26 countries that submitted state-of-the-art reports, 16 made some reference to the use of the SPT. These 16 countries together with Brazil, form the basis of Table III, in which has been recorded the adjustments, correlations and applications described in the state-of-the-art reports, augmented with a few relatively recent technical contributions. It is appreciated that such a review has severe limitations for a number of reasons but it is believed that it is of interest in relation to the consideration of the content of an International Reference Procedure. A more comprehensive review is beyond the scope of this document.

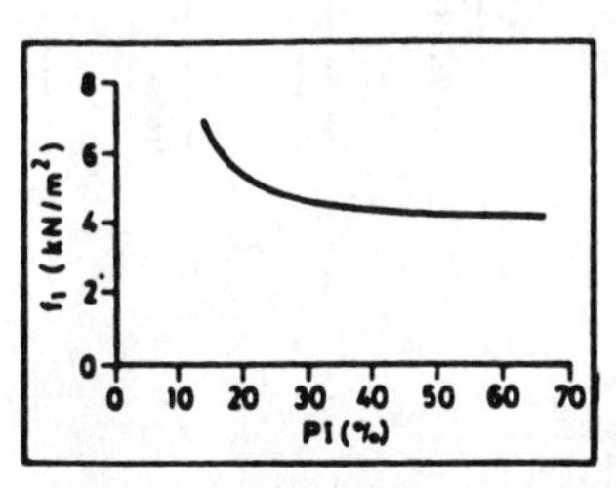

Fig. 5. Variation of f_1 with plasticity index.

5.1 *Materials in which used*

Summarised information with comments from the state-of-the-art reports to ESOPT I is shown in Table II. Beyond the familiar testing of fine grained cohesive and non-cohesive soils there was a fairly general reference to gravels, linked occasionally with the use of the solid cone, but with little caution of its application in such coarse material. Tests in silt were rarely mentioned. Its use in hard material, including weak rocks and hard residual soils, was considered of positive advantage by Australia, Israel and the United Kingdom.

5.2 *Traditional adjustments to N*

Notwithstanding the exclusion of the critical voids ratio adjustment to high values of resistance in submerged fine or silty sands from the second edition, 1967, of 'Soil Mechanics in Engineering Practice'. Table III shows five countries still taking this precaution. On the other hand, there were fewer references to adjustments for the position of the water table.

Gibbs & Holtz (1957) adjustment for the overburden effect continues to enjoy wide usage. It is worthy of note however that a free fall hammer was not used in their laboratory tests and that the energy transfer at impact was probably variable and low since two turns of the rope around the cathead were used.

5.3 *Correlations of SPT data with other soil parameters*

Such references as were made in the state-of-the-art reports have been included in Table III, but clearly many more exist in conjunction with the examination of local conditions. Referring to the table, mention may be made of correlation tests in Canada with the elastic properties of dense glacial till (Trow 1965), in Western Germany with the modulus of compressibility and density of sands (Schultze & Melzer 1965), in Israel both with the properties of dune and beach sands and with the influence of stress history (Zolkov & Wiseman 1965). In Japan correlations have been made with seismic measurements in relation to earthquake engineering (Japanese Society 1981).

Of particular importance in the context of correlations are the difficulties that arise due to the considerable variability that has also existed in the SPT equipment and procedure within, as well as between, different countries. Development of a calibration method, the subject of Document No.4 must be an advantage, and one of the first examples of the application of such a method is contained in a study at a site near Vancouver International Airport where a considerable improvement in uniformity of SPT results occurred after adjusting energy levels of the hammer blows (Robertson et al 1983).

5.4 *Application of SPT data*

Surprisingly there were few references in the state-of-the-art reports emphasising the limitations in application of the results of the SPT due to variability in procedures and equipment, although in certain cases positive statements were made concerning 'crudeness'

Table III. Adjustments, correlations and applications used within some member countries.

Country	Adjustments			Correlations			Applications						
	N′ in saturated, fine & silty sands	Water table influence	Relative density	Other penetration tests	Stress/strength	Deformation	Shallow foundation sands	Shallow foundation clays	Shallow foundation gravels	Piles	Use in weak rock	Compaction control	Liquefaction potential
Australia	●	●	GH57			Al64 D'A70	TP48 M56	H57		M59 N63	●		
Brazil			Me71			Me71	Me71			De82			
Canada		●	TR73	RW83		Tr70	M56 PT74			Ta71 M59			
Czechoslavakia		Dr74		local dynamic tests									
Germany, Federal Republic			SM65			SM65 SS73							
Greece	●		GH57 Th63		TP48		TA74	TP48					
India	●		TP48 GH57				PT73	PT73		Indian C.O.P.			

Israel					residual stress		Zo74					sandy below water table	
Italy		○	GH57 SM65	local dynamic tests		SM65	TP67						
Japan			JS81 MM74	MM74 ram sounding		JS81 MM74	JS81	JS84		JS81			JS81 MM74
Norway				Se74 ram & CPT		○	○	○	● piles	Se74			
Portugal							TP48 M56	● where sandy		M59			
South Africa	●	PB69	GH57			SM61 We69	PT74						
Spain							TP48 M56			M59		US74	
Turkey			TP48				M56			M59			
United Kingdom	●		GH57 Th63	● CPT	●	BM77	TP48		TP48	RT74	GS76		
U.S.A.			GH57	Sc79			PT74						Se79
17. Uses noted	**6**	**4**	**13**	**7**	**3**	**9**	**14**	**5**	**2**	**10**	**3**	**2**	**3**

Legend ● = use is made of technique ○ = some against use Code letters/numbers – see references

and 'lack of reliability', and pleas for better standardisation.

Nothwithstanding these deficiencies, it is clear that widespread use is made of SPT results in design throughout the world, sometimes still using the procedures given when the test method was first publicly announced in 1948 (Terzaghi & Peck 1948). Although this success has been due, at least in part, to a very conservative approach, it undoubtedly included sufficient compensation generally for shortcomings in the test techniques, and a number of changes in the design procedure have since been proposed, e.g. (Terzaghi & Peck 1967, de Mello 1971) with the object of achieving greater economy in foundation costs. In this connection mention should be made of the recognition in Table III to the significant contributions made by G.G.Meyerhof (Meyerhof 1956 and 1959).

Applications in design are most common for determining the bearing capacity and settlement of footings in sands. The results of tests in cohesive soils for shallow foundation design are treated cautiously.

Applications for pile design has been the other subject of most interest. Again it is more fine grained granular soils rather than clays. After the proposals by Meyerhof (1959) many separate studies have been carried out, and reference is made to a selection of these in Table III. The subject was reviewed again in the general report to ESOPT I (Meyerhof 1974).

The convenience and robustness of the test has led to its application to uses for construction and to assess specialist problems. For some time predictions of the driving resistance of piles have been made on the basis of the test, but a natural extension has been for compaction control and a number of cases have been described in the literature where this has been successfully carried out, particularly in marine situations.

One special problem has led almost to a unique reliance upon the test at the present time. This is for estimating the liquefaction potential of saturated cohesionless soils under undrained conditions. The effects of earthquakes on such soils is usually the concern, but explosions and even the operation of machinery may cause unstable conditions to occur. The opportunity to study results of SPT tests before and after earthquakes in Japan has provided experience on which to base predictions (Seed 1979). Research on this application has been carried out in the USA, Japan and the Republic of China.

REFERENCES

AL64: Alpan, I. 1964. Estimating settlement of foundations on sand. *Civ. Eng. & Pub. Wks. Review* 59: 1415.

ASTM D 1586, 1958. *Tentative method for penetration test and split-barrel sampling of soils.* ASTM Philadelphia, Pa.

ASTM D 1588, 1967. *Standard method for penetration tests and split-barrel sampling of soils.* ASTM Philadelphia, Pa.

BM77: Burland, J.B., B.B.Broms and V.F.B.de Mello 1977. Behaviour of foundations and structures. State of Art Review. *IX ICSMFE* 3: 495-546.

D'A70: D'Appolonia, D.F. & E.D'Appolonia 1970. Use of the SPT to estimate settlement of footings on sand. *P.Symposium on Foundations on Interbedded Sands, Perth*: 16.

De82: Decourt, L. 1982. Prediction of the bearing capacity of piles based exclusively on N values of the SPT. In: A.Verruijt, F.L.Beringen & E.H.de Leeuw (eds), *Penetration testing*; *Proc. ESOPT II* 1: 29-34. Balkema, Rotterdam.

Dr74: Drozd, K. 1974. The influence of moisture content in sand on penetration results. *Proc. ESOPT* I 2: 162-164.

1974. Standards and Comments on Standards. *Proc. ESOPT* I 2: 219-259.

Fletcher, G.F.A. 1965. Standard Penetration Test, its uses and abuses. *ASCE J. SMFD* 91(SMA): 67-75.

GH57: Gibbs, H.J. & W.G.Holtz 1957. *Research and determining the density of sands by spoon penetration testing*. I ICSMFE, London.

GS76: *Geotechnique Symposium in Print on piles in weak rocks* 1976.

H57: Hough, B.K. 1957. *Basic soils engineering*.Ronald Press, New York.

Hvorslev, M.J. 1949. *Sub-surface exploration and sampling of soils for civil engineering purposes.* The Engineering Foundation, New York.

JS81: Japanese Society SMFE 1981. Present state and future trend of penetration testing in Japan. Separate report at X ICSMFE, Stockholm.

Japanese Society, SMFE, 1983. Standard Penetration Testing in Japan. Separate report presented at Regional Conference, Helsinki.

Kovacs, W.D., L.A.Salomone & F.Y.Yokel 1983. *Comparisons of energy measurements in the Standard Penetration Test using the Cathead and Rope Method, by Geot. Eng.Gp.* Nat. Bureau of Standards for US Nuclear Regulatory Commission.

Me71: de Mello, V.F.B. 1971. The penetration test. *4th Pan Am Conference, SMFE*, 1: 1-86.

M56: Meyerhof, G.G. 1956. Penetration tests and bearing capacity of cohesionless soils. *Proc. ASCE.* 82(SMI): paper 866.

M59: Meyerhof, G.G. 1959. Compaction of sands and bearing capacity of piles. *Proc. ASCE.* 85(SM6): 1-29.

Mohr, H.A. 1937. Exploration of soil conditions and sampling operations, *SM Series* 21. Harvard University.

Mohr, H.A. 1966. Discussion on Standard Penetration Test, its uses and abuses. *ASCE, J. SMFD* 92(SM1) :196-199.

MM74: Muromachi, T., Ogural & T.Miyashita 1974. Penetration Testing in Japan. *Proc. ESOPT* I 1: 193-196.

Nixon, I.K. 1982. Standard Penetration Test, State of the Art Report. In: A.Verruijt, F.L.Beringen & E.H.de Leeuw (eds), *Penetration Testing*; *ESOPT* II 1: 3-24. Balkema, Rotterdam

N63: Nordlund, R.L. 1963. Bearing Capacity of Piles in Cohesionless Soils. *Proc. ASCE.* 89(SM3): 1.

PB69: Peck, R.B. & A.R.S.Bazaraa 1969. Discussion of settlement of spread footings on sand. *J.SMFE.ASCE* 95(SM3): 905.

PT53: Peck, R.B., W.E.Hanson & T.H.Thornburn 1953. *Foundation Engineering*. Wiley, New York.
PT73: Peck, R.B., W.E.Hanson & T.H.Thornburn 1973. *Foundation Engineering* (2nd edition).
PT74: Peck, R.B., W.E.Hanson & T.H.Thornburn 1974. *Foundation Engineering* (3rd edition).
Riggs, C.O., N.O.Schmidt & C.L.Rassiur 1983. Reproducible SPT Hammer Impact Force with an Automatic Free Fall SPT Hammer System. *Geot. Test J.* 6(3): 201-209.
RW83: Robertson, P.K., R.G.Campanella & A.Wightman 1983. SPT-CPT Correlations. *ASCE. Geotech. Eng.* 109(11): 1449-1459.
RT74: Rodin, S., B.O.Corbett, & S.Thorburn 1974. Penetration Testing in United Kingdom. *Proc. ESOPT* I 1: 139-146.
Schmertmann, J.H. 1975. Measurement of insitu shear strength, state-of-the-art report. *ASCE Spec. Conf. GED. insitu measurement of soil properties, Raleigh* 2: 57-138.
SC79: Schmertmann, J.H. 1979. Statics of SPT. *ASCE.J.GED*. 105(GT5) 655-670.
SM61: Schultze, E. & E.Menzenbach 1961. Standard Penetration Test and Compressibility of Soils. *V ICSMFE* 1: 249-255.
SM65: Schultze, E. & K.J.Melzer 1965. The determination of the density and the modulus of compressibility of non-cohesive soils by soundings. *VI ICSMFE Montreal* 1: 354-358.
SS73: Schultze, E. & G.Sherif 1973. Predictions of settlements from evaluated settlement observations for sand. *VIII ICSMFE Moscow* 1(3): 225-230.
Se79: Seed, H.B. 1979. Soil Liquefaction and cyclic mobility evaluation for level ground during earthquakes. State-of-the-Art. *ASCE J.GED* GT2: 201-255.
Se74: Senneset, K. 1974. Penetration Testing in Norway. *Proc. ESOPT I* 1: 85-93.
Stroud, M.A. 1974. The Standard Penetration Test in insensitive clays and soft rocks. *Proc. ESOPT I* 2(2): 367-375.
TA74: Tassios, T.P. & A.G.Anagnostopoulos 1974. Penetration Testing in Greece. *Proc. ESOPT I* 1: 65-68.
Ta71: Tavenas, F. 1971. Discussion to Session 1: The Standard Penetration Test. *Proc. 4th Panam. CISSMFE*, III: 64-70.
TR73: Tavenos, F.A., R.S.Ladd & P.La Rochelle 1973. Accuracy of relative density measurements: results of a comparative test program. *ASTM Spec. Symposium LA*: 18-60.
TP48: Terzaghi, K. & R.B.Peck 1948. *Soil Mechanics in Engineering Practice*.
TP67: Terzaghi, K. & R.B.Peck 1967. *Soil Mechanics in Engineering Practice* (2nd edition).
Th63: Thorburn, S. 1963. Tentative correction chart for the standard penetration test in non-cohesive soils. *Civ. Eng. Pub. Works Review* 58(6): 752-753.
Tr70: Trow, W.A. 1965. Discussion of elastic properties of a dense glacial till deposit by E.J.Klohn. *Canadian Geotech. J.* 2(2):132-139
US74: Uriel, A.O., J.Rodriguez Sanchez & M.Sanchez Morales 1974. Test results concerning the influence of the cone angle in the dynamic penetration resistance. *Proc. ESOPT I* 2(2): 401-406.
We69: Webb, D.L. 1969. Settlement of structures on deep alluvial sandy sediments in Durban, S.A. *Proc. BGS Conf. insitu investigations in soils and rocks*: 181-188.
ZW65: Zolkov, E. & G.Wiseman, 1965. Engineering Properties of Dune and Beach Sands and the Influence of Stress History. *Proc. VI ICSMFE* 1: 134-138.
Zo72: Zolkov, E. 1972. Standard Penetration Test and Foundation Practice in Fine Sands in Israel. *J. Mat. JMSLA (ASTM)* 7(3): 336-344.
Zo74: Zolkov, E. 1974. The Nature of a Sand Deposit and the Settlements of Shallow Foundations. *Proc. ESOPT I* 2(2): 421-431.

DOCUMENT 3

INTERNATIONAL REFERENCE TEST PROCEDURES

1 SCOPE

1.1 This specification describes the principles constituting acceptable test procedures for the SPT from which results are comparable.

1.2 The SPT determines the resistance of soils in a borehole to the penetration of a tubular steel sampler, and obtains a disturbed sample for identification. The penetration resistance can be related to soil characteristics and variability.

The basis of the test consists of dropping a hammer weighing 63.5 kg onto a drive head from a height of 760 mm. The number of blows (N) necessary to achieve a penetration by the steel tube of 300 mm (after its penetration under gravity and below a seating drive) is regarded as the penetration resistance (N).

2 BORING METHODS AND EQUIPMENT

2.1 *Boring methods*

2.1.1 The boring equipment shall be capable of providing a reasonably clean hole to ensure that the penetration test is performed on relatively undisturbed soil.

2.1.2 When wash boring, a side-discharge bit should be used and not a bottom-discharge bit. The process of jetting through an open tube sampler and then testing when the desired depth is reached shall not be permitted.

2.1.3 When shell and auger boring with temporary casing, the drilling tools shall have diameters not more than 90% of the internal diameter of the casing.

2.1.4 When boring in soil that will not allow a hole to remain stable, casing and/or mud shall be used.

2.1.5 The diameter of the borehole should be between 63.5 and 150 mm.

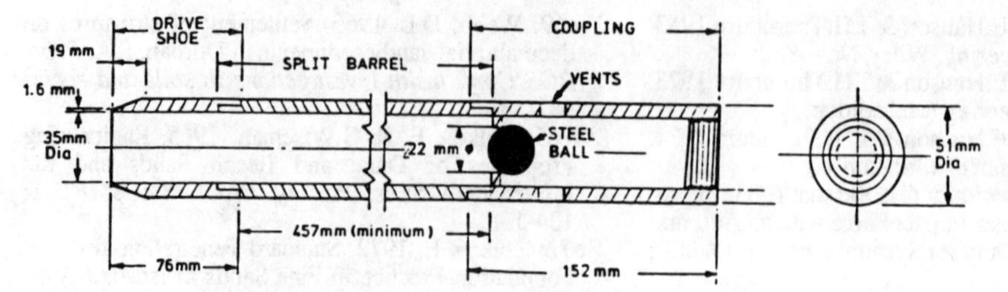

Fig. 1. Cross Section of SPT Sampler.

2.2 *Tubular steel sampler assembly*

The tube of the sampler shall be made of hardened steel with a smooth surface externally and internally. The external diameter shall be 51 mm plus or minus 1 mm and the internal diameter throughout shall be 35 mm plus or minus 1 mm. Its length shall be 457 mm minimum.

The lower end of the tube shall have a driving shoe 76 mm long plus or minus 1 mm having the same bore and external diameter as the tube. Over the lowermost 19 mm it will taper uniformly inwards to reach the bore at the bottom edge. The material shall be the same as the tube. The drive shoe shall be replaced when it becomes damaged or distorted.

At the upper end of the tube a steel coupling shall be fitted to connect with the drive rods. Inside shall be a non-return valve with wide vents in the coupling wall, which are of sufficient size, to permit unimpeded escape of air or water on entry of the sample. The valve should provide a watertight seal when withdrawing the tubular steel sampler.

One acceptable form of the sampler assembly is shown in Fig 1.

2.3 *Steel drive rods*

2.3.1 The steel drive rods, connecting the sampler assembly to the hammer assembly. shall have a section modulus appropriate to their total length and lateral restraint.

Appropriate section properties are:

Rod diameter (mm)	Section modulus ($x 10^{-6}m^3$)	Rod weight (kg/m)
40.5	4.28	4.33
50	8.59	7.23
60	12.95	10.03

Rods heavier than 10.03 kg/m shall not be used.

2.3.2 Only straight rods shall be used and periodic checks shall be made on site. When measured over the whole length of each rod the relative deflection shall not be greater than 1 in 1000.

2.3.3 The rods should be tightly coupled by screw joints.

2.4 *Hammer assembly*

2.4.1 The hammer assembly shall comprise:

a. A steel drive head tightly screwed to the top ofthe drive rods. The energy transferred on impact shall be maximised by a suitable design of drive head.

b. A steel hammer of 63.5 kg (plus or minus 0.5 kg) weight.

c. A release mechanism which will ensure that the hammer has a free fall of 760 mm.

2.4.2 The guide arrangement shall permit the hammer to drop with minimal resistance.

2.4.3 The overall weight of the hammer assembly shall not exceed 115 kg.

2.4.4 In situations where comparisons of SPT results are important calibrations will be made to evaluate the efficiency of the equipment in terms of energy transfer.

3 TEST PROCEDURE

3.1 *Preparing the borehole*

3.1.1 The borehole shall be carefully cleaned out to the test elevation using equipment that will ensure the soil to be tested is not disturbed.

3.1.2 When boring below the groundwater table or in sub-artesian conditions the water or mud level in the borehole shall at all times be maintained at a sufficient distance above the groundwater level to minimise disturbance. The water or mud level in the borehole shall be maintained to ensure hydraulic balance at the test elevation.

3.1.3 The drilling tools shall be withdrawn slowly to prevent loosening of the soil to be tested.

3.1.4 When casing is used, it shall not be driven below the level at which the test is to commence.

3.2 *Executing the test*

3.2.1 The sampler assembly shall be lowered to the bottom of the borehole on the drive rods with the hammer assembly on top. The initial penetration under this total deadweight shall be recorded. Where this penetration exceeds 450 mm the test drive will be omitted and the 'N' value taken as zero.

After the initial penetration, the test will be executed in two stages:

Seating drive: A penetration of 150 mm. If the 150 mm penetration cannot be achieved in 50 blows, the latter shall be taken as the seating drive.

Test drive: A further penetration of 300 mm. The number of blows required for this 300 mm penetration is termed the *penetration resistance*(N). If the 300 mm penetration cannot be achieved in 100 blows the test drive shall be terminated.

The rate of application of hammer blows should not be excessive such that there is the possibility of not achieving the standard drop or preventing equilibrium conditions prevailing between successive blows. Typically, the maximum rate of application of blows is 30 per minute. The number of blows required to effect each 150 mm of penetration shall be recorded. If the seating or test drive is terminated before the full penetration, the record should state the depth of penetration for the corresponding 50 blows.

3.3 *Recovery of soil sample and labelling*

3.3.1 The sampler shall be raised to the surface and opened. The representative sample or samples of the soil in the sampler shall be placed in an air-tight container.

3.3.2 Labels shall be fixed to the containers with the following information:

a) Site;
b) Borehole number;
c) Sample number;
d) Depth of penetration;
e) Length of recovery;
f) Date of sampling;
g) Standard penetration resistance (N).

4 REPORTING OF RESULTS

The following information shall be reported:

1) Site;
2) Date of boring to test elevation;
3) Date and time of commencement and end of test;
4) Borehole number;
5) Boring method and dimensions of temporary casing, if used;
6) Dimensions and weight of drive rods used for the penetration tests;
7) Type of hammer and release mechanism or method;
8) Height of free fall;
9) Depth to bottom of borehole (before test);
10) Depth to base of casing;
11) Information on the groundwater level and the water or mud level in the borehole at the start of each test;
12) The depth of initial penetration and the depths between which the penetration resistances (seating and test drives) were measured;
13) Penetration resistances (seating and test drives);
14) The descriptions of soils as identified from the samples in the sampler;
15) Observations concerning the stability of strata tested or obstructions encountered during the tests etc., which will assist the interpretation of the test results;
16) Calibration results, where appropriate.

EXPLANATORY NOTE

Calibrations of drive rods and hammer assemblies, where appropriate, would normally be carried out for each rig and separately from the investigation work. They would be applicable to a particular project, based upon the personnel and equipment involved.

DOCUMENT 4

CALIBRATION METHODS

1 SCOPE

The method described herein is the procedure for measuring that part of the drive weight kinetic energy that enters the connector rod column during the 'Standard Penetration Test' (SPT). The method uses a formula allowing optional adjustment of the number of hammer blows 'N' (N Values) in the SPT in accordance with different magnitudes of driving energy.

It requires the determination of stress wave energy ratio from a record of the force time of the first compression wave produced by standard hammering. This record is obtained from a load cell capable of sensing impact energy to the connector rod column with sufficient sensitivity.

2 APPLICATIONS

The method has particular application to the comparative evaluation of hammer systems and, to a lesser degree of accuracy, hammer-operator systems, that are used for the SPT.

It may be applied to a means of quality control of the SPT, when using N values as determined in conformance with the International Reference Test Procedure of the SPT in important test, research, analysis and design.

3 DEFINITION OF SYMBOLS AND TERMS

3.1 E^*

The normal kinetic energy in a drive weight of stipulated mass of 63.5 kg after a gravitational free fall from a stipulated fall height of 0.76 m. (the Newtonian kinetic energy at impact)

3.2 E_i

The energy content of the initial (first) compression wave that is produced by a hammer impact. (ideal case rod length L = infinity)

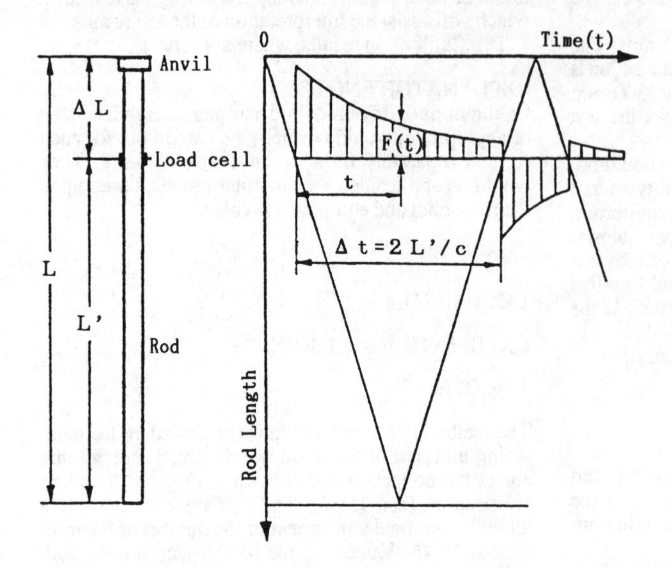

Fig. 1. Principle of measurement of E_i.

Note: Some hammer energy is lost at impact during the transfer of kinetic energy in the hammer to compression wave energy in the connector rods. This impact-transfer loss can vary significantly with the shape of the hammer or the anvil, or both, with the mass of the anvil, and with the properties of any impact cushion at the anvil.

3.3 ER_i (=E_i/E^*) in %

Measured stress wave energy ratio, this value takes the value of rod length L to be infinite.

3.4 L (see Fig. 1)

Length between the hammer impact surface on the anvil and the bottom of the sampler.

3.5 L' (see Fig. 1)

Length between the load cell and the bottom of the sampler.

3.6 ΔL (= L - L', see Fig. 1)

Length between the top of the anvil and the load cell.

3.7 L_e(= M_h/m_r)

Equivalent rod length; the ratio of mass of standard hammer, M_h, to the unit mass of used rod, m_r. (see 7.)

3.8 *Connector rods*

The drill rods that connect the drive weight system above the ground surface to the sampler below the surface (except guide rod).

3.9 *Load cell*

Any instrument placed around, on, or within a continuous column of connector rods for the purpose of sensing that part of the drive weight kinetic energy that is transmitted into the connector rods.

3.10 *Processing instrument*

An instrument which receives the signal from the load cell after a hammer blow and processes it to produce computed E_i or ER_i values.

3.11 *Engineer*

The engineer, geologist, or other responsible professional who are responsible for energy measurement.

3.12 *Operator*

The person responsible for conducting the SPT.

4 SIGNIFICANCE AND USE

4.1 The penetration resistance of soil depends not only upon the mass of the hammer and the soil characteristics, but upon the continuity and geometry of system components and upon system operation factors that affect resistance to free-fall of the drive weight.

The SPT is only standardized with respect to the

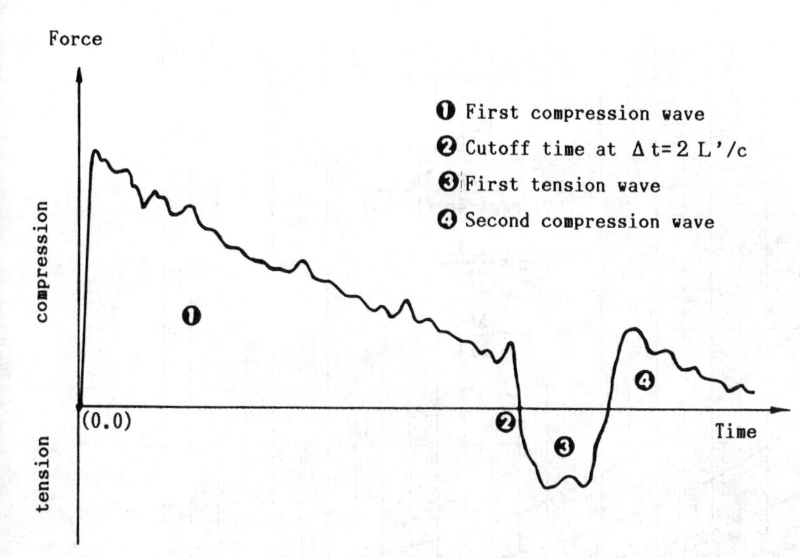

Fig. 2. Idealized example of the force-time waveform.

configuration of the penetrometer, the mass of the drive weight and the height of free fall. The stress wave integration procedure is a method of evaluating variations from differences in system geometry and operation.

4.2 The incremental penetration of the sampler from a hammer blow is directly related to the magnitude of the irregularly shaped force pulse within the first compressive wave from that blow. A processing instrument integrates the forces within the time limits of the first compressive pulse or wave (idealized example in Fig. 2) thus computing the stress wave energy of the primary force pulse.

The integrated stress wave energy, therefore, provides an approximate measure of the effective driving force.

4.3 The integrated stress wave method has particular application to evaluation of various hammer systems of the SPT.

There is an approximate, linear relationship between the incremental advance of the sampler and the stress wave energy that enters the connector rods. Therefore, there is also an approximate inverse relationship between the number of hammer blows, N, and the stress wave energy, E_i (or ER_i).

4.4 Stress wave measurements may be used to evaluate both operator-dependent cathead and rope-hammer drop system and the relatively operator-independent mechanized systems.

When operator-dependent hammer systems are tested, the possible inability of the operator to reproduce environmental influences must be considered.

4.5 Equation (a-1) provides an accurate method of comparing N values obtained by drill rig source 2 and reference source 1, based energy measurements obtained at these sources and using this method.

Equation (a-1) adjusts N for energy differences only and assumes all the other factors affecting N remain constant.

$$N'_2 = N_2\left(\frac{ER_{i2}}{ER_{i1}}\right) \qquad (a-1)$$

where N' = N after adjustment to a reference ER_i value.

5 PRINCIPLE OF ANALYSIS (STRESS WAVE INTEGRATION METHOD)

The following is one of a number of recognized theories that provides a basis for making the energy measurements. The stress wave integration method uses the following formula:

$$E_i = \frac{cK_1K_2K_c}{AE}\int_0^{\Delta t}[F(t)]^2dt \qquad (a-2)$$

$$K_1 = \frac{1-\exp(-4m)}{1-\exp\{-4m(1-d)\}} \qquad (a-3)$$

$$K_2 = \frac{1}{1-\exp(-4m)} \qquad (a-4)$$

$$K_c = \frac{c_a}{c} \qquad (a-5)$$

where:

F(t) = dynamic compressive force in the connector rods as a function of time, t.

Δt = time duration of the first compression pulse starting at t=0.

A = cross-sectional area of the connector rods above and below the load cell.

E = Young's modulus of the connector rods.

c = theoretical sound velocity of the compression wave in the connector rods (=5,120 m/s).

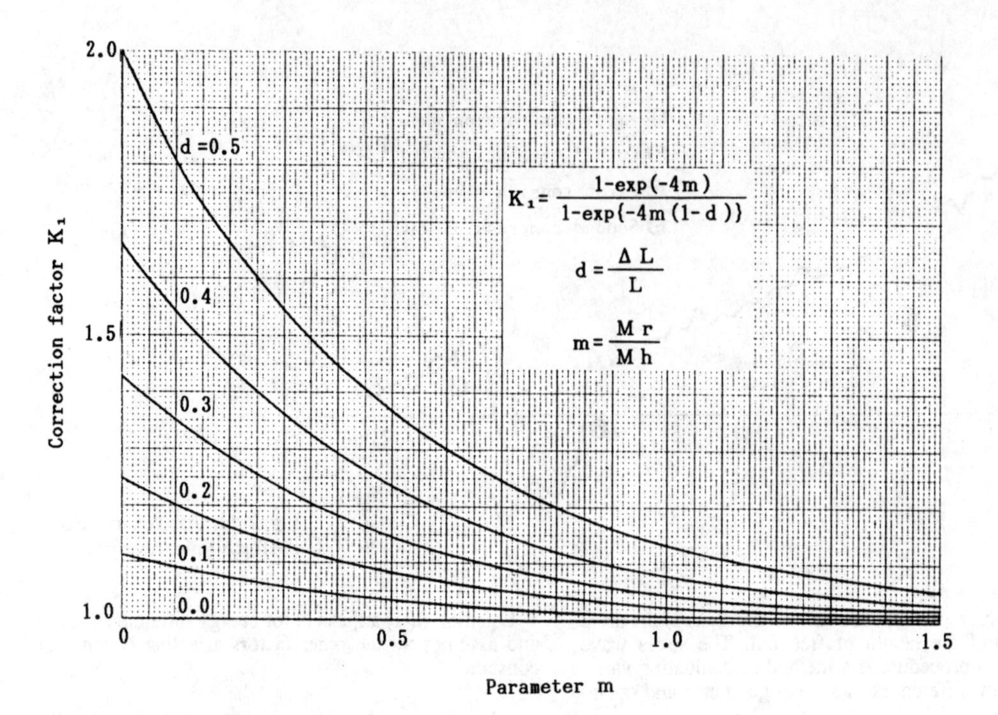

Fig. 3. Correction factor K_1.

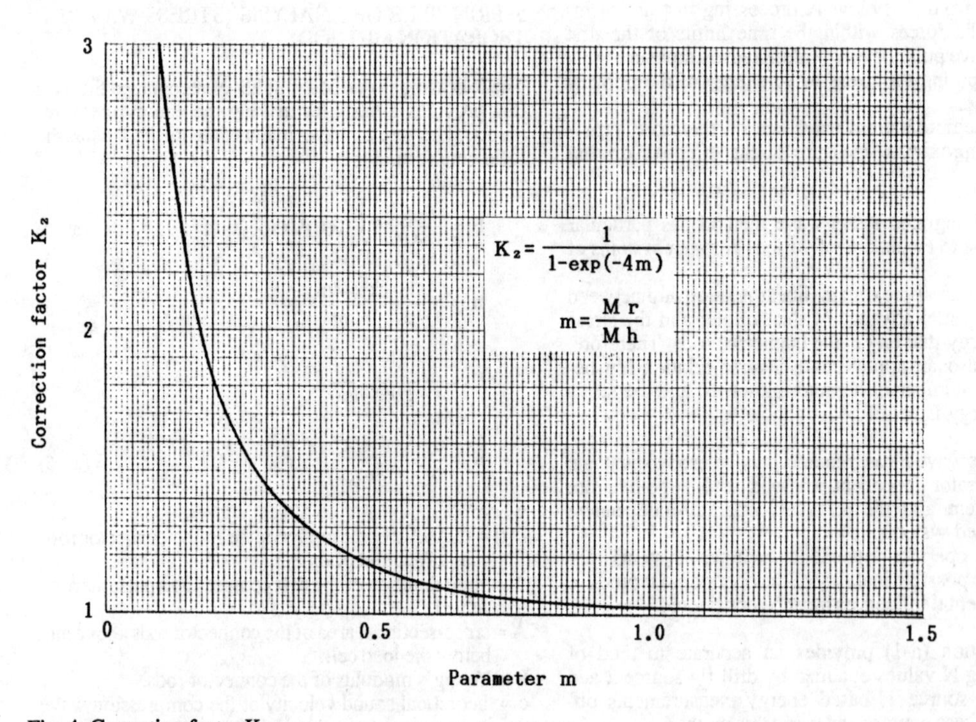

Fig. 4. Correction factor K_2.

m = ratio of total mass of rod for L (= M_r) to mass of standard hammer (= M_h). (= M_r/M_h)
$d = \Delta L/L$, defined in 3.4 and 3.6.
K_1 = a correction factor to account for the compression wave energy not sensed by the load cell due to the length, ΔL, of the connector rod between the hammer impact surface and the load cell. K_1 for used rod is calculated in expression (a-3). (see Fig. 3)
K_2 = a theoretical correction factor needed when the rod length, L, is less than the equivalent rod length, L_e. (see Note 3 in 7) K_2 for any used rod length, L, is calculated in expression (a-4). (see Fig. 4)
K_c = a factor to correct the theoretical sound velocity, c, to the actual (or measured) sound velocity, c_a. K_c is calculated in expression (a-5).

Note: Total energy, E_i, of that part of rod having an impact stress consists of equal measures of strain energy (potential energy) and kinetic energy. In general, it will be easier to measure strain in that part of rod than its displacement velocity. Then, E_i may be obtained by measuring changes in strain for Δt, calculating strain energy and doubling the result. Eq. (a-2) is expressed in the form of doubled strain energy.

6 MEASURING APPARATUS

6.1 *Sensor*

A load cell (defined in 3.9) with as little dynamic impedance as possible.

6.2 *Processing instrument (defined in 3.10)*

Normally it consists of an amplifier, an integration circuit that displays ER_i after each blow, and an integration timer measuring Δt. This system must be compatible with the sensor.

6.3 *Oscilloscope*

An oscilloscope can be a useful adjunct, and a digital oscilloscope hooked to a computer can act as a combination instrument and timer.

7 EQUIVALENT ROD LENGTH

To use this principle without a correction for rod length, the rod used should have a length greater than the equivalent rod length, L_e, as determined by the following formula:

$$L_e = M_h/m_r \qquad \text{(a-6)}$$

where
M_h = mass of standard hammer (= 63.5 kg)
m_r = unit mass of used rod = M_r/L (kg/m)

When circumstances make measurements at depth less than L_e unavoidable, then a correction for rod length is necessary.

Note: If the engineer wishes to adjust E_i value entering actually into the rods for a case when $L < L_e$ to that of infinite L case, then adjust by multiplying the measured E_i value by k_2. (see Fig. 4)

8 MEASURING PROCEDURE

8.1 *Preparations*

The operator first prepares a borehole to keep the conditions $L > L_e$. The engineer calibrates the load cell using the processing instrument.

8.1.1 After the sampler and rod assembly have been connected together for insertion into the borehole, connect the load cell no closer than 0.6 m to the base of the anvil. Use load cell adaptors that minimize abrupt changes in cross-sectional area with respect to the connector rods. The cross-section of the rod above and below the load cell must be the same for a distance of at least 0.3 m in both directions.

8.1.2 Follow the processing instrument manufacturer's calibration procedures to ensure the load cell and processing instrument are operating properly. Match the processing instrument setting to the type of rod used above and below the load cell.

8.1.3 If the instrument cannot be preset to the type of rod used, then convert the results from that set on or built into the instrument to that actually used by adjusting instrument E_i or ER_i values by the relative rod cross-sectional areas in accordance with Eq. (a-2). All connector rods to include the short section between the load cell and the anvil must have the same section diameter and area.

8.1.4 Do not perform measurements of E_i or ER_i until several preparatory sequences of blows obtained under the same conditions of operator, equipment, soil conditions, and the like, have been performed by the operator. Record the measurements by any suitable means.

Note: Preparatory sequences of blows have the objective of bringing the equipment and operator to their normal functioning conditions. They can accomplish such things as repolishing the cathead, drying wet or damp rope, providing fresh lubrication for mechanical parts, and providing refamiliarization practice for all the personnel concerned with these measurements.

8.2 *Measurement*

The engineer continues to measure E_i as the operator conducts the SPT hammer drops. The amount of penetration per hammer blow is determined by an automatic recording device. However, if that is not available, total penetration in mm is measured per blow, using an appropriate scale.

8.2.1 Perform a minimum of 30 measurements to determine the E_i or ER_i performance of a drive weight system.

Note: It is preferable to make as many measurements as practical so as to reduce the statistical sampling error. It is also preferable to make the measurements when

Table 1. General information sheet on energy measurement [SPT].

Item		Information
1. Date		2. Weather:
3. Engineer		Name; Position; Affiliation
4. Operator		Name; 5. Project: Name
6. Location		Place name; Country
7. Bore hole		No. of bore hole; Elevation(m); Depth tested(m)
8. Drill rig	Machine	Type of machine; Type of bit; Bit diameter(mm); Manufacturer
		Capacity, Special feature ☆
	Rod	Type & No.; O.D./I.D.(mm); Special feature
		Unit mass(kg/m); Le(m); Section modulus(X10⁻⁶ m³)
9. Hammer	Donut type	Mass(kg); Height(mm); Top:O.D./I.D.(mm); Bottom:O.D./I.D.(mm)
	Others	Type; Net hammer mass(kg); Total mass(kg); Special feature ☆
10. Anvil		Diameter(mm); Height(mm); Mass(kg); Special feature
11. Guide rod		Mass(kg); Length(m); O.D./I.D.(mm); Connection to anvil
12. Trigger		Type; Special feature; Rate of hammer blow per min. ☆
13. Cathead		Shape; Min.diameter(mm); Rate of revolution per min. ☆
14. Rope		Kind; Diameter(mm); Number of turns; New or used(years); Condition
15. Sampler		Length(mm); O.D./I.D.(mm); Mass(kg); Special feature ☆

16. Bore hole protection	Casing	Diameter (mm)	Distance between bottom and end of casing (m)
	Mud	Kind	Density
17. Water level		Depth from G.L. (m)	Distribution of water pressure (if measured)
18. Soil profile		Outline with N values ☆	
19. Sensor		Name, Type, Manufacturer ☆	
20. Processing instrument		Name, Type, Manufacturer ☆	
21. Comment of engineer		on all other factors that might influence the general validity of the energy measurement	
☆ N.B.		Drawing or picture in separate sheets are requested	

using the operator-dependent drive weight system in as near a routine manner as practical.

8.2.2 It is necessary to measure K_c in Eq. (a-2) to assure the proper cutoff time for Δt. If a processing instrument without oscilloscope is used, then the engineer needs to make as many measurements of the Eq. (a-2) integration time during each sequence of hammer blows as needed to assure that the instrument makes the proper time cutoff. Compare these measurements with the theoretical time to cutoff, 2L'/c.

Note: To conform to the theory in 5, the processing instrument should perform the integration until the first time the load cell senses a zero load. This should happen when the compression wave reflects at the sampler and returns to the load cell as a tension wave, at which time the net force at the load cell crosses zero force. For some circumstances, particularly near penetration refusal, the reflected wave arriving at the load cell may be a compression wave. Under these circumstances the force at the load cell may not cross zero until significant time after 2L'/c. The instrument will then continue the integration and can give a stress wave energy value much greater than that actually in the first compression wave.

8.2.3 When the recorded time of integration is within the range of (0.9 ~ 1.2) x the theoretical 2L'/c, the E_i or ER_i values displayed by the processing instrument should be multiplied by a factor:

$$K_c = c_a/c_i$$

where:
c_a = actual, and
c_i = the value of c built into the instrument and is usually the theoretical sound velocity c.

8.2.4 If the measured trip time of the compression wave is greater than 1.2 x 2L'/c, the data should not be used. Either use a processing instrument that also allows a visual check of the integration time of the stress wave or make the measurements at another ground location with a lower N value. When the time ratio is less than 0.9, something has produced a premature electrical signal indicating zero force. The operator needs to discontinue testing until discovering and correcting the cause.

9 REPORT

All energy measurement reports shall include two sets of information – general and special. For general information, the engineer makes necessary entries on the general information sheet.

9.1 *General information* (see Table 1)

9.2 *Special information for E_i and ER_i*

1) A record of all the energy measurements made together with their average and standard deviation.
2) Correct length of L, L' and ΔL.
3) Special connector rod size below the anvil and above and below the load cell.
4) Information on the calibration of the energy processing instrument and its matching load cell.
5) Comparisons between measured and theoretical 2L'/c integration times, spaced to help assure proper cutoff times over the range of ER_i values obtained.
6) Oscilloscope records of typical stress waves.

Penetration Testing 1988, ISOPT-1, De Ruiter (ed.)
© 1988 Balkema, Rotterdam, ISBN 90 6191 801 4

Cone penetration test (CPT): International reference test procedure

ISSMFE Technical Committee on Penetration Testing
CPT Working Party:
E.E.De Beer, Chairman (Belgium)
E.Goelen (Belgium)
W.J.Heynen (Netherlands)
K.Joustra (Netherlands)

Contents :

I. INTRODUCTION AND OBJECTIVES

I.1 The Cone Penetration Test originated in the Netherlands in the early thirties, and it spread first out to the foreign countries under the name of "Dutch sounding Test". At first the purpose was to have some quantitative data about the thickness and the consistency of very soft layers, by means of hand operated equipment. Soon it appeared that also very interesting results could be obtained concerning the bearing capacity and shear resistance of the deeper sand-layers, in which piles find their bearing. The results can be used to estimate the bearing capacity of these piles. Of course much stronger mechanically operated thrust machines were to be used to obtain sufficient penetration.

Since the early applications the static sounding technique has known a tremendous expansion. Many changes and improvements were and are still accomplished.

As an entrance to this report a thankfull tribute should be paid to the Dutch pioneers, remembering that their first aim was not to make theory, but to obtain quantitative and directly useful information for solving practical problems in geotechnics.

Of course the Dutch quasi static sounding technique, is only one out of many possible techniques. For instance the dynamic penetration techniques, in their most elementary forms, are as old as human building activities are.

During the presidential term of Prof. Fukuoka an European subcommittee was set up for the standardization of penetration testing in Europe, including, among others, the Dutch sounding test.

Why only Europe ? The reason for that was connected with practice in Europe where most of the penetration testing methods were already widely applied and accepted.

This European Subcommittee reported at the IXth International Conference on Soil Mechanics and Foundation Engineering held in Tokyo in 1977. The report, containing recommended procedures for the Cone Penetration Test or Dutch sounding, the Standard Penetration Test, the different Dynamic Probing Tests and the Weight Sounding Test, has been published in the Proceedings of the Conference.

After the Xth Int. Conf. in Stockholm the new president, Prof. de Mello installed the Committee on Penetration Testing.This committee has been very active under the chairmanship of Prof. Broms. At the meeting at Amsterdam at the occasion of the second European Symposium on Penetration Testing (ESOPT-II) in 1982 it was decided to include the piezo-cone in the recommended procedure for the Cone Penetration Test (CPT).

It appeared to be impossible to arrive at a final agreement on the draft reports during the XIth ICSMFE at San Francisco in 1985.

The actual president, Prof. Broms decided that the work should proceed. A new committee was nominated under the chairmanship of Mr. Bergdahl. The terms of reference are as follows :

1. To present reference test procedures for the different penetration methods ; in the chapter concerning the CPT the piezocone should be included.
2. The tentative drafts should be pre-

sented for a general discussion at the first International Symposium on Penetration Testing in Florida in 1988.

3. On the basis of comments made at ISOPT I a final draft will be made, submitted in due time to all member societies. This final report will then be presented for agreement to the XIIth ICSMFE to be held in Rio de Janeiro in 1989.

Already before the San Francisco Conference a team has worked out a preliminary draft for the Reference CPT Procedure.

In course of time this preliminary draft has been commented by the different members of the Committee on penetration testing and also by some of the member societies. Consequently several objections and proposals of modifications have been made. Step by step three tentative drafts have been presented to the members of the Committee.

After the San Francisco Conference the Working Party has drafted the 4th draft trying to fullfill the following conditions :

1°) respect as much as possible the text of the European subcommittee (Tokyo 1977).

2°) include the piezo-cone.

3°) describe a Reference Procedure but not as an obligatory method.

This 4th draft has been submitted to all members of the new committee appointed by the president Prof. Broms. The suggestions received before 15th September 1987, have again been analysed. Finally this has resulted in the 5th tentative draft, which is included in this report.

This draft will be supplied to the attendants of ISOPT I in time so that they can present comments and suggestions at the relevant session of this symposium.

After ISOPT I the Committee on Penetration Testing will finalise the document concerning the Reference Procedure. It will then be sent to the member societies. It will be submitted for approval to the meeting of the Executive Committee of ISSMFE at its meeting in Rio de Janeiro 1989.

I.2 The main objectives of establishing a Reference Test Procedure for the Cone Penetration Test (CPT) are as follows :

- Improvement of the quality of the exchange of information about investigations in the field of geotechnics between the members of ISSMFE.
- Provide a harmonised system of testing methods which will be beneficial for all parties involved in projects in the field of geotechnics.
- Prevention of more CPT-methods which deviate from the reference method.
- Provide a basis for further standardization on an international level of field testing techniques in geotechnical engineering (e.g. ISO).

The first item is of vital importance for a proper understanding of the results of studies and tests between research workers in the various countries. The standardized CPT-procedure offers an excellent basis of comparison in this respect.

The second item is aimed at the improvement of the understanding between parties involved in the building process. Reference to the Reference procedure - or to special deviating methods - in terms of reference and tender documents, will certainly be of great value.

The third item is important because the acceptance of a reference procedure and a limited number of deviations certainly will help to prevent the appearance of new deviating CPT-methods on the market.

The fourth item is connected with the widespread tendency to obtain standardization on an international basis. The ISO Technical Committee TC 182 "Geotechnics" is already active in this respect. The same applies to the Eurocode system of the European Community.

ISSMFE Reference Procedure will certainly be a good basis for this international standardization activity.

In the meeting of the Committee on Penetration Testing in Helsinki in May, 1983 the decision was taken to base the international CPT reference procedure on the already issued European Reference Procedure. The Working Party was also asked to consider the possibility of including the piezo-cone in the international reference procedure and to investigate whether deviating methods could be left out of the document.

To obtain a better idea about opinions and current practice with respect to CPT-testing in the various countries, the Working Party has sent out a questionnaire to the members of the former ISSMFE subcommittee on Penetration Testing, in the beginning of 1983.

II. REVIEW OF CURRENT PRACTICE

Practice with CPT differs considerably between the various countries in which the method is used.

In some countries, like Belgium and the

Netherlands, results of CPT's, expressed in terms of cone resistance and local friction, are applied directly for the prediction of bearing capacity of piles, foundation pads and other types of geotechnical structures. The CPT results frequently are also used for checking the effect of compaction works and other soil improvement methods.

Besides such applications it is also quite common to transform cone resistance values into basic soil parameters, like angle of internal friction of sand, undrained shear strength of cohesive soils and deformation parameters (Young's modulus and modulus of subgrade reaction).

Much emphasis is also laid on the assessment of composition and stratification of the ground on the basis of the so-called friction ratio, which is the ratio between local friction and cone resistance.

In most of the other countries emphasis is laid on the last types of applications. Relative density is derived from empirically established relations between the course of cone resistance with depth in sands and the relative density. This type of application appears to become more and more important for the evaluation of the risk of liquefaction under the influence of ground motions evoked by earthquakes.

The appearance of the piezo-cone has opened many new perspectives for the CPT method. In many countries extensive research programmes are going on aimed at a still better interpretation of CPT results.

To obtain adequate and up-to-date information concerning the CPT practice in the various countries a questionnaire was sent out, early 1983, to the members of the committee on penetration testing.

Their answers are summarized in table 1. The numbers of the columns refer to the following questions :

1. What is the number of CPT's yearly performed in your country
2. What type of equipment is used
3. If several types of equipment are used, what number of CPT's is performed with each type.
4. What is the relation between the number of CPT's and the volume of other types of soil investigation in your country.
5. Is there a reference standard for CPT, is such standard under discussion or is the European standard used for this purpose.
6. What is your opinion on the suggestion of the Steering Committee to leave out the cone types deviating from the European Standard.
7. Do you agree with the opinion of the Steering Committee to consider all penetrometer tips with a diameter in excess of 35.7-36.0 mm as deviations from the Standard.
8. Should friction sleeve geometry be as given in figure 1, page 101, Volume 3 of the Proceedings of the Conference of ISSMFE, Tokyo 1977.
9. Piezo-cone.

a) Geometry of piezo-cone ; place of measuring water pressures
b) Should the piezo-cone be considered as a Standard
c) What requirements should be imposed regarding the stiffness of the piezometer.

10. Should electrical equipment be more standardized and should stiffness requirements be included.
11. Do you have any further suggestions.

The answers give a rather clear picture of the current practice with the CPT method throughout the world. - They also show that in addition to the standard procedure deviating penetrometer tips are applied frequently, in particular the dutch mechanical cone types.

The viewpoints with respect to restricting the International Recommended Reference CPT method differ considerably. In countries where the CPT-method came in use recently and practice is still limited, preference is given to exclude all deviating penetrometer tips from the document. Countries or regions where the CPT method is used intensively since several decades and many thousand CPT's are performed yearly, emphasize that, apart from the reference testing procedure, also those deviating methods should be included which can boast on a long and extensive practice. Long practical experience in Geotechnical engineering is often based on the results of tests with the mechanical cone types.

It is therefore reasonable that cone types included in the European Standard should only be deleted when the country in which they are extensively used makes a proposal in this respect.

However no such proposals were brought forward. Neither proposals were received from countries outside Europe to include new cone types in the document on the International Recommended CPT procedure.

There are certainly more reasons for maintaining deviating methods in the document. Some of them are already mentioned earlier.

But also the following considerations are of importance :

a) Electrical cones are more vulnerable

Table 1 : Compilation of answers to the CPT questionnaire of 4.02.83.

Reporter \ Question	1	2	3	4	5	6	7	8	9	10	11
Jamiolkowski Italy	1500-2000	S M2	S - 400 M2-1500	very important	no	all excl. M2	tip ø 35.7 mm	agree discussion on location	S-type piezo-cone no friction sleeve	no	-
Melzer Germany	1.500 (30.000m)	S-E2 (M1/M2 ?)	S-1000 E2-500	equal to DPT or somewhat less	DIN 4094	all excl. M2-E2	tip ø 35.7 mm	agree	to early to standardize	no	-
Muromachi Japan	?	M1/M2	?	approx. 2 %	JIS-A1220-1976	no comment	tip ø 35.7 mm	agree	be careful with standardization	no	-
Trofimenkov USSR	50.000	M3 E1 A3-350	M3-45000 E1 A3-350 : - 5000	very important	TOCT 20 069-81	exclude M5.1 ; M5.2 ; H1.1 ; H1.2	tip ø 35.7 mm	agree	no experience	no	-
Bergdahl Sweden	500	S	S : 500	3rd most important	yes	all, except S and M	tip ø 35.7 mm	in case of piezo-cone friction sleeve has to be more upwards	location of filter directly above cone (see remark point 8)	no	-
Holden Australia	2000-5000	S M1/M2/M4	mostly S	5 % of all soil investig.	yes	all, best gradually	tip ø 35.7 mm	agree	S-geometry filter directly above cone base ; height filter 2.5 mm	no	special attention to gap-sealing
Amar France	approx. 700	M1 E1.1 ø 45 E1.1 ø 35.7	M1 : 260 E1 35.7:80 E1 45 : 340	20 %	Eur. Standard	none	tip ø 35.7 mm allow : 45 mm	agree	use only E1 ø 45 with piezo-element above cone base	no	-
Parez France	?	H1.1 ; H1.2 M1 M5.1 : M5.2	?	30 %	Eur. Standard	exclude M1, M2 and M3	tip ø 35.7 mm allow : 44 mm	agree	only with E1 ø 45: or S without friction sleeve : stiffness to be described	no	-

Question / Reporter	1	2	3	4	5	6	7	8	9	10	11
Sanglerat France	2000	M5.2/M4	M5.2-1300 M4 - 700	important 30 %	Eur. Standard	none	tip ø 35.7 mm	agree	still in research phase	no	-
Goelen Belgium	10.000	M1/M2/M4 S /E1.2	M1 - 5000 M4 - 3000 M2 - 2000	60 %	no	none	tip ø 35.7 mm	agree	no objection against standardization no experience	no	for normal use apply M1/M2 and M4
Joustra/ Heijnen Netherlands	80.000	S M1/M2	S - 20.000 M1/M2 - 60.000	90 %	NEN 3680	all except M1 ; M2 ; M4	tip ø 35.7 mm	enlarge e_2	location : directly above cone: with friction sleeve e_2 to be increased	no	-
Stefanoff Bulgaria	300-500	M1	M1 : 300-500	10-15 %	BDS-1002 . 82	excl. M4 (?)	tip ø 35.7 mm	agree	no experience	no	-
Thorburn/ B.G.Soc. U.K.	3000	S1 M1/M2 E1.1 ; E1.2	S : 2400 M1 : 150 M2 : 300 E1.1/E1.2: 150	increasing	no Eur. Standard	all	tip ø 35.7 mm	agree	no standardization of piezo-cone	no	-
Jones South Africa	500	M2 S (piezo-cone)	M2 - 400 S(piezo) - 100	increasing	no Eur. Standard	excl. : M4	tip ø 35.7 mm	agree	S-type : filter directly above cone	yes better indications should be given	-
Campanella Canada	increasing	S M2	?	increasing	no ASTM D 3441-79	excl. all div. except M2	tip ø 35.7 mm	increase e_1 ; e_2 because of filter	location directly above cone base e_1/e_2 to be increased S-type	no	include inclino-meter in S-type.

The symbols S, M, E and H with numbers refer to the abbreviations for the various penetrometer tips applied in the European Recommended Standard for the Cone Penetration Test.

to defects than mechanical types. Good back service is of paramount importance.

b) In continuous testing the total capacity of the thrust machine is more rapidly reached than in discontinuous testing.

c) Electrical cones are much more expensive than the mechanical cones.

d) In continuous testing with the standard penetrometer tip the cone resistance is influenced by the side friction along the part of the shaft above the cone. The effect strongly depends on the ambient soil conditions.

e) Operating an electrical cone on a spot not accessible to a truck is often not possible.

In consequence of these considerations the Working Party has chosen in consonance with the wishes of the ISSMFE to base the International Recommended Procedure on the already existing European Recommended Standard Procedure for Cone Penetration Testing, including also a section concerning deviations.

The opinion of the respondants about the piezo-cone was generally not to standardize the piezo-cone now. Many research programmes are in progress and no definite conclusions can be drawn yet with regard to the location of the filter element and other important features. In line with the wishes of the ISSMFE presidency and the members of the Committee on Penetration Testing, the Working Party has choosen to include the piezo-cone in the international document, without further details about location and size of filter, stiffness of the measuring system and so on.

III. REFERENCE TEST PROCEDURE FOR THE CONE PENETRATION TEST (CPT)

Contents

11. EXPLANATORY NOTES AND COMMENTS

1. SCOPE

The cone penetration test consists of pushing into the soil, at a sufficiently slow rate, a series of cylindrical rods with a cone at the base, and measuring in a continuous manner or at selected depth intervals the penetration resistance on the cone and, if required, the total penetration resistance and/or the local side friction resistance on a friction sleeve. In addition, the pore-water pressure present at the interface between penetrometer tip and soil can be measured during penetration by means of a pressure sensor in the cone. This pore-water pressure includes the pore-water pressure increase or decrease due to compression and dilation of the saturated soil around the cone arising from the penetration of the cone and the push rods into the ground.

Cone penetration tests are performed in order to obtain data on one or more of the following subjects :

1. the stratigraphy of the layers, and their homogeneity over the site,
2. the depth to firm layers, the location of cavities, voids and other discontinuities,
3. soil identification,
4. mechanical soil characteristics,
5. driveability and bearing capacity of piles.

2. DEFINITIONS

2.1 CPT stands for Cone Penetration Test and includes what has been variously called Static Penetration Test, Quasi-Static Penetration Test and Dutch Sounding Test.

2.2 Penetrometer : an apparatus consisting of a series of cylindrical rods with a terminal body, called the penetrometer tip, and the measuring devices for the determination of the cone resistance, the local side friction resistance, the total resistance, and/or the pore-water pressure present in the immediate vicinity of the cone during penetration.

2.3 Penetrometer tip : the terminal body with a length of 1000 mm at the end of the series of push rods, which comprises the active elements that sense the cone resistance, the local side-friction resistance and/or the pore-water pressure present at the interface of cone and soil during penetration.

2.3.1 Shaft : the cylindrical part of the penetrometer tip above the cone and/or the friction sleeve.

2.4 Cone : the cone-shaped end piece of the penetrometer tip on which the end bearing is developed.

2.4.1 According to the degree of freedom of the cone the following types are defined :
- fixed cone penetrometer tip : where the cone can only be subjected to micro-displacements relative to the other elements of the tip.
- free cone penetrometer tip : where the cone can move freely with respect to the other elements of the tip.

2.4.2 According to the shape of the cone the following are defined :
- Simple cone : a cone having a cylindrical extension above the conical part with a length considerably smaller than the diameter of the cone.
- Mantle cone : a cone which is extended with a fixed more or less cylindrical sleeve of which the diameter is smaller than the diameter of the cone and with a length 1 to 3 times the cone diameter : this sleeve is called the mantle.

2.4.3 Piezo-cone : a cone with a filter inserted in, or in the immediate vicinity of the cone, to measure the pore-water pressure present in the soil during penetration by means of a pore-water pressure sensor.

2.5 Friction sleeve : the section of the penetrometer tip upon which the local side friction resistance to be measured is developed.

2.6 System of measurement : the system includes the measuring devices themselves and the means of transmitting information from the tip to where it can be seen or recorded. For example, the following can be defined :

2.6.1 Electric penetrometer : which uses electrical devices such as strain gauges, vibrating wires, etc. built into the tip.

2.6.2 Mechanical penetrometer : which uses a set of inner rods to operate the penetrometer tip.

2.6.3 Hydraulic and pneumatic penetrometer : which uses hydraulic or pneumatic devices built into the tip.

2.7 Push rods : the thick-walled tubes or solid rods, preferably with a length of 1 metre, used for advancing the penetrometer tip.

2.8 Inner rods : solid rods which slide inside the push rods to extend the tip of a mechanical penetrometer.

2.9 Thrust machine : the equipment that pushes the penetrometer into the soil. The necessary reaction for this machine is obtained by dead weight and/or anchors.

2.10 Friction reducer : narrow local protuberances outside the push rod surface, placed above the penetrometer tip, and provided to reduce the total friction on the push rods.

2.11 Continuous and discontinuous tests (see Note 1).

2.11.1 Continuous penetration test : a penetration test in which the cone resistance is measured, while all elements of the penetrometer have the same rate of penetration (see Note 1).

2.11.2 Discontinuous penetration test : a penetration test in which the cone resistance is measured, while the other parts of the penetrometer tip remain stationary. When a friction sleeve is also included, the sum of the cone resistance and local side friction resistance is measured when both cone and friction sleeve are pushed down together, while the other parts of the penetrometer tip remain stationary.

2.12 Cone resistance q_c : the cone resistance q_c is obtained by dividing the ultimate axial force acting on the cone Q_c by the area of the base of the cone A_c

$$q_c = Q_c : A_c$$

This resistance is expressed in MPa or kPa.

2.13 Local unit side friction resistance f_s : the local unit side friction resistance f_s is obtained by dividing the ultimate frictional force Q_s acting on the sleeve, by its surface area, A_s

$$f_s = Q_s : A_s$$

The local unit side friction resistance f_s is expressed in Pa, kPa or MPa.

2.14 Total force Q_t : the force needed to push cone and push rods together into the soil. Q_t is expressed in kN.

2.15 Total side friction resistance Q_{st}: this is generally obtained by subtracting the ultimate force on the cone Q_c from the total force Q_t

$$Q_{st} = Q_t - Q_c$$

Q_{st} is expressed in kN, as are Q_t and Q_c.

Certain penetrometers allow Q_{st} to be measured directly.

2.16 Friction Ratio R_f and Friction Index I_f (see Note 2)

2.16.1 Friction Ratio R_f : the ratio of the local unit side friction resistance f_s to the cone resistance q_c measured at the same depth, expressed as a percentage.

2.16.2 Friction Index I_f : the ratio of the cone resistance q_c to the local unit side friction resistance f_s measured at the same depth.

3. REFERENCE TEST PENETROMETER AND EQUIPMENT

3.1 General geometry of the penetrometer tip : in the reference penetration test, penetrometer tips with or without a friction sleeve and with or without a pore-water pressure meter may be used. If a gap between the cone and the other elements of the penetrometer tip exists, it shall be kept to the minimum necessary for the operation of the sensing devices and designed and constructed in such a way as to prevent the entry of particles (see Note 3). This shall also apply to the gaps at either end of the friction sleeve, if one is included, and to the other elements of the penetrometer tip. The axes of the cone, the friction sleeve, if included, and the body of the penetrometer tip shall be coincident.

The diameter of the shaft of the penetrometer tip shall nowhere be less than 0.3 mm smaller nor more than 1 mm greater than the nominal diameter of 35.7 mm of the reference cone.

In addition, in the case of a penetrometer tip with a friction sleeve, no part of the penetrometer tip shall project beyond the sleeve diameter.

An example of a reference penetrometer is presented in Fig. 1 (a) and 1 (b).

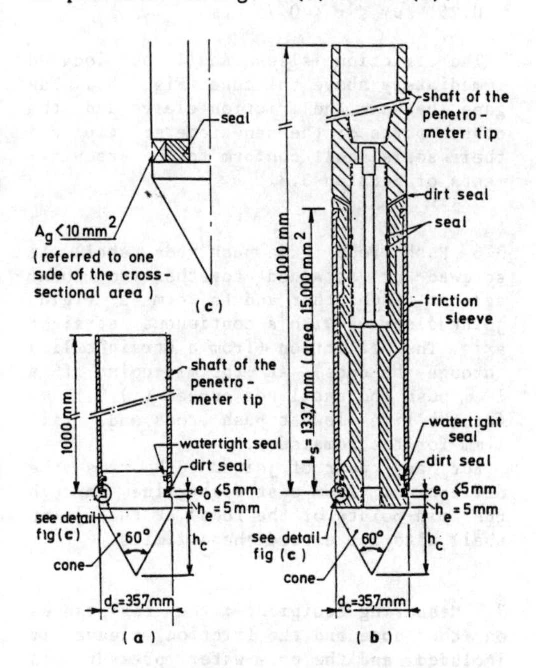

Fig.1 - Example of a reference penetrometer with a fixed cone and (a) without or (b) with friction sleeve. Detail of gap : (c).

3.2 Cone : The cone shall consist of a conical part and a cylindrical extension (Fig. 2) ; the apex angle of the cone shall be 60°.

For the reference cone, when there is no filter element, the length h_e of the cylindrical part directly above the conical part shall not exceed 5 mm.

In case of a piezo-cone and when the filter element is located directly above the conical part, than the length of the cylindrical extension must be sufficient to contain the filter element but in any way shall not exceed 15 mm (Fig. 3).

The area A_c of the base of the cone shall be 1000 mm^2 giving a cone diameter of 35.7 mm. Cone diameter is defined as the diameter of the cylindrical extension.

The surface roughness in the longitudinal direction of the cone shall not exceed 1 m which is equivalent to the roughness produced by the friction of the soil.

Tolerance on the dimensions (see Note 4) :

(a) On the area of the base of the cone A_c

$$A_c = 1000\ mm^2 - 5\ \%, + 2\ \%$$

giving a cone diameter d_c

$$34.8\ mm \leq d_c \leq 36.0\ mm$$

For a worn tip, the cone diameter shall be measured across the top section of the cylindrical extension.

(b) on the height h_c of the conical part of the cone

$$24.0\ mm \leq h_c \leq 31.2\ mm$$

(c) on the height h_e of the cylindrical extension :

$$2\ mm \leq h_e \leq 5\ mm$$

If a filter element is placed in the cylindrical extension :

$h_e \leq 15$ mm and wear on h_e shall not exceed 3 mm

Cones with a visible asymmetrical wear are to be rejected.

3.3 Gap and seal above the cone (Fig. 1) : the gap between the cone and the other elements of the penetrometer shall not be greater than 5 mm.

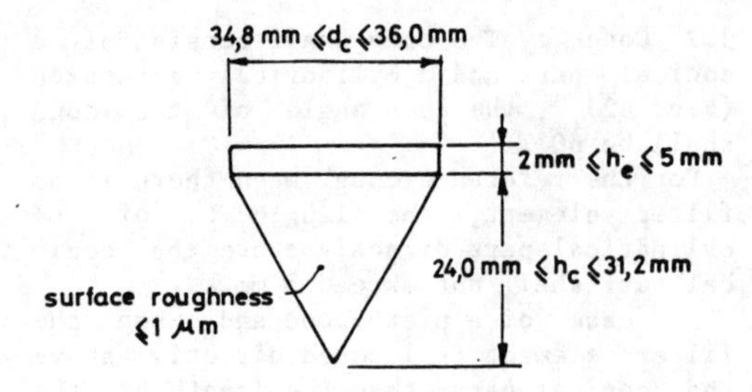

Fig.2 - Tolerances on the dimensions of the reference cone (without filter element in the cylindrical extension).

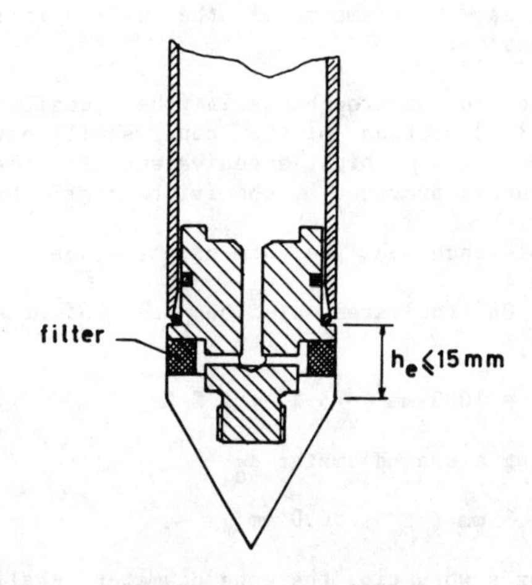

Fig.3 - Piezo-cone with a filter element in the cylindrical extension.

The outer limits of the gap shall be shaped in such a way that the measurements cannot be affected by possible bridging by soil particles.

The seal placed in the gap shall be properly designed and manufactured to prevent the entry of soil particles into the penetrometer tip. It shall have a deformability many times larger than that of the sensing devices inside the tip. The cross-sectional area A_g of the gap, remaining after deduction of the area occupied by the seal, shall be less than 10 mm^2 (Fig. 1).

3.4 Sensing devices : the sensing devices for measuring the cone resistance and the friction resistance shall be designed in such a way that eccentricity of these resistances cannot influence the readings.

The sensing device for the local friction resistance shall operate in such a way that only the shear stresses, and not normal stresses, acting on the friction sleeve are recorded.

3.5 Friction sleeve (Fig. 1b) : the diameter of the friction sleeve shall not be less than the actual diameter of the base of the cone. The surface area of the friction sleeve shall be 1.5 x 10^4 mm^2.

Tolerances on the dimensions :
(a) on the diameter d_s of the friction sleeve

$$d_c \leqslant d_s \leqslant d_c + 0.35 \text{ mm}$$

d_c being the actual diameter of the base of the cone.
(b) on the surface area A_s of the friction sleeve

$$A_s = 1.5\text{x}10^4 \text{ mm}^2 \text{ +2 \%, -2 \%, i.e.}$$

$$1.47\text{x}10^4\text{mm}^2 \leqslant A_s \leqslant 1.53\text{x}10^4\text{mm}^2$$

on the surface roughness r of the friction sleeve in the direction of the longitudinal axis (see Note 5)

$$0.25 \quad \mu\text{m} \leqslant r \leqslant 0.75 \quad \mu\text{m}$$

The friction sleeve shall be located immediately above the cone (Fig. 1b). The gaps between the friction sleeve and the other parts of the penetrometer tip and their seals shall conform to the requirements of Section 3.3.

3.6 Push rods : the push rods shall be screwed or attached together to bear against each other and to form a rigid-jointed series with a continuous straight axis. The deflection (from a straight line through the ends) at the mid-point of a 1 m push rod shall not exceed (i) 0.5 mm for the five lowest push rods and (ii) 1 mm for the remainder.

For any pair of joined push rods the deflection (from a straight line through the mid-points of the rods) at the joint shall also not exceed these limits.

3.7 Measuring equipment : the resistances on the cone and the friction sleeve if included and the pore-water pressure in case of a piezo-cone shall be measured by suitable devices and the signals transmitted by a suitable method to a data recording system.

Recording test data exclusively on a tape, which does not permit direct accessibility during the test, is not recommended.

3.8 Thrust machine : the machine shall have a stroke of at least one metre, and shall push the rods into the soil at a constant rate of penetration. The thrust machine shall be anchored and/or ballasted such that it does not move relative to the soil surface during the pushing action.

3.9 Friction reducer : if a friction reducer is included, it shall be located at least 1000 mm above the base of the cone.

4. TESTING PROCEDURE

4.1 Continuous test : the testing procedure shall be that of continuous penetration testing, in which the measurements are made while all elements of the penetrometer tip have the same rate of penetration.

4.2 Verticality : the thrust machine shall be set up so as to obtain a thrust direction as near vertical as practicable. The deviation from vertical of the thrust direction shall not exceed 2 %. The axis of the push rods shall coincide with the thrust direction.

4.3 Rate of penetration : the rate of penetration shall be 20 mm/sec with a tolerance of ±5 mm/sec. In case of a piezo-cone this tolerance shall be narrowed. Between these tolerances a constant rate shall be maintained during the entire stroke, even if readings are taken only at intervals.

4.4 Interval of readings : a continuous reading is recommended. In no case shall the interval between the readings be more than 0.2 m.

4.5 Measurement of the depth : the depths shall be measured with an accuracy of at least 0.1 m.

5. PRECISION OF THE MEASUREMENTS

Taking into account all possible sources of error (parasitic frictions, errors of the measuring devices, eccentricity of the load on the cone with respect to the sleeve, temperature effects, etc...), the precision of measurement shall not be worse than the following whichever is the greater

5 % of the measured value

1 % of the maximum value of the measured resistance in the layer under consideration.

The precision shall be verified in the laboratory or in the field taking into account all possible disturbing influences (see Note 6).

6. PRECAUTIONS, CHECKS AND VERIFICATIONS

6.1 Straightness of push rods : regular checks shall be made on the straightness of the push rods and their joints, particularly for the lowest five rods of the series (see par. 3.6 and Note 7).

6.2 Wear : regular inspections shall be made for wear of the cone, friction sleeve and shaft of the penetrometer tip.

6.3 Distance to other tests : the CPT shall not be performed too close to existing boreholes or other penetration tests. It is recommended that reference deep CPT's shall not be performed closer than 25 boring diameters from boreholes, or at least 2 m from previously performed CPT's (see Note 8).

6.4 Seals : the seals between the different elements of a penetrometer tip shall be regularly inspected to determine their condition. Prior to use, the seals shall be checked for the presence of soil particles and cleaned.

6.5 Temperature compensation : electric penetrometer tips shall be temperature compensated. If the shift observed after extracting the tip is so large that the precision defined under Section 5 is no longer met, the test shall be discarded.

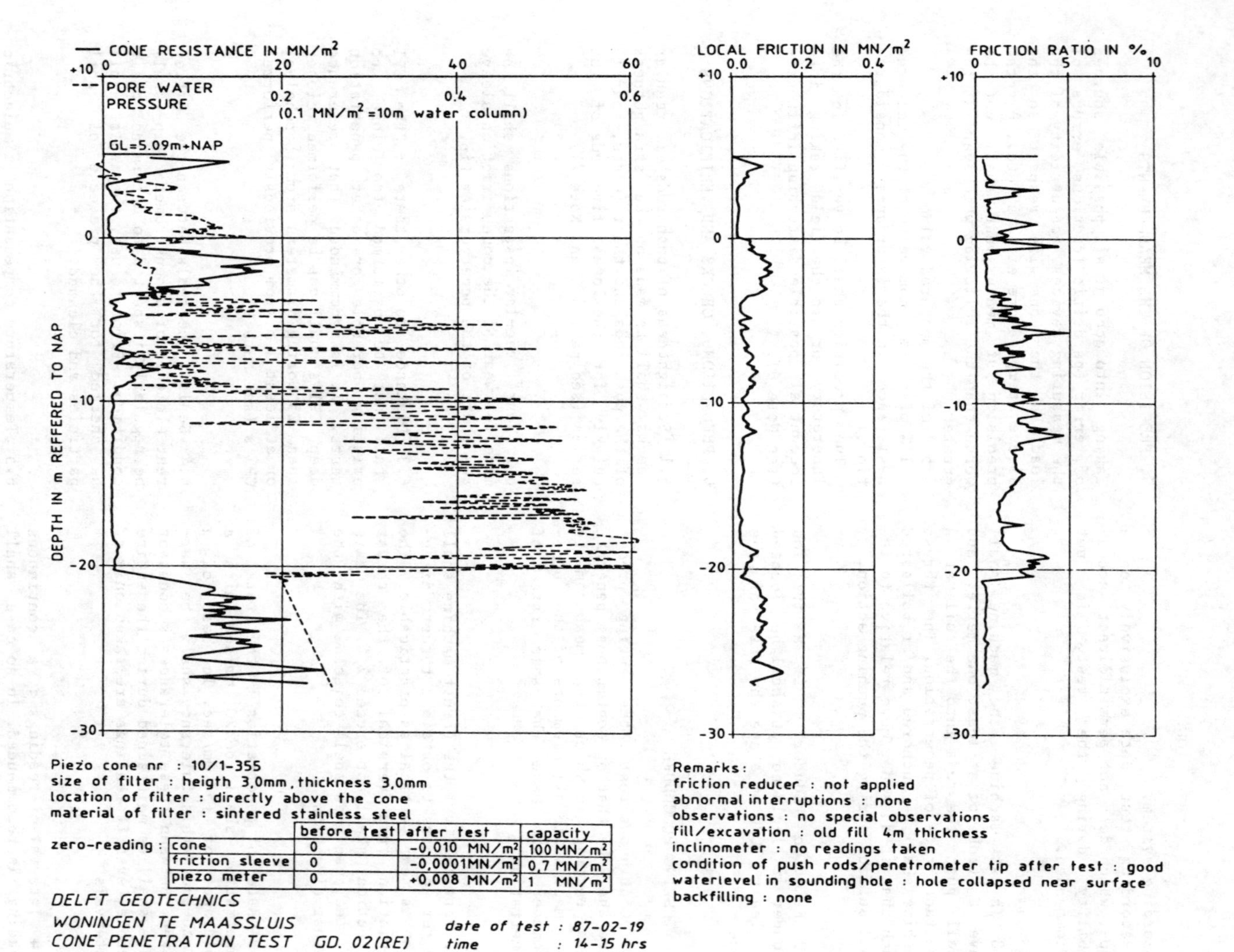

Piezo cone nr : 10/1-355
size of filter : heigth 3,0mm, thickness 3,0mm
location of filter : directly above the cone
material of filter : sintered stainless steel

zero-reading :	before test	after test	capacity
cone	0	-0,010 MN/m²	100 MN/m²
friction sleeve	0	-0,0001MN/m²	0,7 MN/m²
piezo meter	0	+0,008 MN/m²	1 MN/m²

DELFT GEOTECHNICS
WONINGEN TE MAASSLUIS
CONE PENETRATION TEST GD. 02(RE)

date of test : 87-02-19
time : 14-15 hrs

Remarks:
friction reducer : not applied
abnormal interruptions : none
observations : no special observations
fill/excavation : old fill 4m thickness
inclinometer : no readings taken
condition of push rods/penetrometer tip after test : good
waterlevel in sounding hole : hole collapsed near surface
backfilling : none

Fig. 4 - Example of the presentation of the test results on a graph.

7. CALIBRATION

7.1 Manometers : manometers shall be calibrated at least every 6 months.

For each type of manometer, there shall be two identical units, each with its own calibration, available with the machine. At regular intervals, the manometer used in the tests shall be checked against the reserve manometer.

7.2 Load cells/proving rings : load cells or proving rings shall be calibrated at least every 3 months.

Regular checks on the site with an appropriate field control unit are recommended.

8. SPECIAL FEATURES

8.1 Push rod guides : in order to prevent buckling, guides shall be provided for the part of the push rods protruding above the soil and for the push rod length in water.

8.2 Inclinometers : in order to obtain more precise information on the drift of the push rods in the soil, inclinometers may be built into the penetrometer tip.

The need for such information depends on the soil conditions, and increases with increasing depth of the test.

8.3 Push rods with smaller diameters : in order to decrease the skin friction on the rods, use may be made of push rods with a smaller diameter than that of the penetrometer tip. The distance between the smaller diameter push rods and the cone base shall be at least 1000 mm.

8.4 Piezo-cone : the tip may be equipped with a pore-water pressure sensing device connected to a filter placed in, or in the immediate vicinity of, the cone (see Note 9). The filter and all fluid spaces of the piezo-cone shall be filled with water or another suitable fluid and thoroughly deaired before each set of tests. Precautions shall be taken to maintain full saturation of the filter and the other spaces of the measuring system during transport to the test site and during the execution of the piezo-cone penetration tests, especially when the upper layers of the ground are not saturated.

9. REPORTING OF RESULTS

9.1 The following information shall be reported on the graphs of the test results :

1. Where the penetrometer and the test procedure are completely in agreement with the reference procedure each graph shall be marked with the letter R (Reference test procedure). One of the following letters shall be added to indicate the system of measurement :

M = mechanical
E = electrical
H = hydraulic

The capacities of the different measuring devices of the penetrometer tip shall be reported.

2. The date and time of the test and the name of the firm.

3. The identification of the CPT and the location of the site.

4. The depth over which a friction reducer, or push rods with a reduced diameter, has been used. The depth at which the push rods have been partly withdrawn in order to reduce the side friction resistance, and thus achieve greater penetration.

5. Any abnormal interruption of the reference procedure of the CPT even as all interruptions in case of a piezo-cone.

6. Observations made by the operator such as soil type, sounds from the push rods, indications of stones, disturbances, etc.

7. Data concerning the existence and thickness of fill, or existence and depth of an excavation, and starting level of the CPT with respect to the original or modified soil surface.

8. The elevation of the soil surface at the location of the test.

9. If a piezo-cone is used, the position, the size and the material of the filter shall be clearly indicated.

10. The readings of the inclinometer, if taken.

11. The zero readings of all sensors before and after the test.

12. All checks made after extracting the push rods, the condition of the push rods and the penetrometer tip.

13. The depth of the water in the hole remaining after withdrawal of the penetrometer, or the depth at which the hole collapsed.

14. Whether or not the test-hole has been backfilled and, if so, by which method.

9.2 Besides the information indicated in Section 9.1, the internal files shall also record :

1. The identification of the penetrometer tip used.
2. The name of the operator in charge of the crew which performed the test.
3. The dates and reference numbers of the calibration certificates for the measuring devices.

9.3 The following scales are recommended for the presentation of the test results on graphs :

Depth scale : vertical axis
1 unit length (arbitrary) for 1 m

Cone resistance q_c : horizontal axis, the same unit length for 2 MPa.

Local side friction resistance f_s : horizontal axis, the same unit length for 50 kPa.

Total penetration force Q_t : horizontal axis, the same unit length for 5 kN.

Total friction Q_{st} : horizontal axis the same unit length for 5 kN.

Pore-water pressure : horizontal axis, the same unit length for 20 kPa.

Provided that the recommended relationship between the scales on the vertical and horizontal axes is maintained, the scales can be chosen arbitrarily in such a way that standard sheets can be used (see Fig. 4).

9.4 Site plan

For every investigation which is carried out, a clear site plan shall be drawn, with clear reference points in order that the locations of the penetrometer tests can be plotted accurately.
Also, when made in conjunction with borings, the time sequence of the borings and CPT's shall be indicated.

10. DEVIATIONS FROM THE REFERENCE TEST

10.1 General

All deviations from the Reference Procedure shall be described explicitely and completely on the graphs with the test results.
The different possible deviations are :

10.2 deviation of the cone dimension

10.3 deviation of the apex angle of the cone

10.4 deviation of the friction sleeve dimensions

10.5 deviation of the location of the friction sleeve

10.6 deviation related to the shape of the penetrometer tip

10.7 deviation related to the possibility of relative movements of the cone with respect to the push rods (so called free penetrometer tips).

10.2 Deviation of the cone dimension

In certain circumstances it may be necessary to deviate from the Reference Test Procedure by adapting smaller or larger cone diameters.
All tolerances specified for the Reference Test shall be adapted in direct proportion to the diameter.

10.3 Deviation of the apex angle

It may be necessary to deviate from the apex angle of the cone, smaller or larger than 60°.
Examples of penetrometers using cones with deviating diameters and deviating apex angle are given in Fig.5 and Fig.6, assuming for Fig. 6 that the test is run in a continuous testing manner.

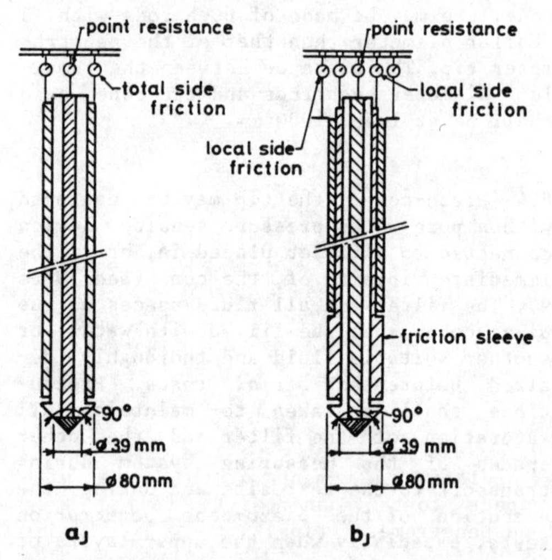

Fig.5-(a) the Andina cone penetrometer tip and (b) the friction sleeve cone penetrometer tip.

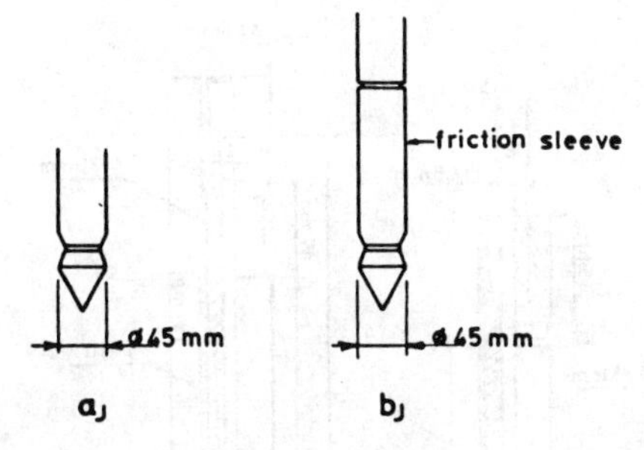

Fig.6-(a) the Parez hydraulic penetrometer tip and (b) the friction sleeve hydraulic penetrometer tip

10.4 Deviation of the friction sleeve dimensions

If the length of the friction sleeve of a penetrometer having a cone diameter of 35.7 mm deviates from that of the Reference tip then the surface area of the sleeve A_s should not be larger than 3.5×10^4 mm^2 and not smaller than 1×10^4 mm^2.

In the case of a cone diameter different from 35.7 mm the surface area of the sleeve shall be adjusted proportionally with respect to the cross sectional area of the cone.

10.5 Deviation of the location of the friction sleeve

The distance between the base of the cone and the lower end of the friction sleeve can be larger than that corresponding to the reference tip.

An example of penetrometer tip with a location of the sleeve which deviates from the reference is given in Fig. 7.

10.6 Deviation related to the shape of the penetrometer tip

In certain circumstances special electrical penetrometer tips with a shape deviating from the Reference Test can be used. An example is given in Fig. 8.

10.7 Deviation related to the possibility of relative movements of the cone with respect to the push rods (so called free cone penetrometer tips)

With a free cone penetrometer tip, testing in either continuous or discontinuous manner is possible : (see note 10).

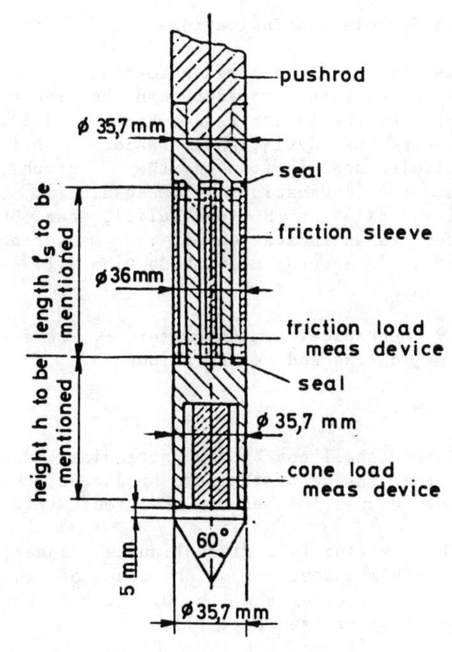

Fig.7 - The Degebo friction sleeve electric penetrometer tip.

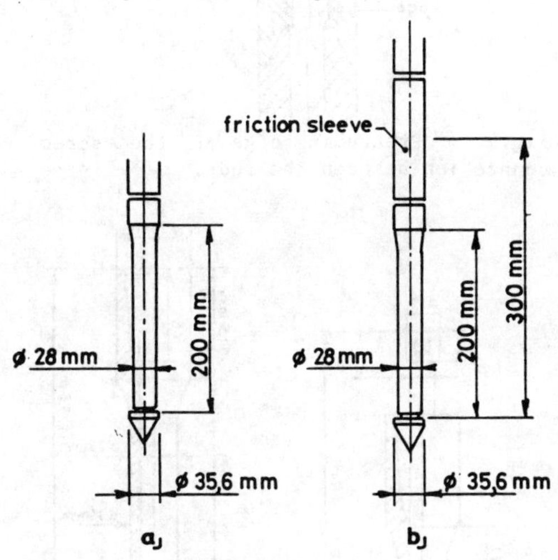

Fig.8 - (a) the Delft electric penetrometer tip and (b) the friction sleeve electric penetrometer tip.

In the case of a test run in a discontinuous manner, although the rate of downward movement due to the thrust machine is known, the rate of penetration of the cone at the point of rupture of the soil can be different from that of the movement due to the thrust machine. They correspond only when there is continuous downward movements of the push rods.

10.8 Symbols and indications

Only the test performed according to the Reference Test Procedure can be represented by the letter R (see Section 9.1). In case of deviations, besides their explicit description on the graphs, also the indications M (Mechanical), E (Electrical) or H (Hydraulic), can be added as an indication of the measuring system, in accordance with Section 9.1.

10.9 Mechanical penetrometer - Precautions, checks and verifications

Push rods

There shall not be any protruding edge on the inside of the push rods at the screw connection between the rods (Fig. 9).

When testing in a discontinuous manner, the minimum movement of the cone or the friction sleeve shall be 0.5 times the cone diameter (see Note 11).

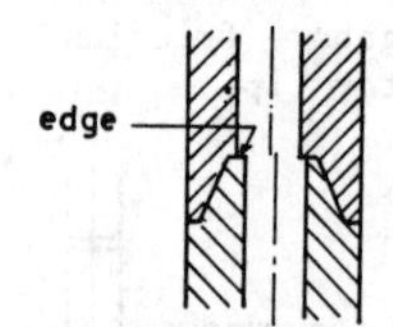

Fig.9 - Protruding edge at the screw connection between the rods.

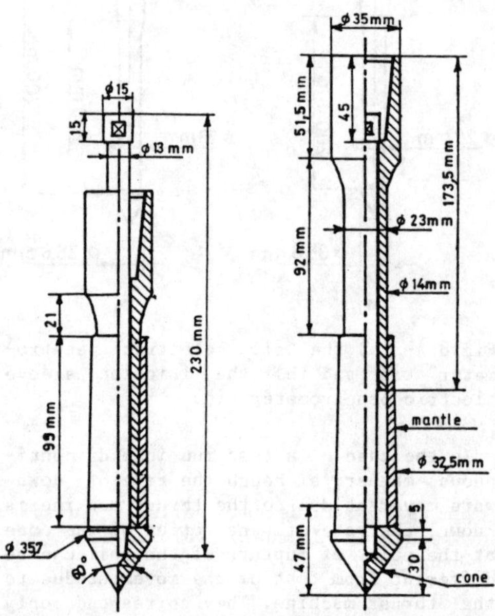

Fig.10-The Dutch mantle cone penetrometer tip.

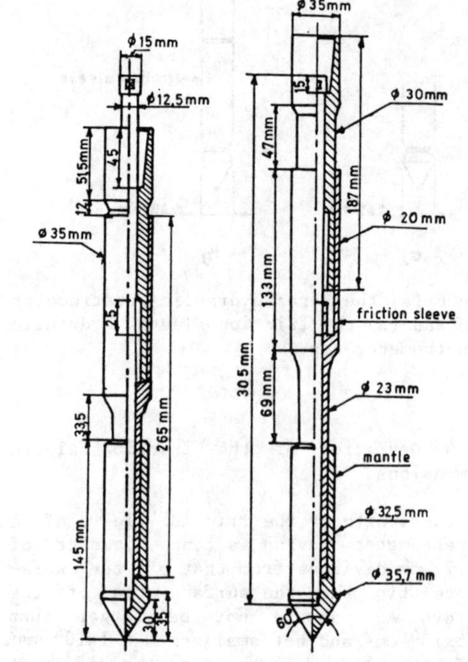

Fig.11-The Dutch friction sleeve penetrometer tip.

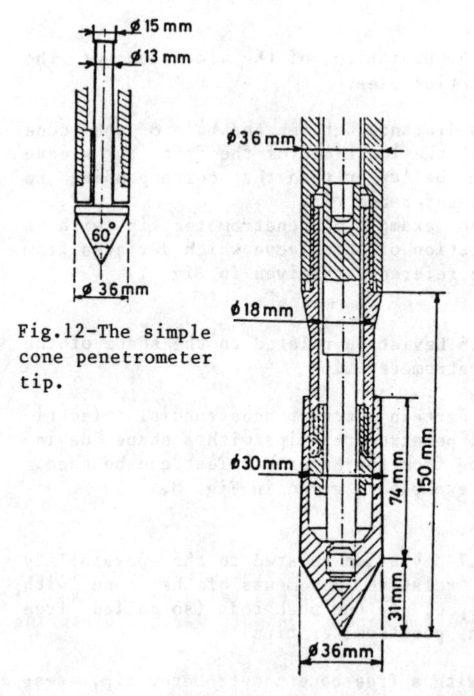

Fig.12-The simple cone penetrometer tip.

Fig.13-The U.S.S.R. mantle cone penetrometer tip.

Examples of free cone penetrometer tips in use in many countries are presented in Fig. 10, 11, 12,13 and 5 if used in a discontinuous testing manner (see Note 12).

Inner rods

The diameter of the inner rods shall be 0.5 to 1 mm less than the internal diameter of the push rods. The inner rods shall slide very easily through the push rods.

The ends of the inner rods shall be exactly at right angles to the axis of the rod and be machined to a smooth surface.

In order to maintain the degree of freedom of the inner rods, they shall not screw together or be joined in any way, as this has been found to increase the parasitic friction between the inner rods and the push rods. Before and after the test, a check shall be made that the inner rods slide very easily in the push rods, also that the cone and the friction sleeve move easily with respect to the body of the penetrometer tip. For improved accuracy at low values of the resistances, the thrust data registered at the surface shall be corrected for the total weight of the inner rods in the case of the cone resistance, and for that of the push rods and inner rods in the case of the total resistance.

10.10 Precision of the measurements : when testing deviates from the Reference Test two classes of precision are defined :

Normal precision : see Section 5

Lower precision : the precision obtained shall not be worse than the following whichever is the greater

10 % of the measured value

2 % of the maximum value of the measured resistance in the layer under consideration.

In all such cases, the class of precision of the tests shall be indicated in the report and on the test graphs.

10.11 Static/dynamic penetrometers and preboring cone penetrometers : penetration may be increased by the use of static/dynamic penetrometers, also by the use of penetrometers equipped with preboring tools.

It shall be clearly indicated in the report and on the test graphs when such equipment has been used.

11. EXPLANATORY NOTES AND COMMENTS

Note 1 : Definitions (2.11)

The terms "continuous" and "discontinuous" applied to penetration tests are not quite correct. The respective phrases "with simultaneous pushing of cone and push rods" and "with non-simultaneous pushing of cone and push rods" are more explicit. However, the terms "continuous" and "discontinuous" have been maintained, as they are already well established.

Note 2 : Definitions (2.16)

The friction ratio R_f (the ratio of the local side friction resistance f_s to the cone resistance q_c) shall be expressed as a percentage in order to obtain a figure larger than one. Although in the past the friction ratio has been mostly used, one may also use the friction index I_f(the ratio of the cone resistance q_c to the local side friction resistance f_s) which gives directly a figure larger than one.

Care shall be taken to calculate the friction ratio and/or the friction index for measurements of cone resistance and local side friction resistance at the same depth. Although this is clearly stated in the definition given in 2.16, attention is drawn to the fact that, because of the depth discrepancy between the cone and the friction sleeve, these parameters shall not be calculated from measurements made at the same time.

Note 3 : General geometry of the penetrometer tip (3.1),cone (3.2), gap and seal above the cone (3.3), and friction sleeve (3.5)

During penetration, water pressures act in opposite directions on the end areas of the cone. When these are unequal, this can result in a difference in the measured cone resistance which depends on the value of the water pressure. This holds also for the friction sleeve. To take account of this influence, the geometry can further be defined, namely by the area ratios as defined by R.G.Campanella, P.K. Robertson, D. Gillespie : Cone penetration testing in deltaic soils. - Canadian Geotechnical Journal, volume 20, Nol, February 1983, pp.23-35.
More data are needed to define quantitatively this influence.

Note 4 : Tolerance on the cone dimensions (3.2)

In the rare cases where a new cone is worn beyond the specified limits during a single sounding, the test results need not be discarded.

Note 5 : Friction sleeve (3.5)

The roughness is defined as the mean deviation of the real surface of a body from the mean plane. The roughness is expressed in micrometers (μm).

Note 6 : Possible errors (5, 10.10)

After completion of the test, a non-return to zero is sometimes noted. Such an error, when combined with the other possible errors, shall not exceed the precision limits quoted in Sections 5 and 10.10.

Note 7 : Checks (6.1)

As a minimum the push rods shall be checked after either a number of soundings, say 15 or 20, or after a total length of soundings, say 500m.

They shall be checked after every sounding in certain ground conditions that are known to be conducive to rod bending, for example :

(a) large depths of very soft soils over hard layers,

(b) residual soils containing floaters, and

(c) soils containing large gravels and boulders.

Note 8 : Distance to other tests (6.3)

When a CPT and a boring are to be executed close the one to the other it is recommended that whenever possible the CPT shall be executed before executing the boring. This has further the advantage that the levels where soil-specimens are to be taken can be selected according to the soil condition.

As a guidance : a deep CPT is a test which reaches to a depth of 10 m or more.

Note 9 : Piezo-cone penetrometer (8.4)

With the piezo-cone, the pore-water pressure present in the soil adjacent to the cone is measured continuously during penetration. The measured pore pressure consists of the sum of the pore pressure before penetration and the positive or negative pore pressure which is caused by tendency for soil compression and dilation due to the penetrating cone.

The experience and the state of knowledge regarding the piezo-cone is still rather limited. Therefore it seems premature to present recommendations regarding its features. Nevertheless, the subcommittee decided to give some general recommendations with regard to the location of the filter and some other important aspects, in order to obtain a reasonable state of uniformity in publications.

In selecting the location of the filter immediately above the base of the cone, in the cylindrical extension h_e (Fig.3), the majority of applications in practice was followed. However, this must be considered as a tentative recommendation. Due to the selection of this location, the recommended maximum length of h_e is increased to 15 mm. The influence of this change on the cone resistance is mostly small and it is possible to correct for this change.

The material of the filter shall be wear resistant. Its structure shall be such that blocking by soil particles is very unlikely. The device for the pore-water pressure measurement, which includes the sensor, the housing of the sensor, the filter and the connections, shall have very stiff characteristics because of the great influence on a) the transmission of the pore-water pressure from the soil to the sensor and b) the prevention of penetration of small soil particles into the filter.

Further developments have to be awaited.

Note 10 : Free cone penetrometer tips (10.7)

Continuous testing with a free cone penetrometer tip is not recommended for high accuracy, as the movement of the inner rods relative to the push rods can change its sense at different depths, increasing the margin of error due to the parasitic internal friction. Furthermore, it is necessary to check at least every metre during the test that the inner rods are still free to move relative to the push rods.

Note 11 : Discontinuous test (10.7)

In the case of mechanical penetrometer tips, in order to be certain that the cone and the friction sleeve move sufficiently with respect to the push rods, due account shall be taken of the elastic shortening of the inner rods. Therefore, at the surface the movement of the inner rods relative to the push rods shall be at least equal to the sum of the minimum imposed movement of the cone (see Section 10.9) plus the shortening of the inner rods.

Note 12 : Simple cone (10.7, Fig. 12)

In the case of the simple cone, special precautions shall be taken against soil entering the sliding mechanism and affecting the resistance measurements. After extracting the penetrometer tip, a check shall be made, in order to be certain that the cone stem still moves completely freely relative to the bush.

IV. DISCUSSION CONCERNING RECEIVED COMMENTS

1. Scope

A CPT test consists of the complete set of readings of cone resistance, mantle friction and pore-water pressure measured along a vertical line in the ground.

The abreviation CPT (cone penetration test) has been introduced in place of "Dutch sounding test". A sounding is a continuous observation along a given vertical. Consequently a CPT test is the whole of all measures of cone resistance, mantle friction, pore-water pressure, taken along a given vertical.

This clarification is necessary, because some define a CPT as one measurement on a given level.

2. Definitions

Some propose to limit the definition to the items related to the Reference testing procedure. In section II are given the practical reasons to give a more complete list of definitions, including also those belonging to the deviation.

2.14 and 2.15 Some propose to delete the definition of total force Q_t and total side friction resistance Q_{st}. The total force Q_t can easily be measured. Some consider it as a useful indication.

2.16.2. Although the friction ratio R_f is more used, the friction index I_f has the advantage not to be a fraction ; its use increases in some countries.

3. Reference test penetrometer and equipment

3.2. Cone : the heigh h_e of the cylindrical extension is limited to 5 mm when there is no filter element and 15 mm when a filter element is located in the cylindrical extension. Muromachi and Parez object because of the parasitic friction on the cylindrical extension the measured cone resistance q_c when using a cone with a cylindrical extension larger than 5 mm, which could be the case for a piezocone with a filter in the cylindrical part, will be larger than in the normal cone. However it can be stated that this difference will remain between acceptable limits. Some advocate to impose for normal cones and piezo-cones a uniform maximum length of the cylindrical extension of 10 mm.

3.5. Friction sleeve

The diameter of the friction sleeve shall always be at least the same as the actual diameter of the base of the cone. One shall ensure that the cone is never larger than the sleeve. Tests performed by Fugro (de Ruiter 1982) have shown that a sleeve diameter slightly less than the cone produced a marked reduction in friction resistance. Slightly increasing the sleeve diameter above the cone diameter did not significantly result in an increase of the measured friction.

8. Special Features

8.2. Inclinometers

Campanella insists that verticality should be measured wherever possible. The British Geotechnical Society asks that for reference testing the tip should be fitted with a device for measuring deviation from the vertical. However in order not to severely restrict practical applications, it is felt that such an obligation for all cases is too restrictive.

Electrical Equipment

Jones among others is of the opinion that the electrical equipment should be standardized and stiffness requirements included. The British Geotechnical Society and Jamiolkowski consider that there is no need to standardize the electrical equipment, as the developments in the electrical and electronic fields are very rapid. As furthermore the reference test procedure includes also the mechanical and hydraulic operated equipment, no specifications concerning the electrical equipment are included.

10. Deviations from the reference test

Among the cones deviating from the reference testing procedure, cones in frequent and standing use are the Dutch mantle cone penetrometer tip (fig. 10) ; the Dutch friction sleeve penetrometer tip (fig.11) ; the simple cone penetrometer tip (Belgium) (fig.12) the U.S.S.R. mantle cone penetrometer tip (fig.13), the Delft electric penetrometer tip and the friction sleeve electric penetrometer tip (fig.8), the Degebo friction sleeve electric penetrometer tip (Germany) (fig.7) ; the Andina cone and Andina friction sleeve cone penetrometer tip (fig.5), and the Parez hydraulic penetrometer tip and friction sleeve hydraulic penetrometer tip (fig.6). They are also included in the European recommendation (1977).

The CPT's realized with all these deviations outweigh the reference tests with a decuple.

Among them the Dutch mantle cone and Dutch friction sleeve penetrometer tip, (fig.10 and 11) are used in many countries, inside and outside Europe.

Because of their generalized use, most are in favour to refer to them in the deviations.

A general feeling is that the simple cone with closing nut (fig.12) is less reliable than the mantle cone, because of the possible friction caused by the introduction of soil particles between stem and closing nut.

Experience at the Belgian Geotechnical Institute for more than 30 years shows that when closing nut and stem are regularly checked and replaced the friction error can be minimized. Systematic comparisons between tests with simple cones and mantle cones show that the readings with the simple cone are always smaller than those with the mantle cone. The measurements with the mantle cones are burdened with an error due to the friction on the mantle, and which may increase the resistance with 30 % in stiff clays, and 10 % in sands (Parez,Muromachi). It appears that the possible errors due to a parasitic friction between stem and closing nut, in a test performed with a unit in good shape, is always smaller than the error introduced by the friction on the mantle of the mantle cone.

Consequently the readings with the simple cone with closing nut (fig.12) are more correct than those with the mantle cone (fig.10 and 11). As furthermore the simple cones (fig.12) even when regularly replaced, are cheaper than the mantle cones (fig.10 and 11), in Belgium preference is generally given to the use of the simple cones (fig.12). The simple cone with closing nut is furthermore the original shape, and is still used in the hand sounding equipment. As thousands of tests have been performed in the course of years with the simple cone with closing nut, a large experience has been gathered with this type of CPT. Consequently there are sufficiently reasons to mention also this cone type in the deviations.

The Piezo-cone

1. In the meetings of the Committee in Amsterdam 1982, and in San Francisco 1985, it was agreed that the piezo-cone should be included in the Reference Test Procedure.

As the piezo-cone technique is still under development, some advocate (Melzer, British Geotechnical Society) that it is premature to include the piezo-cone. However some (Jones) strongly insist on such an introduction.

2. There is not yet agreement concerning the geometry of the piezo-cone penetrometer tip to be adopted as reference. Several members strongly support the measuring element of the pore-water pressures being placed at the apex of the cone primarely for practical reasons (Jones). At the contrary Campanella and Thorburn recommend to locate the porous element in the cylindrical extension directly above the conical part of the cone. Jamiolkowski states that there is no unique optimum position of the porous stone.

3. Campanella feels that a 4 channel cone 1)bearing 2) friction 3) pore pressures 4) slope should be or become the standard reference electronic cone.

Jamiolkowski expresses the view that by advent of the piezo-cone the usefulness of the friction sleeve could be questioned. Indeed the measurement of penetration pore pressure gives a better indication of the soil types than the friction ratio, and furthermore the sleeve friction measurements are quite unreliable. Also other members (Jones) are of the opinion that when pore pressure data are available, the need for a friction sleeve is very much reduced.

4. Some suggest that in the Reference Test Procedure not only the geometry of the cone but also a description of the measures to check the deairing of the pore-water pressure measuring devices should be included. Also operational procedures for detecting the smearing of the filter element should be given. Regarding the stiffness of the piezometers, some suggest a definition of system stiffness, but for establishing requirements more data of experimental work are needed (Jones). Also Jamiolkowski insists on the necessity to set up criteria concerning the theoretical stiffness and the deairing procedure.

5. In case of the use of a piezo-cone Thorburn proposes to add a figure in order to explain the corrections to be applied because of the waterpressures on cone and sleeve (Fig. T1). The corresponding formula's are given in the figure.

More complicated corrections are proposed by Senneset. The formulas are based on the assumption that the waterpressure on the gap is the same as the waterpressure on the tip of the cone. However depending on the stress field, above the cone can exist a zone of dilation, while at the tip of the cone exists a zone of compression. With the piezocone one knows only the waterpressure in a specific location.

In each case such corrections seem too delicate to be introduced as a rule in a reference test procedure.

6. Jones proposes to represent a piezocone test by CUPT. If a different symbol should be needed, it is perhaps preferable to use CPTU.

Conclusions

In the course of years CPT tests in several countries have delivered valuable data to practicing engineers for solving their soil and foundation problems. Successive improvements have been brought to equipment and procedures. However it must be reminded that CPT tests are field tests, they must remain simple and cheap. It is of course worthwhile to have all over the world an unique Reference Test Procedure, in order to be able to compare correctly the results and exchange obtained experience. But at the same time the reference procedure must be such that it

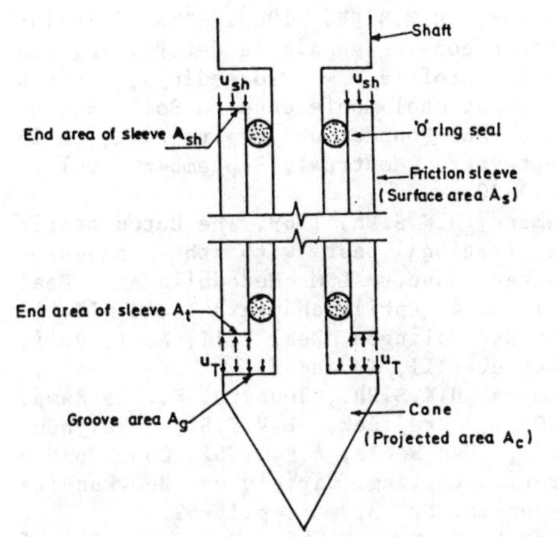

If the grooves are completely filled the following corrections apply

$$q_T = q_c(\text{corrected}) = q_c(\text{measured}) + \frac{u_T A_g}{A_c}$$

$$f_s(\text{correted}) = f_s(\text{measured}) + \frac{u_{sh} A_{sh} - u_T A_t}{A_s}$$

Fig. T1 - Correction of penetrometer readings (Thorburn).

does not constitute an hindrance, but at the contrary an incentive for an increased use in practice.

This is the double aim which the restricted team has tried to respect in the tentative draft yet presented for discussion.

Literature.

Aas, G., Lacasse, S., Lunne, T. and Madshus, C. 1984. In situ testing : new developments. - Publication No 153 of the Norwegian Geotechnical Institute, p.1-7.

Amar, S., Baguelin, F., Jezequel, J.F., Nazaret, J.P. 1974. The use of the static penetrometer in the laboratoires des Ponts et Chaussées. - Proceedings of the European Symposium on Penetration Testing, Stockholm, June 5-7, Vol. 2:2, p.7-12.

Baligh, M.M. 1975. Theory of deep site static cone penetration resistance. - Massachusetts Institute of Technology. Department of Civil Engineering. Publication No 75-56, September.

Begemann, H.K.S.Ph. 1963. The use of the static soil penetrometer in Holland. -North Zealand Engineering, February 15, p.41-49.

Begemann, H.K.S.Ph. 1965. The friction jacket cone as an aid in determining the soil profile. - Proceedings, Sixth International Conference on Soil Mechanics and Foundation Engineering, 8-15 september, Montreal, September, Vol.I, p.17-20.

Begemann, H.K.S.Ph. 1969. The Dutch static penetration test with the adhesion jacket cone. - LGM Mededelingen, Deel XII No 4, April, chapters I and II. - LGM Mededelingen, Deel XIII, No 1, Juli, chapters III, IV and V.

Begemann, H.K.S.Ph., Joustra, K., Te Kamp, W.G.B., Krajiček, P.V.F.S., Heijnen, W.J., Van Weele, A.F. 1982 Cone penetration testing. Civiele en Bouwkundige techniek. Nr. 3, mei, p.15-59.

Berezantsev, V.G. 1966. Determination of the limiting resistance of sand soils under the point of a pile by means of static penetration. - Soil Mechanics and Foundation Engineering, No 4, July-August, p.229-234.

Beringen, F. L., Windle, D. and van Hooydonk, W.R. 1979. Results of loading tests on driven piles in sand. - Proceedings of the Conference on recent developments in the design and construction of piles. I.C.E. London, March, p.153-165.

Beringen, F.L., Kolk, H.J. and Windle, D. 1981. Cone penetration and laboratory testing in marine calcareous sediments. Fugro. Leidschendam.

Bogdanovic, L. 1961. The use of penetration tests for determining the bearing capacity of piles. - Proceedings, Fifth International Conference on Soil Mechanics and Foundation Engineering, Paris, Vol. II, p.17-22.

Broug, N.W.A. 1982. The quasi-static penetration test for pile design. The Institution of Engineers, Australia, 13 january, p.31-37.

Broug, N.W.A. 1982. The analysis of cone resistance q_c and sleeve friction f_s as interactive stresses, resulting in a new pile bearing capacity design method. - Proceedings, Second European Symposium on Penetration Testing, Amsterdam, 24-27 May, p.19-27.

Bustamante, M., Gianeselli, L. 1983. Calcul de la capacité portante des pieux à partir des essais au pénétromètre statique.- Bulletin de Liaison des laboratoires des ponts et chaussées N°127, septembre-octobre 1983, p.73-80.

Campanella, R .G., Berzins, W.E. and Shields, D.H. 1979. A premilinary evaluation of Menard pressuremeter, cone penetrometer and standard penetration tests in the lower Mainland British Columbia. University of British Columbia, Soil Mechanics, Series No 40, Department of Civil Engineering.

Campanella, R.G. and Gillespie, D. 1981. Consolidation characteristics from pore pressure dissipation after piezometer cone penetration. University of British Columbia, Soil Mechanics, Series No 47, Department of Civil Engineering.

Campanella, R.G., Gillespie, D. and Robertson, P.K. 1981. Pore pressures during cone penetration testing. - Proceedings, Second European Symposium on Penetration Testing. Amsterdam, May 24-27, Vol. 2, p.507-512.

Campanella, R.G., Robertson, P.K. 1981. Applied cone research. - Proceedings of a session at the ASCE National Convention, St.Louis, Missouri, October 26-30, p.343-362.

Campanella, R.G. and Robertson, P.K. 1982. State of the art in situ testing of soils : developments since 1978. University of British Columbia, Soil Mechanics, Series No 56, Department of Civil Engineering.

Campanella, R.G., Gillespie, D., and Robertson, P.K. 1982. Pore pressures during cone penetration testing. - Proceedings Second European Symposium on Penetration Testing. Amsterdam. May 24-27, Vol. 2, p.507-512.

Campanella, R.G., Robertson, P.K. and Wightman, A. 1983. SPT-CPT correlations. - Proceedings ASCE, GT11, Vol. 109, November, p.1449-1459.

Campanella, R.G. and Robertson, P.K. (et al.). 1984. Piezometer-friction cone investigation at a tailings dam. Canadian Geotechnical Journal, Vol. 21, No 3, p.551-562.

Campanella, R.G. and Robertson, P.K. 1985. Liquefaction potential of sands using the CPT. - Proceedings, ASCE, Vol. 111, March, p.384-403.

Campanella, R.G. and Robertson, P.K. (et al.). 1985. Design of axially and laterally loaded piles using in situ tests : a case history. - Canadian Geotechnical Journal, Vol. 22, N° 4, November, p.518-527.

Carpentier, R. 1970. Vergelijking van de resultaten van enkele vinproeven en van de overeenkomstige weerstanden opgemeten in sonderingen. - Tijdschrift der Openbare Werken van België, Nr.3, 1970, blz.179-186.

Carpentier, R. 1982. Relationship between the cone resistance and the undrained shear strength of stiff fissured clays. - Proceedings, Second European Symposium on Penetration Testing, Amsterdam 24-27 May, Vol. 2, p.519-528.

De Beer, E., 1945. Etude des fondations sur pilotis et des fondations directes. L'appareil de pénétration en profondeur. - Annales des Travaux Publics de Belgique, Avril, juin et août 1945.

De Beer, E. 1948. Données concernant la résistance au cisaillement déduite des essais de pénétration en profondeur. - Géotechnique, Vol. I. N° 1, June, p.22-39.

De Beer, E. 1971. Méthodes de déduction de la capacité portante d'un pieu à partir des résultats des essais de pénétration. Annales des Travaux Publics de Belgique, N° 4, août, p.191-266.

De Ruiter, J. 1971. Electric Penetrometer for Site Investigations. - Proceedings ASCE, SM-2, February, p.457-472.

De Ruiter, J. 1981. Current penetrometer practice. Cone penetration testing and experience. - State of the Art Report. - Proceedings of a session at the ASCE National Convention, st. Louis, Missouri, October 26-30, p.1-48.

De Ruiter, J. 1982. The static cone penetration test. State-of-the-art report. - Proceedings, Second European Symposium on Penetration Testing, Amsterdam, May 24-27, Vol. 2, p.389-405.

Fugro. 1981. How geotechnical engineers conquered the Beafort Sa. Offshore Services and Technologies, Vol. 14, No 6, June, p.2111-24.

Geuze, E.C.W.A. Résultats d'essais de pénétration en profondeur et de mise en charge de pieux-modèles. - Annales de l'Institut Technique du Bâtiment et des Travaux Publics, Paris, N°s 63-64, mars-avril 1953, p.313-319.

Gielly, J, Lareal, P., Sanglerat, G. 1969. Correlations between in situ penetrometer tests and the compressibility characteristics of soils. - Conference on situ investigations in soils and Rock, 13-16 May, London.

Hartikainen, J. 1976. On the estimation of the bearing capacity of friction piles from the sounding resistance. - Proceedings, Fifth, Budapest Conference on Soil Mechanics and Foundation Engineering, October 12-15, p.265-285.

Heijnen, W.J. 1973. The Dutch cone test : Study of the shape of the electrical cone. - Proceedings, Eight International Conference on Soil Mechanics and Foundation Engineering, Moscou, August, Vol. 1.1, p.181-184.

Heijnen, W.J. and Janse, E. 1985. Case studies of the Second European Symposium on Penetration Testing. - LGM-Mededelingen Nr. 91, augustus, p.1-33.

Holden, J.C., 1976. The determination of deformation and shear strength parameter for sands using the electrical friction-cone penetrometer. Publication No 110 of the Norwegian Geotechnical Institute. p.55-59.

Jamiolkowski, M., Baldi, G. (et al.) 1985. Penetration resistance and liquefaction of sand. Proceedings, Eleventh International Conference on Soil Mechanics and Foundation Engineering, San Francisco, August, Vol. 4, p.1891-1986.

Jones, G.A. 1975. Deep sounding - its value as a general investigation technique with particular reference to friction ratios and their accurate determination. - Proceedings, Sixth Regional Conference for Africa on Soil Mechanics and Foundation Engineering, Durban, September, Vol.1, p.167-175.

Joustra, K. 1970. Modern sonderen. T.B.V. Grondonderzoek in het terrein. -Polytechnisch tijdschrift editie Bouwkunde water- en wegenbouw, Nr. 9.

Joustra, K. and Fugro N.V. 1973. New developments of the Dutch cone penetration test. - Proceedings, Eight International Conference on Soil Mechanics and Foundation Engineering, Moscou, August, Vol. 1.1, p.199-201.

Joustra, K. and de Gijt, J.G. 1982. Results and interpretation of cone penetration tests in soils of different mineralogic composition. - Proceedings, Second European Symposium on Penetration Testing, Amsterdam, May 24-27, Vol. 2, p.615-626.

Kahl, H. und Muhs, H. 1952. Uber die Untersuchung des Baugrundes mit einer Spitzendrucksonde. - Die Bautechnik, April, Heft 4, S.81-88.

Kantey, B.A. 1951. Significant development in sub-surface explorations for piled foundations. - The Transactions of the South African Institution of Civil Engineers, Vol. I, No 6, August, Discussion Vol. 2, No 12, 1951.

König, G. 1969. Die Ermittlung der Tragfähigkeit von Pfählen und die Bestimmung des Reibungswinkels nichtbindiger Böden mit Hilfe von Drucksondierungen. - Der Bauingenieur, Heft 9, S.343-346.

Lacasse, S., Jamiolkowski, M., Lancellotta, R. and Lunne, T. 1981. In situ characteristics of two Norwegian clays. Proceedings, Tenth International Conference on Soil Mechanics and Foundation Engineering, Stockholm, June 5-7, Vol. 2, p.7-22.

Lunne, T., Eide, O. and De Ruiter, J. 1976. Correlations between cone resistance and vane shear strength in some Scandinavian soft to medium stiff

clays. - Canadian Geotechnical Journal, Vol. 13, No 4, p.430-441.
Lunne, T., Christoffersen, H.P. 1985. Interpretation of cone penetrometer data for offshore sands. - Publication No 156 of the Norwegian Geotechnical Institute. p.1-11.
Marsland, A. 1974. Comparisons of the results from static penetration tests and large in-situ plate tests in London Clay. - Proceedings of the European Symposium on Penetration Testing, Stockholm, June 5-7, Vol. 2:2, p.245-252.
May, R.E. 1982. Second European Symposium on Penetration Testing. A report on the British Geotechnical Society's meeting at the Institution of Civil Engineers on June 23, 1982. - Ground Engineering, Vol. 15, No 6, September, p.6-7.
Menzenbach, E. 1961. The determination of the permissible point load of piles by means of static penetration tests. -Proceedings, Fifth International Conference on Soil Mechanics and Foundation Engineering, 17-22 July, Paris, Vol. II, p.99-104.
Meyerhof, G.G. 1956. Penetration tests and bearing capacity of cohesionless soils. - Proceedings, ASCE, SM1, January. paper 866 - p.1-19.
Meyerhof, G.G. 1982. Bearing capacity and settlement of foundations in sand based on static cone penetration tests. Amici et Alumni, Em. Prof. Dr ir E.E. De Beer. Komitee ter huldiging van Professor E. De Beer, Brussel, p.217-222.
Muromachi, T. 1974. Experimental study on application of static cone penetrometer to subsurface investigation of weak cohesive soils. - Proceedings of the European Symposium on Penetration Testing. Stockholm, June 5-7, Vol. 2:2, p.285-292.
Muromachi, T. 1981. Cone penetration testing in Japan. - Proceedings of a session at the ASCE National Convention, St. Louis, Missouri, October 26-30, p.76-107.
Muromachi, T. and Kobayaschi S. 1982. Comparative study of static and dynamic penetration tests currently in use in Japan. - Proceedings, Second European Symposium on Penetration Testing, Amsterdam, 24-27 May, Vol. 1, p.297-302.
Muromachi, T., Tsuchiva, H., Sakai, Y. and Sakai, K. 1982. Development of multi-sensor cone penetrometers.- Proceedings, Second European Symposium on Penetration Testing, Amsterdam 24-27 May, Vol. 2, p.727-732.
Parez, L. 1974. Static penetrometer : The importance of the skin friction associated with the point resistance. - Proceedings of the European Symposium on Penetration Testing. Stockholm, June 5-7, Vol. 2:2, p.293-300.
Parez, L., Bachelier, M. et Sechet, B. 1976. Pression interstitielle développée au fonçage des pénétromètres. - Proceedings, Sixth European Conference on Soil Mechanics and Foundation Engineering, Vienna, 22-24 March 1976, Vol. 1:2, p.533-538.
Plantema, G. 1957. Influence of density on sounding results in dry, moist and saturated sands. - Proceedings, Fourth International Conference on Soil Mechanics and Foundation Engineering, London, August 12-24, Vol.I, p.237-240.
Robertson, P.K., Campanella, R.G. 1983. Interpretation of cone penetration tests. Part I : sand. Part II : clay - Canadian Geotechnical Journal, Vol. 20, No 4, November, p.718-733, p.734-745.
Sanglerat. G. 1965. Le pénétromètre et la reconnaissance des sols. Interprétation des diagrammes de pénétration. - Théorie et pratique. - Ed. : Dunod, Paris.
Sanglerat, G. 1972. The penetrometer and soil exploration. - Elsevier, Amsterdam.
Sanglerat, G. 1974. The determination of soil profile. - General discussions groupe 4 : Interpretation of results of static penetration test. - Proceedings of the European Symposium on Penetration Testing, Stockholm, June 5-7, Vol. 2:1, p.144-145.
Sanglerat, G. and Nhiem, T.V., Sejourne, M., Andina, R. 1974. Direct soil classification by Static penetrometers with special friction sleeve. - Proceedings of the European Symposium on Penetration Testing. Stockholm, June 5-7, Vol. 2:2, p.337-344.
Sanglerat, G., Girousse, L. et Bardot, F. 1979. Controle in situ des prévisions de tassements basées sur les essais de pénétration statiques pour 79 ouvrages sur 17 sites différents. - Annales de l'Institut Technique du Bâtiment et des Travaux Publics, N° 369 février, p.29-50.
Sanglerat, G. and Mlynarek, Z.B. 1981. Bearing capacity equations of static sounding of pliocene clay. - Proceedings, Tenth International Conference on Soil Mechanics and Foundation Engineering, June 15-19, Stockholm, Vol. 2, p.523-526.
Sanglerat, G., Mlynarek, Z. B. and Sanglerat, T.R.A. 1982. The statistical analysis of certain factors influencing cone resistance during static sounding of cohesive soils. - Proceedings, Second European Symposium on Penetration

Testing, Amsterdam, 24-27 May, Vol. 2, p.827-834.

Sanglerat, G., Olivari, G. and Cambou, B. 1984. Practical problems in soil mechanics and foundation engineering. - Elsevier, Amsterdam.

Schmertmann, J.H. 1974. Penetration pore pressure effects on quasi-static cone bearing q_c. - Proceedings of the European Symposium on Penetration Testing, Stockholm, June 5-7, Vol. 2:2, p.345-352.

Schmertmann, J.H. 1982. A method for determining the friction angle in sands from the Marchetti dilatometer test (DMT). - Proceedings, Second European Symposium on Penetration Testing. Amsterdam, May 24-27, Vol. 2, p.853-861.

Swedish Geotechnical Society. - Proceedings of the European Symposium on Penetration Testing in Stockholm, June 5-7, 1974.

Thorburn, S., Rigden, W.J., Marsland, A., Quatermain, A; 1982. A dual load range cone penetrometer. - Proceedings, Second European Symposium on Penetration Testing, Amsterdam, May 24-27, Vol. 2, p.787-796.

Trofimenkov J.G. 1974. Penetration testing in USSR. State-of-the-art report. Proceedings of the European Symposium on Penetration Testing, Stockholm, June 5-7, Vol. 1, p.147-154.

Trofimenkov J.G. 1974. Penetration testing in Eastern Europe. - General Report 2. - Proceedings of the European Symposium on Penetration Testing, Stockholm, June 5-7, Vol. 2:1, p.24-28.

van den Berg, A.P. 1982. Latest developments in static cone penetrometers and other soil-testing equipment (on land-offshore). - Proceedings, Second European Symposium on Penetration Testing, Amsterdam, May 24-27, Vol. 2, p.447-455.

van den Berg, A.P. 1984. Nieuwste ontwikkelingen in sondeerapparatuur voor zeebodemonderzoek. - Land Water - Vol. 24 N° 1, p.35-39.

Van Der Veen, C. and Boersma, L. 1957. The bearing capacity of a pile, predetermined by a cone penetration test. - Proceedings, Fourth International Conference on Soil Mechanics and Foundation Engineering, London, 12-24 August, Vol. II. p.72-75.

Van Wambeke, A., d'Hemricourt, J., 1982. Correlation between the results of static or dynamic probings and pressuremeter tests. - Proceedings, Second European Symposium on Penetration Testing, Amsterdam, May 24-27, Vol.2, p.941-944.

Van Weele, A. 1961. Deep sounding tests in relation to the driving resistance of piles. - Proceedings, Fifth International Conference on Soil Mechanics and Foundation Engineering, 17-22 July, Paris, Vol. II, p.165-169.

Verhandelingen Fugro - Sondeersymposium 1972. p.1-62.
Broug, N.W.A. Historisch overzicht v/h sonderen.
Van Ree, W.F. Mechanische en elektrische sondeersystemen.
Heijnen, W.J. De vorm van de elecktrische sondeerconus.
Loof, W.H. Mogelijke foutenbronnen bij de uitvoering van sonderingen.
Schmertmann, J.H. Effects of in situ lateral stress on friction cone penetrometer data in sands.
Vlas, G. Bouw- en woningtoezicht en het sonderen.
De Ruiter, J. Normalisatie van de sondeermethode.
Zuidberg, H.M. Sonderen op zee.

Verruijt, A., Beringen, F.L., De Leeuw, E.H. 1982. Penetration Testing - Proceedings, Second European Symposium on Penetration Testing, Amsterdam 24-27 May, Vol. 1 & 2.

Vlasblom, A. 1985. The electrical penetrometer ; a historical account of its development. - LGM-Mededelingen No 92, december.

Zuidberg, H.M. 1975. A submersible cone-penetrometer rig. Marine Geotechnology No 1, Vol. 1, p.15-32.

Zuidberg, H.M., Schaap, L.H.J., Beringen, F.L. 1982. A penetrometer for simultaneously measuring of cone resistance, sleeve friction and dynamic pore pressure. - Proceedings, Second European Symposium on Penetration Testing, Amsterdam, May 24-27, Vol. 2, p.963-970.

Penetration Testing 1988, ISOPT-1, De Ruiter (ed.)
© 1988 Balkema, Rotterdam, ISBN 90 6191 801 4

Dynamic probing (DP): International reference test procedure

ISSMFE Technical Committee on Penetration Testing
DP Working Party:
G.Stefanoff (Bulgaria)
G.Sanglerat (France)
U.Bergdahl (Sweden)
K.-J.Melzer, Chairman (FR Germany)

ABSTRACT: The Dynamic Probing (DP) Working Party of the ISSMFE committee on penetration testing presents the results of its work as a three-part report. Document no. 1: Introduction and objectives; Document no. 2: Review of present practice; Document no. 3: International reference test procedures.

DOCUMENT NO. 1

INTRODUCTION AND OBJECTIVES

Dynamic probing (DP) or dynamic penetration testing is probably the oldest method of penetration testing for the purpose of subsoil exploration in the field of foundation engineering. Because of the simplicity of the test, a variety of equipment types have been in use for the last fifty years worldwide. The following main facts have led to such a wide variety of equipment:

- Type of field application
  - -- Preliminary field investigations
  - -- Final field investigation
  - -- Supervision of construction
- Topography (accessibility of the area of investigation)
- Geology (availability of general information about soil types, etc.).

Reproducibility, accuracy and compatibility of the results of DP are important requirements as for any other field test. Because of the urgent demand for comparability of DP test results on an international level in the areas of both industrial application and research, the need for reference test procedures has become evident.

In 1977, the European Sub-Committee on Penetration Testing published recommendations for a European Standard for DP at the IXX ISSMFE, Tokyo, basically presenting a new one-type reference test. In 1981, a second Standard for the light equipment (DPL) was added.

In 1982, the decision was taken at the second European Symposium on Penetration Testing, Amsterdam, to form an ISSMFE Technical Committee on Penetration Testing because of the necessity for international cooperation in the preparation of reference test procedures, the task established for the Technical Committee to be completed by 1985.

As part of the Technical Committee, the DP Working Party in general agrees to the following principles as established by the SPT Working Party in order to permit comparability of the test results from different member countries and within those countries:

- A better communication of information between the ISSMFE member countries
- A better understanding of the differences in current test procedures
- Better correlations between the work of site personnel, organizations, and countries.

The DP Working Party considers the establishment of an International Standard as premature and not acceptable at the present time. A strict standard could mean a loss of vast knowledge and experience gained on the national and international levels during the past fifty years, besides the possible negative economic consequences.

The Working Party considers International Reference Test Procedures, however, a useful and necessary step towards complying with the principles outlined above, and having a realistic chance to be accepted by the international community.

Before this background, the Working Party has chosen the following main criteria in establishing the Reference Test Procedures (RTP):

- number of countries in which a certain type of DP is used,
- frequency of use (at least qualitatively),
- compatibility of equipment dimensions with those of equipment that is in use or even standardized nationally,
- selection of more than one RTP covering the range from light to very heavy equipment in order to encourage the use of RTP world wide.

The actions that have been taken can be summarized as follows:

- May 1982, Amsterdam: First meeting of the DP Working Party
- November 1982: Distribution of a questionnaire about equipment in use, etc., to 25 member countries (Distribution list: Appendix A)
- April/May 1983: Distribution of RTP, first draft (distribution list: Appendix A)
- May 1983, Helsinki: 2nd meeting of the DP Working Party
- November 1983: Distribution of RTP, second draft (distribution list: Appendix A)
- June 1984: Distribution of RTP, third (final) draft, to DP Working Party
- August 1985: Delivery of the final document to the ISSMFE Committee on Penetration Testing.

The final version of the DP Reference Test Procedures was to be written in a rather simplistic manner, following the principles and criteria outlined above. Accordingly, it is structured into the following main chapters:

- Scope
- Definition
- Apparatus
- Test Procedure
- Measurements
- Reporting of Test Results
- Explanatory Notes.

The DP Working Party has summarized its efforts in a three-part documentation in order to give the participating international community at least an idea of its considerations and decisions:

1 Introduction and Objectives
2 Review of Present Practice
3 International Reference Test Procedures

DOCUMENT NO. 2

REVIEW OF PRESENT PRACTICE

1 Introduction

The purpose of this document is to give the ISSMFE community a summary of the information that was available to the DP Working Party and that was used as the basis for establishing the Reference Test Procedures (RTP; DOCUMENT NO. 3), taking into account the basic considerations set forth in DOCUMENT NO. 1. This document is not intended to provide a classical state-of-the-art report, an almost inconceivable task considering the fact that some kind of dynamic probing was mentioned first in Europe by Goldmann as early as in 1699. It is rather focused on existing compiled literature and information obtained through direct communication with ISSMFE member countries. It is hoped that communication within the member countries has also taken place. Therefore, our list of references is:

1. Proceedings of the 1st European Symposium on Penetration Testing, Stockholm, 1974 (ESOPT I)
2. Proceedings of the 2nd European Symposium on Penetration Testing, Amsterdam, 1982 (ESCOPT II) especially the State-of-Art Report on Dynamic Penetration Testing
3. Some additional literature references, including:

3a. the questionnaire distributed to 25 member countries (see DOCUMENT NO. 1), and

3b. private communications, especially among the members of the Working Party.

The results of the evaluation of the references are summarized in Tables 1 to 3, and will be described in greater detail in the following Chapters. The headings of these Chapters correspond to the main headings of the questionnaire (Ref. 3a):

- Equipment
- Test procedures
- Measurements and reporting of results

- Field application
- Use of DP in member countries
- Interpretation of test results
- Standardization.

In the final Chapter the reasons for the selection of the individual Reference Test Procedures are summarized.

The following basic definitions in regard to dynamic probing have been followed throughout the evaluation and the establishment of the RTP: only the penetrometers that are used for continuous dynamic penetration testing have been considered. Because of the great variety of penetrometers in use, the following classification according to the hammer masses applied was chosen (Ref. 2):

Type	Abbreviation	Mass, kg
Light	DPL	≤10
Medium	DPM	>10<40
Heavy	DPH	≥40≤60
Super-heavy	DPSH	>60

2 Equipment

Dynamic probing actually emerged as a side product: engineers conducting pile driving or sheet piling jobs started to estimate the necessary depth of penetration, at which the bearing stratum supposedly was to be reached, according to the increase in driving resistance by counting the number of blows necessary to achieve a certain depth of penetration. Therefore, it is not surprising that the first source reporting this type of "dynamic probing" goes back to the late 17th century. Between the two world wars, dynamic probing became known besides the traditional boring methods as a means of subsoil exploration in the field of foundation engineering, especially in Europe.

After 1945, the known use of dynamic probing equipment became widespread within and beyond Europe (Tables 1 and 2). There is no doubt, however, that the use outside Europe, at least partially, has been substantially influenced by European consultants and by the construction industry operating beyond their home borders.

Considering the historic development of dynamic probing, it is not surprising at all that the present equipment parameters vary considerably, at least in part. According to the classification of the test equipment given in Chapter 1, the following can be stated in summary with regard to the equipment parameters such as mass of hammer, heigth of fall, cone, anvil and rod dimensions, etc. (Table 1).

The equipment parameters are rather consistent for DPL and DPH, one of the possible reasons being that most of this equipment is being designed closely to the requirements of the German standard DIN 4094, at least for the DPL. The equipment parameters for DPM and DPSH vary more.

Especially for the case of the DPSH, it is observed that the equipment dimensions of one group (see Nos. 8, 17, 18, 20 and 21 in Table 1) are similar to those of the Standard Penetration Test (SPT) equipment, whereas others (see Nos. 9, 10, 20, 21 in Table 1) deviate from these, in some cases even considerably.

The importance of understanding and considering the equipment-related parameters has considerably increased during the past decade or more, mainly for two reasons:

- The reliability of the results of dynamic penetration tests had to be improved, compared with static cone penetration testing (CPT)
- Establishing relations among the results of various dynamic tests and between the results of dynamic tests and those of the static test has become more and more important with the increasing drive for standardization at the national and international levels.

At ESOPT I, a French contribution tackled the old controversy that results from dynamic probing are not reliable and cannot be used for the design of foundations. It has been proven that dynamic probing using properly designed equipment and an adequate procedure, based on full understanding of the mechanism of driving a cone into the soil, allows measurements to be made that are as reliable as those performed with static equipment. This one example stands only for various others and remains to be disproven. Before this background it is not surprising that equipment specifications including

Table 1: Summary of questionaire evaluation

No.	Country	Reply	Penetrometers in Use				Recomm. or Standard	Cone Dimensions Diameter mm	Apex Angle deg.	Rod Dimensions Inside Ø mm	Outside Ø mm	Length m	Mass of Hammer kg	Height of Fall cm
			DPL	DPM	DPH	DPSH								
1	Argentine	No	-	-	-	-	-	-	-	-	-	-	-	-
2	Belgium	Yes	x	-	-	-	-	22.5;25.2;35.7	60	0	20	1.0	10.0	50
3	Brazil	No	-	-	-	-	-	-	-	-	-	-	-	-
4	Bulgaria	No	-	-	-	-	-	-	-	-	-	-	-	-
5	Canada	No	-	-	-	-	-	-	-	-	-	-	-	-
6	China	No	-	-	-	-	-	-	-	-	-	-	-	-
7	CSSR	Yes	x -	- -	- x	- -	DIN 4094 DIN 4094	35.7 43.7	90 90	6 9	22 32	1.0 2.0	10.0 50.0	50 50
8	Denmark	Yes	-	-	-	x	DPB [1]	51.0	90	-	32	1.0-2.0	63.5	75
9	Finland	Yes	x -	- -	- -	- x	DPL [1] Finnish Standard	35.7 45.0	90 90	- -	22 32	1.0 1.0	10.0 63.5	50 50
10	France	Yes	- - - - -	x x - x -	- - - - x	x x x - x	- - - - -	62.0 70.0 45.0 35.7 43.7	90 90 90 90 90	- - - - -	40 40 32 22 32	1.0 1.0 1.0 1.0 1.0	32; 64; 96 120 30; 60; 90 63.5 30.0 50.0-100.0	76 40 50 20 50
11	FRG	Yes	x - -	- x -	- - x	- - -	DIN 4094 DIN 4094 DIN 4094	25.2; 35.7 35.7 35.7; 43.7	90 90 90	6 6; 9 9	22 22; 32 32	1.0 1.0-2.0 2.0	10.0 30.0 50.0	50 20; 50 50
12	Greece	Yes	x -	- -	- x	- -	DIN 4094 DIN 4094	35.7 43.7	90 90	6 9	22 32	1.0 2.0	10.0 50.0	50 50
13	India	No	-	-	-	-	-	-	-	-	-	-	-	-
14	Israel	Yes	-	-	-	-	-	-	-	-	-	-	-	-
15	Italy	Yes	-	-	-	-	-	-	-	-	-	-	-	-
16	Japan	No	-	-	-	-	-	-	-	-	-	-	-	-
17	Norway	Yes	-	-	-	x	DPB	51.0	90	-	32	1.0-2.0	63.5	75
18	South Africa	Yes	x -	- -	- -	- x	- -	20.0 50.0	60 60	- 29	16 41	1.0 1.5	8.0 63.5	58 75
19	Spain	Yes	x - -	- - -	- - -	- x x	DPL - DPA/DPB	35.7 40x40 or Ø 45 62/51	90 90 90	- - -	22 32 40-45/32	1.0 1.0 1.0-2.0	10.0 63.5 63.5	50 50 75
20	Sweden	Yes	- -	- -	- -	x x	HfA DPB	45.0 51.0	90 90	- -	32 32	1.0-2.0 1.0-2.0	63.5 63 5	50 75
21	Switzerland	Yes	- - -	x - -	- x -	- - x	- - -	35.7 43.7 62.5; 60.0	30; 90 30 90	- 15 26	22; 25 32 42; 38	1.0-2.0 1.0-2.0 1.5-2.0	30.0;20.0 50.0 63.0;60.0	20; 50 50 76; 50
22	Taiwan	No	-	-	-	-	-	-	-	-	-	-	-	-
23	UK	No	-	-	-	-	-	-	-	-	-	-	-	-
24	USA	No	-	-	-	-	-	-	-	-	-	-	-	-
25	USSR	No	-	-	-	-	-	-	-	-	-	-	-	-

1) former European Recommended Standard

tolerance levels today appear in more detail in standards and recommendations.

3 Test Procedures

The selection of the DP equipment to be used for a given job normaly depends on the local conditions and the purpose of the particular test. The number of tests and the necessary depth of penetration are determined by the type and the size of the planned structure. Key borings are being made next to some DP locations in order to enable the analysis of soil types, layer stratification and strength distribution with depth.

Table 2: Overview of the use of dynamic probing in additional countries

No.	Country	Penetrometer Type DPL	DPM	DPH	DPSH
1	Algeria	X	X	-	X
2	Australia	X	-	-	X
3	Benin	X	X	-	-
4	Burundi	-	X	-	-
5	Cameroon	X	-	-	-
6	Central African Republic	X	-	-	-
7	Gabon	X	-	-	-
8	Guinea	X	-	-	-
9	India	X	-	-	X
10	Ivory Coast	X	-	-	-
11	Japan	-	-	-	X
12	Madagascar	X	-	-	-
13	Mali	X	X	-	-
14	Morocco	X	X	-	X
15	Martinique	X	X	-	-
16	New Caledonia	-	X	-	X
17	Nigeria	-	X	-	-
18	Réunion	-	X	-	X
19	Romania	X	X	X	-
20	Tunisia	-	X	-	-
21	Turkey	-	-	-	X
22	United Kingdom, Overseas Territories	X	-	-	-
23	Upper Volta	X	-	-	-
24	Zaire	X	-	-	-

Recent standardization attempts at the European level (Eurocode 7) include a geotechnical categorization of building structures which in turn requires specified subsoil explorations. While these factors concerning the selection of test types do not influence the results of a specific test, but rather the interpretation of the test results, there are the following test-procedure-related factors that mainly influence the test results:

- deformation of driving rods
- skin friction
- driving rate
- driving interruptions
- application of driving energy.

Our survey shows that it has become common practice to give special attention to the fact that deformation of the driving rods has to be avoided. Normally, probing is conducted vertically, the probing equipment being firmly supported. Especially at the beginning of the test, the rods are carefully guided to keep them straight. In some instances, pre-boring is used. Any bending of the rods may cause divergency of the driving energy and additional skin friction along the driving rods which may falsify the results.

Avoiding skin friction is one of the major concerns in dynamic probing, because the cone resistance is the only result that allows interpretation of the test results. Four measures are being taken that are to help reduce or avoid skin friction:

- the cone diameter is larger than the rod diameter
- the rods are being turned at certain depths of penetration
- drilling mud is injected
- push rods similar to the CPT separate skin friction from cone resistance.

Using cones larger in diameter than the rods may be considered as being mandatory practically worldwide. Test equipment with cone/rod diameter ratios exceeding about 1.3 leads to results in cohesionless and in many cohesive soils that are little or not at all seriously influenced by skin friction. At smaller ratios (as mentioned, e.g., for one of the light penetrometers according to the German standard DIN 4094, which is also used

Table 3: Summary of use, standardization and interpretation of test results

No.	Country	Penetrometer in Use				Recomm. or Standard	Qualitatively	Interpretation of Test Results: Eng. Propert. of Soils		Bearing Capacity	
		DPL	DPM	DPH	DPSH			Cohesionless	Cohesive	Shall.Found.	Piles
1	Belgium	x	–	x	–	x	x	3	3	2	–
2	Bulgaria	–	x	x	x	x	x	1,2	1,2	1	3
3	CSSR	x	–	x	–	x	x	1	–	1	3
4	Denmark	x	–	x	x	x	x	–	–	–	3
5	Finland	x	–	–	x	x	x	1	–	2	1
6	France	–	x	x	x	–	x	1,2	1	–	1
7	FRG	x	x	x	x	x	x	1,2	2	–	1,2
8	GDR	x	–	x	x	–	x	1,2	1,2	–	–
9	Greece	x	–	x	–	x	x	1	1	–	–
10	Hungary	–	–	x	–	x	x	1	–	–	–
11	Italy	x	–	–	x	–	x	1,2	–	1	–
12	Norway	–	–	–	x	x	x	–	–	–	1
13	Poland	x	–	–	x	x	x	2	1	–	–
14	Portugal	x	–	–	–	–	x	–	–	–	–
15	South Africa	x	–	–	x	–	x	2	3	3	3
16	Spain	x	–	–	x	x	x	3	3	1	1
17	Sweden	–	–	–	x	x	x	3	–	3	3
18	Switzerland	x	x	x	x	–	x	1	1	3	3
19	UK	–	–	–	x	–	x	3	–	–	–
20	USSR		x		x	x	x	1,2		–	3

1 = Ranges (estimates)
2 = Relations
3 = Not specified

in Belgium), specific caution is necessary in interpreting results obtained from dynamic probing in cohesive soils if no additional measures are being taken to avoid skin friction.

Turning of the rods is one of the measures that is supposed to reduce skin friction. Skin friction is avoided practically completely if drilling mud is used during the performance of the test. This method is most effective if the drilling mud is injected through holes in the hollow rods near the cone that are directed horizontally or slightly upwards. The use of push rods (casings) similar to the method used in the CPT is as effective as, but less used than the drilling-mud method.

It is being recognized that the driving rate influences the test results. Driving rates of 15 to 30 blows per minute are commonly used. In pervious soils such as sands and gravels, the influence of the driving rate is lower; thus higher rates, e.g. 60 blows per minute, are used.

Extensive driving interruptions may lead to a build-up of skin friction especially when using DP in cohesive soils. Therefore, under normal circumstances, the penetrometer is being continuously driven into the subsoil allowing for interruptions only while the driving rods are being extended. These interruptions are generally kept as short as possible.

Application of driving energy is being realized as a very critical factor influencing the test results. Besides ensuring that the driving

rods are being kept straight (as mentioned above) and the rod couplings are tight, the following two facts are being considered as most important:
- guarantee of free fall of the hammer
- guarantee of constant height of fall of the hammer.

These facts become even more important if one considers that dynamic penetrometers are either hand- or machine-operated. Recent investigations made in Germany - which compare the classical hand-operated DPL showing the expected consistent results with test equipment using an air-pressure-powered hammer - revealed dramatic differences in the test results that lead to serious questioning of the results obtained with this specific machine-operated equipment in some soil types.

In these and all other respects, the users seem to be aware of the influences of the test procedures; for this reason, test procedures appear to be relatively consistent.

4 Measurements and Reporting of Results

Results are generally reported as penetration resistance measurements versus penetration depth diagrams. Measurements are mainly N_{10} (number of blows per 10 cm penetration) for the lighter equipment (DPL, DPM) and the DPH, whereas N_{20} is preferred for the DPSH. Results are also reported using driving resistance formulae. Also, the driving work is often used, especially in relating the results of different DPs to one another; the driving work is defined as

$$R_d = \frac{MgH}{A} \times \frac{N}{D'}$$

where:
M is the mass of the hammer
H is the height of fall
A is the base area of the cone
g is the acceleration of gravity
D' is the defined depth interval of penetration (e.g. 10 cm)
N is the number of blows per defined depth interval of penetration.

The general site and equipment descriptions are rather consistent and contain mainly:
- location of probing, type, date, number
- elevation, groundwater level
- deviations from the normal procedure such as interruptions or damage to rods.

5 Field Application

According to Reference 2, the kinds of subsoil exploration for which dynamic penetration tests are mostly used can be divided into the following three groups:
a) Preliminary field investigations
b) Final field investigations
c) Construction supervision.

For the different types of investigation (Table 4), an analysis of our questionnaire (Ref. 3a) concerning the use of DP has given the following results for some member countries:

Table 4: DP field application

Country	Type of investigation		
	Preliminary	Final	Construction supervision
Australia	X	-	X
Belgium	X	-	X
Bulgaria	X	X	X
CSSR	X	X	X
Finland	X	X	X
France	X	X	X
FRG	X	X	X
Greece	-	X	X
Norway	X	X	-
South Africa	X	X	X
Spain	X	-	X
Sweden	-	X	-
Switzerland	X	X	X

Which type of penetrometer should be used for the various investigations is strongly influenced by the following factors:
- Topography (accessibility of the area of investigation)
- Geology (general information about soil types, etc.)
- Necessary depth of penetration
- Necessity of key penetration tests in connection with borings.

Considering these tasks and the influencing factors, the process of selecting the right equipment has to be optimized under economic and technical aspects. Specific guidelines that would be valid worldwide cannot be given. Optimization can only be regional.

Usually, an investigation (especially one of Group a or b) will be started with the simplest equipment available that is suitable for the specific job in hand. General rules concerning, for example, the best suited penetration equipment depths (DPL: about 8 m; DPSH: larger than 25 m, etc.) serve as guidelines. Additional influences on the results of DP such as skin friction, groundwater level, critical depth, etc., are being considered when selecting the equipment which is first to be used. When first results of the site investigation are availabe, the use of other, maybe more sophisticated equipment which is more suitable for the problems encountered will be taken into consideration.

The general trend is that in jobs of construction supervision the lighter equipment (especially DPL) is preferred to the heavier equipment. More or less all penetrometers are used for preliminary and final field investigations.

Another aspect should be taken into account which has become increasingly important in recent years: the necessity of applying statistics and probability calculations to results of field and laboratory tests. This may already influence the design even of preliminary field investigation programs (e.g. the spacing of DP points). Equally important is that the test results - which normally are subject to considerable scattering - should be analyzed using statistical approaches; however, care must be taken that basic physical principles are not violated and valuable detailed information does not get lost using, for example, curve-fitting filtering techniques.

6 Use of DP in Member Countries

Looking at our worldwide survey (Tables 1 to 3), the general statement is permitted that the use of the various types of dynamic penetrometers is widespread:

	DPL	DPM	DPH	DPSH
Number of countries	21	16	10	23

Concerning the frequency of use, an attempt was made to get at least a qualitative picture of the overall situation. This picture may be qualitative - and more or less subjective - only because it is hardly possible to obtain comprehensive numbers of jobs per year covering all countries under consideration. However, the main purpose was not to obtain exact numbers but a trend of the use of the various types of penetrometers. The frequency of use was classed into "occasionally" (1 point), "frequently" (2 points) and "very often" (3 points). Table 5 contains the results for the countries that have answered our questionnaire; the evaluation shows the following sequence for the various types (starting with the type primarily in use): DPL, DPSH, DPH, DPM. Based on existing literature (mainly Refs. 1 and 2), a similar evaluation (but even more qualitative) was made for additional countries (Table 6).

Summing up the two evaluations resulted in the following trend confirming the basic sequence in Table 5: DPL (29 points), DPSH (27 points), DPH (18 points), DPM (13 points). Considering the additional countries of Table 3 with occasional use only, the sequence is: DPL (42 points), DPSH (31 points), DPM (23 points). DPH (18 points).

Table 5: Frequency of use*) in countries having replied to the questionnaire (Table 1)

No.	Country	DPL	DPM	DPH	DSH
2	Belgium	3	-	-	-
7	CSSR	2	-	2	-
8	Denmark	-	-	-	1
9	Finland	1	-	-	2
10	France	1	2	2	2
11	FRG	3	2	3	-
12	Greece	2	-	1	-
17	Norway	-	-	-	1
18	South Africa	2	-	-	1
19	Spain	1	-	-	2
20	Sweden	-	-	-	2
21	Switzerland	-	3	2	2
	Sum	15	7	10	13

*) 1 - occasionally
2 - frequently
3 - very often

Table 6: Frequency of use*) in countries not included in Table 5

Country	DPL	DPM	DPH	DPSH
Australia	2	-	-	2
Bulgaria	-	2	2	1
GDR	2	-	2	2
Hungary	-	-	2	-
India	2	-	-	2
Italy	1	-	-	1
Japan	-	-	-	1
Poland	3	-	-	1
Portugal	1	-	-	-
Romania	2	2	2	-
Turkey	-	-	-	1
UK	1	-	-	1
USSR	-	2	-	2
Sum	14	6	8	14

*) 1 - occasionally
2 - frequently
3 - very often

7 Interpretation of Test Results

Although the interpretation of the test results is beyond the scope of the RTP, an attempt is made to summarize the current state of practice.

Besides the factors related to test equipment and procedures (Chapters 2 and 3), at least the following soil-related parameters can additionally influence the penetration resistance:

- Soil type
- Grain characteristics
- Critical depth
- Degree of saturation (e.g. groundwater, porewater pressure)
- Density
- Consistency
- Shear characteristics
- Compressibility
- Degree of consolidation.

Concerning the effect of these parameters on the penetration resistance, the following qualitative statements can be made that generally have to be observed in interpreting the test results:

- In cohessionless soils, the penetration resistance is affected by the following parameters besides relative density: grain size distribution, grain shape and roughness, mineral type, and soil cementation. If all the other parameters are kept constant, the following general rules are valid:
  -- With increasing relative density, the penetration resistance will increase more than linearily
  -- Soils possessing grains with rough surfaces yield a higher penetration resistance than soils with round and smooth grains
  -- Stones, especially those having a diameter close to that of the cone, may result in a considerable increase in penetration resistance
  -- At the same relative density, the penetration resistance is higher in uniform than in non-uniform soils.
- Soil cementation causes an increase in penetration resistance
- The same penetration resistance does not necessarily indicate the same soil type or - in the case of cohesionless soils - the same relative density
- The penetration resistance in cohesive soils is mainly influenced by consistency, plasticity and structure, which in turn is dependent on the geological history of the soil
- Low penetration resistances, especially in man-made deposits, may be due to loose layers or cavities
- In general, the penetration resistance will scatter more in sands and gravels (especially if light DP equipment is used) than in cohesive soils. The scatter band depends on changes in soil strength and on the contents of gravel and stones
- The penetration resistance in organic soils (e.g. peat) is essentially influenced by their structure, geological history and presence of other soil types
- If DP equipment without push rods or drilling mud is used, the skin friction normally is negligible in medium- to high-density cohesionless soils - especially above the groundwater level - because the cones have larger diameters than the driving rods. In many soils, especially in soft cohesive and in organic soils, the skin friction can have a substantial effect on the penetration resistance; at the same consistency, the penetration resistance increases with depth in these cases. Also, cavities will

not be detected sufficiently clearly
- Until a certain critical depth is reached by the penetrometer, the penetration resistance - at constant relative density - increases with depth; below the critical depth, the penetration resistance remains almost constant or increases at a lower rate if all the other soil characteristics remain the same. The critical depth increases with relative density and cone diameter. The overburden pressure on a soil layer that is penetrated or an additional load (e.g. filling or foundation loads) close to the DP location may increase the penetration resistance
- In cohesionless soils, DP results in lower penetration resistances at low depths below the groundwater level than above if all the other soil characteristics remain constant. This influence is especially strong in clean sands and at low relative densities; it is less pronounced at high relative densities and very large depths
- Cohesive soils can be saturated also above the groundwater level; therefore, there is no recognizable influence of the groundwater on the penetration resistance even at low depths of penetration. On the other hand, a low degree of saturation, which is due,e.g., to dehydration, may increase the penetration resistance. Porewater or groundwater pressures may influence the penetration test results especially in fine sands; vertical groundwater pressure, e.g. in a borehole, reduces the penetration resistance
- The penetration resistance close to the border lines of a given soil layer will be influenced by the soil types above and below that layer being penetrated. Compressibility and inclination of the layer below the penetration location will be of influence; if the layer has a higher strength than the layer that is being penetrated, the penetration resistance will increase; at lower strength, the resistance will decrease.

The DP results are interpreted qualitatively and/or quantitatively. In both cases the knowledge about layering and soil types is deemed necessary in addition to DP, e.g. from key borings.

Qualitative interpretation of the DP results is in general used for the following purposes:
- Assessment of the uniformity or non-uniformity of the subsoil or of a fill
- Exploration of very loose or of very dense layers, e.g. fill zones or rock surfaces
- Assessment of a compaction effort, e.g. by comparing the penetration resistance before and after compaction
- General assessment of layering and types of subsoil and of its horizontal extension if the soil type distribution with depth is known, e.g., from key borings.

Quantitative interpretation (see also Table 3) of DP results is generally used for the following purposes:
- Determination of the engineering properties of cohesionless and cohesive soils
- Determination of the bearing capacity of soils for shallow foundations and piles.

For quantitative interpretation, it is definitely mandatory that the soil types of the layers penetrated be known, e.g., from the results of key borings.

Quantitative interpretation is mainly used for the results from DP obtained in cohesionless soils (Table 3). Relative density, shear parameters and compressibility are mostly evaluated from the test results. Relations between the engineering properties and the DP results are established directly and/or - as in the cases of shear parameters and compressibility - evaluated indirectly using correlations between density and penetration resistance.

While in the case of the interpretation of the DP results obtained in cohesionless soils the engineering properties are given as functions or at least as ranges of penetration resistance, establishing relations for the evaluation of the results obtained in cohesive soils is not as widely known. Evaluation procedures for compressibility and consistency have become known only lately (Ref. 2). Ranges and estimates still prevail (Table 3).

Reports about relations to determine the bearing capacity of shallow foundations directly from the results of dynamic penetration tests remain scarce (Table 3). Most of the time, the bearing capacity is determined indirectly through relative density or shear strength predictions.

Because of the similarity of penetrometers and piles, numerous attempts have been made to determine the pile bearing capacity on the basis of penetrometer results (see also Table 3). Most of the commonly known methods are described in Reference 1. Essential improvements in the evaluation of the corresponding test results were made using statistical approaches (Ref. 2).

Theoretical considerations still have not resulted in closed analytical solutions (Ref. 2). However, theoretical findings help to understand the basic principles involved, and thus to improve the quantitative interpretation.

Table 3 summarizes the application of qualitative and quantitative interpretation of DP results in some member countries. The results of dynamic penetration tests are being interpreted mainly qualitatively; however, the tendency toward quantitative interpretation - especially of results obtained in cohesionless soils - has increased considerably within the last ten years.

8 Standardization

Today, thirteen out of the twenty member countries listed in Table 3 and Australia are using national standards or recommendations of some kind; in 1974 this number was as low as eight (Ref. 1). In addition, it should be noted that national standards, e.g. DIN 4094 (DPL, DPH), and the former European Recommended Standard (DPB) are used across borders (Table 1).

At this point some general remarks may be repeated about the push for international standardization of dynamic penetration testing:

The dynamic penetrometer is the oldest device of its kind. As a consequence, a large variety of equipment is used in different countries, which is in some cases defined by national standards, in others not. Any international standard defining only one or two types of this equipment will give rise to severe difficulties, e.g.
- existing national standards will be overruled
- private industry and consultants as well as research institutes will be forced to buy new equipment, which is inconceivable in the present economic situation.

Therefore, international standardization should be pursued rather cautiously and appears to be premature at the present time.

9 Selection of International Reference Test Procedures (RTP)

Based on the principles and the main criteria for RTP selection, on which the Working Party had agreed (DOCUMENT NO. 1), and based on the results of the review of the present practice (Chapters 1 to 8), we agreed upon establishing RTPs for the following methods of dynamic probing:
- Dynamic probing, light (DPL), according to the former European Recommended Standard (ERS)
- Dynamic probing, medium (DPM)
- Dynamic probing, heavy (DPH)
- Dynamic probing, superheavy (DPSH), DPB type of the former ERS.

Additional reasons for this selection are:
- DPL is widely used in quality control during construction supervision besides the regular site investigations. Its international use seems to have increased since the publication of the ERS
- DPM is widely used and fills the gap of dimensions between DPL and DPSH
- DPH is widely used and also fills the gap of dimensions between the two extremes DPL and DPSH; the dimensions of the equipment are rather consistent in the countries in which it is used; therefore, it is predestined for use as RTP
- Dynamic probing of the superheavy type (DPSH) shows widespread use; however, the range of the existing dimensions is rather wide. The DPB of the former ERS was selected in favour of the former DPA because its dimensions simulate the SPT (that has a different test procedure). Furthermore, the former DPB's use appears to have increased in at least a few countries since the introduction of the ERS, in contrast to the DPA. Therefore, the DPB represents a better example of DPSH-type penetrometers.

DOCUMENT NO. 3

INTERNATIONAL REFERENCE TEST PROCEDURES

CONTENTS

1 Scope

The expression probing is used to indicate that a continuous record is obtained from the test in contrast to, for example, the Standard Penetration Test (SPT). The aim of dynamic probing is to measure the effort required to drive a cone through the soil and so obtain resistance values which correspond to the mechanical properties of the soil. Four procedures are recommended:

Dynamic probing light (DPL) representing the lower end of the mass range of dynamic penetrometers used worldwide; the investigation depth usually is not larger than about 8 m if reliable results are to be obtained.

Dynamic probing medium (DPM) representing the medium mass range; the investigation depth usually is not larger than about 20 to 25 m.

Dynamic probing heavy (DPH) representing the medium to very heavy mass range; the investigation depth usually is not larger than about 25 m.

Dynamic probing superheavy (DPSH) representing the upper end of the mass range of dynamic penetrometers and simulating closely the dimensions of the SPT; the investigation depth can be larger than 25 m.

2 Definition

2.1 General Principles and Nomenclature

A hammer of mass M and a height of fall H is used to drive a pointed probe (cone). The hammer strikes an anvil which is rigidly attached to extension rods. The penetration resistance is defined as the number of blows required to drive the penetrometer a defined distance. The energy of a blow is the mass of the hammer times the acceleration of gravity and times the height of the fall (MxgxH).

The results of different types of dynamic probing may be presented (and/or compared) as resistance values q_d or r_d (see Note 1, Section 7).

Dynamic probing is mainly used in cohesionless soils. In interpreting the test results obtained in cohesive soils and in soils at great depth, caution has to be taken when friction along the extension rod is significant*). Dynamic probing can be used to detect soft layers and to locate strong layers as, for example, in cohesionless soils for end-bearing piles (DPH, DPSH). In connection with key borings, soil type and cobble and boulder contents can be evaluated under favorable conditions. The results of the DPL can also be used to evaluate trafficability and workability of soils.

After proper calibration, the results of dynamic probing can be used to get an indication of such engineering properties as, e.g.,
- relative density
- compressibility
- shear strength and
- consistency.

For the time being, quantitative interpretation of the results including predictions of bearing capacity remains restricted mainly to cohesionless soils; it has to be taken into account that the type of cohesionless soil (grain size distribution, etc.) may influence the test results.

---

*) for skin friction see Note 2, Section 7

2.2 Classification

Four different probing methods, types DPL, DPM, DPH and DPSH, are recommended to fit different topographic and geological conditions and various purposes of investigation.

Apparatus, test procedures, measurements and recording are described in the following Sections. Technical data are summarized in Table 1.

Other types of equipment may be required for special purposes or different cone dimensions, e.g. for DPL (see Note 3, Section 7).

3 Apparatus

3.1 Driving Device

The driving device consists of the hammer, the anvil and the guide rod. Dimensions and masses are given in Table 1.

The hammer shall be provided with an axial hole with a diameter which is about 3-4 mm larger than the diameter of the guide rod. The ratio of the length to the diameter of the cylindrical hammer shall be between 1 and 2. The hammer shall fall freely and not be connected to any object which may influence the acceleration and deceleration of the hammer. The velocity shall be negligible when the hammer is released in its supper position (Note 4, Section 7).

The anvil shall be rigidly fixed to the extension rods. The diameter of the anvil shall not be less than 100 mm and not more than half the diameter of the hammer. The axis of anvil, guide rod and extension rod shall be straight with a maximum deviation of 5 mm per meter.

3.2 Extension Rods

Dimensions and masses of the extension rods are given in Table 1. The rod material shall be of high-strength steel with high resistance to wear, have a high toughness at low temperatures and a high fatique strength. Permanent deformations must be capable of being corrected. The rods shall be straight. Solid rods can be used; hollow rods should be preferred in order to reduce the weight (Note 5, Section 7). The rod joints shall be flush with the rods (Note 6, Section 7). The deflection (from a straight line through the ends) at the mid point of a 1-m push rod shall not exceed 0.5 mm for the five lowest push rods and 1 mm for the remainder.

3.3 Cones

Dimensions of cones are given in Table 1. The cone consists of a conical part (tip), a cylindrical extension, and a conical transition with a length equal to the diameter of the cone between the cylindrical extension and the rod (Fig. 1). The cones when new shall have a tip with an apex angle of 90°. The tip of the cone may be cut (e.g. by wear) about less than 10 % of the diameter from the theoretical tip of the cone.

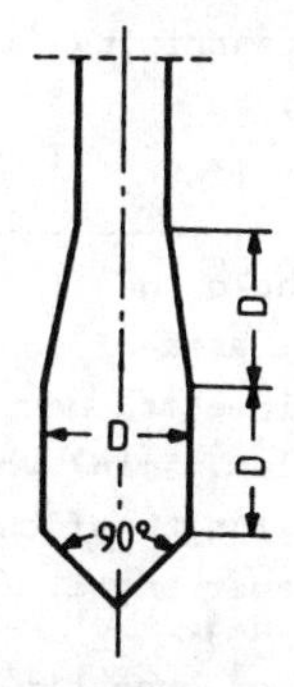

Fig. 1: Scheme of cones and rods (for dimensions see Table 1)

The maximum permissible wear of the cone is given in Table 1. The cone shall be attached to the rod in such a manner that it does not loosen during the driving. Fixed or "lost" (detachable) cones can be used.

4 Test Procedure

4.1 General

Criteria for the determination of a test should be specified in advance. The depth required will depend on the local conditions and the purpose of the particular test.

4.2 Probing Equipment

Probing shall be effected vertically unless specified otherwise. The maximum devitation of the test rig is

Table 1: Technical data of the equipment

Factor	Reference Test Procedure DPL	DPM	DPH	DPSH
Hammer mass, kg	10 ± 0.1	30 ± 0.3	50 ± 0.5	63.5 ± 0.5
Height of fall, m	0.5 ± 0.01	0,5 ± 0.01	0.5 ± 0.01	0.75 ± 0.02
Mass of anvil and guide rod (max), kg	6	18	18	30
Rebound (max), %	50	50	50	50
Length to diameter (D) ratio (hammer)	$\geq 1 \leq 2$	$\geq 1 \leq 2$	$\geq 1 \leq 2$	$\geq 1 \leq 2$
Diameter of anvil (d), mm	$100 < d < 0.5D$	$100 < d < 0.5D$	$100 < d < 0.5D$	$100 < d < 0.5D$
Rod length, m	1 ± 0.1 %	1-2 ± 0.1%	1-2 ± 0.1 %	1-2 ± 0.1 %
Maximum mass of rod, kg/m	3	6	6	8
Rod deviation (max), first 5 m, %	1.0	1.0	1.0	1.0
Rod deviation (max), below 5 m, %	2.0	2.0	2.0	2.0
Rod eccentricity (max),mm	0.2	0.2	0.2	0.2
Rod OD, mm	22 ± 0.2	32 ± 0.3	32 ± 0.3	32 ± 0.3
Rod ID, mm	6 ± 0.2	9 ± 0.2	9 ± 0.2	-
Apex angle, deg.	90	90	90	90
Nominal area of cone, cm^2	10	10	15	20
Cone diameter, new, mm	35.7 ± 0.3	35.7 ± 0.3	43.7 ± 0.3	50.5 ± 0.5
Cone diam. (min),worn, mm	34	34	42	49
Mantle length of cone, mm	35.7 ± 1	35.7 ± 1	43.7 ± 1	50.5 ± 2
Cone taper angle, upper, deg.	11	11	11	11
Length of cone tip, mm	17.9 ± 0.1	17.9 ± 0.1	21.9 ± 0.1	25.3 ± 0.4
max wear of cone tip length, mm	3	3	4	5
Number of blows per cm penetration	10 cm; N_{10}	10cm; N_{10}	10 cm; N_{10}	20 cm; N_{20}
Standard range of blows	3 - 50	3 - 50	3 - 50	5 - 100
Specific work per blow: Mgh/A, kJ/m^2	50	150	167	238

2 % : 1 (horizontal) to 50 (vertical). The probing equipment shall be firmly supported. The cone and the extension rods must be guided at the beginning of a test to keep the rods straight. Pre-boring may required.

The diameter of the bore hole shall be slightly larger than that of the cone. The test rig shall be positioned in such a way that the extension rods cannot be bent above the ground surface.

4.3 Driving

The penetrometer shall be continuously driven into the subsoil at a rate of 15 to 30 blows per minute. In pervious sands and gravels, the driving rate has a minor influence on the results; consequently, the driving rate can be increased up to 60 blows per minute in such soils.

All interuptions shall be recorded in the site log. All factors which may influence the penetration resistance (e.g. tightness of the rod couplings, straightness of extension rods) should be checked regularly. Any deviations from the recommended test procedures shall be recorded. The rods shall be rotated one and a half turn every meter to keep the hole straight and vertical, and to reduce skin friction (see Note 2, Section 7). When the depth exceeds 10 m, the rods shall be rotated more often, e.g. every 0.2 m. It is recommended to use a mechanized rotating device for large depths.

5 Measurements

The number of blows should be recorded every 0.1 m for DPL, DPM and DPH (N_{10}) and every 0.2 m for DPSH (N_{20}) (see Note 7, Section 7). The blows can easily be measured by marking the defined penetration depth (0.1 or 0.2 m) on the rods. The normal range of blows - especially in view of any quantitative interpretation of the test results - is between N_{10} = 3 and 50 for DPL, DPM and DPH and between N_{20} = 5 and 100 for DPSH. The rebound per blow should be less than 50 % of the penetration per blow. In exceptional cases (outside these ranges), when the penetration resistance is low, e.g. in soft clays, the penetration depth per blow can be recorded. In hard soils, where the penetration resistance is very high, the penetration for a certain number of blows can be recorded.

It is recommended to measure the torque required for rotating the extension rods to estimate skin friction. The skin friction can also be measured by means of a slip coupling close to the cone.

6 Reporting of Test Results

The following information shall be reported:

a) Location of probing
 Type of application
 Purpose of probing
 Date of probing
 Number of probing

b) Probing number, evaluation and location of probing and of the borehole (in case of reference boring). Position of the test rig with respect to the ground surface. Elevation or depth of the ground water table

c) Equipment used. Type of penetrometer, cone, rod, casing, bentonite etc.

d) Mass of hammer, height of fall and number of blows required per defined penetration

e) Elevation or depth at which the rods were rotated

f) Deviations from the normal procedure such as interruptions or damage to rods

g) Observations made by the operator such as soil type, sounds in the extension rods, indication of stones, disturbances, etc.

An example of a site log is shown in Fig. 2.

The probing results shall be presented in diagrams which show the N_{10} or N_{20} values on the horizontal axis and the depth on the vertical axis. An example is given in Fig. 3. If other measurements are taken such as the settlement per blow or the penetration per a certain number of blows, these values should be transformed to N_{10}, N_{20} or r_d, q_d values before drawing or numbering the diagram. Alternatively, it may be advantageous to transform the number of blows per defined penetration into resistance values r_d or q_d. The resistance values shall be plotted on the horizontal axis.

PROJECT LOCATION SECTION / HOLE-NO				SHEET-NO DATE OPERATOR
ELEVATIONS GROUND SURFACE		GROUND WATER TABLE		REFERENCE LEVEL
PURPOSE OF TEST				TYPE
EQUIPMENT: DYNAMIC PROBING ROD	CASING		POINT	DRILLING FLUID
DEPTH BELOW REFERENCE LEVEL m	HAMMER kg	HEIGHT OF FALL m	NUMBER OF BLOWS PER 0.2 m	NOTATIONS: INTERRUPTIONS, ROTATION, SOUNDS REASON FOR TERMINATING, ROD SHAPE

Fig. 2: Example of site log from dynamic probing

If the test was conducted according to the RTP, the letter R should appear on site logs, graphs etc. followed by the abbreviation of the penetrometer type (see Fig. 3). All divergences from the RTP must described completely on logs and graphs containing test results.

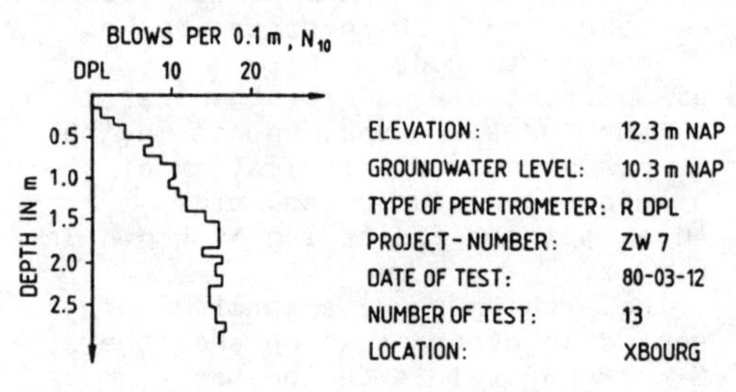

Fig. 3: Example of the presentation of the test results from dynamic probing (DPL)

7 Explanatory Notes

Note 1

Equations for r_d and q_d are:

$$r_d = \frac{MgH}{Ae}; \quad q_d = \frac{M}{M + M'} \times \frac{MgH}{Ae}$$

where:

r_d and q_d are resistance values in Pa, kPa or MPa
M is the total mass of the extension rods, the anvil and the guide rods
H is the height of fall
e is the average penetration per blow
A is the base area of the cone
g is the acceleration of gravity.

r_d- and q_d-resistances do not relate to strength ranges that the corresponding mechanical device can sustain. Especially, at high resistances r_d- and q_d-diagrams should be analysed with caution.

Note 2

To eliminate skin friction, drilling mud shall be injected through holes in the hollow rods near the cone. The holes have to be directed horizontally or slightly upwards. The injection pressure should be sufficient that the drilling mud fills the annular space between the soil and the rod. A casing can be used in addition.

Note 3

Some light penetrometers have hammers of 20 kg mass (e.g. Bulgarian State Standard 8994-70); in some countries 5 cm^2 cones are used (e.g. Belgium, German Standard DIN 4094). Some medium penetrometers have hammers of 20 kg mass, and heights of fall of 20 cm are being used in some countries (e.g. DIN 4094 of FRG, and Switzerland). Also, a height of fall of 50 cm is used in some instances of the DPSH (e.g. Finland).

Note 4

The free-falling hammer should be raised slowly to ensure that the inertia of the hammer does not carry it above the defined height. Also the pickup assembly should be lowered slowly to avoid significant impact on the hammer.

Note 5

Instead of the hollow extension rods (OD = 22 mm) of the DPL, massive rods of OD = 20 mm are being used. In regard to the extension rods of DPSH, it is recommended to increase the OD from 32 to 36 mm (this suggestion comes from France, Spain and Sweden).

Note 6

Curvature and eccentricity are best measured by coupling a rod together with a straight rod and holding the straight rod in contact with a plane surface.

Note 7

In the case of DPL, DPM and DPH, the numbers of blows are occasionally counted per 0.2 m depth interval of penetration.

Appendix A

Technical Committee (May 1983)

Dr. J.C. Holden
25 Marbary Drive
GLEN WAVERLY VIX 3150
Australia

Prof. E. de Beer
Rijkinstituut voor
Grondmechanica
Tramstraat 44

9710 ZWIJNAARDE
Belgium

Mr. L. Décourt
Avenue Brigadeiro Faria Lime 1857

01451 SAO PAULO, SP
Brazil

Prof. G. Stefanoff
University of Civil Engineering
Dept. of Soil Mechanics
and Foundation Engineering
Ul. Latinka 35

SOFIA 1113
Bulgaria

Dr. R. G. Campanella
Department of Civil Engineering
University of British Columbia
2324 Main Mall

VANCOUVER B.C. V6T, 1W5
Canada

Ing. V. Bures
Institute of Geology for
Civil Engineering
Ladbabská 4

160 00 PRAHA 6
Czechoslovakia

Civil Eng. H. Denver
Danish Geotechnical Institute
Maglebjergvej 1

DK 2800 LYNGBY
Denmark

Prof. Abdel Gawad A. Basaara
Faculty of Engineering
University of Cairo

GIZA
Egypt

Prof. Markku Tammirinne
Technical Research Centre
of Finland
Geotechnical Laboratory

SF-02150 ESPOO 15
Finland

Prof. Guy Sanglerat
Soils Investigation
182 Ave Félix Faure

69003 LYON
France

Dr. Klaus-J. Melzer
Battelle-Institut e.V.
Am Römerhof 35
Postfach 900 160

D-6000 FRANKFURT AM MAIN 90
Federal Republic of Germany

Mr. B. P. Papadopoulos
Foundations Engineering Dept.
National Technical University
42 Patission Street

ATHENS
Greece

Mr. Eli Zolkov
Eng. E. Zolkov Ltd.
P.O. Box 1005

RAMAT HASHARON
Israel

Prof. Giorgio Berardi
Facolta di Ingegneria
Via Montallegro 1

16145 GENOVA
Italy

Dr. Todahiko Muromachi
Kiso-Jiban Consultants Co. Ltd.
11-5 Kudan-Kita 1 chome

Chiyoda-ku, TOKYO 102
Japan

Mr. W. J. Heijnen
Delft Soil Mechanics Lab.
Stieltjesweg 2
Postbus 69

DELFT
Netherlands

Civ. ing. L. I. Finborud
Division of Soil Mechanics
NTH
Högskoleringen 7

7034 TRONDHEIM
Norway

Mr. G. A. Jones
16th Floor, 20 Anderson Street

JOHANNESBURG 2001
South Africa

Ing. V. Escario
Laboratorio de Carreteras
y Geotecnica
Alfonso XII 3

MADRID 7
Spain

Prof. B. Broms
Nanyan Technological Institute
Upper Jurong Road

SINGAPORE 2263
Singapore

Mr. S. Thorburn
Thorburn and Partners
145 West Regent Street

GLASGOW
Scotland

Prof. J. H. Schmertmann
Schmertmann & Crapps Inc.
4509 NW, 23rd Ave, Suite 19

GAINSVILLE, FL 32601
U.S.A.

Prof. YU. G. Trofimenkov
Gosstroy USSR
Prospekt Marx, 12

MOSCOW
USSR

Ing. Civ. M. L. Galavis
Apartado 68.651
Altamira

CARACAS 106
Venezuela

Mr. Ulf Bergdahl,
Swedish Geotechnical Institute

S-581 01 LINKÖPING
Sweden

S e c r e t a r y

Corresponding members, Working Party DP

Prof. Ir. A. van Wambeke
Ecole Royale Militaire
30 Avenue de la Renaissance

B-1040 BRUXELLES
Belgium

Ing. K. Drozd
Institute of Geology
for Civil Engineering
Mylynská 7

160 00 PRAHA 6
Czechoslovakia

Dr. A. G. Anagnostopululos
Dept. Foundation Engineering
Nat. Techn. University of Athens
113 Patission Street

ATHENS 813
Greece

Mr. Hans Zeindler
Juraweg

CH-3110 Münzingen
Switzerland

Dr. S. Ohya
Oyo Corporation 2-1
Otsuka 3 chome

Bunkyo-ku, TOKYO 112
Japan

Dr. V. S. Aggarwal
Soil Engineering Div.
Central Building Research
Institute

ROORKEE (UP)
India

Mr. Nils Rygg
Veglaboratoriet
Ganstadalleen 25

OSLO 3
Norway

Mr. Eero Slunga
Technical Research Centre
of Finland
Geotechnical Laboratory

SF-02150 ESPOO 15
Finland

Mr. A. Hansen
Danish Geotechnical Institute
Maglebjergvej 1

DK-2800 LYNGBY
Denmark

Mr. S. J. Trevisan
Calle 53, No. 390
Subsuelo

1900 LA PLATA
Argentine

Associacao Brasiliéra de
Mocanica dos Solos
SP-Caixa Postal 7141

0100 SAO PAULO
Brazil

Prof. G. G. Meyerhof
889 Beaufort Avenue

HALIFAX, N.S. B3H 3X7
Canada

Chinese Society of Soil
Mechanics and Foundation
Engineering
CCES, 10 Fluxing Road
P.O.Box 2500

BEIJING
China

Dr. S. Frydman
Geotechnical and Mineral
Engineering Area
Faculty of Civil Engineering
Technion City
Israel Institute of Tech.

HAIFA
Israel

Dr. S. Marchetti
Via Braciano 38

00189 ROMA
Italy

Dr. Z.-C. Moh
Moh & Associates
11th Floor, 75 Nanking E. Road
Section 4

TAIPEI
Taiwan

British Geotechnical Society,
Attn. Mr. B. A. Leach
Great George Street
Westminster

LONDON SW1P 3AA
United Kingdom

Penetration Testing 1988, ISOPT-1, De Ruiter (ed.)
© 1988 Balkema, Rotterdam, ISBN 90 6191 801 4

Weight sounding test (WST): International reference test procedure

ISSMFE Technical Committee on Penetration Testing
WST Working Party:
U.Bergdahl, Chairman (Sweden)
B.B.Broms (Sweden, Singapore)
T.Muromachi (Japan)

ABSTRACT: The Working Party on Weight Sounding Test of the ISSMFE Committee on Penetration Testing of Soils presents the results of its work in a three-part report. Document No 1: Introduction and objectives; Document No 2: Development and current use of the WST; Document No 3: International Reference Test Procedure of the WST.

DOCUMENT NO 1

INTRODUCTION AND OBJECTIVES

1. The Weight Sounding Test (WST) is extensively used in Sweden, Norway, Finland and Denmark. During the last few decades the WST method has also been used in Poland, Hungary, Czechoslovakia, Japan, Singapore, the Philippines as well as in Algeria. The method was recommended in Sweden for the first time in 1917.

The Weight Sounding Test is a simple inexpensive and fast method. The penetrometer can either be operated manually or mechanically which makes it suitable for different applications. Its sensitivity in soft soil is rather good but it can also penetrate very dense sand and gravel due the the rotation of the screw-shaped point.

Thanks to these advantages, the existing correlations with other penetration methods and experiences with the interpretation of the results it is believed that the method will stay and its use be extended.

The WST is primarily utilized to obtain a continous soil profile and an indication of the layer sequence and of the lateral extent of different soil layers. The results can also be used to get an indication of the relative density of cohesionless soils and the shear strength of cohesive soils. The bearing capacity of friction piles and spread footings in cohesionless soils can be determined as well.

It has been found that the results for silty, gravelly or stony soils must be handled with caution since the measured resistance might be higher than that which corresponds to the actual shear strength or compressibility of the soil.

2. In 1977, the ISSMFE European Sub-Committee on Penetration Testing proposed a European Recommended Standard for the Weight Sounding Test. The proposed reference standard was adopted by the ISSMFE Executive Committee at its meeting in Tokyo in 1977.

In 1982, at the Second European Symposium on Penetration Testing in Amsterdam, an ISSMFE Technical Committee on Penetration Testing was appointed. At the meeting in Amsterdam it was agreed that the Committee should work out proposals for Reference Test Procedures for the following penetration testing methods: Standard Penetration Test (SPT), Cone Penetration Test (CPT), Dynamic probing (DPA, DPB, DPL) and Weight Sounding Test (WST).

Such Reference Test Procedures will facilitate the correlation between the different penetration testing methods used in the various countries and with the bearing capacity and driving resistance of piles, plate load tests, settlements of piles, spread footings and rafts etc. It will also give a better understanding of the influence of the differences in current test procedures and facilitate the selection of various testing methods suitable for different tasks.

At the committee meeting in Amsterdam it was also decided to use the already adopted Recommended European Standard as a base for the Reference Test Procedures.

A questionnaire together with the Recommended European Standard and the State-of-the-art report on WST from Amsterdam was sent to the following countries: Czechoslovakia, Denmark, Finland, Hungary, Japan, Norway and Poland to determine if any modifications were required. The following questions were raised.

1. Are the descriptions of equipment, test procedure, recording and presentation of test results adequate or do you think something has to be added or can be omitted?

2. Are there details in the equipment or test procedure you think are necessary to change in a forthcoming Reference test procedure? E.g. it has been discussed to use 25 mm diameter rods instead of 22 mm. There is also a desire to limit the worn-down of the sharp edge of the lower pyramidal-shaped part of the point.

3. What are your experiences from manual compared to mechanized sounding tests?

4. Are there other applications of this test method than those indicated in the Introduction to the Recommended Standard or do you think something has to be omitted?

5. Do you know of any other methods of interpretation of the WST-results than those indicated in the State-of-the-art report -82? N.B. Interpretations will not be included in the Reference test procedure but information on existing methods will be reported later on.

6. Can you make an estimate of the number of WST-equipment in use in your country.

7. Can you make an estimate of the number of projects in your country where WST-test is part of the site investigations.

8. We will also ask you to give us a reference for WST-standard or Recommendation existing in your country. If possible please enclose a copy.

9. Are there other points of view that have to be considered in the forthcoming work on an International Reference test procedure for the WST?

The result of this questionnaire indicated that only minor changes are required compared to the European Recommended Standard. These changes were:

i. The detailed description of the joint can be omitted.

ii. When the penetrometer is used as a static penetrometer and not rotated the results should be indicated by a diagram and not only with numbers.

iii. The rod diameter should be 19 to 25 mm and not only 22 mm. However, 22 mm diameter rods should be recommended.

At the meeting of the committee in Helsinki in May 1983 it was agreed that the committee should also include review of presently used penetration testing methods and the prediction of bearing capacity and settlements of footings, rafts and piles based on the results from WST. Only minor modifications have been made since the meeting in 1977 in Tokyo and 1982 in Amsterdam of the Weight Sounding Test and its application. Therefore only a revised summary of previous reports on the WST has been included in this report.

DOCUMENT No 2

DEVELOPMENT AND CURRENT USE OF THE WST

1. Introduction

The most common penetration testing method in the Scandinavian countries and in Finland is by far the weight sounding method (Aas, 1969, Broms, 1974). The method is frequently used also in Poland and Hungary. It is used in Czechoslovakia, Japan and Algeria as well.

For this method a screw shaped point, rods with normally 22 mm diameter and a number of weights (5, 10, 10, 25, 25 and 25 kg) are used as illustrated in Fig. 1. The point is manufactured of a steel bar with a square cross-section (⌀ 25 mm) which is twisted one turn to the left. Details of the Swedish standard rod is given in Fig. 2.

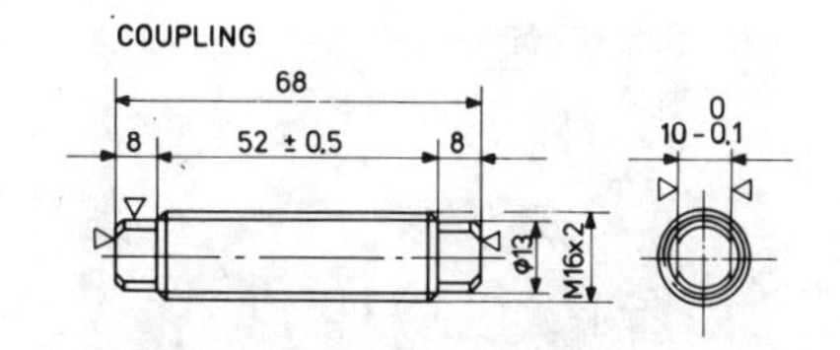

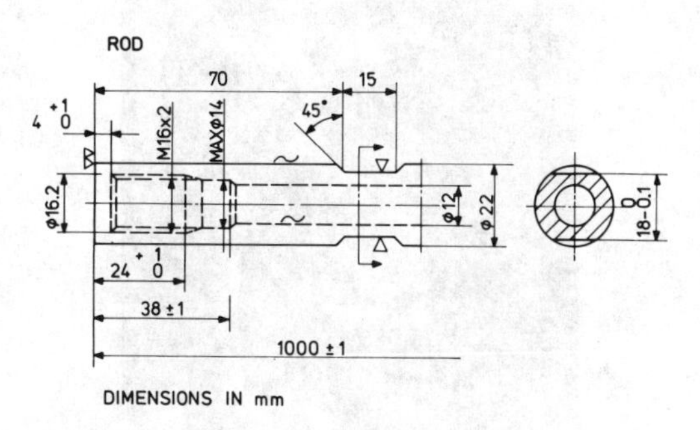

Fig. 2. Recommended tolerances for manufacture of 22 mm weight penetrometer rods and couplings.

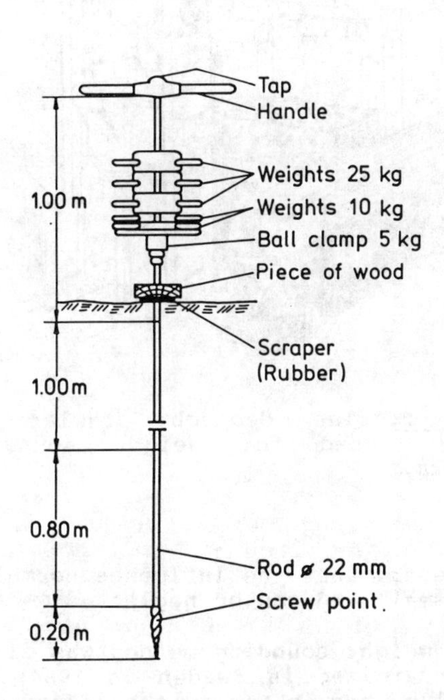

Fig. 1. Details on the manually operated weight penetrometer.

For a weight sounding test the load on the penetrometer is gradually increased to 0.05, 0.15, 0.25, 0.50, 0.75 and 1.0 kN without rotating the penetrometer by increasing the number of weights. The load is adjusted to keep the penetration rate constant, about 50 mm/s (3.0 m/min). When the penetrometer does not penetrate further when loaded to 1.0 kN it is rotated and the required number of halfturns every 0.2 m of penetration is recorded (N_{WST}, halfturns/0.2 m) Fig. 3.

A gasoline driven engine type Borros, Fig. 4, is often used to rotate the rods. The required number of halfturns for the penetrometer to penetrate 0.2 m (N_{WST}) is registered automatically by a counter. The penetrometer is usually pushed down manually by the two operators and the applied load is read on a separate force indicator (dynamometer). When the weight penetrometer was first developed it was rotated by hand. Today it is common to use a crawler tractor for the weight sounding tests as illustrated in Fig. 5.

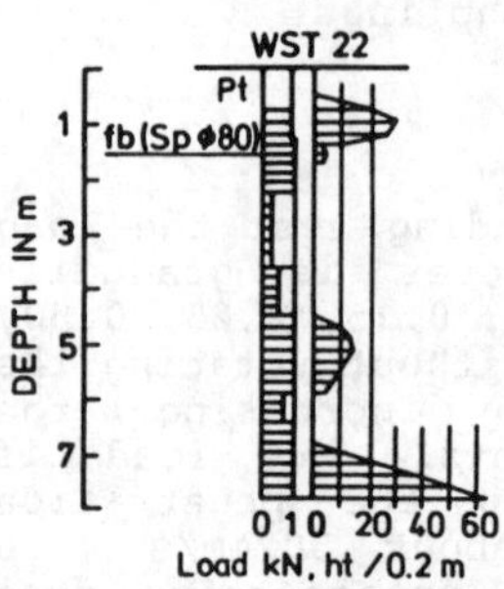

WST 22 — WEIGHT SOUNDING TEST, 22 mm RODS
ht/0.2 m — NUMBER OF HALFTURNS PER 0.2 m OF PENETRATION
Pt — DRY CRUST OF CLAY
fb(Sp φ80) — PREBORING TO THIS LEVEL WITH 80 mm DIAM AUGER

DIAGRAM TO THE LEFT INDICATE LOADS APPLIED IN kN

Fig. 3. Manually operated weight penetrometer with an example of test results.

Fig. 4. The Borro power unit in use for rotating the weight penetrometer.

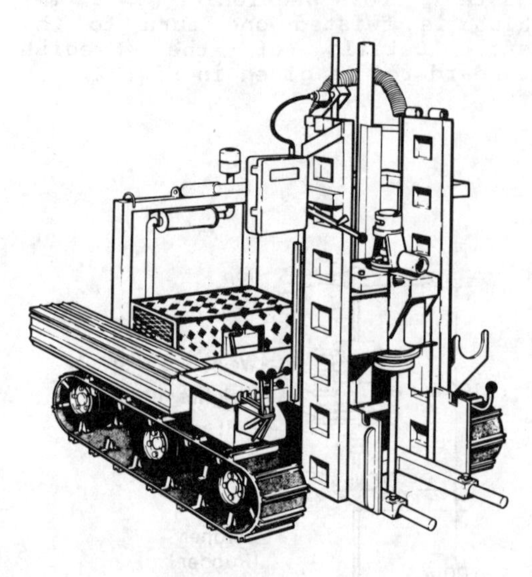

Fig. 5. The Geotech crawler is often used for weight sounding tests.

If an engine is used for the rotation it is difficult to determine the type of soil (clay, silt, sand or gravel) at the screw shaped point from the sounds and the vibrations of the rods during the rotation. The vibrations from the engine can also affect the results. Investigations in Sweden (Bergdahl, 1969) and in Norway (Senneset, 1974) indicated, however, that the influence normally is small and can be neglected.

The weight sounding method was first standardized in Sweden in 1964. In Finland and in Norway the method was standardized in 1968 and in 1973 respectively (NGS, 1982). The proposed European standard, for the weight sounding test was approved at the meeting of the Executive Committee of the ISSMFE in Tokyo in 1977. This recommended standard corresponds closely to the standards in the Scandinavian countries and Poland.

2. Development of the weight sounding method

The weight sounding test is an old method. At the end of the 19th century it was common in Sweden to push or to drive square or circular steel rods with 15 to 30 mm diameter through the soil to determine the depth to a till layer or to bedrock.

The first modern weight sounding device is that described by O. Olsson in 1915 in a manual published by the Swedish State Railway. One metre long steel rods with 15 mm diameter were then used. The penetrometer was provided with a twisted square steel point. The penetration resistance was expressed in the number of men required to push the penetrometer through the soil. If it was not possible to push the penetrometer by hand a 90 kg weight was added and the penetrometer was rotated. The penetration every 25 halfturns was recorded.

This early weight penetrometer was later modified (The Swedish Geotechnical Commission 1917, 1922). The force required to push down the penetrometer was measured using a series of weights. The soil was classified with respect to the required force.

A soil was for example classified as a "50 kg clay" when a force of 0.5 kN (50 kg) was required to push down the penetrometer. When the penetrometer stopped when loaded to 1 kN (100 kg) it was rotated and the penetration every 25 halfturns was measured.

The weight sounding method was used extensively by the Swedish Geotechnical Commission (1922) to determine the layer sequence and the thickness of the different strata when the stability of railway embankments were investigated. At that time several large landslides had occurred in Sweden e.g. at Lake Aspen (1913) and at Getå (1918).

Additional developments have been made to increase the strength of the different parts (rod diameter has increased from 19 to 20, 22 and now 25 mm) and to mechanize the method, Fig. 4-5. Weight soundings are often combined with percussion drilling so that also very dense strata can be penetrated. Extensive investigations of the interpretation of the test results for different application have been carried out as well. These investigations indicate also that it is necessary to standardize the penetrometer and the test procedure.

3. Factors affecting the penetration resistance

The need to standardize different sounding methods was discussed intensively in Sweden in the 1950th. The Swedish Geotechnical Society (SGF) appointed in 1958 a committee for standardization and mechanization of existing penetration testing methods. The committee investigated the influence of different factors on the penetration resistance. The effects of the wear of the penetrometer point, straightness of the rods, overburden pressure, method of rotation and friction along the rods were studied.

Senneset (1973) investigated for example the effect of the shape and the wear of the penetrometer point in different soils. He found for example that in soft clay the twist (1/2, 3/4 or 1 turn) did not significantly affect the results. The penetration resistance was reduced in clay if a worn point was used compared with a new point. In silt the difference was small. In dense sand the penetration resistance was higher for a worn point than for a new point when the depth was large. Close to the surface, the penetration resistance was lower for the worn point.

Similar results have been reported by Bergdahl (1969). He found that the penetration resistance in a sand was reduced somewhat close to the ground surface when a worn point was used. Below 7 m depth the penetration resistance was about 25 % higher for a worn point compared with a new point. In a silt the resistance was slightly reduced when a worn point was used.

The penetration resistance is affected when the penetrometer is rotated by machine. An increase of 60 to 70 % has e.g. been reported by Senneset (1973) for a clay when the weight penetrometer was rotated by a petrol driven engine compared to the resistance (halfturns/0.2 m) when the penetrometer was rotated by hand. Bergdahl (1969) found an increase of 30 % for a soft clay from Sweden. Natukka (1969) has reported an increase of the penetration resistance of 14 % for a clay in Finland. For a silt the penetration resistance was reduced by 7 to 17 % when the penetrometer was rotated by machine.

Senneset (1973) investigated also the effect of the straightness of the rods. The measured penetration resistance was hardly effected except in clay close to the ground surface. The penetration resistance was larger for the bent rods than for the straight rods. For large depths the penetration resistance was slightly smaller for the bent rods than for the straight rods.

Also the rotation speed affects the results. For a soft clay Bergdahl (1969) found that the penetration resistance increased 22 % when the speed was 125 rpm compared to a manually rotated penetrometer. When the speed was 50 rpm the increase was 16 %. For a fine sand the difference was small. For sand and sandy gravel the increase was 10 %. It is recommended that the speed of rotation should be between 15 and 40 rpm as stated in the European recommended standard. The rate should not exceed 50 rpm.

The effect of the diameter of the sounding rods on the penetration resistance has been investigated by Bergdahl (1969). In clay the effect was found to be small while in sand the penetration resistance was 7 to 8 % larger for 22 mm rods compared with 19 mm rods due to the difference in surface area. In silt the difference was small.

4. Relative density

Weight soundings are commonly used in Sweden and Finland to determine the relative density of cohesionless soils. Also the degree of compaction can be investigated (Hellman et al, 1979). The increase of the penetration resistance with depth close to the surface is then utilized. Below a certain critical depth the total penetration resistance is used. However, the penetration resistance is also affected by the particle size and by the gradation of the soil. It is therefore necessary to calibrate each material separately when the relative density is determined. Also the friction along the uncased rods can affect the results.

According to the Swedish Building Code (SBN, 1980) sand is classified as dense when the penetration resistance exceeds 15 halfturns/0.2 m and as medium to loose between 1 and 15 halfturns/0.2 m penetration. A somewhat different classification is used by the Swedish Road Administration (1976). Sand is then classified as middle dense when N_{WST} is between 10 and 30 halfturns/0.2 m and as dense when N_{WST} > 30 halfturns/0.2 m penetration.

In order to compare the results from weight soundings with other methods Bergdahl and Sundqvist (1974) and Helenelund (1966) have proposed the following classification for non-cohesive soils.

Classification	Sweden (Bergdahl and Sundqvist 1974) Halfturns/ /0.2 m	Finland (Helenelund 1966) Halfturns/ /0.2 m
Very loose	<8	<10
Loose	8- 20	10- 30
Medium	20- 60	30- 60
Dense	60-100	60-100
Very dense	>100	>100

Recent investigations by Bergdahl and Ottosson (1984) have shown that minor modifications of these limits ought to be done if comparisons are made with the point resistance

Classification	CPT, Point resistance q_c MPa	WST-resistance[1)] ht/0.2 m
Very loose	0- 2.5	0-10
Loose	2.5- 5.0	10-30
Medium	5.0-10.0	20-50
Dense	10.0-20.0	40-90
Very dense	>20.0	>80

1) In silt and silty sands the weight sounding resistance should be reduced by a factor of 1.3 before classification.

values, q_c from cone penetration tests.

In Norway the penetration resistance of a sand is classified as very low, low, medium, dense and very dense when the number of halfturns/0.2 m penetration is 0, <7, 7-25, 25-50 and >50, respectively (Senneset, 1974).

Test data indicate that the penetration resistance of cohesionless soils is not affected when the material is either dry or saturated (Bergdahl, 1973). The penetration resistance is, however, affected by the degree of saturation due to the false cohesion as found by Bergdahl (1973). The relative density can then be overestimated.

The size and the gradation of the soil also affect the results as pointed out by Bergdahl (1973). He also pointed out that there is no unique relation between relative density and the penetration resistance. An investigation in Finland (Gardemeister et al, 1974) has also indicated that the particle size affects the penetration resistance. At a given dry unit weight the penetration resistance was found to increase with decreasing particle size. Also the water content affects the results as well as the location of the ground water level (Tammirinne, 1973).

Dahlberg has (1974) investigated the effect of the overburden pressure on the penetration resistance in a fine sand. The tests were made in an excavation where the depth was gradually increased and the overburden pressure was reduced. The investigation indicated that the penetration resistance was affected below the bottom of the excavation down to a depth of 0.9 to 2.6 m where the effective overburden pressure was 25 to 30 kPa. Below this critical depth the penetration resistance did not change. These tests indicated that the penetration resistance at weight soundings is mainly governed by the lateral pressure in the ground. The lateral pressure is uneffected by a reduction of the overburden pressure except close to the bottom of the excavation.

5. Compressibility

The weight sounding method can also be used to estimate the settlement of spread footings and rafts. The compression modulus (M) is according to Helenelund (1966) approximately 10 to 20 MPa at 10 to 30 halfturns/0.2 m, 20 to 50 MPa at 30 to 60 halfturns/0.2 m and 50 to 80 MPa at 60 to 100 halfturns/0.2 m. The investigations by Bergdahl and Ottosson (1984) have indicated quite similar results as those indicated by Helenelund when considering the long term settlements in sands. They propose the following relation.

WST resistance ht/0.2 m	Modulus of elasticity E, MPa
0-10	<10
10-30	10-20
20-50	20-30
40-90	30-60
>80	>60

Before evaluation of the modulus in silt and silty sands the weight sounding resistance should be reduced by the factor 1.3 as mentioned above.

6. Shear strength and sensitivity

With weight soundings it is normally possible to separate the stiff surface crust of a deep clay layer from the underlying soft clay. The penetrometer must normally be rotated in the stiff surface crust while in the underlying soft clay the penetrometer often can be pushed down with a force which is less than 1 kN.

The friction along the penetrometer rods is normally low in the Scandinavian countries due to the often high sensitivity of the soil. The friction is mainly governed by the remoulded shear strength rather than the undisturbed strength. The sensitivity of Scandinavian clays is normally 10 to 20 while in quick clay it is higher than 50.

In areas where the sensitivity of the clay is low and where filling material occur close to the surface the skin friction resistance along the shaft of the penetrometer rods can be large and appreciably influence the results.

Several attempts have been made to correlate the results from weight soundings with the undrained shear strength for clays. However, the proposed relationships are very uncertain. Fukuoka (1974) has correlated the penetration resistance with the unconfined compression strength (Fig. 6). It can be seen that the penetration resistance increased with increasing shear strength. When the penetration resistance is less than 1 kN the undrained shear strength of the clay normally does not exceed 20 to 30 kPa. Similar results have been reported by Saarelainen (1979) for soft clays from Helsinki in Finland. When the penetration resistance was 0.5 kN the average undrained shear strength was about 13 kPa. The scatter of the results was large particularly close to the ground surface.

Möller (1980) has utilized the increase of the penetration resistance with depth as an indication of the sensitivity of clays. When the penetration resistance is constant with depth the sensitivity of the clay is often high because the shaft resistance is small. When the sensitivity is low the shaft resistance will be high and the penetration resistance will

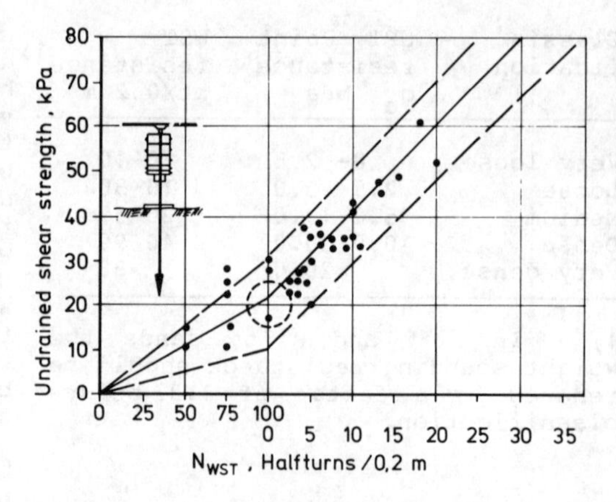

Fig. 6. Relationship between undrained strength and weight sounding resistante N_{WST} (Fukuoka 1974)

increase normally with depth. Broms and Bergdahl (1982) also show the influence of the sensitivity of the clay. Two places were compared one with medium sensitive clay and one with quick clay. At e.g. 17 m depth the shear strength of the clay was in both places about 40 kPa. In the medium sensitive clay the weight penetrometer resistance was 5-7 ht/0.2 m of penetration while it in the quick clay was only 0.25 kN.

7. Comparisons with other penetration testing methods

Comparisons with the weight sounding method and the standard penetration test (SPT) have been carried out in Japan where the weight sounding method is frequently used. The following relationship has been reported by e.g. Miki for cohesionless soils (Muromachi et al, 1974):

$$N_{SPT} = 0.42\ N_{WST}$$

where N_{SPT} is the standard penetration resistance (blow/ft) and N_{WST} is the penetration resistance (halfturns/0.2 m).

In Fig. 7 is shown a comparison between the results from weight soundings and standard penetration tests carried out in a preloaded deposit of medium to coarse sand.

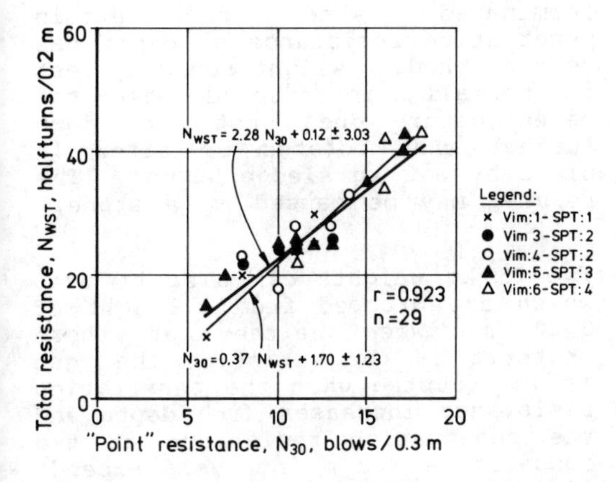

Fig. 7. Relationship between weight sounding resistance N_{WST} and SPT-values (Dahlberg, 1975)

The relationship $N_{SPT} = 0.37\ N_{WST} + 1.70$ corresponds closely to that proposed by Miki.

Also Muromachi and Kabayashi (1982) and Bergdahl and Ottosson (1984) have presented similar correlations for cohesionless soils as can be seen from Fig. 8.

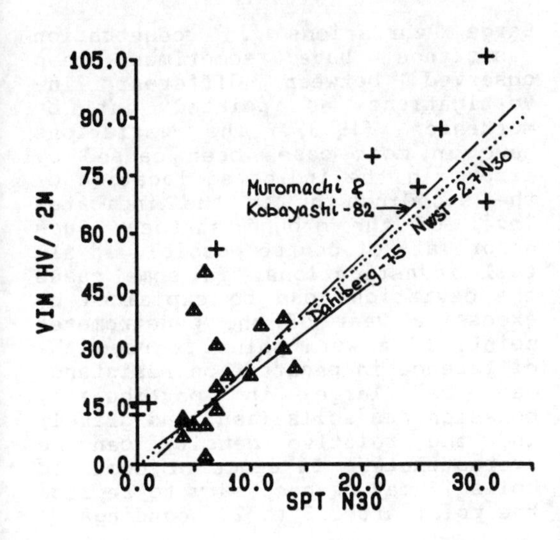

Fig. 8. Correlations between weight sounding resistance N_{WST} ht/0.2 m (VIM HV/.2 M) and N_{30} from SPT according to Bergdahl and Ottosson (1982).

Tammirinne (1974) has compared the penetration resistance of weight soundings and cone penetration tests (CPT) in coarse and fine sand using the Dutch cone penetrometer. For coarse sand the following relationship was obtained

$$q_c\ (\text{MPa}) = 0.5\ N_{WST}\ (\text{halfturns}/0.2\ \text{m})$$

For fine sand

$$q_c\ (\text{MPa}) = 0.2\ N_{WST}\ (\text{halfturns}/0.2\ \text{m})$$

where q_c is the cone penetration resistance.

Similar results have been reported by Bergdahl (1973), Muromachi and Kabayashi (1982) and Bergdahl and Ottosson (1984), Fig. 9.

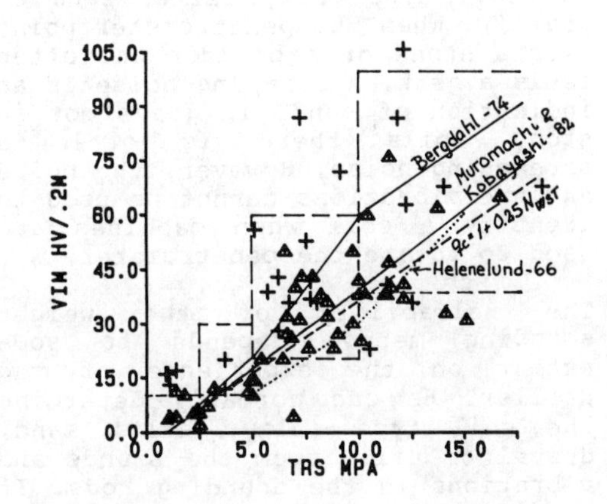

Fig. 9. Correlations between weight sounding resistance N_{WST} in ht/0.2 m (VIM HV/.2 M) and the point resistance q_c (TRS MPA) from CPT-test according to Bergdahl and Ottosson (1984).

8. Soil exploration

Weight soundings can be used in most soils except very dense sand and gravel or in stony soils and very stiff clays. It can be used in inaccessable areas where it is difficult to bring in the equipment for e.g. cone or standard penetration test or for dynamic probings.

Weight soundings are used in Sweden primarily in the exploratory phase of an investigation to determine the stratification, location and thickness of the different strata, the depth to "firm bottom", the depth where soil samples should be taken, the suitable location of e.g. field vane tests, cone penetration tests or dynamic probings. Weight soundings are also used to estimate the length and the bearing capacity of point bearing and friction piles in sand as well as the bearing capacity and settlement of spread footings and rafts. Weight soundings have been used as well to check the relative density of compacted fills and the digability of clay, silt, sand and gravel (Korhonen and Gardemeister, 1972).

The vibrations and noise in the sounding rods during a penetration test can also be useful to identify the soil type (clay, silt, sand or gravel). When the penetrometer point hits a stone or a boulder one often feels a jerk. A scraping noise is an indication of sand. In gravel or in stony soils there is often a screeching noise. However, the noise and the vibrations cannot be used to identify a soil when machines are used to rotate the penetrometer.

The reliability of the weight sounding method depends to some extent on the experience of the driller. He can normally determine the soil type (clay, silt, sand, gravel or till) from the sounds and vibrations in the sounding rods. If a fill or a sand or gravel layer is located close to the ground surface the determination becomes much more difficult because of the friction along the rods which affects the penetration resistance. Predrilling or precoring will than be required.

Predrilling or precoring should be used when e.g. a hard layer is located close to the surface since the friction in this layer may affect the results. Casing is sometimes used when the depth exceeds 20 to 25 m (Natukka, 1969) to reduce the friction along the rods or when the density of cohesionless soils is estimated at great depths (>10 m).

A weight sounding test is normally terminated when a certain penetration resistance or depth has been reached. A weight sounding test is normally interrupted when the penetrometer does not go down further when rotated or after 10 blows by a 3 kg sledge hammer. (The refusal may be caused by a stone.)

Two 25 kg weights can also be used which are dropped from a height of 0.20 m. When neither of these criteria can be satisfied the test is interrupted when the penetration resistance increases with depth and the penetration resistance at two consecutive 0.2 m intervals exceeds 40 halfturns/0.2 m and the penetration is less than 10 mm/blow at 5 blows by a sledge hammer or by the two 25 kilogram weights. If the penetration exceeds 50 mm/5 blows or the penetration resistance decreases with depth the test should be continued.

The maximum depth of all weight soundings should be recorded even if the tests have been interrupted close to the ground surface. There are probably stones or boulders in the soil if the depth of refusal varies which may affect for example the installation of piles or sheet piles.

Large variations in penetration resistance have sometimes been observed between different investigations as pointed out by Mortensen (1973). The variations have in most cases been caused by errors in the indicated location of the soundings or in the indicated level of the ground surface. Such errors are of course possible in all soil investigations. In some cases the deviations can be explained by excessive wear of the penetrometer point. If a worn point is used the difference in penetration resistance can be large in particularly cohesionless soils (sand and gravel) and the relative density can be overestimated. In dense or cemented soils it may be necessary to replace the point after 5 to 25 soundings.

The weight sounding is a flexible and inexpensive investigation method. Because of the relatively low cost it is possible to carry out a large number of soundings for every boring. The costs today (1987) in Sweden is about 90 $/sounding for a 10 m deep hole in soft clay or in loose to medium dense sand.

Soil investigations and weight soundings are in many cases carried out at a fixed price for each borehole or per metre of penetration regardless of the soil conditions. The disadvantage with this method of payment is that it is often difficult to change the investigation program to fit the actual conditions in the field. The advantage with this system is that the owner knows in advance the total cost of the investigation.

9. Shallow foundations

The design of spread footings and rafts in Sweden is primarily based on the results of weight soundings. The weight sounding tests are normally supplemented by auger borings where disturbed but representative samples are obtained so that the soil can be classified. However, there are examples where shallow foundations have been designed on the basis of sounding alone. In one case a building was founded just above a peat layer (Mortensen, 1973). It was not possible to detect this layer from the weight soundings due to the high friction along the penetrometer rods in the peat.

Different rules of thumb exist which are based on local experience. For example in Sweden bridge abutments are founded on spread footings on sand if the penetration resistance (N_{WST}) exceeds 10 halfturns/0.2 m. In Denmark two storey buildings are founded on spread footings in boulder clay when the penetration resistance is 10 to 15 halfturns/0.2 m.

Weight soundings are normally used in Sweden to estimate the allowable bearing capacity of spread footings on gravel, sand or sandy silt. According to the present Swedish Building Code (SBN 1980) sand and silt are classified as dense when the penetration resistance exceeds 15 halfturns/0.2 m and as medium to loose when the penetration resistance is between 1 and 15 halfturns/0.2 m penetration. Spread footings and rafts are not used when the penetration resistance is less than 1 halfturn/0.2 m penetration, because of large expected differential settlements.

The Swedish Road Administration (1976) classifies sand and gravel as dense when the penetration resistance N_{WST} exceeds 30 halfturns/0.2 m and as medium dense between 10 and 30 halfturns/0.2 m. It is worth mentioning that the Swedish classification limits do not agree with international practice as can be seen from the second table in chapter 4. However, a new classification is being worked out also including the N_{WST} resistance.

The allowable bearing capacity is calculated according to the Swedish Building Code from a special bearing capacity formula based on the ultimate resistance of the soil. The allowable soil pressure increases linearly with increasing size of the footing. In order to keep the settlements small the allowable soil pressure is limited. The maximum allowable soil pressure is 0.5 and 0.3 MPa for medium to coarse sand when the soil is dense and loose, respectively. The corresponding maximum soil pressure for a fine sand is 0.4 and 0.2 MPa, respectively. In gravel the maximum soil pressure is 0.6 MPa.

The allowable soil pressure should be calculated in a similar way according to the Swedish Road Administration. The factor of safety F_{ϕ} should be at least 1.5. The angle of internal friction can be determined from the following table:

Soil type	Penetration resistance halfturns/0.2 m 10-30	>30
Fine sand	31°	35°
Medium to coarse sand	35°	38°
Gravel	38°	42°

The indicated allowable soil pressures should not be used without a settlement calculation.

A more detailed evaluation of the angle of internal friction Ø from penetration resistance has recently been proposed by the Swedish Committee on Penetration Testing. This proposal indicate the q_c-values from CPT-tests to be the most reliable results for evaluation of soil characteristics. However, correlations with weight sounding resistances are also given. These correlations have been obtained both from international practice and from Swedish experiences and compilations. The following values have been obtained.

Classification	WST-resistance[1)] N_{WST} ht/0.2 m	Angle of internal friction, Ø °2)
Very loose	0-10	29-32
Loose	10-30	32-35
Medium	20-50	35-37
Dense	40-90	37-40
Very dense	>80	>40

1) In silt and silty sands the weight sounding resistance should be reduced by a factor of 1.3 before evaluation of the angle of internal friction.

2) The values given in the table is valid for sands. In silty soils the angle of internal friction shall be reduced by 3°. In gravel Ø can be increased by 2°.

10. Deep foundations

The results from weight soundings are also used to predict the driving resistance of piles as illustrated in Fig. 10 (Helenelund, 1974). It can be seen that the driving resistance of precast concrete piles (0.25 x 0.25 m) driven by a 3 to 4 ton drop hammer increased with increasing penetration resistance. However, the scatter of the results is large particularly for glacial till.

The driving resistance of fine sand and silt is often high probably due to the high negative pore water pressures that develop in the soil during the driving when the relative density is high. In glacial till stones and boulders in the soil have often a large influence on the penetration resistance as well as on the driving of timber and concrete piles and of sheet piles.

Experience in Finland (Eklund, 1970) indicates that the length of precast concrete piles corresponds approximately to the maximum depth that can be reached with weight soundings. In Sweden the length of end-bearing piles often correspond to the penetration depth plus 1 to 2 m. In Denmark (Winkel, 1969) it has been found that point bearing piles in boulder clay can be driven 0.5 to 1.0 m deeper than the depth where the penetration resistance (N_{WST}) is 60 to 80 halfturns/0.2 m. Dynamic probing is however a more accurate way of determining the length of end-bearing piles.

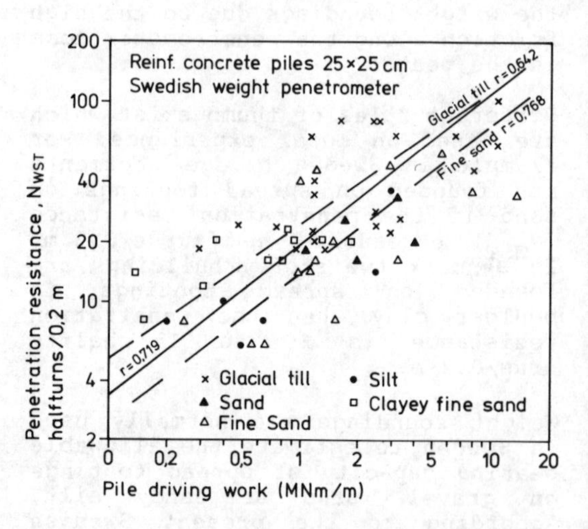

Fig. 10. Driving resistance of precast concrete piles in relation to the weight sounding resistance in different soils. (Helenelund 1974).

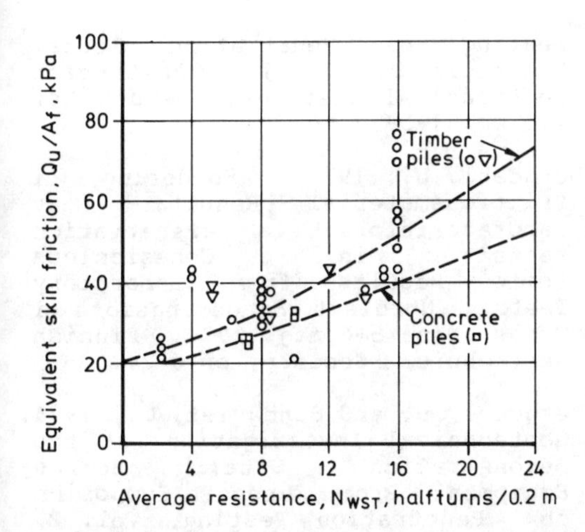

Fig. 11. Relationship between the equivalent skin friction resistance and the average penetration resistance from weight soundings. (Norwegian Pile Committe, 1973).

Results from weight soundings can also be used to estimate the bearing capacity of friction concrete and timber piles in sand. In Fig. 11 is shown the results from a number of pile load tests in sand (Norwegian Pile Committe, 1973). The ultimate bearing capacity of the piles (Q_u) divided with the total surface area of the piles (Q_u/A_f) neglecting the point resistance has been plotted as a function of the average penetration resistance from weight soundings. It can be seen that the bearing capacity of timber piles due to the taper of the piles is normally higher than that of concrete piles at the same penetration resistance. The ultimate bearing capacity per unit shaft area for timber piles increases in general with increasing penetration resistance of the soil from about 20 kPa when the penetration resistance is low (N_{WST}>24). The results are applicable to piles with a length of 12 to 20 m.

According to the Recommendations on Design of Bored Piles published by the Swedish Commission on Pile Research the angle of internal friction in cohesionless soils can be determined within the range 15 to 50 halfturns/0.2 m from the following table.

Soil type	Penetration resistance halfturns/0.2 m 15	20	30	50
Silt	-	-	-	27°
Sand	31°	32°	34°	38°
Gravel	32°	34°	37°	40°

It should bo noted that these values also include the installation effect on a specific soil. Therefore these values cannot be directly compared to these given in the previous table.

11. Compaction control

Weight soundings are frequently used in Sweden to check the relative density of fine and medium sand below the ground water level where other methods cannot be used or are difficult to apply. Lagging and Eresund (1974) have for example described a case where weight soundings have been used to check the uniformity of a hydraulically placed fill compacted by a 3 ton vibratory roller. The thickness of the fill varied between 8.3 and 9.0 m.

The penetration resistance of different soils can vary considerably at the same relative density due to a variation in grain size or the gradation of the material as reported for example by Bergdahl (1973) and Tammirinne (1973). Therefore the weight penetrometer must be calibrated for each particular soil before it can be used. The accuracy of different control methods have been compared by e.g. Hellman et al (1979).

Weight soundings have also been used to check the compaction caused by pile driving using short tapered timber or concrete piles (compaction piles). Compaction piles have been used in Sweden to compact sand and gravel for e.g. bridge abutments as reported by Sandegren (1974). The length of the piles was 6 to 8 m. The compaction was small when the spacing of the 235 x 235 mm precast concrete piles was 1.5 m. The compaction increased considerably when the number of piles was doubled.

12. Summary and conclusions

The most common penetration testing method in Sweden, Finland, Norway and Denmark is by far the weight sounding method (WST). It is estimated that more than 20.000 weight soundings are carried out anually (1981). It is an inexpensive method which can be used in almost all soil types and in areas which are difficult to reach with other types of penetrometers. The main application of the method is in soft to medium stiff clay and silt and in loose to dense sand. The main limitation of the method is that layers of very dense sand or gravel or layers of glacial till are very difficult to penetrate. Predrilling may then be required.

Weight soundings are primarily used during the exploratory phase of a soil investigation to determine the depth and the thickness of the different strata. The method has also been used to estimate the length and the bearing capacity of friction and point bearing piles in cohesionless soils as well as the allowable bearing capacity and settlements of spread footings and rafts. Weight soundings can also be used to check the compaction of fills and the diggability of soil.

References

Aas, G., 1969. Sonderingsmetoder i Norge (Penetration testing methods in Norway). Nordic Meeting on Penetration Tests in Stockholm Oct 5-6, 1967. Swedish Geotechnical Institute, Report No. 31, pp 19-22.

Andresen, A. and Rygg, N., 1975a. Rotary Pressure Sounding, Proc, European Symp. on Penetration Testing, Vol. 2:2, Stockholm. pp 15-18.

Andresen, A. and Rygg, N., 1975b. Borrigg og metoder for vegterace undersökelser (Boring Equipment and Methods for Road Investigations) Nordisk Geoteknikermöde i Köpenhavn, pp 357-366.

Bergdahl, U., 1969. Resultat av försök med viktsond. (Results from weight soundings). Scandinavian meeting on Penetration Tests, Stockholm Oct. 5-6, 1967. Swedish Geotechnical Institute, Report No. 21, pp 51-59.

Bergdahl, U., 1973. Sondering i friktionsmaterial - Resultat av laboratorieförsök (Penetration tests in Cohesionless soils - Results from Laboratory Tests). Nordiskt Sonderingsmöte i Otnäs den 5-6 maj 1971. Finnish Geotechnical Society, pp 6-24.

Bergdahl, U. and Sundqvist, O., 1974. Geotechnical Investigations at the Demonstration Site, Borros Equipment. Proc. European Symposium on Penetration Testing, Vol. 2, pp 193-198.

Bergdahl, U. and Ottosson, E., 1984. Jordegenskaper ur sonderingsresultat - En jämförelse mellan olika undersökningsmetoder i friktionsjord. Nordiskt Geoteknikermöte, Linköping. (Soil characteristics from penetration test results - A comparison between different investigation methods in cohesionless soils.)

Broms, B., 1974. General Report, Scandinavia, Proc. European Symposium on Penetration Testing, ESOPT, Vol. 2.1, pp 14-23.

Broms, B.B. and Bergdahl, U., 1982. The Weight Sounding Test, State-of-the-art Report. ESOPT II.

Dahlberg, G., 1913. Skred och sättningar eller s k rad vid järnvägsbyggnader och liknande arbeten.

Dahlberg, R., 1973. Bestämning av friktionsjordars kompressionsegenskaper med hjälp av sondering. (Determination of the Deformation Properties of Cohesionless Soils with Penetration Tests). Nordiskt sonderingsmöte i Otnäs den 5-6 maj 1971, pp 25-48. Finnish Geotechnical Society.

Dahlberg, R., 1974a. Penetration Testing in Sweden, Proc. European Symposium on Penetration Testing (ESOPT), Vol. 1, pp 115-131.

Dahlberg, R., 1974b. A Comparison Between the Results from Swedish Penetrometers and Standard

Penetration Test, Results in Sand, Proc. European Symposium on Penetration Testing, ESOPT, Vol. 2.2, pp 67-68.

Dahlberg, R., 1974c. The effect of overburden pressure on the penetration resistance in a pre-loaded natural fine sand deposit. Proc. European Symposium on Penetration Testing, ESOPT, Vol. 2.2, pp 89-91.

Gardemeister, R., and Tammirinne, M., 1974. Penetration Testing in Finland. Proc. European Symposium on Penetration Testing (ESOPT), Vol. 1, pp 35-46.

Hansen, Å., 1969. Sonderingsmetoder i Danmark (Penetration testing methods in Denmark). Nordic Meeting on Penetration Tests in Stockholm Oct. 5-6, 1967. Swedish Geotechnical Institute, Report No. 31, p 13.

Hartmark, H., 1969. Norska erfarenhetern av dreiesondering (Experience in Norway with Weight Soundings). Nordic Meeting on Penetration Testing in Stockholm Oct. 5-6 1967. Swedish Geotechnical Institute, Reprints and Preliminary Reports No. 31, pp 45-49.

Helenelund, K.V., 1966. Kitkamaalajien Kantavuusominaissukusta Valtion (On the bearing capacity of frictional soils) Teknillinen Tutkimus laitos, Tideotus. Sarja III, Rakennus 97, Helsinki.

Helenelund, K.V. 1974. Prediction of Pile Driving Resistance from Penetration Tests. Proc. European Symposium on Penetration Testing, Vol. 2.2, pp 169-175.

Hellman, O., Pramborg, B.O. and Svensson, G., 1979. Kontroll av packad friktionsjord. Kontrollmetoder för bestämning av deformations- och brottbärighetsegenskaper (Control of the Compaction of Cohesionless Soils. Methods to check the deformation properties and bearing capacity). Swedish Council for Building Research, Report R 102:1979, Stockholm, 146 pp.

Korhonen, K-H. and Gardemeister R., 1972. Ett nytt system för klassificering av schaktbarhet. (A new classification system with respect to diggability). Väg- och Vattenbyggaren nr 3. Also Finnish Geotechnical Society, Nordiskt Sonderingsmöte i Otnäs den 5-6 maj 1971. pp 139-145.

Lagging, L.B. and Eresund, S., 1974. Test Loading at a Hydraulic Sand Fill. Proc. European Symposium on Penetration Testing, Vol. 2.2, pp 221-227.

Möller, B., 1980. Bedömning av lerors sensitivitet ur vikt- och trycksonderingsresultat (Estimate of the sensitivity of clays from weight and ram soundings). Swedish Geotechnical Institute, Report 1-206/79, 15 pp.

Mortensen, K., 1973. Der begås mange fejl ved utførelse og anvendelse af drejesonderinger (Many mistakes are made when weight soundings are used). Nordiskt Sonderingsmöte Otnäs den 5-6 maj 1971. pp 131-133. Also Ingeniørens Ugeblad No. 23, June 3, 1966. Finnish Geotechnical Society.

Muromachi, T., Oguro, I. and Miyashita, T., 1974. Penetration testing in Japan, Proc. European Symposium on Penetration Testing (ESOPT), Vol. 1, pp 193-200.

Muromachi, T., Kobayashi, S., 1982. Comparative study of static and dynamic penetration tests currently in use in Japan. ESOPT II, Vol. 1, Amsterdam.

Natukka, A., 1969. Finska Sonderingskommitténs rekommendationer för viktsonderingsstandard (Recommendations for Weight Soundings by the Finnish Penetration Committee on Testing.) Nordic Meeting on Penetration Testing in Stockholm Oct. 5-6, 1967. Swedish Geotechnical Institute, Reprints and Preliminary Reports No. 31, pp 39-43.

Nordiskt sonderingsmöte i Otnäs 5-6 maj, 1971. Föredrag och diskussioner. Finlands Geotekniska Förening, 1973.

Norwegian Geotechnical Society, Sounding Committee, Dreiesondering - Forslag till Standard (Weight Sounding - Proposed Standard), Oslo, 1972.

Norwegian Geotechnical Society, 1982. Veiledning for utførelse av dreiesondring. Melding Nr 3, 9 pp. (Guide for Weight Soundings, Guide No. 3.

Norwegian Pile Committee, 1973. Veiledning ved pelefundamentering (Guide for Pile Foundations) Norwegian Geotechnical Institute, Veiledning nr 1, (Guide No. 1) 108 pp.

Olsson, H., 1915. Banlära. Järnvägars byggnad och underhåll (Railway Construction Guide, Construction and Maintenance of Railways). Stockholm, Sweden, Vol. 1, 512 pp, Vol. 2, 421 pp.

Pitkänen, R., 1971. Daalujen tunketuminen eri maalajiolo suhteissa sekä korrelaatio heijariä ja painokairangvastusten kanassa. M. Sc Thesis, Helsinki University of Technology.

Saarelainen, S., 1979. Grovt sättningsestimat på grundval av viktsonderingsmotstånd. (A rough estimate of settlements from weight soundings) Nordiska Geoteknikermötet 1979 NGM -79, Finnish Geotechnical Society, pp 718-727, Esbo, Finland.

SBN, 1980. Svensk Byggnorm (Swedish Building Code) National Board of Physical Planning and Building) Liberförlag, Stockholm.

Sandegren, E., 1974. Swedish Weight Sounding as an Aid to Control Compaction of Loose Sand by mean of Piles. Proc. European Symposium on Penetration Testing, Vol. 2.2, pp 331-336.

Senneset, K., 1973. En undersøkelse vedrørende dreieborutstyr (An investigation of weight soundings equipment) Nordiskt sonderingsmöte i Otnäs den 5-6 maj 1971. Finnish Geotechnical Society.

Senneset, K., 1974. Penetration Testing in Norway, Proc. European Symposium on Penetration Testing (ESOPT) Vol. 1, pp 85-95.

Svensson, M., 1899. Beteckningssätt för lös mark å undersökningsprofiler (The recording of Soft Soils in Boring Profiles) Tekn, Tidskrift Vol. 29, No. 3, pp 55-56.

Swedish Committee on Pile Research, 1979. Recommendations on Bored Piles, Design construction and supervision. Report No. 58, Stockholm.

Swedish Geotechnical Commission, 1917. Vägledningar vid jordborrningar för järnvägsändamål (Guide to Soil Borings for Railways). Stat. Järnv. Geotekn. Medd. No. 1, 37 pp.

Swedish Geotechnical Commission, 1922. Statens Järnvägars Geotekniska Kommission. 1914-22. Slutbetänkande (Geotechnical Commission on the Swedish State Railways, Final Report). Stat. Järnv. Geotekn. Medd. No. 2, Stockholm, 180 pp.

Swedish Geotechnical Commission, 1922. Statens järnvägars geotekniska kommission. 1914-22. Slutbetänkande (Geotechnical Commission of the Swedish State Railways, Final report). Stat. Järnv. Geot. Medd. No. 2, Stockholm, 180 pp.

Swedish Road Board, 1976. Bronormer, Statens Vägverk TB 103, 1976-09 Stockholm, 200 pp.

Tammirinne, M., 1972. Karakearakeisten maalajien ja karkeiden silttimaalajien tiiveyden määvittäminen kairausvastuksen perus teella. Valtian taknillinen tut kimuskeskus. Rakennus - ja yhdryskintateknikka) Jalkaisu 2. Helsinki.

Tammirinne, M., 1973. Bestämning av torrvolymvikt i grus, sand och grov silt på grund av sonderingsmotstånd (determination of the Dry Unit Weight of Gravel, sand and coarse silt from the penetration resistance) Nordiskt Sonderingsmöte i Otnäs den 5-6 maj 1971, pp 49-62. Finnish Geotechnical Society.

Tammirinne, M., 1974. Relation between Swedish Weight Sounding and Static Penetration Test, Resistance of Two Sands, Proc. European

Symposium on Penetration Testing ESOPT, Vol. 2.1, pp 154-156.

Winkel, C.T., 1969. Erfaringer vedrørende sondeboret i Danmark (Experience with soundings in Denmark). Nordic Meeting on Penetration Testing in Stockholm, Oct. 5-6, 1967, Swedish Geotechnical Institute, Reprints and Preliminary Reports No. 21, pp 35-37.

DOCUMENT No 3

INTERNATIONAL REFERENCE TEST PROCEDURE OF THE WEIGHT SOUNDING TEST

1. Introduction

1.1 The weight-penetrometer consists of a screw-shaped point, rods, weights and a handle. It is used as a static penetrometer in soft soils when the penetration resistance is less than 1 kN. When the resistance exceeds 1 kN the penetrometer is rotated and the number of rotations for a given settlement is noted. Its ability to penetrate even stiff clays and dense sands is good. The penetrometer is primarily used to give a continuous soil profile and an indication of the layer sequence and to determine the lateral extent of different soil layers. It is also used to determine whether cohesionless soils are loose, medium-dense or dense and to estimate the relative strenghts of cohesive soils. The results obtained in cohesionless soils are also used to get an indication of the bearing capacity of spread footings and piles.

2. Apparatus

2.1 Weights

2.1.1 These comprise one 5 kg clamp, two 10 kg weights, three 25 kg weights. Total 100 kg. The weights can be replaced by a dynamometer when the penetrometer is pushed in manually or mechanically.

2.1.2 The maximum allowable deviation for the weight and the dynamometer scale is ± 5 %.

2.2 Rod and coupling

2.2.1 The diameter of the rod should be 19-25 mm, preferably 22 mm. Regarding material and the influence of rod diameter, see note 1 para. 6.

2.2.2 The deviation from the straight axis should not exceed 4 °/₀₀*) for the lowest 5 m of the rod and 8 °/₀₀*) for the remainder. Determination of the deviation for the rods see Fig. 1. Maximum allowable eccentricity for the coupling is 0.1 mm. Maximum angular deviation for a joint between two straight rods is 0.005 rad.

2.2.3 Flush joint should be used.

2.3 Point

2.3.1 Manufactured from a 25 mm square steel bar with a total length of 0.2 m. The bar has a 80 mm long pyramidal tip. The point is twisted one turn to the left over a length of 130 mm as shown in Fig. 2. If the point should be attached to other rod diameters than 22 mm as shown in Fig. 2 the upper conical part should end with the same diameter as the rod.

2.3.2 The diameter of the circumscribed circle of the point shall not exceed 35.0 ±0.2 mm for a new point and shall not be less than 32.0 ±0.2 mm for a worn point. The diameter shall be checked by circular gauges with different inner diameters.

Maximum allowable shortening of the length of the point is 15 mm due to wear. The tip of the point shall not be bent or broken.

---

*) These deviations correspond in case of an even curvature to a deflexion of 1 and 2 mm in 1 m length respectively.

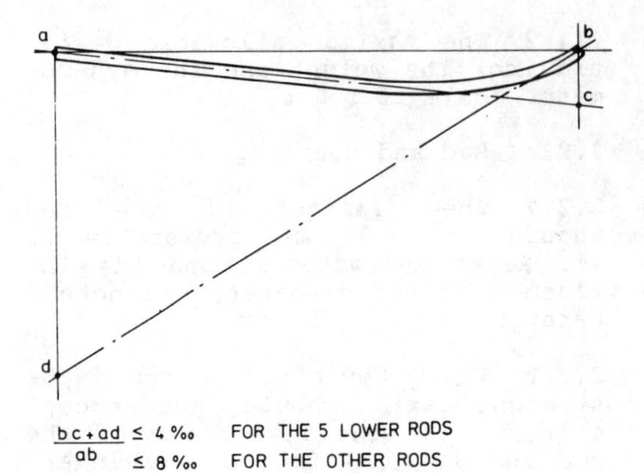

Fig. 1. Determination of the deviation from the straight axis for rods.

2.4 Additional tools

Two fixed wrenches, a handle, extraction device and augers for preboring.

3. Test procedure

3.1 Manual weight sounding*)

When the penetrometer is used as a static penetrometer in soft soils the test should be made in accordance with clause 3.1.1 and 3.1.2. In stiffer soils the penetrometer should be rotated as described in clause 3.1.3.

3.1.1 The rod is loaded in steps using the following reference loads.

loads in kN	mass in kg
0	0
0.05	5
0.05 + 0.10 = 0.15	5 + 10 = 15
0.15 + 0.10 = 0.25	15 + 10 = 25
0.25 + 0.25 = 0.50	25 + 25 = 50
0.50 + 0.25 = 0.75	50 + 25 = 75
0.75 + 0.25 = 1.00	75 + 25 = 100

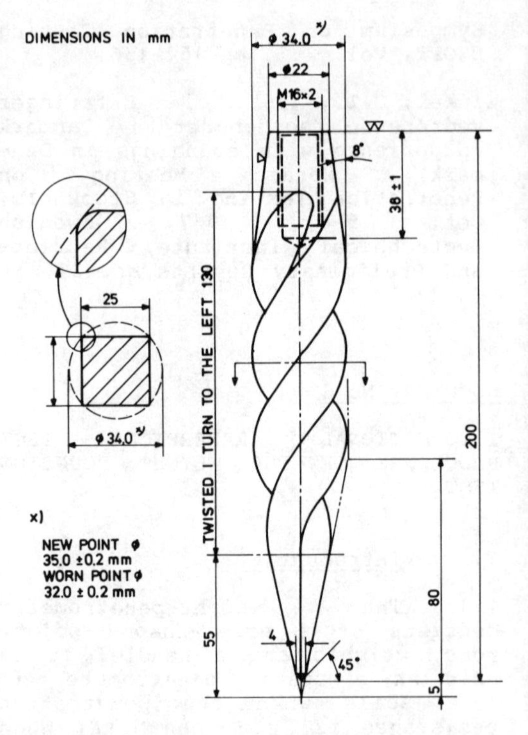

Fig. 2. Tolerances for the manufacture of weight penetrometer points, applicable to 22 mm rods.

3.1.2 The load shall be adjusted to give a rate of penetration of about 50 mm/sec. This means that the rod must be partly unloaded when a layer of stiff soil, such as a dried crust, has been penetrated.

3.1.3 If the penetration resistance exceeds 1 kN or the penetration rate at 1 kN is less than 20 mm/sec the rod should be rotated. The load 1 kN is maintained and the number of half turns required to give 0.2 m of penetration is measured. The rod must not be rotated when the penetration resistance is less than 1 kN.

3.2 Mechanized weight sounding

3.2.1 Tests are carried out in a similar manner as for the manual soundings. The rod is rotated mechanically in stiff soils.

3.2.2 The applied load is measured with a dynamometer or a measuring cell attached to the machine.

*) When manually operated vibrations and sounds from the rod gives a better indication of the soil penetrated. Such indications are often lost when the penetrometer is mechanically operated.

3.2.3 When the penetration resistance is less than 1 kN and rotation is not required, the engine must be stopped to prevent the vibrations from the engine to affect the measured penetration resistance. The rate of rotation should be between 15 and 40 rpm and should not exceed 50 rpm. The average rate of rotation should be 30 rpm. (See note 2 para. 6.)

3.3 General considerations

3.3.1 The possible need to prebore through the upper soil layers shall be estimated in each case. (See note 3 para. 6.)

3.3.2 The criteria to be used for the termination of a WST test shall be stated for each investigation, e.g. exceed the minimum penetration resistance or reach a minimum depth. (See note 4 para. 6.)

4. Recording

4.1 Penetration resistance

4.1.1 When the penetration resistance is less than 1 kN the reference load required to give a rate of penetration of about 50 mm/sec shall be recorded against the depth (Fig. 3). It should be noted in the boring log and on drawings whether weights or a dynamometer have been used. (See note 5 para. 6.)

4.1.2 When the penetration resistance exceeds 1 kN, the number of half turns required for every 0.2 m of penetration shall be recorded (Fig. 3).

4.1.3 When the penetrometer is driven by blows of a hammer or some of the weights, the depths penetrated during driving shall be recorded.

4.2 General notes

4.2.1 All observations which may help in the interpretation of the test results shall be noted in the boring log. e.g. diameter of rods, sounds and vibrations in the rods when the point penetrates cohesionless soils (stones, gravel and sand). Also interruptions etc. shall be recorded.

4.2.2 The type of rotating equipment and the rate of rotation shall be noted in the boring log.

5. Presentation of test results

5.1 The form of presenting the results of weight sounding tests is shown in Fig. 3.

6. Explanatory notes

Note 1. Clause 2.2.1.

The rods and couplings should be made of high tensile steel. It has to be considered that 25 mm rods may give higher total resistance than 22 mm rods especially in cohesive soils.

Note 2. Clause 3.2.3.

Differences between manually and mechanically performed tests sometimes occur. Where this may be the case, for example when estimating the relative density of loose cohesionless soils, comparisons between manually and mechanically performed tests are recommended.

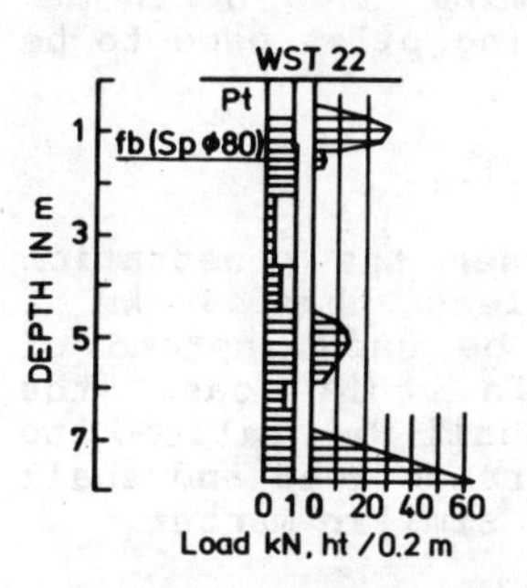

WST 22 WEIGHT SOUNDING TEST, 22 mm RODS
ht/0.2 m NUMBER OF HALFTURNS PER 0.2 m OF PENETRATION
Pt DRY CRUST OF CLAY
fb(Spφ80) PREBORING TO THIS LEVEL WITH 80 mm DIAM AUGER

DIAGRAM TO THE LEFT INDICATE LOADS APPLIED IN kN

Fig. 3. An example of the presentation of test results from weight sounding test (WST).

Generally the penetration resistance at rotation is higher for mechanically than for manually operated penetrometers.

Note 3. Clause 3.3.1.

In case where the skin friction resistance along the upper parts of the rod can significantly influence the results, a comparison should be made with a test in a prebored hole. Preboring is normally required through a dry crust or through a fill. When the difference in penetration resistance is large between the two tests, preboring is necessary for all tests within the area. The preboring shall be made using an auger with a minimum diameter of 50 mm. To estimate the thickness of the dry crust, however, the sounding test is performed directly from the ground surface.

Note 4. Clause 3.3.2.

A penetration test to "firm bottom" shall be terminated by striking the rod with a hammer or by dropping some of the weight onto the clamp in order to check that the refusal is not temporary. If it is possible to penetrate the stiff layer the test shall be continued.

Sometimes the weight penetrometer test is followed by percussion boring to deeper levels, e.g. in order to determine the depth to which point-bearing piles need to be driven.

Note 5. Clause 4.1.1.

In soft soils when the penetration resistance is less than 1 kN a dynamometer can be used instead of the weights. In this case the recorded load shall be related to the closest reference load and shall be recorded in a similar manner.

Special lectures

Penetration Testing 1988, ISOPT-1, De Ruiter (ed.)
© 1988 Balkema, Rotterdam, ISBN 90 6191 801 4

Current status of the piezocone test

R.G.Campanella & P.K.Robertson
Civil Engineering Department, University of British Columbia, Vancouver, Canada

ABSTRACT: The piezocone test was first introduced in the early 1970's and has shown rapid development in the last 12 years. This paper discusses the current knowledge and opinions related to areas of equipment, procedures and interpretations. Detailed discussion is provided on the influence of pore pressure element location and the importance of saturation. Current methods of interpretation for soil type, undrained shear strength, OCR and flow characteristics (c_H) are presented and discussed. A brief review of standards of practise is also presented.

KEY WORDS: Piezocone, equipment, procedures, interpretation.

INTRODUCTION

The measurement of pore water pressures during the penetration of a probe into soil was first introduced in the early 1970's (Wissa et al., 1975; Torstensson, 1975). In the late 1970's pore pressure measurements were made using probes of the same shape and dimensions as the standard 60 degree, 10 cm^2 cone, but only pore pressure measurements could be made. In the early 1980's piezometer elements were incorporated into standard electric cone penetrometers with the measurement of cone resistance (q_c) and, in some cases, sleeve friction (f_s) and cone inclination (i) (Baligh et al., 1981; Campanella and Robertson, 1981; de Ruiter, 1982; Muromachi, 1981; Smits, 1982).
In recent years developments in electronics have resulted in the introduction of electronic cone penetrometers capable of measuring q_c, f_s, i, temperature and pore pressure (u) at several locations on the cone.
Figure 1 presents examples of previous and current piezocone designs.

Today the piezocone test is generally regarded as a standard cone penetration test (CPT) with pore pressure measurements (CPTU).
The main advantages of the CPTU over the conventional CPT are:

- ability to distinguish between drained, partially drained and undrained penetration,
- ability to correct measured cone data to account for unbalanced water forces due to unequal end areas in cone design,
- ability to evaluate flow and consolidation characteristics,
- ability to assess equilibrium groundwater conditions,
- improved soil profiling and identification,
- improved evaluation of geotechnical parameters.

Due to these capabilities the CPTU has a wide range of applications in geotechnical engineering.
The objective of this paper is to discuss items of current knowledge and informed opinions related to CPTU equipment, procedures and interpretation. The results of a

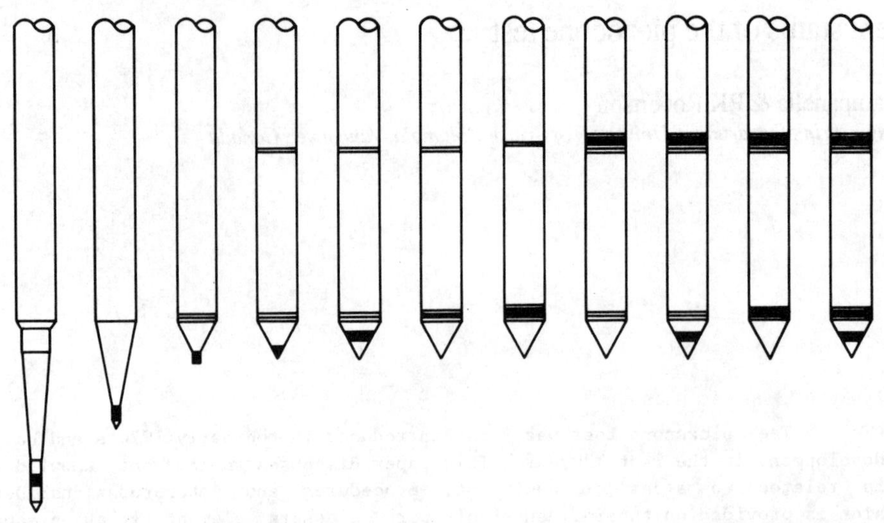

Fig.1 Examples of different piezocone designs illustrating various pore pressure element locations

world wide survey of users of the CPTU are also included.

EQUIPMENT

Pore Pressure Element Location

Significant advances have been made in the last 12 years related to the design of piezocones. However, during that period no agreed standard has emerged concerning the location of the pore pressure element. Figure 1 illustrates some of the previous and current suggested locations for the pore pressure element.
Several modern cones have the capability to record pore pressures at several locations simultaneously (Figure 1). It is generally agreed that no single location can provide information for all possible applications.

Figure 2 shows the theoretical pore pressure distribution developed by Baligh et al. (1980) and data collected by the writers from University of British Columbia (UBC) research sites. The UBC data have been normalized using the excess pore pressure measured behind the friction sleeve (ie. 10 radii behind the tip). Figure 2 illustrates the large range in response possible if pore pressures are recorded at different locations. Campanella et al. (1985), Lunne et al., (1986) and Jamiolkowski et al., (1985), presented additional data to illustrate the large gradient in pore pressure around the tip in stiff overconsolidated clays and dense silty sands. This gradient can result in a variation in measured pore pressure for slightly different measurement locations in the area behind the tip.

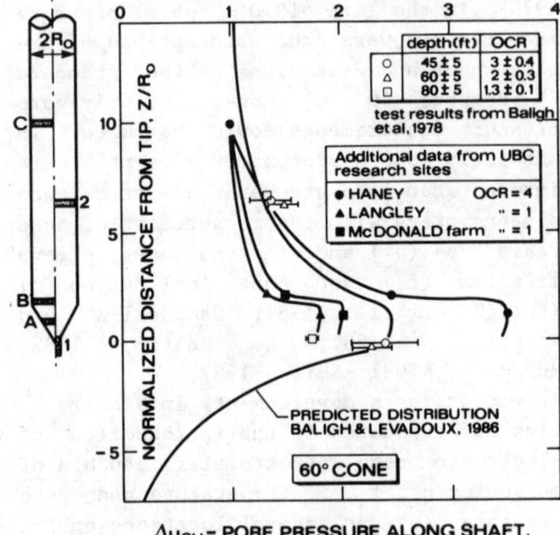

Fig.2 Comparison between measured and predicted pore pressures on a penetrating cone

Figure 3 presents an example of CPTU data from the Imperial Valley, California, showing pore pressure measurements at 3 different locations on a cone during penetration in an overconsolidated stiff clay from a depth of 6.5 m to 11.5 m. The pore pressure measured on the face of the tip with a 5 mm thick element ("A") is approximately 3 times larger than the equilibrium pore pressure, u_o.

The pore pressure measured with a 5 mm thick element located 5 mm behind the shoulder (edge of shoulder to center of element, "C") measures a pore pressure slightly less than u_o. However, a 2.5 mm thick element located only 2.5 mm behind the shoulder ("B") measures a pore pressure almost twice u_o.

The date in Figure 3 illustrate the high pore pressure gradient that can exist around the tip in stiff soils.

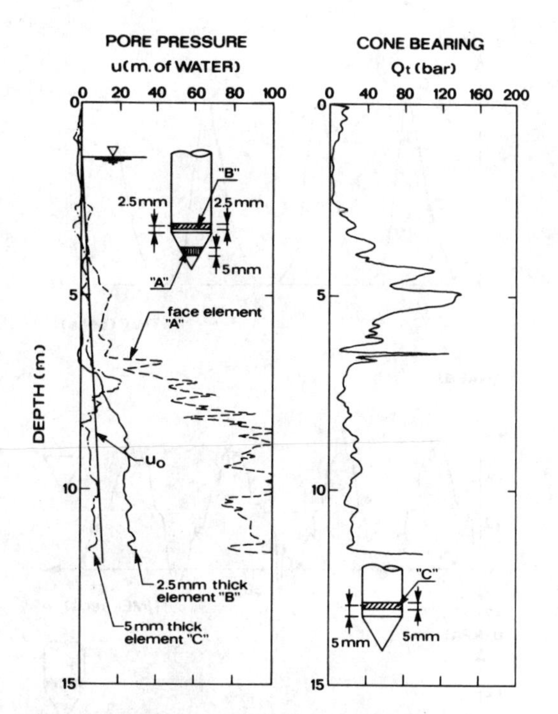

Fig.3 Example of CPTU pore pressures at different locations on the cone

Figures 2 and 3 illustrate some of the difficulty in selecting a suitable location to measure pore pressure during cone penetration. Alternatively, cones that can measure pore pressures at 2 or 3 locations (as shown in Figure 1) show considerable promise. However, <u>all pore pressure measurements from CPTU must clearly identify the location and size of the sensing element.</u> Further discussion regarding pore pressure location will be made in the sections concerning interpretation.

Piezocone Design

For any given location of sensing element there are several factors that influence the reliability and accuracy of the pore pressure results. These factors relate to the design features of the cone.

Measuring dynamic pore pressures with the piezocone requires careful consideration of probe design, choice of porous element and probe saturation.

Mechanical Design

The mechanical design of the cone must ensure that when the cone tip is stressed, no load is transferred to the pore pressure transducer, porous element or fluid volume. This problem can be checked by loading the tip of a fully assembled, saturated piezocone and observing the pore pressure response. If no mechanical transfer of load is taking place, there should be no pore pressure response.

An example of such a test was presented by Battaglio et al., (1986) and is shown in Figure 4, which shows that, for the particular cone tested, the filter element located on the tip of the cone gave a relatively large positive load transfer to the pore pressure transducer. In a dense sand the cone with the filter element on the tip, in Figure 4, may produce incorrect large positive pore pressures. Also the particular cone in Figure 4 may produce slightly negative pore pressures for the filter element behind the tip.

It is interesting to note from Figure 4 that in soft clays where q_c is generally less than about 1 MPa and where the total pore pressures generated during penetration, either on or behind the tip, are large (>700 kPa) the error due to mechanical load transfer would be negligible.

However, in very stiff soils, where q_c can be large, mechanical load transfer effects can produce misleading pore pressure data unless the cone is carefully designed.

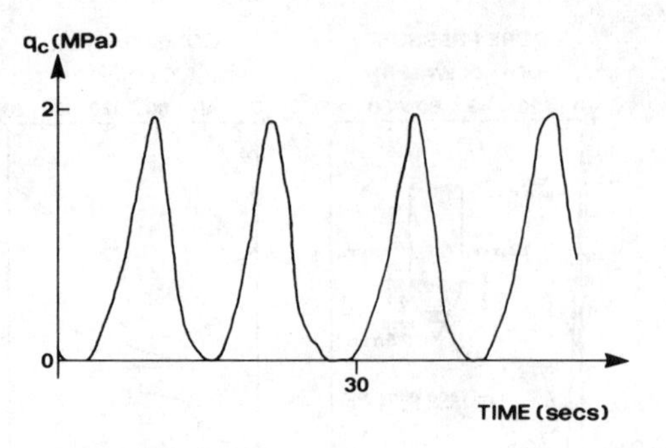

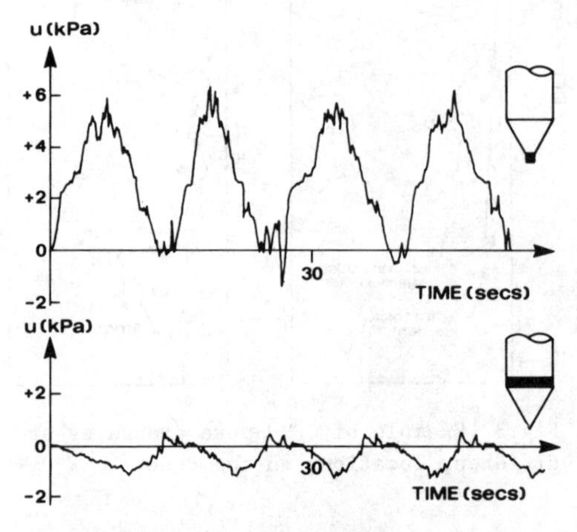

Fig.4 Influence of mechanical load transfer on measured pore pressure (Adapted from Battaglio et al., 1986)

Porous Element

For a high frequency response (i.e. fast response time), Smits (1982) suggested that the design must aim at: a small fluid filled cavity, low compressibilty and viscosity of fluid, a high permeability of the porous filter and a large area to wall thickness ratio of the filter. To measure dynamic pore pressures rather than filter compression effects, the filter should be rigid.
However, to maintain saturation the filter should have a high air entry resistance, which requires a finely graded filter and/or high viscosity of the fluid. Clearly, not all of these requirements can be combined.
Essential requirements are to incorporate a small fluid cavity, a rigid pressure transducer and a low compressibility of saturating fluid. A compromise is required between a high permeability of the porous filter to maintain fast response time and a low permeability to have a high air entry resistance to maintain saturation.
Many different materials, such as; stainless steel, ceramic, sintered bronze, carborundum, aerolith-10, cemented quartz sand, stone, teflon and polypropylene; have been used for porous filter elements. Problems of abrasion and smearing have been reported (Smits, 1982) with some ceramic and stainless steel elements, although many of these problems have now been overcome with special hardened stainless steel and ceramic elements.
Of practical importance, the filter element must survive penetration through dense sands and gravelly soils. Polypropylene is a relatively hard and rugged plastic with high resistance to abrasion and toughness.
Current piezocone filter designs appear to fall into two groups;

- low permeability, rigid filters (ceramic, stainless steel, etc.)
- higher permeability, flexible filters (polypropylene - porous plastic)

Both designs generally use a low compressibility, low viscosity silicon oil or glycerin as a saturating fluid, although water is still sometimes used. The rigid filters are sometimes considered to act as "sealed" systems with very little flow. The more flexible filters (porous plastic) allow rapid movements of extremely small volumes of water needed to activate the pressure sensor while preventing soil ingress or blockage. Good results have been published using both systems (Campanella et al., 1986, Smits, 1982, Battaglio et al., 1986).
The advantage of using a fluid with a viscosity higher than water is that high air entry values can be achieved, even with relatively permeable polypropylene filters. The high air entry values help maintain saturation during penetration through some unsaturated soils. However, penetration through unsaturated clays can generate very large suctions and saturation of the porous element may not be maintained.
Filter element squeeze or compression can be important for cones that measure pore pressures at or on the tip. During

penetration into stiff layers with high cone resistance, the filter element can become compressed and generate high positive pore pressures. This will occur unless the filter element has a very low compressibility (rigid) or if filter and soil are of sufficient permeability to rapidly dissipate the pore pressure due to filter element compression. Experience at UBC comparing a relatively compressible porous plastic filter with rigid ceramic filters on the face has shown no evidence of induced pore pressures due to filter squeeze. This is likely due to the relatively high permeability (0.01 cm/sec) of the porous plastic element. However, problems may occur with these elements in very stiff soils with permeabilities considerably smaller than that of the porous element.

Filter squeeze is generally only critical for pore pressure measurements on the tip during initial penetration into dense fine, silty sands and very stiff clays or glacial silts.

An example of filter compressibility and mechanical load transfer was presented by Battaglio et al. (1986) and reproduced in Fig.5.

Figure 5a shows pore pressure measurements in the soft silty clay at Porto Tolle site ($q_c / \sigma'_{vo} \approx 5$), using two CPTU tips both with the filter elements located at the cone apex. Despite this, the difference in measured pore pressure values are about 30%, reflecting probably a different influence of filter compressibility and load transfer (Battaglio et al., 1986). A similar comparison is given in Figure 5b, showing CPTU results obtained in the heavily overconsolidated hard clay at the Taranto site ($q_c/\sigma'_{vo} \approx 35$). In this case the pore pressure measured with the conical filter is greatly affected by the compression of the filter itself, making the measured values of u meaningless. In both cases in Figure 5 the filter elements were stainless steel on the tip and caborundum on the face and both saturated with silicon oil.

When the filter element is located just behind the tip it is important that the diameter of the element not be smaller than the diameter of the cone, because the existance of a slight gap can produce erroneous data.

The importance of the factors discussed above, such as, mechanical transfer of load, compression, abrasion and smear of filter elements and rapid response time are all generally recognised. Selection of filter element and saturating fluid will depend on filter element location and the type of soils anticipated during penetration. The cost of the filter elements can also become an important factor since the writers believe that filter elements should be changed following each sounding because abrasion,

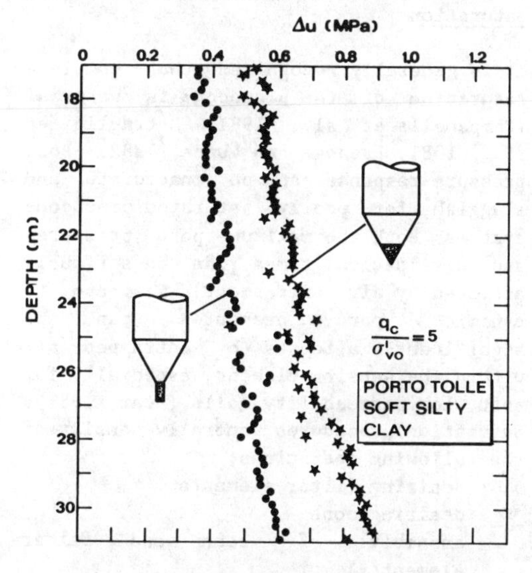

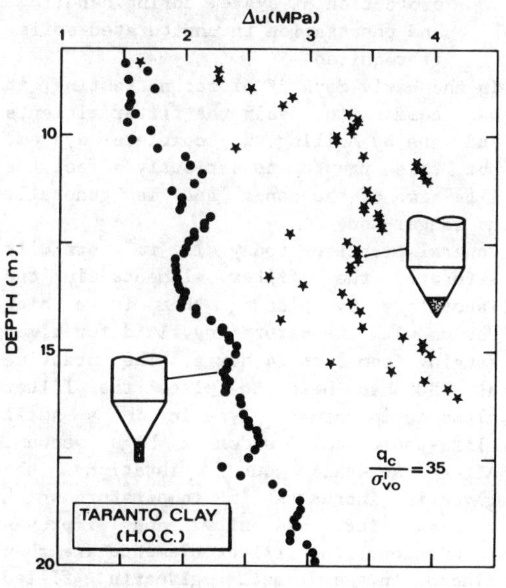

Fig.5 Comparison of pore pressure response (After Battaglio et al., 1986)

smearing and clogging of filter elements may restrict their reuse.

PROCEDURES

There are generally few differences in field test procedures between standard CPT and CPTU soundings, except those related to, saturation, penetration and pauses in penetration.

Saturation

It is generally recognised that complete saturation of the piezocone is essential (Campanella et al., 1981; Battaglio et al., 1981; Lacasse and Lunne, 1982). Pore pressure response can be inaccurate and sluggish for poorly saturated piezocone systems. Both the maximum pore pressures and dissipation times can be seriously affected by air entrapment. Response to dynamic pore pressures can be significantly affected by entrapped air within the sensing element, especially for soft, low permeability soils (Acar, 1981). Saturation procedures generally consist of the following operations;

- dealring filter elements
- dealring cone
- assembly of cone and filter element(s)
- protection of system during handling and penetration in unsaturated soils, if required.

In the early days of piezocone testing, it was common to deair the filter elements and cone by boiling the complete system, but this proved to seriously affect the life-time of the cones, and is generally no longer done.

General practice today is to carefully saturate the filter elements in the laboratory by placing them in a high vacuum with the saturating fluid for times ranging from 3 to 24 hours. The practice at UBC has been to place the filter elements in warmed glycerin in a small ultra-sonic bath under a high vacuum. After several hours vibration, the glycerin increases in temperature which reduces its viscosity and improves saturation. The filter elements are then placed in a small glycerin filled container for transportation into the field. Similar procedures are adopted by many CPTU operators using different saturating fluids.

The voids in the cone itself should be deaired by flushing with a suitable fluid, usually the same fluid used for saturation of the elements. Techniques for saturating piezocones will depend on individual cone design but, where possible, it is suggested that flushing the voids within the cone be performed with some type of hypodermic needle. The cone and filter elements should be assembled while submerged in the saturating fluid.

The next step after cone preparation and assembly is the lowering of the string of cone rods and penetration. A thin protective rubber sleeve or container is sometimes placed over the cone. To avoid premature rupture of the protective sleeve, a small hole can be pushed using a "dummy" cone of a larger diameter than the piezocone. Sometimes a hand dug or predrilled hole is made depending on circumstances and soil stratigraphy. Predrilling to the water level is not always necessary if the filter element and saturating fluid develop a high air entry value to prevent loss of saturation. However, soil suctions can be very large in unsaturated clays where predrilling may be necessary.

The authors generally recommend that the entire saturation procedure be repeated after each sounding, including a change of filter element.

One of the major difficulties with piezocone testing can be the evaluation of saturation.

Frequently it has been suggested to check saturation before penetration. Unfortunately, it is very difficult to check saturation and it is questionable if such a check would be reliable since even a small amount of entrapped air produced during the handling or early penetration can drastically increase the system compressibility. Generally, saturation is evaluated by careful review of the pore pressure data during the sounding. Several examples have been published to illustrate the pore pressure response from poorly saturated piezocones (Battaglio et al., 1981; Campanella and Robertson, 1981;

Lacasse and Lunne, 1982). In general, the pore pressure response for poorly saturated piezocones is sluggish and detailed macro-structure and soil stratigraphy is subdued.
The requirement for complete saturation is very important at shallow depth where equilibrium pore pressures (u_o) are very low. Once significant penetration below the water table has been achieved (≥5 m) the resulting equilibrium water pressure is often sufficient to ensure saturation. Penetration at shallow depth in saturated sands can produce negative pore pressures behind the tip which may cause temporary cavitation if u drops below about -100 kPa.

Rate of Penetration

The standard rate of penetration for CPTU is 2 cm/sec. For penetration in medium grained clean sands and coarser materials, these pore pressures dissipate almost as fast as they are generated and penetration takes places under drained conditions. For penetration in fine grained soils, such as clays and clayey silts, significant excess pore pressures can be generated because of their relatively low permeability, and penetration takes place under predominantly undrained conditions. Penetration into fine sands and silty sands can generate excess pore pressures, but penetration may be taking place under partially drained conditions.

Correct interpretation of the pore pressure data requires some knowledge that the penetration is predominantly undrained. Radial consolidation theory can be used to obtain a plausible estimate of the upper limit to soil permeability for which the piezo element will observe undrained pore pressures. The estimate of the upper limit to permeability for undrained penetration depends on the following; soil compressibility and stiffness, size of the cone and size and location of porous element. For standard 10 cm^2 base area cones and a 5 mm thick porous element located on or just behind the tip, a plausible upper limit to soil permeability for undrained penetration (at 2 cm/sec) is in the order of $1 \cdot 10^{-7}$ m/s. A partially drained CPTU response may be observed for soils with a permeability in the range of $1 \cdot 10^{-4}$ m/s to $1 \cdot 10^{-7}$ m/s, that is, soils such as fine sands and silts. For permeabilities greater than about $1 \cdot 10^{-4}$ m/s penetration is most likely fully drained. These values are approximate but have been generally confirmed by field observations. It has often been suggested that if penetration is partially drained at 2 cm/sec, the rate of penetration could be increased or decreased to produce an undrained or drained penetration, respectively. However, since the permeability of soils varies by orders of magnitude, the change in penetration rate required to significantly change the drainage process would generally also vary by orders of magnitude (Campanella et al., 1983). Penetration rates of 20 cm/sec or faster and 0.2 cm/sec or slower become impractical and also introduce further strain rate effects.

Pore Pressure Dissipation

Penetration is generally performed in 1 m strokes since most push-rods are generally 1 m in length. This produces pauses in the penetration process which last from about 15 to 90 seconds depending on the individual pushing assembly. During these pauses in penetration any excess pore pressures start to dissipate.
The rate of dissipation depends upon the coefficient of consolidation which, in turn, depends on the compressibility and permeability of the soil. When penetration is resumed some movement is required in order to regain the original penetration pore pressure values. The amount of movement appears to vary with soil type and can range from about 2 cm to 50 cm. No clear explaination has been proposed to clarify these large differences. When presenting CPTU data it is important to either remove or clearly identify the pauses in the penetration.

To avoid the problem of regular pauses in the penetration process several pushing assemblies have been developed to allow truly continuous penetration with no pauses. However, these have generally been restricted to offshore applications.

The dissipation of excess pore pressures

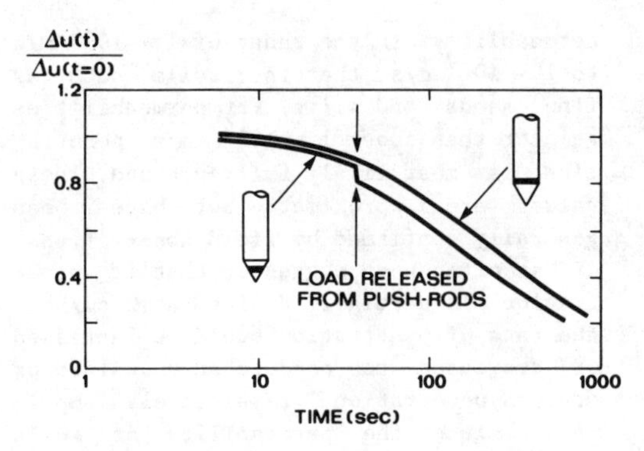

Fig.6 Example of influence of load release from push-rods during CPTU dissipation test

during a pause in penetration can provide valuable additional information regarding; drainage conditions, soil permeability and, if allowed to fully dissipate, equilibrium piezometric pressures, u_o.
A dissipation test can be performed at any depth. In a dissipation test the rate of dissipation of excess pore pressure to a certain percentage of the equilibrium pore pressure is measured.
Sometimes the push rods are clamped to the pushing rig during the dissipation test. Athough this stops the movement of the push rods the cone tip will continue to move very slightly as the elastic strain energy in the rods releases and the tip load reduces. The longer the push rods, and the greater the tendency for the soil to creep, the more significant this movement may be. This movement alters the total stresses in the soil around the conical tip and may influence the measured decay of pore pressures with time. It is generally agreed that this is only significant with the piezo element located on the tip. With the piezo element behind the tip, it has not been necessary to clamp the rods. An example of the effect of load release on the dissipation record in a soft silt is shown in Figure 6.

Sometimes a fixed period of dissipation for all soil layers is used and sometimes dissipation is continued to a predetermined percentage of the assumed hydrostatic pressure.

A common period is t_{50} or the time to record a dissipation to half the excess pore pressure.
The pore pressure is recorded in a time base mode and the measurement of equilibrium pressures can provide important hydro-geologic information.

Because of the importance of good saturation and the need for careful procedures the CPTU is often more operator dependant than the standard CPT. Therefore it is essential that the operators have training and experience in the use of the CPTU.

INTERPRETATION

Before interpretation of any CPTU data it is important to realize and account for the potential errors that each element of data may contain. Specific aspects that relate to equipment design and test procedures have been discussed above. One area of advance that has resulted from piezocone data is the ability to correct measured cone data to account for unbalanced water forces due to unequal end areas in cone design.

Corrections

Water pressures can act on the exposed surfaces behind the cone tip and on the ends of the friction sleeve (see Figure 7). These water forces result in measured tip resistance (q_c) and sleeve friction (f_s) values that do not represent true total stress resistances of the soil. This error introduced in the measurement can be overcome by correcting the measured q_c for unequal pore pressure effects using the following relationship (Baligh et al., 1981; Campanella et al., 1982):

$$q_T = q_c + u\,(1 - a) \qquad \ldots (1)$$

where:
q_T = corrected total tip resistance
u = pore pressure generated immediately behind the cone tip
a = net area ratio

An example of the determination of the net area ratio using a simple calibration vessel is shown in Figure 8. The calibration vessel is designed to contain the cone and to apply an all around air or water pressure. Many cones have values of net area ratio ranging from 0.90 to 0.6, but sometimes this ratio may be as low as 0.38 (Figure 8; Battaglio and Maniscalco, 1983). The correction cannot be eliminated except with a unitized, jointless cone design.
In offshore practice the correction can be evaluated during the lowering of the cone through the water.
The importance of this correction is especially significant in soft clays, where high values of pore pressure and low cone resistance may lead to the physically incorrect situation of $u > q_c$.

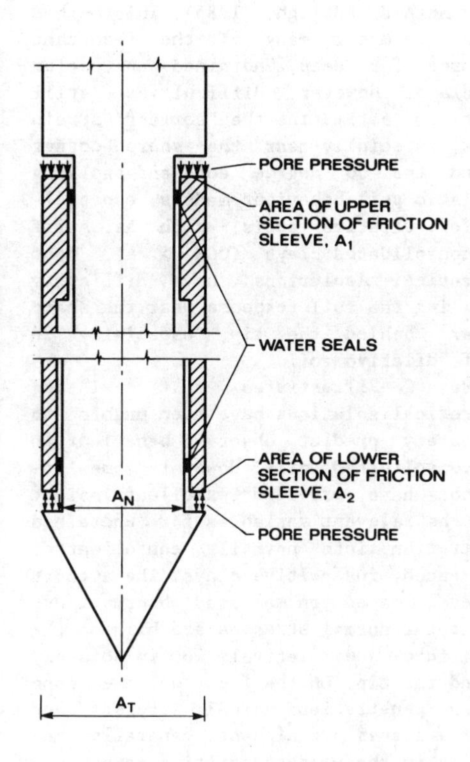

Fig.7 Unequal end area correction

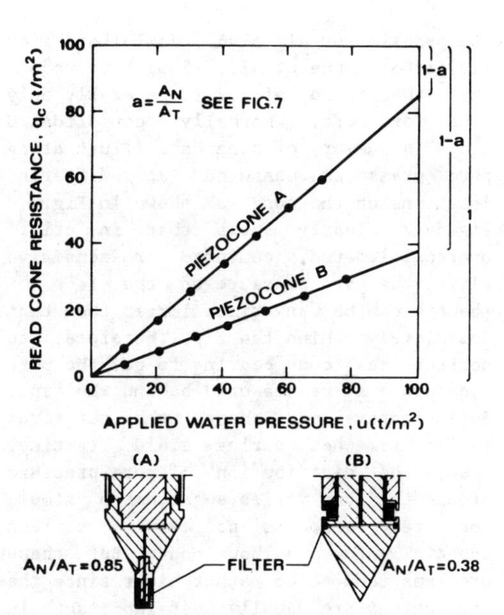

Fig.8 Determination of unequal end area correction for two types of CPTU probes (After Battaglio and Maniscaldo, 1983)

Also, previous correlations developed to obtain soil properties, such as undrained shear strength (s_u), from q_c measurements incorporate systematic errors, depending on cone design.
A similar correction is required for sleeve friction data. However, information is required of the pore pressures at both ends of the friction sleeve. The importance of the sleeve friction correction can be significantly reduced using a cone design with an equal end area friction sleeve.

Location of Porous Element

Several CPTU operators and researchers who use cones that record the pore pressure on the face of the cone tip have suggested correction factors to convert the measured pore pressures on the face to those that are assumed to exist immediately behind the tip. The assumed ratio of the pore pressure on the face to the pore pressure behind the tip is generally taken to be about 1.2 (i.e. the pore pressure on the face is assumed to be 20% larger than that immediately behind the tip). Measurements

(Campanella et al., 1985; Jamiolkowski et al., 1985; Lunne et al., 1986) have shown that the ratio of 1.2 is generally only true for soft, normally consolidated clays. A summary of some data illustrating pore pressures measured at different locations on the cone was shown in Fig.2.
Figure 2 clearly shows that in stiff, overconsolidated, cemented or sensitive clays, the pore pressure on the face of the tip can be many times larger than that immediately behind the tip. Therefore, to correct the cone bearing to q_T, the pore pressure must be measured behind the tip.
Soil ingress may change the net area ratios somewhat during field testing. Also, the distribution of pore pressure around the cone varies such that a simple net area ratio is not always correct especially for a bulbous cone. But these problems tend to be rather minor since the corrections are usually most important in soft cohesive soils where the variation in pore pressures around the cone are generally small. The potential error due to these problems are significantly less than the error if no correction is applied.
The location of the pore pressure element is very important with regard to data interpretation. Pore pressures can be generated in saturated soils due to changes in both mean normal stresses and shear stresses. When saturated soils are subjected to increases in mean normal stresses, positive pore pressures are generated. When saturated soils are subjected to only shear stresses, pore pressures generated can be either positive or negative depending on the contractive or dilative response of the soil. During cone penetration the soil elements adjacent to the penetrating cone experience changes in both mean normal stresses and shear stresses.
The cone penetration process represents a very complex problem for theoretical solutions. Most solutions have been developed for undrained penetration into saturated clays or drained penetration into clean sands. The solutions can be categorised as follows (with typical examples);

1. Bearing Capacity Methods
   (Meyerhof, 1961; Begemann, 1965; Durgunoglu and Mitchell, 1975).
2. Cavity Expansion Methods
   (Gibson, 1950; Vésic, 1972; Al Awkati, 1975)
3. Steady Penetration
   (Baligh, 1975; Tumay et al. 1985; Acar and Tumay, 1986)
4. Finite Element Methods
   (Borst and Vermeer, 1982)
5. Stress Characteristics Methods
   (Houlsby and Wroth, 1982; Luger et al., 1982)
6. Strain Path Method
   (Baligh, 1985)

It is generally agreed that the bearing capacity methods do not adequately represent the steady penetration problem. The other methods, generally have the advantage of using more realistic soil behaviour models. The advantage of the cavity expansion methods is in their simple closed-form solutions. One of the most promising solutions is the Strain Path Method (Baligh, 1985). This method appears to model many of the important features for deep, undrained penetration in clays. However, difficulties still exist in estimating the "correct" strain paths, especially near the sharp corner behind the 60 degree cone and applying realistic soil behaviour models, especially for dilative soils, such as, stiff overconsolidated clays (OCR > 4). Most theoretical solutions have difficulty modelling the soil response near the sharp corner behind the tip, especially in stiff, dilative soils.
Figure 2 illustrates that existing theoretical solutions have been unable to adequately predict observed behaviour in overconsolidated soils. However, some solutions have provided excellent insight into the relevant variables for undrained penetration into normally consolidated, uncemented, insensitive clays. The authors believe, based on measured observations, that total normal stresses are high on the cone face and relatively low immediately behind the tip. On the face of the cone during penetration normal stresses and shear stresses are highest, generally resulting in the highest positive pore pressures in fine grained soils (Jamiolkowski et al., 1985). However, as an element of soil passes behind the tip, there appears to be a small strain decrease resulting in a drop in total normal stresses. This

stress relief can be very large in sandy soils (Hughes and Robertson, 1985) and stiff clays, due to arching effects, and less in soft clayey soils. Therefore, pore pressures measured on the face of a penetrating cone are influenced by both large normal and shear stresses, whereas, pore pressures measured immediately behind the 60 degree tip tend to be influenced more by high shear stresses.

Since theories to truly model the penetration process for all soils are still in the process of development, much disagreement still exists over the distribution of stresses around a penetrating cone.

Soil Profile

The major application of CPTU data has been for soil profiling.

Traditionally, soil classification from CPT data has been related to cone bearing, q_c, and friction ratio, FR = $(f_s/q_c)\cdot 100\%$. Several charts have been developed that use this basic CPT data. All the charts are similar in that sandy soils generally have high cone bearing and low friction ratios whereas, clayey soils generally have low cone bearing and high friction ratios. However, the measurement of sleeve friction is sometimes less accurate and reliable than the cone resistance. Also cones of different designs will often produce variable friction sleeve measurements (Lunne et al., 1986). This can be caused by variations in mechanical and electrical design features of the friction sleeve as well as unequal end areas.

Initially, it was considered that the pore pressures measured on the apex of the cone or along the face of the cone tip were better for soil profiling, since at these locations the pore pressure response was a maximum. However, there have been reasonable arguments (Tavenas et al., 1982; Campanella et al., 1982; Jamiolkowski et al., 1985) to locate the filter element just behind the cone tip for the following reasons:

- porous element is much less subject to damage and abrasion
- measurements are less influenced by element compressibility
- position is appropriate for correction due to unequal end areas
- good stratigraphic detail still possible.

For these reasons the published charts for soil profiling incorporate the pore pressure measurement immediately behind the cone tip.

To overcome the problems associated with sleeve friction measurements, several classification charts have been proposed based on q_c or q_T and pore pressures (Jones and Rust, 1982; Baligh et al., 1980; Sennest and Janbu, 1984). The chart by Senneset and Janbu (1984) uses the pore pressure parameter ratio, B_q, defined as;

$$B_q = \frac{\Delta u}{q_T - \sigma_{vo}} \qquad \ldots (2)$$

where

Δu = excess pore pressure $(u - u_o)$

q_T = cone resistance corrected for pore pressure effects

σ_{vo} = total overburden stress

The original chart by Senneset and Janbu (1984) uses q_c. However, it is generally agreed that the chart and B_q should use the corrected cone bearing, q_T. The correction is usually only significant in soft, fine grained soils where q_c can be small and Δu can be very large.

The chart proposed by Senneset and Janbu (1984) and modified to use q_T is shown in Fig.9. The chart is based on pore pressures measured immediately behind the cone tip, as shown in Fig.9. Recent experience has shown that the measured pore pressures are influenced by factors, such as, stress history, sensitivity and stiffness to strength ratio (G/s_u). Experience has also shown that it is possible to record pore pressures behind the tip that are less than the static equilibrium pressure (u_o) in some overconsolidated and dilative soils (Campanella et al., 1983; Robertson et al., 1986; Lunne et al., 1986). Therefore Δu can be negative in some soils. The possibility of a negative Δu was incorporated into the classification chart proposed by Jones and Rust (1982), which is reproduced in Fig.10. It is interesting to note that the slope of the boundaries defining the zones of various soil types in Fig.10 are comprised of the same parameters used to define B_q.

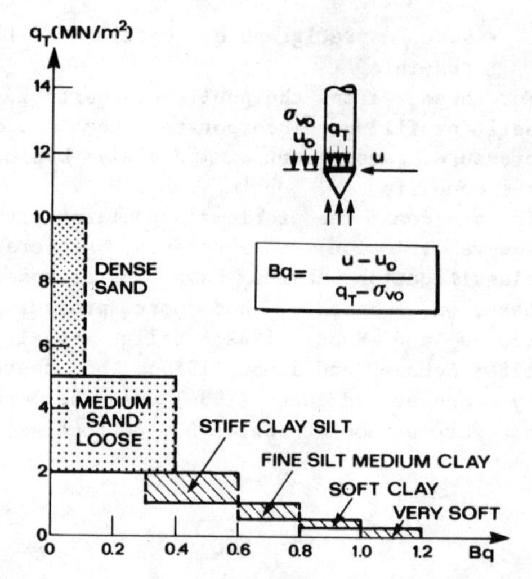

Fig.9 Soil classification chart from CPTU data proposed by Senneset and Janbu (1984)

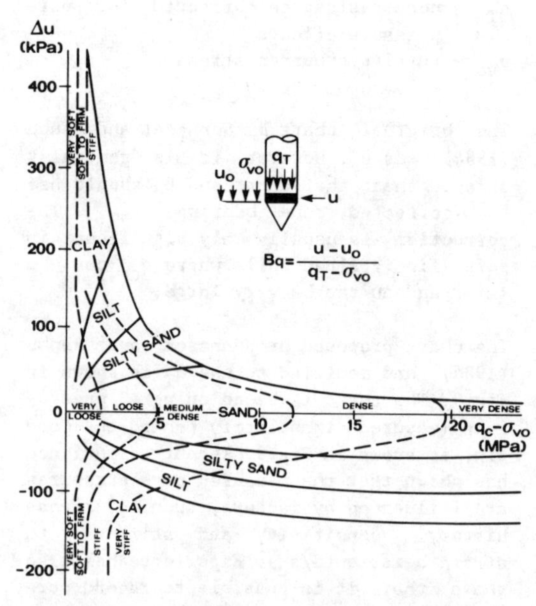

Fig.10 Soil classification chart from CPTU data proposed by Jones and Rust (1982)

Certainly the measurement of u adds an extra dimension to soil classification. However, classification cannot always be reliably based on q_T (or q_c) and B_q alone, in the same way that q_T (or q_c) and FR is not always universally reliable. The best practice is to use all three pieces of data (q_T, u, f_s) in the form of q_T, B_q and FR as suggested by Robertson et al. (1986) and shown in Figure 11.

The charts in Figures 9, 10 and 11 are global in nature and should be used as a guide to define soil behaviour type based on CPTU data. In addition dissipation data and samples are highly desireable. It is strongly recommended that local correlations or regionally adjusted classification charts be developed. Factors such as changes in, stress history, sensitivity, stiffness and void ratio, compressibility, fissuring and cementation will influence the classification using either the FR or the B_q chart.

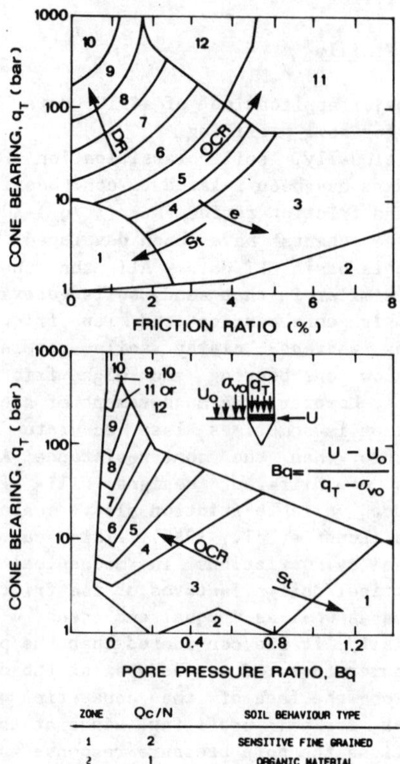

ZONE	Qc/N	SOIL BEHAVIOUR TYPE
1	2	SENSITIVE FINE GRAINED
2	1	ORGANIC MATERIAL
3	1	CLAY
4	1.5	SILTY CLAY TO CLAY
5	2	CLAYEY SILT TO SILTY CLAY
6	2.5	SANDY SILT TO CLAYEY SILT
7	3	SILTY SAND TO SANDY SILT
8	4	SAND TO SILTY SAND
9	5	SAND
10	6	GRAVELLY SAND TO SAND
11	1	VERY STIFF FINE GRAINED (*)
12	2	SAND TO CLAYEY SAND (*)

(*) Overconsolidated or cemented

Fig.11 Soil behaviour type chart from CPTU data proposed by Robertson et al. (1986)

Occasionally soils will fall within different zones on each chart, in these cases judgment is required to classify correctly the soil behaviour type. Often the rate and manner in which the excess pore pressures dissipate during a pause in the cone penetration will aid in the classification. For example, a soil may have the following CPTU parameters; q_T = 10 bars (1 MPa), FR = 4%, B_q = 0.1. It would classify as a clay on the FR chart and as a clayey silt to silty clay on the B_q chart. However, if the rate of pore pressure dissipation were very slow this would add confidence to the classification of a clay. If the dissipation were rapid (t_{50} < 60 sec) the soil may be more like a clayey silt or possibly a clayey sand. The manner of the dissipation can also be important. In stiff, overconsolidated clay soils, the pore pressure behind the tip can be very low in comparison to the high pore pressures on the face. When penetration is stopped, pore pressures recorded immediately behind the tip may initially rise before dropping to the equilibrium pressure. The rise can be caused by local equilization of the high pore pressures on the nearby cone face, although poor saturation can also cause a similar response (Campanella et al., 1983).

A further problem associated with existing CPT classification charts is that soils can gradually change in their apparent classification as cone penetration increases in depth. This is due to the fact that q_T, u and f_s all tend to increase with increasing overburden pressure. For example, in a thick deposit of normally consolidated clay, the cone bearing will increase linearly with depth resulting in an apparent change in CPT classification. Existing classification charts are based predominantly on data obtained from CPT profiles extending to a depth of less than 30 m. Therefore, for CPT data obtained at depths significantly greater than 30 m or at very shallow depth, some error can be expected when using the standard global CPT classification charts.

Soil macrofabric, such as fissuring, and soil stabilization procedures, such as, vibroreplacement and dynamic compaction, will often change the classification of a given soil.

Attempts have been made to normalize cone data with the effective overburden stress, σ'_{vo} (Robertson and Campanella, 1985; Olsen, 1984; Douglas et al., 1985; Olsen and Farr, 1986). However, it is not clear how CPT data in general should be normalized. Olsen and Farr (1986) use different normalization methods for clayey and sandy soils, but this produces a somewhat complex iterative interpretation that requires a computer program.

In theory, any normalization to account for increasing stress should also account for changes in horizontal stresses. This could be achieved by using a parameter such as the octahedral stress, σ'_m; where:

$$\sigma'_m = \frac{1}{3} \sigma'_{vo} (1 + 2 K_o) \qquad \ldots (3)$$

However, at present, this has little practical benefit without a prior knowledge of the in-situ horizontal stresses (K_o at rest).

Normalization of the CPT data would avoid some of the problems associated with variations in q_T with soil density. At present, a very loose clean sand may be classified as a sandy silt to silty sand because of the low q_T. Attempts to normalise the q_T-B_q chart (Figure 11) by the writers, has met with little success since it becomes impossible to distinguish between, for example, a heavily overconsolidated clay and a normally consolidated sandy silt based only on q_T and B_q. It is important to understand that the charts in Fig.s 9, 10 and 11 are <u>not</u> "grain size" classification charts, but are "soil behaviour type" charts.

Soil Layering

A potential problem for identification of soil stratigraphy is the difference in response of q_c and u due to soil layering. The tip resistance is influenced by soil properties ahead and behind the tip. Chamber studies (Schmertmann, 1978, Treadwell, 1975) show that the tip senses an interface between 5 to 10 cone diameters ahead and behind. The distance over which the cone tip senses an interface increases with increasing soil stiffness. For interbedded deposits, the thinnest stiff layer the cone bearing can respond fully (i.e. q_c to reach full value within the layer) is about 10 to 20 diameters. For the standard 10 cm^2 electric cone, the minimum stiff layer thickness to ensure full tip resistance is therefore between 36 cm to 72 cm. The tip may however, respond fully for soft layers considerably thinner than 36 cm in thickness.

Since the cone tip is advanced continuously, the tip resistance will sense much thinner stiff layers, but not fully. This has significant implications when interpreting cone bearing for relative density determination in sand.

A major advantage of the CPTU is that the pore pressure responds to the soil type in the immediate area of the pore pressure element. To aid in the identification of very thin silt or sand layers within clay deposits, Torstensson (1982) proposed and successfully used thin (2.5 mm) pore pressure elements located immediately behind the cone tip to identify thin layers of soil.

The frequency response of a fully saturated piezometer cone is usually fast enough to observe changes in pore pressure with a period of 0.25 seconds or less. This corresponds to layer thickness of about 5 mm or less at the standard penetration rate of 2 cm/sec. Whether or not such thin layers are observed in practice depends on the response of the soil to the advancing cone, the response of the recording equipment and the depth interval of recording. For thin sand layers within a body of clay the drainage characteristics of the sand become very important.

Undrained Soil Strength

Interpretation of soil strength from CPTU is dependent on drainage conditions. Generally undrained shear strength (s_u) is determined from undrained penetration. Comprehensive reviews of s_u evaluation from CPT and CPTU data have been presented by Baligh et al. (1980), Lunne and Kleven, (1981), Jamiolkowski et al., (1982) and Robertson et al., (1986). Unfortunately the evaluation is complicated by the fact that s_u is not a unique parameter and depends on type of test (stress path followed), rate of strain and orientation of the failure planes (Wroth, 1984).

As discussed briefly above, undrained cone penetration represents a very complex problem for which there is no generally accepted theoretical solution for the determination of s_u. Therefore, CPT and CPTU data is generally interpreted based on empirical or approximate theoretical solutions. Unfortunately, the reference s_u for many empirical methods have been different. The authors believe that the reference s_u should be taken as the field vane (FVT) value and should be clearly stated.

In the past s_u has been estimated from the measured cone bearing, q_c, using the following bearing capacity equation:

$$s_u = \frac{q_c - \sigma_o}{N_k} \quad \ldots (4)$$

where N_k is an empirical or semi-empirical cone factor and σ_o is generally taken to be the total overburden pressure (σ_{vo}). With the corrected cone resistance, q_T, the cone factor has been expressed (Lunne et al., 1985) as:

$$s_u = \frac{q_T - \sigma_{vo}}{N_{KT}} \quad \ldots (5)$$

Unfortunately, N_{KT} varies between 4 and 30, depending on factors, such as; sensitivity, stress history, stiffness, fissuring (macrofabric) and definition of s_u.

Senneset et al. (1982) have suggested the use of CPTU data using the effective cone resistance, q_E, to determine s_u. Where q_E is defined as follows;

$$q_E = q_c - u \quad \ldots (6)$$

and u = total penetration pore pressure measured immediately behind the cone tip.

Robertson and Campanella (1983) redefined the effective cone bearing using the corrected cone resistance, q_T.

One major drawback using the effective cone resistance, q_E, is the reliability to which q_E can be determined. In soft normally consolidated clays, the total pore pressure, u, generated immediately behind the tip is often approximately 90 percent or more of the measured cone resistance, q_c. Even when q_c is corrected to q_T, the difference between q_T and u is often very small. Thus, q_E is often an extremely small quantity and is therefore sensitive to small errors in q_c measurements.

The problems of accuracy and pore pressure effects associated with the measured cone resistance in soft clays may explain some of the large scatter in published data concerning the cone factors N_K and N_{KE}.

Several relationships have been proposed between excess pore pressure (Δu) and s_u based on theoretical or semi-theoretical approaches using cavity expansion theory (Vésic, 1972; Battaglio et al., 1981; Randolph and Wroth, 1979; Henkel and Wade, 1966; Massarch and Broms, 1981; Campanella et al., 1985) using,

$$s_u = \frac{\Delta u}{N_{\Delta u}} \quad \ldots (7)$$

where $N_{\Delta u}$ varies between 2 and 20.

These methods have the advantage of increased accuracy in the measurement of Δu, especially in soft clays where Δu can be very large. In soft clays, q_c is very small and typically the cone tip load cell may be required to record loads less than 1% of rated capacity with an associated inaccuracy of up to 50% of the measured values.

The semi-empirical solution proposed by Massarch and Broms (1981) based on cavity expansion theories and including the effects of overconsolidation and sensitivity by using Skempton's pore pressure parameter at failure (A_f) is given in Fig.12. Approximate values for A_f can be estimated from Table 1. Clearly a knowledge of the shear modulus (G) or plasticity index (PI) would assist in the estimate of s_u. The addition of shear wave

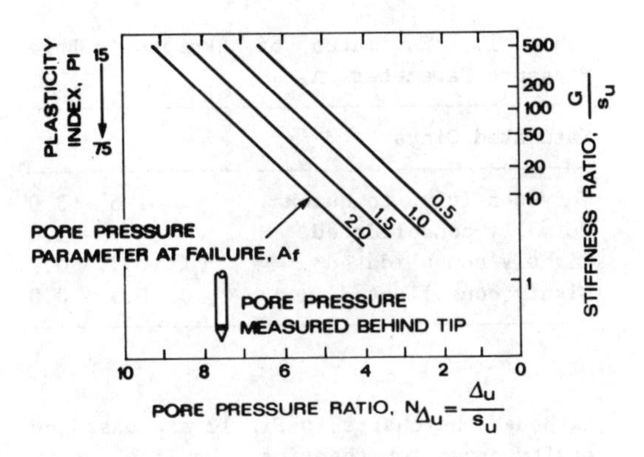

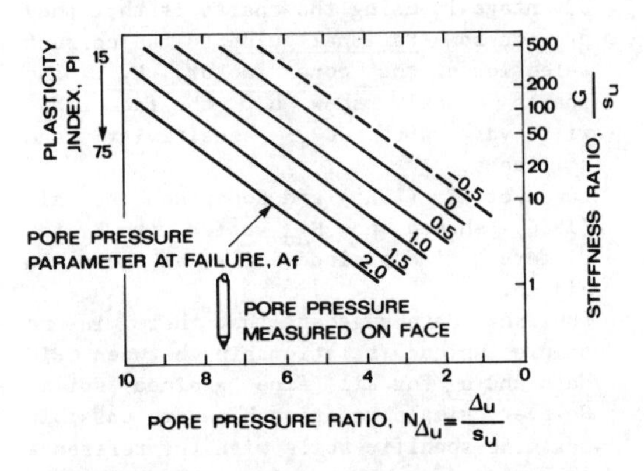

Fig.12 Charts to obtain s_u from excess pore pressure, Δu, measured during CPTU

velocity measurements during seismic CPTU is a promising method to obtain an independent measure of the shear modulus, (Robertson et al., 1986).

If pore pressures are measured immediately behind the cone tip, the measured values may not have reached the cylindrical cavity expansion value. Therefore s_u estimated from the chart with the pore pressures behind the tip may be slightly overestimated. Also because of the tendency for low or negative pore pressures measured behind the tip in insensitive, overconsolidated clays (see Fig.2), the chart in Fig.12 is not recommended for highly overconsolidated clays ($-0.5 < A_f < 0$).

Table 1. Estimates of Skempton's Pore Pressure Parameter, A_f

Saturated Clays	A_f
Very sensitive to quick	1.5 - 3.0
Normally consolidated	0.7 - 1.3
Lightly consolidated	0.3 - 0.7
Highly consolidated	-0.5 - 0.0

Although the charts in Fig.12 are based on cavity expansion theories, they are basically semi-empirical in nature. The advantage in using the charts is that they provide some rational guide to the correct selection of the cone factor, $N_{\Delta u}$. The charts clearly show how the factor $N_{\Delta u}$ will vary with OCR, sensitivity and stiffness.

Lunne et al. (1985) and Robertson et al. (1986) showed how $N_{\Delta u}$ varied with B_q for different fine grained soils, as shown in Fig.13.

Published data suggests that there is no simple unique relationship between CPTU data and s_u for all fine grained soils. However, simple relationships are possible for site specific soils with the reference s_u related to the design method. The use of semi-empirical charts, such as those in Figure 12, provide a useful rational guide for the estimate of s_u. It is also important to be aware that the use of Δu for estimating s_u can also be complicated by; poor saturation, partial drainage and soil macrofabric.

It is also important to note that CPTU data provides a continuous profile of soil strength which can be extremely valuable for design. CPTU data can also be used for direct correlation to foundation performance.

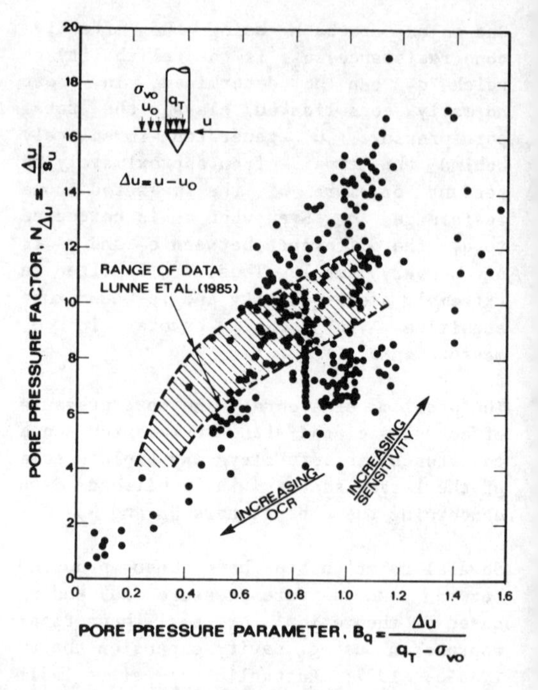

Fig.13 Pore pressure factor, $N_{\Delta u}$, versus pore pressure parameter, B_q (After Robertson et al., 1986).

Drained Soil Strength

Senneset et al. (1982) have suggested using CPTU data to determine the drained effective stress strength parameters (c', ϕ'). However, this method is based on bearing capacity theory and, as with any method for determining drained parameters from undrained cone penetration, can be subject to serious problems. An important problem, which was not addressed by Senneset et al. (1982), is the location of the porous element, since different locations give different measured total pore pressures.

Keaveny and Mitchell (1986) suggested an alternate method to determine effective stress strength parameters from CPTU data. Their method uses empirical correlations to estimate OCR, A_f and K_o and Vésic's (1972) cavity expansion method to estimate s_u. These are then combined to estimate the effective stress at failure, which when combined with s_u provides an estimate of the effective stress strength parameters (c', ϕ'). Keaveny and Mitchell (1986) report good results for silts and overconsolidated clays but rather poor results for normally consolidated clays. This approach does account for different pore pressure element locations but relies on simplified empirical correlations to estimate OCR, A_f and rigidity index, I_r. The authors believe that the current state of the art has not reached a level where reliable estimates of drained strength parameters can be made from undrained CPTU data.

Stress History

Baligh et al. (1980) suggested that the pore pressure measured during undrained cone penetration may reflect the stress history of a deposit. Since then several methods have been suggested to correlate various pore pressure parameters to OCR. A summary of the main pore pressure parameters are as follows;

	Parameter	Reference
(i)	$\frac{u}{q_c}$	Baligh et al. (1981)
(ii)	$\frac{\Delta u}{q_T}$	Campanella and Robertson (1981)
(iii)	$\frac{\Delta u}{q_c - u_o}$	Smits (1982)
(iv)	$\frac{\Delta u}{q_c - \sigma_{vo}}$	Senneset et al. (1982); Jones and Rust (1982); Jefferies and Funegard (1983); Wroth (1984);
(v)	$\frac{q_T - \Delta u - \sigma_{vo}}{\sigma'_{vo}}$	Jamiolkowski et al. (1985)
(vi)	$\frac{\Delta u}{\sigma'_{vo}}$	Azzouz et al. (1983), Mayne (1986)
(vii)	$\frac{u_T}{u_o} - \frac{u_B}{u_o}$	Sully et al. (1987)

where;
u_T = pore pressure on tip
u_B = pore pressure behind tip
u_o = equilibrium pore pressure

It is generally agreed that q_c should always be corrected to q_T. Therefore (iv) becomes B_q, as shown in Figures 11 and 13. Battaglio et al. (1986) presented several examples of the parameters (iv) and (v) for different Italian clays.

Wroth (1984) correctly pointed out that only the shear induced excess pore pressure reveals the nature of the soil behaviour and depends on stress history. Unfortunately, because of the complex nature of cone penetration it is not possible to isolate the shear induced pore pressures. However, as suggested earlier, the pore pressures measured immediately behind the cone tip appear to be influenced by shear stresses, although changes in octahedral stresses complicate any quantitative interpretation.

A review of published correlations shows that no unique relationship exists between the above pore pressure ratios and OCR, because pore pressures measured at any one location are influenced by clay sensitivity, preconsolidation mechanism, soil type and local heterogeneity (Robertson et al., 1986; Battaglio et al., 1986).

Methods have also been suggested to relate CPTU data to preconsolidation stress (p'_c) (Konrad and Law, 1987). However, these methods suffer from many of the same problems as those given above.

Since the shear induced pore pressures cannot be isolated with measurements at any one location on the cone, Campanella et al. (1985) suggested that the difference between pore pressures measured on the face and somewhere behind the tip may correlate better with OCR (Sully et al., 1987).

At present any empirical relationship should be used to obtain only qualitative information on the variation of OCR within the same relatively homogeneous deposit.

Flow Characteristics

In the last 10 years, much attention has been devoted to the analysis of dissipation tests with the CPTU (Torstensson, 1977; Randolph and Wroth, 1979; Baligh and Levadoux, 1980; Acar et al., 1982; Gupta and Davidson, 1986). A dissipation test consists of stopping cone penetration and monitoring the decay of excess pore pressures (Δu) with time. From these data an approximate value of the coefficient of consolidation in the horizontal direction (c_h) can be obtained.

A comprehensive study and review of this topic was recently published by Baligh and Levadoux (1986).

The most relevant conclusions from this study are:

1. The simple uncoupled solutions provide reasonably accurate predictions of the dissipation process.
2. Dissipation is controlled by the horizontal c_h.

3. Consolidation is taking place predominantly in the recompression mode, especially for dissipation less than 50%.
4. Initial distribution of excess pore pressures around the cone have a significant influence on the dissipation process.

Based on these findings the following procedure for evaluating c_h from CPTU dissipation tests is recommended:

a. Plot the normalized excess pore pressure with log time, as shown in Figure 14.
b. Compare the measured dissipation curve with theoretical curves.
c. If the curves are similar in shape compute c_h from;

$$c_h = \frac{R^2 T}{t} \quad \ldots (8)$$

where:

T - theoretical time factor for given tip geometry, porous element location and soil rigidity index, $I_r = G/S_u$

t - time to reach given value of $\Delta u(t)/\Delta u$

R - radius of cone

A recent comprehensive study by C-I. Teh (1987) has shown the importance of the rigidity index, I_r, and has suggested the use of a normalized theoretical time factor to account for variations in I_r. Examples of a range of recorded dissipation results compared with the theoretical rate of decay (Baligh and Levadoux, 1986) is shown in Figure 14. The data in Figure 14 was obtained at the McDonalds Farm site where the OCR ≈1.0. However, Figure 14 shows that, for all the pore pressure element locations evaluated, the measured rates of decay diverge from the theoretical values. Thus, the calculated values of c_h depend on the degree of dissipation selected and on the location of the pore pressure element.
At present, because of the difficulties in predicting the initial distribution of excess pore pressures around a cone in stiff, overconsolidated clays (OCR > 4), the theoretical solutions for estimating c_h from dissipation tests is limited to normally to lightly overconsolidated clays (OCR < 4).

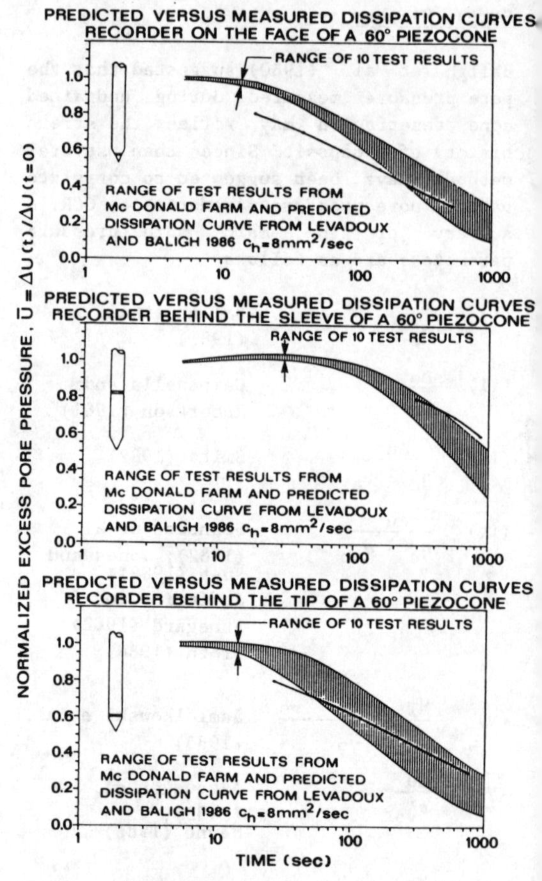

Fig.14 Comparison of predicted and measured dissipation curves

The applicability and meaning of the solutions is also complicated by several phenomena, such as;

- importance of vertical as well as horizontal dissipation,
- effect of soil disturbance,
- uncertainty over distribution, level and changes in total stresses,
- soil anisotropy and nonlinearity,
- non-homogeneity due to soil layering or nearness to a layer boundary,
- influence of macrofabric, such as fissuring,
- influence of clogging and smearing of filter element.

In spite of the above limitations the authors believe that the dissipation test provides a useful means of evaluating <u>approximate</u> (i.e. within one order of magnitude) consolidation properties, soil macrofabric and related drainage paths of

natural fine grained deposits. The test also appears to provide very important information for the design of vertical drains (Battaglio et al., 1981; Robertson et al., 1986).
Baligh and Levadoux (1986) also suggested an approximate procedure to estimate the horizontal coefficient of permeability, k_h, from c_h.

APPLICATIONS OF CPTU DATA

Despite the considerable interest shown in the CPTU the major area of application in geotechnical engineering has been in soil profiling. Interpretation of the pore pressure data in a quantitative manner has been slow to gain acceptance by the practising profession. This may be due, in part, to the confusion in much of the published literature regarding pore pressure element location combined with the complexity of the data in many "real" soils.
The continuous nature of the CPTU data has encouraged many operators to develop direct correlations to foundation performance. However, a full discussion is beyond the scope of this paper.
Further discussion on the application of CPTU data is deferred to the paper on "New Applications of Penetration Tests in Design Practice" by Jamiolkowski et al. presented in this conference.

REVIEW OF CPTU PRACTISE

A questionnaire was sent out to over 80 people around the world to obtain information on the current status of CPTU practise. The main results of the questionnaire can be summarized as follows,

- There are over 20 different types of CPTU equipment in use around the world.
- Most of the equipment meets the basic standard of a 60 degree, 10 cm^2 cone with a shaft of equal diameter. Several cones have been manufactured to be 15 cm^2, especially in offshore practise.
- The pore pressure element location varies as shown on Figure 1 (60 degree cones). Although a slight majority of systems have the element located immediately behind the tip (>60%). Several cone designs can have the element at different locations.
- Saturation of porous elements is generally performed under vacuum using either water, glycerin or silicon oil. Some countries still boil the elements. Water appears to be the least common saturation fluid and silicon oil or glycerin the more common.
- Saturation of the cone assembly varies considerably. Generally saturation is performed in the field using either; a flushing technique, specially designed devices or submerging the system under water.
- Procedures to maintain saturation vary depending on local geological conditions, but generally a special sheath is placed over the assembled and saturated cone.
- Generally no special tests are performed to evaluate saturation before or after a CPTU. Evaluation is generally based on a careful review of the CPTU data.
- Dissipation tests are generally not performed on a routine basis, but most people perform dissipations when required.
- The primary purpose of the CPTU is for improved, continuous stratigraphic logging.
- Most users make corrections to q_c for unequal end area effects.
- In most countries the CPTU generally represents less than 10% of all CPT work. Exceptions are; the offshore industry and some countries (Norway, and Canada) were the CPTU represents greater than 80% of all CPT work.
- The current level of supervision provided for CPTU varies considerably. In some areas no supervision is provided whereas, in others (notably offshore) high levels of supervision is provided.
- The major stated advantages of the CPTU are;
  - continuous data
  - improved stratigraphic logging
  - improved interpretation of soil parameters.

• The major stated disadvantages of the CPTU are;
  . lack of standardization, especially for element location,
  . difficult and time consuming saturation procedures,
  . complex interpretation in some soils,
  . limited to soft, saturated soils.

The level of experience stated by the people who responded to the questionnaire varied from as little as 1 to 2 years to up to 17 years, although the average was about 5 years.

SUMMARY

The CPTU has many advantages over the conventional CPT. The main advantages are:
. ability to distinguish between drained, partially drained and undrained penetration,
. ability to correct measured cone data to account for unequal end areas in cone design,
. ability to evaluate flow and consolidation characteristics,
. ability to assess equilibrium groundwater conditions,
. improved soil profiling and identification, improved evaluation of geotechnical parameters.

The importance of the pore pressure elemnt location is clearly understood and all CPTU data should clearly state the size of the element and its location. New cone designs that incorporate 2 or more porous elements show considerable promise. For general CPTU the authors suggest that the pore pressure element should be located immediately behind the tip for the following reasons;
• porous element is much less subject to damage and abrasion
• measurements are less influenced by element compressibility
• position is appropriate for correction due to unequal end areas
• good stratigraphic detail is still possible.

However, the element should not be closer than 2.5 mm from the shoulder of the tip and should never be smaller than the diameter of the probe.

The importance of cone design related to mechanical load transfer and filter compression is also clearly recognised. However, no standard has emerged regarding selection of the filter material and its associated saturating fluid. The authors suggest that the filter material should be resistant to abrasion and smearing and should have compatible rigidity and permeability consistent with the element location and the soils to be penetrated. Filter elements should also be replaced after each sounding.

Complete saturation of the piezocone is very important. Methods to assess the degree of saturation are difficult and unreliable. Generally saturation is evaluated by careful review of the data. However, this procedure is somewhat subjective.

Interpretation of CPTU data is often complex and influenced by many variables. Correct interpretation of the pore pressure data requires some knowledge that the penetration is predominantly undrained. Dissipation data can provide valuable additional information concerning the drainage conditions.

The major application of CPTU data has been for soil profiling. Several classification charts have been developed using CPTU data. These charts are global in nature and should be used only as a guide to define soil behaviour type. In specific geological areas the charts can be adjusted for local experience to provide excellent local correlations.

Methods have been developed to estimate s_u from undrained CPTU data. No unique relationship appears to exist between CPTU data and s_u, since factors such as stress history, sensitivity, soil stiffness and macrofabric complicate the pore pressure response. However, simple relationships are possible for site specific soils.

New methods have been developed to estimate effective stress strength parameters from undrained cone penetration data. However, these methods incorporate many simplified assumptions and require further evaluation.

Correlations have been developed relating various pore pressure parameters to stress history (OCR). However, the pore pressure measured at one location on the cone is influenced by other factors, such as, sensitivity, soil type, preconsolidation mechanism and local heterogeneity. Therefore, the existing empirical relationships should be used to obtain only qualitative information on the variation of OCR within the same relatively homogeneous deposit. An attractive alternative is to correlate OCR with the difference in Δu measured at two different locations on the cone (Sully et al., 1987).

Comprehensive studies have been published concerning the analyses of pore pressure dissipations during a pause in cone penetration. In general, these theoretical solutions for estimating c_h from dissipation CPTU data is limited to normally to lightly overconsolidated clays (OCR < 4). Many limitations exist in the solutions, especially related to the initial distribution of excess pore pressures around the cone before dissipation. However, in spite of these limitations the dissipation test provides a useful means of evaluating approximate consolidation properties, soil macrofabric and relative drainage conditions in natural soils. If the excess pore pressures are allowed to completely dissipate the resulting equilibrium piezometric pressures provide important hydro-geologic information.

The continuous profiles of CPTU data provide excellent information on soil stratigraphy. CPTU data can also be used for direct correlation to foundation performance.

In spite of the complexities in interpretation of CPTU data, the test provides valuable additional information, especially related to estimates of; drainage conditions during penetration, soil profiling, soil hydraulic characteristics and equilibrium groundwater conditions. With increasing use of the CPTU in geotechnical practise continued improvements can be expected in the use and interpretation of the test.

ACKNOWLEDGEMENTS

The assistance of the Natural Sciences and Engineering Research Council and the technical staff of the Civil Engineering Department, University of British Columbia is much appreciated.
The valuable work and assistance of the graduate students at UBC, especially D. Gillespie and J. Greig is also appreciated.
The support and review comments of Prof. M. Jamiolkowski during the stabbatical leave of P. Robertson is also much appreciated.

REFERENCES

Acar, Y., (1981), "Piezocone Penetrating Testing in Soft Cohesive Soils", Louisiana State University, Dept. of Civil Eng., Fugro Postdoctoral Fellowship, Activity Report No.4.

Acar, Y.B. and Tumay, M.T., (1986). "Strain Field around Cones in Steady Penetration", ASCE Journal of Geotechnical Engineering Division, Vol.112, No.2, pp.207-213.

Al-Awkati, Z.A., (1975), "On Problems of Soil Bearing Capacity at Depth", Ph.D. Thesis, Duke University, Department of Civil Engineering.

Azzouz, A.S., Baligh, M.M., Ladd, C.C., (1983), "Cone Penetration and Engineering Properties of the Soft Orinoco Clay". Proc. of the BOSS 83, Cambridge, Mass.

Baligh, M.M., (1975), "Theory of Deep Site Static Cone Penetration Resistance", Report No. R.75-76 Massachusetts Institute of Technology, Cambridge, Mass., 02139.

Baligh, M.M., (1985). "The Strain Path Method", ASCE, JGED, No.9, Vol.III.

Baligh, M.M. and Levadoux, J.N., (1980), "Pore Pressure Dissipation After Cone Penetration", Massachusetts Institute of Technology, Department of Civil Engineering, Construction Facilities Division, Cambridge, Massachusetts 02139.

Baligh, M.M., Azzouz, A.S., Wissa, A.Z.E., Martin, R.T. and Morrison, M.J., (1981), "The Piezocone Penetrometer", ASCE, Geotechnical Divsion, Symposium on Cone

Penetration Testing and Experience, St. Louis, pp.247-263.
Baligh, M.M., Vivatrat, V., and Ladd, C.C., (1980), "Cone Penetration in Soil Profiling", ASCE, Journal of Geotechnical Engineering Division, Vol.106, GT4, April, pp.447-461.
Battaglio, M., Bruzzi, D., Jamiolkowski, M., and Lancellotta R., (1986), "Interpretation of CPT's and CPTU's - Undrained Penetration of Saturated Clays", Proceedings 4th International Geotechnical Seminar, Singapore.
Battaglio, M., Jamiolkowski, M., Lancellotta, R., Maniscalco, R., (1981), "Piezometer Probe Test in Cohesive Deposits", ASCE, Geotechnical Division, Symposium on Cone Penetration Testing and Experience, pp. 264-302.
Battaglio, M., and Maniscalco, R., (1983), "Il Piezocone. Esecuzione ed Interpretazione." Scienza delle Costruzioni Politecnico di Torino, No.607.
Begemann, H.K.S., (1965). "The Friction Jacket Cone as an Aid in Determining the Soil Profile". Proc. VI ICSMFE, Montreal, Vol.I.
Borst, R. de., and Vermeer, P.A., (1982), "Finite Element Analysis of Static Penetration Test", Proceedings of 2nd European Symposium on Penetration Testing, Amsterdam.
Campanella, R.G. and Robertson, P.K., (1981), "Applied Cone Research", Symposium on Cone Penetration Testing and Experience, Geotechnical Engineering Division, ASCE, Oct. 1981, pp. 343-362.
Campanella, R.G., Gillespie, D., and Robertson, P.K., (1982), "Pore Pressures during Cone Penetration Testing", Proc. of 2nd European Symposium on Penetration Testing, ESOPT II, pp. 507-512.
Campanella, R.G., Robertson, P.K. and Gillespie, D., (1986), "Factors Affecting the Pore Water Pressures and its Measurement around a Penetrating Cone", Proceedings 39th Canadian Geotechnical Conference Ottawa.
Campanella, R.G., Robertson, P.K. and Gillespie, D., (1983), "Cone Penetration Testing in Deltaic Soils", Canadian Geotechnical Journal, Vol.20, No.1, February, pp.23-35.
Campanella, R.G., Robertson, P.K., Gillespie, D.G. and Greig, J., (1985), "Recent Developments in In-Situ Testing of Soils, Proceedings of XI ICSMFG, San Francisco.
Cone Penetration Testing and Experience, (1981), Edited by G.M. Norris and R.D. Holtz, Proceedings of a Sympoisum sponsored by the Geotechnical Division of ASCE, St. Louis, Missouri, Oct., 1981.
de Ruiter, J., (1982), "The Static Cone Penetration Test State-of-the-Art Report", Proceedings of the Second European Symposium on Penetration Testing, ESOPT II, Amsterdam, May 1982, Vol.2, pp.389-405.
Douglas, B.J., Strutynsky, A.I., Mahar, L.J. and Weaver, J., (1985), "Soil Strength Determinations from the Cone Penetration Test", Proceedings of Civil Engineering in the Artic Offshore, San Francisco.
Durgunoglu, H.T. and Mitchell, J.K., (1975), "Static Penetration Resistance of Soils: I-ANALYSIS", ASCE Specialty Conference on In-situ Measurement of Soil Parameters, Raleigh, Vol.I.
Gibson, R.E., (1950), Discussion of G. Wilson, "The Bearing Capacity of Screw Piles and Screw-crete Cylinders", Journal of the Institution of Civil Engineers, Vol.34, No.4, pp.382.
Gupta, R.C. and Davidson, J.L., (1986). "Piezoprobe Determined Coefficient of Consolidation", Soil and Foundations, Vol.26, No.3, Japanese Society of Soil Mechanics and Foundation Engineering.
Henkel, D.J. and Wade, N.H., (1966), "Plane Strain on a Saturated Remoulded Clay", ASCE-J, Vol.92, No.6.
Hughes, J.M.O. and Robertson, P.K., (1985), "Full-displacement Pressuremeter Testing in Sand", Canadian Geotechnical Journal, Vol.22, No.3.
Houlsby, G.T., and Wroth, P., (1982), "Determination of Undrained Strengths by Cone Penetration Tests", Proceedings of 2nd European Symposium on Penetration Testing, Amsterdam.
Jamiolkowski, M., Ladd, C.C., Germaine, J.T. and Lancellotta, R., (1985), "New Developments in Field and Laboratory Testing of Soils", State-of-the-Art Paper at the 11th International Conference Society for Soil Mechanics and Foundation Engineering (ICSMFE), San Francisco.
Jamiolkowski, M., Lancellotta, R., Tordel-

la, L. and Battaglio, M., (1982), "Undrained Strength from CPT", Proceedings of the European Symposium on Penetration Testing, ESOPT II, Amsterdam, May 1982, Vol.2, pp.599-606.

Jefferies, M.G. and Funegard, E., (1983), "Cone Penetration Testing in the Beaufort Sea", ASCE Specialty Conference, Geotechnical Practice in Offshore Engineering, Austin, Texas, pp.220-243.

Jones, G.A. and Rust, E.A., (1982), "Piezometer Penetration Testing CUPT", Proceedings of the 2nd European Symposium on Penetration Testing, ESOPT II, Amsterdam, Vol.2, pp.607-613.

Keaveny, J.M. and Mitchell, J.K., (1986), "Strength of Fine Grained Soils Using the Piezcone", Proceeding of In-Situ '86, ASCE Specialty Conference, Blacksburg, Virginia.

Konrad, J.-M. and Law, T., (1987). "Preconsolidation Pressure for Piezocone Tests in Marine Clays", Geotechnique, Vol.37, No.2, pp.177-190.

Lacasse, S., and Lunne, T., (1982), "Penetration Tests in Two Norwegian Clays", Proceeding of the 2nd European Symposium on Penetration Testing, Amsterdam, Vol. II.

Levadoux, J.N. and Baligh, M.M., (1986), "Consolidation After Undrained Piezocone Penetration. I: Prediction", Journal of Geotechnical Division, ASCE, Vol.112, No.7, pp.707-726.

Luger, H.J., Lubking, P., and Nieuwenhnis, J.D., (1982), "Aspects of Penetration Tests in Clay", Proceedings of 2nd European Symposium on Penetration Testing, Amsterdam.

Lunne, T. and Kleven, A., (1981), "Role of CPT in North Sea Foundation Engineering", Symposium on Cone Penetration Testing and Experience, Geotechnical Engineering Division, ASCE, Oct. 1981, pp. 49-75.

Lunne, T., Christoffersen, H.P., and Tjelta, T.I., (1985), "Engingeering Use of Piezocone Data in North Sea Clays", Proceedings XI ICSMFE, San Francisco.

Lunne, T., Eidsmoen, T.E., Gillespie, D. and Howland, J.D., (1986), "Laboratory and Field Evaluation of Cone Penetrometers", Proceedings of In-Situ 86, Specialty Conference, ASCE, Blacksburg, Virginia.

Lunne, T., Eidsmoen, T.E., Powell, J.J.M. and Quatermann, R.S.T., (1986). "Piezocone Testing in Overconsolidated Clays", XXXIX Canadian Geotechnical Conference, Ottawa, pps 209-218.

Massarsch, K.R., and Broms, B.B., (1981), "Pile Driving in Clay Slopes", Proceeding ICSMFE, Stockholm.

Mayne, P.W., (1986). "CPT Indexing of In-situ OCR in Clays", Proceedings of In-Situ 86, ASCE Specialty Conference, Blacksburg, Virginia, pps.780-789

Meyerhof, G.G., (1961), "The Ultimate Bearing Capacity of Wedge Shaped Foundations", Proceedings of 5th International Conference on Soil Mechanics and Foundation Engineering, Paris, Vol.2.

Muromachi, T., (1981), "Cone Penetration Testing in Japan", Symposium on Cone Penetration Testing and Experience, Geotechnical Engineering Division, ASCE, October, St. Louis, pp.76-107.

Olsen, R.S., (1984), "Liquefaction Analysis Using the Cone Penetration Test", Proceedings of the 8th World Conf. on Earthquake Engineering, San Francisco.

Olsen, R.S., and Farr, J.V., (1986), "Site Characterization Using the Cone Penetration Test", Proceeding of In-Situ 86, ASCE Specialty Conference, Blacksburg, Virginia.

Randolph, M.F. and Wroth, C.P., (1979), "An Analytical Solution for the Consolidation Around a Driven Pile", International Journal for Numerical and Analytical Methods in Geomechanics, Vol.3, pp.217-229.

Robertson, P.K. and Campanella, R.G., (1985), "Evaluation of Liquefaction Potential of Sands Using the CPT", Journal of Geotechnical Division, ASCE, Vol.III, No.3, Mar., pp.384-407.

Robertson, P.K. and Campanella, R.G., (1983), "Interpretation of Cone Penetration Tests - Part II (Clay)", Canadian Geotechnical Journal, Vol.20, No.4.

Robertson, P.K., Campanella, R.G., Gillespie, D. and Grieg, J., (1986), "Use of Piezometer Cone Data", Proceedings of In-Situ '86, ASCE, Specialty Conference, Blacksburg, Virginia.

Schmertmann, J.H., (1978a), "Guidelines for Cone Penetration Test, Performance and Design", Federal Highway Administration, Report FHWA-TS-78-209, Washington, July 1978, 145 pgs.

Senneset, K., Janbu, N. and Svanø, G., (1982), "Strength and Deformation Parameters from Cone Penetrations Tests", Proceedings of the European Symposim on Penetration Testing, ESOPT II, Amsterdam, May 1982, pp. 863-870.

Senneset, K. and Janbu, N. (1984), "Shear Strength Parameters Obtained from Static Cone Penetration Tests", ASTM STP 883, Symposium, San Diego.

Skempton, A.W., (1957), "Discussion: The Planning and Design of the New Hong Kong Airport", Proceedings, Institutiuon of Civil Engineers, London 7, pp.305-307.

Smits, F.P., (1982), "Penetration Pore Pressure Measured with Piezometer Cones", Proceedings of the Second European Symposium on Penetration Testing, ESOPT II, Amsterdam, Vol.2, pp.887-881.

Sugawara, N. and Chikaraishi, M., (1982), "On Estimation of ϕ' for Normally Consolidated Mine Tailings using the Pore Pressure Cone Penetrometers", ESOPT II, 883-888.

Sully, J.P., Campanella, R.G. and Robertson, P.K., (1987). "Overconsolidation Ratio of Clays from Penetration Pore pressures". ASCE, Journal of Geotechnical Division (to be published).

Tavenas, F., Leroueil, S. and Roy, M., (1982) "The Piezcone Test in Clays: Use and Limitations", Proceedings of the Second European Symposium on Penetration Testing, ESOPT II, Amsterdam, pp.889-894.

Teh, Cee-Ing (1987). "An Analytical Study of the Cone Penetration Test", Ph.D. Thesis, University of Oxford.

Torstensson, B.A., (1975), "Pore Pressure Sounding Instrument", Proceedings, ASCE Spec. Conf. on In-Situ Measurement of Soil Properties", Vol. II, Raleigh, N.C., pp.48-54.

Torstensson, B.A., (1977), "The Pore Pressure Probe", Nordiske Geotekniske Møte, Oslo, Paper No.34.1-34.15.

Torstensson, B.A., (1982), "A combined Pore Pressure and Point Resistance Probe", Proceedings of the Second European Symposim on Penetration Testing, ESOPT II, Amsterdam, Vol.2, pp.903-908.

Treadwell, D.D., (1975), "The Influence of Gravity, Prestress, Compressibility, and Layering on Soil Resistance to Static Penetration", Ph.D. Dissertation, Graduate Division of the University of California, Berkeley, 210 pgs.

Tumay, M.T., Acar, Y.B., Cekirge, M. and Ramesh, N., (1985). "Flow Field Around Cones in Steady Penetration", ASCE Journal of Geotechnical Engineering Division, GT2 pp.325-342.

Vésic, A.S., (1972), "Expansion of Cavities in Infinite Soil Masses", Journal of the Soil Mechanics and Foundation Division, ASCE, Vol.98, SM3, pp.265-290.

Wissa, A.E.Z., Martin, R.T. and Garlanger, J.E., (1975), "The Piezometer Probe", Proceedings ASCE Spec. Conf. on In-situ Measurement of Soil Properties, Raleigh, N.C., Vol.1, pp.536-545.

Wroth, C.P., (1984). "The Interpetation of In-Situ Soil Tests." Rankine Lecture, Geotechnique No.4.

Penetration Testing 1988, ISOPT-1, De Ruiter (ed.)
© 1988 Balkema, Rotterdam, ISBN 90 6191 801 4

Penetration tests for dynamic problems

Kohji Tokimatsu
Tokyo Institute of Technology, Japan

ABSTRACT: The current status of penetration type testing for determining dynamic soil properties is presented and discussed with special emphasis placed on liquefaction related problems of cohesionless soils. The major penetration tests discussed include the standard penetration test and the static cone penetration test. Discussion is also extended to other penetration type tests (e. g., the dilatometer test and large versions of the standard penetration test), and newly developed tests that combine some of the good features of penetration tests and specific tests (e. g., seismic cone and vibrocone penetration tests). The applicability and limitation of each in-situ testing are evaluated, if any, based on field performance data and laboratory test results on high-quality undisturbed samples obtained by the in-situ freezing method.

INTRODUCTION

The interest in soil dynamics and geotechnical earthquake engineering covers the deformation and stability problems in soil deposits subjected to various kinds of cyclic loadings including those caused by earthquakes, ocean waves, machines, and traffic. However, in view of the serious damage reported in every significant earthquake, the main problems in this area are 1) the seismic stability of soil deposits, primarily comprising saturated cohesionless soils, and 2) the seismic response of the deposits during earthquakes.

The dynamic soil characteristics of primary interest corresponding to the problems are 1) the liquefaction resistance of cohesionless soils, and 2) the shear moduli and damping factors of the soils, respectively.

The dynamic characteristics have generally been determined by one or a combination of 1) laboratory tests on undisturbed samples, and 2) in-situ tests. Significant advantages of the in-situ tests for dynamic problems as compared to laboratory tests are as follows (e. g., Seed and De Alba, 1986, and Robertson, 1986):

1. It is fast and cost effective for obtaining soil characteristics of the deposits which are inherently erratic. Laboratory tests on limited samples may miss such erratic nature and thus may not serve a sufficient basis for design.
2. It can circumvent some difficulties associated with sampling and laboratory testing, such as sample disturbance and imperfect simulation of in-situ stresses.
3. It has a potential to determine properties of soils, such as cohesionless soils and offshore deposits, that cannot be easily sampled in the undisturbed state.

At the same time one should also keep in mind the following disadvantages of in-situ tests (e. g., Jamiolkowski et al., 1985a).

1. Boundary and drainage conditions cannot be controlled independently.
2. Modes of deformation and failure imposed on the soil are generally different from those caused by earthquake shaking.
3. Measured values may not correspond directly to the dynamic soil properties. Determination of such properties usually requires some empirical correlations.

However, these limitations tend to be outweighed by the aforementioned advantages of in-situ testing. Accordingly, in-situ testing has played an important role in evaluating dynamic soil properties.

The in-situ tests can be classified into: 1) penetration type tests, and 2) specific tests (Robertson, 1986). Compared to specific tests, penetration type tests are fast and economical, and thus extensively used primarily for stratigraghic profiling. They can also provide some indices to empirically correlate dynamic soil properties such as shear modulus and, in particular, liquefaction resistance. However, there currently exists no in-situ testing that enables straightforward interpretation of dynamic soil properties. The only exception is the shear wave velocity measurement, classified as a specific test, that provides elastic shear modulus without any empirical correlation. In these respects, while shear wave velocity measurements are preferred in the direct estimation of shear modulus, penetration type tests are used for the evaluation of dynamic soil strength including liquefaction resistance.

The object of this paper is to summarize the current status of penetration type tests in relation to dynamic soil properties with special emphasis on liquefaction related problems. The major penetration tests discussed in this paper include the standard penetration test, and the cone penetration test. Discussion is also extended to other penetration tests, and newly developed tests that combine features of penetration type tests and specific tests.

INFLUENCE OF SAMPLE DISTURBANCE EFFECTS ON DYNAMIC PROPERTIES

It has been generally believed that the mechanical properties of soils may seriously deteriorate as a result of sampling disturbance. Besides, recent studies have indicated that the dynamic characteristics are more sensitive to sample disturbance than the static strength (Tokimatsu and Hosaka, 1986, and Kokusho, 1987).

Dramatic evidence of this is shown in Figs. 1 to 3. Fig. 1 compares elastic shear moduli measured in the laboratory for sands obtained by conventional sampling, with those from shear wave velocity measurements in-situ. It seems that the laboratory measurements yield lower moduli than those measured in-situ, and that this difference increases as the shear modulus increases. Note that the laboratory shear moduli can be as low as only one fifth of the in-situ values for very stiff sands.

Figs. 2 and 3 respectively compare liquefaction resistance of sands obtained by conventional sampling and by in-situ freezing. The latter is currently considered one of the best methods to obtain high quality undisturbed samples of sand, though expensive.

Because of the disturbance during conventional sampling, the tube sample of medium dense sand becomes dense (Fig. 2) and the

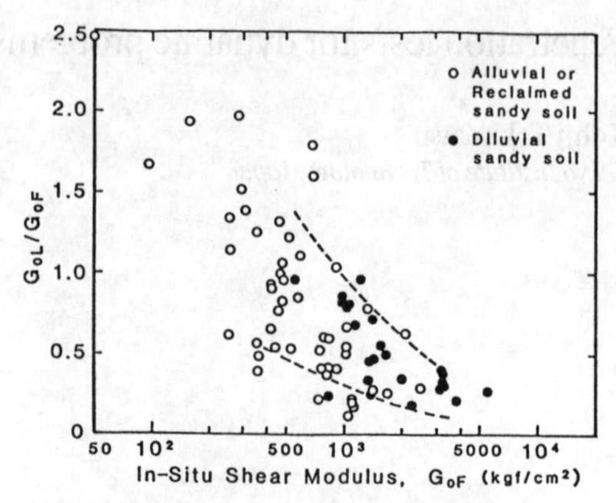

Fig. 1 Effects of sample disturbance on shear moduli of sands (after Yasuda and Yamaguchi, 1984)

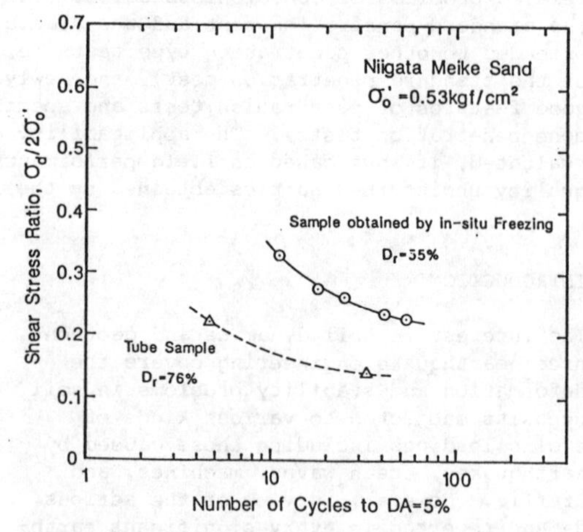

Fig. 2 Effects of sample disturbance on liquefaction resistance of medium dense sand

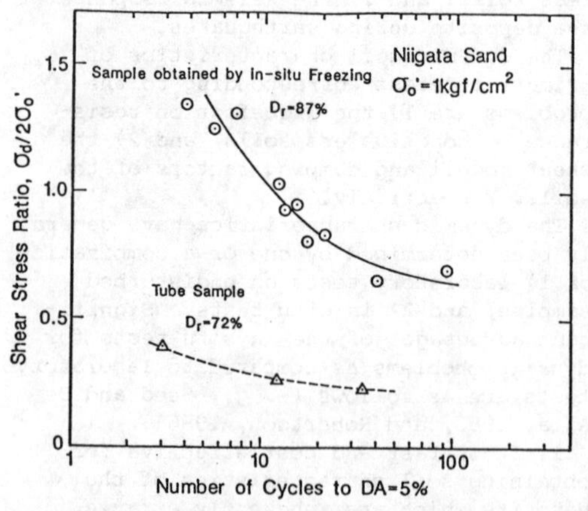

Fig. 3 Effects of sample disturbance on liquefaction resistance of dense sand

tube sample of very dense sand gets loose (Fig. 3) as compared to the in-situ frozen samples. But in any case, the tube sample provides significantly lower liquefaction resistance than does the sample obtained by in-situ freezing.

This is due to the fact that sample disturbance has changed not only soil density but soil fabric which governs the negative dilatancy of sand. Previous studies have shown that it is the negative dilatancy that controls pore pressure generation in undrained loading, i. e., liquefaction resistance (Seed, 1979).

The key to evaluating dynamic characteristics from the laboratory testing is therefore recovering good quality undisturbed samples that retain soil fabric and density in-situ. However, relatively high expenses involved in the in-situ freezing method, plus sample disturbance in conventional sampling, emphasize the significant role of in-situ tests, and thereby leading to the renewed attention to the development of these test methods.

FACTORS AFFECTING PENETRATION RESISTANCE

There probably are several basic factors that govern the mechanical properties of a specific soil:

1. Soil Density
2. Fabric of Soil
3. Effective Confining Pressures
4. Drainage Condition

Laboratory results have clarified the individual influence of these factors on the dynamic properties of sand to a certain extent. Their general effects on the dynamic properties and penetration resistance were summarized by Seed (1979) and Tokimatsu et al. (1986), and are shown in Table 1. The detailed factors in the table may correspond to more then one basic factor cited above, but are listed generally in the same order. Individual contribution to the measured penetration resistance is, however, only qualitatively evaluated. It is therefore required to quantify each of these influences on the measured resistance for better understanding and development of in-situ penetration tests for dynamic problems.

Overburden Effects

Both static and dynamic penetration resistances are significantly affected by the effective stresses at the depth where the measurements are made. In consequence, penetration resistance increases with depth even though both soil density and fabric are the same. This requires the correction of the measured resistances in terms of effective stress, if soil characteristics at different depths are compared. In this regard, penetration resistances are generally corrected to an effective overburden pressure of 1 kgf/cm^2 or 1 $tonf/ft^2$ as:

$$N_1 = C_N N \quad (1)$$

$$q_{c1} = C_q q_c \quad (2)$$

in which N = measured standard penetration resistance, q_c = measured cone penetration resistance, N_1 and q_{c1} are the corrected penetration resistances, and C_N and C_q are correction coefficients in terms of the vertical effective stress as shown in Fig. 4.

Based on a review of previous studies, Liao and Whitman (1985) suggested the following formula:

$$C = 1/(\sigma'_{vo})^n \quad (3)$$

Table 1 Factors affecting liquefaction resistance, shear modulus and penetration resistances

Factor	Effects on stress ratio to cause liquefaction or cyclic mobility	Effects on shear modulus	Effects on SPT N-values	Effects on CPT q_c-values
Increased relative density	+	+	+	+
Increased stability of structure	+	Probably +	+	Probably +
Increased time under pressure	+	Probably +	+	+
Prior seismic strains	+	+	+	Probably +
Increased K_o	+	+	+	+
Increased effective stress		+	+	+
Drainage condition	Undrained	Undrained	Undrained	Drained

+ means positive effect

where σ'_{vo} is the effective overburden pressure in kgf/cm^2, and n is a function of such factors as soil density, soil gradation, soil type, stress history and type of test. Based on a statistical study on available experimental results for normally consolidated sands, Jamiolkowski et al. (1985b) obtained n = 0.56 and n = 0.72 for the SPT and CPT resistances, respectively. This means that the CPT is more sensitive to effective stress than the SPT. Liao and Whitman (1985) proposed n = 0.5 for the standard penetration resistance.

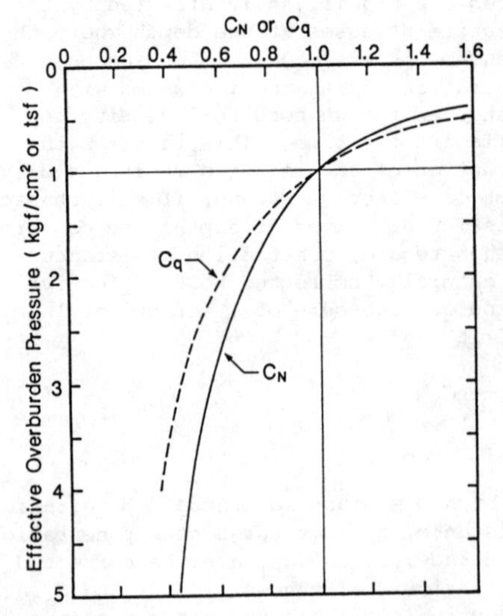

Fig. 4 Chart for values of C_N and C_q

It appears that not only the vertical effective stress but the lateral effective stress or the coefficient of earth pressure at rest, K_o, influences the measured resistance. Replacing σ'_{vo} with $(1+2K_o)/3\sigma'_{vo}$ in Eq. (3) may differentiate the effects of lateral stress from the others as stated by Skempton (1986). This implies that the penetration resistances for any K_o condition, N and q_c, would be defined by:

$$N/N_{nc},\ q_c/q_{cnc} = \left(\frac{1 + 2K_o}{1 + 2K_{onc}}\sigma'_{vo}\right)^n \quad (4)$$

where K_{onc} is the coefficient of earth pressure at rest for normally consolidated state, and N_{nc} and q_{cnc} are the corresponding resistances.

Possible increase in resistance as a result of lateral stress increase can readily be estimated from Eq. (4) and is

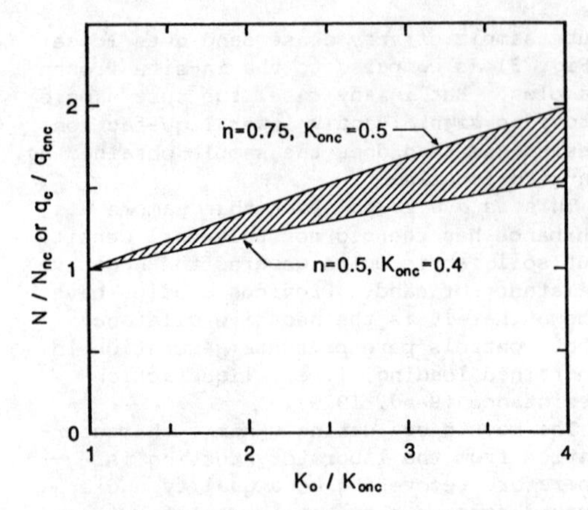

Fig. 5 Effects of lateral stress on penetration resistance

shown in Fig. 5. The trend in which penetration resistance increases with increasing K_o is in good accord with the previous findings by Baldi et al. (1981) and Jamiolkowski et al. (1985b).

For liquefaction evaluation, however, penetration resistances are not corrected in terms of mean effective stress but vertical effective stress. This is consistent with the fact that liquefaction resistance is usually normalized with respect to vertical effective stress [see Eq. (11)]. For density evaluation, on the other hand, the effects of lateral stress on penetration resistance should be accounted for as suggested by Skempton (1986).

Drainage Effects

Unlike clayey soils for which undrained conditions prevail in any penetration test, such conditions for cohesionless soils can vary depending on the type of test. It is apparent that drainage conditions can have an enormous effect on the penetration resistance as well as soil strength. Besides, drained strength may not uniquely correlate with undrained strength. It is therefore reasonable to anticipate that any static penetration test made under essentially drained conditions, may not provide a good index for liquefaction evaluations. However, among penetration tests conventionally used, only the SPT is considered as a dynamic test. Further research is therefore required to study this important issue.

STANDARD PENETRATION TEST FOR LIQUEFACTION EVALUATIONS

Advantages and Disadvantages of SPT

Kovacs and Salomone (1982) estimated that up to 80-90% of the routine foundation designs in the U.S.A. have been accomplished using the SPT N-value. The ratio is even higher in Japan where sand and gravelly soil frequently predominate. Practicing engineers therefore prefer the SPT which has a capability to penetrate these layer and to explore the underlying deposits. This is one of the main reasons for the extensive use of the SPT in these countries.

The following conditions seem to provide additional and essential reasons for using the SPT to determine the liquefaction resistance of sands (e. g., Schmertmann, 1977; Seed, 1979; and Tokimatsu and Yoshimi, 1983):

1. The SPT is an in situ test which reflects soil density, soil fabric, stress and strain history effects, and horizontal effective stress all of which are known to influence the liquefaction resistance but are difficult to retain in most so-called undisturbed samples.
2. The SPT is primarily a shear strength test under essentially undrained conditions in all but very coarse soils.
3. Numerous case histories of soil liquefaction during past earthquakes are available with SPT N-values. The method based on these case histories can reflect actual soil behavior during earthquakes which cannot be adequately simulated in the laboratory.
4. The SPT yields soil samples for the determination of grain size characteristics which are required for any investigation.

It has been apparent, however, that the SPT has the following limitations.

1. It does not provide continuous information of soil resistance with depth, which may miss thin weak layers within otherwise a relatively dense deposit.
2. Its apparatus and procedure have not been standardized enough to be the "standard."

Because of considerable differences in the test apparatus and procedures used in different countries and even within a country, there is a significant variability of measured resistance. It is therefore important to understand and correct for these differences when establishing SPT based design correlations. Such understanding is vital when applying such results to liquefaction evaluation of a site where the SPT procedure is different from the standard.

There are several basic causes of the problem to consider:

1. Change in effective stress at the bottom of borehole
2. Dynamic energy reaching sampler
3. Sampler design
4. Interval of impact
5. Penetration resistance count

Detailed causes and their effects on measured blowcounts were first evaluated by Schmertmann (1978a), and are resummarized in Table 2 in the light of recent studies. Note that the key factors influencing SPT blowcount are the effective stress at the bottom of the borehole and the dynamic energy reaching the sampler. However, most of these effects can be minimized by standardizing the test procedures suggested by Seed et al. (1985) as listed in Table 3. Recent studies by Oh-oka and Tatsui (1987) show that no apparent difference exists in the measured blowcounts in sands for borehole diameters between 66 mm and 116 mm.

Energy Corrections to SPT N-values

Among the factors listed in Table 2, the energy actually delivered to the drilling rods is especially important, since the blowcount for a given soil is inversely

Table 2 Major factors influencing SPT N-values of sand

Cause		Estimated % change in N
Basic	Detailed	
Effective Stress at Bottom of Borehole	1. Casing and Water vs. Drilling Mud	-50%
	2. Allow Head Imbalance	+100%
	3. Large vs. Small Borehole	-35%
Dynamic Energy Reaching Sampler	4. Rope-Cathead vs. Free Drop	+100%
	5. Large vs. Small Anvil	+50%
	6. Short vs. Long Rod	+30%
Sampler Design	7. Large ID for Liners, but Noliners vs. Standard	±10-20%
Blowcount Rate	8. Slow vs. Standard	+10%
Penetration Resistance Count	9. 0-12 in. vs. 6-18 in.	-15%
	10. 12-24 in. vs. 6-18 in.	+15%

+ :increase in N, - :decrease in N

Table 3 Recommended SPT procedure for liquefaction evaluations (after Seed et al., 1985)

1. Borehole: 4 to 5 in. diam rotary borehole with bentonite drilling mud for borehole stability
2. Drill Bit: Upward deflection of drilling mud
3. Sampler: O.D. = 2.00 in.; I.D. = 1.38 in.-Constant (i. e., no room for liners in barrel)
4. Drill Rods: A or AW for depths less than 50ft; N or NW for greater depths
5. Energy Delivered to Sampler: 60% of theoretical maximum
6. Blowcount Rate: 30-40 blows per minute
7. Penetration Resistance Count: Measures over range of 6-18 in., of penetration into the ground

proportional to this energy (Schmertmann and Paracious, 1979). The major factors influencing actual energy reaching the sampler are:

1. Method of releasing SPT hammer
2. Hammer and anvil type
3. Rod length

First, rope-cathead and hammer-rod frictions during hammer release make the kinetic energy of the hammer just before impact less than the theoretical free-fall energy. The theoretical free-fall energy for the standard weight (63.5 kg or 140 lb), and the height of drop (76.4 cm or 30 in.) of the hammer can be defined by

$$E^* = \frac{1}{2}\frac{w}{g}(v^*)^2 = wh = 48.5 \text{ kg-m} \quad (5)$$

in which v^* is theoretical free-fall velocity of the hammer just before impact and 3.87 m/s. Thus the actual kinetic energy of the hammer just before impact, E_v, may be defined by

$$E_v = e_1 E^* \quad (6)$$

in which e_1 is the efficiency depending on the method of hammer releasing procedure and hammer-rod friction.

Kovacs et al. (1978), Kovacs and Salomone (1982), and Oh-oka (1984) measured impact velocities for various hammer release methods. Based on these studies, typical values of e_1 were determined and are shown in Fig. 6. Note that even free-fall methods such as Tombi would not seem to provide the theoretical free-fall energy, probably due to friction between hammer and rod.

Second, shape of the anvil controls the energy transfer from the hammer to the rod

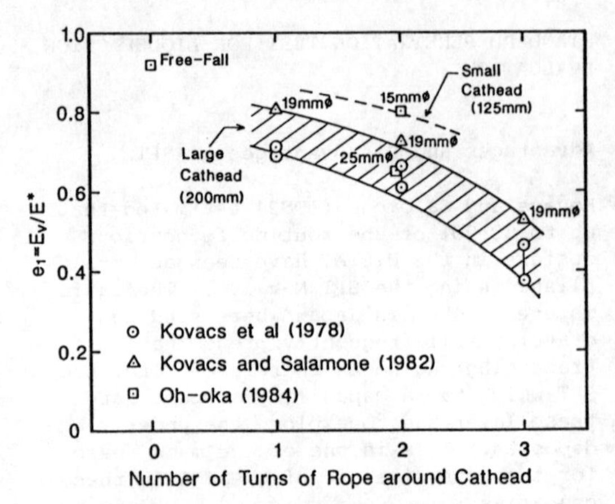

Fig. 6 Effects of hammer release procedure on energy efficiency just before impact

(Schmertmann, 1978a, and Skempton, 1986). The actual energy delivered to the rod of considerable length, E_i, can be defined by

$$E_i = e_2 E_v = e_1 e_2 E^* = ER_i E^* \quad (7)$$

in which $ER_i = e_1 e_2$ and e_2 is the efficiency of hammer and anvil system in transferring energy to the rods, and strongly depends on the weight of the anvil as listed in Table 4.

Table 4 Effect of anvil weight on energy transfer to the rod

Hammer	Anvil	e_2
Donut	Small (2kg)	0.85
	Large (12kg)	0.70
Pilcon	Large (19kg)	0.65
Safety	Small (2.5kg)	0.90

Third, theoretical studies by Schmertmann and Palacios (1979) indicated that reflection of energy in the rods may further reduce the energy actually reaching the sampler, E_{ri}, as

$$E_{ri} = e_3 E_i = e_1 e_2 e_3 E^* \quad (8)$$

in which e_3 is rod transfer efficiency depending on the rod length. Typical values of e_3 are listed in Table 5.

Table 6 summarizes the results from studies on the fractions of theoretical energy actually delivered to the rod for the current practice in several countries. The energy ratio of 60% is considered to be an average worldwide (Seed et al., 1985). The higher energy efficiency in

Table 5 Effect of rod length on rod transfer efficiency

Rod Length	e_3	
	Seed et al.(1985)	Skempton (1986)
less than 3m	0.75	
3- 4m	1.0	0.75
4- 6m	1.0	0.85
6-10m	1.0	0.95
more than 10m	1.0	1.0

Japanese practice is due to more efficient energy transfer by the use of smaller anvils, ropes and pulleys. Possible effective energies provided by other combinations of hammer-release procedure and anvil system may be estimated by Eq. (7) based on Fig. 6 and Table 4.

Table 6 Summary of energy ratios for SPT procedures

Country	Release	e_1	Hammer	Anvil	e_2	ER_i
Japan	Free-Fall	0.92	Donut	Small	0.85	0.78
	Rope-Cathead	0.80	Donut	Small	0.85	0.68
USA	Rope-Cathead	0.65	Donut	Large	0.70	0.45
	Rope-Cathead	0.65	Safety	small	0.90	0.60
UK & China	Free-Fall	0.92	Pilcon	Large	0.65	0.60

Japanese SPT results have additional corrections for borehole diameter and frequency effects.

The above discussion requires either to select the combination of SPT apparatus and procedure that produces an energy ratio of 60 %, or to correct measured N-values to an average energy ratio of 60 % using the following equation.

$$N_{60} = (ER_m/60)N_m \qquad (9)$$

where N_m is the measured blowcount and ER_m the corresponding energy ratio in percent delivered to the rods. Thus the combined correction in terms of overburden pressure and energy efficiency is given by:

$$(N_1)_{60} = C_N(ER_m/60)N_m \qquad (10)$$

Finally, it is of course essential to develop a SPT system which consistently produces a energy ratio of 60% with little operator variability. An automatic hammer-release method appears suitable for this purpose.

Liquefaction Criteria Based on Field Performance

In establishing SPT based design correlation, the index to determine the liquefaction resistance of soil is the cyclic stress ratio defined as:

$$\frac{\tau_{av}}{\sigma'_o} = 0.65\,\frac{a_{max}}{g}\,\frac{\sigma_o}{\sigma'_o}\,r_d \qquad (11)$$

where a_{max} = maximum acceleration at the ground surface, σ_o = total overburden pressure at depth under consideration, σ'_o = effective overburden pressure at the same depth, r_d = a stress reduction factor decreasing from a value of 1 at the ground surface to about 0.9 at a depth of 10 m.

The procedure used to develop the correlations has been fully described elsewhere (Seed et al., 1985). The relationship between the stress ratio and the SPT blowcount normalized to an effective overburden pressure of 1 kgf/cm^2 and and SPT energy ratio of 60 % is shown in Fig. 7 for sites which have or have not liquefied during earthquakes with magnitudes of about 7.5 for clean sands with fines content less than 5%.

Soils known to have liquefied are plotted as solid symbols, soils which did not liquefy are plotted as open symbols.

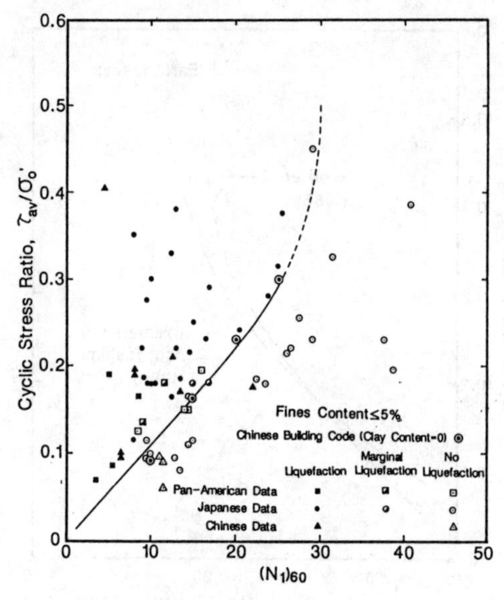

Fig. 7 Relationship between stress ratios causing liquefaction and N_1-values for clean sands for M=7-1/2 (after Seed et al., 1985)

Boundary line is to separate liquefaction and non-liquefaction conditions. Supported by a significant body of field performance data, the boundary curve can be drawn and thus used for design with a sufficient degree of confidence. The boundary curve shown in Fig. 7 appears generally consistent with that proposed by Tokimatsu and Yoshimi (1983) as shown in Fig. 8 when the difference in the effective energy between the two countries is taken into account. The correlation proposed by Seed et al., however, appears to result in a conservative estimate for loose sand.

Test Results on High Quality Undisturbed Samples Obtained by In-Situ Freezing Method

In the most favorable situation it should be possible to obtain reasonable agreement on the potential for liquefaction and cyclic mobility between the results of the available two approaches, e. i., one based on in-situ testing and the other based on laboratory testing. The need for such confirmation cannot be overemphsized, considering the ambiguousness caused by the disadvantages of the empirical approach based on in-situ tests. For instance, the field performance data are often ambiguous as to the quantitative estimation of the earthquake intensity, and as to the identification of the soil stratum that did liquefy. It is particularly difficult to identify the possible sign of cyclic mobility behavior of medium dense to dense sands.

There is a need therefore for determining the in-situ liquefaction resistance of sands using laboratory tests on high quality undisturbed samples. In this connection, Yoshimi et al. (1984, 1988) conducted laboratory tests on high quality undisturbed samples obtained by in-situ freezing method. The results are summarized in Fig. 9 together with the empirical boundary curves separating liquefiable from nonliquefiable conditions. The detailed procedure to predict the in-situ strength based on laboratory liquefaction tests were described elsewhere (Yoshimi et al., 1984, 1988).

It seems that the laboratory data fall near the critical boundary curves based on field performance. The good agreement between the two approaches confirms the overall capability of N_1-value for liquefaction and cyclic mobility evaluation for natural deposits of clean sand.

It is important and noteworthy that the SPT approach is equally effective for artificially compacted soils. Open circles in the figure correspond to the data from natural deposits, and open squares indicate those from artifically

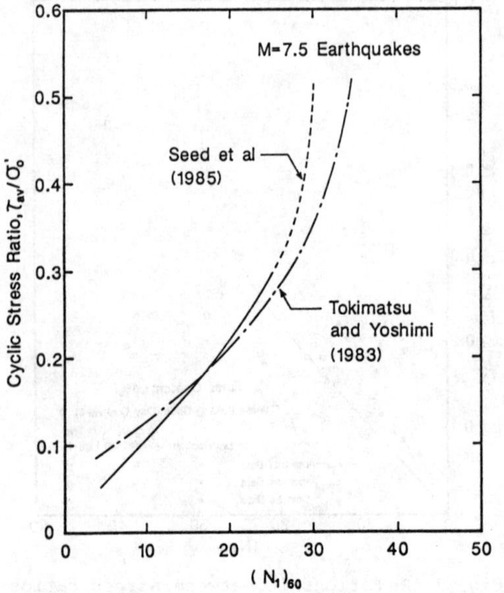

Fig. 8 Comparison of proposed boundary curves for liquefaction resistance in terms of SPT

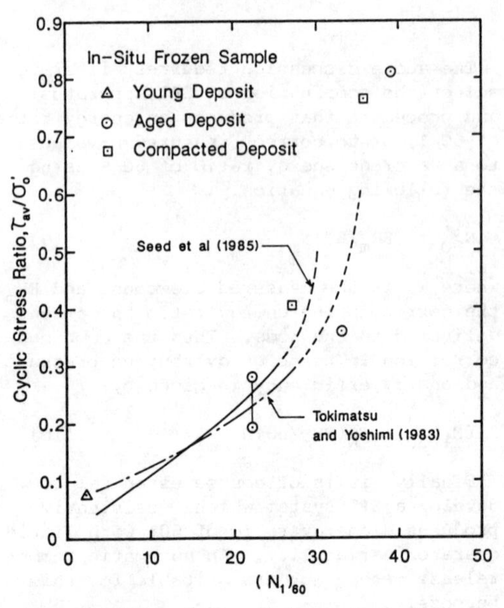

Fig. 9 Comparison of laboratory test results on in-situ frozen samples with field performance data

compacted deposits. The sands in these environments can have different soil fabrics and may involve different coefficients of earth pressure at rest. Nevertheless, there exists a fairly well defined correlation between N_1-value and liquefaction resistance, irrespective of their different geological histories.

It should be remembered that the criteria shown in Fig. 9 are for M = 7.5 earthquakes. The conversion of the boundary curve for other earthquake magnitudes can readily be made if the number of shear stress cycles is assumed to increase with increasing earthquake magnitude. Such a design chart has already been proposed by Seed (1979), but based on laboratory tests on reconstituted sands. The availability of high quality undisturbed samples can provide a similar chart as shown in Fig. 10, indicating that the chart proposed by Seed is reasonably satisfactory.

Effects of Fines

It has been recognized that the presence of fine reduces the SPT blowcount without significantly changing liquefaction resistance (Tokimatsu and Yoshimi, 1983, and Seed et al., 1985). This is probably because the excess pore pressure generated during sampler penetration is easy to be accumulated as the grain size gets small,

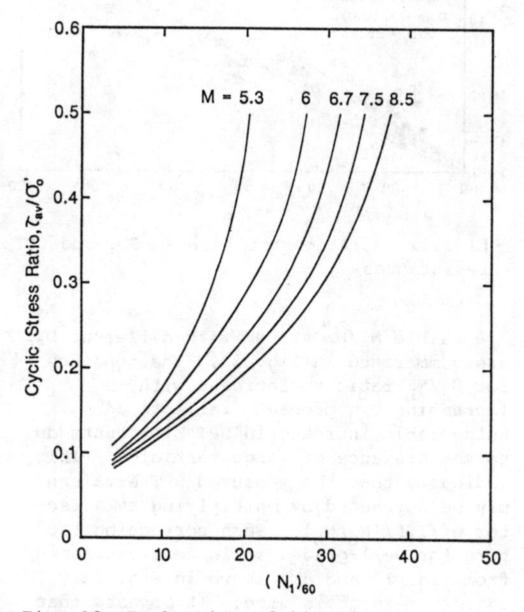

Fig. 10 Evaluation of liquefaction potential for sands for different magnitude earthquakes

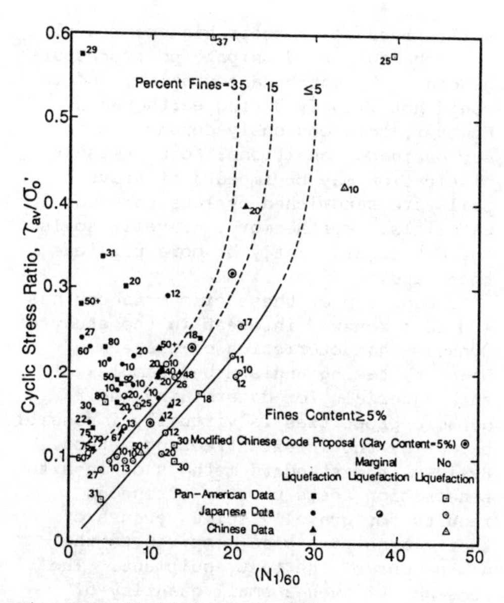

Fig. 11 Relationship between stress ratios causing liquefaction and N_1-values for silty sands for M=7-1/2 (after Seed et al., 1985)

which tends to lessen the soil resistance to further penetration. Thus the SPT blowcounts cannot stand by itself but with grain size information.

Fig. 11 shows such evidence in which the critical boundary shifts leftward as the fines content increases. The number beside each data point indicates fines content. It appears that boundary lines are somewhat obscure because of a limited number of field performance data and because of the disregard of such factors as soil plasticity that might be a more essential factor controlling liquefaction resistance of these soils. Further compilation of case histories for silty sands is therefore required to establish more reliable design criteria.

LARGE PENETRATION TESTS FOR LIQUEFACTION EVALUATION OF GRAVELLY SOILS

Liquefaction of Gravelly Soils

There have been a number of cases in recent years where the liquefaction of gravelly soils has been observed to occur during earthquakes, with associated detrimental consequences. These cases are summarized by Harder and Seed (1986). It was generally believed that gravelly soils, because of their high permeability, would be able to dissipate pore pressure generated by earthquake shaking, and thus would not liquefy during earthquakes. However, this obviously depends on boundary drainage conditions; pore pressure dissipation may be impeded if gravelly soils are sandwiched by less pervious materials. Furthermore, gravelly soils may not significantly be more pervious than sand.

Recognition of these circumstances has led to a renewed interest in the study of dynamic characteristics of gravels. However, taking undisturbed samples of such materials for determining their dynamic properties is virtually impossible, except by the in-situ freezing method. Besides, the standard methods of in-situ penetration tests provide erroneous results for gravelly soils because of their large particles compared to the dimensions of the test equipment. The presence of even a small quantity of gravel tends to increase the penetration resistance, without significantly reducing the susceptibility of soil to liquefaction.

Such an error may be eliminated to some extent by measuring penetration resistance on an inch-by-inch basis and then interpreting this information to determine N-value for sand zone only (Vallee and Skryness, 1979, and Seed and De Alba, 1986).

It seems desirable, however, to investigate the possibility of evaluating liquefaction characteristics using a large scale version of in-situ tests which would be less affected by the presence of large particles. An advantage of such an approach is that it can be correlated with extensive body of field performance data, through the development of correlations between large-scale penetration test and the standard penetration test. The large-scale version of penetration tests currently available for this purpose are:

1. Large diameter penetration test (LPT)
2. Becker Drill Hammer

The geometry of the equipment and the test procedure adopted in these tests are compared in Table 7, together with those of the SPT.

Large Penetration Test

The LPT proposed by Kaito et al (1971), adopts a larger hammer, rods and sampler than the SPT equipment, but follows the SPT procedures except for the falling height. The great similarity in the test procedure and equipment enables one to

Table 7 Comparison of dynamic penetration test procedure and apparatus

	Standard Penetration Test	Large Penetration Test	Becker Drill Hammer
Drive Method	Fall Weight	Fall Weight	Diesel Hammer
Weight	63.5kg	100kg	
Drop Height	76.4cm	150cm	
Impact Energy	48.5kgf-m	150kgf-m	35-664kgf-m
Drive length	30cm	30cm	30cm
Drill Bit OD	5.1cm	7.3cm	16.8cm
ID	3.5cm	5.0cm	closed

establish a correlation between the test results of the two methods. Provided that the energies delivered to the unit surface of the two samplers are the same, the following equation can be derived:

$$N_S/N_L = 1.5 \tag{12}$$

in which N_S and N_L are blowcounts for the SPT and LPT, respectively. This relation appears generally satisfactory for soils without gravel as shown in Fig. 12. The correlation factor, however, tends to become greater than 1.5 for gravelly soils.

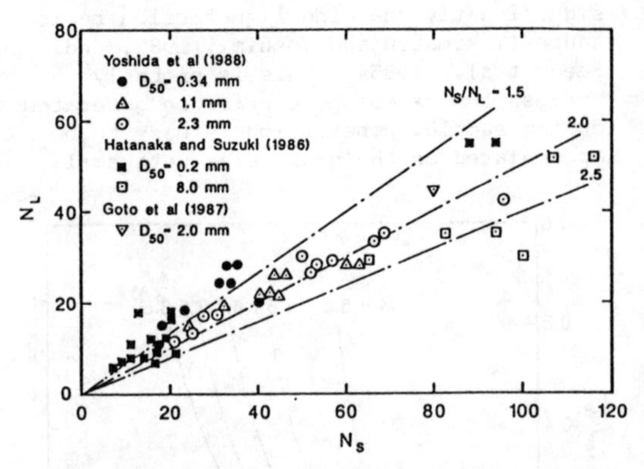

Fig. 12 Relationship between SPT and LPT resistances

Possible N_S/N_L values with different D_{50} are summarized in Fig. 13. The tendency for N_S/N_L ratio to increase with increasing D_{50} probably reflects an unfavorable increase in SPT blowcounts due to the presence of large particles. This indicates that the measured SPT N-values may be corrected by multiplying by a factor of $1.5/(N_S/N_L)$. Such correction factors for well-graded soils were estimated from Fig. 13 and are shown in Fig. 14 against mean grain size. It appears that the correction factor decreases with increasing mean grain size. In fact, Skempton (1986) suggested a correction

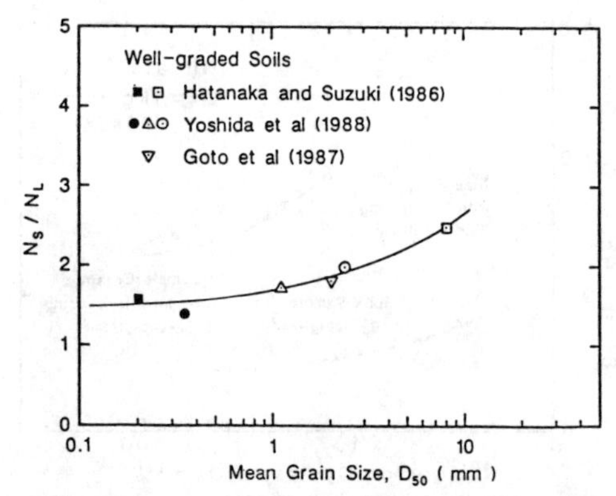

Fig. 13 Variation of N_S/N_L ratio with mean grain size for well-graded soils

factor of 0.9 for coarse sand, and the review of the study by Tokimatsu and Yoshimi (1983) indicates a correction factor of about 0.75 for sands containing 20% gravel (D_{50} = 0.8 mm).

Although the effects are less significant, unfavorable increase in blowcount may also occur for the LPT as the soil particles get large. This tendency and some degree of engineering judgment lead to tentative corrections to convert measured blowcounts into equivalent blowcounts for sands, N, in terms of liquefaction resistance as follows.

$$N = C_{Sg}N_S \quad (13)$$

$$N = C_{Lg}N_L \quad (14)$$

in which C_{Sg} and C_{Lg} are correction factors as shown in Fig. 14, for the SPT and the LPT, respectively. Thus the liquefaction evaluation for gravelly soils may be made by: first correcting penetration resistance by Eqs. (13) or (14), and then entering the conventional chart using SPT N-values.

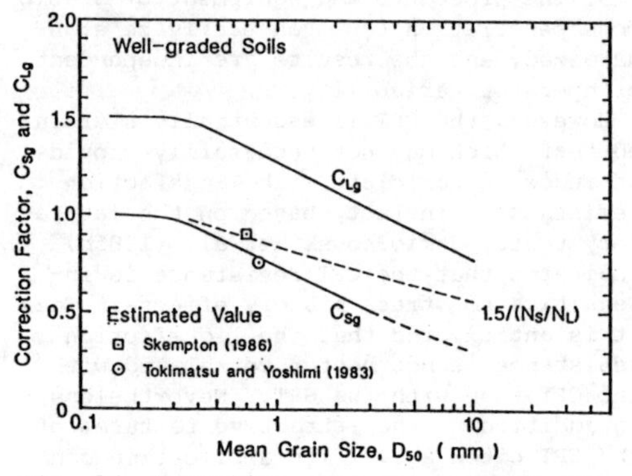

Fig. 14 Tentative correction factor for penetration resistances of gravelly soils

Liquefaction and Cyclic Mobility Evaluations of Gravelly Soils by In-Situ Frozen Samples

Recently, Hatanaka and Suzuki (1986) and Goto et al. (1987) extended the application of the in-situ freezing method to gravelly soils. Gravelly soils were frozen and then cored into 300 mm diameter specimens in-situ for laboratory liquefaction tests. The results of in-situ and laboratory tests are summarized in Table 8. The N-values were corrected according to Eqs. (13) and (14), and are plotted in Fig. 15 against the measured liquefaction resistance in the laboratory.

For comparison purposes, the SPT based correlation is also shown in the figure. Although the laboratory data are quite limited, the fairly good agreement between

Table 8 In-situ and laboratory test results on gravelly soils

	Mandano Gravel	Tokyo Gravel
D_{50} (mm)	2	8
U_c	10	65
N_L	45	40
N_{S78}	80	100
σ'_v (kgf/cm^2)	1.2	3
$(N_1)_{60}$	57	26
Reference	Goto et al. (1987)	Hatanaka & Suzuki (1986)

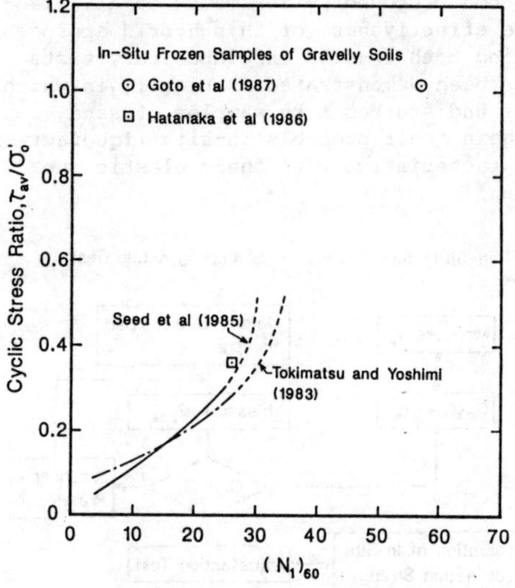

Fig. 15 Comparison of laboratory test results on in-situ frozen samples of gravelly soils with SPT based design correlation

the two approaches indicates that the proposed method may be used as a tentative guideline for the evaluation of liquefaction potential of gravelly soils.

Similarly, Harder and Seed (1986) have established the chart correcting Becker Drill blowcounts into SPT blowcounts. However, the kinetic energy delivered to the drill rod varies considerably depending on the nature of the diesel hammer, and in this respect, substantial corrections are required on the measured blowcounts.

In the light of the variability of the correlation between the large-scale version of penetration test and the standard penetration test, special care and engineering judgment are crucial in using these correlation charts in practice.

Other In-Situ Testing for Gravelly Soils

Using other in-situ tests such as shear wave velocity measurements, though beyond the scope of this paper, shows promise, considering its potential applicability and the inability of using penetration type tests for gravelly soils. Possible application of this approach was described by Seed et al. (1983) and Tokimatsu et al. (1986). While the Seed method is based on the concept of threshold strain, Tokimatsu et al. take a full advantage of shear wave velocity that can be measured both in the field and in the laboratory.

Fig. 16 illustrates the flow of the method proposed by Tokimatsu et al. (1986). The effectiveness of this hybrid approach using both in-situ and laboratory tests has been demonstrated in Fig. 17 in which the undisturbed tube samples of sand regain their probable in-situ liquefaction characteristics when their elastic shear

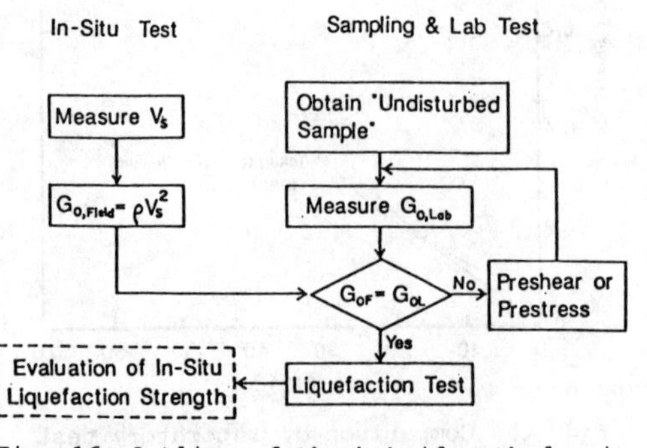

Fig. 16 Outline of the hybrid method using both in-situ and laboratory tests (after Tokimatsu et al., 1986)

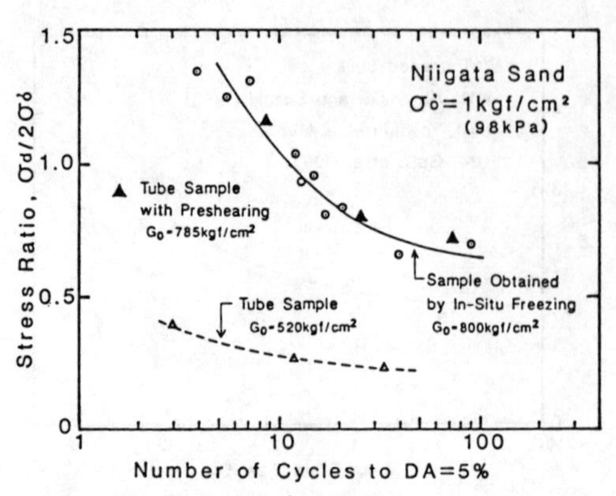

Fig. 17 Verification of the method proposed by Tokimatsu et al.

moduli, G_o, are adjusted to the in-situ values. If penetration type tests provide erroneous results or cannot be performed for some reason, this type of approach would be an important role in liquefaction evaluations.

STATIC CONE PENETRATION TEST FOR LIQUEFACTION EVALUATIONS

It is generally believed that a cone penetration test (CPT) provides a more consistent evaluation of the strength and stiffness of soils than does the SPT, and hence there is considerable interest in using the CPT to evaluate liquefaction susceptibility. There are important potential advantages to the use of the CPT.

1. It is fast and economical compared to the SPT.

2. It provides a continuous record of penetration resistance, and thus it makes a more thorough interpretation of a soil profile than the SPT.

3. The procedure and equipment of static cone penetration test can easily be standardized, and the results are independent of operator variability.

However, the CPT is essentially a drained test which may not necessarily provide an index to correlate with liquefaction resistance. In fact, based on the laboratory tests, Jamiolkowski et al. (1985b) indicated that the CPT resistance is insensitive to stress history effects. Thus it is anticipated that the liquefaction resistance is not better correlated with the CPT than with the SPT. Nevertheless, in addition to the attractive features of the CPT cited above, its application can be extended to offshore investigation where the application of the SPT is severely restricted. Described hereafter is the current status of the CPT for the evaluation of liquefaction resistance.

CPT-SPT Correlation

There exist few case histories in which the occurrence or nonoccurrence of liquefaction during actual earthquakes is directly correlated with cone penetration resistance, q_c. In consequence, it is not possible directly to establish charts such as those developed using the SPT. The usual approach is therefore to use the SPT based design correlation together with a correlation between CPT tip resistance and SPT blowcount. There is a clear need therefore to establish a good SPT-CPT correlation so that the CPT can be used in conjunction with the existing SPT based design correlation.

There are two possibilities to obtain such correlation:

1. To establish a site-specific CPT-SPT correlation based on preliminary investigation of the two tests at each site
2. To use an empirical correlation between SPT and CPT test data

There have been a number of empirical correlations between cone tip resistance and SPT blowcount (e. g., Muromachi and Kobayashi, 1982, and Robertson et al, 1983). It has been generally believed that q_c/N ratios vary depending on soil type, and test apparatus and procedure. The most up-to-date version to define such relation is shown in Figs. 18 and 19 in which SPT N-values are corrected in terms of an effective energy of 60%. It can be

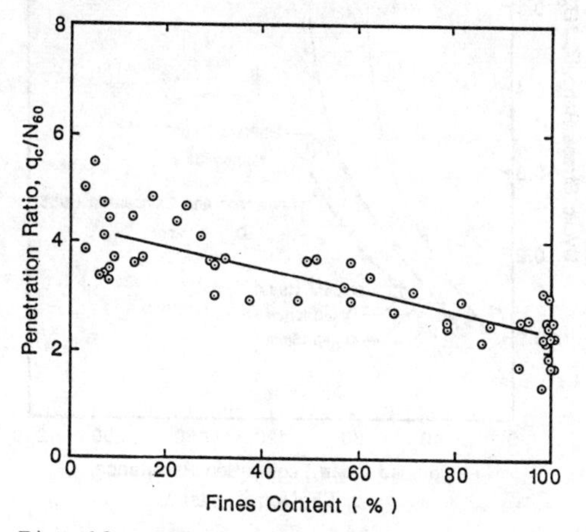

Fig. 18 Variation of q_c/N_{60} ratio with fines content (after Muromachi and Kobayashi, 1982)

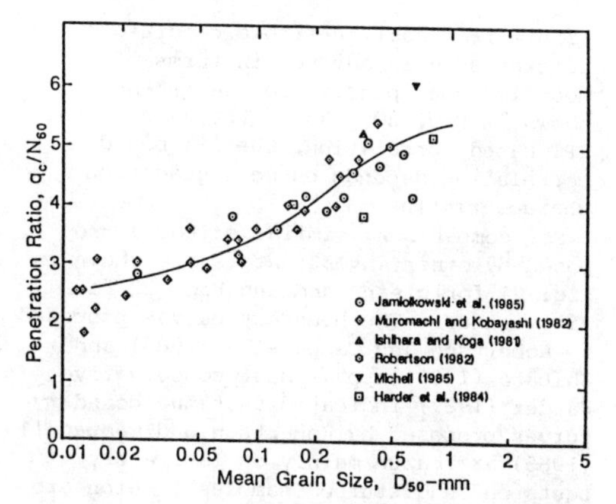

Fig. 19 Variation of q_c/N_{60} ratio with mean grain size (after Seed and De Alba, 1986)

seen that the ratio of cone tip resistance to SPT blowcount decreases with decreasing fines content (Fig. 18) or increasing mean grain size (Fig. 19).

As shown in the figures, there can be a considerable scatter in the correlation even after the SPT energy correction was made. This obviously indicates that neither the fines content nor the mean grain size can by itself be uniquely correlated with q_c/N ratio. This is probably because soil material, soil density, effective stress and geometry of the cone, (Schmertmann, 1978b) have some influence on the relation.

Jamiolkowski et al. (1985b) indicated that each data point shown in Figs. 18 and 19 usually corresponds to an average of several data with considerable variations. Thus compilation of case histories with CPT data is badly needed for directly establishing CPT based design criteria and then for determining the applicability and limitation of the CPT for liquefaction evaluation. Until then, it seems desirable, for important projects, to establish a site-specific correlation between SPT and CPT or to use a conservative value resulting from either Fig. 18 or 19.

Liquefaction Evaluation Chart based on CPT

Using Fig. 18 or 19, cone penetration resistances normalized to an overburden pressure of 1 kgf/cm^2, can readily be converted to equivalent N_1-values and then used with Figs. 7 and 11 for the evaluation of liquefaction potential. Instead, by using the CPT-SPT correlation (Fig. 18) and the SPT based correlation (Fig. 11), the critical boundaries separating liquefiable from nonliquefiable conditions can be expressed in advance in terms of modified cone penetration resistance as shown in Fig. 20. As is the case of the SPT based correlation, the CPT based correlation depends on soil gradation characteristics.

For comparison, similar criteria proposed by other investigators are shown in Fig. 21 for clean sand and Fig. 22 for silty sands. The boundary curves proposed by Robertson and Campanella (1985) and Shibata (1987) appear most conservative. Harder (1987) indicated that the boundary curves proposed by Robertson and Campanella (1985) are based mainly on laboratory tests on "undisturbed samples", which are inevitably affected by sample disturbance. In view of the variability of q_c/N ratio and the lack of reliability of the SPT correlation, the CPT correlation for silty sands appears highly questionable and require further verification.

Comparison with Available Case Histories

As stated previously, few CPT studies have been reported for sites whose performance during past earthquakes has been known. However, valuable field data have been provided by Ohsaki (1970), Ishihara et al. (1981), Ishihara and Koga (1981), Ishihara and Perlea (1984), Youd and Bennett (1983), and Zhou (1981).

The data from these studies are plotted

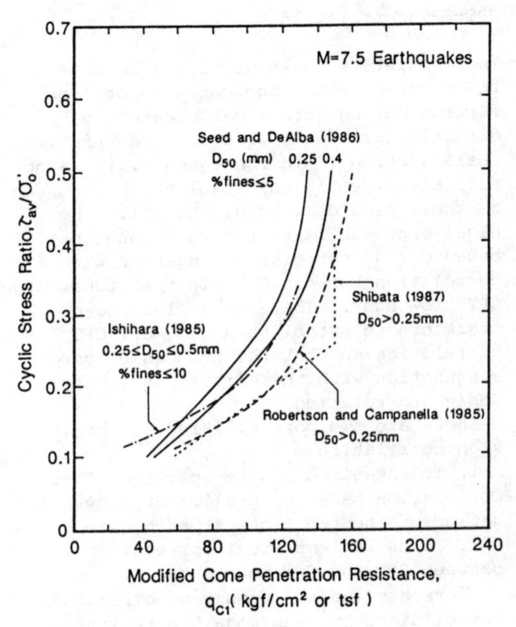

Fig. 21 Comparison of proposed boundary curves for liquefaction resistance evaluations for clean sands in terms of CPT

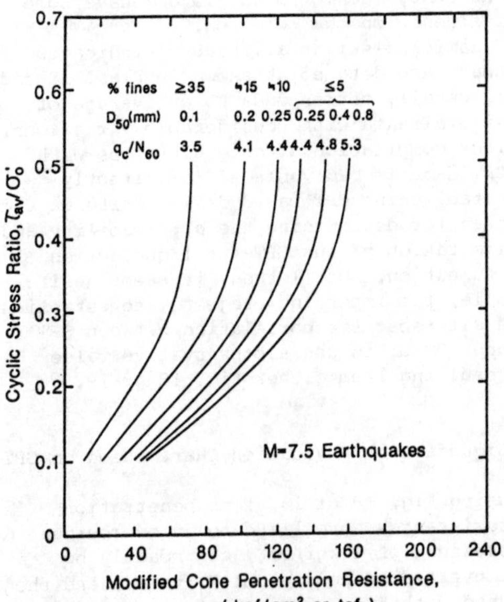

Fig. 20 Relationship between stress ratio causing liquefaction and cone tip resistance for sands and silty sands (after Seed and De Alba, 1986)

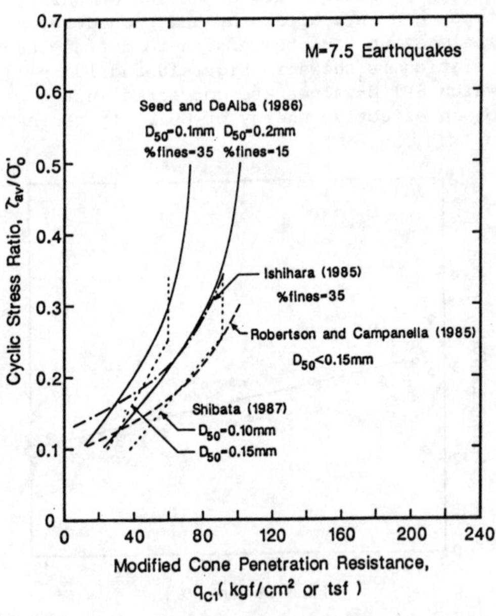

Fig. 22 Comparison of proposed boundary curves for liquefaction resistance evaluations for silty sands in terms of CPT

in Fig. 23 with the critical boundaries shown in Fig. 20. Each data point represents one case history with one CPT logging data, as is the case of the SPT correlation shown in Fig. 7. The number besides each data point indicates fines content. It appears that the field data are generally consistent with the critical boundaries. However, more field data are obviously needed before using this chart for design with sufficient degree of confidence.

Disadvantages of CPT

The availability of correlation charts for liquefaction resistance in terms of the CPT as well as the SPT requires that one should keep in mind the significant disadvantages of the CPT approach (e. g., Seed and De Alba, 1986).

1. Whereas liquefaction strength depends not only on penetration resistance but also on grain size distribution for a soil, the CPT does not provide samples for determining grain size characteristics. Using sleeve friction data cannot be sufficiently reliable to detect grain size parameter for this purpose. Thus, unless the grain size parameter is reliably known from other investigation, the method poses considerable uncertainty in evaluating liquefaction potential.

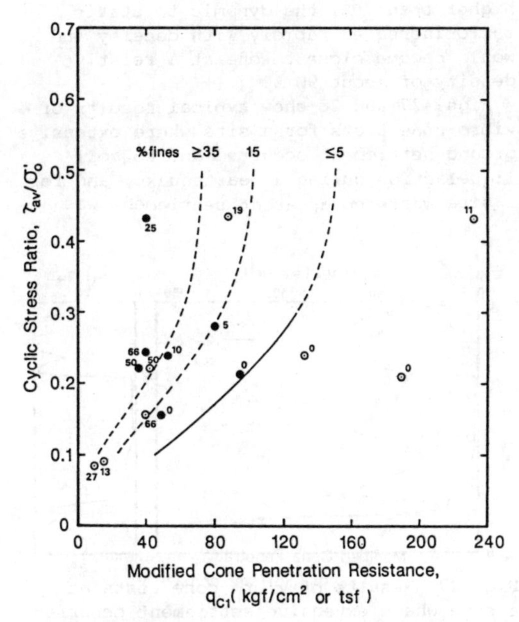

Fig. 23 Comparison of proposed boundary curves and field performance data

2. There is only limited field case histories to establish a correlation between liquefaction strength and CPT resistance directly.

3. The CPT can sometimes not be performed at large depths because of the inability to penetrate even medium dense sands under high confining pressure.

4. The reliability of the CPT data becomes questionable when dealing with coarse sands and gravelly soils.

However, the CPT has potential capabilities to increase its usefullness. For example, the introduction of the piezocone penetrometer has allowed the pore pressure measurements in addition to tip resistance and sleeve friction, which somewhat improves the interpretation of the soil profile. Robertson et al. (1986a) proposed a chart classifying the soil type based on these parameters.

The development of seismic cone penetrometer (Robertson et al., 1986b) and lateral stress sensing cone penetrometer (Huntsman et al., 1986) provides additional valuable information without losing the economical advantage of the CPT. Of particular significance is the seismic cone penetrometer that provides information concerning soil profiling, modulus and strength in one sounding.

OTHER METHODS FOR LIQUEFACTION EVALUATIONS

Vibro-Cone Penetrometer

Recently a vibro-cone penetrometer has been developed (Sasaki et al, 1984, 1985) primarily for liquefaction potential evaluations. The apparatus is similar to the static cone penetrometer except somewhat larger diameter for a built-in vibrator as shown in Fig. 24. The centrifugal force and frequency of a typical vibrator are 80 kgf and 200 Hz, respectively.

The tip resistances with and without axial vibration, q_{cd} and q_{cs}, are measured during quasi-static penetration. If the soil is contractive, the penetration resistance with vibration would be considerably lower than the static resistance due to the induced excess pore pressure. For dilative soil in which no excess pore pressure is generated by vibration, both resistances may not be significantly different. For clayey soils other than sensitive one, measured resistances are also similar to each other since the drainage conditions of the two tests are the same.

Comparing resistances with and without

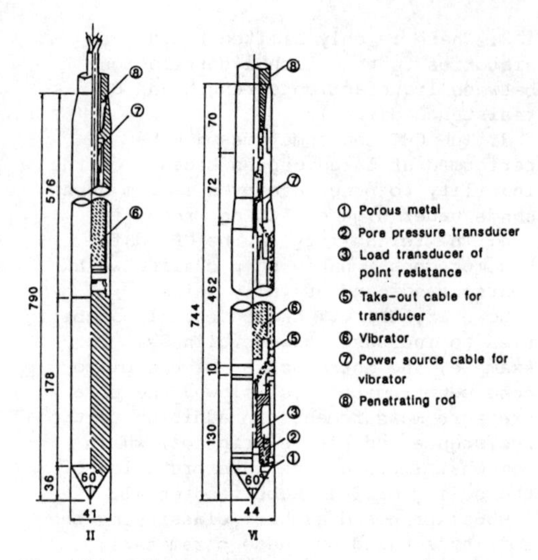

Fig. 24 Vibro-cone penetrometer (after Sasaki et al., 1985)

vibration can therefore serve in detecting soils susceptible to liquefaction. For this purpose, dynamic-static ratio of cone penetration resistance can be defined as:

$$R_{DS} = q_{cd}/q_{cs} \quad (15)$$

The dynamic to static ratio defined herein is somewhat different from that used in the original paper. However, direct comparison between dynamic and static resistances appears more appealing for characterizing soil properties, considering a trend in which the liquefaction resistance decreases as the dynamic-static ratio decreases.

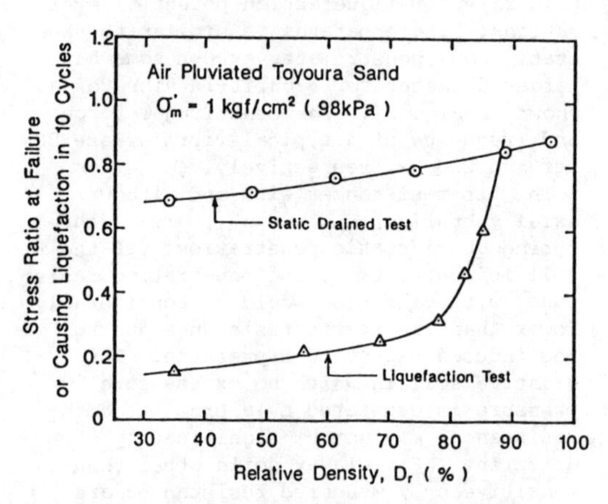

Fig. 25 Variation of static and dynamic strengths for air-pluviated Toyoura sand with relative density (after Tatsuoka et al., 1982)

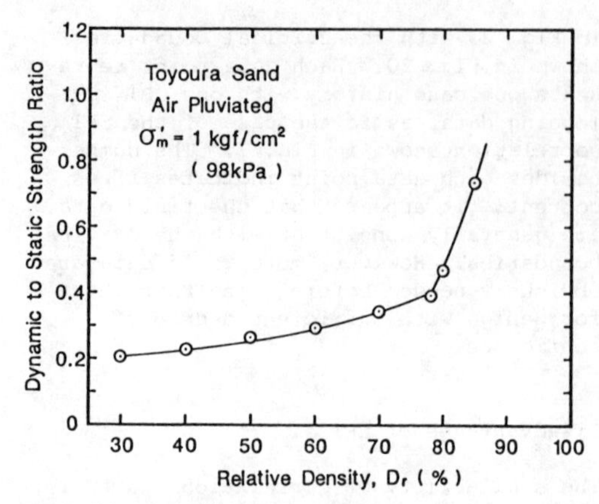

Fig. 26 Relationship between dynamic to static strength ratio and relative density for air-pluviated Toyoura sand

Fig. 25 shows such evidence in which dynamic and static strengths of a sand are compared from laboratory tests by Tatsuoka et al. (1982). Unlike the static strength, the dynamic strength is sensitive to the change in soil density.

The dynamic strength was normalized with respect to static strength and are shown in Fig. 26. The dynamic to static ratio for relative densities less than 60%, which is considered highly liquefiable, is less than about 0.25. For relative densities higher than 75%, the dynamic to static ratio increases rapidly with density and would become close to one at a relative density of about 90 %.

Figs. 27 and 28 show typical results of vibro-cone tests for a site where extensive ground settlement occurred due to soil liquefaction during an earthquake, and for a site where no apparent settlement was

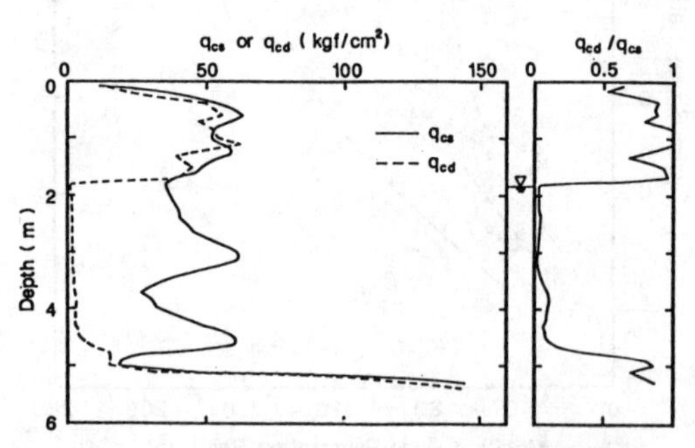

Fig. 27 Results of vibro-cone tests at a site where extensive settlement occurred due to soil liquefaction (Sasaki et al., 1984)

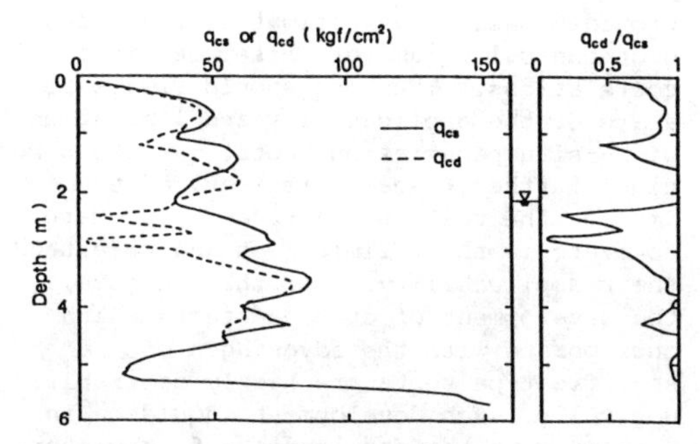

Fig. 28 Results of vibro-cone tests at a site where no apparent settlement was observed (Sasaki et al., 1984)

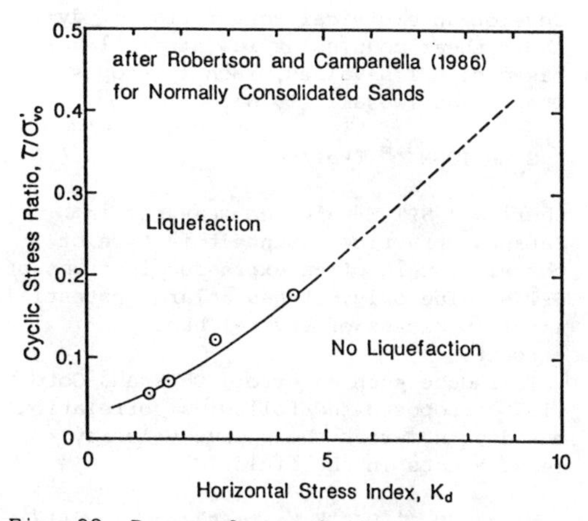

Fig. 29 Proposed correlation between liquefaction resistance under level ground conditions and dilatometer horizontal stress index for sands (after Robertson and Campanella, 1986)

observed. Despite similar characteristics of static penetration resistance for the two sites, the difference in the dynamic penetration resistances are significant. As a result, the dynamic to static ratios of penetration test for the liquefied site are considerably lower than those for the nonliquefied site.

It appears from Figs. 27 and 28 that the sand with dynamic to static ratios less than about 0.2 has high liquefaction potential, which is consistent with the laboratory finding mentioned above. Considering the difference in the test conditions between laboratory and field, the fairly good agreement in the dynamic to static ratio separating liquefiable from nonliquefiable conditions appears fortuitous. Obviously, the dynamic penetration resistance depends on the centrifugal force, frequency of vibrator, and effective stress. Nevertheless, the vibrocone seems to offer a promising means for liquefaction evaluation through dynamic-static ratio. Further refinement of the apparatus and accumulation of test data in this direction should be encouraged.

Flat Plate Dilatometer

Flat plate dilatometer testing (DMT) developed by Marchetti is a simple, inexpensive test which can rapidly provide soil profile, and index parameters for liquefaction assessment, though it is considered to be a static test.

The dilatometer test provides three parameters for the estimation of soil characteristics. They are material index, I_d, horizontal stress index, K_d, and dilatometer modulus, E_d. The material index can be used for identifying soil type, and the horizontal stress index may be used for estimating liquefaction resistance. In fact, many of the factors that influence liquefaction resistance are likely to increase K_d to some extent as in the case of the CPT and the SPT (Marchetti, 1982), although the individual contribution of each factor is unknown.

Robertson and Campanella (1986) have proposed a tentative relation between K_d and liquefaction resistance of clean sand as shown in Fig. 29. Because the chart is based on a K_d-D_r relation and a D_r-liquefaction correlation for normally consolidated sands, direct use of the correlation for aged sands may result in underestimation of liquefaction resistance. Since its effectiveness has not been proven by field performance, the application of DMT for liquefaction evaluation requires considerable care and engineering judgment. Note that the chart shown in Fig. 29 is only applicable to clean sands and may not be used for silty sands and silts (Robertson and Campanella, 1986).

PENETRATION TESTS FOR SHEAR MODULUS EVALUATIONS

Dynamic shear modulus of soil depends not only on stress level but also on strain level. However, the shear modulus at low strain level, G_o, is only correlated with penetration test results, and the strain dependent shear modulus is usually estimated based on test results on undisturbed samples.

There have been numerous studies to

develop an empirical correlation of dynamic shear modulus at low strain level based on SPT N-values, such as proposed by Ohsaki and Iwasaki (1973):

$$G_o = 120N^{0.8} \text{ (kgf/cm}^2\text{)} \quad (16)$$

where N = SPT N-value as measured in Japanese practice. Since this type of correlation is often expressed in terms of SPT N-value only, it has a large potential error in excess of several hundred percent.

To reduce such an error, Ohta and Goto (1976) proposed the following correlation based on numerous shear wave velocity measurements in the field.

$$V_s = 69N^{0.17}D^{0.2}F_1F_2 \text{ (m/s)} \quad (17)$$

where D = depth of soil below ground surface; F_1 = a factor, depending on the nature of the soil, having a value of 1 for alluvial deposits and 1.3 for diluvial deposits; and F_2 = a factor, depending on the nature of the soil as shown in Table 9. The introduction of these factors in addition to SPT N-value has significantly improved the reliability of the empirical approach (Seed et al., 1986).

Table 9 F2 factors for various soil types

Soil Type	F_2
Clay	1.0
Fine sand	1.09
Medium sand	1.07
Coarse sand	1.14
Sandy gravel	1.15
Gravel	1.45

Nevertheless, it is recommended, for dynamic analyses, to measure shear wave velocity directly by means of specific type tests or combined tests (e. g., seismic cone penetration tests) that can provide more consistent evaluations of shear modulus.

CONCLUDING REMARKS

Without thorough understanding of what are reflected in the measured values, penetration type tests have been and will be extensively used in practice for determining dynamic characteristics of soils as well as soil profiling, because of their significant advantages over laboratory tests conducted on undisturbed samples. However, penetration type tests can only provide qualitative estimates of various geotechnical parameters based on empirical correlations. Thus one should always be aware of the applicability and limitation of in-situ penetration tests. At the same time, further research is required to improve the test apparatus and procedure to overcome their limitation and to extend their applicability. For this purpose, the development of combined tests which incorporate with the advantages of specific type tests are highly desirable. Hopefully, such development should be in harmony with the development of laboratory test together with the field verification of the problem, since they are complementary in some respects. The verification using in-situ frozen samples demonstrated in Figs. 9 and 15, and the hybrid approach using both in-situ and laboratory tests illustrated in Fig. 16 are typical examples along these lines.

The most important topics the author considers for the future research are listed in the following:

1. Better recognition of what is being measured by penetration tests or what is reflected in the measured values
2. Better understanding of the effects of lateral stress on the penetration resistance
3. Collection of field performance data with CPT and direct development of a CPT based correlation for liquefaction evaluations
4. Collection of well documented case histories concerning gravelly soils for a better recognition of dynamic characteristics of these soils
5. Development of a new in-situ test procedure which is more directly correlated with liquefaction resistance

ACKNOWLEDGMENTS

The author is grateful to Professor Y. Yoshimi for his critical review of the manuscript and valuable comments. The author also wishes to acknowledge the valuable advice provided by Dr. T. Kokusho, Mr. Y. Sasaki, Mr. K. Tamaoki, Dr. H. Oh-oka, and Dr. M. Hatanaka.

REFERENCES

Baldi, G., Bellotti, R., Ghionna, V., Jamiolkowski, M., and Pasqualini, E. (1981) "Cone resistance in dry N.C. and O.C. sands," Proceedings, a session on Cone Penetration Testing and Experience, ASCE National Convention, St. Louis, pp.

145-177.
Goto, S., Shamoto, Y., and Tamaoki, K. (1987) "Dynamic properties of undisturbed gravel sample by in-situ frozen," Proceedings, 8th ARCSMFE, Kyoto, Vol. 1, pp. 233-236.
Harder, Jr., L. F. (1987) "Discussion", Journal of Geotechnical Engineering, ASCE, Vol. 113, No. 6, pp. 673-676.
Harder, Jr., L. F. and Seed, H. B. (1986) "Determination of penetration resistance for coarse-grained soils using the becker hammer drill," Report, Earthquake Engineering Research Center, Report No. EERC-86/06.
Hatanaka, M. and Suzuki, Y. (1986) "Dynamic properties of undisturbed Tokyo gravel obtained by freezing," Proceedings, 7th Japan Earthquake Engineering Symposium, Tokyo, pp. 649-654 (in Japanese).
Huntsman, S. R., Mitchell, J. K., Klejbuk, L. W., and Shinde, S. B. (1986) "Lateral Stress Measurement During Cone Penetration," Proceedinds of In Situ '86, ASCE, pp. 617-634.
Ishihara, K. (1985) "Stability of natural deposits during earthquakes," Proceedings, 11th International Conference on Soil Mechanics and Foundation Engineering, Vol.1, pp. 321-376.
Ishihara, K. and Koga, Y. (1981) "Case studies of liquefaction in the 1964 Niigata earthquake," Soils and Foundations, Vol. 21, No. 3, pp. 35-52.
Ishihara, K., Shimizu, K. and Yamada, Y. (1981) "Pore water pressures measured in sand deposits during an earthquake," Soils and Foundations, Vol. 21, No. 4, pp. 85-100.
Ishihara, K. and Perlea, V. (1984) "Liquefaction-associated ground damage during the Vrancea earthquake of March 4, 1977," Soils and Foundations, Vol. 24, No. 1, pp. 90-112.
Jamiolkowski, M., Ladd, C. C., Germaine, J. T., and Lancellotta, R. (1985a) "New developments in field and laboratory testing of soils," Proceedings, 11th ICSMFE, Vol. 1, pp. 57-153.
Jamiolkowski, M., Baldi, G., Bellotti, R., Ghionna, V., and Pasqualini, E. (1985b) "Penetration resistance and liquefaction of sands," Proceedings, 11th ICSMFE, Vol. 4, pp. 1891-1896.
Kaito, T., Sakaguchi, S., Nishigaki, Y., Miki, K., and Yukami, H. (1971) "Large penetration Test," Tsuchi-to-Kiso, No. 629, pp. 15-21.
Kokusho, T. (1987) "In-situ dynamic soil properties and their evaluations," Proceedings, 8th Asian Regional Conference on Soil Mechanics and Foundation Engineering.
Kovacs, W. D., Griffith, A. H., and Evans, J. C. (1978) " An alternative to the cathead and rope for the standard penetration test," Geotechnical Testing Journal, Vol. 1, No. 2, pp. 72-81.
Kovacs, W. D. and Salomone, L. A. (1982) "SPT hammer energy measurement," Journal of the Geotechnical Engineering Division, Vol. 108, No. GT4, pp. 599-620.
Liao, S. and Whitman, R. V. (1985) "Overburden correction factors for SPT in sand," Journal of Geotechnical Engineering, Vol. 112, No. 3, pp. 373-377.
Marchetti, S. (1982) "Detection of liquefiable sand layers by mean of quasi-static penetration tests, " Proceedings, 2nd European Symposium on Penetration Testing, Vol. 2, pp. 689-695.
Muromachi, T. and Kobayashi, S. (1982) "Comparative study of static and dynamic penetration tests currently in use in Japan," Proceedings, 2nd European Symposium on Penetration Testing, pp. 297-302.
Oh-oka, H. (1984) "Instantaneous velocity measurements just before impact for SPT hammer," Proc., Annual Meeting of JIA (in Japanese).
Oh-oka, H. and Tatsui, T. (1987) "Effects of borehole diameter on N-values and energy efficiency measurements just before hammer impact," Annual Meeting of JSSMFE (in Japanese).
Ohsaki, Y. (1970) "Effects of sand compaction on liquefaction during the Tokachioki Earthquake," Soils and Foundations, Vol. 10, No.2, pp. 112-128.
Ohsaki, Y. and Iwasaki, R. (1973) "On dynamic shear moduli and Poisson's ratios of soil deposits," Soils and Foundations, Vol. 13, No. 4, pp. 61-73.
Ohta, Y. and Goto, N. (1976) "Estimation of S-wave velocity in terms of characteristic indices of soil," Butsuri-Tanko, Vol. 29, No.4, pp.31-41 (in Japanese).
Robertson, P. K. (1986) "In Situ Testing and Its Application to Foundation Engineering," Canadian Geotechnical Journal, Vol.23, No. 4, pp. 573-594.
Robertson, P. K., Campanella, R. G., and Wightman, A. (1983) "SPT-CPT correlations," Journal of Geotechnical Engineering, ASCE, Vol. 109, No. 11, pp. 1449-1459.
Robertson, P. K. and Campanella, R. G. (1985) "Liquefaction potential of sands using the CPT," Journal of Geotechnical Engineering, ASCE, Vol. 111, No. 3.
Robertson, P. K. and Campanella, R. G. (1986) "Estimating liquefaction potential of sands using the flat plate dilatometer," Geotechnical Testing

Journal, Vol. 9, No. 1, pp.38-40.
Robertson, P. K., Campanella, R. G., Gillespie, D. and Greig, J. (1986a) "Use of piezometer cone data," Proceedings of In-Situ'86, pp. 1263-1280.
Robertson, P. K., Campanella, R. G., Gillespie, D. and Rice, A. (1986b) "Seismic CPT to measure in situ shear wave velocity," Journal of Geotechnical Enginnering, ASCE, Vol. 112, No. 8, pp. 791-803.
Sasaki, Y., Koga, Y., Itoh, Y., Shimazu, T., and Kondo, M. (1985) "In-situ test assessing liquefaction potential using vibratory cone penetrometer," Paper presented at 17th Joint Meeting, UJNR, Tukuba, May, 18pp.
Sasaki, Y., Itoh, Y. and Shimazu, T. (1984) "A study on the relationship between the results of vibratory cone penetration tests and earthquake-induced settlement of embankments," Proceedings, 19th Annual Meeting of JSSMFE.
Schmertmann, J. H. (1978a) "Use the SPT to measure dynamic soil properties? - yes, but..!," Dynamic Geotechnical Testing, ASTM SPT 654, pp. 341-355.
Schmertmann, J. H. (1978b) "Guidelines for cone penetration test -performance and design," Report No. FHWA-TS-78-209, U. S. Department of Transportation, Federal Highway, Administration, Washington, D. C.
Schmertmann, J. H. and Palacios, A. (1979) "Energy dynamics of SPT," Journal of the Geotechnical Engineering Division, ASCE, Vol. 105, No. GT8, pp. 909-926.
Seed, H. B. (1979) "Soil liquefaction and cyclic mobility evaluation for level ground during earthquakes," Journal of the Geotechnical Engineering Division, ASCE, Vol. 105, No. GT2, pp. 201-255.
Seed, H. B., Idriss, I. M., and Arango, I. (1983) " Evaluation of liquefaction potential using field performance data," Journal of Geotechnical Engineering, ASCE, Vol. 109, No. 3.
Seed, H. B., Tokimatsu, K., Harder, L. F., and Chung, R. M. (1985) "Influence of SPT procedures in soil liquefaction evaluations," Journal of Geotechnical Engineering, ASCE, Vol. 111, No. 12, pp. 1425-1445.
Seed, H. B. and De Alba, P. (1986) "Use of SPT and CPT tests for evaluating the liquefaction resistance of sands," Proceedings of In Situ '86, ASCE, pp. 281-302.
Seed, H. B., Wong, R. T., Idriss, I. M., and Tokimatsu, K. (1986) "Moduli and damping factors for dynamic analyses of cohesionless soils," Journal of Geotechnical Engineering, ASCE, Vol. 112, No. 11, pp. 1016-1032.
Shibata, T. (1987) "Discussion," Journal of Geotechnical Engineering, ASCE, Vol. 113, No. 6, pp.676-678.
Skempton, A. W. (1986) "Standard penetration test procedures and the effects in sands of overburden pressure, relative density, particle size, aging and overconsolidation," Geotechnique, Vol. 36, No. 3, pp. 425-447.
Tatsuoka, F., Muramatsu, M., and Sasaki, T. (1982) "Cyclic undrained stress strain behavior of dense sands by torsional simple shear test," Soils and Foundations, Vol. 22, No. 2, pp. 55-70.
Tokimatsu, K. and Hosaka, Y. (1986) "Effects of sample disturbance on dynamic properties of sand," Soils and Foundations, Vol. 26, No. 1, pp. 53-64.
Tokimatsu, K., Yamazaki, T., and Yoshimi, Y. (1986) "Soil liquefaction evaluations by elastic shear moduli," Soils and Foundations, Vol. 26, No. 1, pp. 25-35.
Tokimatsu, K. and Yoshimi, Y. (1983) "Empirical correlation of soil liquefaction based on SPT N-value and fines content," Soils and Foundations, Vol. 23, No. 4.
Vallee, R. P. and Skryness, R. S. (1979) "Sampling and in-situ density of a saturated gravel deposit," Geotechnical Testing Journal, Vol. 2, No. 3, 136-142.
Yasuda, S. and Yamaguchi, I. (1984) "Dynamic shear moduli measured in the laboratory and the field," Proceedings, Symposium on Evaluations of Deformation and Strength Characteristics of Sandy Soils and Sand Deposits, JSSMFE, pp.115-118.
Yoshimi, Y., Tokimatsu, K., Kaneko, O., and Makihara, Y. (1984) "Undrained cyclic shear strength of an dense Niigata sand," Soils and Foundations, Vol. 24, No. 4, pp. 131-145.
Yoshimi, Y., Tokimatsu, K. and Hosaka, Y. (1988) "Evaluation of liquefaction resistance of clean sands based on high-quality undisturbed samples," to be published.
Yoshida, Y., Kokusho, T. and Ikemi, M. (1988) "Empirical formulas of SPT blow-counts for gravelly soils," Proceedings, ISOPT'88.
Youd, T. L. and Bennett, M. J. (1983) "Liquefaction sites, Imperial Valley, California," Journal of Geotechnical Engineering, ASCE, Vol. 109, No. 3, pp. 440-459.
Zhou, S. G. (1981) "Influence of fines on evaluating liquefaction of sand by CPT," Proceedings, International Conference on Recent Advances in Geotechnical Earthquake Engineering and Soil Dynamics, St. Louis, Vol. 1, pp. 167-172.

Penetration Testing 1988, ISOPT-1, De Ruiter (ed.)
© 1988 Balkema, Rotterdam, ISBN 90 6191 801 4

Current status of the Marchetti dilatometer test

Alan J.Lutenegger
Clarkson University, USA

ABSTRACT: The use of the Marchetti Dilatometer Test is rapidly expanding worldwide such that the test is becoming one of the premier in situ tests available to the geotechnical profession. The current state-of-the-practice of the test is described and a summary of current usage is presented. A review of the procedures for determining conventional design parameters of soils and an assessment of the quality of predicted field performance is presented and discussed.

1 INTRODUCTION

The Marchetti Dilatometer Test (DMT) has rapidly grown in the past decade to one of the more popular in situ tests available to researchers and practicing engineers. This rapid growth seems to reflect a need in the geotechnical profession for simple, rapid, and cost-effective tools to characterize sites for geotechnical projects. The DMT seems to possess most of the preferred qualitites of in situ tests, i.e., it is simple to operate, rugged, non-electronic, can be used with a wide variety of practical insertion equipment and appears to give very reproducible results.

In situ tests fulfill a very real need of many practicing engineers; namely, they provide rapid estimates of soil parameters for predicting field performance of real structures. This is true of both long-standing tests, such as the Standard Penetration Test and Vane Shear Test, and relatively newer tests, such as the Dilatometer Test. However, even tests with a long history of use by the profession are not without problems, e.g., the high variability of the SPT and the need for a shear strength correction obtained from the field vane.

Like all soil tests, the DMT may have limitations and may not apply to all geotechnical materials and problems. The purpose of this paper is to provide a review of the current status of the DMT with respect to both predicting conventional geotechnical parameters and predicting field performance of geotechnical works. In providing such a review, it is appropriate to briefly look at where the test has been, what changes have occurred and finally what new developments have recently taken place.

At the preparation of this paper, it is evident that the worldwide use of the DMT has expanded significantly in the past 10 years. This is made obvious by the wide range of materials wherein the test has seen use. Using predominantly information available in the open literature, Table 1 summarizes reported usage by the profession not including the original work presented by Marchetti (1980). Quite obviously, practicing engineers, who are by nature keen to expand any technique to new areas, probably have encompassed an even wider range of materials.

Even so, Table 1 clearly demonstrates that the DMT has been used in a broad range of materials: cohesive and cohesionless; saturated and partially saturated; normally consolidated and overconsolidated; "quick" and very stiff; natural and artificial. There appear to be only minor limitations in regard to use of the DMT in natural geologic materials, namely, bouldery glacial sediments or gravelly deposits, both of which resist penetration and may damage the blade and/or diaphragm. Offshore use of the DMT has been reported by Marchetti (1980), Burgess (1983), Sonnenfeld et al. (1985) and Lacasse and Lunne (1988). A specially designed DMT for offshore use is currently under development at NGI.

Table 1. DMT tested materials.

Material	Reference
Sensitive marine clay	Lacasse and Lunne (1983) Fabius (1985) Bechai et al. (1986) Hayes (1986) Lutenegger and Timian (1986) Lutenegger (1987)
Soft non-sensitive clays	Minkov et al. (1984) Ming-Fang (1986) Saye and Lutenegger (1988 a)
Lacustrine clay	Chan and Morgenstern (1986)
Glacial tills and/ or very stiff over-consolidated clays	Davidson and Boghrat (1983) Schmertmann and Crapps (1983) Powell and Uglow (1986)
Sand	Schmertmann (1982) Baldi et al. (1986) Clough and Goeke (1986) Lacasse and Lunne (1986) Schmertmann et al. (1986)
Deltaic silt	Campanella and Robertson (1983) Konrad et al. (1985)
Loess	Lutenegger and Donchev (1983) Hammandshiev and Lutenegger (1985) Lutenegger (1986)
Peat	Hayes (1983) Kaderabek et al. (1986)
Compacted fill	Borden et al. (1985)
Industrial slime	GPE (1984)
Soft/medium rock	Sonnenfeld et al. (1985)

2 CONDUCTING THE TEST

The DMT consists of a flat-plate penetrometer which is instrumented with a flexible, circular diaphragm mounted on one face of the blade and a console to provide operational control. The dimensions and geometry of the blade are shown in Fig. 1. A detailed recommended procedure for conducting the test has been presented by ASTM Subcommittee D18.02 (Schmertmann, 1986a)and will only briefly be summarized here.

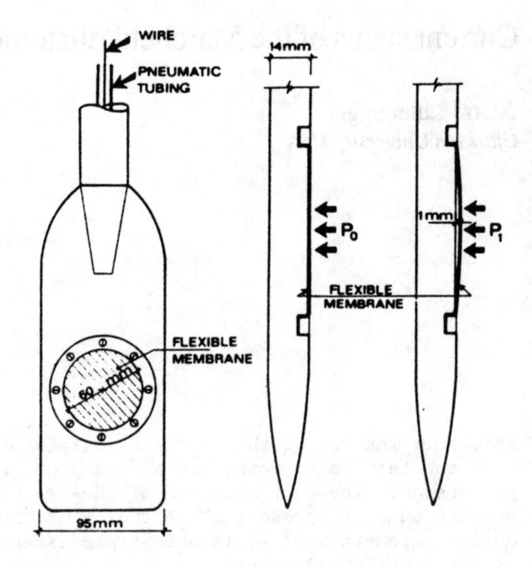

Figure 1. Marchetti Dilatometer

Immediately after the blade is forced into the ground to a desired test depth, preferably by quasi-static penetration, the flexible diaphragm is expanded by compressed gas. As gas pressure is slowly increased and the membrane moves outward against the soil, an electric signal identifies the pressure required for the diaphragm to lift off the plane of the blade. As diaphragm expansion continues, a second electric signal denotes when a central diaphragm expansion of about 1 mm is reached. The two pressures are denoted as the A- and B-Reading, respectively. These pressures are corrected for diaphragm inertial resistance via a simple calibration procedure such that:

$$p_o = A + A \text{ correction} \quad (1)$$

$$p_1 = B - B \text{ correction} \quad (2)$$

The pressures p_o and p_1 are used along with an estimate of the vertical effective stress, σ'_{vo}, and in situ pore pressure, u_o, at the elevation of the test, to provide three indices denoted as:

$$\text{Material Index; } I_D = \frac{p_1-p_o}{p_o-u_o} \quad (3)$$

$$\text{Horizontal Stress Index; } K_D = \frac{p_o-u_o}{\sigma'_{vo}} \quad (4)$$

$$\text{Dilatometer Modulus; } E_D = 34.7\ (p_1-p_o) \quad (5)$$

The use of these indices for predicting a variety of soil parameters was proposed by Marchetti (1980) who presented a series of empirical correlations based on more conventional laboratory and field test data. The interrelationships between these index values and soil parameters proposed by Marchetti (1980) and others are summarized in Table 2.

A third pressure reading designated as C may also be obtained by controlled gas deflation after obtaining the B-Reading and denotes the pressure at which the diaphragm recontacts the plane of the blade. The pressure p_2 is obtained from the C-Reading as:

$$p_2 = C - A \text{ correction} \quad (6)$$

This pressure reading has only recently been introduced and its use has not been fully investigated. An additional DMT index has been proposed (Lutenegger and Kabir, 1988) which has the form:

$$U_D = \frac{p_2 - u_o}{p_o - u_o} \quad (7)$$

The use of U_D will be discussed further in a later section.

3 ESTIMATING SOIL PARAMETERS

One of the main uses of the DMT is to provide estimates of a number of conventional soil parameters, e.g., undrained strength, K_o, compressibility, etc. Since the test may be conducted at intervals of about 20 cm, for many projects a large amount of data is obtained. This allows the use of statistical analyses for probabilistic designs. Since tests may be initiated at depths as shallow as 20 cm, this may be particularly advantageous for pavement subgrades, shallow foundations and laterally loaded pile problems. As with most in situ penetration tests, the values obtained are estimates; the use of which is often to give the range of actual values. Quite often the use of penetration tests is in the form of preliminary site investigations and the tool is used for rapid identification of problem layers where more detailed in situ or laboratory tests may be required. However, the DMT has often provided accurate estimates of soil properties and therefore it is appropriate to review these predictions.

Table 2. Interrelationships between soil parameters and DMT Indices.

Soil Parameter	DMT Index	Reference
s_u (clays)	I_D, K_D	Marchetti (1980)
ϕ' (sands)	I_D, K_D, thrust or adjacent q_c	Schmertmann (1982) Marchetti (1985)
K_o (clays)	I_D, K_D	Marchetti (1980) Marchetti (1986)
K_o (sands)	K_D, thrust	Schmertmann (1982)
OCR (clays)	I_D, K_D	Marchetti (1980)
OCR (sands)	K_D, thrust	GPE (1983)
M	I_D, E_D, K_D	Marchetti (1980)
E_i	I_D, E_D	Robertson et al. (1988)
E_{25}	E_D	Campanella and Robertson (1983) Baldi et al. (1986)
Cyclic stress ratio to cause liquefaction	K_D	Robertson and Campanella (1986)
k_h (subgrade reaction modulus)	p_o, K_D	Schmertmann and Crapps (1983) Robertson et al. (1988)
CBR	E_D	Borden et al. (1985)

3.1 Site Stratigraphy

The Material Index, I_D, given in Eq. [3] was used by Marchetti (1980) as a simple means of identifying soil type. Although no specific values could be related to grain-size distribution, the value of I_D appeared to be reproducible in similar materials and had a range of only about 2 magnitudes roughly 0.1 to 10.0. Marchetti proposed a simple classification scheme, shown in Table 3, so that the engineer would have some means of identifying soil type. This is particularly appealing to engineers since no sample is obtained. Similar schemes are used in other penetration tests such as the CPT and CUPT.

I_D is a measure of the relative change from p_o to p_1 with respect to the original

Table 3. Proposed Soil Classification Based on I_D.

Peat or Sensitive clays	CLAY		SILT			SAND	
		Silty	Clayey		Sandy	Silty	
I_D values	0.10	0.35	0.6	0.9	1.2	1.8	3.3

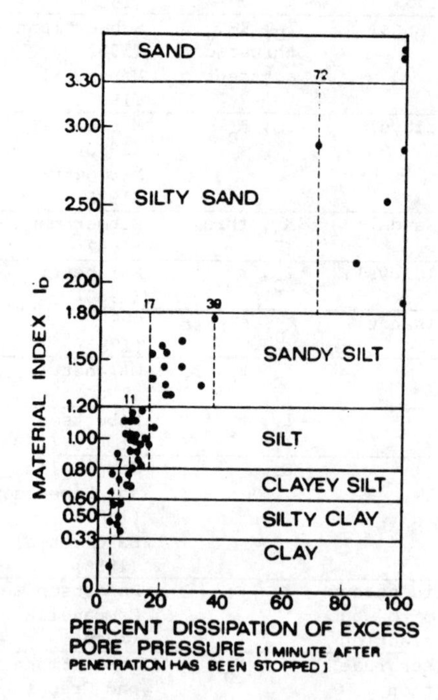

Figure 2. Degree of Dissipation of Excess Pore Water Pressure 1 min. After Penetration as a Function of I_D (Davidson and Boghrat, 1983)

p_o, corrected for static pore water pressures. Using an instrumented DMT, Campanella et al. (1985) showed that in soft clays, the increase in pressure during the expansion from p_o to p_1 was equally matched by an increase in pore water pressure; while in sands, the pressure increase from p_o to p_1 occurs without generating pore water pressures, and is a real measure of soil response. Therefore, one might logically conclude that I_D is a measure of the response of the soil to the increased cavity expansion. As will be shown later, in soft clays, p_o should closely predict P_L from a pressuremeter, while in overconsolidated clays and sands, P_L is closer to p_1. I_D also reflects the degree of drainage which takes place in a particular soil and therefore should be related to permeability or the coefficient of consolidation, Fig. 2.

Is I_D a material property or is it dependent on other parameters, such as degree of saturation, etc? The writer conducted a series of tests at a site in which partially saturated silts were artificially wetted. After each successive wetting, a DMT profile was conducted. Results are shown in Fig. 3. With increasing saturation there is a systematic decrease in K_D and E_D as the material becomes softer, however, the Material Index I_D, remains essentially constant, indicating the same material type. Similar results have been reported by Schmertmann (1982) and Lacasse and Lunne (1986) in compacted sands.

It is the writer's opinion that I_D may be potentially useful as a measure of other soil properties such as sensitivity in cohesive soils. However, at the current time it is used predominantly as an approximate indication of soil type and as a qualifier for application of empirical correlations.

3.2 Undrained Shear Strength

Marchetti (1980) suggested estimating s_u based on:

$$s_u = f(K_D) \qquad [8]$$

This relationship is based on the suggestion by Ladd et al. (1977) that:

$$\frac{\left(\frac{s_u}{\sigma'_{vo}}\right)_{OC}}{\left(\frac{s_u}{\sigma'_{vo}}\right)_{NC}} = OCR^m \qquad [9]$$

i.e., many clays exhibit normalized undrained shear strength behavior. Thus, the estimate of undrained strength provided by the DMT is via OCR through the empirical relation to K_D, which is in turn linked directly to p_o as given by Eq. [4]. However, as reviewed herein it can be shown that s_u also correlates to the DMT via limit pressure or pore pressure behavior concep

Several investigations (e.g., Lacasse and Lunne, 1983; Fabius, 1985; Greig et al., 1986; Lutenegger and Timian, 1986; Ming-Fang, 1986) have shown that the DMT prediction of undrained shear strength in soft saturated clays compares very well with <u>uncorrected</u> field vane results. In stiff overconsolidated clays, the current correlation appears less accurate. Why does the DMT provide accurate results of s_u in soft clays and how accurate is it?

The quality of the correlation between $s_{u(DMT)}$ and $s_{u(reference)}$ of course depends

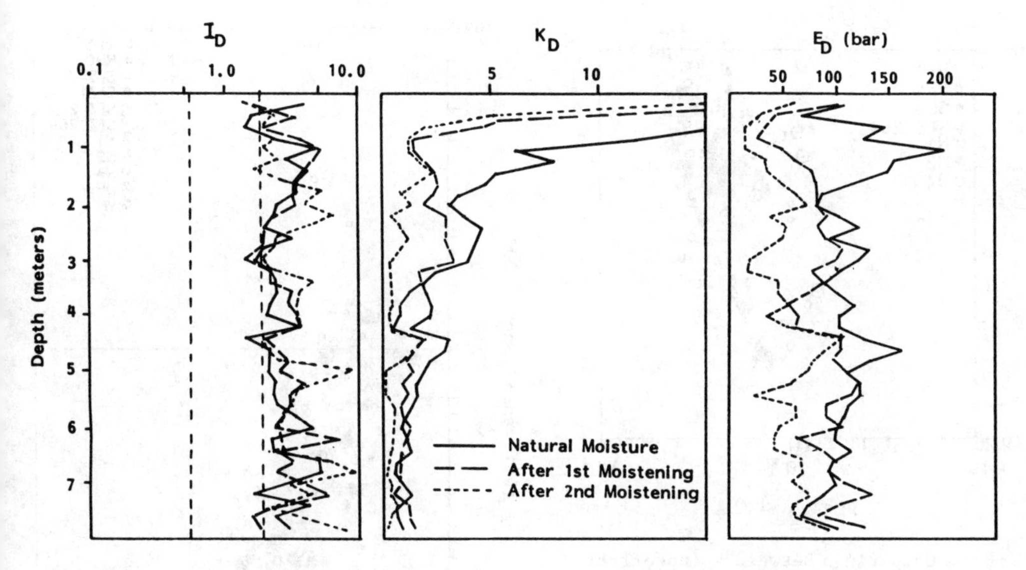

Figure 3. Insensitivity of I_D to Changes in Saturation in a Silt

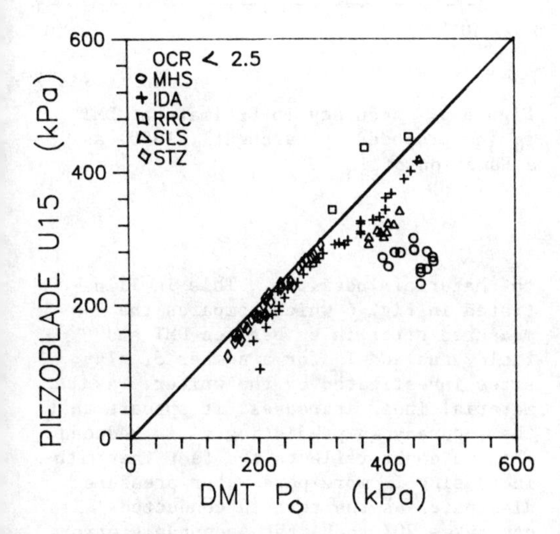

Figure 4. Comparison Between DMT p_o and Penetration Pore Pressures from Piezoblade in Normally Consolidated and Lightly Overconsolidated Clays

on the quality of the reference measurement. This discussion will assume that the uncorrected s_u from the field vane represents the "most likely" value of undrained shear strength. Problems associated with misuse of the vane in materials which are too stiff or do not remain undrained during testing or problems of soil variability will be neglected. Additionally, it should be left up to the engineer's discretion whether or not to apply a vane correction depending on the design problem.

As noted, the current estimate of s_u indirectly makes use of the p_o pressure reading, which is normally obtained within 15 to 30 sec. after penetration. Several studies (Davidson and Boghrat, 1983; Campanella et al., 1985; Lutenegger and Kabir, 1988) have shown that in soft saturated clays (OCR<2) the DMT is essentially an undrained test and p_o is dominated by penetration pore water pressures. Figure 4 shows a comparison between p_o and u_{excess} measured with a DMT Piezoblade 15 sec. after penetration. It can be seen that in these materials the p_o reading primarily reflects penetration pore water pressures.

Cavity expansion theory (Vesic, 1972; Ladanyi, 1963) and strain path analysis (Baligh, 1985) for cylindrical penetrometers accurately predict undrained shear strength from penetration pore water pressures for normally consolidated and lightly overconsolidated (non-dilating) clays. Therefore, it should not be too surprising that the DMT is able to accurately predict s_u in these materials provided that substantial pore water pressure dissipation does not occur in the time required to obtain p_o. Mayne (1987) has shown that despite the differences in geometry, the DMT

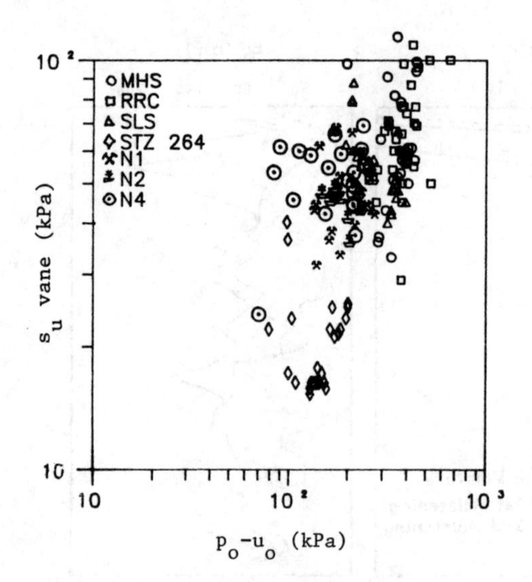

Figure 5. Comparison Between s_u (uncorrected field vane) and Corrected DMT p_o

p_o is nearly identical to pore water pressures obtained behind the base of a piezocone. In the case of a cylindrical piezocone Robertson and Campanella (1983) suggest that, based on cavity expansion theory the undrained shear strength for normally consolidated clays can be estimated as:

$$3 < \frac{\Delta u}{s_u} < 5 \qquad [10]$$

for a cylindrical cavity with $\Delta u = (u_{measured} - u_{equilibrium})$ measured behind the cone. Since the penetration of the DMT represents some form of cavity expansion and p_o in soft saturated clays is predominantly excess pore water presures, one might expect a relationship similar to [10] to exist for the DMT.

A preliminary check on this approach is shown in Fig. 5, which presents data obtained in normally consolidated clays by the writer. It appears that this approach indeed has some merit with the values generally ranging from:

$$3 < \frac{p_o - u_o}{s_u} < 9 \qquad [11]$$

The range in values no doubt reflects differences in rigidity index and other factors relating to cavity expansion.

The accuracy of the prediction of s_u appears to be strongly linked to proper soil identification through the use of

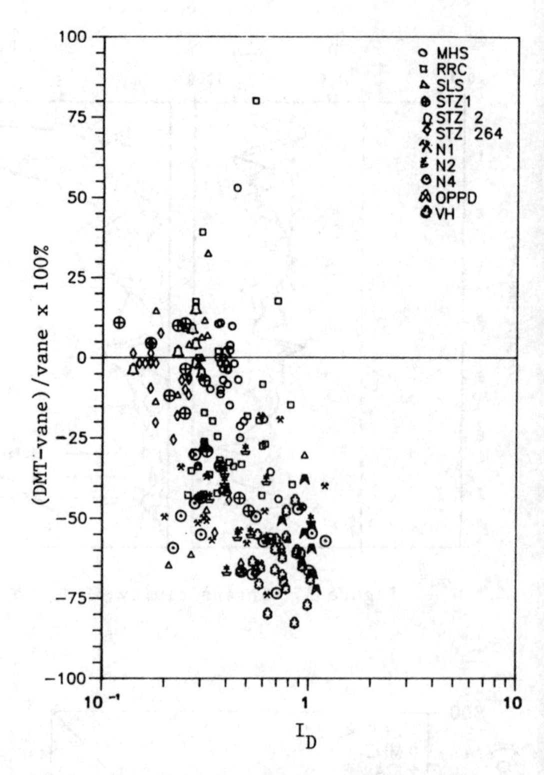

Figure 6. Accuracy in Estimate of DMT s_u (as proposed by Marchetti, 1980) as a Function of I_D

the Material Index, I_D. This is illustrated in Fig. 6 which compares the measured error in s_u between DMT and field vane and I_D for a number of clay sites investigated by the writer. As the material index increases, it appears that the accuracy in predicting s_u is reduced. This no doubt reflects the fact that with increasing I_D more pore water pressure dissipates as the test in conducted. If one takes 20% to be the acceptable error, Fig. 6 suggests that s_u estimates should be restricted to I_D values less than about 0.33. This value roughly coincides with the limit shown by Davidson and Boghrat (1983) for approximately 5% penetration pore pressure dissipation after 1 min. One encouraging point noted in Fig. 6 is that the DMT consistently underpredicts s_u.

The data shown in Fig. 6 suggest that for various values of I_D, a simple correction may be applied to the current DMT empirical correlation to bring the estimate of s_u to within acceptable limits. However this correction should be used with caution considering the degree of scatter indicated in Fig. 6.

Results obtained in stiff overconsolidated clays (e.g., Powell and Uglow, 1986) seem to indicate that the current relationship for estimating s_u from the DMT is not as accurate, sometimes overpredicting and sometimes underpredicting s_u. However, the comparisons have not been with field vane, but with other field tests (e.g., PMT, Plate Load) or laboratory tests (e.g., UU triaxial).

Powell and Uglow suggested that in highly overconsolidated clays and tills the normalized undrained strength could be expressed as a function of K_D. This is essentially the same as comparing s_u directly to p_o as previously shown in Fig. 5 since both are normalized with respect to σ'_{vo}. Figure 7 shows pore pressure measurements obtained from Piezoblade and DMT p_o values in overconsolidated (2.5 < OCR < 10) soils. In contrast to results obtained in more normally consolidated clays and shown in Fig. 4, data from overconsolidated clays clearly show that p_o is predominantly greater than u_{excess}. Therefore, in overconsolidated clays there is a component of p_o which is attributed to soil resistance, which may be regarded as the "overstress" created by penetration of the blade. In light of these data, it is not surprising that current empirical correlations do not predict s_u as accurately as in normally consolidated materials.

3.3 Stress History

The estimation of soil stress history using the DMT was proposed by Marchetti (1980) by correlating K_D with OCR from oedometer tests. Marchetti restricted the use of the correlation to "materials free of cementation, attraction etc. in simple unloading" and for I_D < 1.2. Unfortunately, many natural soils do not fit this description for one reason or another and therefore attempts have been made to apply the correlation outside the original scope. This is only natural, since engineers deal with a multitude of materials and wish to extend the use of any tool to its practical limits.

Recently, Mayne (1987) has reviewed the existing literature and proposed a simple relationship between the "effective" DMT p_o (p_o-u_o) and vertical preconsolidation stress obtained from stress controlled incremental oedometer tests, p'_c. These data are shown in Fig. 8 and include an additional 32 data points obtained by the

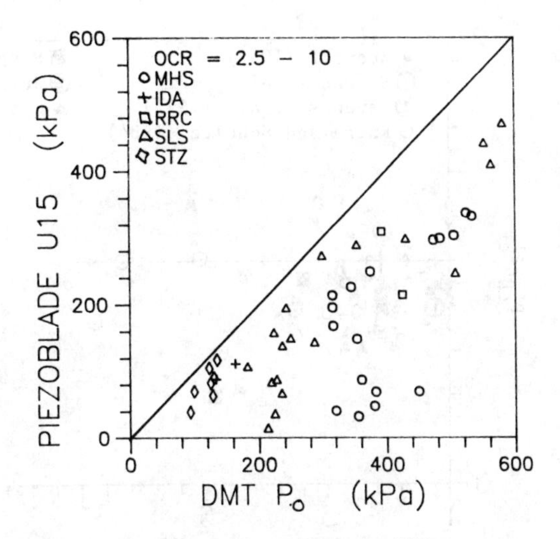

Figure 7. Comparison Between DMT p_o and Penetration Pore Pressures from Piezoblade in Overconsolidated Clays

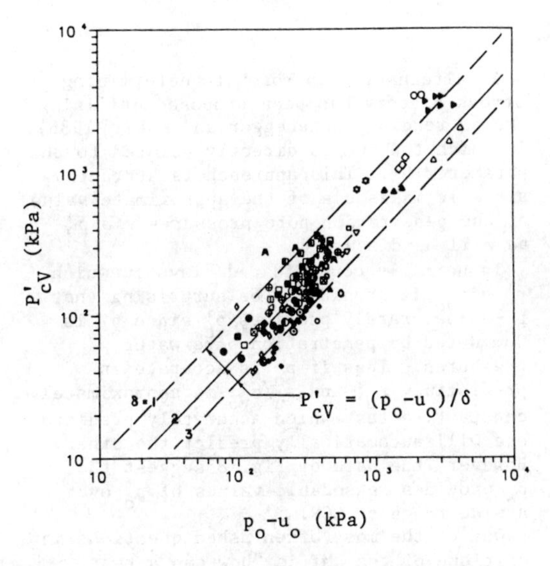

Figure 8. Relationship Between Corrected p_o and Vertical Preconsolidation Stress from Oedometer (modified from Mayne, 1987)

writer in a variety of materials. The simple relationship between p'_c and p_o indicated in Fig. 8 is similar to that presented by Marchetti (1980) since both OCR and K_D are normalized with respect to the effective overburden stress.

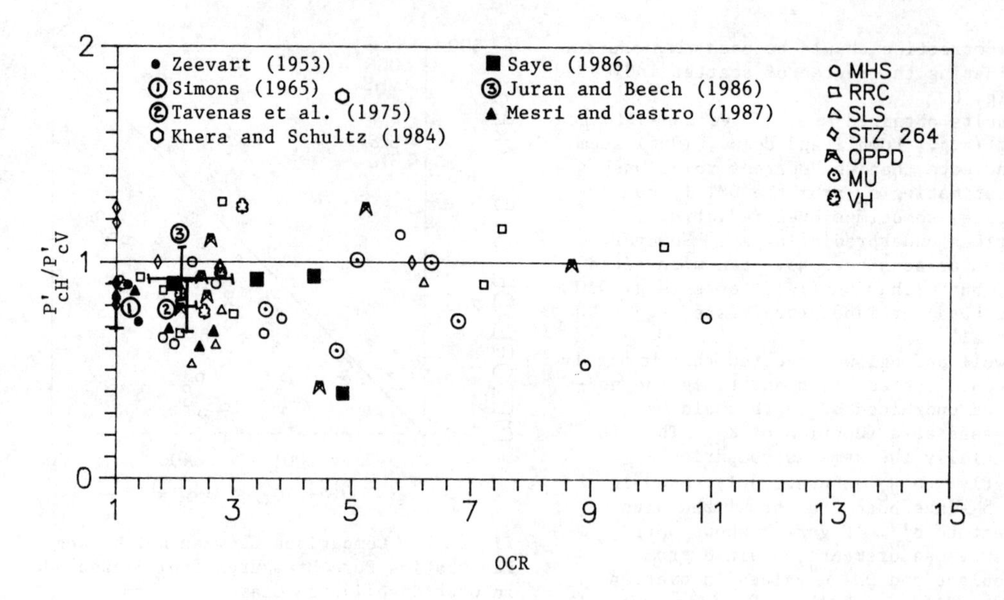

Figure 9. Variation in p'_{cH}/p'_{cV} with OCR

An alternative approach to determining stress history has been proposed utilizing the p_2 reading (Lutenegger and Kabir, 1988), in which the OCR is directly related to the parameter U_D. This approach is attractive since it makes use of the approximate value of the penetration pore pressures via p_2 as will be discussed.

In normally consolidated clays in which $p'_c = \sigma'_{vo}$ it should not be surprising that $p_o - u_o$ accurately predicts p'_c since p_o is dominated by penetration pore water pressures. Thus if p_o is accurate in predicting s_u; and s_u/σ'_{vo} is approximately constant, a test which accurately predicts one will automatically predict the other. However, the data of Fig. 8 suggest that p_o provides reasonable values of p'_c over a wide range of OCR.

One of the most often asked questions in critique of the DMT is "how can a test which presumably measures soil response in the lateral direction predict vertical soil properties"? The answer to this question lies partially in the fact that in softer more normally consolidated soils, the p_o reading is dominated by pore water pressures, as previously discussed. In these materials, if s_u/p'_c is approximately constant, then the p_o value reflects both s_u and p'_c. However, it may also be that a large percentage of naturally occurring sediments are not highly anisotropic with respect to oedometric yield stress and compressibility, both of which are controlled in large part by soil fabric.

During the course of research work connected with the DMT and other projects, the writer has conducted a number of oedometer tests on undisturbed Shelby tube and piston samples with samples oriented in the horizontal direction, such that the preconsolidation stress normal to the ground surface and therefore parallel to the DMT may be determined. A comparison between the ratio of horizontal to vertical preconsolidation stress and vertical OCR for all tests is shown in Fig. 9. Also shown are available tests from the literature.

These results indicate (quite unexpectedly to the writer) that there is no significant trend in p'_{cH}/p'_{cV} with changing OCR. This is probably due to the fact that the majority of the samples represent geologic materials which <u>have not</u> been physically preloaded, and thus the apparent preconsolidation pressure obtained from oedometer tests is a result of some other phenomenon, such as, freeze-thaw, shrink-swell, fluctuations in water level, etc. There is not sufficient evidence to indicate that such materials should be anisotropic with respect to yield stress in K_o loading. Some of the scatter indicated in Fig. 9 may also derive from sample disturbance and other problems associated with determining p'_c, e.g., graphical

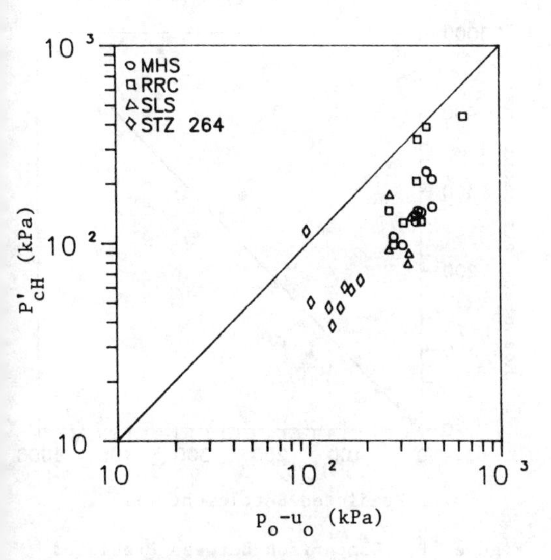

Figure 10. Comparison Between p_o and p'_{cH} from Oedometer Tests.

interpretation.

In light of these data, the relationship between p_o and p'_c presented in Fig. 8 may be reasonable, irrespective of OCR. In order to further investigate the response of p_o to horizontal yield stress, Fig. 10 presents a comparison between corrected p_o and p'_{cH} for tests conducted by the writer. These data fall within the correlation range suggested by Mayne (1987) for vertical p'_c as previously shown in Fig. 8.

3.4 Lateral Stress Ratio, K_o

Because of its geometry, the DMT largely records the horizontal response to penetration when placed vertically. Thus, a measure of horizontal total stress is obtained, which Marchetti used to define a horizontal stress index K_D, Eq. 4. In recent years, engineers have become increasingly aware of the influence that horizontal ground stresses have on engineering behavior. Schmertmann (1985) recently summarized a number of engineering problems wherein the lateral stress invokes a significant influence, Table 4.

The value of K_D was directly related to K_o by Marchetti (1980) primarily using the empirical relation to OCR presented by Brooker and Ireland (1965). The initial correlation appeared to be independent of soil type (excluding sands) and stress history and therefore has been used by a number of investigators.

Table 4. Qualitative Effects of Changing K on Engineering Behavior (Schmertmann, 1985). (arrows show direction of usual less conservative behavior)

Engineering Behavior	In Situ Low K — High K
Bearing Capacity	safety decreases ←
Slope Stability	safety decreases →
Fracture of Dams	safety decreases ←
Pressure on Walls	increases →
Pile Friction	decreases ←
Settlement/Deformation	increases ←
Liquefaction	safety decreases ←
Ground Treatment Improvement	more difficult →

One of the difficulties in establishing a direct relationship between K_D and K_o is that a reference value of K_o is difficult to obtain. Unlike other soil parameters such as undrained strength or compressibility, which may be reasonably determined by acceptable methods, there is no specific technique which is agreed upon by the profession as the preferred method for determining K_o. Several recent investigations have made use of other field or laboratory tests to compare the K_o value obtained using the DMT: e.g., push-in spade cells (Chan and Morgenstern, 1986); K_o-Stepped Blade (Lutenegger and Timian, 1986); prebored pressuremeter (Powell and Uglow, 1986); hydraulic fracturing and laboratory correlations (Lacasse and Lunne, 1983); and self-boring pressuremeter (Clough and Goeke, 1986).

These and other studies indicate that K_o values derived from the DMT K_D correlation are nearly all within a factor of about 1.5 for a wide range of materials. Many geotechnical engineers are now comfortable with horizontal stress measurements obtained with the self-boring pressuremeter as a reference test. The writer suspects that within the next few years, a number of studies will be available to combine SBPMT and DMT data to help refine the K_D prediction of K_o.

Based on calibration chamber tests in sands, Baldi et al. (1986) have suggested that K_o be determined using both DMT K_o values and adjacent CPT penetration data. The proposed equation for predicting K_o has the form:

$$K_o = 0.376+0.095K_D-0.00172\ q_c/\sigma'_{vo} \quad [12]$$

Of course the drawback is in having to assess the CPT value q_c.

One area in which the DMT appears particularly useful is in the assessment of changes in horizontal stress which have occurred as a result of induced loading, excavation, landslides, or other changes. In this case, the DMT offers the ability to rapidly and inexpensively obtain data to compare the relative changes in horizontal stress, e.g., as a result of preloading under known stress history (Saye and Lutenegger, 1988a).

3.5 Deformation Characteristics

3.5.1 Constrained Modulus

Expansion of the DMT diaphragm from p_o to p_1 produces a known displacement which was used by Marchetti to define the Dilatometer Modulus, E_D, as given in Eq. [5]. Marchetti (1975) had previously suggested that the DMT expansion could be used to define a lateral subgrade reaction modulus value, however engineers are more often in need of a deformation parameter for settlement estimates. Marchetti thus proposed a correlation between E_D, K_D, and M, the local oedometric constrained modulus at the effective in situ overburden stress, σ'_{vo}. More specifically, the correlation is for the local reload modulus. Therefore, in normally consolidated soil, the DMT should not be expected to provide information about the reload modulus and in overconsolidated soil the DMT will provide no direct measurement of the virgin compression modulus.

Schmertmann (1981) had shown that the values of M estimated by the DMT were generally within a factor of about 2 when compared with laboratory oedometer tests. Other studies have generally indicated about the same accuracy in both clays and sands (e.g., Lacasse and Lunne, 1982; Lutenegger and Timian, 1986; Ming Fang, 1986; Lacasse and Lunne, 1986; Borden et al., 1986). Baldi et al. (1986) note that, in the case of calibration chamber tests on overconsolidated sands, the Marchetti correlation results in a pronounced underestimate of M.

There is a distinct advantage in estimating M, which is particularly appealing to practitioners, since the Janbu (1963, 1967) technique for estimating settlement may be used. The relative accuracy of the estimate of M was clearly demonstrated by Schmertmann (1986a) who compared DMT settlement estimates with measured settlements over a wide range in magnitude.

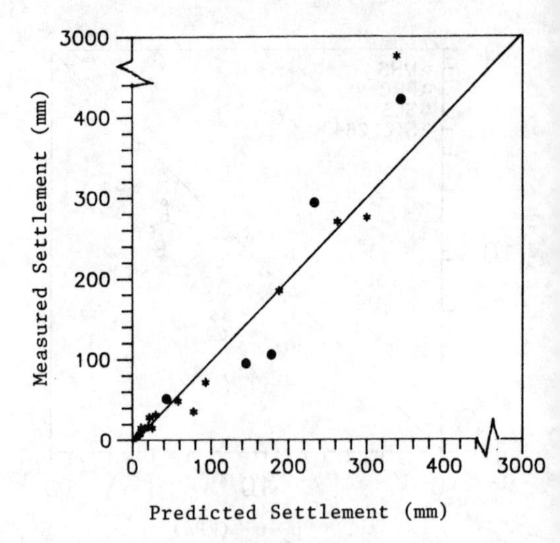

Figure 11. Comparison Between Predicted and Measured Settlements Using M and DMT. (modified from Schmertmann, (1986b)

The real excitement in the data is that if one is interested in a rapid estimate of settlement for preliminary design the DMT appears particularly accurate! The writer has replotted settlement values obtained by Schmertmann (1986b) and included six additional case histories which have been independently investigated using this procedure, Fig. 11.

It is obvious to the writer that constrained modulus values for comparison with the DMT should be obtained from back calculated settlement records on projects wherein an exact determination of footing or other loading stress may be made. It should also be kept in mind that the accuracy of predictions with respect to field performance is also dependent upon the variability associated with the use of the mathematical model involved. This applies to any test.

3.5.2 Elastic Modulus

In many design situations, the engineer may need an estimate of elastic modulus, E, e.g., for use in drilled shaft design or immediate settlement estimates. Depending on the design problem, a different value of E may be required, i.e., E_i, E_{25}, etc.

Davidson and Boghrat (1983) suggested that in highly overconsolidated clays, the value of E_i, obtained from unconsolidated undrained triaxial compression tests could

Table 5. Suggested Correction Factor F For Use in Eq. [13].

Soil Type	Modulus	F	Ref.
Cohesive soils	E_i	10	Robertson et al. (1988)
Sand	E_i	2	Robertson et al. (1988)
Sand	E_{25}	1	Campanella et al. (1985)
NC Sand	E_{25}	0.85	Baldi et al. (1986)
OC Sand	E_{25}	3.5	Baldi et al. (1986)

be related directly to the DMT E_D using a factor of about 1.4. Recently, Robertson et al. (1988) have suggested factors of 10 and 2 for clays and sands, respectively, for use in laterally loaded pile design. All of these results suggest a simple correction factor of the form:

$$E = F E_D \quad [13]$$

Table 5 presents suggested factors of F for different materials.

Baldi et al. (1986) presented results of calibration chamber tests on pluviated sands over a wide range of relative densities and compared E_D to E_{25} obtained from CK_oD triaxial compression tests for both the normally consolidated and overconsolidated condition. Their results follow the form of Eq. [13] and are included in Table 5. For normally consolidated sands, a value of F=1 was also suggested by Campanella, et al. (1985) to estimate E_{25}.

Borden et al. (1985) suggested a relationship between the initial tangent modulus from unconfined compression tests on partially saturated compacted A-6 soil and E_D which had the form:

$$E_i = 0.142\, E_D^{1.298} \quad [14]$$

The writer reanalyzed these data to fit into the form of Eq. [13] and found that the value of F ranged from about 0.4 to 1.1.

The writer was curious about the variation in factor F for estimating the initial tangent modulus. If I_D relates to soil stiffness, then it should be suspected that the value of F is related to I_D. Figure 12 presents a comparison

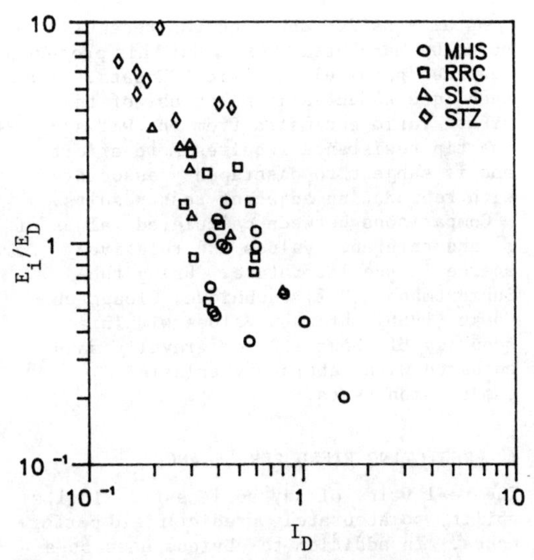

Figure 12. E_i/E_D in Cohesive Soils as Related to I_D (E_i from UU triaxial compression)

of E_i/E_D as a function of I_D. E_i values were obtained from UU triaxial compression tests at confining stresses equal to total horizontal stress estimated from the DMT. The trend of decreasing F with increasing I_D is in agreement with Table 5 and again indicates the importance of I_D.

3.6 Drained Friction Angle in Cohesionless Soils

The penetration of the DMT blade in sands and other freely draining soils represents a drained bearing capacity failure approximating a plane-strain condition. Since the failure condition is controlled by the frictional strength component and in situ state of stress in granular materials, it is reasonable to expect that the results from the DMT may be used to determine ϕ'.

A procedure for obtaining the effective axi-symmetric friction angle in sands was presented by Schmertmann (1982) using the wedge penetration theory of Durgunoglu and Mitchell (1975) and is attractive since it incorporates the horizontal effective stress which is estimated during the DMT. This technique requires measurement of the pushing thrust to advance the DMT, such that an estimate of the tip resistance may be obtained. This measurement may be made using an appropriate load cell which ideally should be located immediately behind the blade.

An alternative approach was recently proposed by Marchetti (1985), but this procedure requires parallel electric CPT data. The technique essentially makes use of the stress ratio estimated from the DMT and the tip resistance requires more effort and is subject to discrepancy associated with reproducing adjacent test results.

Comparisons between calculated values of ϕ' and reference values are relatively scarce in the literature. Using the Schmertmann (1982) technique, Clough and Goeke (1986) obtained values within an accuracy of about 15% for gravelly sand compared with laboratory triaxial compression tests.

4 PREDICTING FIELD PERFORMANCE

The real value of any soil test is in its ability to accurately predict field performance. In addition to obvious uses as a site profiling tool and in obtaining predictions of conventional soil properties, several more direct applications of the DMT to specific engineering problems have been reported. Table 6 presents a current summary. The list is no doubt larger since the application of the DMT is rapidly expanding and new uses are continually being investigated.

Most of the current applications of the DMT to real engineering design problems make use of conventional soil parameters predicted by the DMT. Therefore the accuracy of the predictions generally indicate the ability of the DMT to accurately predict properties.

In the case of quality control wherein the DMT has been used to investigate changes in parameters, the DMT may find a new use in investigating the effect that changing in situ lateral stress has on other in situ tests, e.g., SPT, CPT, VST, etc.

The success of the DMT to accurately predict performance has generally been linked to a design approach based on an accepted engineering practice. Thus, at the current time, conventional design procedures using DMT derived conventional soil engineering properties are generally being used. This is in contrast to a hybrid design approach which is often used with other in situ tests; for example, the pressuremeter approach to foundation design based on E_m or P_L. The writer considers this a distinct advantage of the DMT and an attribute which should make the test more appealing to practicing engineers.

Table 6. Current Reported Application of DMT for Design.

Application	Reference
Settlement Prediction	Schmertmann(1986b) Hayes (1986) Saye and Lutenegger(1988b)
Laterally Loaded Driven Pile Design	Schmertmann and Crapps (1983) Robertson et al. (1988)
Skin Friction of Axial Loaded Piles	Marchetti et al. (1986)
Load on Buried Pipe	Schmertmann
Liquefaction Potential of Sands	Marchetti (1982) Robertson and Campanella(1986)
Compaction Control/ Verification	Schmertmann (1982) Schmertmann et al. (1986) Lutenegger (1986) Lacasse and Lunne (1986)
Ultimate Uplift of Anchor Foundations	Lutenegger et al. (1988)
Transmission Tower Foundation Design	Bechai et al. (1986)
End Bearing, Side Friction and Settlement of Drilled Shafts	Schmertmann and Crapps (1983)
Earth Pressures on Existing Retaining Walls in Distress	Schmertmann (pers. comm. 1987)

5 RECENT DEVELOPMENTS

5.1 Penetration Pore Pressure

Davidson and Boghrat (1983) had clearly shown that in some cases large penetration pore water pressures could be generated when a flat plate with identical geometry to the DMT was forced into saturated soils. Campanella et al. (1985) demonstrated that the corrected DMT closure pressure, p_2, obtained by deflation after obtaining the p_1 pressure could be used to estimate such

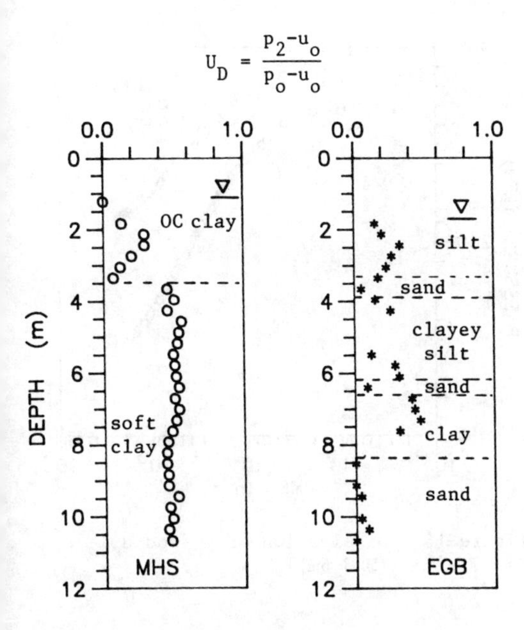

Figure 13. Use of U_D for Site Stratigraphy

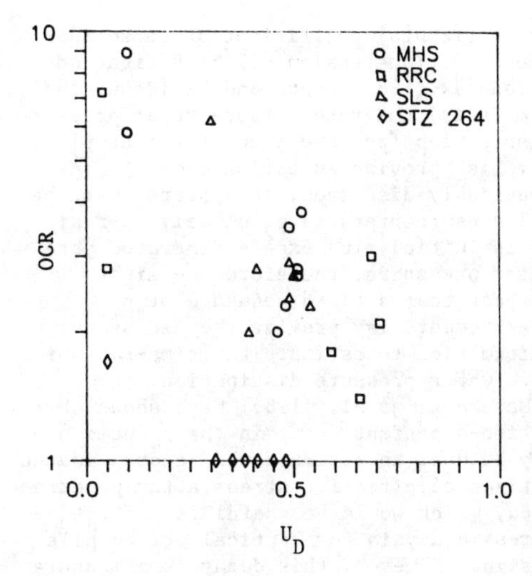

Figure 14. Variation in U_D with OCR

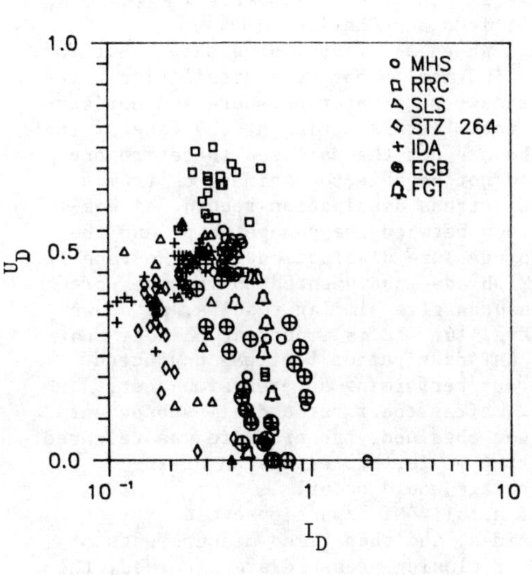

Figure 15. Relationship Between U_D and I_D

pore water pressures. In sands, where the test measures predominantly drained behavior, the p_2 reading reflects the initial in situ water pressure u_o. In saturated clays, in which little drainage occurs, the pressure p_2 reflects both initial and excess pore water pressure generated during penetration. The generated pore water pressures may be as high as seven times the initial pore water pressure in soft clays (Lutenegger and Kabir, 1987).

The parameter U_D, defined by Eq. [7] has shown to be useful in determining site stratigraphy, as shown in Fig. 13. Variations in U_D reflect drainage conditions and, for a given soil, the tendency for generating positive pore pressures, which one can expect to vary with the stress history of the soil. Therefore, one should logically expect a relationship between U_D and OCR. Available data from oedometer tests for a number of sites are shown in Fig. 14. Additionally, if both U_D and I_D are an indication of soil type, one should expect a strong relation between these two parameters. Combined data from several sites are shown in Fig. 15 and verify this.

It is clear to the writer that the p_2 pressure and U_D provide additional insight into the drainage conditions and pore pressure surrounding the DMT. At this time, the full implications of p_2 are not known, however, the writer recommends that the deflation C-Reading should be obtained as a routine part of every DMT.

5.2 Pore Pressure Dissipation

The dissipation record of excess pore water pressures following penetration of a cylindrical probe, i.e., piezocone, into saturated soils may be used to estimate the horizontal coefficient of consolidation, c_h (Torstenssen, 1975; Baligh and Levadeaux, 1986, Gupta and Davidson, 1986). Similarly, the rate of pore water pressure dissipation from the face of a flat-plate may also provide an estimate of c_h. As previously discussed, it appears that the DMT closure pressure p_2 closely approximates initial plus excess generated pore water pressures; therefore one might suspect that a timed sequence of p_2 measurements may provide the necessary information to estimate the time-rate of pore water pressure dissipation.

Marchetti et al. (1986) have shown that a time-dependent decay in the p_o pressure may be used to estimate the reconsolidated horizontal effective stress after penetration, which would be useful for effective stress analysis for vertical driven pile design. However, this decay is a measure of the rate of total horizontal stress dissipation, which would be expected to be similar to that obtained by others using total stress spade cells (e.g., Massarsch, 1975; Tedd and Charles, 1981).

Baligh et al. (1985) have shown that for a cylindrical probe, the dissipation rates of excess pore water pressure and horizontal total stress acting at the face of the probe are not the same and therefore one would not be able to estimate c_h from a total stress dissipation record. A comparison between the change in p_o and the pore pressure dissipation from the face of a DMT blade instrumented to measure pore pressures give similar results, as shown in Fig. 16. It is important to note that the DMT dissipation test was conducted without performing an expansion test; i.e., after the first and all subsequent p_o was obtained, the pressure was released so that no further expansion of the diaphragm would occur.

If a full DMT expansion cycle is obtained, and then timed measurements of the p_2 closing pressure are recorded, the pressure dissipation curve more closely matches the Piezoblade dissipation disspation curve, Fig. 17. In this case it is of interest to note that the reexpansion p_o values are nearly identical to the p_2 closure pressures. This may indicate that the p_1 expansion to 1 mm has created a cavity between the blade and the soil.

At the present time, no theoretical solution is available to calculate the time-rate of consolidation using pore water pressure dissipation rates from flat plates. However, based on the theoretical time factors for a cylindrical probe, Robertson et al. (1988) have proposed a technique to estimate c_h from the DMT.

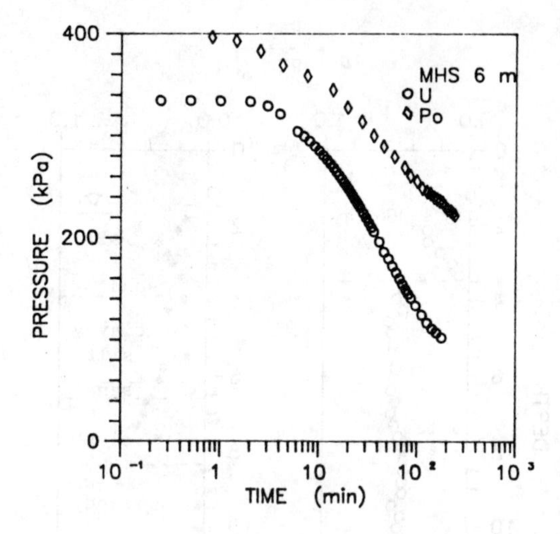

Figure 16. Dissipation of p_o and u (MHS 6m)

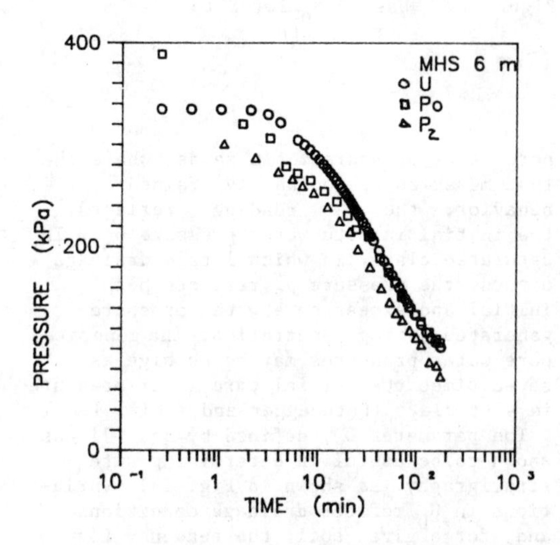

Figure 17. Dissipation of p_o, p_2, and u (MHS 6m)

5.3 Comparison with Other In Situ Tests-PMT

Campanella et. al (1985) noted that the DMT penetration might be sufficient to produce a limit pressure in soft clays, keeping in line with cavity expansion theory. However, because of time-rate effects, location of the membrane in relation to the tip, etc., the p_o may actually be less than the limit pressure, P_L. In a prebored pressuremeter (PMT) test engineers normally define the limit pressure at two times the initial cavity volume for practical reasons, in contrast to the infinite volume ratio which occurs in the DMT which expands from zero thickness to the thickness of the blade. This would make P_L less than p_o. Additionally, pore pressure dissipation may occur during the PMT expansion test which normally takes about 20 minutes compared with the approximately 30 seconds required to obtain p_o. This would also reduce P_L vs. p_o. As noted by Campanella et al. (1985) the DMT expansion to obtain p_1 may reestablish P_L.

Powell and Uglow (1986) compared the limit pressures obtained from pressuremeter tests (Menard, self-boring and push-in types) to the p_1 DMT pressures at three stiff clay sites in the U.K. The bulk of their data fell within the range of $p_1 = P_L$ to $p_1 = 1.4\ P_L$.

Field tests conducted by the writer using a prebored Menard PMT to obtain limit pressures in soils ranging from very soft clays to dense silts and sands indicate a similar relationship to p_1 as shown in Fig. 18. These data are generally within the ranges indicated by Powell and Uglow (1986). If one compares the DMT p_o with P_L, the data indicate a trend slightly skewed to $p_o < P_L$. Thus, as an approximation for a wide range of conditions it appears that one could use the average pressure $((p_o+p_1)/2)$ to estimate P_L.

This may be justified if we consider the following. In very soft clays, the increase from p_o to p_1 is small and consists mainly of pore water pressure, hence the low values of I_D. As stiffness of the material increases, the difference between p_o and p_1 likewise increases which indicates a component of soil stiffness, and thus I_D increases. Therefore, there is a tendency for p_1 to underpredict P_L in soft soils where limit pressures have already been reached and for p_o to overpredict P_L in stiff soils. Hence, the average value for a wide range of materials may be appropriate.

A comparison between p_o and the limit pressure obtained from a full-displacement (Pencel) pressuremeter, which creates a plane strain cylindrical cavity expansion from zero to infinity, is shown in Fig. 19. In softer materials (OCR < 2.5) the values of p_o and P_L are nearly identical, while in stiffer materials (OCR 2.5-10) P_L is significantly underpredicted by p_o. While some of the difference relates to obvious probe geometric differences, rates of testing, creep, drainage etc., these results help explain why the p_o prediction of undrained strength is accurate in

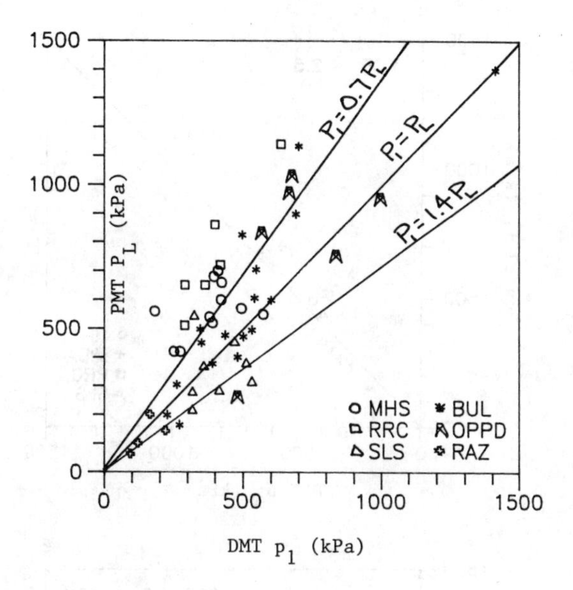

Figure 18. Comparison Between P_L (prebored PMT) and p_1

softer materials, i.e., the penetration of the blade creates a cavity expansion failure. In stiffer materials no limit pressure failure has taken place and further membrane expansion is required. The data are consistent with Figs. 4 and 7.

Elastic modulus values obtained from a prebored DMT may be compared with E_D from the DMT as shown in Fig. 20. These data indicate considerable scatter which may be in part related to variations in borehole drilling techniques used for PMT testing (e.g., augered vs. thin-walled tube sampler). However, because of the displacement created by the DMT, one should suspect the E_D is really a reload modulus and therefore it might be more appropriate to compare the PMT reload modulus with E_D. Unfortunately, these results are not available for the tests shown in Fig. 20. However, since the PMT reload modulus is higher than the initial modulus, usually by a factor of about 2 to 5, the tendency for E_D to overpredict E_M in many of the tests indicates the correct trend.

6 SUMMARY AND CONCLUSIONS

The DMT is quickly earning a place in the geotechnical profession as a cost-effective tool for conducting <u>routine</u> site investigations. The device has developed to a

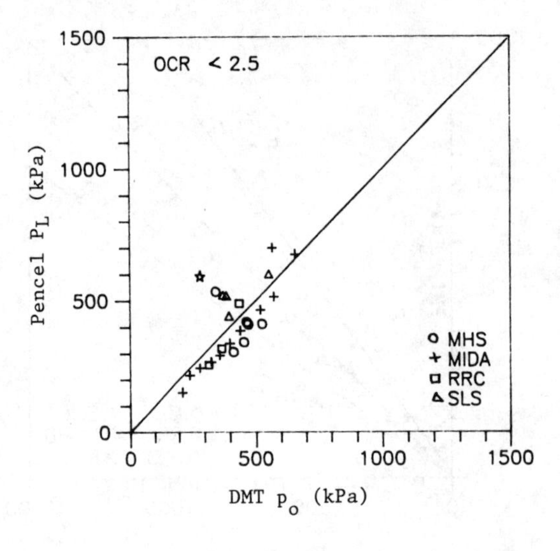

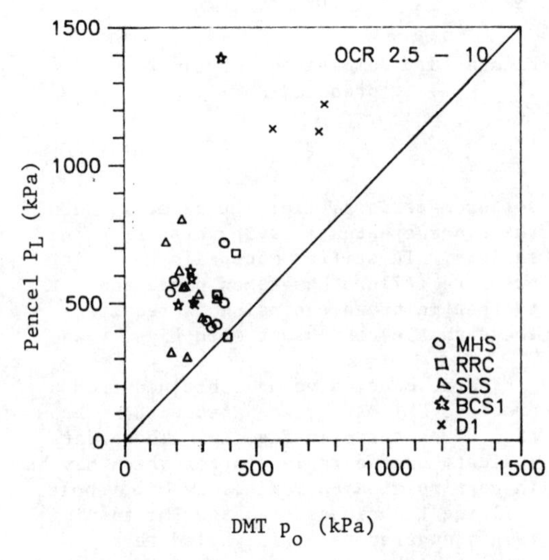

Figure 19. Comparison Between P_L (full displacement PMT) and p_o

point where there now appears to be substantial theoretical justification for many of the empirical correlations proposed by Marchetti. Certainly, the use of the instrument will continue to expand into new areas with more general use. The standardization of the test by ASTM, which is currently underway, comes at a timely point and will help strengthen the development and accuracy of correlations by reducing test variability.

The writer would like to offer the following concluding remarks about the test and interpretation:

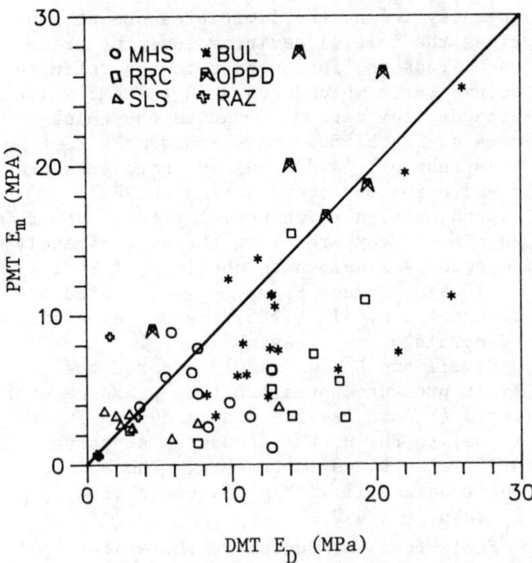

Figure 20. Comparison Between E_M (prebored PMT) and E_D

(1) The DMT is a simple and efficient test for estimating a number of soil engineering properties, and may also be used as an extremely useful logging device since it allows closely spaced vertical test points. In specific applications the test may be even more efficient if the engineer chooses to test at a few preselected depths.
(2) Use of the DMT in a wide range of earth materials has been demonstrated with nearly all materials providing reasonable test results.
(3) In soft, saturated, cohesive materials, the penetration of the blade creates a cavity expansion failure condition leading to the generation of large excess positive pore water pressures. Therefore, estimates of undrained shear strength in these materials using p_o are very accurate. In stiffer overconsolidated materials further expansion of the DMT diaphragm is required to create a failure, therefore estimates of undrained strength, based on p_o, are somewhat lower than predicted by other tests.
(4) Prediction of soil stress history relies on the p_o measurement and appears to be more accurate in normally consolidated and lightly overconsolidated materials for similar reasons given in (3) above.

(5) Estimates of other soil parameters e.g., deformation modulus, K_o and drained friction angle of sands, are generally very good for most materials.
(6) The use of DMT results for design is well established and encompasses a wide range of common geotechnical problems. Predictions for both deformation and limit equilibrium problems, which rely not only on the quality of input parameters but also on a mathematical model for a given problem are excellent.
(7) The recent addition of a controlled deflection C-Reading following the conventional A and B pressure readings allows for determination of an additional index parameter , U_D. This parameter which approximates total pore pressure around the DMT blade appears to have significant merit in determining site stratigraphy and stress history and may have other uses, e.g., determining undrained strength. Therefore, U_D may provide an alternative assessment of soil engineering properties to compare with current procedures. Because of it's apparent usefulness, the writer recommends that it become a routine measurement.
(8) It appears that by using different testing procedures, it is possible to conduct both total stress and pore pressure dissipation tests using the DMT. The results of these tests are useful for practical problems such as estimating effective stress axial pile capacity and designs involving horizontal drainage.

7 ACKNOWLEDGEMENTS

The work conducted by the writer with the DMT during the past five years has primarily been funded by the Geotechnical Research Lab, Clarkson University. A number of coworkers have contributed to the work in various stages. The writer wishes to thank them for their efforts; J. Dickson, M. Kabir, B. Saber, S. Saye, B. Smith, K. Tierney, D. Timian. The writer would also like to thank Dr. J. Schmertmann for his helpful suggestions on this manuscript and various discussions concerning the DMT.

REFERENCES

Baldi, G., Bellotti, R., Ghionna, V., Jamiolkowski, M., Marchetti, S. and Pasqualini, E. 1986. Flat Dilatometer Tests in Calibration Chambers. Use of In Situ Tests in Geotechnical Engineering, ASCE: 431-446.

Baligh, M.M. 1985. The Strain Path Method. Jour. of the Geotech. Div., ASCE, Vol. 111:1108-1136.

Baligh, M.M. and Levadous, J.N. 1986. Consolidation After Undrained Piezocone Penetration. II: Interpretation. Jour. of the Geotech. Div., ASCE, Vol. 112: 727-745.

Baligh, M.M., Martin, R.T., Azzouz, A.S. and Morrison, M.J. 1985. The Piezo-Lateral Stress Cell. Proc., 11th Int. Conf. on Soil Mech. and Found. Engr., Vol. 2:841-844.

Bechai, M., Law, K.T., Cragg, C.B.H. and Konrad, J.M. 1986. In Situ Testing of Marine Clay for Towerline Foundatons. Proc., 39th Can. Geotech. Conf.:115-119.

Borden, R.H., Aziz, C.N., Lowder, W.M. and Khosla, N.P. 1985. Evaluation of Pavement Subgrade Support Characteristics by Dilatometer Test. Trans. Res. Rec. No. 1022:120-127.

Burgess, N. 1983. Use of the Flat Dilatometer in the Beaufort Sea. Proc., 1st Int. Conf. on the Flat Dilatometer, Edmonton, Mobile Augers and Research Ltd.

Campanella, R. and Robertson, P.K. 1983. Flat Plate DMT: Research at UBC. Proc., 1st Int. Conf. on the Flat Dilatometer, Edmonton, Mobile Augers and Research Ltd.

Campanella, R., Robertson, P., Gillespie, D. and Grieg, J. 1985. Recent Developments in In Situ Testing of Soils. Proc., 11th Int. Conf. on Soil Mech and Found. Engr., Vol. 2:849-854.

Chan, A.C.Y. and Morgenstern, N.R. 1986. Measurement of Lateral Stresses in a Lucustrine Clay Deposit. Proc., 39th Can. Geotech. Conf.:285-290.

Clough, G.W. and Goeke, P.M. 1986. In Situ Testing for Lock and Dam 26 Cellular Cofferdam. Use of In Situ Tests in Geotechnical Engineering, ASCE:131-145.

Davidson, J.L. and Boghrat, A. 1983. Flat Dilatometer Testing in Florida. Proc., Inter. Symp. on In Situ Testing of Soil and Rock, Paris, Vol. 2:251-255.

Durgunoglu, H.T. and Mitchell, J.K. 1975. Static Penetration Resistance of Soils: II - Evaluation of Theory and Implications for Practice. In Situ Measurement of Soil Properties, ASCE, Vol. 1:172-189.

Fabius, M. 1985. Experience with the Dilatometer in Routine Geotechnical Design. Proc., 38th Can. Geotech. Conf.

GPE, Inc. 1983. Dilatometer Digest No. 1.

GPE, Inc. 1984. Dilatometer Digest No. 4.

Greig, J.W., Campanella, R.G. and Robertson, P.K. 1986. Comparison of Field Vane

Results with Other In Situ Test Results. Soil Mech. Series No. 106, Dept. of C.E., Univ. of British Columbia.

Gupta, R.C. and Davidson, J.L. 1986. Piezoprobe Determined Coefficient of Consolidation. Soils and Foundations, Vol. 26:12-22.

Hammandshiev, K.B. and Lutenegger, A.J. 1985. Study of OCR of Loess by Flat Dilatometer. Proc., 12th Int. Conf. on Soil Mech. and Found. Engr., Vol. 4: 2409-2414.

Hayes, J.A. 1983. Case Histories Involving the Flat Dilatometer. Proc., 1st Int. Conf. on the Flat Dilatometer, Edmonton, Mobile Augers and Research, Ltd.

Hayes, J.A. 1986. Comparison of Dilatometer Test Results with Observed Settlement of Structures and Earthwork. Proc., 39th Can. Geotech. Conf.:311-316.

Janbu, N. 1963. Soil Compressibility as Determined by Oedometer and Triaxial Tests. Proc., 3rd Europ. Conf. on Soil Mech. and Found. Engr.:19-25.

Janbu, N. 1967. Settlement Calculations Based on the Tangent Modulus Concept. Three guest lectures at Moscow State Univ., Bulletin No. 2, Soil Mechanics, Norwegian Institute of Tech.:1-57.

Juran, I. and Beech 1986. Effective Stress Analysis of Soil Response in a Pressuremeter Test. The Pressuremeter and Its Marine Applications, STP 950 ASTM:150-168.

Kaderabek, T.J., Barreiro, D. and Call, M.A. 1986. In Situ Tests on a Florida Peat. Use of In Situ Tests in Geotechnical Engineering, ASCE:649-667.

Khera, R.P. and Schultz, H. 1984. Past Consolidation Stress Estimates in Cretaceous Clay. Journ. Geotech. Engr., ASCE, Vol. 110:189-202.

Konrad, J.-M., Bozozuk, M., and Law, K.T. 1985. Study of In-situ Test Methods in Deltaic Silt. Proc., 11th Int. Conf. on Soil Mech. and Found. Engr., Vol. 2: 879-886.

Lacasse, S. and Lunne, T. 1983. Dilatometer Tests in Two Soft Marine Clays. Norwegian Geotechnical Institute Publ. No. 146:1-8.

Lacasse, S. and Lunne, T. 1986. Dilatometer Tests in Sand. Use of In Situ Tests in Geotechnical Engineering, ASCE:686-699.

Lacasse, S. and Lunne, T. 1988. Offshore Dilatometer Test. (paper submitted to 1st ISOPT).

Ladanyi, B. 1963. Expansion of a Cavity in a Saturated Clay Medium. Jour. of The Soil Mech. and Found. Div., ASCE, Vol. 89:127-161.

Ladd, C.C., Foott, R., Ishihara, K., Schlosser, F. and Poulos, H.G. 1977. Stress-Deformation and Strength Characteristics. Proc., 9th Int. Conf. on Soil Mech. and Found. Engr., Vol. 2:421-494.

Lutenegger, A.J. 1986. Application of Dynamic Compaction in Friable Loess. Jour. of Geotech. Engr., ASCE, Vol. 112: 663-667.

Lutenegger, A.J. 1987. Flat Dilatometer Tests in Leda Clay. Dept. of CEE Report No. 87-2, Clarkson University, Potsdam, NY.

Lutenegger, A.J. and Donchev, P. 1983. Flat Dilatometer Testing in Some Meta-Stable Loess Soils. Proc., Inter. Symp. on In Situ Testing of Soil and Rock, Paris, Vol. 2:337-340.

Lutenegger, A.J. and Timian, D.A. 1986. Flat-Plate Penetrometer Tests in Marine Clays. Proc., 39th Can. Geotech. Conf.: 301-309.

Lutenegger, A.J. and Kabir, M.G. 1987. Pore Pressure Generated by Two Penetrometers in Clay. Dept. of CEE Report No. 87-2, Clarkson University, Potsdam, NY.

Lutenegger, A.J. and Kabir, M. 1988. Use of Dilatometer C-Reading (paper submitted to 1st ISOPT).

Lutenegger, A.J., Smith, B.L. and Kabir, M.G. 1988. Use of In Situ Tests to Predict Uplift Performance of Multihelix Anchors. (paper submitted to ASCE Symp. on Special Topics in Foundations).

Marchetti, S. 1975. A In Situ Test for the Measurement of Horizonatl Soil Deformability. In Situ Measurement of Soil Properties, ASCE, Vol. 2:255-259.

Marchetti, S. 1980. In Situ Tests by Flat Dilatometer. Jour. of the Geotech. Engr. Div. ASCE, Vol. 106: 299-321.

Marchetti, S. 1982. Detection of Liquefiable Sand Layers by Means of Quasi-Static Penetration Tests. Proc., 2nd ESOPT, Vol. 2:689-695.

Marchetti, S. 1985. On the Field Determination of K_o in Sand. Panel Presentation Session: In Situ Testing Techniques, Proc., 11th Int. Conf. on Soil Mech. and Found Engr.

Marchetti, S., Totani, G., Campanella, R.G., Robertson, P.K. and Taddei, B. 1986. The DMT-σ_{HC} Method for Piles Driven in Clay. Use of In Situ Tests in Geotechnical Engineering, ASCE:765-779.

Massarsch, K.R. 1975. New Method for Measurement of Lateral Earth Pressures in Cohesive Soils. Canadian Geotech. Jour., Vol. 12:142-146.

Mayne, P.W. 1987. Determining Preconsolidation Stress and Penetration Pore Pressures from DMT Contact Pressures. Geotech. Testing Journal. GTJODJ, Vol. 10:146-150.

Mesri, G. and Castro, A. 1987. C_α/C_c Concept and K_o During Secondary Compassion.

Jour. of Geotech. Engr. ASCE, Vol. 113: 230-247.

Ming-Fang, C. 1986. The Flat Dilatometer Test and Its Application to Two Singapore Clays. Proc., 4th Int. Geotech. Seminar on Field Instrumentatino and In Situ Measurements, Singapore:85-101.

Minkov, M., Karachorov, P., Donchev, P. and Genov, R. 1984. Field Tests of Soft Saturated Soils. Proc., 6th Conf. on Soil Mech. and Found. Engr., Budapest: 205-212.

Powell, J.J.M. and Uglow, I.M. 1986. Dilatometer Testing in Stiff Overconsolidated Clays. Proc., 39th Can. Geotech. Conf.:317-326.

Robertson, P., and Campanella, R.G. 1983. Interpretation of Cone Penetration Tests. Part II: Clay. Can. Geotech. Jour., Vol. 20:734-745.

Robertson, P.K. and Campanella, R.G. 1986. Estimating Liquefaction Potential of Sands Using the Flat Plate Dilatometer. Geotech. Test. Jour. GTODJ, Vol. 9:38-40.

Robertson, P.K., Campanella, R.G., Lunne, T. and Tronda 1988. Excess Pore Pressures and the DMT. (paper submitted to 1st ISOPT).

Robertson, P.K., Davies, M.P. and Campanella, R.G. 1988. Design of Laterally Loaded Driven Piles Using the Flat Plate Dilatometer. (paper submitted to Geotechnical Testing Journal, ASTM).

Saye, S.R. 1986. In Situ Testing at Westside Reservoir, Omaha, Nebraska. Unpubl. report presented to Omaha Section-ASCE Annual Geotech. Conf.

Saye, S.R. and Lutenegger, A.J. 1988a. Performance of Two Grain Bins Founded on Compressible Alluvium. (paper submitted to ASCE Symp. Special Topics in Foundations).

Saye, S.R. and Lutenegger, A.J. 1988b. Stress History Evaluation Beneath Surcharge Fills in Stiff Alluvium Using the DMT (paper submitted to 1st ISOPT).

Schmertmann, J.H. 1981. Discussion of In Situ Tests by Flat Dilatometer. Jour. of the Geotech. Engr. Div., ASCE, Vol. 107:831-832.

Schmertmann, J.H. 1982. A Method for Determining the Friction Angle in Sands from the Marchetti Dilatometer Test. Proc., 2nd ESOPT, Vol. 2:853-861.

Schmertmann, J.H. 1985. Measure and Use the In Situ Lateral Stress. in The Practice of Foundation Engineering, Dept. of Civil Engr., Northwestern Univ.: 189-213.

Schmertmann, J.H. 1986a. Suggested Method for Performing the Flat Dilatometer Test. Geotechnical Testing Journal, GTJODJ, Vol. 9:93-101.

Schmertmann, J.H. 1986b. Dilatometer to Compute Foundation Settlement. Use of In Situ Tests in Geotechnical Engineering, ASCE:303-321.

Schmertmann, J.H. 1986c. Some 1985-6 Developments in Dilatometer Testing and Analysis. Proc., Innovations in Geotechnical Engineering, Harrisburg, PA.

Schmertmann, J.H. and Crapps, D.K. 1983. Use of In Situ Penetration Tests to Aid Pile Design and Installation. Proc., Geopile '83, Associated Pile and Fitting Corp.:27-47.

Schmertmann, J.H., Baker, W., Gupta, R. and Kessler, K. 1986. CPT/DMT QC of Ground Modification at a Power Plant. Use of In Situ Tests in Geotechnical Engineering, ASCE:985-1001.

Simons, N.E. 1965. Consolidation Investigation on Undisturbed Fornebu Clay. Norwegian Geotech. Inst. Publ. No. 62.

Sonnenfeld, S., Schmertmann, J. and Williams, R. 1985. A Bridge Site Investigation Using SPT's, MPMT's and DMT's from Barges. Strength Testing of Marine Sediments, STP 883, ASTM: 515-535.

Tavenas, F., Blanchette, G., Leroueil, S., Roy, M. and LaRochelle, P. 1975. Difficulties in the In Situ Determination of K_o in Soft Senstive Clays. In Situ Measurement of Soil Properties, ASCE, Vol. 1:450-476.

Tedd, P. and Charles, J.A. 1981. In Situ Measurement of Horizontal Stress in Overconsolidated Clay Using Push-in Space-Shaped Pressure Cells. Geotechnique, Vol. 31:554-558.

Torstensson, B.A. 1975. Pore Pressure Sounding Instrument. Proc., ASCE Spec. Conf. on In Situ Measurement of Soil Properties, Vol. 2:48-54.

Vesic, A. 1972. Expansion of Cavities in Infinite Soil Masses. Jour. of Soil Mech. and Found. Div., ASCE, Vol. 98: 265-290.

Zeevart, L. 1953. Theories and Hypotheses of General Character, Soil Properties, Soil Classification, Engineering Geology. Discussion, Proc. 3rd Int. Conf. on Soil Mech. and Found. Engr., Vol. 3:113-114.

Penetration Testing 1988, ISOPT-1, De Ruiter (ed.)
© 1988 Balkema, Rotterdam, ISBN 90 6191 801 4

History of soil penetration testing

Bengt B.Broms
Nanyang Technological Institute, Singapore
Nils Flodin
Royal Institute of Technology, Stockholm, Sweden

ABSTRACT: The development of dynamic and static penetration testing methods has been reviewed. A large number of different tools are now used for the investigation in-situ of the shear strength and the deformation properties of both soils and rocks as described in the paper.

1 INTRODUCTION

1.1 Penetration Testing Methods

A large number of different static and dynamic penetration tests are used today as described by Sanglerat (1972). The most common are :
- o Standard penetration test (SPT)
- o Cone penetration test (CPT)
- o Weight sounding test (WST)
- o Ram sounding test (DPA and DPB)

European reference standards have been adopted by ISSMFE in 1977 for these methods.

Penetrometers are generally used in Europe during the exploratory phase of a soil investigation to determine the soil conditions in general such as the depth, thickness and lateral extent of the various strata so that an evaluation of different possible foundation methods can be made such as spread footings, rafts, piles or caissons or if it is possible to improve the soil conditions e.g. by preloading, with excavation and replacement, stone columns, pressure berms, lime or cement columns or with embankment piles. In the Scandinavian countries and Finland the weight penetrometer (WST) is common. This method is very fast and inexpensive and a large number of soundings can be carried out within a short time. In Holland, the Dutch Cone Penetrometer (CPT) is used for the same purpose while in U.K., Germany, Spain, Portugal, France, Italy and Greece different types of dynamic penetrometers are often utilized (SPT and DP) because of the limited penetration depth of CPT in dense or hard soils.

Penetration tests are also very valuable during the detailed exploration phase especially in silt, sand and gravel so that the compressibility of the soil in the different strata can be estimated. It is also possible to get an indication of the shear strength so that the ultimate bearing capacity of footings and piles can be assessed. In most cases the settlements will govern the design rather than the ultimate bearing capacity of the soil.

Electrical cone penetrometers and pore pressure probes (piezocones) are mainly used during the detailed exploration phase. These penetrometers are relatively delicate and can easily be damaged by stones or boulders in the soil. The maximum capacity and the maximum depth of the electrical cone penetrometers are limited. It is difficult or not possible in most cases to penetrate very dense or cemented layers. Predrilling or precoring may be required.

1.2 Design Values

Dynamic penetrometers are generally used to estimate the ultimate bearing capacity of piles and of caissons. With CPT or WST it is often not possible to reach the required depth. Therefore Standard penetration tests (SPT) or ram soundings (DPA or DPB) are used.

Design parameters cannot as a rule be determined from the penetration resistance alone. In most cases, the results from penetration tests had to be supplemented by borings and laboratory tests using undisturbed samples. Field vane tests, fall-cone tests, unconfined compression tests, direct shear tests or undrained triaxial tests are usually required to determine the undrained shear strength.

The results from penetration tests can according to the proposed new Eurocode (1986) be used to evaluate the design values required in either :

- o Ultimate limit state design or
- o Serviceability limit state design

Design values are generally chosen such that the probability of lower values is very unlikely. However, some of the factors that affect the design values are difficult to evaluate from static or dynamic penetration tests such as the location of the ground water level, the stress history and the stratificaiton of the soil. Also such factors as the experience and care of the operator can affect the results. The accuracy of the calculation method which is used to evaluate the bearing capacity or the settlement and the consequences when the limit state is exceeded should also be considered as well as any warnings before the limit state has been reached e.g. large deformations, so that precautionary measures can be taken in time.

The ultimate bearing capacity of footings on granular soils is generally estimated from the following equation when the applied load is vertical and the footing is concentrically loaded

$$Q_{ult} = D\gamma N_q + 0.5 B\gamma N_\gamma \quad (1)$$

where N_γ and N_q are bearing capacity factors, D is the depth to the foundation level of the footing, B is the width and γ is the unit weight of the soil. The unit weight is affected by the location of the ground water level.

The settlement of a spread footing sand and gravel is in general estimated from :

- o the <u>modulus of elasticity</u> or an estimated <u>shear modulus</u> from field measurements or laboratory tests using elastic theory or from
- o the <u>penetration resistance</u> as measured by cone penetration tests (CPT), Standard penetration tests (SPT) or weight soundings (WST).

The settlement of a footing is often calculated from the general equation :

$$s = \frac{q\,B\,f}{E_m} \quad (2)$$

where q is the average serviceability limit state bearing pressure, B is the width, E_m is a general stiffness parameter (equivalent modulus of elasticity or compression modulus) and f is a coefficient that depends on the shape and dimensions of the loaded area, the thickness of the compressible layers and on Poissons ratio. This equation has been used by e.g. Parry (1971) and D'Appolonia et al (1970) to estimate the settlement of footings on sand. Values on E_m should preferably be obtained from settlement observations of adjacent structures. When settlement observations are not available then the results from plate load tests, pressuremeter tests and from penetration tests such as CPT and STP are used. Local experience is important.

Penetration tests (CPT, SPT, WST) are commonly used to check the compaction of fills especially below the ground water level (see e.g. Choa et al, 1979, Mitchell, 1986).

There have been two specialty conferences on penetration testing (ESOPT 1 in 1974 and ESOPT II in 1982) where applications and recent advances were discussed. The development up to 1972 have been summarized by Sanglerat (1972). An international symposium is planned for 1988 in Florida, USA (ISOPT 1).

1.3 Standardization

The interest in standardization of commonly used penetration testing methods dates back to the 4th International Conference on Soil Mechanics and Foundation Engineering which was held in London in 1957. A subcommittee was then appointed with the task to study both static and dynamic penetration testing methods with respect to possible standardization. In Paris in 1961, Prof M Vargas, then the Chairman of the Sub-committee, reported that the Standard penetration test (SPT) as it was used at that time was far from being a standard method and that the procedure to carry out the test varied between different countries. The committee felt that there was a real need for standardization but could not agree on any recommendations.

The work of the committee was summarized in three reports. They suggested that the work should continue on a regional basis. A European subcommittee with Dr H Zweck, as Chairman, was therefore appointed at the 6th International Conference in Montreal in 1965 with the task to continue with the work on standards. A report on the work of the committee was published as well as recommendations for static and dynamic penetration tests.

The ISSMFE European Sub-committee on Standardization of Penetration Testing presented its final report at 9th ICSMFE in Tokyo 1977 including reference standards for CPT, SPT, WST and DP. The proposal was approved by the ISSMFE Council with the recommendation that it also should be used outside of Europe.

At the 10th ICSMFE in Stockholm in 1981 the Committee repeated its recommendations that "papers to international conferences or journals presenting results from penetration tests should also include results from at least one recommended standard penetration testing method." The Sub-committee also recommended that "comparisons between the different recommended standard testing methods should be made in different soils to facilitate the evaluation of soil characteristics from different penetration tests."

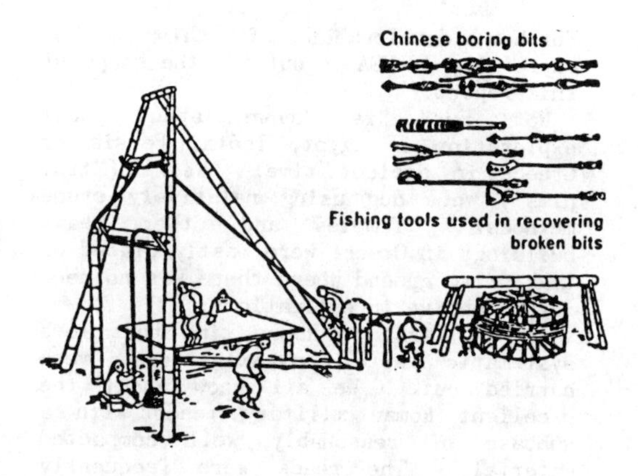

Fig 1a Deep boring 2000 years ago.

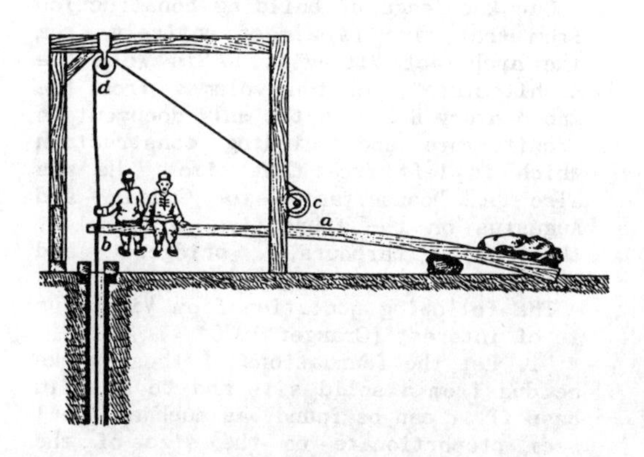

Fig 1b Simple shallow boring, 15th century

2 EARLY HISTORY OF SOIL PENETRATION TESTING

2.1 Ancient Time

The Chinese were undoubtedly pioneers with respect to drilling of soil and rock. Already 2,000 years ago. they introduced the so-called cable-tool method in their search for salt, water and minerals. Boreholes with a diameter of 120 to 150 mm down to 500 a 600 m depth were common, using heavy chisels fixed to bamboo-fibre ropes of varying lengths. The upper part of the ropes was connected to a large drum with a diameter of up to 5 m. Depths exceeding 1000 m could be reached.

The chiesel was lifted by a boring crew in different ways, as shown in Fig 1 (Gaskell 1969, Tecklenburg 1886/1896, Nachmanson and Sundberg 1936, respectively). At shallow depths the hoisting was made by hand while a draught animal was used when the depth was large (Fig 1a). In Fig 1b the crew consisting of up to 10 men, is sitting on a resilent wooden beam. They had to be relieved every ten minutes. The percussion rate was 50 blows/min and the height of fall about 100 mm. For the other methods, 12 to 15 blows/min have been mentioned.

The cleaning of the borehole and the sampling was done with a bailer.. Kerisel (1985) has mentioned that a spoon-sampler known as a Loyang spoon was used very early. The boreholes were stabilized with a casing of bamboo which was wrapped in canvas as shown in Fig 1d.

The cable-tool method, which was brought to Europe in the first half of the 18th century, has gradually been improved. In the 19th century, the human power was replaced by steam engines. The further development of this method especially in USA is outside the scope of this report.

Not much is known about soil exploration in Egypt, India, Persia or Greece in ancient time. Wells ("test pits") were dug using relatively crude methods. Temples and other heavy buildings in Greece were mostly placed on high, firm ground where there was no need for extensive investigations.

It was first in Roman time that more systematic site investigations were carried out. We all know about the excellent Roman military roads with a subbase of reasonably well compacted material. The roads were frequently constructed in marshy areas, far from the homeland and bridges must often be supported on wooden piles which still in many cases are in good conditions.

Our knowledge of building construction from that time is almost entirely from the architect Vitruvius. His work "De Architectura", in ten volumes from the 2nd century B.C., is the only document on architecture and building construction which is left from that time. He was also a "consultant" to Caesar and Augustus on the foundation of temples, theatres, harbours, bridges and cofferdams.

The following quotation from Vitruvius is of interest (Granger 1970) :

"1. Let the foundations of those works be dug from a solid site and to a solid base if it can be found, as much as shall seem proportionate to the size of the work; and let the whole site be worked into a structure as solid as possible. And let walls be built, upon the ground under the columns, one-half thicker than the columns are to be, so that the lower portions are stronger than the higher;...

2. But if a solid foundation is not found, and the site is loose earth right down, or marshy, then it is to be excavated and cleared and re-made with piles of alder or of olive or charred oak, and the piles are to be driven close together by machinery, and the intervals between are to be filled with charcoal."

This is still good advice. In fact, many centuries would pass before any kind of soil exploratory tool and soil investigation methods appeared.

2.2 Development from the 15th century to about 1700

Many churches, cathedrals and other heavy buildings were built before the 15th century even on weak soils. These buildings were constructed by the "trial and error-method" and with a "calculated risk" and experience was gained very slowly. The hand-written documents before Gutenberg did not deal very much with such "down to earth problems" as the foundation of buildings and soil investigations.

Fig 1c Boring using a light chisel

Fig 1d Bamboo-casing tube wrapped in canvas

An excessive number of piles were probably used in many cases to be safe. The first few were usually "test piles" to determine the required length. Pile driving can be regarded as an early type of penetration testing. Anyhow, it is surprising to see that so many heavy buildings are still standing, apparently in "good order". However, there are some leaning towers from that time!

In the 15th century, at the end of the "dark" medieval time, there were some new developments in Europe with respect to soil exploration, as illustrated by the peculiar screw auger shown in Fig 2 (Cambefort 1955). However, the purpose of this tool is not clear. (The tool has been attributed to the German military engineer Conrad Keyser aus Eichstatt (1366-1405?). It was reproduced in a work by B Gilles in 1964, as mentioned by J P Daxelhofer, 1987).

The genius of that time, Leonardo da Vinci (1452-1519), was also involved in "soil engineering" and has presented a sketch of an early soil auger as shown in Fig 3 (Kerisel 1985). The "instructions" of Leonardo de Vinci reads in translation as follows. "If you want to make a hole in the ground easily, you can do this with the device illustrated; the screw may be advanced into the ground by means of turning the lever m-n in a right hand manner. Having reached the desired depth, with m-n fixed, turn the lever f-g in the opposite sense : this will make the screw rise without turning, bringing the soil with it." This screw auger was probably not used during the lifetime of the inventor. The model shown in Fig 3 is probably of a later date.

The Italian architect Anarea Palladio (1508-1580) later took over Vitruvius all-embracing role. He presented four textbooks in 1570 and many of his architectural ideas are still of great interest (Atterbom and Branzell, 1928). He was, among other things, familiar with piling and other foundation problems and had some concept of what today is called "differential settlements" and how to prevent buildings from cracking. However, most of his ideas about soil exploration seems to be peculiar. He mentions, however, that certain species of grass would only grow where the ground

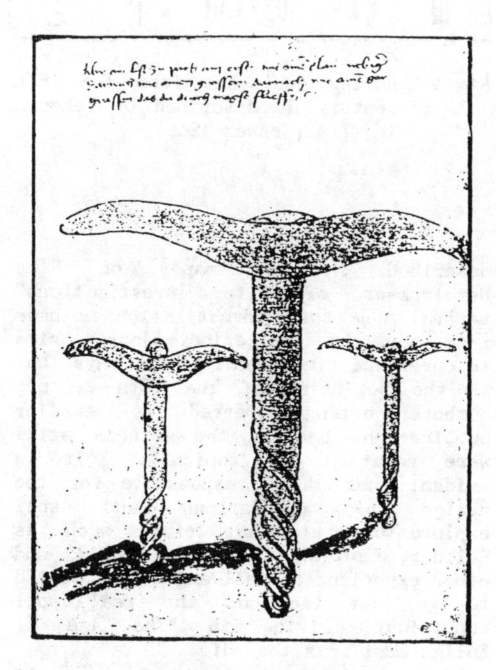

Fig 2 Screw auger from the 15th century (Cambefort 1955)

Fig 3 Sketch of a soil auger device by Leonardo da Vinci. Below a model of later design (Kerisel 1985)

is firm. Other indications of firm ground are "if ground does not respond or tremble if something heavy is dropped. In order to ascertain this, one can observe whether some drum-skins placed on the ground vibrate, give off a weak sound or whether the water in a vessel placed on the ground gets into motion."

At the end of the 17th century, a ram penetrometer was developed in Germany to evaluate the soil resistance as described by Zweck (1969) in his book "Baugrunduntersuchungen durch Sonden". He refers there to Nicolaus Goldmann (1699), a German "theorist of architecture" who lived between 1611 and 1665! Jensen (1969) also mentions Goldmann. (Most of the work by Goldmann was edited and published by his younger colleague L. C. Sturm. This means that Goldmann's penetrometer was developed earlier, probably between 1650 and 1660.) Zweck (1987) has kindly translated Goldmann's description of his method from an old German text: "Hereto on the site at each place, a pointed rod can be driven, and one can notice the penetration depth for each blow, and in this manner one can find differences in the subsoil."

This very interesting early contribution (a "solitaire") to the history of soil penetration testing was forgotten until Künzel, Paproth and others in Germany developed the light-ram penetration method in the 1930's and 1940's.

2.3 Development during the 18th century

Germany dominated the development of new soil exploratory methods in the 18th century. There were two "schools"

1. "The Continental School", i.e. countries with comparatively favourable soil conditions (friction soils), and
2. "The Nordic School", except Denmark but including the Netherlands and Canada, i.e., countries with deep deposits of soft cohesive soils.

The "Continental School" worked mainly on the "auger line" in the 18th and most of the 19th centuries while the "Nordic School" followed the more simple "rod line" where special equipment was used for soil exploration and sampling. Later, the two schools would merge.

In the 18th century several interesting penetrometers were developed as described by Jensen (1969) in his valuable work "Civil Engineering around 1700. With special reference to equipment and methods." Also Whyte (1976) has

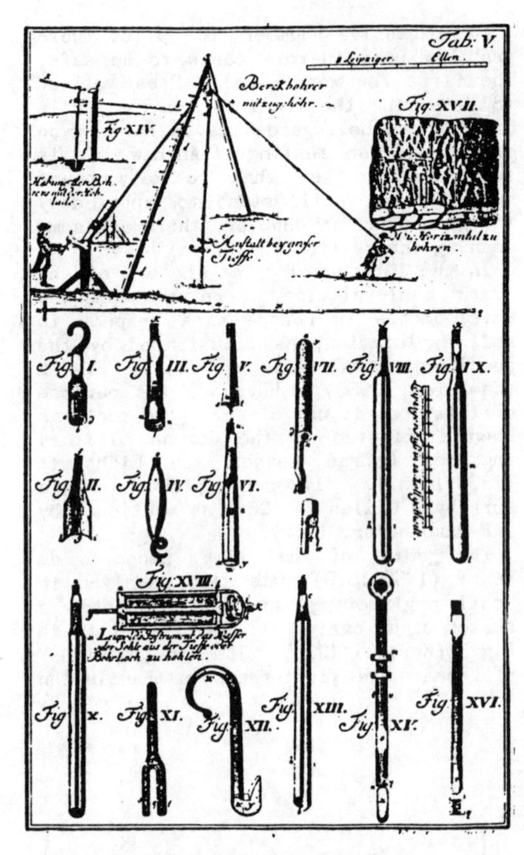

Fig 4 Boring equipment from the 18th century as described by Lehmann in 1714 (Jensen 1969)

described in his paper on "The Development of Site Investigations" several new soil investigation methods from about the same period. Jensen tries to cover the time from 1650 and says that in the beginning of the 18th century authors often referred to earlier publications because the methods still were regarded as "topical". It is evident that those responsible for the design of foundations and soil explorations at that time, such as Belidor, Gautier, Labelye, Leupold, and other experienced "soil engineers", tried to get an idea of the real soil conditions, living in "The Age of Enlightment", as they did.

A popular tool from around 1700 is shown in Fig 4. It was used in, e.g., England and a set was brought to Sweden by Triewald (1740). This tool is known in Sweden as "The English Earth Auger". However, a better name would be "The German Earth Auger" since it was already introduced in 1714 by the German physician and professor in Leipzig J. E. Lehmann (Jensen 1969). This soil exploration method, probably after some improvements, has been described by another German, namely J Leupold (1724), director of the mines in Saxony. Also the water sampler in Fig 4, XVIII, carries his signature (Leupold's name can be found e.g. on many drawings of machines, foundation constructions, piling and hydrology from the 18th century). He was also engaged on a project in Amsterdam in the 1720's concerned with the sinking of a well down to 76 m depth. An excellent record of the soil profile was then obtained.

H Gauthier (1660-1737), a French engineer who was active at about the same time mentions in his book on bridge construction (1716) that after a plan of the region has been prepared and sounding of the water depth has been made, the ground should be investigated with an iron probe (!). There are different soil sampling methods, he adds and continues: "Sometimes when the ground consists of such large stones that the probe cannot be got through, a 3 to 6 inches thick oak pile is used. An iron shoe with a long spike, to guide the pile between the stones, is fixed at the lower end, while at the upper end the pile is strengthened by an iron ring. The pile is driven down with a ram which can have one, two or three handles." This is, thus, a combination of probing and test piling. "No efforts must be spared in carrying out the investigation with the utmost diligence, but then there is the satisfaction of being able to draw in the profile of the ground, in which the bridge pillars will be founded."

Jensen (1969) has described the investigations of the river bottom for the Westminster Bridge in London 1738. Responsible for the work was Labelye (1705-1781). He was born in Switzerland but was active in England from about 1725. "A hard 3 to 4 feet thick gravel made it difficult to penetrate the river bed. The sharpest boring tool could only with difficulty penetrate down to 14 feet depth. The gravel particles increased in size with depth as well as the compactness."

A large number of borings was carried out about 100 feet upstream as described by Labelye where also "sand and mud quite a few feet in thickness were found, but there is a reason to suppose that these sand and mud deposits rest on the hard gravel". He was criticized for not taking samples, so that the character and the colour of the soil could be determined.

In a book with the title "A Short Account of the Method made use of in Laying the Foundations of Westminster Bridge", Labelye (1739) have some very interesting comments on the "reaction" of the drill rods during the investigation. "It has always been understood that the purpose of the borings is to estimate whether the nature and consistency of the ground (no matter what colour) allows the direct foundation of a substantial bridge. The sharp bore used gave perfect information in that respect. The nature and consistency of the soil could be determined by the resistance to the penetration of the boring rod. Vibrations in the rod, and the noise it gave, made it easy to decide whether the bore was in mud, sand, clay or gravel. This was checked by boring in other places and taking up samples. It transpired that these samples always confirmed what had been estimated from the sound (noise) and feel of the boring rod."

Labelye's comments could have been part of a manual on the Swedish Weight Sounding method which was developed almost two hundred years later. It should be added that one of the piers failed in 1746 when the bridge was almost completed. The bridge was opened for traffic in 1750 and has been admired for many years since then.

B.F. de Belidor (1697-1761), a French engineer and author of several books on, i.a., building techniques and hydraulic engineering, investigated the bearing capacity of soils in connection with "test piling". An iron rod was fixed to the foot of each pile by three or four fishplates. The shape of the lower end was such that a sample of soil could by brought up when the test pile was extracted. It is uncertain, however, if this method really can be classified as a penetration testing method.

Belidor indicates that a so-called "sound" is an iron rod which at the lower end is provided with a worm blade which would take a sample when turned. However, it should be called an auger rather than a "sound". A drawing by Diderot (1768), which speaks for itself, (Fig 5) concludes the developments in the 18th century in Central Europe.

Some notes on penetration testing were

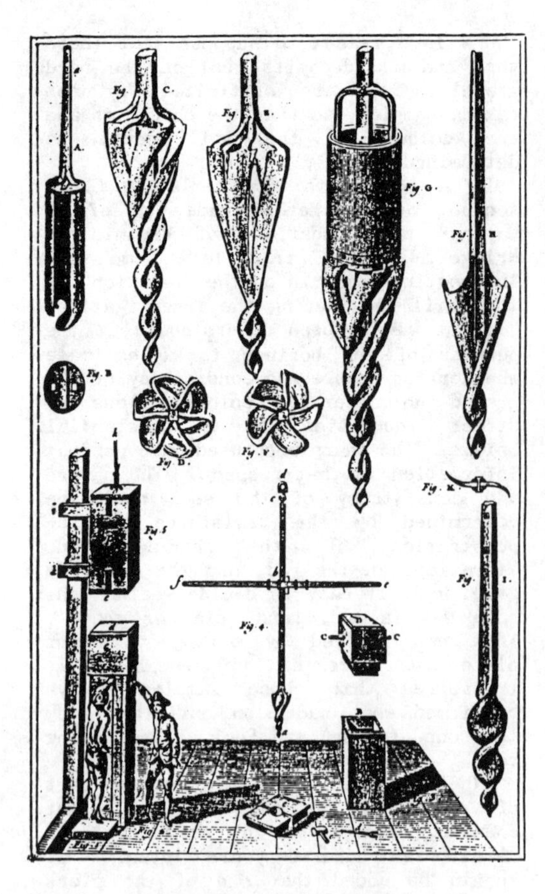

Fig 5 Soil penetration equipment from the 18th century according to Diderot (1768)

Fig 6 Sounding equipment used at the construction of the Gota Kanal in Sweden about 1810 (Flodin 1984)

published in Scandinavia in the 18th century. Nygård (1944) mentions that after the landslides at Sarpfossen in Norway 1722 and 1726, 14 to 16 "alen" (about 7 m) long rods were used at the site investigation without reaching a firm layer. Senneset (1974) describes in another example from Norway how 10 to 12 m long rods were forced into the ground in 1736 in an area near the River Glomma. The rods were either circular with 15 to 30 mm diameter or square with 28 mm side. Hexagonal and octagonal rods were also used. A commission consisting of three officers was in charge of the investigations. They characterized the ground conditions from the penetration resistance of the rods. Extensive soundings were also carried out in Drammen, Norway, in 1838 to investigate a deposit with soft clay along the unstable river bank (Senneset 1974).

Penetration tests were also carried out by Thomas Telford, the "Father of British Civil Engineering" in connection with the construction of the Caledonian Canal in England in 1804. He mentions that the soil consisted of soft mud into which an iron rod could be thrust 55 feet (about 18 m) (Telford 1830; Tomlinson 1956).

Telford was for a short time (20 days) also engaged in 1804 as a consultant for the Göta Kanal in Sweden. There he stressed the need for extensive site investigations. The soil conditions were investigated with 347 "test pits" down to a maximum depth of 4.2 m. A spit was used to probe the bottom. All results were carefully recorded. Sounding rods were also used as can be seen from Fig 6 (Flodin 1981).

2.4 Handbooks from the middle of the 19th century

In the middle of the 19th century several handbooks appeared in Europe, mainly in Germany, which included chapters on soil, rock and foundation engineering. The material in the Scandinavian handbooks was to a large extent "borrowed" from German handbooks. Thus the continental "firm-soil philosophy" prevailed in Scandinavia for a long time.

One of the first handbooks was written by Broch (1848), a captain in the Norway-Brigade. His book which is typical for its time was mainly concerned with military applications and "by translations and excerpts from works by the best and most recent authors". However, only a few augers were described in this book.

A remarkable handbook in six volumes on mainly rock and soil boring techniques was published in 1886/1896 by Tecklenburg, at that time chief mining engineer in Darmstadt. The title of the work is "Handbuch der Tiefbohrkunde". He describes in detail a large number of tools and methods, their origin and history. Also methods used outside of Europe are included. However, most of the work is concerned with the drilling of rock.

Sounding rods ("Sondiereisen" or "Visitireisen") were described for the first time in detail by Tecklenburg, as well as soundings as a method to determine the properties of both soil and rock, although vaguely. The sounding rod in its most simple version was provided with a point at the bottom and an eye at the top for a wooden cross bar intended for the driving and the withdrawal of the rod. The maximum length was about 4 m. It was recommended to drill some small inclined "pockets" at the lower end of the rod to get some soil samples from the bottom (cf. Strukel and Fig 8). It is indicated that one can get an impression of the strength and the bearing capacity of the various soil strata from the driving resistance.

The rods were generally of solid iron (steel?). Wooden rods were provided with iron fittings and special joints to reduce the wear (cf. Fig 8). Square rods, 25 to 30 mm wide were generally used for shallow soundings while for deeper soundings up to 40 to 60 mm diameter round bars were common. Three 8 to 12 m long rods which were coupled together were also used. In exceptional cases the rods could be up to 24 m long. Two boring kits are shown in Fig 7, the upper one is known as the Tecklenburg kit (1886/1887).

Two Nordic handbooks which were more directly concerned with soil exploration and foundation engineering soon followed in the Nordic countries, one by Kolderup (1894), captain-engineer and teacher at the military university in Kristiania (Oslo), and one by Strukel (1895), visiting teacher and later professor at the Polytechnical Institute in Helsinki. It is obvious that both have studied Tecklenburg's "guide book" in detail. Especially Strukel referred to Tecklenburg's book.

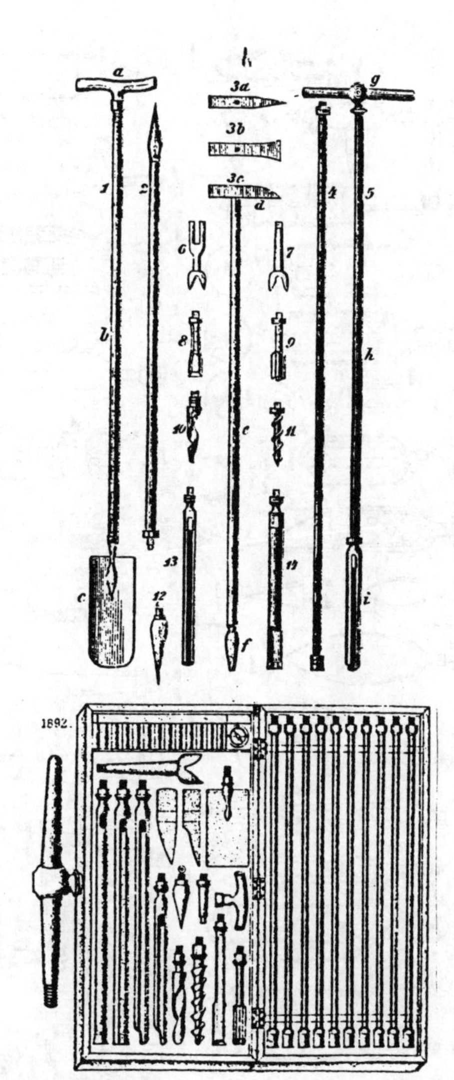

Fig 7 Two kits of hand-equipment for soil borings (Tecklenburg 1886/1896)

Kolderup has some interesting comments in his book about soundings, some of

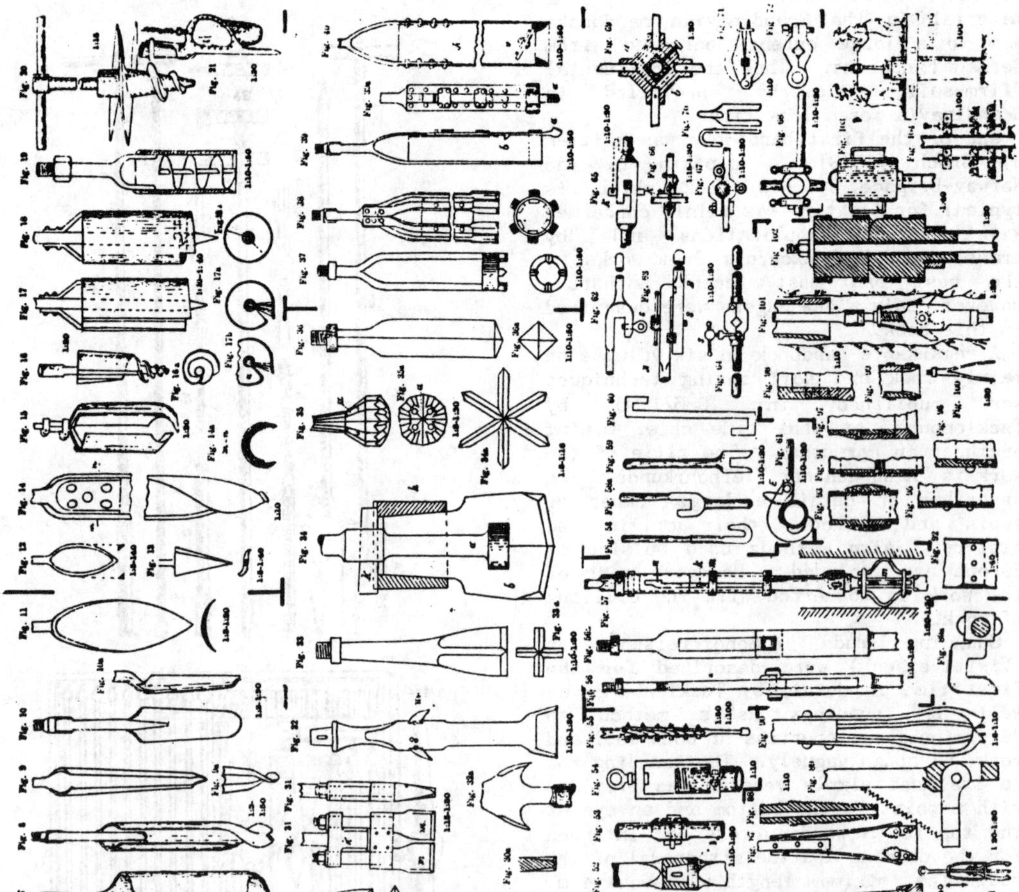

Fig 8 Collection of boring equipment used during the 19th century (Strukel, 1895)

which have already been mentioned. He states that some experience is required to judge the character of the soil and its bearing capacity and that these factors can be evaluated from the driving resistance. He, too, stressed the importance of the sound (noise) and the vibrations observed during the driving. The author also points out the importance of the trace of soil which is left on the rods after the withdrawal and the colour.

Strukel's handbook is, like the one by Tecklenburg, very detailed with respect to both text and illustrations. It can be seen in Fig 8 that this handbook contains a fascinating collection of different equipment for drilling and sampling of both soil and rock. Many of these can also be found in the six volumes by Tecklenburg. The collection is divided into several main groups. Of greatest interest is here 1-4: Sounding equipment; 49-53: Rods and couplings; and 62-69: Handles to be attached to the rods.

Strukel indicates that the lifting eye in (1) (cf. Tecklenburg above) is only for short rods. At greater lengths a movable handle is used as can be seen in (2) and 62-69. The diameter of the rods, according to Strukel, is 25 to 30 mm for shallow depths and 40 to 60 mm or more for large depths. It also mentioned that 8 to 10 m long rods can be used for deep soundings. In the second edition of Strukel's book published in 1906 a long rod consisting of several sections of different lengths is shown. The bottom rod is provided with an enlarged point. A total length of up to 24 m is indicated, together with scaffolding. The inclined holes for soil "sampling" mentioned above are also indicated (2a). Of special interest in this connection is the fact that the diameter of all couplings is larger than the diameter of the rods (49-51).

Several other handbooks from the 19th century present more or less the same equipment as those above, although not so much in detail. The well known "Hütte. Des Ingenieurs Taschenbuch" from 1872 does not have anything about field investigations. In the 1889 edition one can read: "For small projects ("Bauwerk") it is sufficient to use sounding rods ("Visitireisens") at the investigation, 1 to 2.5 m long, ca 3 cm thick and with a pointed iron rod, which for longer rods has an eye in the blunt end for the withdrawal using a cross bar". A few literature references were added.

2.5 Penetration testing

The "soft-clay countries" outside Europe, for example, Canada and USA should not be forgotten. There wash borings were for a long time the most commonly used drilling method, as described by Mohr (1943/44) and by Terzaghi (1953). A geotechnical "boom" started during the reconstruction of Chicago after the Great Fire in 1871. See, e.g. Baron (1941), Peck (1948, 1963) and Flodin (1985). In 1902 C.R. Gow introduced the so-called the dry sampling method, the embryo of the Standard Penetration Test, SPT. More about this method later.

Some remarkable penetration tests were carried out in soft soil in Canada in 1872 in connection with the construction of the Miramachi railway bridge in New Brunswick, 1871-1875 (Legget and Peckover 1973). It was found by the chief engineer Sir Stanford Fleming, "a notable Canadian civil engineer", that earlier site investigations were incorrect. He proposed a new soil investigation method where a steel rod was pushed down into the soil and where the required force was measured. The friction along the rod was eliminated by a 125 mm diameter steel casing. The rod used for the testing had a diameter of 75 mm and the end was blunt. The tests were carried out in water and the rod was loaded axially using weights. The time of the loading was varied. This was probably the first modern static penetrometer.

With the new method Fleming could determine the bearing capacity of the various strata and the design of the bridge was changed accordingly. However, Fleming's advanced penetration testing method was forgotten for almost 100 years, a fate which is not rare in geotechnical engineering.

In Scandinavia, in the 1890's, almost everyone in charge of a site investigation used his own penetration testing method and had his own method to evaluate the results, based on his own experience. The shape and the dimensions of the sounding rods varied greatly.

In Norway, several large landslides took place at that time in the Trondheim area (Trondelag). Among these was the huge landslide in Vaerdalen where 112 people lost their lives (Reusch 1901a). Friis (1898), a geologist, was given the task to carry out a thorough general "terrain investigation". A large number of boreholes were made, some up to 100 m deep using a "rather simple tool". The soil conditions varied from sand, gravel, stones and hard clay to very soft clay. Frequently quick clays and more or less remoulded materials were encountered. The consistency of the soil was described as that of soup. Other similar expressions were used to describe the different strata. Remarkably detailed records of the different boreholes were given. For one borehole, it was indicated that the rod sank only 10 cm after 600 revolutions! The soil conditions for another borehole at Vaerdalen (Borehole XXXVI), were described as follows :

4.00 m Gravel and hard clay alternatingly, then
1.00 m Coarse sand upon :
28.93 m of soft clay that the rod sank by its own weight. Thereafter followed
8.80 m Stiff clay strongly mixed with sand, and
2.24 m Coarse sand upon hard rock, 44.97 m from the ground surface

In an other report on a landslide at Morsil, Stjørdalen Valley in the Trondelag, Reusch (1901b) describes the boring equipment as follows :

1. A clumsy, heavy apparatus consisting of solid iron rods which could be coupled together and had a bottom rod formed like an extended pyramide, and
2. a slender hand tool.

When classifying the clay, Reusch used the following terms : soup, thin porridge, thick porridge, doughy bread (i.e. bread not well baked), stiff clay, and hard clay. He was also the first to use the English term "quick clay" (Reusch, 1901a).

Around the turn of this century, there was a hectic time in Sweden when the rail road along the West coast of Sweden was built and Gothenburg harbour with its deep deposits of soft clay and peat was enlarged. The many landslides and the failures of several quay structures and embankments that occurred at that time caused a public outcry for remedial measures and better site investigations.

Svensson (1899), a railway engineer, complained that the strength and the bearing capacity of soils could not be predicted scientifically, unlike other materials. He suggested that the penetration resistance should be measured more systematically and be expressed in terms of the number of men required to push down a sounding rod. Grade 0 was used to indicate when the rod sunk by its own weight, Grade 1 when one man was required, Grade 2 when two men were required etc. Half grades were also used.

Dahlberg (1913), a railway chief engineer proposed the following relationshhip between bearing capacity and the results from soundings using 25 mm rods :

Table 1. Relationship between Penetration Resistance and Bearing Capacity (after Dahlberg, 1913)

Penetration resistance (number of men)	Maximum height of an embankment in metres
< 1.5	< 0.5
1.5 - 1	0.5 - 1
1 - 2	1 - 2
2 - 3	2 - 5
3 - 4	5 - 15

Samuelsson (1926), another railway engineer, described how a 20 m deep deposit with soft clay below a railway embankment about 50 km south of Stockholm was investigated about 1912. It was specified :

<u>that</u> the total depths of the soft clay should be determined and
<u>that</u> some kind of resistance should be recorded during the penetration.

A round steel rod with 15 mm diameter was used during the preliminary phase. In the second final phase the soil conditions were investigated using a 25 mm octagonal rod. The resistance was indicated by the number of men required to push down the rod.

For the sake of completeness some words should be added about the "Gothenburg school". Wendel (1900), who at that time was head of the Harbour and River Works in Gothenburg, complained about the square sounding rods which were used to investigate soft cohesive soils. He pointed out that the protruding couplings gave trouble in, expecially, stony soils. He recommended a "new" probe which in some respects was similar to the one shown in Fig 7. The probe, called the "needle probe", was composed of several short pipe sections which were connected using inside couplings in order to give a smooth outside surface. Instead of expressing the soil resistance in terms of number of men, Wendel proposed to drive down the probe using a weight which was dropped from a predominated height and to measure the penetration for each blow.

A drawing signed by Lidén, a geologist, in 1914 shows a penetrometer consisting of 12.5 or 19 mm diameter pipes with inside couplings and with an extended smooth point as proposed by Wendel. In an accompanying letter Lidén mentions that he had used this tool in Gothenburg. Up to six men were required to reach 30 m depth.

Returning to the Swedish State Railways, a drawing from 1911, signed by Wolmar Fellenius, is an important link in the development of the Swedish weight sounding method. This drawing (Fig 9) shows a penetrometer consisting of 1.00 m long solid steel rods, 19 mm in diameter with outside couplings and a 0.80 m long lower rod which is provided with a twisted screw point, 0.20 m in length. A handle and a clamp and some other auxiliaries are also included (not shown here).

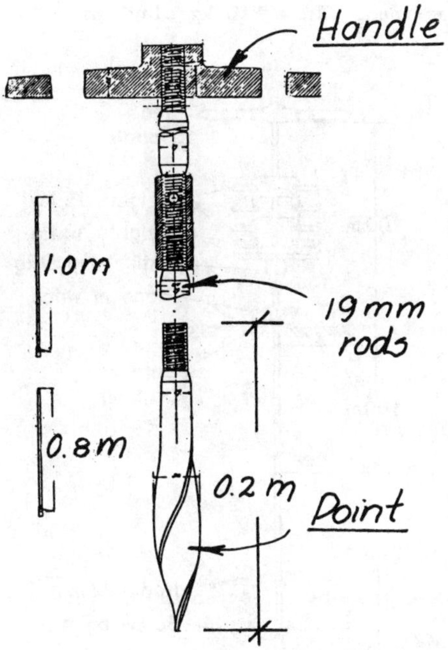

Fig 9 Swedish sounding tool from 1911 (After a drawing signed by Wolmar Fellenius)

A drawing in a handbook by Olsson, (1915) at the Swedish State Railways indicated some further developments. A penetration test was described as follows:

"The rod is first pressed down by the force of one man and the penetrated depth recorded, then by two men and the depth recorded. After that the 90 kg weight is fastened to the handle (total mass about 100 kg), the rod is rotated and the penetration for each 25 turns (half turns?) is recorded." The boring record in Fig 10 from Gothenburg exemplifies the method.

In another part of this handbook it was pointed out how an experienced person can get an impression of the character of the soil from the noise and the penetration resistance. We are now approaching the time when the Swedish weight sounding method as it is known today appeared.

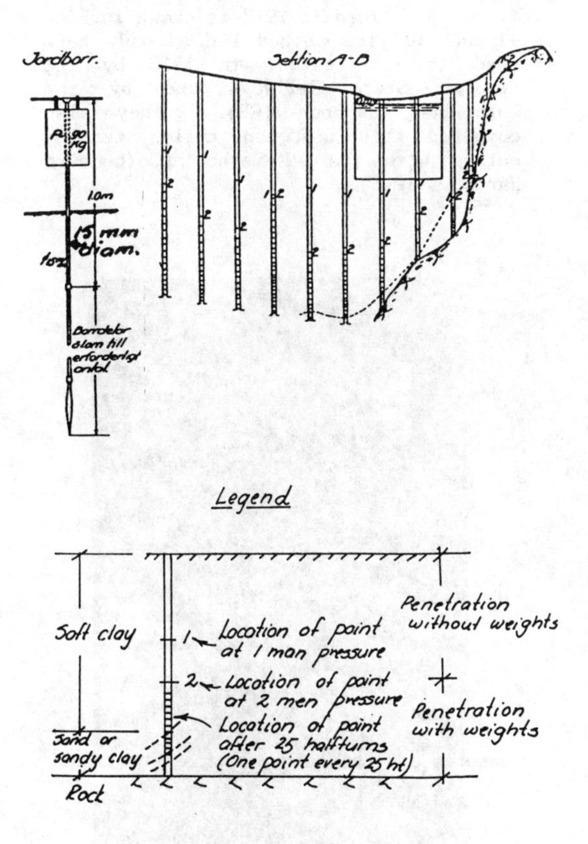

Fig 10 Swedish sounding tool about 1915 and example of recording the results (After H Olsson 1915)

3 THE WEIGHT SOUNDING METHOD (WST)

3.1 The Swedish weight sounding method of 1917/1922

One of the first tasks of the Swedish Geotechnical Commission in 1914 was to develop inexpensive and accurate field and laboratory soil investigation methods, among them a simple exploratory sounding method. In 1917, the Commission presented its first report with the title (in translation) "Soil Investigations for Railways" (Swedish Geotechnical Commission, 1917). John Olsson, as Secretary of the Commission, was the main author. Wolmar Fellenius, the father of the Swedish "Slip Circle" method, was a member of the Commission from the beginning and its Chairman from 1919.

The weight sounding method was described in detail in the 1922 Final Report of the Commission. The method as it was practised in 1917 is shown in Figs 11 and 12 (The method had already been used to some extent in 1916 by the Swedish State Railways and by the Gothenburg Harbour (GH). They also compared this new penetration testing method with the GH method (Gothenburg Harbour, 1916).

The diameter of the 1.0 m long steel rods had now been increased to 19 mm and a 0.2 m long screw shaped point was attached to the 0.8 m long bottom rod. The cross-section of the screw shaped point was square, 25 x 25 mm. It had been twisted one turn as can be seen in Fig 13. Threaded pins were used to splice the rods internally. This was a significant advancement since the surface of the penetrometer rods was smooth.

It can be seen from Figs 11 and 12 that the single 90 kg weight had been replaced by cast iron plates (two 10 kg weights and three 25 kg weights). The total mass including the 5 kg clamp was 100 kg. The penetrometer was gradually loaded during a test starting with the clamp, and then with the 10 kg and the 25 kg weights. The penetration after each load increment 0, 0.15, 0.25, 0.50, 0.75 and 1.00 kN, was measured and recorded. (0 kN means that the rods sank under its own weight). The load was adjusted in order to keep the penetration rate constant, about 20 mm/s. In the case when there was a soft layer below a stiffer layer, the probe had to be unloaded until the prescribed penetration rate was reached (20 mm/s). The consistancy of the different clay layers could now be defined. In a "15 kg clay" or a "75 kg

Fig 11 Weight sounding penetrometer in practice, 1917

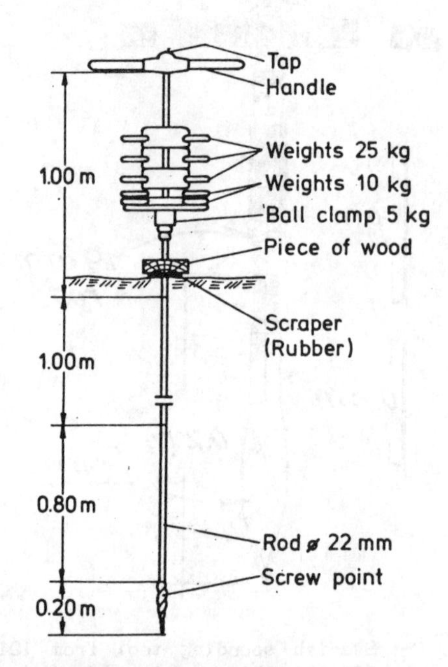

Fig 12 Details of recent weight sounding penetrometer (Today a vertically movable handle is used)

clay", etc an axial load of 0.15 kN and 0.75 kN, respectively was required to obtain the required penetration rate (20 mm/s). This was a significant step forward.

The penetrometer was rotated when the rod did not sink any further when the applied load was 100 kg. The penetration every 25 halfturns was then measured and recorded. Several series with 25 halfturns, for example 100, 200 halfturns could be required to penetrate a hard layer as indicated in Fig 13b. The rotation was normally continued until the penetration after 100 halfturns was only 10 to 20 mm.

It was indicated in the report that one had to be very careful when the depth to firm bottom was determined. Refusal was checked by striking the penetrometer a few times with a sledge hammer as discussed by Dahlberg (1974) and by Broms and Bergdahl (1982). Even then it might not be possible to penetrate a layer with dense sand. Ram soundings might be required.

The character of the different layers and the stratification was determined from the noise generated by the twisted screw point. (For this reason the screw point should not be worn). A weak squeeking noise was an indication of sand while a strong squeeking noise was typical for gravel. Stones generated a "hewing" noise and the rotation became jerky. The record of a complete soil profile is shown in Fig 14 (Note the inclined soundings!). Samples of e.g. organic soils "gyttja" were obtained with helical or post-hole augers or with an open drive sampler.

Predrilling was required through a hard surface layer in order to reduce the friction along the rods. The penetration test was then carried out from the bottom of the borehole. In deep water, the weight of the rods, 2.0 kg/m, had to be added to the recorded penetration resistance.

The Commission stated that the main purpose of weight soundings was to determine the strength and the thickness of the different soil layers. Also, embankments that have failed were investigated by the new method. Weight soundings were also used to determine the bearing capacity of piles in sand and gravel (Fig 14). The work of the

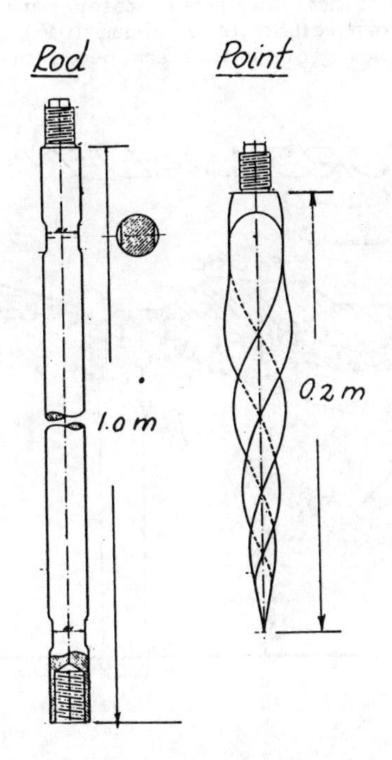

Fig 13a Rod and screw point of the original weight sounding penetrometer

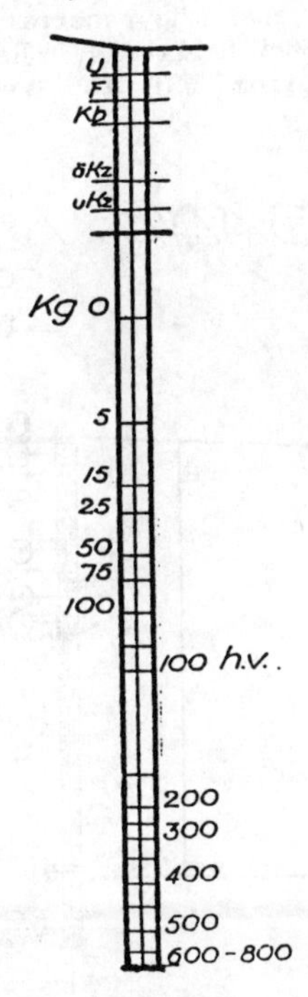

Fig 13b Example of a borehole (Swed Geot Comm 1917/1922)

Commission was of great importance especially in the Nordic countries where the geological conditions and the geotechnical problems were similar.

There was also considerable interest in the weight sounding method in Germany (Hoffmann, 1930). Today (1988), the method is also common in Poland, Hungary and Czechoslovakia, as well as in Japan and Algeria.

3.2 Further developments of WST

The WST became internally known through a paper presented by the Danish geotechnical engineer Godskesen (1936) at the first international geotechnical conference which was held in Harvard, Mass., USA (1. ICSMFE). He had previously described the method in 1930 in the Danish magazine Ingenøren (Godskesen, 1930).

In his 1936 paper, Godskesen mentions that numerous bridge foundations had been designed in Denmark based on the information obtain from WST alone, without sampling! He also indicated that e.g. shallow foundations could be designed for an allowable bearing presure of 0.3 MPa when the penetration was less than 50 cm/25 halfturns. He could also determine from WST the required pile length without driving test piles. Godskesen concluded that, in his opinion, the soil data from WST have never been on the unsafe side. The penetrometer shown in his figure has incorrectly a smooth point instead of the usual screw shaped point. This mistake was later corrected in a special note.

The following year, Godskesen (1937) presented another paper (in German) where he summarized his experiences with WST during the last 10 years. There he repeated his satisfaction with the method. An interesting diagram is presented in the paper where the results from WST have been compared with the bearing capacity of a concrete pile (Fig 15).

In contrast to the favourable conditions in Denmark where sampling according to Godskesen was not necessary, Skaven Haug, a Norwegian geotechnical engineer, was of the opinion that especially quick clays became disturbed in front of the penetrometer point and to some extent also along the penetrometer rod so that the results could not be used directly in design. At great depths, the penetration resistance was often equal to zero since the penetrometer sank under its own weight (Skaven Haug 1946).

A new geotechnical era began in Sweden

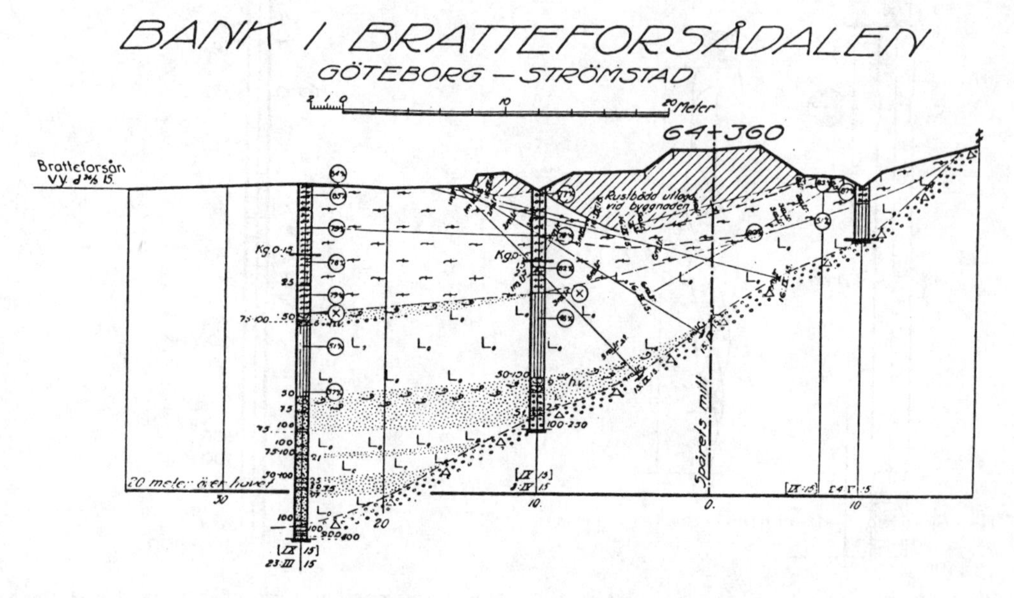

Fig 14 Example of a railway soil profile with boreholes, some inclined to check embank profile. Circle to the right of boreholes indicate samples

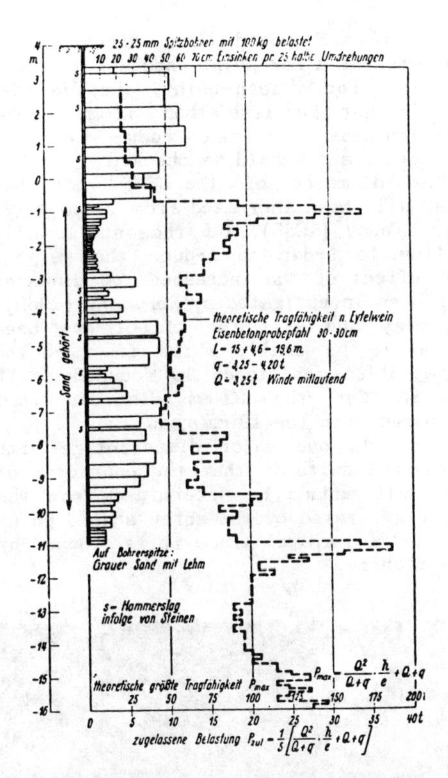

Fig 15 Comparison between results from penetration testing and theoretical bearing capacity of a concrete pile (Godskesen 1937)

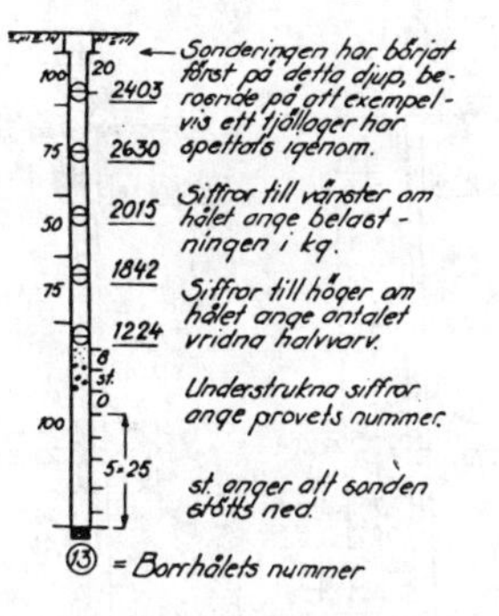

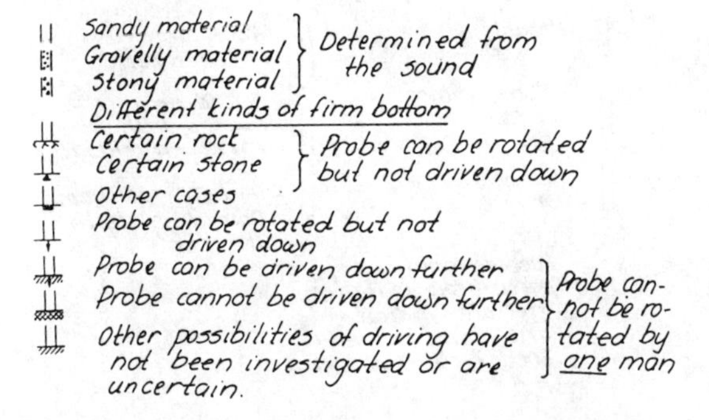

Fig 16 SGF standard 1947. Presentation of results from weight soundings including samples (2403 etc). Numbers to the left of the borehole indicate the applied load

when the Swedish Geotechnical Institute (SGI) was established in 1944 with Walter Kjellman as Director. Kjellman, who was greatly influenced by Terzaghi, developed together with Torsten Kallstenius, a new self-recording penetrometer, the "Fast Sounding Machine" which was mounted on a jeep (Swedish Geotechnical Institute, 1949). This cone penetrometer which was used by SGI on many large projects has gradually been improved as discussed later.

The method of recording the results from WST according to a SGI manual for site investigations from 1947 is shown in Fig 16. The method used today (1988) illustrated in Fig 17 for comparison.

The Swedish Geotechnical Society (SGF) proposed in 1958 to change the method of recording the results from WST to the number of halfturns required to increase the penetration 0.2 m (Figs 16 and 17). This change was adopted in 1965. The first WST standard in Sweden was approved in 1964, with revisions in 1965, 1970 and 1974. National standards for WST were adopted in Finland 1968 and in Norway 1973.

An important event was the introduction by SGI in 1964 of an electrically driven penetrometer. Other similar motorized units would soon follow (see Fig 18). However, the manual method was still used as a reference. The applied load which is regulated by the operators can be read on a separate force indicator (dynamometer). The number of halfturns (N_{wst}) required for 0.2 m penetration was registered by a counter.

The rate of the rotation has been found to affect significantly the results (Bergdahl, 1969). Comparisons with the manual method indicated that the mechanical rotation increased the

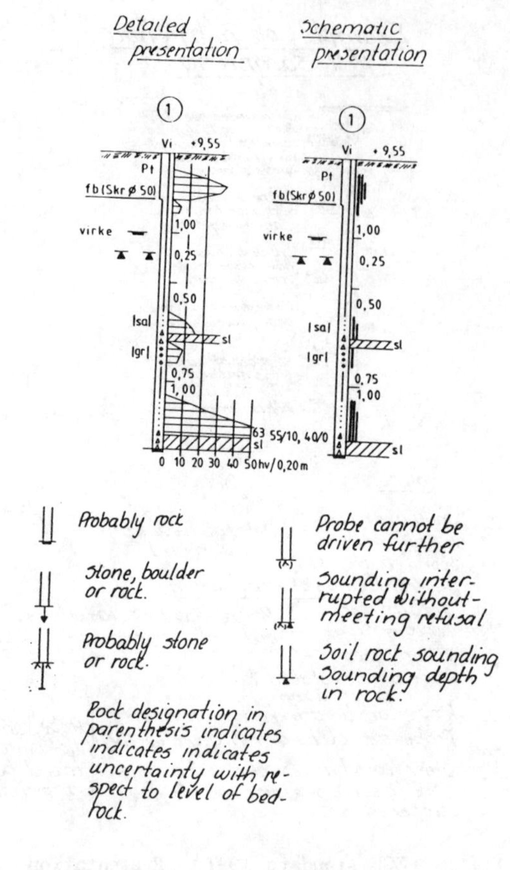

Fig 17 SGF standard 1987 of presenting results from weight soundings, detailed and schematic presentation, respectively

penetration resistance (N_{WST}) for a Norwegian clay by 60% to 70% as reported by Senneset (1974). Bergdahl (1969) found a 30% increase for a clay in Sweden. In Finland, Natukka (1969) reported an increase of 14%. In silt, the penetration resistance was reduced by 7% to 17%.

According to the Swedish National Standard of 1970 the rotation rate at WST should not exceed 50 rpm. It should preferably be between 15 and 40 rpm, with 30 rpm as a recommended value. The motor should be stopped when the penetrometer sinks under its own weight. Otherwise, the vibrations from the machine could affect the results.

Wear of the screw shaped point influences also the measured penetration resistance. According to the Swedish 1970 WST Standard the circumscribed diameter of a new point should not exceed 35 mm. For a worn point, the diameter should not be less than 32 mm. The straightness of the rods is also important and should be checked.

The diameter of the WST rods has gradually been increased from 19 mm to 22 mm. Today (1988), the rods are usually hollow in order to reduce the weight. The effect of the increased rod diameter has been investigated by Bergdahl (1969). In clay and silt the effect has been found to be small, while in sand the penetration resistance is reduced by 7% to 8% for the 22 mm diameter rods compared with the 19 mm rods.

There is one major disadvantage with motorized units in that the character of the soil cannot be determined from the noise generated by the screw shaped point in sand and gravel since it is drowned by the machine.

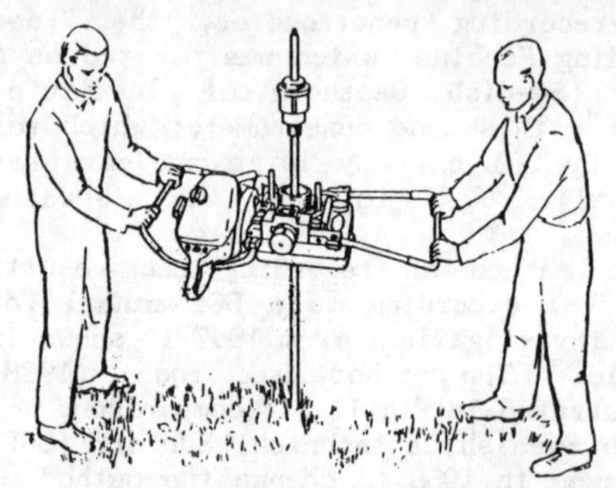

Fig 18 Professor De Beer, Belgium admiring the motor-driven penetration aggregate, type Borro. (The machine can also be used for earth anchors and for helical augering (sampling)

3.3 ESOPT I and II

The WST method was one of the principal methods discussed at ESOPT I in Stockholm 1974. The experience with the method in Sweden, Norway and Finland was then summarized by Dahlberg (1974), Senneseth (1974) and by Gardemeister and Tammirinne (1974), respectively. Muromachi et al (1974) have presented some results from Japan, where the method is also used extensively, on the relationship between the penetration resistance (N_{wst}), the shear strength, the water content and the constrained modulus. Of special interest is the Norwegian experience with the method for offshore investigations (Broms, 1974).

A reference Standard for WST was approved by the ISSMFE Council at the 9th ICSMFE in Tokyo 1977. It included a description of the weights, the rods, the couplings and of the point (Figs 19 and 20). Some explanatory notes were also given.

At ESOPT II in Amsterdam a State-of-the-Art report and two general papers on WST were presented. Broms and Bergdahl (1982) summarized the experience in the Nordic countries and the interpretation of the test results. The WST method was also included in a case study at ESOPT II which was concerned with the prediction of the bearing capacity of a driven reinforced concrete pile (0.35 x 0.35 m) from the results of different penetration tests.

3.4 Evaluation of Soil Properties

The weight sounding test is used extensively mainly in Sweden to estimate the bearing capacity of shallow and deep foundations (Broms 1974). The design methods based on WST have gradually been improved. It has been estimated that more than 20,000 weight soundings are carried out annually in Sweden alone.

In the Nordic countries WST is mainly used to get an indication of the relative density of cohensionless soils and the degree of compaction (Bergdahl 1973; Tammirinne 1973; Gardemeister and Tammirinne 1974; Hellman et al 1979). The method is also utilized to check the compaction of fine to medium coarse sand below the ground water level where other methods cannot be used or are difficult to apply. Lagging and Eresund (1974) have described a case where WST was

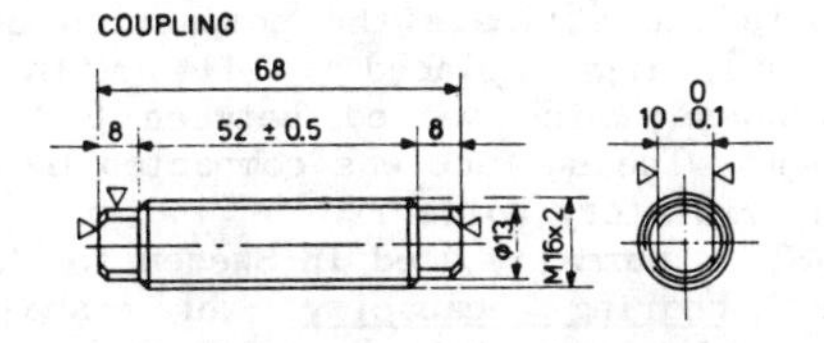

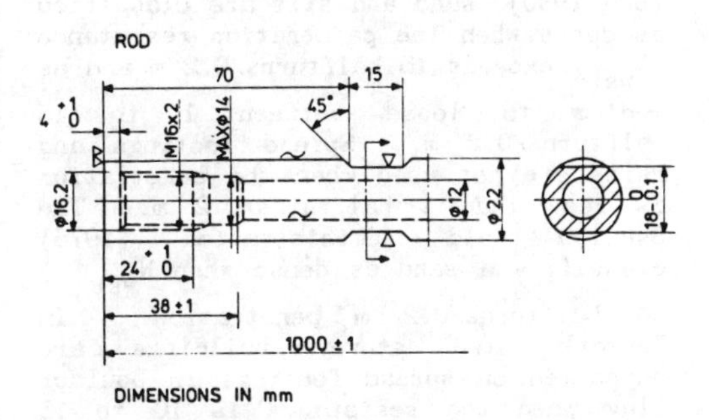

Fig 19 Recommended tolerances for manufacture of weight penetrometer rods and couplings (Subcommittee on standardization of penetration testing 1977)

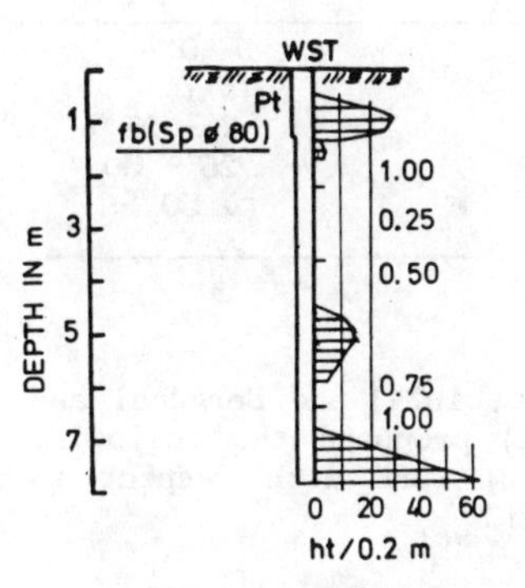

WST — WEIGHT SOUNDING TEST
ht/0.2m — NUMBER OF HALFTURNS PER 0.2m OF PENETRATION
Pt — DRY CRUST OF CLAY
fb(Sp ø 80) — PREBORING TO THIS LEVEL WITH 80mm DIAM AUGER

FIGURES TO THE RIGHT IN THE DIAGRAM INDICATE LOADS APPLIED IN kN

Fig 20 An example of the presentation of test results from weight sounding test (WST) according to the standard proposal 1977

applied to estimate the compaction of a hydraulically placed fill with a thickness which varied between 8.3 and 9.0 m. The surface was compacted by a 3 tons vibratory roller.

WST is normally used in Sweden to check the <u>bearing capacity</u> of shallow foundations on gravel, sand and silt. According to the Swedish Building Code (SBN 1980), sand and silt are classified as dense when the penetration resistance (N_{wst}) exceeds 15 halfturns/0.2 m and as medium to loose between 1 to 15 halfturns/0.2 m. Spread footings and rafts are not used when the penetration is less than 1 halfturns/0.2 m. The Swedish Road Administration (1976) classifies a sand as dense when $N_{wst} \geq$ 30 halfturns/0.2 m penetration. In Denmark, two storey buildings are supported on spread footings in boulder clay when the resistance is 10 to 15 halfturns/0.3 m (Broms and Bergdahl 1982).

In Norway, granular soils are classified as follows :

Table 2. Classification of granular soils in Norway.

Classification	Halfturns/0.2 m
Very low	0
Low	< 7
Medium	7 - 25
Dense	25 - 50
Very dense	> 50

Helenelund (1966, 1974) and Bergdahl and Sundqvist (1974) proposed the following classification of sand with respect to the results from N_{wst}

Table 3. Classification of granular soils in Sweden and Finland after Bergdahl and Sundqvist (1974) and Helenelund (1966)

Classification	Sweden	Finland
	Halfturns/0.2 m	
Very loose	< 8	< 10
Loose	8 - 20	10 - 30
Medium	20 - 60	30 - 60
Dense	60 - 100	60 - 100
Very dense	> 100	> 100

The difference between the two classification systems is small. Dahlberg (1974) has also investigated the effect of the overburden pressure on the penetration resitance of a fine sand. Test results indicate that the penetration resistance was affected down to a depth of 0.9 to 2.6 m below the bottom of the excavation where the overburden pressure was about 25 kPa. Below this depth, the effect on the penetration resistance has been found to be small.

The angle of internal friciton is generally estimated as shown in the following table :

Table 4. Angle of internal friction from WST

Soil type	Penetration resistance, N_{wst}, Halfturns/0.2 m 10 to 30	> 30
Fine sand	31°	35°
Medium to coarse sand	35°	38°
Gravel	38°	42°

The weight sounding method is also used to estimate the settlement of spread footings and rafts. According to Helenelund (1966) the compression modulus (M) is equal to 10 to 20 MPa at 10 to 30 halfturns/0.2 m, 20 to 50 MPa at 30 to 60 halfturns/0.3 m and 50 to 60 MPa at 60 to 100 halfturns/0.2 m. The investigation by Bergdahl and Ottosson (1984) has indicated similar values. For sand they also considered long-term effects. The following values have been proposed.

Table 5. Modulus of elasticity from WST

WST resistance halfturns/0.2 m	Modulus of elasticity E_s, MPa
0 - 10	< 10
10 - 30	10 - 20
20 - 50	20 - 30
40 - 90	30 - 60
> 80	> 60

In silt and silty sands the weight sounding resistance should be reduced by

30%.

It is difficult to obtain consistant results for clays with a high sensitivity ratio (S_t), where the penetrometer will sink under its own weight due to the disturbances of the clay during the testing. This difficulty has also been pointed out by Möller (1980). Also "more normal clays" have been investigated. Fukuoka (1974) and Saarelainen (1979) have correlated the results from weight soundings with the undrained shear strength. However, the reported relationships are uncertain.

3.5 Correlation with other penetration testing methods

Several have compared WST with other penetration test. In Japan, Muromachi et al, (1974) compared the WST with SPT. For cohesionless soil, the following relationships have been proposed :

$N_{30} = 0.42\ N_{wst}$ (Miki)

$N_{30} = 2 + 0.34\ W_{wst}$ (Inada)

Where N_{wst} is the penetration resistance in halfturns/0.2 m at WST and N_{30} is the resistance in blows/0.3 m at SPT. The difference between these expressions is relatively small at $N_{wst} > 10$ halfturns/0.2 m.

For a preloaded medium to coarse sand, Dahlberg (1975) indicated the relationship $N_{30} = 0.37\ N_{wst} + 1.70$ which is about the same as that proposed by Inada.

Tammirinne (1974) has for coarse to fine sand compared the WST with CPT using the Dutch cone penetrometer. For coarse sand the following relationship was obtained

$$q_c \text{ (MPa)} = 0.2\ N_{wst} \quad (3)$$

where q_c is the cone penetration resistance. Similar results have also been reported by Bergdahl (1974). This relationship corresponds closely to that used in Sweden for the classification of cohesionless soils.

The result from WST are used to estimate the <u>driving resistance</u> of piles as illustrated in Fig 21 (Helenelund 1974). Experience in Finland indicates that the length of the driven precast concrete piles corresponds approximately

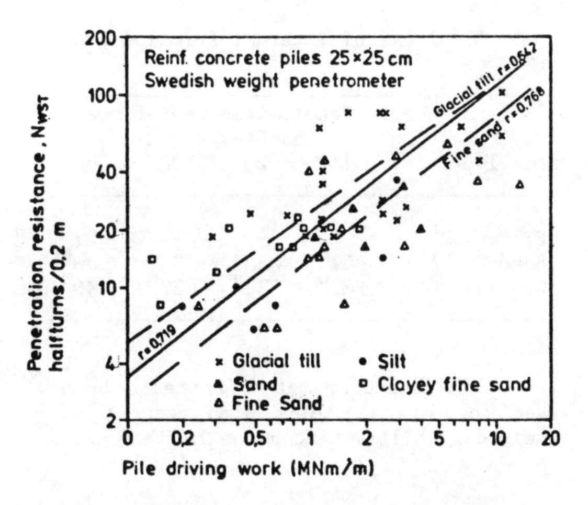

Fig 21 Driving resistance of precast concrete piles in relation to the weight sounding resistance in different soils (Helenelund 1974)

to the maximum depths that can be reached with WST. In Sweden, it is often assumed that the length of end-bearing piles corresponds to the penetrated depth plus 1 to 2 m. Winkel (1969) has reported that point bearing piles in boulder clay in Denmark can be driven 0.5 to 1.0 m deeper than the depth where the penetration resistance (N_{wst}) is 60 to 80 halfturns/0.2 m.

Results from load tests with timber and concrete piles in sand in Norway are shown in Fig 22 (Norwegian Pile Committee, 1973). The ultimate bearing capacity (Q_{ult}) of the piles divided by the total surface area (A_f) of the piles (Q_{ult}/A_f) neglecting the point resistance has been plotted as a function of the penetration resistance as determined by weight soundings. It can be seen that the ultimate bearing capacity of timber piles which are tapered is somewhat higher than that of driven precast concrete piles. The length of the piles is 12 to 20 m.

The Swedish Commission on Pile Research (1979) have proposed for cohesionless soils (silt, sand and gravel) the following values on the angle of internal friction at a penetration resistance of 15 to 50 halfturns/0.2 m as summarized in the following table :

Table 6. Angle of internal friction from WST

Soil Type	Penetration resistance halfturns/0.2 m 15	20	30	50
Silt	-	-	-	27°
Sand	31°	32°	34°	38°
Gravel	32°	34°	37°	40°

The weight sounding method has also been used to classify soils with respect to their diggability (Magnusson 1973).

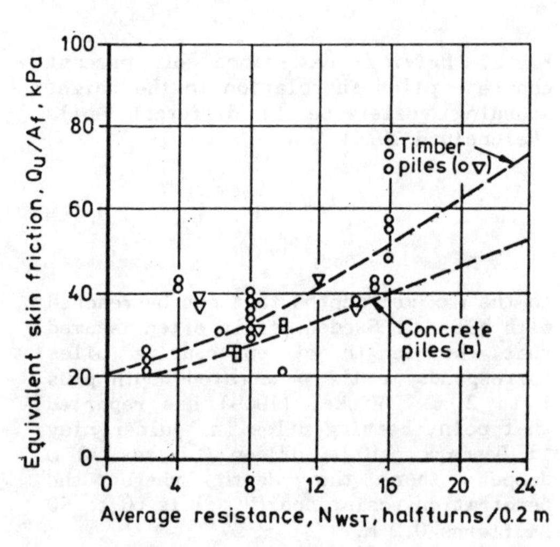

Fig 22 Relationship between the equivalent skin friction resistance and the average penetration resistance from weight soundings (Norwegian Pile Committee, 1973)

4 THE STANDARD PENETRATION TEST (SPT)

4.1 SPT Testing

The Standard penetration test (SPT) as shown in Fig 23 is today (1988) by far the most widely used penetration testing method. It is a "standard" method in most parts of the world to determine the bearing capacity of both shallow and deep foundations. In e.g. Japan, SPT is used during the exploratory phase in more than 90% of all borings. The main advantages with SPT are that the method can be applied in almost all soil types and in weak rock and that samples are obtained

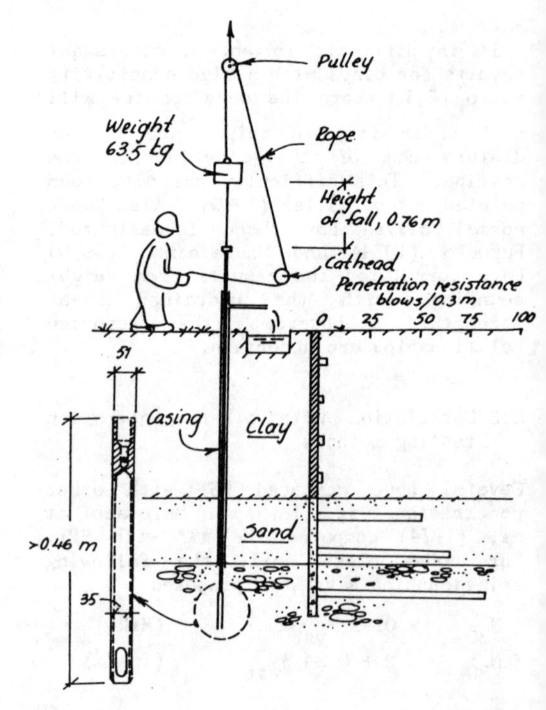

Fig 23 Standard penetration test (SPT)

so that the soil and the rock can be classified.

At the Standard penetration test, a 51 mm diameter thick-walled split spoon sampler is driven down into the bottom of a borehole by a 63.5 kg (140 lb) hammer with a height of fall of 760 mm (2 1/2 ft). The inside diameter of the sampler is 35 mm. The number of blows (N_{30}) required to drive the samplers 0.30 m is registered discounting the first 0.15 m. The testing is normally carried out every 2 to 3 m. The testing is usually terminated when the penetration resistance exceeds about 100 blows/0.3 m. Casing or drilling mud is required when the borehole is not stable. The diameter of the boreholes should normally be between 63.5 mm and 150 mm.

The most controversal aspect of the Standard penetrationt test is the operation of the hammer. In North and South America the procedure in ASTM (1586 - 84) using a rope and a cathead or a free-falling hammer is generally followed while in Europe and Asia the test is often carried out in accordance with the European Reference Standard (ISSMFE, 1977) using a free falling hammer. The difference in the results can be large as discussed in the following.

The Standard penetration test is mainly utilized today to estimate the settlement and the bearing capacity of mainly shallow foundations, to indicate the liquefaction potential of sand and silt and to evaluate the density index of granular soils. It is also used to get an indication of the shear strength of cohesive soils. The main limitation is that SPT is by far a "standard" method. The procedure followed to carry out the test varies greatly in different parts of the world with large differences in the results.

Peck (1971) has summarized the application of STP as follows : "The results can be used for the routine design of footings, at the price of conservatism in most instances. Whenever there is reason to believe that the design should not be routine, or that economy could be achieved by using more sophisticated methods of investigation, such methods should be used. The test is, indeed, of great practical value to those whc appreciate its limitations and who recognize the circumstances under which better information is justified by the requirement of the job including the economic consequences."

de Mello (1971) has pointed out in a very comprehensive State-of-the-Art report on SPT that over the past 25 years thousands of skyscrapers and other important foundations in Brazil have been designed and executed in possibly 99% of the cases on the bases of the penetration resistance alone.

4.2 Early History of SPT

The Standard penetration test (SPT) can be traced back to Colonel Charles R Gow who in 1902 developed a 25 mm diameter sampler which was driven by a 50 kg hammer into the bottom of a borehole. The split spoon sampler as it is used today for SPT is mainly the work of H. A. Mohr, then district manager of the Gow Division in New England, USA of the Raymond Concrete Pile Company and of G.F.A. Fletcher (1927). The Gow Company became in 1922 a subsidiary of the Raymond Concrete Pile Campany. A split spoon sampler was also manufactured by Spraque & Henwood at about the same time.

Fletcher and Mohr standardized in 1930 the method of driving the 50 mm diameter split spoon sampler using a 62.5 kg (140 lb) weight with a height of fall of 760 mm (30 in) as described by H A Mohr in 1937. Mohr (1943) mentions that it provides a "rough idea of the ground conditions and should always be done". A square concrete block with a hardwood inset mounted on a guide spike was used for the driving at that time (Mohr, 1966).

This new testing method has gradually replaced the samples obtained from wash borings which were used almost exclusively at that time for the classification of soils. The "Raymond sampler" developed by Mohr and Fletcher has been described by Hvorslev in 1948 in his classical report "Subsurface Exploration and Sampling". The sampler with a vented head could either have a 0.55 m (22 ins) long split barrel or be solid with a total length of 0.85 m (34 ins). A ball check valve was not used at that time. The inside diameter of the barrel which did not have a liner was the same as that of the shoe (35.1 mm). Drill rods were introduced about 1945.

Hvorslev (1949) mentions that the "Raymond sampler", used by the Raymond Concrete Pile Company, was forced about 150 mm into the soil in order "to decrease the influence of the disturbed zone below the bottom of the borehole". However, the Gow division continued to use a sampler that was simply a section of a 25 mm diameter extra heavy pipe. The boreholes were stabilized by a 50 mm casing. The penetration resistance was measured from the level where the sampler sank under its own weight.

The original split spoon sampler has gradually been improved. It has been provided with a ball valve just above the sampling spoon to prevent loss of the samples during the retrieval.

Terzaghi indicated in 1947 that the inside diameter of the split barrel was 38.1 mm while Terzaghi and Peck in 1948 showed that the inside diameter of the sampler was 35.1 mm. The difference is about 3 mm which corresponds to the thickness of the liners that were used at that time for the sampling. A 150 mm seating drive was recommended.

The term "Standard Penetration Test" was probably used for the first time by Terzaghi in 1947 in a paper "Recent Trends in Subsoil Exploration" which was presented at the 7th Texas Conference on Soil Mechanics and Foundation Engineering in spite of the fact that this testing method was far from being a "standard" method.

The use of the Standard penetration test in the design of both deep and shallow foundations spread rapidly after 1948 when the first edition of Terzaghi and Peck's book on "Soil Mechanics in

Table 7. Relative Density of Sands according to Results of Standard Penetration Test

No of Blows (N)	Relative Density
0 - 4	Very loose
4 - 10	Loose
10 - 30	Medium
30 - 50	Dense
> 50	Very dense

Engineering Practice" was published. Soon thereafter it was adopted as a standard method by the US Corps of Engineers and the US Bureau of Reclamation and by numerous private organizations (Cummings, 1951). Terzaghi and Peck showed in their book, the following correlation between the penetration resistance (N_{30}) and the relative density for granular materials after correction of the results in fine or silty sand when the penetration resistance exceeded 15 blows/0.3 m. This reduction was proposed to compensate for the effects of the negative pore water pressures that develop during the testing. This correlation is still commonly used.

Design charts for spread footings on sand were proposed by Peck et al in 1953 where the allowable bearing pressure was related to the penetration resistance (N_{30}) at a total settlement of 25 mm (1.0 in). These design charts were modified in 1974 (Peck et al, 1974) where the penetration resistance was corrected with respect to the effective overburden pressure as proposed by Bazaraa (1967). A correction of the results from SPT was also included in the 1977 ASCE Manual No 56.

Also the particle size is important as pointed out by Meyerhof (1956). The SPT becomes unreliable when the particle size approaches the diameter of the sampler. This is of course also true for other penetration testing methods.

4.3 Limitations of SPT

The Standard penetration test has been severely criticised by e.g Fletcher (1965) and by Ireland, et al (1970) mainly because it was not a "standard" and that a large number of factors can affect the results. It should also be noted that the practice to carry out the test varies greatly in different parts of the world. Therefore, the results from SPT are often difficult to compare. Some of the factors that can effect the results are shown in Fig 24. Most of these are controlled by the driller. Thus the reliability of the method depends to a large extent on his experience, interest and care.

The importance of the size of the borehole has been discussed by e.g. Fletcher (1965). He proposed that the maximum size should be limited to 100 mm. However, in U.K. casing with a diameter of up to 250 mm is frequently used (Rodin, 1961; Palmer and Stuart, 1957). Sen Gupta and Aggarwal (1965) report that the effect of the size of the borehole on the penetration resistance (N_{30}) is about 30% and that the resistance decreases with increasing diameter as could be expected. In many codes the maximum size of the boreholes is limited to 150 mm.

It should be noted that the samplers used in North America are often different from those in Asia and Europe. The samplers in North America have generally an inside diameter that is 3 mm larger than the diameter of the cutting shoe in accordance with ASTM 1986 - 84 as mentioned above while the samplers used in Asia and in Europe have an inside diameter that is the same as the inside diameter of the shoe in accordance with the 1954 ASTM specification. This difference may affect the penetration resistance as much as 10% to 30% (Seed et al, 1985).

Also the procedure for the seating of the sampler at the bottom of a borehole varies greatly in and between different countries. It is common to discount the penetration resistance during the first 0.15 m to 0.30 m. In the interpretation of the results it has been proposed that the penetration resistance consists of two straight lines and that the penetration resistance should be counted from the intercept of the two lines (Granger, 1963, Sen Gupta and Aggarwal, 1966). This intercept is located about 0.30 m below the bottom of the borehole. However, most codes specify that a seating drive of 0.15 m.

Granger (1963) has pointed out that the use of drilling mud instead of water can increase the penetration resistance significantly. Parsons (1966) found for one case that drilling mud increased the penetration resistance by about 150% from 20 blows/0.3 m when casing was used to 50 blows/0.3 m with drilling mud. However, the suction and the resulting loosening of the bottom of the borehole caused by

Inadequate cleaning of the borehole	SPT is only partially made in original soil. Sludge may be trapped in the sampler and compressed as the sampler is driven, increasing the blow count. (This may also prevent sample recovery.)	Free fall of the drive weight is not attained	Using more than 1-1/2 turns of rope around the drum and/or using wire cable will restrict the fall of the drive weight.
Not seating the sample spoon on undisturbed material	Incorrect "N" values obtained.	Not using correct weight	Driller frequently supplies drive hammers with weights varying from the standard by as much as 10 lbs.
Driving of the sample spoon above the bottom of the casing	"N" values are increased in sands and reduced in cohesive soils.	Weight does not strike the drive cap concentrically	Impact energy is reduced, increasing "N" values.
Failure to maintain sufficient hydrostatic head in boring	The water table in the borehole must be at least equal to the piezometric level in the sand, otherwise the sand at the bottom of the borehole may be transformed into a loose state.	Not using a guide rod	Incorrect "N" value obtained.
Attitude of operators	Blow counts for the same soil using the same rig can vary, depending on who is operating the rig, and perhaps the mood of operator and time of drilling.	Not using a good tip on the sampling spoon	If the tip is damaged and reduces the opening or increases the end area, the "N" value can be increased.
Overdrive sampler	Higher blow counts usually results from overdriven sampler.	Use of drill rods heavier than standard	With heavier rods more energy is absorbed by the rods causing an increase in the blow count.
Sampler plugged by gravel	Higher blow counts result when gravel plugs sampler, resistance of loose sand could be highly overestimated.	Not recording blow counts and penetration accurately	Incorrect "N" values obtained.
Plugged casing	High "N" values may be recorded for loose sand when sampling below groundwater table. Hydrostatic pressure causes sand to rise and plug casing.	Incorrect drilling procedures	The SPT was originally developed from wash boring techniques. Drilling procedure which seriously disturb the soil will affect the "N" value, e.g. drilling with cable tool equipment.
Overwashing ahead of casing	Low blow count may result for dense sand since sand is loosened by overwashing.	Using drill holes that are too large	Holes greater than 10 cm (4 in) in diameter are not recommended. Use of larger diameters may result in decreases in the blow count.
Drilling method	Drilling technique (e.g., cased holes vs. mud stabilized holes) may result in different "N" values for the same soil.	Inadequate supervision	Frequently a sampler will be impeded by gravel or cobbles causing a sudden increase in blow count; this is not recognized by an inexperienced observer. (Accurate recording of drilling, sampling and depth is always required.)
Not using the standard hammer drop	Energy delivered per blow is not uniform. European countries have adopted an automatic trip hammer not currently in use in North America.	Improper logging of soils	Not describing the sample correctly.
		Using too large a pump	Too high a pump capacity will loosen the soil at the base of the hole causing a decrease in blow count.

Fig 24 Factors that may affect the results from SPT (after NAVFAC DM-7, 1971)

the lifting of the drill rods could have contributed to the observed reduction of the penetration resistance.

The size and the weight of the drilling rods (Brown, 1977, Schmertmann, 1978a), the drilling method (wash boring, rotary drilling or auger boring) or the stress history of the soil (Marcuson and Bieganousky, 1977a) have generally only a small effect on the measured penetration resistance. However, a continuous flight hollow-stem auger might cause some loosening of the bottom of the borehole below the ground water level because of the suction during the withdrawal of the auger (Schmertmann, 1975). The SPT is carried out inside the hollow auger stem after the plug that seals the bottom of

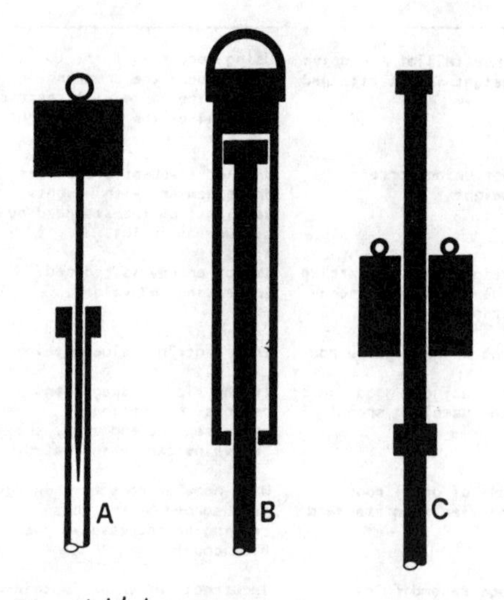

Fig 25 Hammers used in North America at SPT (after Riggs, 1986)

the stem during the drilling has been removed. Often a plug is not used during the drilling. This practice can affect the results significantly.

Skempton (1986) has recognized that aging can significantly affect the results from SPT. The use of correlations obtained from calibration chamber tests can thus lead to an overestimation of the relative density of aged sands.

It should also be pointed out that the compressibility and the crushing strength of the sand particle can have a significant effect on the interpretation of the results from Standard penetration tests. The relative density can be underestimated for e.g. calcareous soils where the particle may be crushed during the testing.

Different types of hammers are used in the United States. The most common is the safety hammer shown in Fig 25. But also pinweight and donut hammers are frequently used (Riggs, 1986). There are some differences in the efficiency between the different hammers.

A significant development in the interpretation of SPT has been the recognition that the energy delivered to the penetrometer is an important parameter as pointed out by e.g. Frydman (1970), Schmertmann and Palacios (1979) and Kovacs and Salomone (1982). Kovacs et al (1977) measured the velocity of the 62.5 kg (140 lb) hammer at the instant of impact while Schmertmann (1978b) investigated the energy delivered to the penetrometer.

Stress wave measurements by Schmertmann (1982) and by Riggs et al (1984) indicate that the transmitted energy is normally 65% to 90% of the theoretical energy when the rope and cathead method is used to lift the hammer with two wraps around the cathead. Values as low as 30% have been reported (Kovacs et al, 1984). The variation of the transfered energy can thus be large. Schmertmann (1976a) suggested that an efficiency of about 55% may be the norm. Also Riggs (1986) has pointed out that the efficiency can be low for some donut hammers and for some types of rotary rigs.

In Fig 26 is shown some results from measurements carried out by Kovacs and Salomone (1984) in Japan using a donut hammer. The measured efficiency ranged from about 60% to about 85% for donut hammers where the rope and cathead method was used to lift the hammer while for trip hammers an efficiency of 80% to 86% was obtained. The donut hammer had a somewhat lower efficiency than the safety hammer. The efficiency of a trip hammer was about 40% higher than that of a donut hammer.

With the rope and cathead method the hammer is often lifted higher than 0.76 m before it is released as reported by Kovacs et al (1984), by Riggs et al (1984) and by Riggs (1986). A height of 0.8 m is typical for the present practice in North America (Kovacs, 1979, Riggs, 1986) as illustrated in Fig 27.

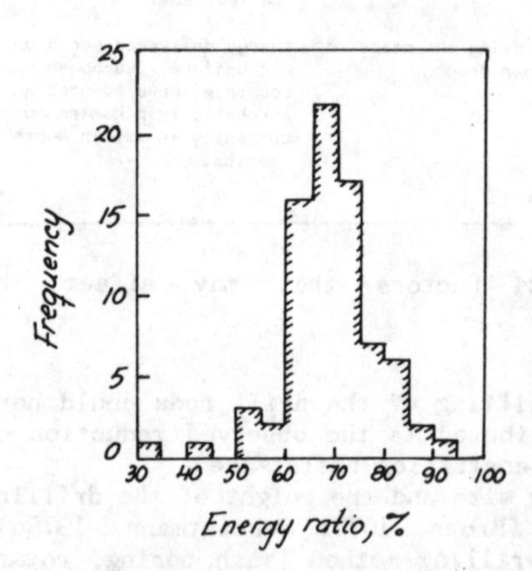

Fig 26 Energy ratio for several cathead and rope operators in Japan (after Kovacs and Salomone, 1984)

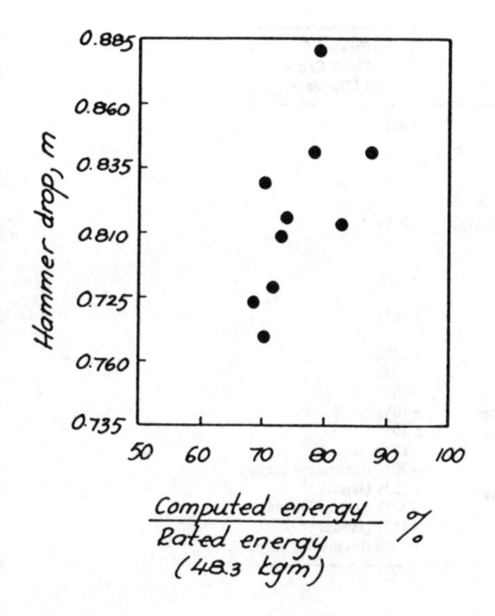

Fig 27 Energy calculated from velocity measurements for ten SPT operators (after Riggs, 1986)

4.4 Standardization of SPT

Standardisation of particularly the lifting is urgently needed for SPT so that the results from different parts of the world can be compared. Schmertmann (1978b) has pointed out that unfortunately, the SPT as practiced in the United States under ASTM Method D1586, suffers from a perhaps fatal or near-fatal flaw. Practicing engineers know all too well that the test and its N-values have a poor reproducibility and great variability between different operators and equipment. Schmertmann (1976a) has listed in Fig 28 what according to his opinions is the main cause of this variability. He suggested that the major differences in the results are those caused by the method used to control the stability of the borehole (drilling mud or casing), the chosen drilling method (e.g. hollow stem augers) and the method for the lifting of the hammer (the rope and cathead method or a free falling hammer). Many have pointed out that standardization of STP is essential so that the results from different investigations can be compared (e.g. Fletcher, 1965, Ireland et al 1970, Serota and Lowther, 1973).

The first ASTM standard (ASTM D1586-54) was that of 1954 "Tentative Method for Penetration Test and Split-barrel Sampling of Soils". The sampler had a constant inside diameter (35.1 mm). It was indicated, however, that the inside diameter of the split-barrel could be 38.1 mm and may contain a liner. Casing or drilling mud was required in weak ground. The maximum size of the borehole was 150 mm (6 in).

The method became an ASTM Standard in 1967 (ASTM D1586-67). Size A drill rods were recommended when the depth of the borehole exceeded 15 m. It was required that the water level in the borehole should be kept at or above the ground water level. The diameter of the split-barrel could be 38.1 mm provided a liner was used with 16 gage wall thickness. The ASTM Standard was revised in 1984.

The penetration resistance between the 1954 standard with two wraps around the cathead and the 1984 standard could differ as much as 25% to 50% as pointed out by Riggs (1986). It is also interesting to note that the penetration resistance (N_{30}) when the rope and cathead method is used to lift the hammer is on the average 40% higher than that with a free falling hammer. Similar results have been reported also by Serota and Lowther (1972) and by Kovacs et al (1977).

There are national standards in many countries which in most cases follow the ASTM Standard where the maximum size of the boreholes is limited to 150 mm (6 in) and where the indicated inside diameter of the sampler is 35.1 mm without a liner and 38.1 mm when a liner is used. In Australia and U.K. a solid cone can be used in gravel or stony soils instead of an open split barrel sampler. In the United States the liner is often omitted (Schmertmann, 1979). The rope and cathead method is commonly specified for lifting the hammer as is the case in Japan but in some codes trip hammers are recommended where the height of fall can be closely controlled. The number of blows required for every 150 mm of penetration is generally recorded.

4.5 Evaluation of Strength and Deformation Properties from SPT

There have been several investigations during the 1960th of the relationship between the penetration resistance (N_{30}) and such properties as relative density, effective angle of internal friction and compressibility for granular soils and the undrained shear strength for cohesive soils.

Cause: Basic	Cause: Detailed	Estimated % by Which Cause Can Change N
Effective stresses at bottom of borehole (sands)	1. use drilling mud versus casing and water	+100%
	2. use hollow-stem auger versus casing and water and allow head imbalance	±100%
	3. Small-diameter hole (3 in.) versus large diameter (18 in.)	50%
Dynamic energy reaching sampler (All Soils)	4. 2 to 3 turn rope-cathead versus free drop	+100%
	5. Large versus small anvil	+50%
	6. Length of rods	
	Less than 10 ft	+50%
	30 to 80 ft	0%
	more than 100 ft	+10%
	7. Variations in height drop	±10%
	8. A-rods versus NW-rods	±10%
Sampler design	9. Larger ID for liners, but no liners	−10% (sands) −30% (insensitive clays)
Penetration interval	10. $N_{0 \text{ to } 12 \text{ in.}}$ instead $N_{6 \text{ to } 18 \text{ in.}}$	−15% (sands) −30% (insensitive clays)
	11. $N_{12 \text{ to } 24 \text{ in.}}$ versus $N_{6 \text{ to } 18 \text{ in.}}$	+15% (sands) +30% (insensitive clays)

Metric conversions: 1 ft = 0.3048 m; 1 in. = 2.54 cm.

Fig 28 Some factors in the variability of standard penetrometer test N (after Schmertmann, 1978b)

One important development has been the correction of the results from SPT with respect to the effective overburden pressure by Gibbs and Holtz (1957) and by Bazaraa (1967) in the United States and by Schultze and Meltzer (1965) in FRG.

A comparison of the different methods is shown in Fig 29. It can be seen that at a certain penetration resistance (N_{30}) the relationsip given by Bazaraa (1967) indicates the lowest relative density. The highest is that obtained from the relationship proposed by Schultze and Melzer (1965). Part of this difference can possibly explained by the method used for the placement and the compaction of the soil, the determination of the maximum and minimum void ratio as indicated by e.g. Tavenas and LaRochelle (1972) and by the method used to lift and to release the hammer.

Zolkov and Weisman (1965) as well as Schmertmann (1970) have pointed out the importance of the overconsolidation ratio and of the lateral stresses at the interpretation of SPT and that the relative density can be overestimated if the sand is overconsolidated. This has been confirmed by Marcuson and Bieganousky in 1977a and 1977b. They found that the penetration resistance increased considerably with increasing overconsolidation ratio.

There have also been attempts by e.g. de Mello (1971) to correlate the <u>undrained shear strength</u> (c_u) of cohesive soils with the penetration resistance (N_{30}) as determined by STP. However, for stiff clays the results are affected by cracks and fissures as reported by e.g. Marsland (1974). Stroud (1974) indicates that the shear strength (kPa) varies between $4N_{30}$ and $6N_{30}$. This relationship corresponds closely to that proposed by Terzaghi and Peck (1948, 1967).

Meyerhof (1956) in Canada has proposed for sand the following conservative relationship between allowable <u>bearing pressure</u> q_a and the penetration resistance N_{30} at a maximum settlement of 25 mm

$$q_a = N_{30} K_d / 80 \text{ (MPa)} \quad (4)$$

for $B \leq 1.2$ m and

$$q_a = N_{30} K_d (1 + 1/3B)^2 / 120 \text{ (MPa)} \quad (5)$$

for $B > 1.2$ m

where B is the width of the loaded area in metres and K_d is a depth factor which is equal to $(1 + D/3B)$ when $D \leq B$.

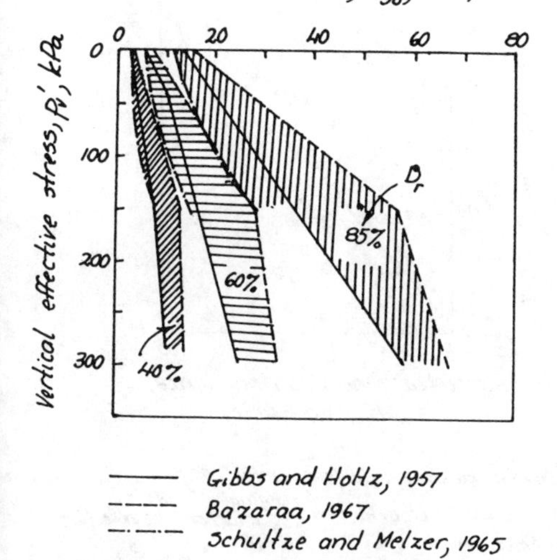

Fig 29 Relationship between N_{30}, P'_{vo} and D_r (after Marcuson and Bieganousky, 1977)

Meyerhof proposed in 1965 that the allowable load as calculated from the equations given above could be increased by 50% without exceeding a maximum settlement of 25 mm.

Meyerhof (1965) suggested also the following relationship between the coefficient of subgrade reaction k_s and N_{30} for a 0.3 m wide plate

$$k_s = q/s = 0.75\ N_{30}\ (MN/m^3) \qquad (6)$$

where q is the applied load (MPa) and s is the settlement (m). The relationship $[B/(B + 0.3)]^2$ has been used by Meyerhof to estimate the settlement of a footing with the width B as proposed by Peck et al (1953) At e.g. $N_{30} = 10$ and q = 0.1 MPa then s = 0.0075 m or 7.5 mm.

Schultze and Menzenbach (1961) found for sands that the compression modulus (M) could be correlated with the penetration resistance N_{30}

$$M = 7.1 + 0.49\ N_{30}\ (MPa) \qquad (7)$$

Stroud and Butler (1975) proposed the relationship

$$M = f\ N_{30}\ (MPa) \qquad (8)$$

The coefficient f was reported to vary between about 0.45 MPa for materials with medium plasticity to about 0.60 MPa at $I_p < 20$. Stroud (1974) proposed a constant value of 0.44 MPa on f.

SPT has also been used to estimate the settlements of footings in sand as well as in overconsolidated clays as summarized by Sutherland (1977).

Simons and Menzies (1977) have suggested for granular soils the following simple relationship

$$s = 3qB/N_{30} \qquad (9)$$

for a footing with the width B.

In USSR (Trofimenkov, 1974) an equivalent modulus of elasticity E_s = (35 to 50) log N_{30} is used to estimate the settlements of footings on sand based on the results from SPT. Parry (1971) proposed the relationship $E_s = 5\ N_{30}$ while Webb (1969) suggested the expression

$$E_s = 0.537\ (N_{30} + 15)\ (MPa) \qquad (10)$$

for saturated fine to medium sand and

$$E_s = 0.358\ (N_{30} + 5)\ (MPa) \qquad (11)$$

for a saturated clayey fine sand.

Correlations have been published for granular soils by e.g. Meyerhof (1956) Peck et al (1974), Muromachi et al (1974), Tassios and Anagnostopolous (1974), Mitchell et al (1978) and by others between the penetration resistance (N_{30}) and the effective angle of internal friction (ϕ'). Muromachi et al, (1974) proposed the following relationship

$$\phi' = 20^o + 3.5\ \sqrt{N_{30}} \qquad (12)$$

where N_{30} (blows/0.3 m) is the measured penetration resistance. The accuracy of this equation is reported to be about ± 5°. The correlation with ϕ is in general more consistant than with the relative density (de Mello, 1971).

The relationship proposed by Mitchell et al (1978) is shown in Fig 30. It can be seen that the effective overburden pressure has a large effect on the interpretation of the results. It should be noted that the scatter of the results is relatively large. It is, therefore, often preferrable to evaluate the bearing capacity of e.g. footings and piles directly from the measured penetration resistance (N_{30}) without using the angle of internal friction. The results may otherwise be misleading.

Meyerhof showed in 1956 that SPT can also be used to evaluate the bearing capacity of piles. The undrained shear

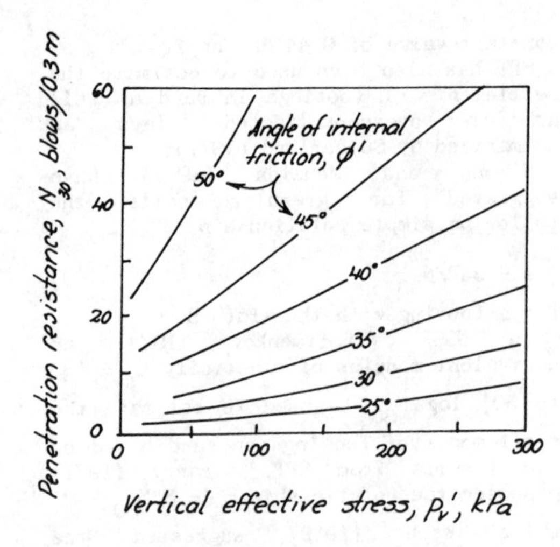

Fig 30 Relationship between N_{30}, P'_v and ϕ' (after Mitchell et al, 1978)

strength when used to calculate the point resistance of piles in stiff clay (9 c_u) was estimated from the relationship

$$c_u = 2.5\ N_{30}\ \text{(kPa)} \tag{13}$$

For the skin friction resistance (f_s) the following relationship was proposed

$$f_s = 2\ N_{30}\ \text{(kPa)} \tag{14}$$

This corresponds to an adhesion factor (α) of about 0.4. However, the maximum skin friction resistance should not exceed 100 kPa.

Meyerhof (1956) also suggested that the point resistance q_c for granular soils can be taken as $4N_{30}$ while the skin friction resistance f_s can be assumed to $N_{30}/50$. A factor of safety of four was recommended.

During the last few years it has become apparent that the bearing capacity of piles especially in calcareous soils can be grossly overestimated when the results from SPT are used (Hagenaar, 1982). Most published relationships are, therefore, only applicable on silica sands where the percentage of fines is low.

The Standard penetration tests (SPT) has also been used to evaluate the liquefaction potential of sand. Seed and Idriss (1974) compared cases where liquefaction has occurred with those with

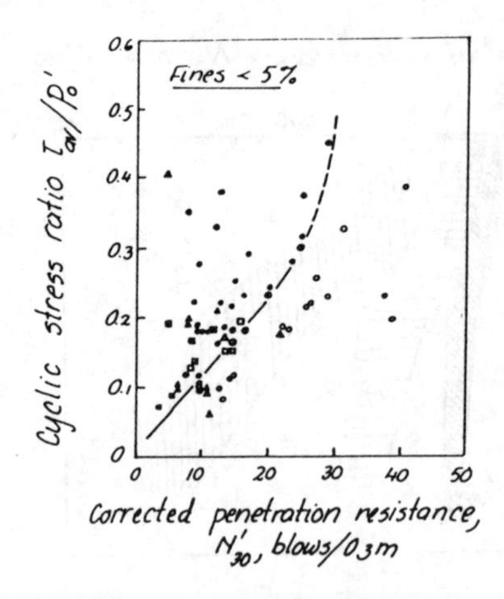

Fig 31 Relationship between stress ratios causing liquefaction and N_1-values for clean sand for M = 7.5 earthquakes (after Seed et al, 1986)

no liquefaction based on the relative density as determined by SPT. The presented charts shows the liquefaction potential as a function of the relative density (density index) and of the depth. This approach has been criticised mainly because of the uncertainty connected with the evaluation of the relative density from SPT. This point has also been discussed by e.g. Peck (1971).

Later the liquefaction potential has been correlated directly with the Standard penetration resistance (N_{30}). Ohsaki (1970) has suggested that liquefaction is not a problem when $N_{30} \geq 2D$ where D is the depth in metres. Seed (1986) has proposed the relationship shown in Fig 31 where the liquefaction potential has been related to the corrected penetration resistance N'_{30} and the cyclic stress ratio τ_{av}/P'_o. The consistancy of the results is remarkably good considering the variation of the procedure used to carry out the Standard penetration test in different parts of the world.

5 THE CONE PENETRATION TEST (CPT)

5.1 CPT Testing

The cone penetration test (CPT) is commonly used in Europe especially during the exploratory phase to determine primarily the layer sequence and the thickness and the lateral extent of the different layers. A mechanical cone penetrometer is mainly utilized during this phase of the investigation. An electrical cone penetrometer is often used during the detailed investigation to evaluate the shear strength and deformation properties of the different layers. The results are used in the design of both shallow and deep foundations. According to Kantey (1951), "The deep sounding apparatus was the most significant development in recent years in the field of piling work".

The main limitation of the cone penetration test is that no samples are recovered during the testing. Therefore, CPT must normally be supplemented by borings, so that the different strata can be classified and for detailed laboratory investigations of the recovered samples. It has been estimated that about 30.000 CPT are carried out every year in Holland alone.

At CPT a 35.7 mm diameter cone (10 cm^2) with a 60 degree apex angle is pushed down into the soil at a constant rate (usually 0.02 m/s). The penetration resistance of the cone as well as the friction resistance along a 150 cm^2 sleeve placed just behind the cone are measured as illustrated in Fig 32. The results are plotted automatically by a strip chart recorder. The diameter of the rods and of the cone are usually the same.

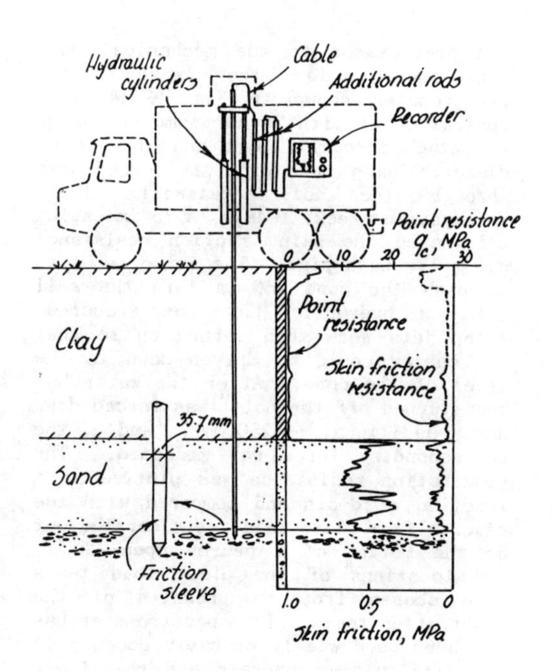

Fig 32 Cone Penetration Test (CPT)

5.2 Early Developments

The idea to determine the shear strength by pushing or dropping a cone into the soil developed very early. Initially the depth of penetration was taken as an indication of the shear strength. This method was used by John Olsson in 1915 to determine the shear strength of very soft clay (Bjerrum and Flodin, 1960). The cone was released when it just touched the surface of the clay. The fall-cone cone test was inspired by the Brinell hardness test which was widely used at that time to determine the yield strength of steel.

The shear strength of the soil was initially expressed in terms of an H-number, which corresponds to the weight in grammes of an equivalent cone to obtain a penetration of 10 mm. Three different H-numbers were used (H_1, H_2 and H_3). The H_1-value indicated the relative strength of the remoulded clay while the H_3-value was used for the undrained shear strength as determined from samples obtained with the piston sampler developed by John Olsson (1925).

He also correlated the H_3-value with the undrained shear strength as determined from pile load tests carried out by Wendel in 1900 and from old landslides using a $\phi = 0$ analysis! He found that an H_3-value of 40 corresponded to an undrained shear strength of about 10 kPa. This was a remarkable achievement at that time.

A pocket penetrometer was developed in Denmark by the Danish State Railroads in 1931 which is based on the principle for the fallcone test as described by Godskesen (1936). This pocket penetrometer (Fig 33) was used extensively to determine in the laboratory the shear strength of stiff clay and the allowable bearing pressure for spread footings.

A predecessor to the mechanical cone penetrometer is the wash point penetrometer shown in Fig 34 where a conical point with 70 mm diameter, which is attached to the lower end of a 50 mm diameter heavy wash pipe, is pushed down through the soil (Paaswell, 1931; Terzaghi and Peck, 1948). A 75 mm casing eliminated the skin friction resistance along the wash pipe. The force required to push the cone 250 mm into the soil using a hydraulic jack was measured. Water jets were then turned on so that the casing could be driven down to the level of the cone. After the water had been turned off the point was forced down an additional 250 mm and the corresponding force was measured. The penetration resistance was plotted as a function of depth and compared with the results from plate load tests carried out at the bottom of a nearby open shaft. The locations of the plate load tests were chosen from the results of the penetration tests. This penetrometer has not been used widely probably because it was difficult to operate and relatively time consuming. However, Terzaghi (1930) utilized this method in New York City in 1929 to determine the bearing capacity of sand for a part of a subway tunnel.

5.3 The Dutch Cone Penetrometer

The Dutch cone penetrometer was initially developed about 1930 by P. Barentsen, a civil engineer at the Department of Public Works ("Rijkswaterstaat") in the Netherlands. The purpose of the penetrometer was to determine the thickness and bearing capacity of 4 m thick hydraulic fill near the town of Vlaardingen in Holland (van der Veen. 1987). The 10 cm^2 cone with a 60 degree apex angle was pushed down by hand (Fig 35) by one to two men as described by

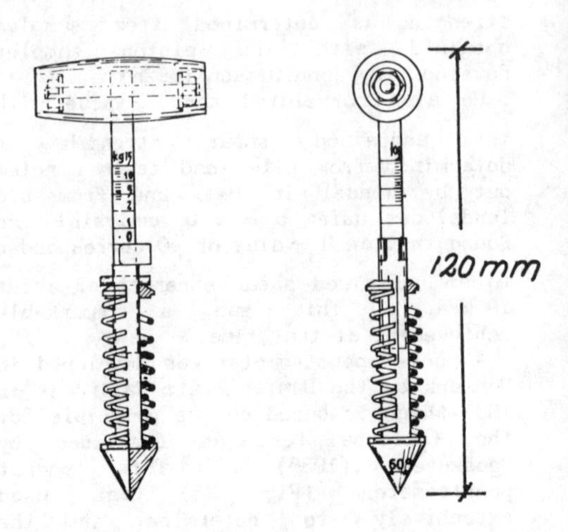

Fig 33 The Godskesen pocket penetrometer (after Godskesen, 1936)

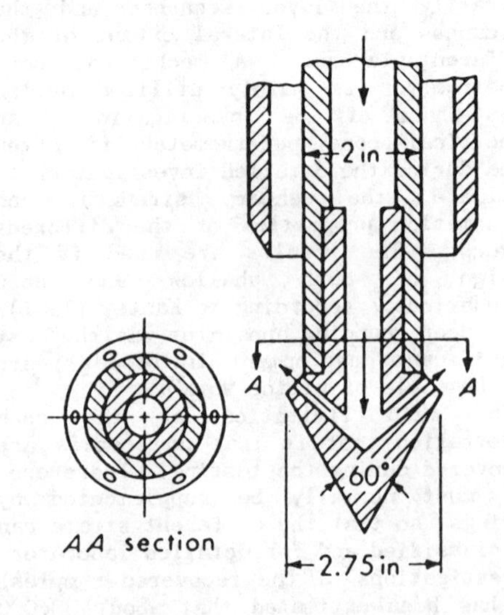

Fig 34 Wash point penetrometer (after Terzaghi and Peck, 1948)

Fig 35 Cone penetrometer developed by Barentsen (1936)

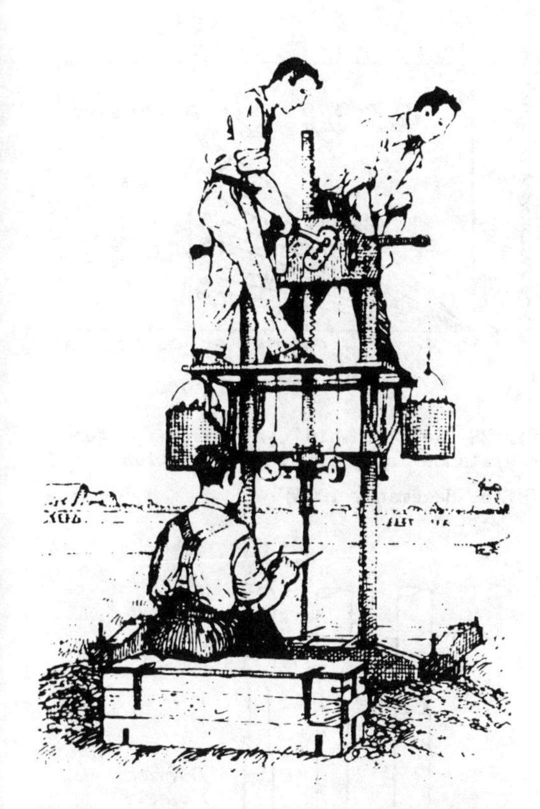

Fig 36 Dutch cone penetrometer about 1936

Barentson (1936). The penetration resistance was read on a manometer. The maximum depth was two to three metres.

A hand-operated cone penetrometer was built by Goudsche Machinefabriek of Gouda, Holland, in cooperation with the Delft Laboratory of Soil Mechanics in 1936 as described by Plantema (1948a, 1948b, 1957a and 1957b). The maximum capacity was 25 kN. The 10 cm^2 cone was pushed down by hand using ratchets and pinions as shown in Fig 36. An outer 19 mm pipe (casing) eliminated the skin friction along the inner rod. After the cone had been pushed down 150 mm, the maximum stroke, the outer pipe was pushed down until it reached the point. Then the outer pipe and the cone were pushed down together until the next level had been reached where the penetration resistance was to be determined. The force was measured with a manometer at the ground surface as shown in the figure every 0.2 m. The length of each rod and of each pipe segment was 1.0 m. This penetrometer was used e.g. in China in the 1930's by the Whangpoo Conservatory

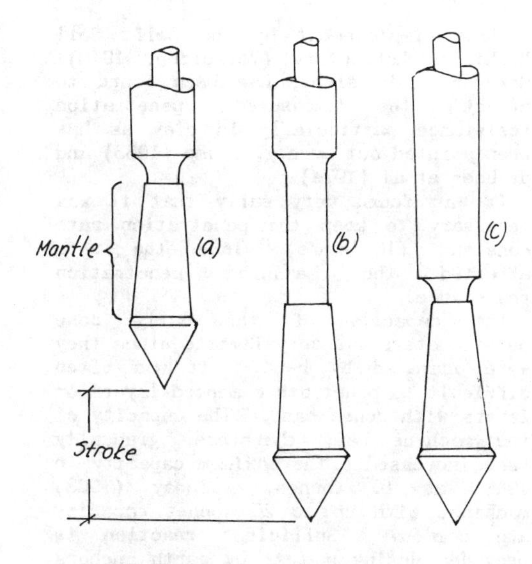

Fig 37 Dutch cone penetrometer with conical sleeve

Board (WCB) to determine the bearing capacity of piles as mentioned by the Engineering Society of China (1937).

Professor A.S.K. Buisman (1935) at the Delft Technological University used this penetrometer to investigate the bearing capacity of piles especially the point resistance. The capacity of the penetrometer was increased so that sand layers located even at 10 to 30 m depth could be reached. Buisman (1940) developed also equations for the prediction of the point resistance of piles from the cohesion and the angle of internal friction of the soil based on Prandtl's work. The important contributions of Terzaghi (1944) and de Beer (1945), Caquot and Kerisel (1953) in this area should also be acknowledged.

The measured cone resistance has been compared with results from pile load tests by e.g. Huizinga (1942). Van der Veen (1952) a civil engineer at the Amsterdam Public Works Department found that the bearing capacity of driven piles corresponded to the average penetration resistance as determined new penetrometer within a depth that extended from four pile diameters above the pile point to one pile diameter below the point.

The original Dutch cone penetrometer was not provided with a conical sleeve just above the cone as shown in Fig 37 to prevent the soil from entering the gap between the casing and the rods which could affect the test results. This was a later improvement by the Delft Soil Mechanics Labroatory (Vermeiden, 1948). However, this sleeve has been found to affect the measured penetration resistance particularly in clay as has been pointed out by e.g. Stump (1953) and de Beer et al (1974).

It was found very early that it was necessary to keep the penetration rate constant (10 mm/s) since the rate affected the measured penetration resistance.

The capacity of the early cone penetrometers was very limited since they were operated by hand. It was often difficult to penetrate cemented layers or layers with dense sand. The capacity of the machine has, therefore, gradually been increased. The maximum capacity in 1948 was 10 tonnes. Today (1988) machines with up to 20 tonnes capacity are common. Sufficient reaction is provided during a test by earth anchors or by a weighted floor. Both the cone resistance as well as the total skin friction resistance along the shaft are measured. The results are often used to estimate the bearing capacity of piles.

A significant advancement was in 1953 when Begemann proposed that the skin friction resistance should be measured every 0.2 m with a separate friction sleeve ("adhesion jacket") located just above the cone (Begemann, 1953, 1957, 1965, 1969). For special purposes the interval could be decreased to 0.1 m. The method was patented in 1955. This improvement was incorporated shortly afterwards by Goudsche Machinefabriek in Holland in their design . The diameter of the sleeve was the same as that of the cone (36 mm). The surface area of the sleeve was 150 cm^2.

Begemann (1965) was also the first to propose that the ratio of the measured friction resistance along the sleeve and the cone resistance, the so-called friction ratio, was affected mainly by the particle size as illustrated in Fig 38 and that the friction ratio could be used to classify the different soil layers. He suggested that the friction ratio (FR) for clays was about 5% and for sand about 1%. (See Fig 38).

The determination of the skin friction resistance using the Begemann friction sleeve is illustrated in Fig 39. First the cone is pushed down alone 40 mm, then the friction sleeve and the cone together an additional 40 mm. The sleeve resistance corresponds to the difference between the two readings. There are some difficulties, however, in the interpretation of the results when the penetration resistance varies rapidly with depth since it is assumed in the interpretation that the cone resistance is constant.

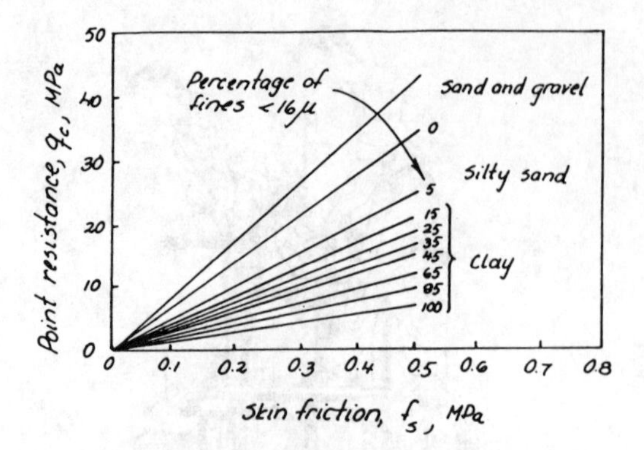

Fig 38 Relationship between cone resistance q_c and skin friction, f_c (after Begemann, 1965)

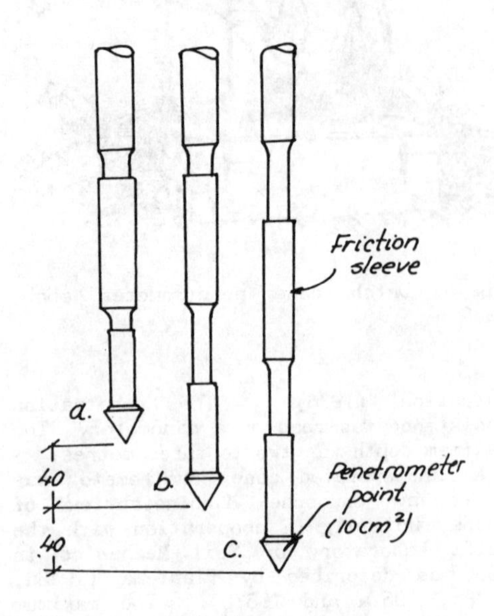

Fig 39 Begemann friction sleeve

5.4 Other Mechanical Cone Penetrometers

Several other mechanical cone penetrometers have been developed with slightly different features. The cone penetrometer developed at the Belgium Geotechnical Institute in Ghent by de Beer (1945) has a fixed cone. The cone

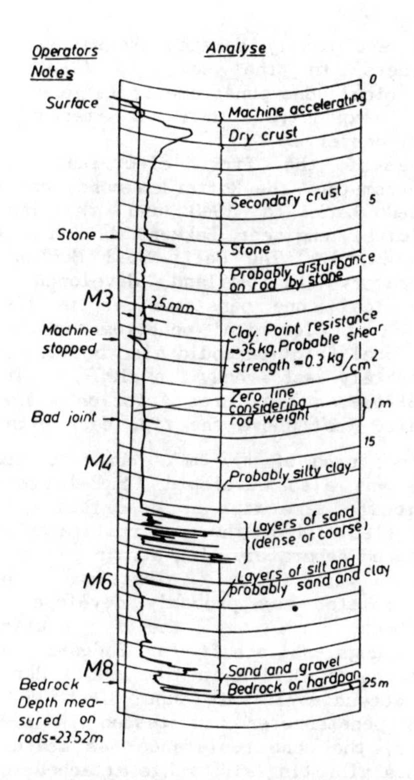

Fig 40 Penetration resistance as determined by the SGI Static Sounding Machine and the evaluation of the test data (Broms, 1974)

and the total skin friction resistances are measured separately and at the same time as the penetrometer is pushed down through the soil.

A. H. Gawith at the Country Roads Board of Victoria built in 1951 the first quasi-static cone penetrometer in Australia (Scala, 1956, Holden, 1987). The capacity of this penetrometer which initially was low has gradually been increased to 20 tonnes (1974).

Another significant development was the SGI cone penetrometer which is mounted on a terrain going vehicle. This penetrometer has a conical point with either 25 mm or 40 mm diameter. A special feature of this penetrometer is that the penetrometer rod is rotated as the point is pushed down through the soil as proposed initially by Kjellman (Flodin, 1986). From the torque measurements it is possible to separate the point resistance from the skin friction resistance. This is done automatically by the machine. A typical recorded is shown in Fig 40 where the

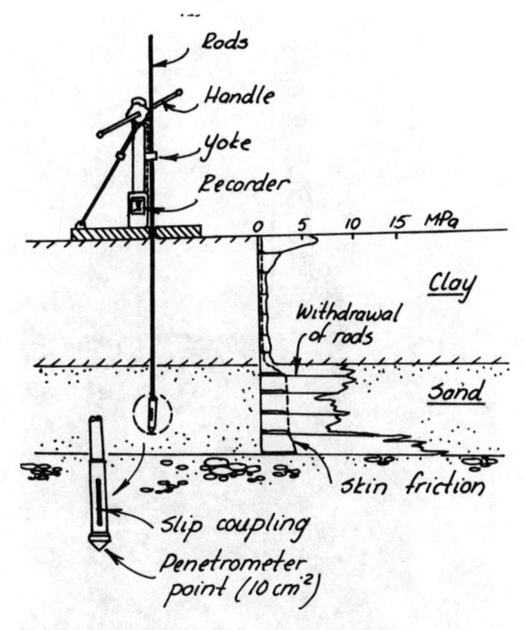

Fig 41 Nilcon penetrometer

cone resistance has been plotted as a function of the depth (Kallstenius, 1960). The maximum axial force that can be applied is 10 kN. The corresponding maximum torque is 0.25 kNm.

Another interesting static penetrometer is the Nilcon penetrometer which was used extensively in Sweden, Finland and Norway in the '60th and 70th. It is a mechanical penetrometer with a 10 cm^2 fixed square point. The point is provided with a slip coupling just above the penetrometer point so that the skin friction resistance can be separated from the point resistance. The skin friction resistance is determined by first retracting the penetrometer 50 mm, every 1 to 2 m, which corresponds to the play of the point. Thereafter, the rod is pushed down alone as illustrated in Fig 41. The force is assumed to correspond to the skin friction resistance until the point is engaged after the rod has been pushed down 50 mm. However, the remoulding of the soil around the rod as it is retracted can affect the skin friction resistance and thus the results.

The first mechanical cone penetrometer USSR was developed in 1953 by the "Hydro-project" Institute. The maximum capacity was 100 kN. The method was standardized in 1974 (GOST-20069-74). This standard which was revised in 1981

Fig 42 Penetrometer type SP-59

is consistant with the European reference standard (Mariupolski and Trofimenkov, 1987). Mechanical cone penetrometers are widely used in USSR to determine the bearing capacity of piles. A hydraulic penetrometer (SP-59) mounted on a tractor is shown in Fig 42.

5.5 Electrical Cone Penetrometers

Several electical cone penetrometers where the cone resistance is measured separately using strain gauges or vibrating wire gauges have been developed since 1950. The main advantage with an electrical cone penetrometer is that a continuous recording of the penetration resistance is obtained as a function of the depth. An electrical cone is more sensitive than a mechanical penetrometer so that it can be used also in loose sand and in soft clay and silt. One disadvantage with electrical cone penetrometers is that they are not as rugged as a mechanical penetrometer. They are easily damaged by stones and boulders in the soil. Also, an electrical cone penetrometer is generally more expensive than a mechanical penetrometer.

One of the first electrical cone penetrometer, the Rotterdam cone, can be traced back to 1948 and the Dutch municipal engineer Bakker who patented the method. The Delft Soil Mechanics Laboratory in Holland developed an electrical cone penetrometers in 1949. The first electrical penetrometer where the local friction could also be measured separately was ready in 1957. This penetrometer had three friction sleeves located just above the tip, each with a surface area of 400 cm^2. An electrical cone was also developed in Belgium at about the same time as described by de Beer (1951) where the cone resistance was measured separately with strain gauges.

The very first electrical cone penetrometer was probably developed by Hoffmann at the Deutsche Forschungsgesellschaft fur Bodenmechanik (Degebo) in Berlin in the 1940's during the second world war (Muhs, 1949, 1978). This penetrometer was tested in 1944. Later, the cone resistance was measured with a vibrating wire gauge attached to a load cell just above the cone. The signals were transmitted to the ground surface through a cable inside the hollow penetrometer rods (Kahl and Muhs, 1952). The total force which was required to push down the penetrometer was also measured. The maximum capacity of the tip of the penetrometer was 50 kN. The penetration rate was kept constant during a test, 20 mm/s.

The most commonly used electrical penetrometer today is that developed by Fugro in 1965 in the Netherlands in collaboration with T.N.O. and the Philips Company. The point and the sleeve resistances of this penetrometer are determined separately with load cells placed just above the tip. The friction sleeve with a surface area of 150 cm^2 is located just above the cone. With the early models only the point resistance was measured. The electrical cone penetrometers does not have a convical extension of the cone that can affect the test results in contrast to the mechanical Dutch cone penetrometer used in Holland. The Fugro cone penetrometer can also be used in very soft clay due to the high sensitivity of the measuring system.

The verticality of the penetrometer

during the testing is important. A deviation of e.g. 1/20, which is not uncommon, can significantly affect the results. The deviation can be checked with an inclinometer located at the lower end of the penetrometer (de Ruiter, 1971).

An electrical cone with a friction sleeve was developed in 1972 at the Delft Soil Mechanics Laboratory (DSML) in Holland. The diameter of the sleeve was only 28 mm, thus 7.6 mm less than the diameter of the cone (35.6 mm) so that similar results as with the mechanical Dutch cone penetrometer could be obtained (Heijnen, 1973). This electrical cone penetrometer was extensively used up to 1980.

A large number of different electrical penetrometers are available today from Sweden, (Dahlberg, 1974), Australia (Gawith and Barlett, 1963, Barlett and Holden 1968, Holden, 1974) and China (Chen, 1987) and Italy (Baldi et al, 1981a, 1981b). An interesting development is the electrical cone penetrometer by Geotech, Gothenburg, Sweden where vibrating wire gauges are used to determine separately the point and sleeve resistances. The signals from the gauges are transferred acoustically through the rods to the ground surface where they are recorded after they have been amplified by a unit located just above the penetrometer point (Dahlberg, 1974; Nilsson, 1975; Jonell, 1975). An hydraulic jack with a 1.1 m stroke is used to force down the penetrometer. The maximum capacity is 30 kN.

Of interest is also the self-recording penetrometer developed in 1959 by Broms and Broussard. The recording unit is located just above the tip of the penetrometer thereby eliminating the need of cables to the ground surface. The point of the penetrometer extends below the drilling bit. The penetrometer unit has been designed to be used with a drilling rig type Failing 1500. The axial thrust provided by the machine is utilized to pushed down the penetrometer through the soil.

An hydraulic operated penetrometer has been designed by Parez in 1953 in France (1974) where the point resistance is measured hydraulically at the surface with a manometer. The penetrometer is mounted on a heavy truck so that the reaction force when the penetrometer is pushed into the soil can be resisted. The maximum capacity is 150 kN. The penetration depth is indicated by a lamp that lights up every 0.1 m of penetration.

Several hydraulic penetrometers were built in 1966 at C.E.B.T.P. (Centre Experimental du Batiment et des Travaux Publics) in Paris, France. The point resistance was measured by load cells located just above the point or hydraulically with manometers at the ground surface. The diameter of the penetrometers was large, up to 320 mm. The purpose was to investigate the behaviour of piles.

Several special penetrometers for exploration of the moon have been developed in USSR and in the United States. The diameter of the Russian penetrometer was 35 mm. The apex angle was 103 degrees. The maximum capacity was 700 kPa.

The use of gravity platforms in the North Sea for the exploration and production of oil has led to the development of several new types of penetrometers so that the bearing capacity of the soil below the platforms can be estimated as well as the penetration resistance of the skirts which are used to protect the structures against erosion (Zuidberg et al, 1986). One example is the NGI offshore penetrometer shown in Fig 43. The maximum capacity is 50 kN which corresponds to the weight of the penetrometer. Other cone penetrometers which are used offshore are the "Seacalf" developed by Fugro (Zuidberg, 1975) and the McClelland "Stingray" (Semple and Johnston, 1979). There are also penetrometers available which are pushed down into the sea bottom from a diving bell (Vermeiden, 1977).

There were requests very early for

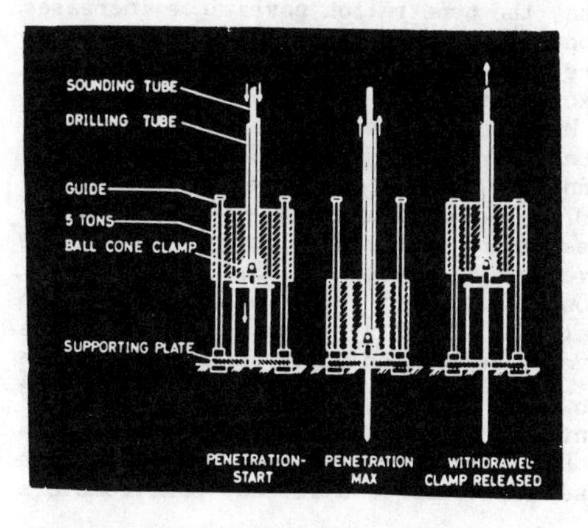

Fig 43 NGI offshore penetrometer

standardization of various penetration testing methods so that the results from penetration tests carried out in different countries could be compared. There is an ASTM standard (D3441-79) for CPT as well as an European Reference Standard (ISSMFE, 1977). In the ASTM standard, the method is referred to as a quasi-static cone penetrometer test.

5.6 Determination of Strength and Deformation Properties from CPT

Several factors affect the penetration resistance at CPT such as the diameter of the cone and of the rods, the particle size, the relative density (density index), the gradation, the compressibility and the degree of saturation of the soil as well as the shape and the compressive strength of the soil particles. According to ASTM standard (D3441-79) and the European Reference Standard and (ISSMFE, 1977) the diameter of the cone and of the rods should be the same.

Kok (1974), for example, has reported that the cone resistance for an electrical cone penetrometer in sand is approximately 30% higher than that of a mechanical penetrometer where the cone is pushed down separately. Even larger differences have been reported by Holm (1983). De Beer (1951) was one of the first to point out this effect for sandy and silty soils. For clays, there is practically no difference between the two methods. According to de Ruiter (1982) the effect of the diameter of the cone is small as long as the diameter of the cone and the rods are the same.

The penetration resistance increases in general with increasing penetration rate. For clay, Muromachi (1974) has reported that the penetration resistance increases about 15% for every tenfold increase of the rate when the penetration rate exceeds the standard rate (10 mm/s).

Much work has been done mainly in Italy (Baldi et al 1981a, 1981b, 1982, 1985) and in the United States (Schmertmann, 1976a) to correlate the cone penetration resistance with the relative density, the overburden pressure, the ground water conditions and the overconsolidation ratio. Tests indicate that the overconsolidation ratio as it affects the K_o-value has an important effect on the interpretation of the results.

Large calibration chambers have been used to evaluate different penetrometers especially the static cone penetrometer. In the calibration chambers, it is possible to control closely the compaction, the stress conditions and the ground water conditions as discussed by e.g. Holden (1971, Veismanis (1974), Marcuson & Bieganousky (1977a and 1977b). However, also the size of the calibration tank is important (Parkin & Lunne, 1982; Jamiolkowski et al, 1985). A correction is generally required.

One limitation of calibration chamber tests is that the effect of aging of the sand cannot be considered. If the correlations derived from the calibration chamber tests are used, then the relative density of the sand may be overestimated (e.g. Parkin and Lunne, 1982; Jamiolkowski et al, 1985).

The relationship between the cone resistance and the relative density was investigated relatively early by e.g. Meyerhof (1956), Schultze and Melzer (1965), Kahl et al (1968) and by Thomas (1968). Later, calibration chambers were used by e.g. Schmertmann (1975), Baldi et al (1981a, 1981b, 1982, 1985) and by Villet and Mitchell (1981) to determine this relationship. The relationship proposed by Schmertmann (1975) is shown in Fig 44. It can be seen that the increase of the penetration resistance

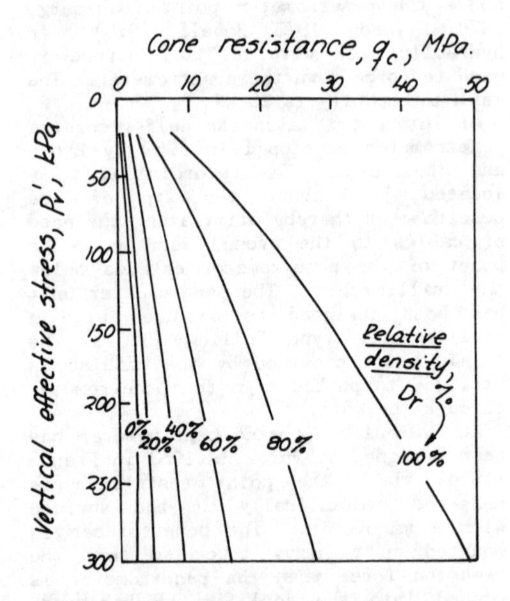

Fig 44 Relationship between q_c, p_v' and D_r (after Schmertmann, 1975)

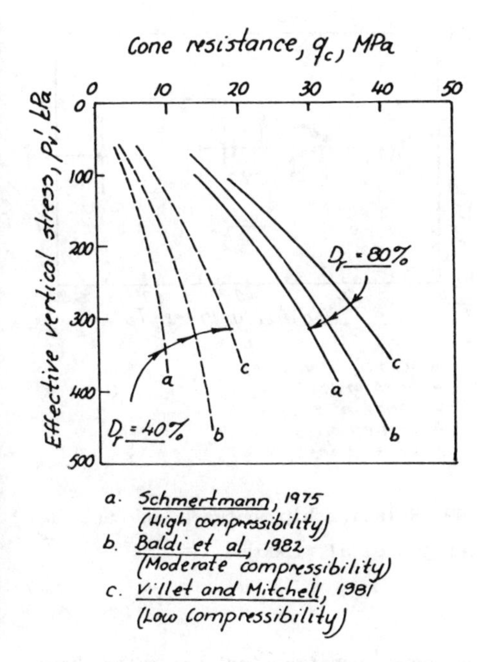

Fig 45 Effect of compressibility on cone penetration resistance (after Robertson and Campanella, 1983a)

with depth is small when the relative density is low while at a high relative density the penetration resistance increases rapidly with increasing penetration depth.

However, the compressibility of the soil also affects the relationship between q_c, p_v' and D_r as illustrated in Fig 45 (Robinson and Campanella, 1983a). It can be seen that the penetration resistance is relatively high when the compressibility of the soil is low. The penetration resistance cannot be used directly to estimate the relative density at shallow depths, without considering the overburden pressure. Since the penetration resistance increases approximately linearly with the depth down to a certain critical depth (5d to about 20d). However, the increase of the penetration resistance with depth can be used to estimate the particle size and to classify the soil as discussed, for example, by Kallstenius (1961).

Relationships between penetration resistance (q_c) and the <u>angle of internal friction</u> ϕ' have been proposed for sand by e.g. de Beer (1945), Meyerhof (1956,

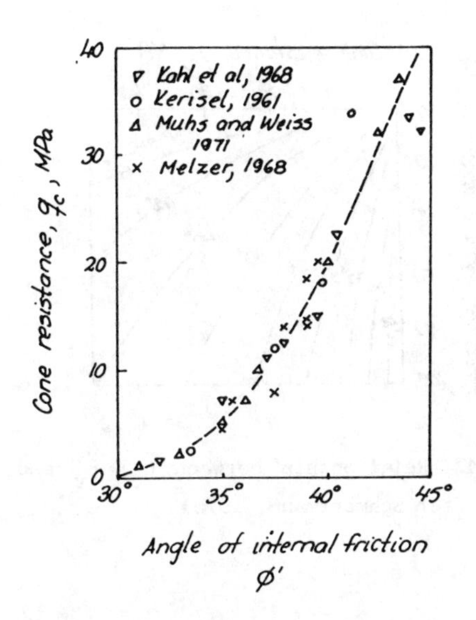

Fig 46 Relationship between q_c and ϕ'

1976), Kahl et al (1968), Kerisel (1961), Vesic (1965, 1967), Meltzer (1968), Muhs and Weiss (1971), Trofimenkov (1974), Janbu and Senneset (1974) and by Schmertmann (1975). Cavity expansion theory has been used to analyze the results (e.g. Baligh, 1976). However, the compressibility of the sand affects the penetration resistance and the interpretation as pointed out by Robertson and Campanella (1983a) as well as the compressive strength of the sand particles.

Meyerhof (1976) has proposed the following relationship

$$\phi' = 29^o + 2.5 \sqrt{q_c} \qquad (15)$$

where q_c is the cone penetration resistance in MPa. The friction angle should be increased by 5 degrees for sandy gravel according to Meyerhof (1976) and be reduced by 5 degrees for silty sand.

A summary of the different proposed relationships is shown in Fig 46. The main reason for the scatter is probably differences in the equipment and in the testing procedures that were followed to carry out the tests. Also the method used to determine the angle of internal friction of the soil with triaxial or direct shear tests has probably also contributed to the scatter of the results.

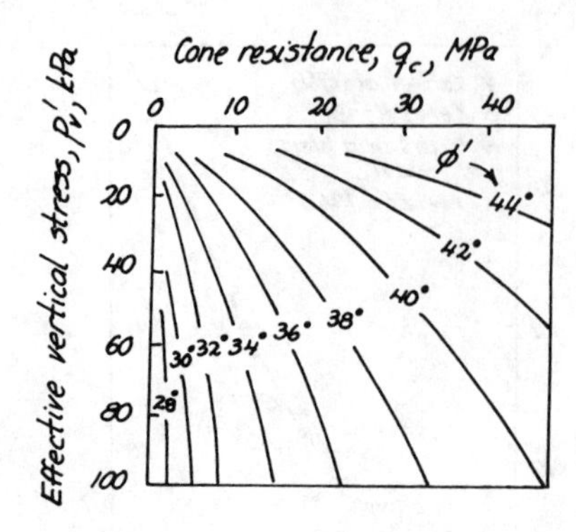

Fig 47 Relationship between q_c, p'_v and ϕ' (after Schmertmann, 1975)

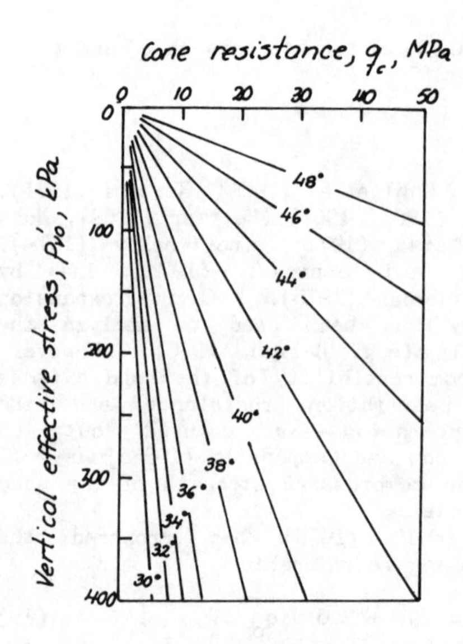

Fig 48 Proposed relationship between q_c, P'_v and ϕ' (after Robertson and Campanella, 1983a)

The results are also affected by the effective vertical stress as shown in Fig 47 (Schmertmann, 1975). The largest effect is obtained when the relative density of the soil is high. Similar results have been reported by Robertson and Campanella (1983a) as shown in Fig 48.

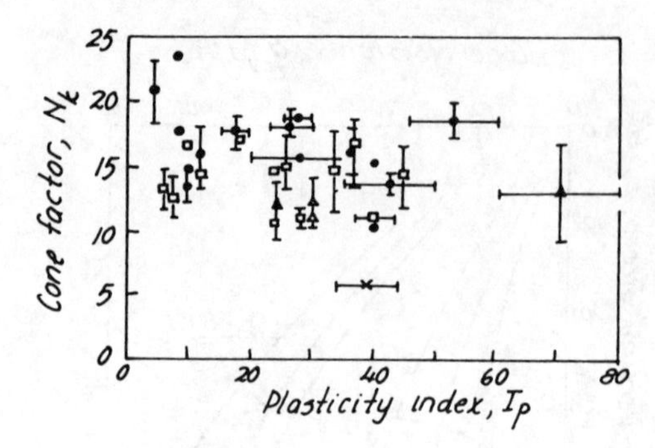

Fig 49 Relationship between N_k and I_p (after Aas et al, 1986)

The cone penetrometer has also been used to estimate for clays the undrained shear strength (Thomas, 1965, Anagnostopoulos, 1974; Brand et al, 1974; Amar et al, 1975; Lunne et al, 1976; Lunne and Kleven, 1981; Jamiolkowski et al; 1982 Aas et al, 1986). The following relationship is normally used

$$q_c = c_u N_k + p_{vo} \tag{16}$$

where q_c is the cone resistance, p_{vo} is the total overburden pressure and N_k is a cone factor. The cone factor has been determined by comparisons with the results from field vane tests. Lunne et al, 1985) later modified this equation.

$$q_T = c_u N_{KT} + p_{vo} \tag{17}$$

where q_T is the corrected cone resistance and N_{KT} is a corrected cone factor.

The cone factor varies in general between 10 and 20. The average value is about 15. When the sensitivity is high, then the cone factor can be as low as 10 (e.g. Ladanyi and Eden, 1969. Barentsen found in 1936 that the cone resistance in clay was 11 times the cohesion while in peat the cone resistance could be up to 21 times the cohesion.

In Fig 49 the cone factor N_k has been plotted as a function of the plasticity index (Aas et al, 1986). It can be seen that the average value on N_k has a tendency to decrease with increasing plasticity index (I_p). There are also several other factors that may affect this relationship such as the sensitivity (Ladanyi and Eden, 1969; and Roy et al, 1974) and the overconsolidation ratio (OCR). Muromachi (1974) found that the effect of the sensitivity ratio (S_t) is proportional to $0.5\ (1 + 1/\sqrt{S_t})$.

The method to determine the undrained shear strength affects the results as well as cracks and fissures in the clay (Marsland, 1974; de Beer et al, 1977) and the type of penetrometer used (de Beer et al, 1974). In stiff fissured clay, the skin friction resistance of a pile can be overestimated from the results of CPT (de Beer et al, 1977). Borings are usually required for the interpretation of the results particularly when the soil is stratified since the results are affected by adjacent layers when the distance is less than about 5d.

The cone penetration test has been used since about 1940 in e.g. Belgium for the design of shallow foundations (De Beer, 1951). The design of all bridges and buildings built by the Ministry of Public Works in Belgium has been based and the results from CPT. In Zair, the method has been used since 1948.

Buisman (1935 and 1940) was the first to propose a relationship between the coefficient of compressibility (C) and the penetration resistance :

$$C = 1.5\ q_c/p'_{vo} \tag{18}$$

where p'_{vo} is the effective overburden pressure. Kerisel (1969) modified this relationship as follows

$$C = \beta q_c/p'_{vo} \tag{19}$$

where β is a coefficient which for sand depends on the relative density as shown in the following table :

Table 8. Proposed values on the coefficient β (after Kerisel, 1969)

Soil	Coefficient β
Very dense sand	≤ 1.0
Medium dense	1.0
Loose sand	1.5

The settlement (s) is often be estimated from the equation:

$$s = 2.3\ \frac{H}{C}\ \log \frac{p'_{vo} + \Delta p}{p'_{vo}} \tag{20}$$

where H is the thickness of the compressible layer and Δp is the load increase.

The settlement can also be calculated from the modulus of elasticity E_s. This modulus and the penetration resistance q_c are usually assumed to be related through the equation :

$$E_s = \alpha q_c \tag{21}$$

where α is a coefficient. Test data indicate that the coefficient α varies considerably partly because soil is not an ideal elastic material and that α depends on the stress level. For plate load tests on sand the coefficient α is normally between 1.5 and 4. These values are often applied in settlement calculations (e.g. de Beer, 1965; Schmertmann, 1970, Mitchell and Gardner, 1975). In USSR, a value of 3.0 on α is used for sand and 7.0 for clay (Trofimenkov, 1974). Vesic (1970) has proposed the following relationsip :

$$\alpha = 2\ (1 + D_r^2) \tag{22}$$

where D_r is the relative density of the sand. However, Dahlberg (1974) found that α increases with increasing penetration resistance q_c, when the sand has been preloaded.

The ultimate bearing capacity of footings on sand is generally calculated from the bearing capacity factors N_γ and N_q, which depends on the effective angle of internal friction of the soil (Muhs, 1969; Muhs and Weiss, 1972; Weiss, 1970). Muhs and Weiss (1971) have proposed the following values on N_γ and N_q.

$$N_\gamma = 6.3\ q_c - 5\ \text{(MPa)} \tag{23}$$

$$N_q = 5.9\ q_c - 1\ \text{(MPa)} \tag{24}$$

where q_c is the cone resistance in MPa. However, the scatter of the results is large particularly when q_c is low ($q_c \leq 1$ MPa). Close to the ground surface, q_c

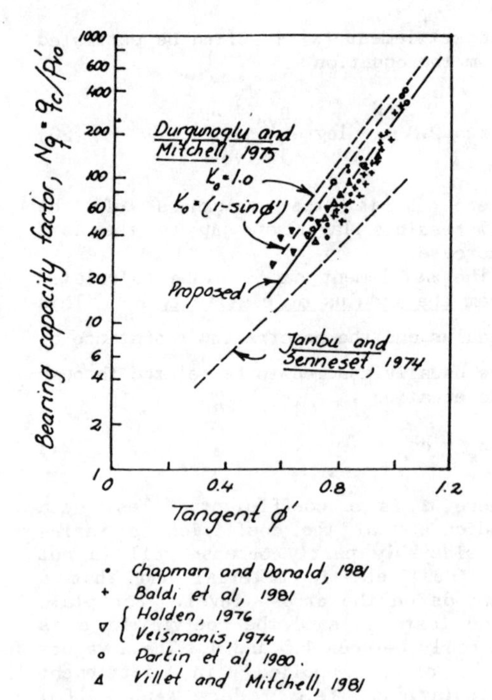

Fig 50 Relationship between N_q and ϕ' (after Robertson and Campanella, 1983a)

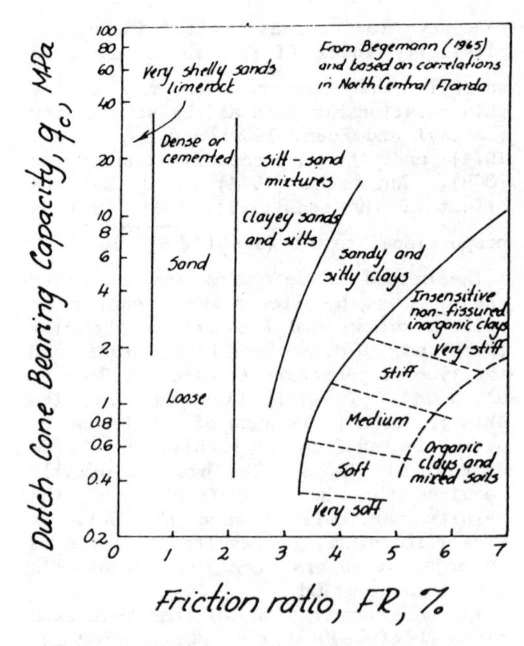

Fig 51 Relationship between q_c, FR and soil type for Dutch cone penetrometer (after Schmertmann, 1978a)

depends also on the effective overburden pressure. Furthermore, the ultimate bearing capacity is a function of size of the loaded area.

The increase of the penetration resistance with depth has been utilized by Janbu and Senneset (1974) and by Durgunoglu and Mitchell (1975a, 1975b) to estimate the bearing capacity factor N_q as illustrated in Fig 50. In the interpretation of the results it should be noted that the bearing capacity of footings or of piles is affected by the distance to the ground surface and by nearby hard strata.

Begemann (1965) was the first to point out that the friction ratio (FR), the ratio of the friction resistance f_s along the sleeve and the point resistance q_c, increases in general with decreasing particle size as mentioned before. Also, Sanglerat (1972) Sanglerat et al (1974), Schmertmann (1978a), Robertson and Companella (1983a) have proposed relationships between the friction ratio and the compostion of the soil. It should be noted that Sanglerat (1972) and Sanglerat et al (1974) used the 80 mm diameter Adena penetrometer in his investigations while Begemann (1965) and Schmertmann (1978a) tested a mechanical penetrometer where the friction sleeve is located some distance above the cone. Robertson and Campanella (1983a) carried out their investigations with an electrical penetrometer where the diameter of the cone and the friction sleeve was the same and where the friction sleeve is located just above the cone. A higher friction ratio is in general obtained for sand with an electrical CPT than with a mechanical cone penetrometer. The type of penetrometer has thus a large influence on the friction ratio (FR) as pointed out by e.g. de Ruiter (1971).

The friction ratio for sand increases in general with increasing compressibility of the soil particles. Carbonate sands may have a friction ratio that can be as high as 3% (Joustra and de Gijt, 1982). Normally the friction ratio for sand is about 1.0% to 0.5%.

The relationships proposed by Schmertmann (1978a) and by Douglas and Olsen (1981) are shown in Figs 51 and 52, respectively. It can be seen that the

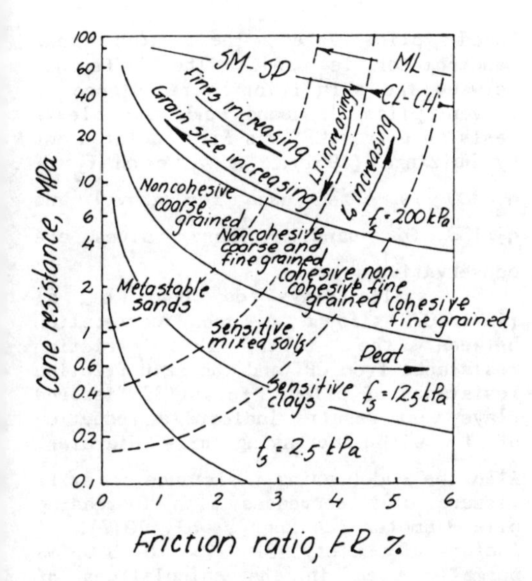

Fig 52 Soil Classification Chart for Standard Electric Friction Cone (Adapted from Douglas and Olsen, 1981)

friction ratio increases with decreasing particle size and with increasing organic content. For granular soils Douglas and Olson (1981) suggested that the friction ratio for an electrical cone penetrometer increases in general with increasing relative density (D_r), with increasing, effective overburden pressure (p_v') and with increasing coefficient of lateral earth pressure at rest (K_o). The friction ratio for very soft clay can be very low according to Douglas and Olsen (1981) especially when the sensitivity of the clay is high.

There are some uncertainties in the determination of the sleeve resistance using a mechanical cone penetrometer as mentioned previously since the cone is first pushed down alone into the soil, then the cone and the friction sleeve together as illustrated in Fig 39. The penetration between each load cycle is 80 to 200 mm. It is assumed in the interpretation of the results that the sleeve resistance corresponds to the difference between the two readings.

It is difficult to analyze the results in soft clay or in loose sand or silt using a mechanical cone penetrometer because of the very low skin friction resistance which is difficult to measure accurately.

In the Canadian Manual on Foundation Engineering (National Research Council of Canada 1975), the allowable load (q_a) for simple structures on clay can be evaluated from the relationship

$$q_a = q_c/10 \text{ (MPa)} \tag{25}$$

where q_c is the cone resistance.

Meyerhof (1956) has recommended the following relationship to estimate the ultimate capacity (q_u) of footings on sand

$$q_u = q_c B\ (1 + D/B)/40 \text{ (MPa)} \tag{26}$$

where B is the width and D is the depth below the ground surface.

Meyerhof (1965) has recommended that the allowable load should be reduced by 50% for clayey sand or when the ground water level is located at or above the foundation level. It might be required to reduce the allowable load still further especially for wide footings or rafts so that the settlements will not be excessive.

The following equations are sometimes used in order to limit the settlements in sand to 25 mm.

$$q_a = q_c/30 \text{ (MPa)} \tag{27}$$

at $B \leq 1.2$ and

$$q_a = q_c\ (1 + 0.3/B)^2/50 \text{ (MPa)} \tag{28}$$

at $B \leq 1.2$ m

where q_c is the average penetration resistance down to a depth which corresponds to the width of the footing.

The method proposed by Schmertmann (1970) is commonly used to estimate the settlement of footings on sand using a value on the equivalent modulus of elasticity (E_s) of 2.5 q_c and 3.5 q_c for square and rectangular footings, respectively. Meyerhof (1965) has proposed the following simple relationship :

$$s = qB/2q_c \text{ (MPa)} \tag{29}$$

where q is the net foundation pressure and q_c is the average penetration resistance down to a depth that corresponds to the width of the loaded area.

Cone penetration tests are also commonly used to estimate the bearing capacity of both driven and bored piles (See e.g. de Beer, 1945; Plantema, 1948a, 1948b, 1953a, Huizinga, 1941, 1951; Franx, 1952; Heijnen, 1974; de Beer et al, 1977). Plantema (1957a) has observed that the changes of the strength and deformation properties of the soil by the pile driving can be large and that it is important to consider these changes in the interpretation of the results. The point resistance q_p of a jacked-down pile in a granular soil is generally assumed to correspond to the cone resistance q_c when the pile diameter is less than 1.0 m as proposed by Plantema (1948b). He also found that the point resistance of driven piles in sand is increased by 50 to 250 percent (Plantema, 1957a). "The effect of compaction due to driving was to give an additional factor of safety to the design based on deep-soundings". Plantema also pointed out that "the pile would have to be driven into the good layer a distance equal to 4 to 6 times the diameter of the pile" to develop the estimated resistance. Plantema was thus the first to recognize the importance of the critical depth for driven piles. In Holland, the average cone resistance (q_c) within a depth equal to 0.7 D to 3.75 D below the pile tip and 8 D above the pile tip is generally used in the calculation of the point resistance (D is the pile diameter). This recommendation corresponds closely to that proposed by van der Veen (1952).

In Belgium, the critical depth which is required to develop the maximum point resistance of a pile is also considered in pile design (de Beer, 1971/1972; van Impe, 1986). For bored piles the point resistance and the skin friction correspond approximately to half the values determined by static cone penetration tests (Meyerhof, 1959; Kerisel, 1961; Vesic, 1967). The zone affected by the piles during loading increases with increasing relative density of the soil as pointed out by e.g. Malyshev and Lavisin (1974). An upper limit of 15 MPa on the point resistance has been proposed by de Ruiter and Beringen (1979) while a somewhat lower value (10 MPa) is recommended by Meyerhof (1976) and by Vesic (1965).

The unit skin friction (f_p) of a pile is generally assumed to $2f_s$ for driven piles in granular soils and to f_s for bored piles when the Dutch cone penetrometer is used. The difference between the skin friction resistance of driven piles in sand and the sleeve resistance from CPT was first pointed out by Huizinga (1951). A value on f_p of $q_c/200$ is often used for gravel and $q_c/150$ for sand. These values are conservative.

For stiff clays de Beer et al (1971/1972) found a good correlation between the total skin friction resistance from CPT and the skin friction resistance of piles. For stiff fissured clays test results indicate a reduction of f_s with increasing pile diameter. Also the end bearing resistance in stiff fissure clay decreases with increasing pile diameter (de Beer et al, 1977). A factor of safety of 2.5 to 3.0 is normally used in the calculations of allowable load. Results from cone penetration tests have also been utilized to evaluate the lateral resistance of piles (e.g. Marche, 1974). The settlements of both single piles and of pile groups are often calculated from elastic theory.

The cone penetration test has also been used to check the compaction of granular soils by e.g. Vibroflotation or pile driving (Sanglerat, 1972). One of the early application of CPT for compaction control is that described by Plantema (1957b).

6 DYNAMIC PROBING (DPT)

6.1 Ram Soundings

Ram soundings are commonly used in the Scandinavian countries, France, Italy and Germany to determine the soil conditions in general during the exploratory phase of an investigation. At a ram sounding test a square or circular point is used which is driven down into the ground and the number of blows (N_{20} or N_{10}) required to drive the penetrometer 0.2 m or 0.1 m is counted. No samples are recovered with this method.

There are essentially four types of dynamic penetrometers depending on the mass of the hammer used: light, medium, heavy and superheavy as shown in the following table :

Table 9. Classificaiton of Dynamic Penetrometer (after Melzer and Smoltczyk, 1982)

Type	Mass of hammer (kg)
Light (DPL)	≤ 10
Medium (DPM)	10 - 40
Heavy (DPH)	40 - 60
Super-heavy (DPSH)	≥ 60

Super-heavy or heavy penetrometers are commonly used in, for example, the Scandinavian countries, FRG, France, Finland, UK, USSR and South Africa to determine the length of piles. The super-heavy penetrometers are similar to the Standard penetration test.

Van Wambeke (1982) has pointed out that the simplicity of the equipment and of the method has make the dynamic penetration test the most economical in-situ testing method and the easiest to use. This is the main reason why it should not be disregarded but the results must be interpretated without going beyond the capability of the method.

Different methods can be used to reduce the skin friction along the rods e.g. by increasing the diameter of the cone so that the diameter becomes larger than the diameter of the rods. It is also possible to reduce the skin friciton with a casing, by rotating the rods every 0.2 m or by injecting drilling mud or water just above the cone during the driving (e.g. Meardi and Meardi, 1974).

Several factors can affect the results especially in cohesive soils such as the driving rate and interruptions of the driving. Interruptions should, therefore, be avoided and the time for the splicing of the rods should be kept as short as possible. Driving rates of 15 to 30 blows/minutes are common.

Dynamic penetration tests are often used to check the uniformity of the soil conditions at a particular site to estimate the location, thickness and lateral extent of the different strata and to determine the depth to bedrock. The method has the advantage compared with the cone penetration test and weight soundings that hard layers can be penetrated. Dynamic penetration tests are used mainly to determine the required length of driven steel or concrete piles and of sheet piles.

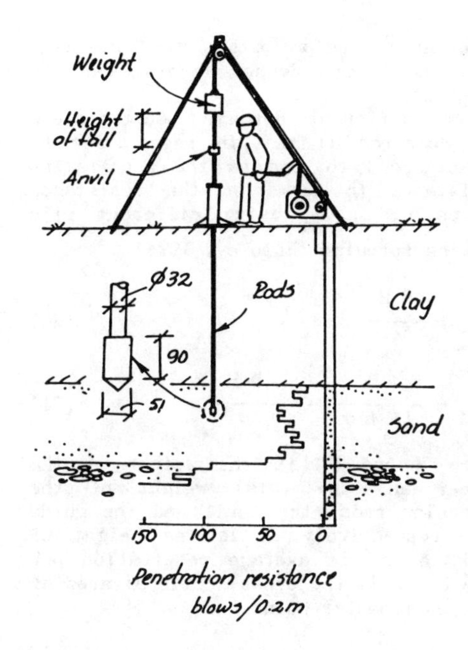

Fig 53 Ram sounding test (DPB)

A reference standard has been proposed for the ram sounding method (DPA and DPB). A 63.5 kg hammer with a height of fall of 0.5 m is used to drive the penetrometer into the ground (Fig 54). The diameter of the point is 62 mm for DPA and 51 mm for DPB. The apex angle is 90 degrees at both methods. Rods with 40 to 45 mm diameter are used at DPA while at DPB the diameter is 32 mm. The skin friction along the rods at DPA is eliminated with drilling mud or with a casing. At DPB the skin friciton resistance along the rods affects the results.

Dynamic penetrometers are mainly used in Europe during the preliminary exploration phase to determine the thickness and the location of the different strata and during the detailed investigation phase to estimate the shear strength and the compressibility of the various strata. Dynamic penetrometers are also used to check the compaction of fills or the loosening of the soil at the bottom of deep excavations.

Light penetrometers are suitable for shallow depth (less than about 8 m) while super-heavy penetrometers are often required when the penetration depth exceeds about 25 m. The heavy and the super-heavy equipment has the advantage that hard or dense layers can be penetrated.

When different dynamic penetrometers are compared it may in some cases be advantageous to express the penetration resistance in terms of the resistance values r_d or q_d using different pile driving formulae (Bolomey, 1974).

$$r_d = \frac{M\,H}{Ae} \tag{30}$$

$$q_d = \frac{M}{(M + M')} \cdot \frac{M\,H}{Ae} \tag{31}$$

where M and M' is the weight of the hammer and the total weight of the extension rods, the anvil and the guide rod, respectively, H is the height of fall, e is the average penetration per blow and A is the cross-sectional area of the penetrometer point.

6.2 Early Developments

In the chapter on the early history of penetration testing, it was mentioned that the driving resistance of a steel rod which is rammed down into the ground can be taken as indication of the bearing capacity of the soil. It was Goldmann (1699) as mentioned by Zweck (1969) that first used the ram penetration method in Germany. The penetrometer was driven down using a sledge hammer and the penetration for each blow was measured. However, the contribution by Goldmann was forgotten until the last half of the 19th century.

In the Swedish National Achives, a record has been found of a ram penetration test (boring log) carried out in 1890 for a project in Stockholm. The method was used to investigate an esker to determine the thickness and the depth of the different layers. The main purpose of the test was to determine the required length of piles. The word "harder" was used to describe the variation of the penetration resistance with depth (Flodin, 1987). The following equipment were used: Rods, 70 feet (about 12 m), probably square (25 x 25 mm), hammer, spits (2), wedge ring, chisels (2) and couplings.

In an investigation in 1910 of the same area, square penetrometer rods, probably 25 x 25 mm, were driven down through the

Fig 54 Ram sounding about 1940 (Wermbergs Mek Verkstad, 1949)

different soil strata by a 60 kg hammer which was probably of wood. The stroke was 0.6 m. Other rod sizes were also used, depending on the geological conditions and on the depth of the borehole. The penetration, which was recorded every 10 blows, varied between 30 and 130 mm/10 blows.

In an interesting leaflet from a mechanical workshop, B. M. Wermbergs Mekaniska Verkstad (1949) in Uddevalla, Sweden, there are some informations about the ram sounding equipment at the turn of the century. The leaflet has some poor but unique photographs of different boring equipment. A ram penetrometer is shown in one of these photographs (Fig 54) with four men lifting a heavy wooden hammer and with a boring foreman inspecting the progress of the testing.

From the list of the equipment offered one can find the following:

20 rods, each 1.0 m long
2 points, 1.0 m long
1 anvil
20 separate splices with conical ends to reduce the friction (!)
1 hammer of oak, with four handles

The testing has been described as follows : "The ram sounding device consists of square steel rods, 25 x 25 mm, coupled together with sleeves. The penetration every 20 blows (= 1 set) is measured to get an idea of the soil resistance". The rods were generally driven to refusal.

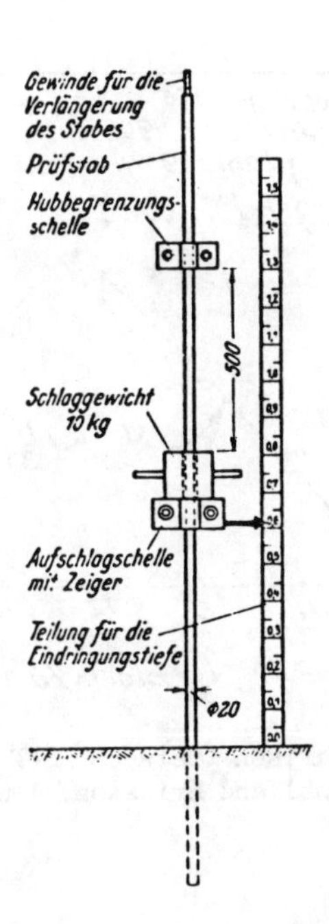

Fig 55 "Prüfstab" Künzel as used by Paproth (1943), with the rod diameter 20 mm and the steel drop hammer weighing 10 kg

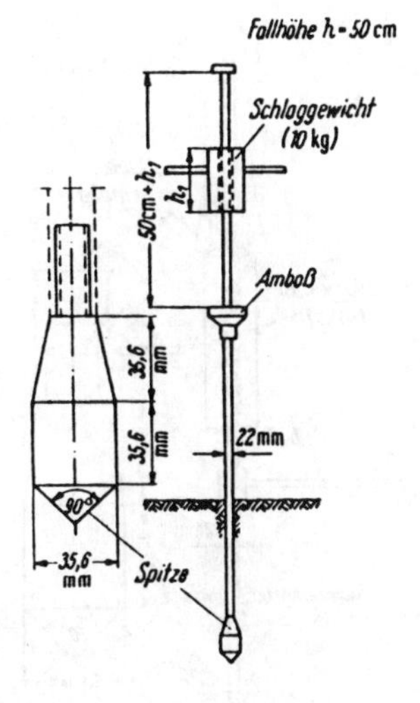

Fig 56 Light penetrometer German Standard (DIN 4094)

6.3 Further Development of the Ram Sounding Method

Goldmann (1699) was the first to determine in Germany the soil resistance with ram soundings. Later a so-called "Prüfstab", as described by Künzel (1936), was used. This is a simple penetrometer where the diameter of the rods is 16 to 20 mm and the length is 5 to 8 m. A 5 to 6 kg hammer with a drop height of 50 cm was used to drive the rods and the penetration every 10 blows was measured, or alternatively the number of blows required to drive the penetrometer a certain distance (100 mm). The results were plotted graphically.

Paproth (1943) used the Künzel penetrometer shown in Fig 55 but the weight of the hammer had been increased to 10 kg. The diameter of the rods was 22 mm. The penetrometer was provided with a conical point with a diameter of 35.6 mm. The skin friction was reduced with a casing. This "light penetrometer" was standardized in 1964 as shown in Fig 56.

The method is frequently used in Central Europe, e.g. in Bulgaria where it also has been approved as a standard method. Light penetrometers have also been developed in Sweden (Massarsch et al, 1976), Australia (Scala 1956; Glick and Clegg, 1965; Harison, 1986) and USSR (Mariupolski and Trofimonkov, 1987). In southeast Asia the so-called Mackintosh probe is extensively used (Chin, 1987).

The "heavy penetrometer" shown in Fig 57 was standardized in Germany 1964 and 1973. The diameter of the rods and of the point is 32 mm and 43.7 mm, respectively. (The cross-sectional area of the point is 15 cm^2). The drop height of the 50 kg hammer is the same as that of the light penetrometer, 0.5 m. The main purpose of the enlarged point was to reduce skin friction resistance along the rods. The number of blows required every 0.1 m is counted.

The first heavy penetrometer in Sweden, the "Borros heavy penetrometer" shown in Fig 58, was developed in 1935 by A Sundberg. It was patented in 1942. This penetrometer has gradually been improved and is used today in many parts of the world (Flodin, 1984).

A heavy ram penetrometers was developed in USSR in 1950. A 60 kg hammer with a height of fall of 0.8 m was used to drive a 74 mm diameter cone. The number of blows required is counted. The method was standardized in 1974 (GOST 19912-74). The standard was revised in 1981 (GOST 19912-81). Heavy dynamic penetrometers are also used in Australia (Holden, 1974) and in France (Sanglerat, 1973).

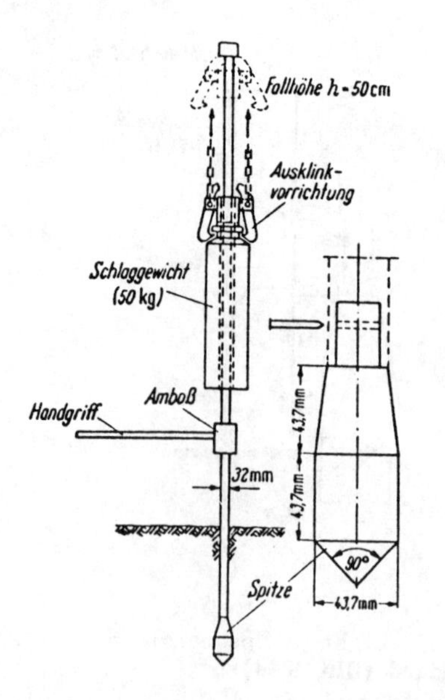

Fig 57 Heavy penetrometer German Standard (DIN 4094)

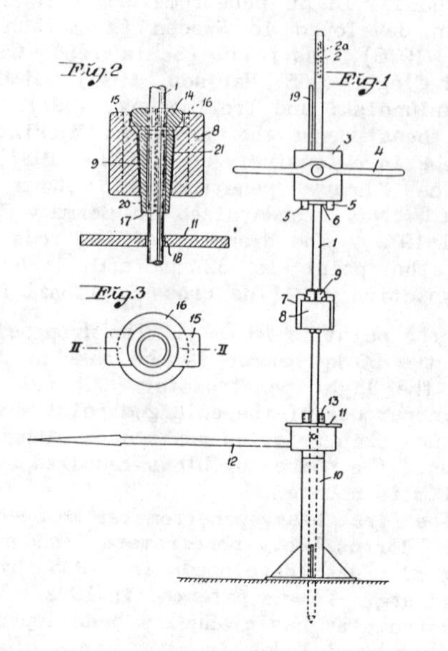

Fig 58 Heavy penetrometer developed by Sundberg in 1935 (Flodin, 1984).

6.4 Comparisons with other Penetration Methods

Comparison have been made for different penetrometers mainly in granular soils

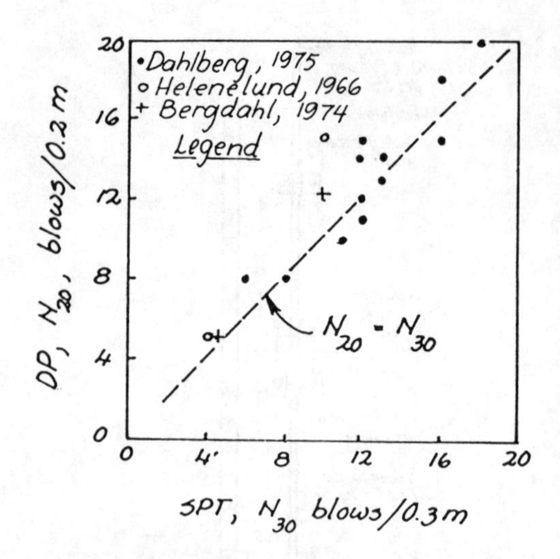

Fig 59 Comparison between SPT and DP (after Bergdahl and Eriksson, 1983)

between the penetration resistance and the relative density or the relative compaction. Calibration of the probe at each site is generally required. The greatest source of error is connected with the operation of the hammer to ensure that the hammer is falling freely and that the height of fall is constant. A large number of factors affect the results such as the dimensions of the point and of the rods, the mass of the hammer and the drop height.

Results from ram soundings (DPA and DPB) have been compared with the Standard penetration test by e.g. Helenelund (1966), Dahlberg and Bergdahl (1974), Dahlberg (1975) and by Bergdahl and Eriksson (1983) as shown in Fig 59. The results, which are consistent, indicate that the penetration resistances from ram soundings (N_{20}) and Standard penetration tests (N_{30}) are about the same :

$$N_{30} = N_{20} \tag{32}$$

Chang and Wong (1986) have reported the following relationship for residual granitic soils in Singapore.

$$N_{30} = 2N_{20} \tag{33}$$

Attempts have also been made to correlate the penetration resistance with the bearing capacity of piles as shown in Fig 60 (e.g. Norwegian Pile Committee,

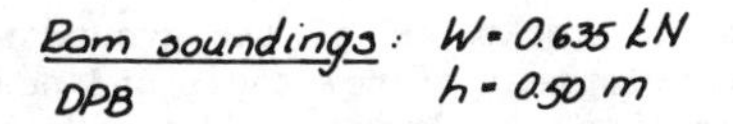

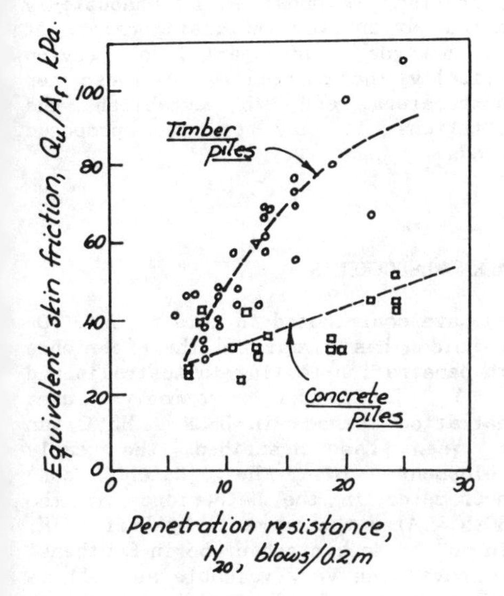

o Timber pile
▼ Timber pile, spliced
□ Concrete pile

Fig 60 Bearing capacity of piles from ram soundings (after Norwegian Pile Committee, 1973)

1973; Rollberg, 1976). In the figure Q_u is the ultimate capacity and A_f is the surface area of the piles. It can be seen that the ultimate bearing capacity is higher for timber than for concrete piles due to the tapered shape of the timber piles.

7 OTHER TYPES OF PENETROMETERS

7.1 In-situ Testing Methods

New tools are constantly being developed for the determination in-situ of various properties of soils and rocks. The capacity of existing machines has been increased so that even dense or cemented strata can be penetrated. This has often been at the expense of the simplicity of the various testing methods and of the costs. However, there is still a need for simple, inexpensive and rugged penetrometers especially during the exploratory phase of an investigation in order to get an indication of the soil conditions in general of the layer sequence and of the thickness of the different strata. Special penetrometers are useful mainly during the detailed investigation when accurate and reliable informations about the strength and deformation characteristics of the different strata are required in order to solve special problems.

The development especially during the last 15 years have been very rapid with respect to new types of penetrometers e.g. the lateral stress cone (Baligh et al, 1985), the vibratory cone (Sasaki et al, 1984) the acoustic cone (Villet et al, 1981, Massarsch, 1986) and the pressio-cone (Hughes et Robertson, 1985). Also the dilatometer can be regarded as a type of penetrometer (Marchetti, 1975, 1980). This new generation of penetrometers have partly been the result of recent advances of different electric sensors and improvements of data acquisition and processing systems.

A significant development is the pore pressure probe (piezo-cone) where the pore water pressure that is generated when a penetrometer is pushed down into the soil is measured. The main advantage with this type of probe is that relatively thin sand or silt seams can be detected which is important when the consolidation rate of clay and silt layers is calculated or estimated. It is also possible to determine the coefficient of consolidation of the soil by stopping the penetration and by measuring the dissipation of the excess pore water pressures with time. This method was developed independently in Sweden by B. A. Torstensson (1975) and in the United States by A. E. Z. Wissa at MIT (Wissa et al, 1975). However, the Delft Soil Mechanics Laboratory (DSML) claims that a piezo-cone penetrometer was built at DSML already in 1962.

Different types of piezo-cones are available today where the location of the porous elements has been varied and where it is possible to measure the pore water pressure at several locations simultaneously. In most penetrometers the porous element is located just above the tip of the penetrometer.

Pore pressure sensors have also been incorporated into electric cone penetrometers so that the measurement of the pore pressures can be combined with the registration of the cone resistance (q_c), the sleeve resistance (f_s) and the inclination of the probe (Baligh et al, 1981; Campanella and Robertson, 1981; de Ruiter, 1982).

There have also been attempts to estimate the shear strength using cavity expansion theory from the excess pore water pressures which are generated with the piezo-cone (Vesic, 1972; Randolph and Wroth, 1979; Massarsch and Broms, 1981). The ratio $\Delta u/c_u$ varies between 3 and 5 when the excess pore water pressure (Δu) is measured immediately above the cone. This ratio increases with decreasing plasticity index of the clay.

The ratio $\Delta u/(q_c - p_{vo})$ has also been used to classify soils (Senneset et al, 1982; Senneset and Janbu, 1984). This ratio decreases with increasing particle size and with increasing relative density of the soil. The change of the pore water pressure can even be negative in some overconsolidated soils.

The coefficient of consolidation can also be estimated with the piezocone from the dissipation rate of the excess pore water pressures (Torstensson, 1977).

8 CONCLUDING REMARKS

It has been a fascinating assignment to trace the development of the different penetration testing methods which are used in various parts of the world (STP, CPT, WST, DP). One had to admire the ingenuity of the civil engineers especially during the 18th and 19th centuries when canals, dams, railroads, harbours were constructed often in areas with very poor soil conditions. Experience was gained very slowly since there was no commonly used method to classify soils and rocks or to determine the consistancy. Terms like loose and dense were used to characterize granular soils while clays were classified as either soft, stiff or hard.

A very large number of different penetration testing methods were used around the turn of the century. Almost every drilling foreman had his own testing method and his own method of intepretation. Often he used the heel of his boot to estimate the consistancy and the allowable bearing capacity of spread footings and of piles.

Penetration testing of soils and rocks and the interpretation of the result have developed very rapidly especially during the last few years. A large number of highly specialized tools are available today to determine various properties of soils and rocks and of the ground water. The development of new tools and new soil exploration methods will undoubtedly continue at an ever increasing rate. It is, however, important to review critically the capability of these new penetrometers and to establish the limitations if any of the proposed methods.

9 ACKNOWLEDGEMENTS

Many have contributed to this report. Dr T C Holden has summarized the experience with penetration testing in Australia and Dr Y Trofimenkov commonly used penetration methods in USSR. Mr C van der Veen has described the early development of the Dutch cone penetrometer in the Netherlands in the 1930th. Also the comments by Prof F. K. Chin on the Mackintosh probe in Southeast Asia have been very valuable as well as the description by Prof F.M. Yu of the development of penetration testing in China. The help by the Librarian at SGI Mrs Birgitta Eldevall who checked the numerous references is especially appreciated. Also the contributions by Dr H Zweck and Professor J. B. Daxelhofer should be mentioned.

10 REFERENCES

Aas, G., Lunne, T. and Høeg K. 1986. Use of In-Situ Tests for Foundation Design in Clay. Proc. of In-Situ '86, Geotechnical Special Publication No 6, ASCE, Blacksburg, pp 1-30.

Amar, S., Baguelin, F., Jezequel, J.F. and Le Mehaute, A. 1975. In-Situ Shear Resistance of Clays. Proc. ASCE Specialty Conf. on In-Situ Measurements of Soil Properties, Raleigh, Vol. 1, p.22-45.

Anagnostopoulos, A.G. 1974. Evaluation of the Undrained Shear Strength from Static Cone Penetration Tests in a Soft Silty Clay in Patras, Greece. Proc. European Conf. on Penetration Testing, Stockholm, Vol 2:2, p.13-14.

ASCE, 1976. Subsurface Investigation for Design and Construction of Foundations of Buildings. ASCE, Manuals and Report on Engineering Practice, No 56. New York, 61 pp.

Atterbom, E. and Branzell, M., 1928 (translation). Andrea Palladio : Fyra böcker om arkitekturen (Four books on Architecture (translated from the Italian original).

Baguelin, F., Jezequel, J.F. and Shields, D.H., 1978. The Pressuremeter and Foundation Engineering. Series on Rock and Soil Mechanics, Vol. 2, No 4, Trans Tech Publications, Switzerland, 617 pp.

Baguelin, F.J., Bustamente, M.G. and Frank, R.A., 1986. The Pressuremeter for Foundations : French Experience. Proc. of In-Situ 86, ASCE Geotechnical Special Publication No 6, Blacksburg, p.31-46.

Baldi, G., Bellotti, R., Ghionna, V., Jamiolkowski, M. and Pasqualini, E. 1981a. Cone Resistance in Dry N.C. and O.C. Sands. Proc. Cone Penetration Testing and Experience. ASCE National Convention, St Louis, p.145-177.

Baldi, G., Bellotti, R., Ghionna, V., Jamiolkowski, M. and Pasqualini, E. 1981b. Cone Resistance of a Dry Medium Sand. Proc. 10th Int. Conf. Soil Mech. a. Found. Engrg., Stockholm, Vol. 2, p.427-432.

Baldi, G., Bellotti, R., Ghionna, V., Jamiolkowski, M. and Pasqualini, E. 1982. Design Parameters for Sands from CPT, Proc. 2nd European Symposium on Penetration Testing, Amsterdam, Vol. 2, p.425-432.

Baldi, G., Belloti, R., Ghionna, V., Jamiolkowski, M. and Pasqualini, E. 1985. Interpretation of CPT's and CPTV's, 2nd Part, Drained Penetration of Sands, Nanyang Technological Institute, 4th Int. Geotechnical Seminar, Field Instrumentation and In-Situ Measurements, Singapore, p.129-156.

Baligh, M.M., 1976. Cavity Expansion in Sand with Curved Envelopes. Journal Geotechnical Engineering Div., ASCE, Vol. 102, NoGT11, p.1131-1146.

Baligh, M.M., Azzous, A.S., Wissa, A.Z.E., Martini, R.T. and Morrison, M.J. 1981. The Piezocone Penetrometer, ASCE, Geotechnical Division, Symposium on Cone Penetration Testing and Experience, St Louis, p.247-263.

Barentsen, P., 1936. Short Description of a Field Testing Method with Cone-shaped Sounding Apparatus, Proc. 1st Int. Conf. Soil Mech. a. Found. Engrg., Cambridge, Vol. 1, B/3, p.6-10.

Baron, F.M. 1911. The Study of Earth-An American Tradition. Civil Engineering (USA) Vol 11, No 8 p.173-176.

Bartlett, A.H. and Holden, J.C., 1968. Sampling and In-Situ Testing Equipment Used by the Country Roads Board of Victoria for Evaluating the Foundations of Bridges and Embankments. Proc. 4th Australian Road Research Board Conf., Vol. 4, Part 2, p.1723-1742.

Bazaraa, A.R.S.S. 1967. Use of the Standard Penetration Test for Estimating Settlement of Shallow Foundations on Sand, Ph.D. Thesis, University of Illinois, Urbana, Ill, USA, 381p.

Begemann, H.K.S. Ph., 1965. The Friction Jacket Cone as an Aid in Determining the Soil Profile, Proc. 6th Int. Conf. Soil Mech. a. Found. Engrg., Montreal, Vol. 1, p.17-20.

Begemann, H.K.S. Ph. 1969. The Dutch Static Penetration Test with the Adhesion Jacket Cone, LGM Mededelingen, Vol. 12, No.4, April 1969, p.69-100.

Bergdahl, U. 1969. Resultat av forsok med viktsond. (Results from Weight Soundings). 1. Nordic Meeting on Penetration Testing. Stockholm, October 5-6, 1967. SGI Repr. a. Prel. Rep. No 31, p.51-79.

Bergdahl, U. 1973. Sondering i friktionsmaterial - Resultat av laboratorieförsök (Penetration Tests in Cohesionless Soils - Results from Laboratory Tests). 2. Nordic Meeting on Penetration Testing in Otnas. 5-6 May 1971. Finnish Geotech Soc, p.6-21.

Bergdahl, U. and Eriksson. U. 1983. Bestämning av jordegenskaper från sondering. En litteraturstudie. (Soil Properties with Penetration Tests - A Literature Study). Swedish Geotechnical Institute, Linkoping Sweden, Report No 22, 96 pp.

Bergdahl, U. and Ottosson, E. 1982. Calculation of Settlements on Sand from Field and Test Results. Proc. 2nd European Symp. on Penetration Testing, Amsterdam, Vol 1 p.229-234.

Bergdahl, U. and Ottoson, E. 1984. International Reference Test Procedure. Weight Sounding Test (WST). Working group. ISSMFE Technical Committee on Penetration Testing. Submitted to XI ISSMFE 1985 (Draft 85-07-18).

Bergdahl, U. and Sundqvist, O. 1974. Geotechnical Investigations at the Demonstration Site, Borros Equipment. Proc. 1st European Symp. on Penetration Testing, Stockholm, Vol 2:1 p.193-198.

Bjerrum, L. and Flodin, N. 1960. Development of Soil Mechanics in Sweden 1900-1925, Geotechnique, Vol. 10, No 1, p.1-18.

Bolomey, H. 1974. Dynamic Penetration - Resistance Formulae. Proc. European Symposium on Penetration Testing, Stockholm, Vol. 2:2, p.39-46.

Brand, E.W., Moh, Z. C. and Wirojanagud, P. 1974. Interpretation of Dutch Cone Tests in Soft Bangkok Clay. Proc. European Conf. on Penetration Testing, Stockholm, Vol. 2:2, p.51-58.

Broch. T. 1848. Laerebog i bygningskunsten (Testbook on the Art of Building Construction). Werner & Co. Christiania (Oslo).

Broms, B.B. 1974. General Report. Scandinavia. Proc. European Symposium on Penetration Testing, Stockholm, Vol. 2:1, p.14-23.

Broms, B.B. and Bergdahl, U. 1982. The Weight Sounding Test (WST) State-of-the-Art Report, Proc. 2nd European Symposium on Penetration Testing, Amsterdam, p.203-214.

Broms, B.B. and Broussard, D.E. (1965). Self-Recording Soil Penetrometer. Journal Soil Mech. a. Found. Div., Proc. ASCE. No SMI, p.53-62.

Brown, R.E. 1977. Drill Rod Influence on Standard Penetration Test. Journal Geotechnical Engineering Div., ASCE, Vol. 103, NoGT11, p.1332-1336.

Buisman, A.S.K. 1935. De weerstand van paalpunkten in zand. De Ingenieur, Vol 50, No. 14, p.28-35.

Buisman, A.S.K. 1940. Grundmechanica, Delft, Holland.

Chang, M.F. and Wong, I.H. 1986. Penetration Testing in Residual Soils of Singapore. Preprint, Specialty Geomechanics Symposium on Interpretation of Field Testing for Design Parameters, Univ. of Adelaide, Australia.

Cambefort, H. 1955. Forages et Sondages. Leur emploi dans les Travaux Publics. Edition Eyrolles, Paris.

Campanella, R.G. and Robertson, P.K. 1981. Applied Cone Research, Symposium on Cone Penetration Testing and Experience, Geotechnical Engineering Div., ASCE, Oct 1981, p.343-362.

Caquot, A. and Kerisel, J. 1953. Courbes de Glissement sous la Pointe des Pieux. Journées de la Mecanique des Sols, Annales I.T.B.T.P., p.341-342.

Chen, Q.H. 1987. Private information.

Chin, F.K. 1987. Private information.

Choa, V., Karunaratne, G.P., Ramaswamy, S.D., Vijiaratnam, A. and Lee, S.L. 1979. Compaction of Sandfill at Changi Airport. Proc. 6th Asian Regional Conf. Soil Mech. a. Found. Engrg., Singapore, Vol. 1, p.137-140.

Cummings, A.E. 1951. Discussion. Trans. South African Institution of Civil Engineers, Vol. 1, No. 6, p.186-189.

Dahlberg, G. 1913. Skred och sättningar eller s.k. ras vid järnvägsbyggnader och liknande arbeten, deras förekommande eller afhjälpande (Slides and Settlements or so-called "Ras" at Railway Constructions or Similar Works, their Prevention or Remedy). Teknisk Tidskrift, V.o.V., Vol 43, No 12, p.129-141.

Dahlberg, R. 1974. Penetration, Pressuremeter and Screw-Plate Tests in a Preloaded Natural Sand Deposit. Proc. European Symposium on Penetration Testing, Stockholm, Vol. 2:2, p.69-87.

Dahlberg, R. 1975. Settlement Characteristics of Preconsolidated Natural Sands. Swedish Foundation for Building Research (BFR), Document D1:1975, Stockholm, Sweden.

Dahlberg, R. and Bergdahl, U. 1974. Investigation on the Swedish Ram-Sounding Method. Proc. European Symposium on Penetration Testing, Stockholm, Vol 2:2, p.93-102.

D'Appolonia, D.J., D'Appolonia, E. and Brissette, R.F. 1970. Discussion on Settlement of Spread Footings on Sand, Journal Soil Mech. a. Found. Div., ASCE, Vol. 96, No.SM2, p.754-762.

Daxelhofer, J.B. 1987. By correspondence.

de Beer E, 1945. Etude des Fondations sur Pilotis et des Fondations directs. L'Appareil de Pénétration en Profondeur. Tijdschrift der Openbare Werken van Belgie, Annales des Travaux Publics de Belgique. April, June and August, 1945, Belgium, p.1-78.

de Beer, E. 1951. Discussion Transactions of the South African Institution of Civil Engineers, Vol. 1, No 6, p.190-199.

de Beer, E. 1965. Bearing Capacity and Settlement of Shallow Foundations on Sand. Proc. Symp. Bearing Capacity and Settlement of Foundations, Duke University, p.15-33.

de Beer, E. 1971/1972. Methodes de Déduction de la Capacité Portant d'un Pieu a partir des Resultats des Essais de Pénétration. Tijdschrift der openbare Werken van Belgie, No 4, 5 and 6.

de Beer, E., Lousberg, E., Wallays, M., Carpentier, R., de Jaeger, J. and Paquay, J. 1974. Scale Effects in Results of Penetration Tests Performed in Stiff Clays. Proc. European Symp on Penetration Testing, Stockholm, Vol.2:2, p.105-114.

de Beer, E., Lousberg, E., Wallays, M., Carpentier, R., de Jaeger, J. and Paquay, J. 1977. Bearing Capacity of Displacement Piles in Stiff Fissured Clays. Camptes Rendus de Recherches, IRSIA and IWONL, No 39, March, 1977, 136 p.

de Mello, V.F.B. 1971. The Penetration Test. Proc. 4th Panamerican Conf. Soil Mech. a. Found. Engrg. Puerto Rico, Vol 1, p.1-86.

de Ruiter, J. 1971. Electric Penetrometer for Site Investigations, Journal Soil Mech. a. Found. Div., ASCE, Vol. 97, No SM2, p.457-472.

de Ruiter, J. 1981. Current Penetrometer Practice, Proc. ASCE, Cone Penetration Testing and Experience, St Louis, Oct 26-30, 1981, p.1-48.

de Ruiter, J. 1982. The Static Cone Penetration Test. State-of-the-Art Report. Proc. 2nd European Symp. on Penetration Testing, Amsterdam, Vol 2, p.389-405.

Delft Laboratory of Soil Mechanics 1936. Pile Loading Tests, Proc. Int. Conf. Soil Mech. a. Found. Engrg, Vol 1, p.181-184.

Diderot, Denis 1768. Encyclopedie, ou Dictionnaire raisonne des Sciences, des Arts et des Metiers, par une Société de Gens de Lettres. Mis en Ordre et Publié par Diderot, et Quant a la Partie Mathématique par d'Alembert. Vol.6.

Douglas, B.J. and Olson, R.S. 1981. Soil Classification using Electric Cone Penetrometer, Symp. on Cone Penetration Testing and Experience. Geotechnical Div., ASCE, St Louis, p.209-227.

Durgunoglu, H.T. and Mitchell, J.K. 1975a. Static Penetration Resistance of Soils. I. Analysis. Proc. ASCE Spec. Conf. In-Situ Meas. Soil Propert., Raleigh, N.C., USA, p.151-171.

Durgunoglu, H.T. and Mitchell, J.K. 1975b. Static Penetration Resistance of Soils. II. Evaluation of Theory and Implications for Practice. Proc. ASCE Spec. Conf. In-situ Meas. Soil Propert., Raleigh, N.C., USA, p.172-189.

Engineering Society of China. 1937. Preliminary Recommendations for the Observation and Recording of Structural and Earth Settlement and Related Data in the Shanghai Area. Foundation Research Committee, Proc. Engineering Society of China, p.1-29.

Fleming, S. 1876. The Intercolonial - A Historic Sketch. Dawson Brothers. Montreal, Quebec, p.200-202.

Fletcher, G.F.A. 1965. Standard Penetration Test : Its Uses and Abuses. Journal Soil Mechanics and Found. Div., ASCE, Vol. 91, No SM4, p.67-75.

Flodin , N. 1981. Göta Kanal. En historisk-teknisk beskrivning (Göta Canal. A Historical - Technical

Description). Swedish Geotechnical Institute. Linkoping, Sweden.

Flodin, N. 1984. Hejare till borrar-eller tvärtom. Lite historik i anslutning till Borros 40-årsjubileum 1983. (Dynamic Penetration Testing - A Short History in Connection with the 40th anniversary of Borros 1983). Vag och Vattenbyggaren, No 1-2, p.43-45.

Flodin, N.O. 1985. Historical Notes on Soil Sampling, with special Reference to Undisturbed Sampling in the United States and Sweden. The Practice of Foundation Engineering. A volume honoring J. O. Osterberg. Dept of Civil Engineering, The Technological Institute, Northwestern University, Evanston. 233-27 (Ed.R.J. Krizek, C.H. Dowding, F. Somogyi).

Flodin, N. 1987. Pålar som vägrade låta sig nedslås (Piles who did not like to be driven). Vag och Vattenbyggaren. No 2 p.78-82.

Flodin, N. and Broms, B. 1981. History of Civil Engineering in Soft Clay. Soft Clay Engineering. Ed. E. Brand and P.E. Brenner, Elsevier Scientific Publishing Co, Amsterdam, p.27-156.

Franx, C., 1952. Sonderen, heien, kalenderen en proefbelasten in Verband met de puntweerstand van heipalen. De Ingenieur, Beuwen Waterbouwkunde, No 10, June 27, 1952, p.95-101, No.11, July 4, 1952, p.102-112.

Friis, J.P. 1898. Terraenundersøgelser og Jordboringer i Stjørdalen, Vaerdalen og Guldalen samt i Trondhjem 1891, 1895, 1896. Norges Geologiske Undersgelse NGU No.27. Kristiania (Oslo).

Frydman, S., 1970. Discussion. Geotechnique, Vol 20, No 4, p.454-455.

Fukuoka, M. 1974. Swedish Weight Sounding and Physical Properties of Soil. Proc. European Symp. on Penetration Testing, Stockholm, Vol 2:2, p.147-148.

Gardemeister, R. and Tammirinne 1974. Penetration Testing in Finland. Proc. 1st European Symp. on Penetration Testing, Stockholm, Vol 1, p.35-76.

Gaskell, T.F. 1969. Turbo-drilling. Endeavour, Vol 28, No 7, p.27-29.

Gautier, H. 1716. Traité de la Construction des Chemins. Paris (Also in Traite des Ponts. Paris 1716).

Gawith, A.H. and Bartlett, A.H. 1963. Deep Sounding Cone Penetrometer, Proc. 4th Australian-New Zealand Conf. on Soil Mech. a. Found. Engrg., Vol. 1, p.8-11, 287-292.

Gibbs, H.J. and Holtz, W.G. 1957. Research on Determining the Density of Sands by Spoon Penetration Testing. Proc. 4th Int. Conf. Soil Mech. a. Found. Engrg., London, Vol. 1, p.35-39.

Glick, G.L. and Clegg, B. 1965. Use of Penetrometer for Site Investigation and Compaction Control at Perth, W.A., Civil Eng. Trans., Institution of Engineers, Australia, Vol. CE 7, No 2, p.114.

Godskesen, O. 1930. Geoteknik i Danmark. (Geotechnical Engineering in Denmark) Ingeniøren No. 44, p.526-538.

Godskesen, O. 1936. Investigation of the Bearing-power of the Subsoil (Especially Moraine) with 25x25 mm Pointed Drill Weighted with 100 kg without Samples. Proc 1. Int. Conf. Soil Mech. a. Found. Engrg, Vol 1 p.311-317.

Godskesen, O. 1937. 10 Jahre Baugrunduntersuchungen bei den Dänischen Staatsbahnen. Bautechnik Vol 15, No 44, p.568-569.

Goldmann, N. 1699. Vollständige Anweisung zu der Civil-Baukunst. Braunschweig. (Ed. by. L.C. Sturm).

Göteborgs Hamn (Gothenburg Harbour) 1916. Redogörelse över jämförande borrningar mellan gamla och nya borrtyper (Report on Comparing Soundings between Old and New Rod Types).

Granger, V.L. 1963. The Standard Penetration Test in Central Africa, Proc. 3rd African Regional Conf. on Soil Mech. a. Found. Engrg, Vol. 1, p.153-156.

Granger, F. 1970. "Ten Books of Architecture" by Vitruvius. 4th ed. (Translated by Frank Granges).

Hagenaar, J. 1982. The Use and Interpretation of SPT Results for the Determination of Axial Bearing

Capacities of Piles Driven into Carbonate Soils and Coral. Proc. 2nd European Symp. on Penetration Testing, Amsterdam, Vol. 1, p.51-55.

Harison, J.A. 1986. Correlation of CBR and Dynamic Cone Penetrometer Strength Measurement of Soils, Australian Road Research, Technical Note No 2, Vol. 16, No 2, p.130.

Heijnen, W.J. 1973. The Dutch Cone Test. Study of the Shape of the Electrical Cone. Proc. 8th Int. Conf. Soil Mech. a. Found. Eng., Moscow, Vol. 1, p.181-184.

Heijnen, W.J. 1974. Penetration Testing in the Netherlands. European Symp. on Penetration Testing, Stockholm, Vol. 1, p.79-83.

Helenelund, K.V. 1966. Kitkamaalajien Kantavuusominaissukusta Valtion (On the Bearing Capacity of Frictional Soils. Teknillinen Tutkimus Laitos, Tideotus, Sarja III. Rakennus 97, Helsinki.

Helenelund, K.V. 1974. Prediciton of Pile Driving Resistance from Penetration Tests. European Sym. on Penetration Testing, Stockholm, Vol. 2:2, p.169-175.

Hellman, O., Pramborg, B.O. and Svensson, G. 1979. Kontroll av packad friktionsjord. Kontrollmetoder för bestämming av deformations-och brottbärighetsegenskaper (Control of the Compaction of Cohesionless Soils. Methods of Checking Deformation Properties and Bearing Capacity). Swedish Council for Building Research. BFR Rep. R102:1979. Stockholm.

Hoffmann, D.R. 1930. Die geotechnischen Arbeitsmethoden der Schwedischen Statsbahnen. Bauingenieur Vol 11, No 41, p.701-705.

Holden, J.C. 1971. Laboratory Research on Static Cone Penetrometers, Dept. of Civil Engineering, University of Florida, Gainsville, Internal Report CE-SM-71-1.

Holden, J.C. 1974. Current Status of Penetration Testing in Australia, State-of-the-Art Report. Proc. European Symp. on Penetration Testing, Stockholm, Vol. 1, p.155-162.

Holden, J.C. 1987. Private Information.

Holm, G. 1983. The Cone Penetration Test. Swedish Geotechnical Institute, SGI Varia 117, 41 p. (in Swedish).

Huizinga, T.K. 1941. Resultaten van diepsondeeringen als oplosing van vele paalproblemen, De Ingenieur, No 56, p.31-37.

Huizinga, T.K. 1942. Grundmechanica, Arend and Son, Amsterdam, Holland, p.79.

Huizinga, T.K. 1951. Application of Results of Deep Penetration Tests to Foundation Piles. Building Research Congress, London, Vol 1.

Hvorslev, M.J. 1948. Subsurface Exploration and Sampling of Soils for Civil Engineering Purposes, Waterways Experiment Station, Vicksburg, Miss, 465p.

Ireland, H.O., Moretto, O. and Vargas, M. 1970. The Dynamic Penetration Test : A Standard that is not Standardized. Geotechnique, Vol 20, p.185-192.

Jamiolkowski, M., Lancellotta, R., Tordella, L. and Battaglio, M., 1982. Undrained Strength from CPT. Proc. 2nd European Symp. on Penetration Testing. Amsterdam, Vol 2, p.599-606.

Jamiolkowski, M., Ladd, C.C. Germaine, J.T., Lancellotta, R. 1985. New Developments in Field and Laboratory Testing of Soils, Theme Lecture, Proc. 11th Int. Conf. Soil Mech. a. Found. Engrg., San Francisco, Vol 1, p.57-154.

Janbu, N. and Senneset, K. 1974. Effective Stress Interpretation of In-Situ Static Penetration Tests. European Symp. on Penetration Testing, Stockholm, Vol 2:2, p.181-194.

Jensen, M. 1969. Civil Engineering around 1700. With Special Reference to Equipment and Methods. Danish Technical Press Copenhagen.

Jonell, P. 1975. Geotek's akustiska penetrometer (Geotek's Acoustic Penetrometer). Swedish Geotechnical Institute. Reprints and Prelim. Reports No 58, p.137-170.

Jostra, K. and de Gijt, J.G. 1982. Results and Interpretation of Cone Penetration Tests in Soil of Different Mineralogic Composition. Proc.

European Symp. on Penetration Testing, Amsterdam, Vol. 2, p.615-626.

Kahl, H. and Muhs, H. 1952. Uber die Untersuchung des Baugrundes mit einer Spitzendrucksonde, Bautechnik, Vol. 29, p.81-88.

Kahl, H., Muhs, H. and Meyer, W. 1968. Ermittlung der Grosse und des Verlaufs des Spitzendrucks bei Drucksondierungen in Ungleichformigen Sand-Kies-Gemischen und im Kies. Mitt. DEGEBO, Berlin, No. 21, p.1-36.

Kallstenius, T. 1961. Development of Two Modern Continuous Sounding Methods. Proc. 5th Int. Conf. Soil Mech. a. Found. Engrg, Paris, Vol 1, p.475-480.

Kallstenius, 1. 1974. Static Penetration Testing in Sweden - Some Experiences. Proc. European Symp. Penetration Testing, Stockholm, Vol. 2:2, p.205-208.

Kantey, B.A. 1951. Significant Developments in Sub-Surface Explorations for Piled Foundations, African Inst. Civil Engrs, 1951, p.159-185.

Kerisel, J. 1961. Fondations Profondes on Milieux Sables : Variation de la Force Portante Limite en Fonction de la Densite, de la Profondeur, du Diametre et de la Vitesse D'Enforcement. Proc. 5th Int. Conf. Soil Mech. a. Found. Engrg., Paris, Vol 2, p.73-83.

Kerisel, J. 1985. The History of Geotechnical Engineering up until 1700. Proc. 11th Int. Conf. Soil Mech. a. Found. Engrg., San Francisco. Golden Jubilee Volume, p.1-121.

Kok, L. 1974. The Effect of the Penetration Speed and the Cone Shape on the Dutch Static Cone Penetration Test Results. Proc. European Symp. on Penetration Testing, Stockholm, Vol 2:2, p.216-220.

Kolderup, E. 1894. Grundundersøgelser af funderingsarbeider paa land og under vand for alle slags byggearbeider. (Site Investigations and Foundations on Land and under Water for all Kinds of Construction works). F.H. Aschehoug & Co. Forlag, Christiania (Oslo).

Kovacs, W.D. 1979. Velocity Measurement of Free-Fall SPT Hammer. Journal Geotechnical Engineering Div., ASCE, Vol 105, No GT1, p.1-10.

Kovacs, W.D. 1981. Results and Interpretation of SPT Practice Study. Geotechnical Testing Journal, ASTM, Vol 4, No 3, p.126-129.

Kovacs, W.D., Evans, J.C. and Griffith, A.H. 1977. Towards a More Standardised SPT. Proc. 9th Int. Conf. Soil Mech. a. Found. Engrg., Tokyo, Vol 2, p.269 - 276.

Kovacs, W.D. and Salomone, 1982. SPT Hammer Energy Measurements, Journal Geotechnical Engineering Div., ASCE, Vol 108, No GT4. April 1982.

Kovacs, W.D. and Salomone, L.A. 1984. Field Evaluations of SPT Energy Equipment and Methods in Japan compared with the SPT in the United States. NBSIR 84 - 2910, National Burau of Standards, Gaithersburg, MD., USA., 55p.

Künzel, E. 1936. Der "Prüfstab", ein einfaches Mittel zur Bodenprüfung. 10 lehrreiche Beispiele der Bodenprüfung-15 Möglichkeiten weiterer Andwendung. Bauwelt No.14, p.327-329.

Labelye, C. 1739. A short Account of the Methods made use of in Laying the Foundations of the Piles of Westminster Bridge, London.

Ladanyi, B. and Eden, W.J. 1969. Use of the Deep Penetration Test in Sensitive Clays. Proc. 7th Int. Conf. Soil Mech. a. Found. Engrg., Mexico City, Vol. 1, p.225-231.

Lagging, L.B. and Eresund, S. 1974. Test loading of a Hydraulic Sandfill. 1st European Symp. on Penetration Testing, Stockholm, Vol 2:2 p.221-227.

Legget, R.F. and Peckover, F.L. 1973. Foundation Performance of a 100-year Old Bridge. Canadian. Geot. Journal Vol 10, No 3, p.507-519.

Leupold, J. 1724. Theatrum Machinarum, Hydrotechnicarum. Quellen und Brunnen... Pfähle und Rammel.... Leipzig.

Lunne, T., Eide, O. and de Ruiter, J. 1976. Correlations between Cone Resistance and Vane Shear Strength in some Scandinavian Soft to Medium Stiff

Clays. Canadian Geotechnical Journal, Vol. 13, p.430-441.

Lunne, T. and Kleven, A. 1981. Role of CPT in North Sea Foundation Engineering, Symp. on Cone Penetration Testing and Experience, Geotechnical Engineering Div., ASCE, Oct. 1981, p.49-75.

Magnusson, O. 1973. Jordars schaktbarhet. beräkningsmetod och förslag till indelning av jord i schaktbarhetsklasser (Diggability of Earth. A Computation Method and a Proposal for Division of Soils into Classes of Diggability). Byggforskningen BFR. Report R51:1973.

Malyshev, M.V. and Lavisin, A.A. 1974. Certain Results obtained in Cone Penetration of a Sand Base Proc. European Symp. on Penetration Testing, Stockholm, Vol 2:2, p.237-239.

March, R. 1974. Penetration Resistance and Soil-Pile Interaction Parameters in View of Bending Moments Evaluation. Proc. European Symp. on Penetration Testing, Stockholm, Vol 2:2, p.241-243.

Marchetti, S. 1975. A New In-Situ Test for the Measurement of Horizontal Soil Deformability. Proc. Conf. on In-Situ Measurement of Soil Properties. ASCE Specialty Conf., Raleigh, N.C., Vol 2, p.255-259.

Marchetti, S. 1980. In-situ Tests by Flat Dilatometer. Journal Geotechnical Engineering Div., ASCE, Vol 106, NoGT3, p.299-321.

Marcuson, W.F. and Bieganousky, W.A. 1977a. Laboratory Standard Penetration Tests on Fine Sand. Journal Geotechnical Engineering Div., ASCE, Vol 103, No GT6, p.565-588.

Marcuson, W.F. and Bieganousky, W.A. 1977b. SPT and Relative Density in Coarse Sand. Journal Geotechnical Engineering Div., ASCE, Vol. 103, No GT11, p.1295-1309.

Mariupolski and Trofimenkov, Y. 1987. Private information.

Marsland, A. 1974. Comparisons of the Results from Static Penetration Tests and large In-situ Plate Tests in London Clay. Proc. European Symp. on Penetration Testing, Stockholm, Vol 2:2, p.245-252.

Massarsch, K.R. 1986. Acoustic Penetration Testing. 4th NTI Int. Geotechn. Seminar, Field Instrumentation and In-situ Measurements, 25-27 Nov, 1986, p.71-76.

Massarsch, K.R. and Broms, B.B. 1981. Pile Driving in Clay Slopes, Proc. Int. Conf. Soil Mech. a. Found. Engrg., Stockholm, Vol 3, p.469-474.

Massarsch, R., Lindholm, P. and Mårtensson, O. 1976. Ny lätt sonderingsmetod (New Light Penetration Testing Method). Royal Institute of Technology, JOB. Rep. No.3. Stockholm.

Matsumoto, K. and Matsubara, M. 1982. Effects of Rod Diameter in the Standard Penetration Test. Proc. 2nd European Symp. on Penetration Testing, Amsterdam, Vol. 2, p.107-112.

Meardi, G. and Meardi, P. 1974. Employment of the Dynamic Conical Penetrometer in Italy with External Protecting Casing. Proc. European Symp. on Penetration Testing, Stockholm, Vol. 2:2, p.259-261.

Melzer, K.J. 1968. Sondenuntersuchungen im Sand. Mitt. Inst. V.G.B., T.H. Aachen, No 43.

Melzer, K.J. and Smoltczyk, U. 1982. Dynamic Penetration Testing. State-of-the-Art Report. Proc. 2nd European Symp. Penetration Testing, Amsterdam, Vol, 1, p.191-202.

Meyerhof, G. G. 1956. Penetration Tests and Bearing Capacity of Cohesionless Soils. Journal Soil Mech. a. Found. Div., ASCE, Vol 91, No SMI, p.1-19.

Meyerhof, G.G. 1965. Shallow Foundations. Journal Soil Mechanics and Foundation Div., ASCE, Vol 91, No SM2, p.21-31.

Meyerhof, G.G. 1976. Bearing Capacity and Settlement of Pile Foundations, Journal Geotechnical Engineering Div., ASCE, Vol 102, No GT3, p.195-228.

Mitchell, J.K. 1986. Ground Improvement Evaluation by In-situ Tests. Proc. of In-situ 86. ASCE, Geotechnical Special Publication No 6, Blacksburg, p.221-236.

Mitchell, J.K. and Gardner, W.S. 1975. In-situ Measurement of Volume Change Characteristics. State-of-the-Art

Report. Proc. Conf. In-situ Measurement of Soil Properties. Specialty Conf. Geotechnical Div., Raleigh, Vol. 2.

Mitchell, J.K., Guzikowski, F. and Villet, W.C.B. 1978. The Measurement of Soil Properties In-Situ, Present Methods - Their Applicability and Potential. Dept of Civil Eng., Univ of California, Berkeley, Report prepared for U.S. Dept of Energy.

Mohr, H.A. 1943. Exploration of Soil Conditions and Sampling Operations. Harvard Soil Mechanics Series, No 21, 3rd Ed., Graduate School of Engineering, 63p.

Möller, B. 1980. Bedömning av lerors sensitivitet ur vikt och trycksondeningsresultat (Estimation of the Sensitivity of Clay from Weight and Ram Soundings). Swed Geot. Inst. Rep 1-206/79. Linkoping.

Möller, B. and Bergdahl, U. 1982. Estimation of the Sensitivity of Soft Clays from Static and Weight Sounding Tests. Proc. 2nd European Symp. on Penetration Testing, Amsterdam, Vol. 1, p.291-295.

Muhs, H. 1949. Arbeiten der Degebo in den Jahren 1938-1948. Bautechnik-Archiv, No 3, p.20-40.

Muhs, H. 1969. Neue Erkenntnisse über die Tragfähigkeit von flachgegrudeten Fundamenten aus Grossversuchen und ihre Bedeutung für die Berechnung, Bautecknik, Vol 46, p.181-191.

Muhs, H. and Weiss, K. 1972. Der Einfluss von Neigung und Ausmittigkeit der Last auf die Grenztragfähigkeit flack gegrundeter Einzelfundamente, Berichte aus der Bauforschung No 73, p.1-119.

Muhs, H. 1978. 50 Years of Deep Sounding with Static Penetrometers, Berlin Universitat, Deutsche Forschungsgesellschaft fur Boden-Mechanik (Degebo), Mitteilungen No 34, p.45-50.

Muhs, H. and Weiss, K. 1971. Untersuchungen von Grenztragfähigkeit und Setzungsvernalten flachgegrundeten Einzelfundamente im ungleichformigen nichtbindigen Boden. Berichte aus der Bauforschung, Helft 69, Berlin.

Muromchi, T. 1974. Experimental Study on Application of Static Cone Penetrometer to Subsurface Investigation of Weak Cohesive Soils. Proc. European Conf. on Penetration Testing, Stockholm, Vol 2:2, p.285-291.

Muromachi, T., Oguro, I. and Miyashita, T. 1974. Penetration Testing in Japan. Proc. European Symposium on Penetration Testing, Stockholm, Vol. 1, p.193-200.

Muromachi, T. and Kobayashi, S. 1982. Comparative Study of Static and Dynamic Penetration Tests Currently in use in Japan. Proc. 2nd European Symposium on Penetration Testing, Amsterdam, Vol 1, p.297-302.

National Research Council of Canada, 1975. Canadian Manual on Foundation Engineering, Ottawa, Canada, 318 p.

Nachmanson, A. and Sundberg. K. 1936. Sv. Diamantbergborrnings AB 1886-1936. Uppsala. Almqvist & Wicksells Boktr. AB.

Natukka, A. 1969. Finska Sonderingskommittens rekommendationer for viktsonderingsstandard (Recommendations of a Weight Sounding Standard by the Finnish Committee on Penetration Testing. 1. Nordic Meeting on Penetration Testing in Stockholm, Oct 5-6, 1967. SGI Repr. a Preli. Rep. No 31, p.39-43.

NAVFAC DM-7, 1971. Soil Mechanics, Design Manual 7.1, Dept of the Navy, Naval Facilities Engineering Command, Alexandria, Va, USA, 347 p.

Nilsson, S. 1975. Nilcons nya geotechniska fältutrustning. (New Field Equipment from Nilcon). Swed Geot. Inst. Reprints and Preli. Reports. No 58, p.127-136.

Norwegian Pile Committee, 1973. Veiledning ved pelefundamentering (Guide for Pile Foundations). Norwegian Geotechnical Institute, Veiledning Nr 1, Oslo, Norway, 108 p.

Nygård, B. 1944. Det stora jordfallet. Sarpfossen i Dikt og Virkelighet. (The Great Earthfall. Sarpfossen in Poetry and Reality) Tell Forlag, Oslo, p.25-32.

Ohsaki, Y. 1970. Effects of Sand Compaction on Liquefaction during the

Tokachioki Earthquake. Soils and Foundations, Vol. 10, No 2, p.112-128.

Olsson, H. 1915. Banlära. Järnvägars byggnad och underhåll (A Railway Engineering Construction and Maintenance. Swedish State Railways. Stockholm (2 volumes).

Olsson, J. 1925. Kolvborr. Ny borrtyp för upptagning av lerprov. (The Piston Sampler. A New Type of Sampler for Clay). Teknisk Tidskrift, V.o.V. Vol 52, No 2, p.13-16.

Paaswell, G. 1931. Penetration Tests give Bearing Power of Deep Subsurface Soils, Eng. News Record, Vol. 106, No 14, p.570-572.

Palmer, D.J. and Stuart, J.G. 1957. Some Observations on the Standard Penetration Test and the Correlation of the Test In-Situ with a New Penetrometer, Proc. 6th Int. Conf. Soil Mech. a. Found. Engrg. London, Vol. 1, p.231-236.

Paproth, E. 1973. Der Prüfstab Künzel, ein Gerat für Baugrunduntersuchungen. Bautechnik Vol 23, No 52/56 p.327-330.

Paproth, E. 1956. Boden untersuchung mit dem Künzelstab in Standardausführung Bauwelt No 17 p.231.

Parez, L.A. 1974. Static Penetrometer : The importance of the Skin Friction associated with the Point Resistance. Proc. European Symp. on Penetration Testing, Stockholm, Vol 2:2, p.293-299.

Parkin, A.K. and Lunne, T. 1982. Boundary Effects in the Laboratory Calibration of a Cone Penetrometer in Sand. Proc. 2nd European Symp. on Penetration Testing, Amsterdam, Vol. 2, p.761-768.

Parry, R.H.G. 1971. A Direct Method of Estimating Settlements in Sand from SPT Values. Proc. Symp. Interaction of Structures and Foundations, Midlands Soil Mech. a. Found. Engrg., Soc., Birmingham, U.K., p. 29-37.

Parry, R.H.G. 1978. Estimating Foundation Settlements in Sand from Plate Bearing Tests, Geotechnique, Vol. 28, No 1, p.105-118.

Parsons, J.D. 1966. Piling Difficulties in the New York Area, Journ. Soil Mech. a. Found. Div., Proc. ASCE, Vol. 92, No. SM1, p.43-64.

Peck, R.B. 1948/1963. History of Buildings Foundations in the Chicago area. Proc. of Lecture Series, Soil Mech. a. Found. Div. ASCE. (Repr. from Univ Illinois Eng. Exper. Station Bull Series 373:75:39 Jan. 1948).

Peck, R.B., 1971. Discussion Proc. 4th, Panamerican Conf. on Soil Mech. a. Found. Engrg., San Juan, Puerto Rico Vol. 3, p.59-61.

Peck, R.B. and Bazarra, A.R.S.S. 1969. Discussion, Journ. Soil Mech. a. Found. Div., ASCE, Vol. 95, No SM3, p.905-909.

Peck, R.B., Hanson, W.E. and Thornburn, T.H. 1953, 1974. Foundation Engineering. Wiley and Sons, New York, 1st and 2nd Ed., 514 p.

Peckover, F.L. and Legget, R.F. 1973. Canadian Soil Penetration Test of 1872. Canadian Geotechnical Journ., Vol 10, p.528-531.

Plantema, G. 1948a. Construction and Method of Operating a New Deep Sounding Apparatus, Proc. 2nd Int. Conf. Soil Mech. a. Found. Eng., Rotterdam, Vol. 1, p.277-279.

Plantema, G. 1948b. Results of a Special Loading Test on a Reinforced Concrete Pile, a so-called Pile Sounding; Interpretation of the Results of Deep-Soundings, Permissible Pile Loads and Extended Observations, Proc. 2nd Int. Conf. Soil Mech. a. Found. Engrg, Rotterdam, Vol. 1, p.112-118.

Plantema, G. 1957a. Influence of Pile Driving on the Sounding Resistances in a Deep Sand Layer. Proc. 4rd Int. Conf. Soil Mechanics and Found Eng., London, Vol. 2, p.52-55.

Plantema, G. 1957b. Influence of Density on Sounding Results in Dry, Moist andSaturated Sands. Proc. 4th Int. Conf. Soil Mech. a. Found. Engrg, London,Vol 1, p.237-240.

Randolph, M.F. and Wroth, C.P. 1979. An Analytical Solution for the Consolidation around a Driven Pile, Int. Journal for Numerical and Analytical Methods in Geomechanics, Vol. 3, p.217-229.

Reusch, H. 1901(a). Nogle optegnelser fra Vaerdalen (Some Observations from Vaerdalen). Norg. Geot. Unders. Vol 32 p.1-32.

Reusch, H. 1901(b). Jordfallet ved Morsil i Stjørdalen (The Landslide at Morsil in Stjørdalen). Norg. Geol. Unders. Vol 32, p.33-44.

Riggs, C.O. 1986. North American Standard Penetration Test Practice : An Essay. Proc. Use of In-Situ Tests in Geotechnical Engineering, ASCE Geotechnical Special Publication No. 6, Blacksburg, Virginia, p.949-967.

Riggs, C.O., Schmidt, N.O. and Rassieur, C.L. 1983. Reproducible SPT Hammer Impact Force with an Automatic Free Fall SPT Hammer System. Geotechnical Testing Journ, ASTM, Vol 6, No.3, p.201-209.

Riggs, C.O. Mathes, G.M. and Rassieur, C.L. 1984. A Field Study of an Automatic SPT Hammer System Geotechnical Testing Journ., ASTM Vol. 7, No.3, p.158-163.

Robertson, P.K. and Campanella, R.G. 1983a. Interpretation of Cone Penetration Tests. Part I : Sand. Canadian Geotechnical Journ., Vol. 20, No.4, p.718-733.

Robertson, P.K. and Campanella, R.G. 1983b. Interpretation of Cone Penetration Tests. Part II : Clay. Canadian Geotechnical Journ., Vol.20, No 4, p.734-745.

Rodin, S. 1961. Experiences with Penetrometers with Particular Reference to the Standard Penetration Test. Proc. 5th Int. Conf. Soil Mech. a. Found. Engrg., Paris, Vol 1, p.517-521.

Rodin, S., Corbett, B.O., Sherwood, D.E. and Thornburn, S. 1974. Penetration Testing in United Kingdom. Proc. European Symp. on Penetration Testing, Stockholm, Vol. 1, p.139-146.

Roy, M., Michaud, D., Tavenas, F.A., Leroueil, S. and LaRochelle, P. 1974. The Interpretation of Static Cone Penetration Tests in Sensitive Clays. Proc. European Symp. on Penetration Testing, Stockholm, Vol, 2:2, p.323-330.

Saarelainen, S. 1979. Grovt sättningsestimat på grundval av viktsonderingsmotstånd (A Rough Estimate of Settlements from Weight Soundings). Nordic Geotechnical Meeting 1979 in Esbo Finland. Finnish Geotechnical Society, p.718-727.

Samuelsson, G. 1926. Grundförstärkningsarbeten vid statsbanebyggnaden Järna-Enstaberga. Ett försök till systematisk metod for förstärkning av svag undergrund under järnvägsbankar (Earth Strengthening Work at the Construction of the Railway Line Järna-Enstaberga. Use of a Systematic Method to Strengthen Poor Subsoil below Railway Embankments). Teknisk Tidskrift V.a.V. Vol 56, No 4, p.37-42 and No 5, p.54-57.

Sanglerat, G. 1972. The Penetrometer and Soil Exploration. Elsevier, Amsterdam, 464 p.

Sanglerat, G. Nhiem, T.V., Sejourne, M. and Andina, R. 1974. Direct Soil Classification by Static Penetrometers with Special Friction Sleeve. Proc. European Symp. on Penetration Testing, Stockholm, Vol. 2:2, p.337-344.

S.B.N. 1980. Svensk Byggnorm (Swedish Building Code). National Board of Physical Planning and Building. Liberförlag. Stockholm.

Scala, A.J. 1956. Simple Methods of Flexible Pavement Design using Cone Penetrometers, Proc. 2nd Australian-New Zealand Conf. Soil Mech. a. Found. Engrg., p.73.

Schmertmann, J.H. 1970. Static Cone to Compute Static Settlement over Sand. Journ. Soil Mechanics and Found. Div., ASCE, Vol 96, No SM3, p.1011-1014.

Schmertmann, J.H. 1975. Measurement of In-Situ Shear Strength, Proc. ASCE Specialty Conf. on In-Situ Measurement of Soil Properties, Raleigh, Vol. 1, p.57-138.

Schmertmann, J.H. 1976a. An Updated Correlation between Relative Density D_r, and Fugro-Type Electric Cone Bearing, q_c, Waterways Experimental Station, Contract DACW 39-76-M6646.

Schmertmann, J.H. 1976b. Predicting the q_c/N Ratio. Final Report D-636. Engineering and Industrial Experiment Station, Dept of Civil Engrg, Univ of Florida, Gainesville, USA.

Schmertmann, J.H. 1978a. Guidelines for Cone Penetration Test Performance and Design. U.S. Department of Transportation, Federal Highway Admin, Offices of Research and Dev., Repr No FHWA-TS-78-209, Washington, 145 p.

Schmertmann, J.H. 1978b. Use of SPT to Measure Dynamic Soil Properties? Yes, But...." Dynamic Geotechnical Testing, ASTM, STP 654, Philadelphia, p.341-355.

Schmertmann, J.H. 1982. Calibration of 2 FLDOT Drill Riggs for SPT Energy. Bureau of Materials and Research. Florida Dept of Transportation, Gainsville, Florida, USA.

Schmertmann, J.H. and Dalacios, A. 1979. Energy Dynamics of SPT, Journal Geotechnical Engineering Div., ASCE, Vol. 105, No GT8, p.909-926.

Schultze, E. and Melzer, K.J. 1965. The Determination of the Density and the Modulus of Compressibility of Non-Cohesive Soils by Soundings. Proc. 6th Int. Conf. Soil Mech. a. Found. Engrg., Montreal, Vol. 1, p.354-358.

Schultze, E. and Muhs, H. 1967. Bodenuntersuchungen für Ingenieurbauten 2ed. Springer Verlag, Berlin.

Schultze, E. and Menzenbach, H. 1961. Standard Penetration Test and Compressibility of Soils. Proc. 5th Int. Conf. Soil Mech. a. Found. Engrg., Vol. 1, p.527-255.

Seed, H.B. 1979. Soil Liquefaction and Cyclic Mobility Evaluation for Level Ground during Earthquakes. State-of-the-Art. Journ. Geotechnical Engr. Div., ASCE, Vol 105, No GT2, p.201-255.

Seed, H.B. and de Alba, P. 1986. Use of SPT and CPT Tests for Evaluating the Liquefaction Resistance of Sands. Proc. Use of In-situ Tests in Geotechnical Engineering, ASCE, Geotechnical Special Publication No 6, Blacksburg, Virginia, p.281-302.

Seed, H.B. and Idriss, I.M. 1974. Soil Moduli and Damping Factors for Dynamic Response Analysis, Univ. of California, Berkeley, Repr. No. EERC 70-10.

Seed, H.B., Tokimatsu, K., Harder, L.F. and Chung, R.M. 1985. The Influence of SPT Procedures in Soil Liquefaction Resistance Evaluations. Journ. of the Geotechnical Engr Div. ASCE, Vol. 111, No GT12, p.1425-1445.

Semple, R.M. and Johnston, J.W. 1979. Performance of "Stingray" in Soil Sampling and In-Situ Testing. Int. Conf. on Offshore Site Investigations, London, U.K.

Sen Gupta, D.P. and Aggarwal, V.S. 1966. A Study on Dynamic Cone Penetration Tests. Indian Geotechnical Journal, Vol 7, No 1. p.61-76.

Senneset, K. 1974. Penetration Testing in Norway. State-of-the-Art Report. Proc. Proc. 1st European Symp. on Penetration Testing, Stockholm, Vol. 1, p.85-95.

Senneset, K., Janbu, N. and Svano, G. 1982. Strength and Deformation Parameters from Cone Penetration Tests. Proc. 2nd European Symp. on Penetration Testing, Amsterdam, p.863-870.

Senneset, K. and Janbu, N. 1984. Shear Strength Parameters Obtained from Static Cone Penetration Tests, ASTM, STP 883, San Diego, USA.

Serota, S. and Lowther, G. 1973. SPT Practice meets Critical Review, Ground Engr, Vol 6, No 1, p.20-22.

Simons, N.E. and Menzies, B.K. 1977. A Short Course in Foundation Engineering., Butterworth Co. Ltd., London, 1st Ed.

Skempton, A.W. 1986. Standard Penetration Test Procedures and the Effects in Sands of Overburden Pressure, Relative Density, Particle Size, Ageing and Overconsolidation, Geotechnique, Vol 36, No.3, p.425-447.

Skaven-Haug, S. 1976. Geotekniske markundersogelser i Norge (Geotechnical Investigations in Norway). Geoteknik. Foredrag fra kursus i Dansk Ingeniorforening, 2-6 April 1976, No. 20, p.245-258.

Stroud, M.A. 1974. The Standard Penetration Test in Insensitive Clays and Soft Rocks. Proc. European Seminar on Penetration Testing, Stockholm, Vol 2:2, p.366-375.

Stroud, M.A. and Butler, F.G 1975. The Standard Penetration Test and the Engineering Properties of Glacial

Materials, Conf. on the Engineering Behaviour of Glacial Materials, Univ. of Birmingham, U.K., p.124-135.

Strukel, M. 1895. Der Grundbau, dargestellt auf Grundlage einer systematisch geordneten Sammulung zahlreicher, anschaulicher Beispiele aus der Praxis. Vortrage gehalten am Polytechn. Institut Helsingfors. Helsinki and Leipzig. Verlag Wenzel Hagelstam. (2nd rev. ed. 1906).

Stump, S. 1953. Pénétromètre Dynamique avec ou sans Tubage. Proc. 3rd Int. Conf. Soil Mech. a. Found. Engrg., Zurich, Vol. 3, p.148.

Sutherland, H.B. 1974. Granular Materials. Review Paper : Session 1, Proc. Conf. on Settlement of Structures, British Geotechnical Society, Pentech Press, London, p.473-499.

Svensson (Sperling), M. 1899. Beteckningssätt för lös mark å undersökningsprofiler (Recording soft Soils on Profiles of Investigations. Tekn. Tidskrift Byggnadskonst, Vol 29, No 3, p.55-56.

Swedish Commission on Pile Research 1979. Recommendations on Bored Piles. Design, Construction and supervision. Rep No 58. Stockholm.

Swedish Geotechnical Commission 1917. Vägledningar vid jord-borrningar för järnvägsändamål. (Guide for Soil Borings for Railways). Statens Järnv. Geot. Medd. No 1, Stockholm, Sweden, 37 p.

Swedish Geotechnical Commission 1922. Statens Järnvägars Geotekniska Kommission 1914 - 22 Slutbetänkande (Geotechnical Commission of the Swedish State Railways, 1914-22, Final Report), No. 2, Stockholm, Sweden, 180 p.

Swedish Geotechnical Institute 1949. Redogörelse för Statens Geotekniska Instituts verksamhet åren 1977-1949 (Report on the Activities of the Swedish Geotechnical Inst 1944-1949). Swed Geotech. Inst, Medd. No 2.

Swedish Road Admin, 1976. Statens Vägverk, TB103, 1976-09. Stockholm.

Tammirinne, M. 1973. Bestämning av torrvolymvikt i grus, sand och grov silt på grund av sonderingsmotstånd (Determination of the Density of Gravel, Sand and Coarse Silt from Penetration Tests. 2nd. Nordic Meeting on Penetration Testing in Otnäs, 5-6 May 1971. Finnish Geot Soc. p.49-62.

Tammirinne, M. 1974. Relation between Swedish Weight Sounding and Static Penetration Test Resistance of Two Sands. Proc. 1st European Conf. on Penetration Testing, Stockholm, Vol. 2.1, p.154-156.

Tassios, T.P. and Anagnostopoulos, A.G. 1974. Penetration Testing in Greece, Proc. 1st European Symp. on Penetration Testing, Stockholm, Vol. 1, p.65-68.

Tavenas, F.A. and LaRochelle, P. 1972. Accuracy of Relative Density Measurements. Geotechnique, Vol. 22, p.549-562.

Tecklenburg, T. 1886/1896. Handbuch der Tiefbohrkunde. Bd. 1-6. Leipzig. Baumgärtners Buchhandlung.

Telford, T. 1830. Inland Navigation. In: Edinburg Encyclopaedia Vol 15, p.209-315.

Terzaghi, K. 1930. Die Tragfähigkeit von Pfahlgründungen. Bautechnik, Vol. 8, p.517-521.

Terzaghi, K. 1953. Fifty Years of Soil Exploration. Proc. 3rd Int. Conf. Soil Mech. a. Found. Eng., Zurich, Vol.3, p.227-237.

Terzaghi, K. 1944. Theoretical Soil Mechanics, John Wiley and Sons, Inc, New York, 510 p.

Terzaghi, K. and Peck, R.B. 1948, 1967. Soil Mechanics in Engineering Practice, John Wiley & Sons, New York, N.Y., 566 p.

Thomas, D. 1968. Deep-Sounding Test Results and the Settlement of Spread Footings on Normally Consolidated Sand. Geotechnique, Vol 8, No.4 p.472-488.

Thornburn, S. 1963. Tentative Correction Chart for the Standard Penetration Test in Non-Cohesive Soils. Civil Engrg. Public Works Review, Vol. 58, No 6, p.752-753.

Tomlinson, M.J. 1956. Telford and Soil Mechanics, Geotechnique, Vol. 6, p.99-105.

Torstensson, B.A. 1975. Pore Pressure Sounding Instrument Proc., ASCE Specialty Conf. on In-Situ Measurement of Soil Properties, Vol. 2, Raleigh, p.48-54.

Torstensson, B.A. 1977. The Pore Pressure Probe, Nordiske Geotekniske Mote, Oslo, Paper No. 34, p.1-15.

Triewald, M. 1740. Beskrifning om Engelska Jordbåren (Description of the English Earth Auger). Kungl Sv Vetenskapsakad. Handlingar, Vol 1 p.216-227.

Trofimenkov, J.G. 1974. General Report Proc. 1st European Symp. on Penetration Testing, Stockholm, Vol 2:1, p.24-28.

van der Veen C. 1953. The Bearing Capacity of a Pile. Proc. 3rd Int. Conf. Soil Mech. a. Found. Engrg., Zurich, Vol 2, p.84-90.

van der Veen C. 1987. The Early Development of the Cone Penetrometer in the Netherlands (By correspondence).

van Impe, W.F. 1986. Evaluation of Deformation and Bearing Capacity parameters of Foundations, from Static CPT-Results, Proc. 4th NTI Int. Geotechn. Seminar, Field Instrumentation and In-situ Measurements, Singapore 25-27 Nov 1986., p.51-70.

van Wambeke, A. 1982. The Dynamic Penetration Test or ".... ni cet excès d'honneur ni cette indignité". Amici et Alumni, Em. Prof. Dr Ir E.E. De Beer.

Veismanis, A. 1974. Laboratory Investigations of Electrical Friction - Cone Penetrometers in Sauds, Proc. 1st European Symp. on Penetration Testing, Stockholm, Vol 2:2 p.407-20.

Wendel, E. 1900. Om profbelastning på pålar med tillämpning deraf på grundläggningsförhållandena i Göteborg (On the Test Loading of Piles and its Application to Foundation Problems in Gothenburg). Tekn. Samfundets Handlingar Vol 7, p.3-62.

Vermeiden, J. 1948a. Improved Sounding Apparatus as Developed in Holland since 1936. Proc. 2nd Int. Conf. Soil Mech. a. Found. Eng., Vol 1, p.280-287.

Vermedien, J. 1977. Cone Penetration Test. Site Investigations, Delft Soil Mechanics Laboratory, LGM Mededelingen, December, 1977, p.55-68.

Vesic, A.S. 1965. Ultimate Loads and Settlements of Deep Foundations in Sand. Proc. Symposium on Bearing Capacity and Settlement of Foundations, Raleigh, p.53-68.

Vesic, A.S. 1967. A Study of Bearing Capacity of Deep Foundations. Georgia Institute of Technology, Final Report, B-189.

Vesic, A.S. 1970. Tests on Instrumented Piles, Ogeechee River Site. Journ. Soil Mech. a. Found. Div., ASCE, Vol. 96, No SM2, p.561-584.

Vesic, A.S. 1972. Expansion of Cavities in Infinite Soil Masses, Journ. of the Soil Mech. a. Found. Div., ASCE, Vol. 98, No. SM3, p.265-290.

Villet, W.C.B. and Mitchell, J.K. 1981. Cone Resistance, Relative Density and Friction Angle. Symp. on Cone Penetration Testing and Experience. Journ. Geotechnical Engrg Div., ASCE, Vol 107, p.178-208.

Villet, W.C.B., Mitchell, J.K. and Tringale, P.T. 1981. Acoustic Emissions Generated During the Quasi-Static Cone Penetration of Soils, ASTM Spec. Techn. Publ. STP 750, p.174-193.

Webb, D.L. 1969. Settlement of Structures on Deep Alluvial Sand Sediments in Durban, South Africa, British Geotechn. Soc., Conf. In Situ Investigation Soils and Rocks, Session 3, Paper No 16, p.181-188.

Weiss, K. 1970. Der Einfluss der Fundamentform auf die Grenztragfähigkeit flach gegrundeter Fundamente. Berichte aus der Bautechnik, No 65, p.1-69.

Wermbergs Mek Verkstad 1949. Vara standardsatser for grundundersökningar (Our Standard Equipment for Field Investigations)(Leaflet, 4.ed).

Whyte, J.L. 1976. The Development of Site Investigations. Ground Engineering Vol 1, No. 9, p.35-38.

Winkel, C.T. 1969. Erfaringer vedrørende sondeboret i Danmark (Experience with Soundings in Denmark). Nordic Meeting on penetration Testing in Stockholm, Oct 5-6, 1967. Swed Geot. Inst. Repr. a. Prel. Rep, No 21, p.35-37.

Wissa, A.E.Z., Martin, R.T. and Garlanger, J.E. 1975. The Piezometer Probe, Proc. ASCE Specialty Conf. on In-Situ Measurement of Soil Properties, Raleigh, Vol 1, p.536-545.

Zuidberg, H.M. 1975. Seacalf, a Submersible Cone Penetrometer Rig. Marine Geotechnology, Vol 1, No 1, p.15-32.

Zuidberg, H.M., Richards, A.F. and Geise, J.M. 1986. Soil Exploration Offshore, Proc. 4th Int. Geotechnical Seminar, Field Instrumentation and In-Situ Measurements, Singapore, 25-27 Nov, 1986, p.3-11.

Zuidberg, H.M., Schapp, L.H.J. Beringer, F.L. 1982. A penetrometer for Simultaneously Measuring of Cone Resistance, Sleeve Friction and Dynamic Pore Pressure. Proc. 2nd European Symp. on Penetration Testing, Amsterdam, Vol 2, p.963-970.

Zweck, H. 1969. Baugrunduntersuchungen durch Sonden. Ramm- Druck- Dreh- und Flugelsonden. Bauingenieurpraxis, Berlin Munchen, Wilh. Ernst u.Sohn, Heft 71.

Zweck, H. 1987. By correspondence.

Penetration Testing 1988, ISOPT-1, De Ruiter (ed.)
© 1988 Balkema, Rotterdam, ISBN 90 6191 801 4

The calibration of cone penetrometers

A.K.Parkin
Department of Civil Engineering, Monash University, Clayton, Victoria, Australia

ABSTRACT: A review is made of current procedures for the calibration of the CPT in sand and clay, with particular emphasis on the use of large laboratory calibration chambers. An assessment is made of recent laboratory work on sand, and the methods of predicting sand properties from cone resistance. However, laboratory studies have not yet been able to provide data on the important category of fine and silty sands. In the case of clays, calibration is mainly by reference to the field vane or laboratory triaxial tests. Some characteristics are discussed, but because such calibration relates intimately to pore pressure behaviour, the principal coverage will be given in Lecture 1.

1. INTRODUCTION

Since the advent of the original Dutch Cone penetrometer in 1938, inspired by the need for a cheap investigative tool for the soft soils of Holland, a fairly spectacular development of the instrument has taken place. In particular, the introduction of the electric (or Fugro) penetrometer, allowing continuous reading, and later electronic output processing (both analogue and digital), has meant that the cone has become a most sophisticated device for the acquisition and processing of soils data. This is particularly true of the off-shore environment where not all in-situ devices are sufficiently simple or robust for general use. This capability has been even further enhanced to the point where current models extend to as many as six output functions that include cone and friction resistance, pore pressure, inclination, temperature and seismic recorders. Along with this development, however, has come the need for calibration, in order that the full potential of these devices may be realised, and to that end this lecture is devoted.

Whilst a considerable spread of penetration devices are now available, the treatment here will be confined to the electric version, conforming essentially to current standards (ASTM, ISSMFE). This is, firstly, because the electric penetrometer (CPT) is now generally favoured over mechanical ones on account of its greater accuracy and efficiency (to which it should be added that the outputs of these instruments may differ by some 20% because of the clearance that is provided behind the mechanical cone; Heijnen, 1973; de Ruiter, 1982). Secondly, the various dynamic devices (such as SPT and dynamic cone) are generally not deemed to be sufficiently sophisticated to warrant calibration to the same extent, and may only be required to locate, for example, a hard stratum. Thirdly, devices such as the dilatometer (DMT) and piezo-cone, although undoubtedly falling within the relevant scope, have been deemed to warrant special lectures in their own right.

Whilst the penetrometer directly models a driven pile, and perhaps finds its most direct application to the prediction of pile load capacity, it is often the objective of an investigation to determine basic soils parameters, none of which is measured directly, or in isolation, by the CPT. It is, therefore, an empirical test from which required parameters must be back-figured in accordance with some theory, of which there are a variety available. This process cannot, however, be fully trusted, and therefore calibration is used to directly compare cone outputs with data from basic testing methods.

In the case of cohesive soils, the most obvious way to calibrate is to perform field vane tests, or to recover "undisturbed" samples for standard laboratory testing, allowing cone outputs to be compared with field results via theory. However, in the case of cohesionless soils, undisturbed sampling, as is well known, is not possible, so that we are obliged to re-create simulated "natural" samples in the laboratory, with known properties,in which the CPT may be calibrated. There is also one further category of fine (or silty) sandy soils, which are prevalent off-shore (and especially in the North Sea), for which neither of the above approaches is yet possible. This is a matter of some concern to those engaged in such investigations, but to this stage the problems in preparing large saturated laboratory samples of these materials have not been overcome.

2. DEVELOPMENT OF THE CAVITY-WALL CALIBRATION CHAMBER

The current state-of-the-art with respect to cavity-wall testing chambers is represented in Fig. 1, which depicts the ENEL test rig, as described by Bellotti et al. (1982). In principle, the apparatus contains a large triaxial sample of sand, enclosed in a rubber membrane and loaded laterally by a water jacket. In order to have control over the lateral deformation of the sample (for K_o loadings, when required), a very rigid enclosing pressure vessel is necessary, which of course is not possible. However, by using a cavity wall, and by maintaining a cavity pressure equal to the chamber pressure, full rigidity of the inner-wall is effectively established.

The original concept for a test rig of this type is due to Holden, at the Road Construction Authority (then Country Roads Board) of Victoria, Australia, and resulted, in 1969, in the construction of a chamber housing a sample 0.75 m dia. by 0.90 m high. The base piston was inflated by water from an air/water cylinder, with deformations being derived from water level observations. Sample formation was by travelling sand spreader (after the principle of Kolbuszewski and Jones, 1961), and the results of this investigation were reported by Veismanis (1974).

Subsequent developments occurred in various countries, with progressive developments in technology. In 1970, a chamber for samples 1.20 m by 1.20 m was built at the University of Florida, essentially similar to its predecessor but intended to accommodate a larger cone (Holden, 1971 ; Laier et al., 1975). Further developments from the Florida design, and the need for increased travel, led to the construction of a 1.20 m dia. by 1.80 m high chamber at Monash University (Australia), with travelling spreader (Chapman, 1974), followed soon after by a 1.20 m by 1.50 m chamber at the Norwegian Geotechnical Institute, Oslo (to where the RCA chamber had then been shipped).

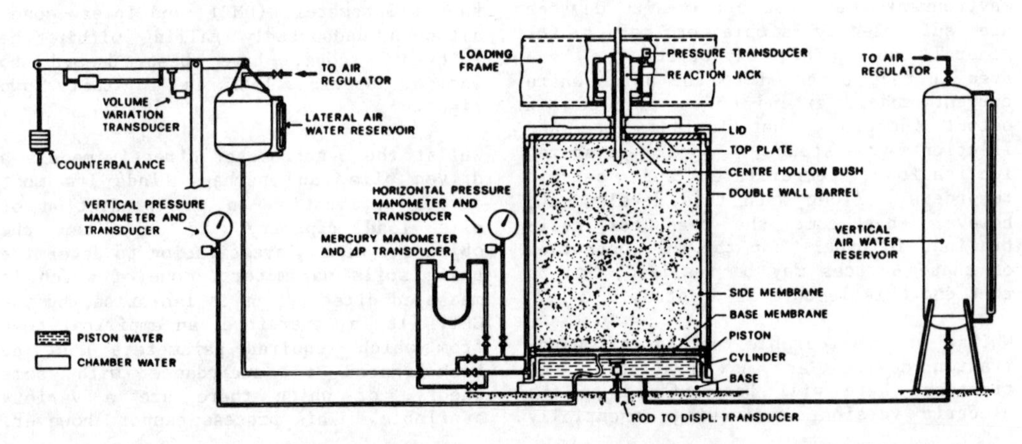

Fig. 1 General arrangement of ENEL Calibration Chamber (Bellotti et al., 1982)

Both these chambers (and the RCA chamber, as modified in Oslo) had air inflation under the base piston, with a transducer provided for the measurement of piston movement (operated from beneath a structural strong floor). Because of the difficulty of achieving sample uniformity, due to the large air currents created by travelling sand spreaders, the latter project also introduced a new static sand rainer (Holden, 1977), developed in association with Jacobsen, at the University of Aalborg (Denmark) (Jacobsen, 1976). This is now considered to be the only satisfactory method of preparing sand samples.

The most recent developments have both occurred in Italy, first at the Italian Electricity Board (ENEL), Milan (1977), and later at the ISMES research institute, Bergamo (1983). Both chambers are of the same size as the NGI chamber, and use similar sand spreaders. In the case of the ENEL chamber (Bellotti, et al., 1982), significant developments were made in respect of a precision servo-controlled mechanical drive for the penetrometer (replacing the hydraulic ram), a highly sensitive device for volume change measurement (Fig. 1), and advanced methods for saturating samples. In the case of the ISMES chamber, the principal development is in modifications to the sand spreader to enable samples to be rained in high vacuum. This has particular value in the preparation of samples at lower densities, and for use with finer materials. (A published description of the ISMES chamber is not yet available, but some observations are reported by Parkin, 1986b.)

3. PRINCIPLES OF CALIBRATION

The operation of the conventional CPT gives two outputs[1] namely cone and friction resistance (q_c and f_s), both of which are stresses and therefore dimensional. One further input, for the purpose of analysis, and which is calculated rather than measured, is the overburden stress. This one just has to presume is correct, although it is probably not unreasonable to do so.

It is then a question of what one can reasonably expect to derive from these outputs. Almost anything can be correlated over a limited range, but the central problem is to determine what, if anything, the cone measures most fundamentally. This may well be quite different when the cone is used in clay, or in sand, and may indeed also vary with density. Is the cone (in sand), for example, causing a bearing failure in accordance with some value of ϕ', or merely expanding a cylindrical cavity against certain parameters of compressibility? From dimensional considerations, it might be reasonable to expect ϕ' to be best estimated from a ratio of stresses, whereas a modulus might be more likely to correlate against a single value of stress.

1 For notes, see Appendix A

3.1 Relative Density

In sand, it is almost universal practice to correlate cone results against relative density, D_r, but this tradition should not be accepted without question. Should CPT's be performed in materials of significantly different specific gravity (eg, mine tailings) then comparisons should undoubtedly be made on the basis of relative density. However, within the usual range of SG, the density limits are not necessarily meaningful in an environment where significant particle crushing is known to occur. The use of relative density is further complicated by the ambiguity in defining limits, depending on the methods used. The latter, particularly, is behind a recent trend, reported by Lunne at al. (1985b) to bypass assessments of D_r and to concentrate on direct predictions of strength and stiffness. In this writer's view, correlations should not be against D_r alone, but also against unit weight or porosity, for the reasons above. Another alternative to D_r may be found in the state parameter (ψ=e-e_{ss}) of Been and Jefferies (1985)

3.2 Overconsolidation in Sand

Whilst overconsolidation in clay has definite and well known consequences, with clearly defined procedures available for establishing stress history, the same is not true of sand. In the calibration chamber at least, the only consequential result of overconsolidation, at zero lateral strain, is an increase in lateral stress, with only minimal disturbance to sand structure. This, upon release of the excess stress, is not detectable by the cone (Parkin, 1977 ; Bellotti et al., 1985).

Sand in-situ may be found to exist under a lateral stress defined by K_o, where K_o is the value of the earth pressure coefficient that may be determined in the laboratory by subjecting that sand to a normally-consolidated loading sequence under rigid confinement. Alternatively, that sand may be found to exist under some other (higher) value of K, but whether this condition has resulted from overconsolidation, and whether this has occurred at zero lateral strain are matters that can be no more than conjecture.

Accordingly, there is some reason to doubt whether overconsolidation (which can only be inferred from geological evidence) has any real meaning for natural sand deposits, and it may be better to regard such materials, quite simply, as sand at a particular value of lateral stress. This is not to disregard the change of modulus induced by a stress history imparted in the laboratory(as discussed by Baldi et al., 1986), but the relevance of this modulus to field conditions has yet to be established (as discussed later).

3.3 Scale Effect

The suggestion is made in various places that the CPT test is subject to a "scale effect", but what this might be is nowhere very clear. Holden (1977), on the one hand, specifically associates scale effect with sands, suggesting that there is some evidence, supported at least by de Beer (priv. comm. to M. Jamiolkowski, 1985), that small penetrometers ($5cm^2$) have a penetration resistance that is different from (higher than) standard penetrometers ($10cm^2$) in normal field situations. Baldi et al. (1982) would also appear to acknowledge this possibility (without detail), but de Ruiter (1982) finds there are no significant differences for cones of 5 to 15 cm^2 area. Battaglio et al. (1986), on the other hand, refer to scale effect in saturated clays, but again without offering details. It is also possible that scale effect may be confused with "chamber size effect" (as referred to by Bellotti et al., 1985, and elsewhere), but which specifically relates to the consequences of the calibration chamber having a less than adequate diameter in relation to the penetrometer diameter. This may cause the larger cones to have either higher or lower penetration resistance, according to the degree of rigidity imposed on the lateral boundary.

In terms of the Terzaghi bearing capacity equation for a circular footing (as quoted by Tomlinson, 1963, for example, with p,p' being total and effective stress at footing level)

$$q_f = 1.3cN_c + p'(N_q-1) + 0.4\gamma B\ N_\gamma + p \quad (1)$$

any scale effect would have to be embodied in the term involving N_γ, which is effectively insignificant at D/B ratios above, say, 10 (ie, all practical situations involving the CPT). Should the CPT output vary with its diameter (B), and particularly to vary inversely with B(as suggested above), there is a difficult conflict with established theories.

Perhaps the only remaining source of a scale effect (as opposed to chamber size effect, discussed elsewhere) is the relationship between cone size and the size of sand grains, but this has not, to this point, been studied in the laboratory situation. It is this writer's opinion, however, that this possibility is unlikely to be one of serious practical consequence.

4. AN ASSESSMENT OF RECENT CHAMBER TESTING OF SAND

Of recent times, the greater bulk of calibration chamber testing has come from the Italian groups (ENEL and ISMES) and the University of Southampton, working with the NGI chamber. A considerable body of new data from these programmes was presented at an informal meeting at Southampton in 1984 (University of Southampton, 1984 and 1985), and an assessment of this data is here presented, along with previous data from Monash University (Chapman and Donald, 1981), NGI (NGI, 1976; Parkin, 1977), and some later contributions.

4.1 The Penetration Test

Because of the finite dimensions of calibration chamber samples, a decision has to be made as to the stress and strain conditions to be imposed on the boundaries. For the simulation of a semi-infinite soil mass, the required boundary condition can be expected to be between the limits of constant stress and zero (average) deformation on both horizontal and vertical boundaries. Accordingly,

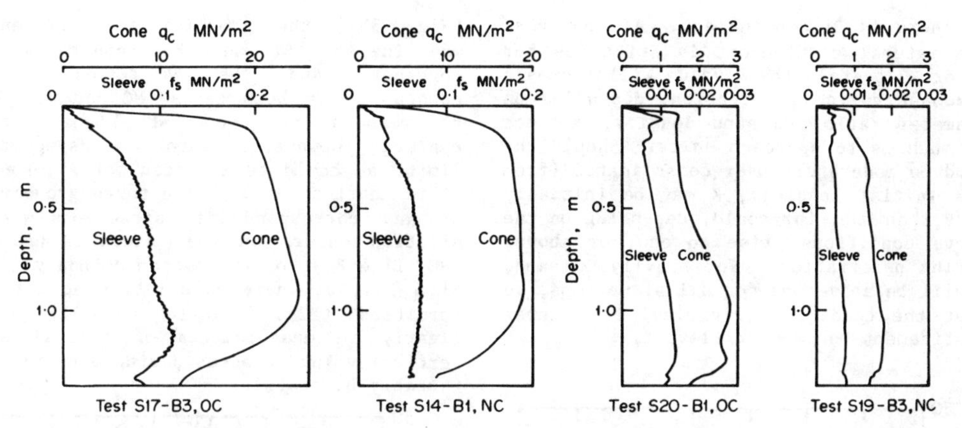

Fig. 2 Typical Penetration Traces, NGI Chamber (Parkin & Lunne, 1982)
(a) Dense sand (b) Loose sand

Holden (1977) identified the following four limiting conditions for the loading of calibration chamber samples, which terminology has been followed by all subsequent workers:

B1 σ_v, σ_h constant

B2 $\varepsilon_v = \varepsilon_h = 0$

B3 σ_v constant, $\varepsilon_h = 0$

B4 σ_h constant, $\varepsilon_v = 0$

Of these possibilities, the most significant are B1 and B3, and testing programmes have mainly adopted these.

After the initial application of the boundary loadings, normally under K_o conditions, and the setting of the selected boundary condition for test, the penetrometer is driven, hydraulically or mechanically, through a hole in the lid, at a constant speed of 20 mm/sec. During penetration, automatic chart records are obtained for cone and friction resistance on the penetrometer, and for lateral pressure and volume change and vertical deformation in the sample.

The chart records of q_c and f_s are found to have characteristic shapes that reflect density and boundary conditions. The typical forms of these curves (Fig. 2) are discussed by Parkin and Lunne (1982), but in particular it should be noted that dense samples under B3 confinement do not reach a "plateau" condition, but have q_c increasing continuously in response to an increasing lateral stress.

For analysis, particular values must be extracted from these charts for correlation against stress and soil parameters. Clearly, there can be no obvious and unique criterion for this that will function satisfactorily for all cases. Peak, or mid-depth (usual) values may be read, or values may be correlated against instantaneous confining stress, where appropriate (NGI), but discretionary judgement may also be required (e.g. Test S20). However, these various options do not have a major influence on final correlations.

4.2 Boundary Conditions

In dense sand (D_r 80 to 90%), all calibration chamber results are affected by boundary conditions, even for the case of a 3 cm² cone, as used by Bellotti et al. (1985)[2] to give a diameter ratio (R_D) of 60. It must be concluded from this that it is impractical to conceive apparatus that would give a value of R_D sufficiently favourable to completely eliminate such boundary effects. But neither would it seem necessary to do so, in view of what can be concluded below. For loose sand, as could be expected, chamber results are independent of boundary conditions, even down to R_D 21 (Parkin, 1977).[3]

The pattern of results analysed herein indicates that the response of the cone to boundary conditions is determined almost entirely by whether the value of K is greater or less than unity. Beyond this, the actual value of K (or overconsolidation ratio, OCR) is irrelevant, as will be discussed later.

In the case of a NC sample, K may be in the region of 0.4 initially (on loading), and will remain at this value thereafter if there is B1 confinement. If, however, the lateral confinement is stiff (as for B2 or B3 tests), the K value will increase somewhat during penetration, depending on diameter ratio and sand density, but not so much as to approach unity. Should the sand be moderately over-consolidated (from the initial loading), K may be initially less than one, but could, depending on the above conditions, rise to one, or above, during penetration. For heavily OC sand, K will be above one for all situations, so that the condition (rigidity) of lateral confinement becomes irrelevant.

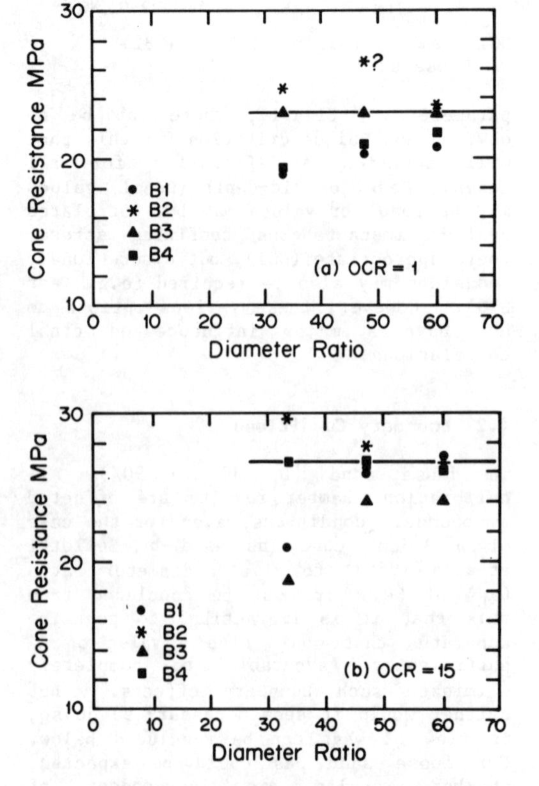

Fig. 3 Boundary effects, very dense Ticino Sand (Bellotti, 1984) (a) Normally consolidated (b) Overconsolidated

These effects are apparent in data presented by Bellotti (1984)[4], reproduced here as Fig. 3, where within the limits of experimental accuracy (and for $R_D > 33$) there is a pronounced grouping of results. For the NC case (Fig. 3a), the B1 and B4 results are effectively identical and vary with R_D, whereas the B3 results are constant. For the OC case (Fig. 3b), the grouping is different, wherein the B1 and B3 results align together[5] and the B4 results are constant. In both cases (NC and OC) the B2 results are somewhat higher, and converge downwards onto an asymptotic limit, as could be expected for all-round rigid confinement[6]. These three groupings of data correspond with situations where dilation occurs on the lateral boundary (NC, B1 & B4), or the bottom boundary (OC, B1 & B3), or where no dilation at all is permitted (B2). These normally show clearly in measurements of lateral and vertical volume changes during penetration (notably B1 tests).

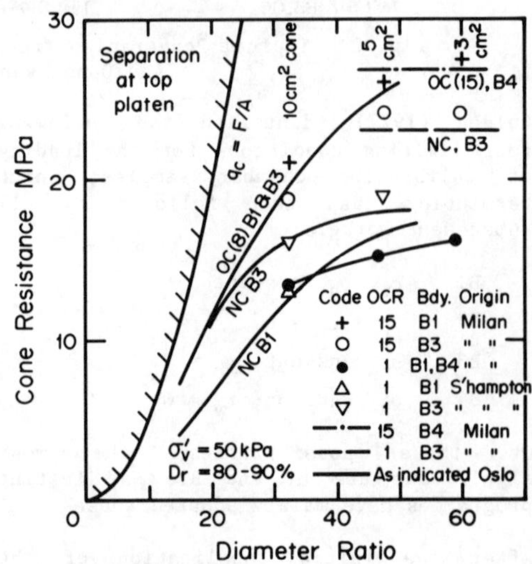

Fig. 4 Boundary Effects in Ticino and Hokksund Sands (Parkin 1986a)

These results (from Bellotti, 1984) can be compared with NGI data on Hokksund Sand (and further tests at Southampton) by superimposing them on a graph of q_c v. R_D from Parkin (1977)[7]. Such a comparison must, however, be made at a common value of σ_v, which, in the case of the NGI tests and most of the Southampton tests, was 50 kPa at the top platen. For the results given by Bellotti (1984), σ_v = 100kPa for NC tests and 50 kPa for OC tests (but not stated therein). These NC results can be correlated back to a common value of σ_v = 50 kPa by means of the square root relation established subsequently.

On the resulting diagram (Fig. 4), the B1 and B3 results for OCR 15 conform closely to the line for Hokksund Sand at OCR 8 (at least in respect of their means, and noting footnote 5). For NC sand, on the other hand, even after correction to a common value of σ_v, there is an appreciable difference of character with respect to the NGI curve for the B1 condition. No explanation is available, but the relatively small response to R_D for these results on Ticino Sand is unexpected by the writer. This is because all results become increasingly affected by the $q_c A_c$ = F boundary for decreasing R_D (see footnote 7), which dependence should, in the writer's opinion, be demonstrated by plotting all results (both NC and OC) on the one diagram, and against the $q_c A_c$ = F boundary. There is also a case for conducting tests at an adverse value of R_D in order to confirm that appropriate trends are being obtained.

4.3 Over-consolidation Ratio (OCR)

The range of OCR studied at NGI (1 and 8) has been significantly extended in a further series of tests on Hokksund Sand at Southampton (Univ. of Southampton, 1985). A set of results for dense sand, B1 boundary and a standard cone (not indicated) is reproduced here as Fig. 5.

In the matter of K_o values determined during the initial loading of samples, it is apparent that the value for NC sand is practically independent of stress level, as indeed it should be (Fig. 5a). In the case of OC samples, the most reasonable conclusion is that K is again independent of stress level, within the limits of experimental error.

During the penetration phase, the cone resistance varied as indicated in Fig. 5b (and sleeve friction in a generally similar manner), from which it was implied that q_c is a function of both OCR and σ_v. This, however, is not the case, as can be demonstrated by plotting the values of σ_h (constant in a B1 test) against each point. The resulting contours then indicate that q_c is a function of lateral stress only (to a sufficient accuracy), at least up to OCR 8. Such behaviour is confirmed by Chapman and Donald (1981), and by Veismanis (1974).

An appearance that q_c depends on OCR is also obtained if these Southampton results are superimposed on the NGI curves of Fig. 4 (Fig. 6). In fact, this variation is not with OCR but with lateral stress, as generated during the initial loading sequence. The cone resistance thus reaches its maximum value at an OCR of around 8, as K approaches unity, and thereafter is independent of further increases in OCR.[8]

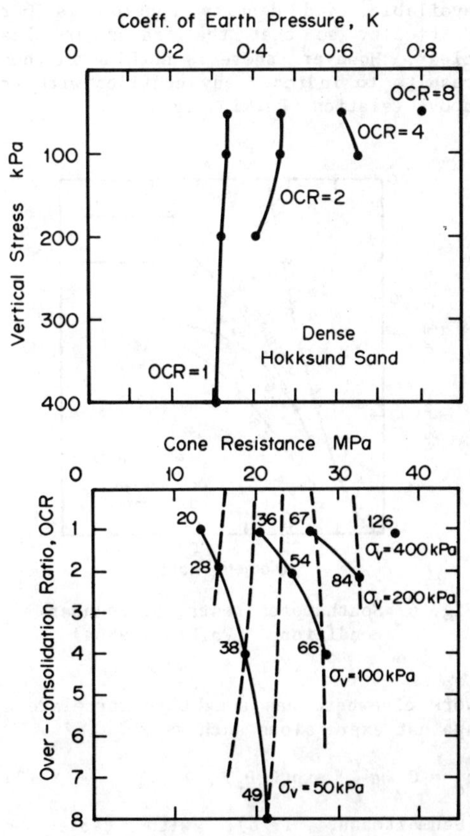

Fig. 5 Southampton test programme (Univ. Southampton, 1985) (a) Variation of K during consolidation (b) Variation of cone resistance

4.4 Stress Level

That cone resistance depends on σ_3, rather than σ_v or σ_h, has been implied in the previous section. This dependence can be further explored in a plot of q_c against σ_h for all available points, including those for B3 tests, wherein σ_h is taken as the developed, rather than the initial value (Fig. 7). Clearly these results (for dense sand and standard cone) conform to a common relationship, linear on log-log scales, and for all practical purposes parabolic, i.e.

$$q_c = \alpha\sqrt{\sigma_h} \qquad (2)$$

For the small cone, an almost identical result holds, but with the line displaced upwards by a factor of about 1.2, reflecting the more favourable diameter ratio. For lower densities, less data is available, and density control is more difficult, so that the trends are less clear. However, there is nothing in these results to indicate any conflict with the above relation (Parkin, 1986a).

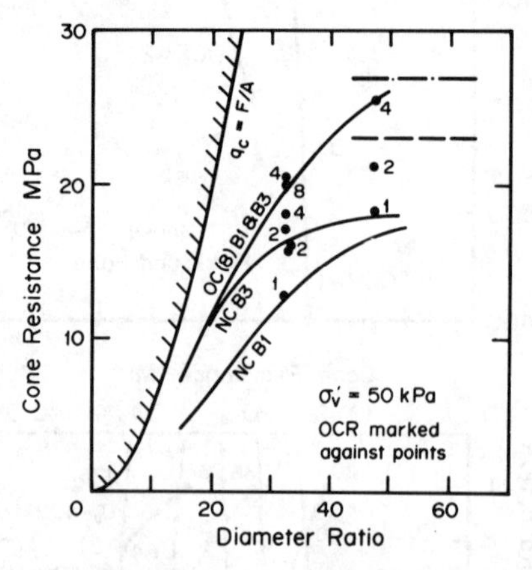

Fig. 6 Southampton tests, B1 Boundary conditions (Parkin, 1986a)

Work elsewhere has sought to correlate q_c against expressions such as

$$q_c = C_o \, \sigma'^{C_1} \exp(C_2 D_r) \qquad (3)$$

(Schmertmann, 1976), with tables of coefficients presented by, for example, Bellotti et al. (1985). These correlations take σ to be, variously, the vertical, lateral, or mean stress, and do not differentiate whether K is greater or less than one, or whether σ_h is changing, as in a B3 test. These factors naturally influence the computed coefficients, with the value of C_1 for all tests performed in Italy being generally in the range 0.53 to 0.63. It is the writer's opinion that the real meaning of such figures is obscure, not only for the above reasons but also because the regression analysis is apt to give undue emphasis to matters of minor significance, in this case the curvature of q_c v. σ, which cannot be supported within experimental accuracy. Therefore, this writer sees little reason to adopt other than C_1 = 0.5, and considers that there is no longer a need for testing at an extensive range of stress levels[9].

In a study of model dynamic cones in a more rudimentary test chamber (taking a sample 0.44 m by 0.66 m, R_D = 22 and B1 boundary), Clayton et al. (1985) note that the increase in penetration resistance with OCR is essentially due to the increase of σ_h, and in proportion to its effect on mean stress. However, OCR values were generally less than 5, and a greater range may be advisable.

4.5 Sleeve Friction

Although readings of cone and friction resistance are obtained quite independently during calibration chamber tests, it can be shown that they are, in fact, closely dependent on each other. A plot of f_s against q_c for the Southampton data (standard cone) shows that all points fall close to a unique relationship that is independent of boundary condition, stress history and stress level (Fig. 8a). A similar result has been obtained previously at NGI (Parkin, 1977), but not by Chapman and Donald (1981).

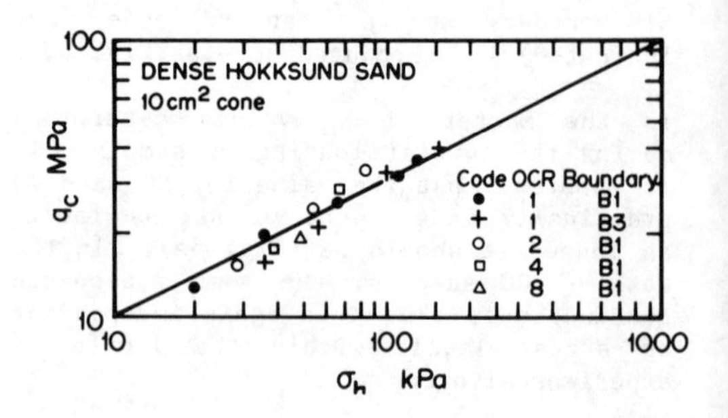

Fig. 7 Cone Resistance v. Lateral Stress (Parkin, 1986a)

Whilst it is tempting to represent this relationship (q_c v. f_s) by a straight line, the trend of individual points is such as to suggest something else. It is, in addition, unreasonable that the correlation should not include the origin. Accordingly, the data has been transferred to a log-log graph, wherein the points correlate closely against a line of gradient 1.6 (Fig. 8b). This, then, indicates a power law relationship, a reason for which needs to be sought in theoretical studies. A similar plot can also be prepared for the small cone, but in this case, the line has a significantly different slope (Parkin, 1986a).

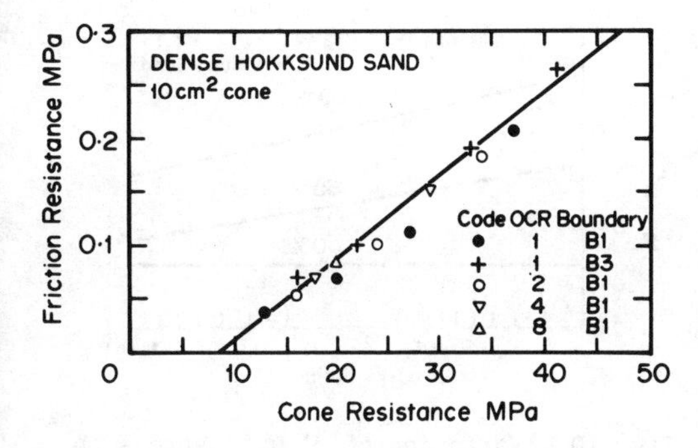

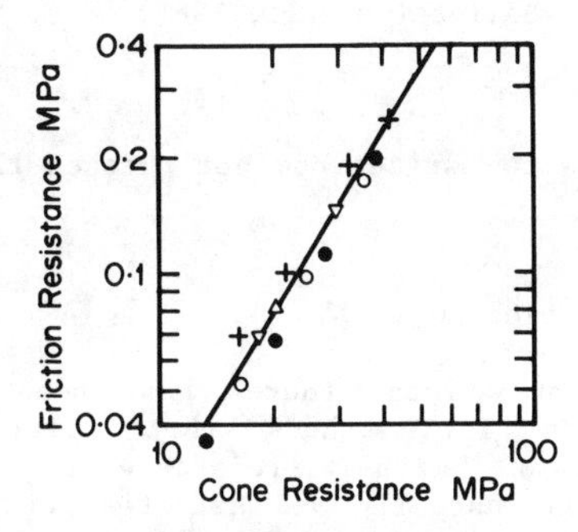

Fig. 8 Cone and Friction Resistance (linear and log scales) (Parkin, 1986a)

It is well known that the friction ratio varies substantially between different soil types, in a manner that readily identifies them. Within any one type of sand, two characteristics emerge from Fig. 8, firstly that f_s is not independent of q_c, and secondly that this relationship is not quite linear. Furthermore, the corresponding plot for the $5cm^2$ cone shows this relationship is unfortunately, also dependent on cone size. Accordingly, sleeve resistance readings would not appear to contribute a lot to the investigation of soil properties.

4.6 Soil Type

It is generally considered that sand type can significantly affect cone response, particularly with respect to differences in hardness, grading, size and roundness. Some light on this matter can be drawn from Fig. 4, wherein the superimposed results for Hokksund and Ticino Sands are indistinguishable, at least for the OC case. Differences do indeed occur for the NC case, the former sand being affected by diameter ratio while the latter apparently is not, as observed by Bellotti et al. (1985), though without explanation. This conflict cannot, however, be dismissed, and because of it these results cannot be taken to indicate differences between the two sands.

The writer therefore sees no strong evidence, at this point, of significantly different behaviour for these sands, which is perhaps fortunate, as the usefulness of the CPT would otherwise be appreciably diminished.

5. PREDICTION OF SAND PARAMETERS

5.1 Relative Density D_r

For the determination of friction angle, normal practice is to first estimate D_r from the measured values of q_c, using correlations determined from calibration chamber testing. The most widely used is that of Schmertmann (1976), as derived for NC sand. With overconsolidation, q_c may be much higher, but provided OCR (or K) is known (which it usually is not, with sufficient accuracy), Schmertmann advises that an equivalent NC value can be calculated as follows:

$$\frac{q_c}{q_{c\ nc}} = 1 + 0.75\left(\frac{K}{K_o} - 1\right) \qquad (4)$$

where $K/K_o = (OCR)^{\beta}$. The same correlations may then be used to establish D_r.

It has been reported that the value of β varies with D_r (Baldi et al., 1981), but Lunne and Christoffersen (1983) advise that $\beta = 0.45$ is adequate for all practical purposes. They further advise that the predicted values of q_c should be reduced by 10 to 15% in coarse sand, with perhaps a similar effect in silty sands (although the influence of grading is still largely unknown).

An updated version of the Schmertmann correlation has been presented by Lunne and Christoffersen (1983). From the curves of Parkin and Lunne (1982) (Fig. 4), correction factors have been derived to allow the prediction of a field cone resistance, free of laboratory size effects. The resulting chart (Fig. 9a) predicts somewhat lower values of D_r than

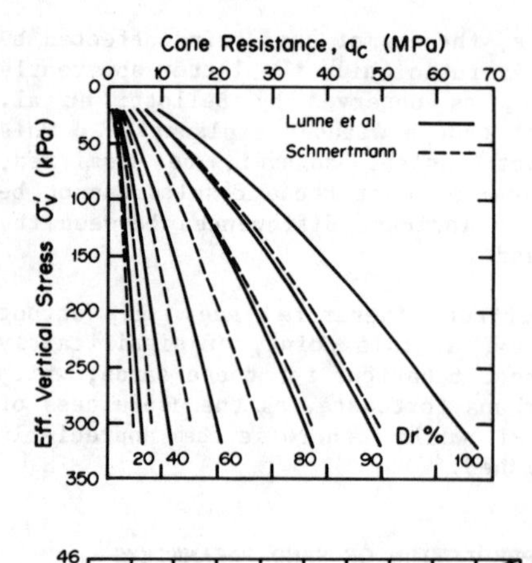

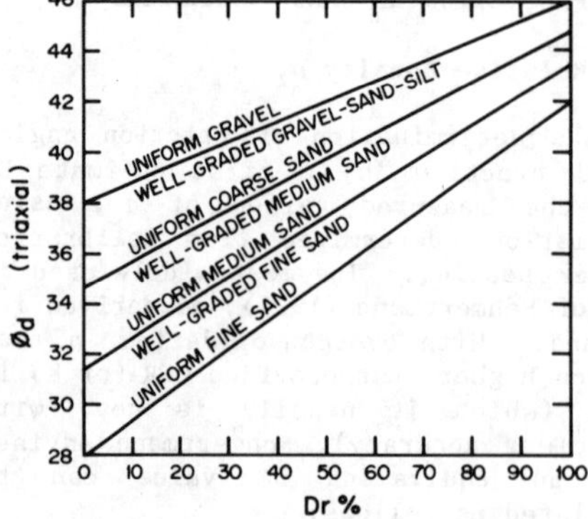

Fig. 9 Cone resistance v. D_r for NC sand (Schmertmann, 1976; Lunne and Christoffersen, 1983)

the original, in line also with Villet and Mitchell (1981), which values may then be used to obtain ϕ' from either triaxial tests or a correlation such as Fig. 9b (Schmertmann, 1976)

The two sands used in the European test programmes have been subjected to an extensive series of triaxial tests in both Norway and Italy, at cell pressures up to about 500 kPa. The data from these tests have been presented as graphs of tan ϕ' v. σ'_{ff}, for various D_r (as in Fig. 10 for Hokksund Sand, with ϕ' for Ticino Sand somewhat less), and as a tabulation of the Baligh (1976) parameters for curved failure envelopes. Whilst failure envelopes may indeed be curved, and enough to significantly affect N_q, the curvature is nevertheless not great, and a curvature factor $\alpha < 5^o$ would be difficult to detect in the usual laboratory investigation. Also, no attempt has yet been made to relate the Baligh parameters to q_c, and

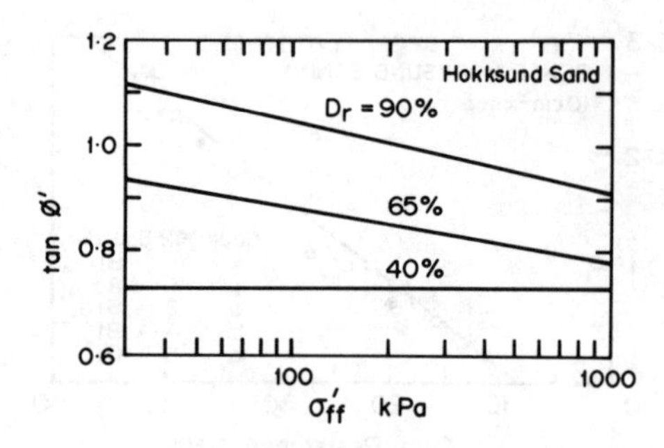

Fig. 10 Variation of ϕ' with stress on failure plane (derived from Bellotti et al., 1985)

therefore the writer has not pursued this aspect.

5.2 Friction Angle ϕ'

Of recent times there has been a preference for methods that directly estimate ϕ', without reference to D_r. This will normally be effected via a bearing capacity factor, such as

$$N_q = q_c/p' \qquad (5)$$

where the value of N_q may be determined experimentally and compared with theory.

Of the various theories that can be applied to the analysis of cone resistance, that of Durgunoglu and Mitchell (1975) is the only one that incorporates a lateral stress. On this basis, Baldi et al. (1982) consider that it deserves preference over other theories but this feature is not an advantage in situations where K may be unknown.

This theory, presented in simplified form in Fig. 11a (with $K_o = 1 - \sin\phi'$), is based on a set of bearing capacity factors $N_{\gamma q}$ for wedges, modified by a shape factor, but the authors advise that $N_{\gamma q}$ may be over-estimated at depths greater than (D/B) critical (the depth just sufficient to contain the full failure mode), which probably means most practical situations. It may also be noted than the K values measured in a calibration chamber test are usually rather greater than (1-sin ϕ') for the NC situation.

Using the D and M theory, but without specifying the K value adopted, Baldi et al., (1982) have calculated ϕ' from values of q_c and have plotted these against triaxial test values at corresponding D_r (Fig. 11b) (on which the writer would suggest confidence limits ±3°). In a similar evaluation, Lunne and Christoffersen (1983) found this theory to work satisfactorily for NC sand, on the assumption of K = 0.4.

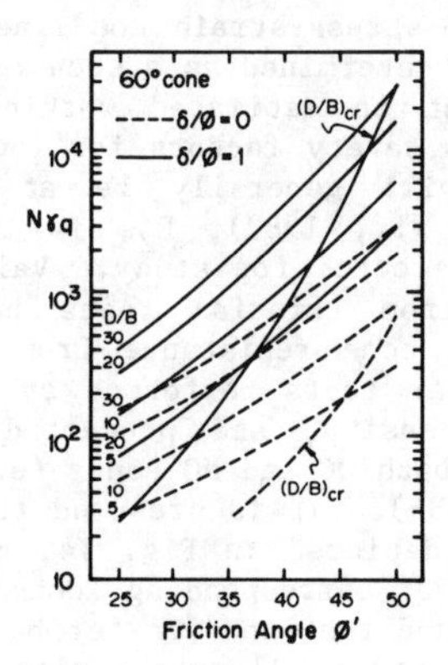

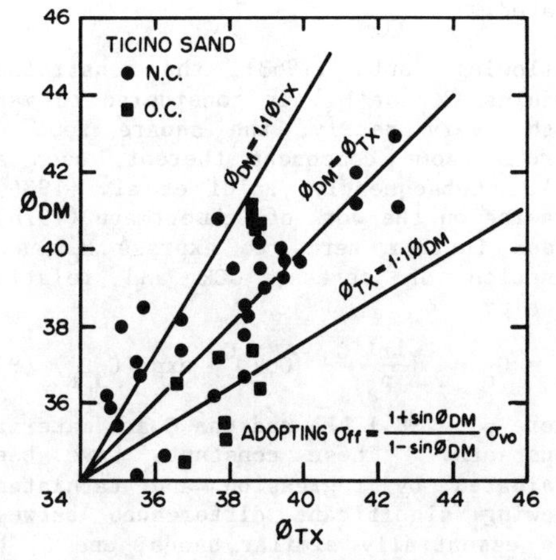

Fig. 11 Friction angle by D and M theory (from Durgunoglu and Mitchell, 1975, and Baldi et al., 1982)

The theory of expanding cavities (Vesic, 1972; Baligh 1976) has also been applied to the prediction of q_c (from triaxial ϕ' values, or vice versa) by Baldi et al. (1982). This work has shown that only the theory for cylindrical cavities is appropriate, and this only for the case of dense to very dense sand. The theory also requires an assessment of the rigidity index

$$I_R = G/(\sigma_h' \tan \phi') \qquad (6)$$

which will, in most cases, not be possible in field use.

Once further theory that would appear to have achieved some success is that of Janbu and Senneset (1974). This theory is a variation of the Prandtl model, recognizing that the failure zone does not close onto the penetrometer shaft, because of compressibility, but terminates on a plane at β (generally 15°) below the horizontal (Fig. 12). It also incorporates a failure criterion written in terms of an attraction a (= c' cot ϕ'), which must be estimated in practice from the q_c v. depth curve (which is only possible if it is linear, or the soil deposit uniform). The formula for cone resistance is then

$$q_c + a = N_q\,(p' + a) \qquad (7)$$

where p' = effective overburden pressure, and N_q is given for various values of β in Fig. 12.

Although the theory is for plane strain only, and does not incorporate a lateral stress, it appears to work well in practice, except that β is found to vary with soil compressibility. Janbu and Senneset therefore give a recommended range (shaded), later revised upwards slightly by Lunne and Christoffersen (1983). It also agrees well with results by Chapman and Donald (1987) (Fig. 13).

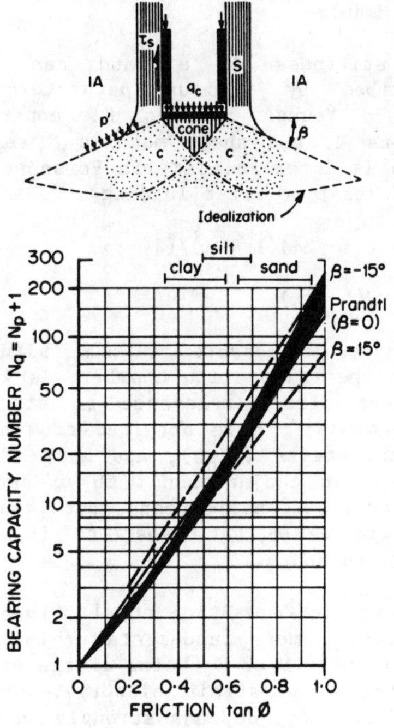

Fig. 12 Bearing capacity theory, Janbu and Senneset, (1974)

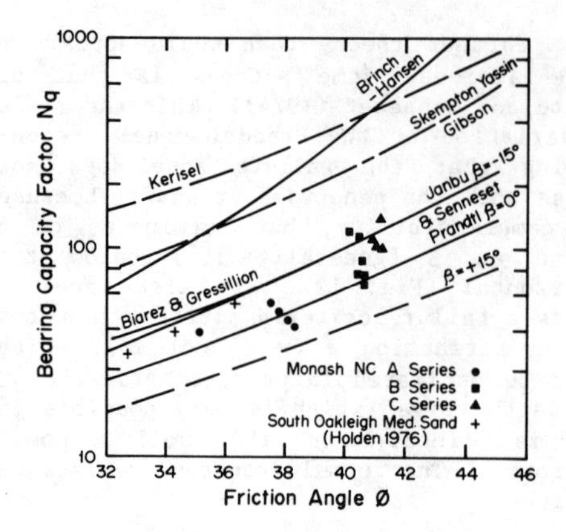

Fig. 13 N_q v. friction angle, Chapman and Donald (1981)

In an analysis of various theories applied to a number of field and laboratory data sources, Mitchell and Lunne (1978) found that the Durgunoglu and Mitchell, Janbu and Senneset, and Schmertmann methods gave generally consistent predictions of ϕ', whilst other methods were mostly conservative. However, they noted that these predictions cannot always be compared with available laboratory results.

5.3 Modulus

The stiffness of a sand can be be described by various parameters that include Youngs modulus E, constrained modulus M, and shear modulus G, each of which is inter-related via Poissons ratio ν (at least on first loading).

$$E/M = (1-2\nu)(1+\nu)/(1-\nu)$$

$$E/G = 2(1+\nu) \quad (8)$$

Of these, M, measured in a K_o situation, as in the calibration chamber, is the one of most direct relevance to structural settlements[10]; E is obtained from drained triaxial tests (CK_oD), and must normally be used in conjunction with ν; and G is derived from dynamic tests, but is usually not applicable, at least to foundation design on-shore.

It is clear from Baldi et al. (1986) that q_c bears a more fundamental relationship to G than to M or E (being little affected by stress and strain history), and they note that G/q_c depends strongly on D_r but only moderately on stress (σ_o'). It is this writer's observation that the product $D_r/100).(G/q_c)$ varies relatively little over the range of variables studied (3.0 to 5.5). However, the derivation of E (from G) is, as explained by Baldi et al., still far from straightforward.

Because of stress-strain non-linearity, E is usually determined as a secant modulus, computed at an estimated working stress level. As safety factors for foundations on sand will generally be at least 4 (Baldi et al., 1982), E_{25}' is the value commonly selected for study. Values thus obtained from triaxial tests have been related to cone resistance from calibration chamber tests performed at the same relative density, and presented graphically for both OC and NC sands (e.g. Baldi et al., 1986). It is presumed that these tests, as depicted in Fig. 14, have been compared at corresponding consolidation stresses, and that the scatter band is for different stress levels, although not stated.

Following Janbu (1963), the constrained modulus (M_o or M_s) is considered to vary with, very nearly, the square root of stress (some component thereof, such as σ_o'). Subsequently, Baldi et al. (1986), drawing on the work of Schmertmann (1976), added further terms to express M_o as a function of stress, OCR and relative density

$$M_o = C_o\, p_a \left(\frac{\sigma'}{p_a}\right)^{C_1} (OCR)^{C_2} \exp(C_3 D_R) \quad (9)$$

where p_a = 98.1 kPa and the C are material constants. These constants have been evaluated by regression and tabulated, showing significant differences between two essentially similar sands, and C_1 by no means always in the region of 0.5.

On the other hand, M may be expressed as a function of cone resistance. Senneset et al. (1982), for example, suggest

$$M_s \approx 40\sqrt{q_n p_a} \quad (10)$$

q_n being net cone resistance (q_c-p), and the coefficient varying ± 10 according to density. This equation would appear to apply for all stress levels and OCR, and could suggest a conflict with data given by Baldi et al. (1986) wherein M_o/q_c is shown to be a function of σ_o', OCR and D_r, as in Fig. 14. However, there are, in fact, not five independent variables, and because of this redundancy it is not possible to determine the M_o - q_c relationship in this case.

In their evaluation of data from various field and laboratory sources, Lunne and Christoffersen (1983) found linear relationships of the form

$$M_o = A\, q_c + B \qquad (11)$$

to be appropriate. However, when these were superimposed graphically, the relationship of Fig. 15 was found to provide the best fit. This form was considered to suggest a parabolic relationship, in line with Eq. 10 (although the writer would suggest other powers of q_c may be more appropriate).

Many empirical relationships of the form

$$M_o = \alpha\, q_c \qquad (12)$$

are to be found in literature, as reviewed for example by Mitchell and Gardner (1975), with α being frequently around 3, but varying with D_r (Vesic, 1970). This type of proportionality is suggested in a result from Parkin (1977), wherein the tangent modulus during initial loading of a chamber specimen is compared with q_c values at a range of stress levels (Fig. 16). Clearly, M_o/q_c takes a range of values that is relatively independent of stress (σ_v or σ_h), and in the case of dense Hokksund Sand, at the most favourable diameter ratio and boundary condition for modelling field conditions (small cone and B3), α = 3 is indicated.

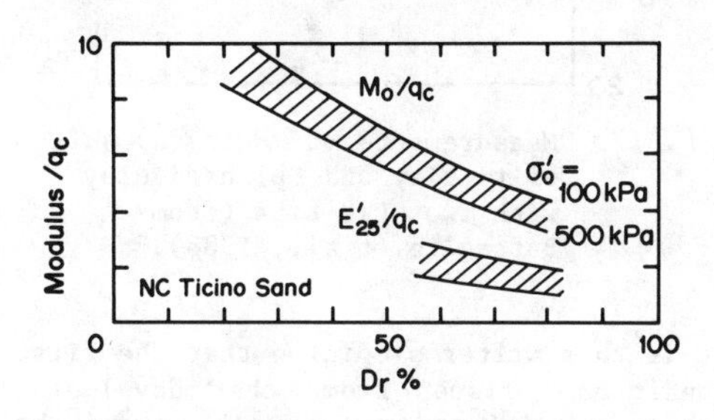

Fig. 14 Modulus - Relative density relationship (derived from Baldi et al., 1986)

There are, however, more ways than one to approximate a set of data, and the various methods above are generally not inconsistent (eg. Senneset et al. predicts α = 3 for q_c = 20 MPa).

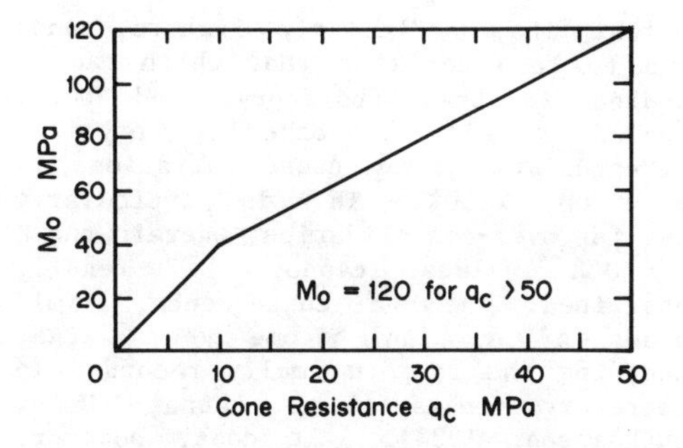

Fig. 15 Recommended relationship, M_o v. q_c (after Lunne and Christoffersen, 1983)

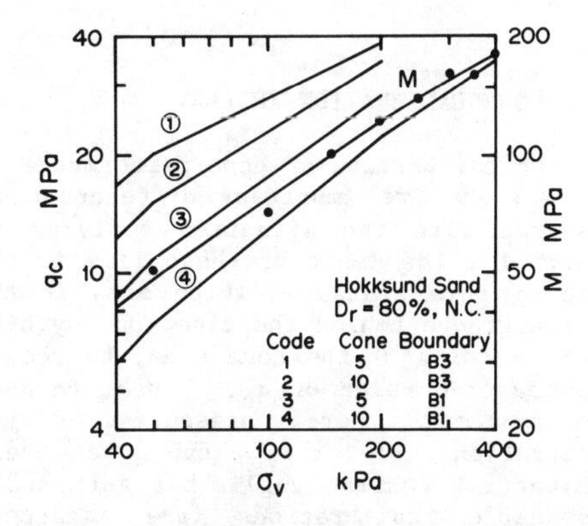

Fig. 16 Variation of M and q_c with σ_v (Parkin, 1977)

Whilst M is known to increase substantially for overconsolidated sand (eg. a value of α = 12 is quoted by Chapman and Donald, 1981, and a multiplier of 1.5 for OC conditions is suggested by Robertson and Campanella, 1983), the definition of these values, and their use in practice, is made difficult by the fact that M changes rapidly on reloading, and may have little meaning until the size of the reload increment is specified.

5.4 "Real" Sands

Of necessity, all the laboratory correlations presented above must be obtained on fairly clean sand of at least medium grain size, because of the requirements of handling in the production of laboratory samples. In most cases, the sand will also be dry, because of the difficulties of saturating and reprocessing large samples if tested wet.

In real life, particularly offshore, sands tend to be finer than that which can be handled in the laboratory, and this, coupled with other physical characteristics, may cause variations in q_c of up to 50%. This is particularly true for over-consolidation, wherein the K or OCR values cannot be readily determined. As a consequence, field values of q_c can become very large, exceeding the range normally recorded in laboratory tests (eg. Lunne, Univ. Southampton, 1984). It does, however, appear that saturation makes very little difference to q_c (less than 10%, Bellotti et al., 1985), but there is a need for laboratory work on very dense and finer sands.

6. CONE CALIBRATION IN CLAY

For the calibration of cone performance in clay, there are important differences as compared with the situation applying to sand. Not the least of these lies in the role of pore pressure, which acts in the O-ring groove behind the cone, on anything up to a third of the cone area, to reduce the measured value of q_c. Hence, because the dynamic pore pressures during penetration of clay can be very substantial (whilst negligible for sand), reasonable calibrations are scarcely possible without pore-pressure measurement (de Ruiter, 1982). This, however, enters upon territory that is the subject of Lecture 1, and therefore will not be examined in depth here.

One consequence of such pore-pressure dependence is that the ultimate determination of clay properties is inevitably influenced by design features in the penetrometer, as discussed by Battaglio et al. (1986), particularly in the location and geometry of the porous stones and O-ring seals. So important is the location of the porous stone that it is possible to record negative pore pressure behind the tip, even when positive pore pressure is being recorded on the tip (Schmertmann, 1974).

Battaglio et al. (1986) do, however, make a further claim, based on piezocone observations in a soft silty clay and a hard clay (Fig. 17), that the pore pressure outputs are also a function of the "mechanical compressibility of the filter stone and cone". In the case of the soft clay, the pore pressure measured at the tip is some 30% higher than that measured ahead of the tip. In the case of the hard clay, the pore pressure is again higher, but very erratic[12], thus contrasting strongly with the measurement ahead of tip.

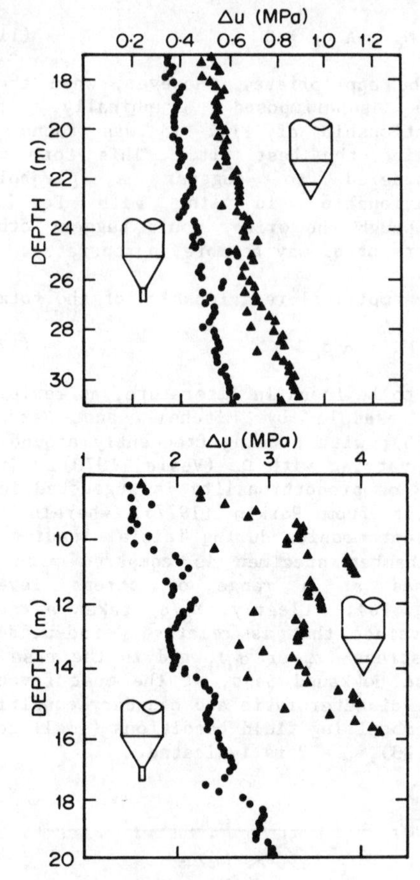

Fig. 17 Measurements of Δu in (a) soft silty clay and (b) hard clay with two CPTU tips (from Battaglio et al., 1986).

It is this writer's opinion that the first condition arises from the developing strain conditions in the soil around the cone tip, and that the second condition in hard clay is the result of fissuring. Therefore there is no sound evidence, at this point, that mechanical compressibility of cone components is involved.

6.1 Drainage State

In clays, q_c is taken to be a measure of the undrained shear strength s_u. This, of course, presumes that penetration takes place undrained, which, if so, should make tip resistance relatively insensitive to variations in penetration rate, as has been shown to be the case for sands. However, apart from the known dependence of s_u on loading rate, the rates of pore-pressure dissipation can be appreciable (eg. of the order of 25% in one minute in soft clay, Battaglio et al., 1986), so that questions of appropriate magnitude and constancy of test rates should not be dismissed too quickly. Some attention has been given to this matter principally in Canada, but studies in Norway on soft marine clays (Lacasse and Lunne, 1982) indicate that u is relatively insensitive to penetration rate, over a modest range, whilst q_c shows a fairly normal dependence on test load rate (5 to 10% increase per 10-fold increase in loading rate). This would suggest that drainage is not significantly affecting the strength of these two clays, but this cannot be relied upon elsewhere.

6.2 Reference Strength

Subject to the limitations of loading rate, calibration becomes a matter of relating observed cone resistance to direct measurements of s_u, from which a cone factor may be evaluated:

$$N_k = \frac{q_c - p_o}{s_u} \qquad (13)$$

where p_o is the total overburden stress.

The choice of a reference standard will normally be between the in situ vane test and laboratory CAUC[13] triaxial tests on undisturbed samples (other in situ arrangements being mostly indirect and often unsuitable for use off-shore). These two procedures do not necessarily agree, and both indeed have their limitations as predictors of structural performance.

The concensus (eg. Lunne et al., 1976) has generally been that field vane strength should be the reference standard, not because it is without limitations, but because it is usually the most convenient method. Additionally, it has been well-correlated against mobilised shear strength in observed failures of embankments and footings on soft clay (principally Bjerrum, 1973; Ladd et al., 1977), wherein safety factor at failure has been computed from field vane strength, and used as the basis of a vane strength correction.

The particular limitations of the vane test (as well as the CPT) have been discussed by Schmertmann (1975). Some apply equally to the triaxial test, as for example the dependence of s_u on strain rate (increasing with plasticity) and possible anisotropy. Others are inherent in the vane, and relate to uncertainties about the shape of the failure surface and the shear stress thereon, and the effects of progressive failure, wherein s_u does not peak simultaneously at all points (Aas, 1967; Donald et al., 1977). However, it seems likely that the extent of available experience overcomes at least part of the difficulty (Aas et al., 1986).

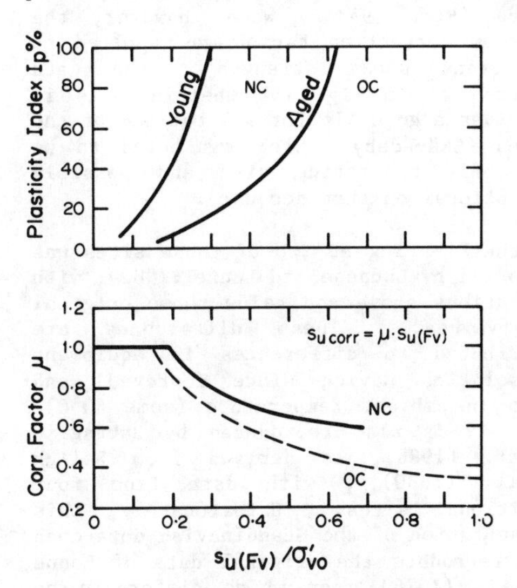

Fig. 18 Correction to field vane strength (after Aas et al., 1986)

For simultaneous strength mobilisation around the vane, the shear strength is calculated as

$$s_{u(FV)} = \frac{6T}{7\pi D^3} \qquad (14)$$

T being torque and D the vane diameter (for a standard 2:1 vane, and standard test rate). This value has traditionally been adjusted according to the Bjerrum (1973) correction curve, to allow for strain rate effects, but because of significant scatter in the original data, this procedure has drawn a measure of criticism, including Lunne et al. (1976). This problem appears to relate to an inadequate consideration of stress history, which matter has been investigated by Aas et al. (1986). In their procedure, s_u/σ'_{vo} and I_p are first used to infer stress history, following which an appropriate curve is selected to obtain μ (Fig. 18).

6.3 Total Cone Factor N_k

From a study of highly uniform deposits of soft clay at six different sites, and comparing cone and field vane results, Lunne et al. (1976) calculated cone factors in accordance with eq. 13. Whilst these values of N_k varied from 7 to 27, a clear correlation with plasticity was noted (Fig. 19a). When, however, the Bjerrum correction factor was applied to the vane shear strength[14], the data plotted as in Fig. 19b wherein $N_k = 17$ provides a good fit for all but one of the sites (Skå-Edeby, which was noted to be the only non-marine site and possibly exceptional on that account).

Further testing at two of these sites was reported by Lacasse and Lunne (1982), with variations above and below these original N_k values. These differences are attributed to differences in equipment (resolution having since improved) and also in ambient temperature (some 40°C). Fig. 19 is also reproduced by Battaglio et al. (1986), as derived from Baligh et al. (1980)[15], with data from four additional sites. Unfortunately, this presentation of the Scandinavian data does not reproduce the original data of Lunne at al. (1976), for which difference no explanation has been offered. The additional data does, however, include marine clays that lower the mean value of N_k to about 14, and which therefore indicates that there may not be any valid distinction between marine and non-marine clays, as suggested by Lunne et al.

Other writers (eg. Schmertmann, 1975) consider sensitivity is also a factor of relevance to N_k, but although this was examined by Lunne et al., no significant trend emerged and sensitivity figures are therefore deleted from Fig. 19.

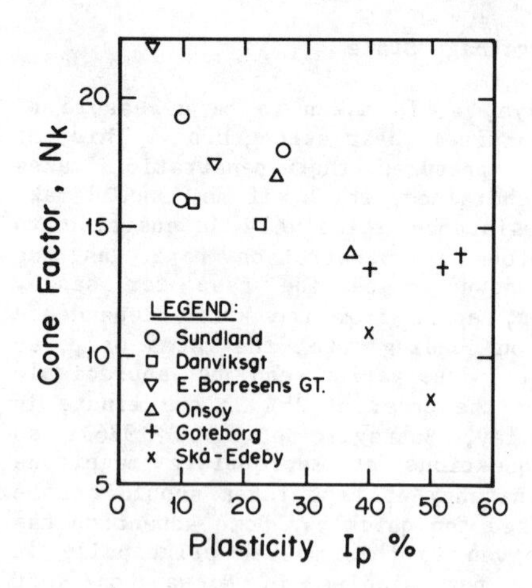

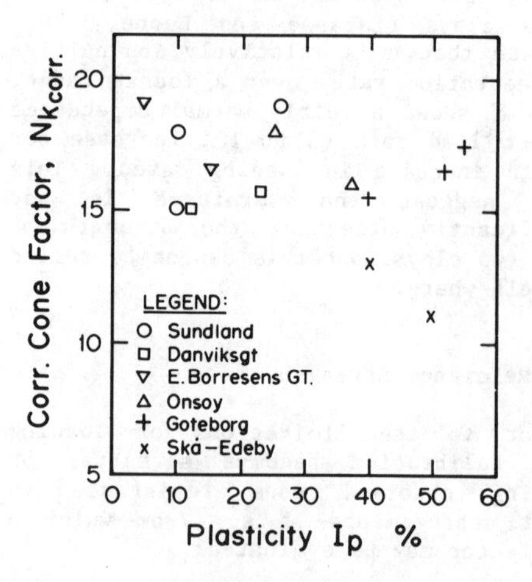

Fig. 19 N_k values from Scandinavian soft clay sites (without and with FV correction) (Lunne et al., 1976)

In a further study from eleven diverse sites, Rad and Lunne (1988) obtained N_k values varying from 11 to 30, with OCR[16] being apparently the principle variable (in a generally logarithmic fashion). Data are correlated against broadly based and overlapping zones of OCR, on which a reasonable degree of separation emerges. Accordingly, they recommend that an estimate of OCR be made from the $q_c - p'_o$ diagram, with s_u being then estimated from q_c and OCR (Fig. 20). It is interesting that a correlation of s_u against cone face pore pressure is virtually independent of OCR[17], but this is considered to be a less attractive method for estimating s_u, for reasons not discussed.

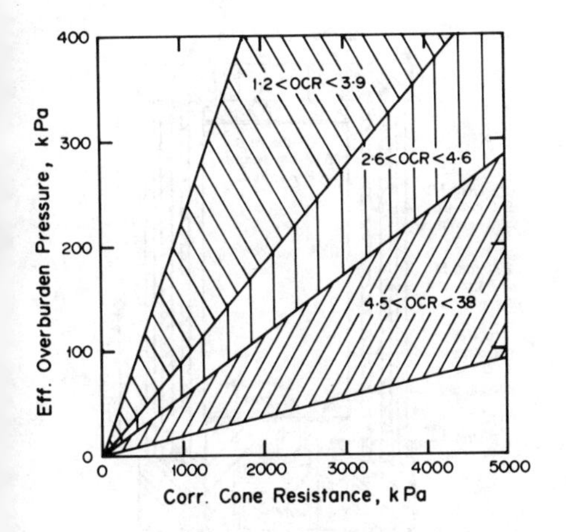

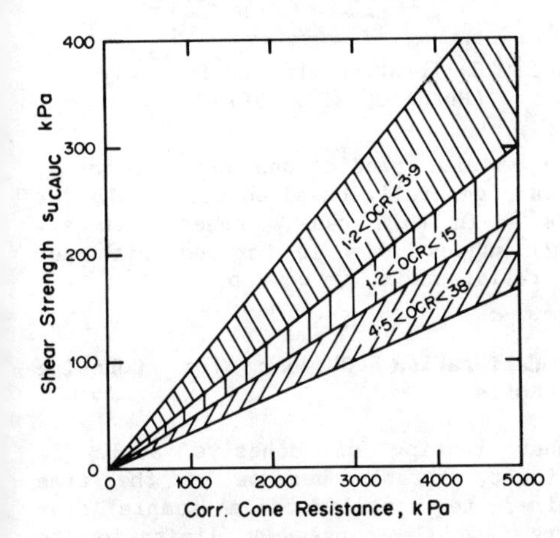

Fig. 20 Effect of OCR on N_k (Rad and Lunne, 1988)

It is not surprising, then, that computations of pore pressure ratio by two different formulations failed to correlate in any useful manner with the OCR-depth curve, as plotted by Battaglio et al. (1986), which writers then concluded that the CPT is unlikely to be a useful predictor of OCR. (They therefore see the oedometer test as the only practicable means for assessing OCR at this time.)

It would, however, seem that the CPT is a somewhat doubtful technique in OC clay, on account of the very small soil volume being tested, and the potential for a very uncertain drainage state on account of fissuring (Battaglio et al., 1986). If anything, N_k values could be slightly higher, with N_k = 17.5 being suggested by Kjekstad et al. (1980)

6.4 Effective Stress Parameters

The effective stress method for the analysis of CPT data was developed by Janbu and Senneset (1974) (as previously applied to sand), and further extended by Senneset and Janbu (1984), for application to clays, by introducing a pore pressure factor N_u. They proposed a bearing capacity formula in terms of the effective overburden stress p_o'

$$q_T - p_o = (N_q - 1)(p_o' + a) - N_u \Delta u \qquad (15)$$

where q_T = cone resistance corrected for pore pressure error, and N_q (for plane strain) is given in Fig. 12. The value of N_u is given as 6 tan ϕ' (1 - tan ϕ'). Equation (15) is then rearranged to

$$q_T - p_o = N_m (p_o' + a) \qquad (16)$$

where

$$N_m = \frac{N_q - 1}{1 + N_u B_q} \quad \text{and} \quad B_q = \frac{\Delta u}{q_T - p_o}$$

Values of N_m have been calculated for β = 0, and the Janbu and Senneset solution is given as the broken lines in Fig. 21. In use, a must be estimated from a graph of (q_T - p) v. p', and the pore pressure parameter B_q obtained from the tip pore pressure excess, Δu. N_m is then calculated and hence ϕ'.

The application of this procedure to several clay sites (mainly off-shore) has been examined by Lunne et al. (1985a) by relating piezocone observations to CAUC triaxial tests. From this, a revised reommendation is given, as the solid lines in Fig. 21. They note that a cannot always be determined adequately , and advise for such situations:

NC and slightly OC clay	a ≈ 2 - 10 kPa
Heavily OC clay	a ≈ 20 - 100 kPa

These writers also define an effective cone resistance as (q_T - u), with appropriate correction factors, and for

$$q_e = \frac{N_{ke}}{s_u} \qquad (17)$$

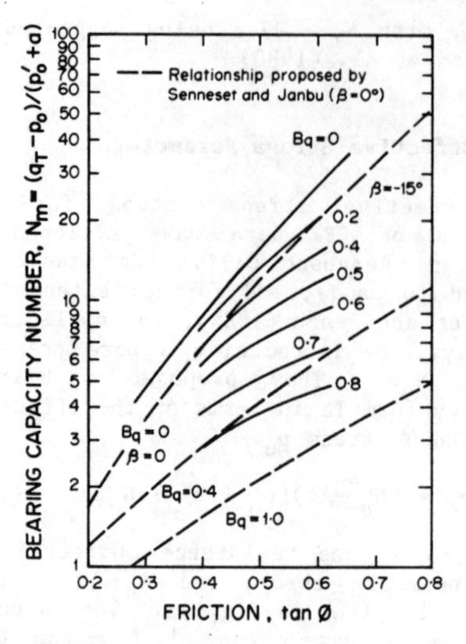

Fig. 21 Effective stress solution (after Lunne et al., 1985a)

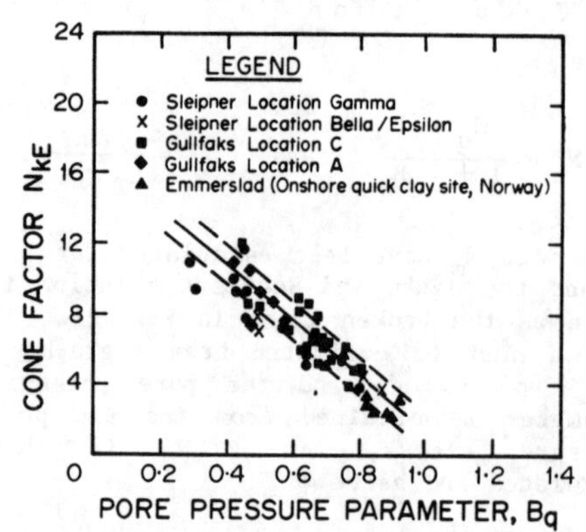

Fig. 22 Effective cone factor v. P.P. parameter (after Lunne et al., 1985a)

they advise that a better prediction of s_u (compensated to some degree for OCR) can be obtained from Fig. 21. However, it might appear that the CPT has not yet found general acceptance as a means of defining effective stress parameters.

6.5 Compression Modulus, M

The CPT is generally considered to be of very limited value as a predictor of modulus for clays. Furthermore, it is often difficult to obtain adequate data for calibration, because the field vane test is unsuitable for this purpose (on account of the flexibility of the rods - Aas et al., 1986).

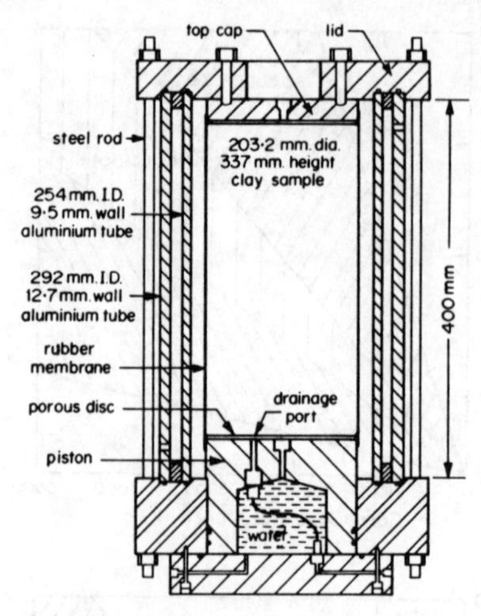

Fig. 23 Calibration chamber for clay (Huang et al., 1987)

Where modulus predictions are attempted, they are generally based on eq. 12, as for sands. In this case, Senneset et al. (1982) suggest α = 10 for use with net cone resistance q_n (= q_c - p).

6.6 Calibration Chamber for Cohesive Soils

Chamber testing of cohesive soils is difficult, firstly because of the time required to consolidate a sample from slurry (and the consequent limitation to much smaller samples than for sand), and secondly because of the large volume change to be accommodated during consolidation (V_i/V_f) up to 2.5). Despite this, Huang et al. (1987) have succeeded in producing Kaolin clay samples, 0.2 m dia. by 0.36 m high, for testing in a calibration chamber that is essentially similar to that used elsewhere for sand (Fig. 23). However, testing so far has been confined to a model pressuremeter (for which R_D = 18)[18][19], and has not yet extended to the CPT.

7 CONCLUSION

The electric friction cone penetrometer is now a sophisticated and robust instrument, capable of operating in diverse and relatively hostile environments and delivering good quality data in both sand and clay.

In sand, the cone tip resistance can be used to predict D_r, ϕ' or M, provided that the OCR (or lateral stress) is known with some confidence. This remains a real problem in practice, even though OCR can be regulated at will and studied in the laboratory. It was shown, from a review of results from laboratory calibration chambers, that the cone responds essentially to the minor principal stress imposed, which may be either σ_v or σ_h according to initial OCR. Consideration is given to the possible presence of a scale effect, wherein cones of different size may give different tip resistance, but it is shown that there is no real evidence for this, and that most references tend to be vague.

Cone tip resistance varies with depth in all cases, but in a manner which can be fitted by various functions, ranging from a power law to a linear function with an "attraction" intercept. Curve fitting has been used extensively to define coefficients, but the writer suggests that this can lead to a measure of confusion in that unwarranted emphasis is often given to minor details. This, for example, often leads to suggestions that CPT results are specific to particular sands, with conclusions that may be over-restrictive.

Of the various theories for predicting ϕ', those of Durgunoglu and Mitchell (1975) and Janbu and Senneset (1974) are the most widely favoured, together with the empirical procedure of Schmertmann (1975), to which only minor modifications have been made since.

The prediction of modulus from q_c is essentially empirical, with a choice of functional relationships being available, with litle in recent data to change the established predictions of, for example, Schmertmann (1975).

In clays, the construction of the cone has a critical bearing on its performance and interpretation as does the drainage state of the soil around the tip (which cannot necessarily be presumed to be undrained). Calibration for shear strength (s_u) is by reference to the field vane or undrained triaxial tests (with no significant role for large chamber tests so far) : the former has generally been preferred on the grounds of convenience and the extent of experience available, but the most recent trend is towards direct use of triaxial tests, consolidated to in situ stress, and sometimes adopting a mean of compression, extension and simple shear tests. Where the field vane is used, a correction for I_p and OCR according to Aas et al. (1986) is recommended. N_k values are shown to be around 14 to 17, varying with I_p and OCR (although the cone is not as reliable in OC clay because of macro-structure).

Methods for deriving effective stress parameters are reviewed, but these do not appear to have the same level of acceptance. It is also used on occasions to predict a modulus, but its value for this purpose is more questionable.

8. ACKNOWLEDGEMENT

The Writer gratefully acknowledges material and comments made available by Mr. T. Lunne (NGI) and Dr. J.C. Holden (RCA, Australia), and also appreciates visits to the Italian installations made possible through Professor M. Jamiolkowski (Univ. of Torino). Much of this material was assembled whilst the Writer was at the Norwegian Geotechnical Institute, Oslo, for which assistance thanks are due to the Director, Dr. K. Høeg.

9. REFERENCES:

Aas, G. (1967). Vane tests for investigation of anistropy of undrained shear strength of clays. Proceedings, Geotech. Conf., Oslo, 1:3-8.

Aas, G., Lacasse, S., Lunne, T. and Høeg K. (1986). Use of in situ tests for foundation design on clay. Proceedings, ASCE Spec. Conf. In Situ 86, Blacksburg, Va. (June 23-25). Also NGI Int. Rep. 52155-41 (29 pp.)

Allersma, H.G.B. (1982). Photoelastic investigation of the stress distribution during penetration. Proceedings, 2nd European Symp. Penetration Testing, Amsterdam, 2:411-418.

American Society for Testing and Materials (1979). Deep quasi-static cone and friction cone penetration test of soil, ASTM D3441-79.

Baldi, G., Bellotti, R., Ghionna, V., Jamiolkowski, M. and Pasqualini, E. (1981) Cone resistance in dry NC and OC sands. Cone penetration testing and experience: Proceedings, ASCE National Convention, St. Louis, Mo., pp. 145-177.

Baldi, G., Bellotti, R., Ghionna, V., Jamiolkowski, M., and Pasqualini E. (1982). Design parameters for sands from CPT. Proceedings, 2nd European Symp. on Penetration Testing, Amsterdam, 2:425-432.

Baldi, G., Bellotti, R., Ghionna, V., Jamiolkowski, M., and Pasqualini, E. (1986). Interpretation of CPT's and CPTU's. Part 2 : Drained penetration of sands. 4th Int. Geotech. Seminar, Singapore.

Baligh, M.M. (1976). Cavity expansion in sand with curved envelopes. Proceedings, ASCE, 102:GT11, pp. 1131-46.

Battaglio, M., Bruzzi, D. Jamiolkowski, M., and Lancellotta, R. (1986). Interpretation of CPT's and CPTU's. Part I : Undrained penetration of saturated clays. 4th Int. Geotech. Seminar, Singapore.

Been, K. and Jefferies, M.G. (1985). A state parameter for sands. Geotechnique 35:2, pp. 99-112.

Bellotti, R., Bizzi, G. and Ghionna, V. (1982). Design, construction and use of a calibration chamber. Proceedings, 2nd European Symp. Penetration Testing, Amsterdam, 2 : 439-446.

Bellotti, R. (1984). Chamber size effects and boundary condition effects. See: Univ. of Southampton (1984), Part 1.

Bellotti, R. et al. (1985). Laboratory validation of in-situ tests. In : Geotechnical Engineering in Italy (Associazione Geotecnica Italiana, published for ISSMFE Golden Jubilee), pp. 251-270.

Bjerrum, L. (1983). SOA Report: Problems of soil mechanics and construction on soft clays and structurally unstable soils. Proceedings, 8th Int. Conf. Soil Mech. and Found. Eng., Moscow, 3:111-159.

Chapman, G. A. (1974). A calibration chamber for field test equipment. Proceedings, European Symp. Penetration Testing, Stockholm, Vol. 2.2, pp.59-65.

Chapman, G.A. and Donald, I.B. (1981). Interpretation of static penetration tests in sand. Proceedings, 10th Int. Conf. Soil Mech. and Found. Eng., Stockholm, 2:455-458.

Clayton, C.R.I., Hababa, M.B. and Simons, N.E. (1985). Dynamic penetration resistance and the prediction of compressibility of a fine-grained sand - a laboratory study. Geotechnique, London 35:1, pp. 19-31.

de Ruiter, J. (1982). The static cone penetration test: SOA Report. Proceedings, 2nd European Symp. Penetration Testing, Amsterdam, 2:339-405.

Donald, I.B., Jordan, D.O., Parker, R.J. and Toh, C.T. (1977). The vane test - a critical appraisal. Proceedings, 9th Int. Conf. Soil Mech. and Found. Eng., Tokyo, 1:81-88.

Durgunoglu, H.T. and Mitchell, J.K. (1975). Static penetration resistance of soil : I-Analysis. Proceedings, ASCE Spec. Conf. on In Situ Measurement of Soil Properties, Rayleigh, N.C., 1:151-171.

Heijnen, W.J. (1973). The Dutch cone test: study of the shape of the electrical cone. Proceedings, 8th Int. Conf. Soil Mech. and Found. Eng., Moscow, 1:1, pp. 181-4.

Holden, J. C. (1971). Laboratory Research on static cone penetrometers. Report No. CE-SM-71-1, Dept. of Civil Eng., Univ. of Florida.

Holden, J.C. (1977). The calibration of electrical penetrometers in sand. Norwegian Geotech. Inst., Int. Rep. 52108-2, (29 pp).

Huang, A.B., Holtz, R.D. and Chameau, J.L. (1987). A calibration chamber for cohesive soils. ASTM Geotech. Testing Journal (in press).

ISSMFE, (1977). Report of the subcommittee on standardization of penetration testing in Europe. App. A: Recommended standard for the cone penetration test (CPT). In: Minutes of Exec. Mtg., App. 5. Proceedings, 9th Int. Conf. Soil Mech. and Found. Eng., Tokyo, 3:99-109.

Jacobsen, M. (1976). On pluvial compaction of sand. Report No. 9, Laboratoriet for Fundamentering, Inst.

of Civil Eng., Univ. of Aalborg, Denmark.

Janbu, N. (1963). Soil compressibility as determined by oedometer and triaxial tests. Proceedings, European Conf. Soil Mech. and Found. Eng., Wiesbaden, 1:19-25.

Janbu, N. and Senneset, K. (1974). Effective stress interpretation of in situ static penetration tests. Proceedings, European Symp. Penetration Testing, Stockholm 2.2, pp. 181-193.

Kjekstad, O., Lunne, T. and Clausen, C.J.F. (1978). Comparison between in situ cone resistance and laboratory strength for overconsolidated North Sea Clays. Marine Geotechnology 3:4, pp. 23-26. (Also in NGI Publ. 124.)

Kolbuszewski, J. and Jones, R. H. (1961). Preparation of sand samples for laboratory testing. Proceedings, Midland Soc. for Soil Mech. and Found. Eng., 4 : 107.

Lacasse, S. and Lunne, T. (1982). Penetration tests in two Norwegian clays. Proceedings, 2nd European Symp. Penetration Testing. Amsterdam, 2:661-669.

Ladd, C.C., Foott, R., Ishihara, K., Schlosser, F. and Poulos, H.G. (1977). SOA Report: Stress-deformation and strength characteristics. Proceedings, 9th Int. Conf. Soil Mech. and Found. Eng., Tokyo, 2: 421-494.

Laier, J.E., Schmertmann, J.H., and Schaub, J.H. (1975). Effect of finite pressuremeter length in dry sand. Proceedings, ASCE Spec. Conf. on In-Situ Measurement of Soil Props., Rayleigh, N.C., 1:241-259.

Lunne, T., Eide, O, and de Ruiter, J. (1976). Correlation between cone resistance and vane shear strength in some Scandinavian soft to medium stiff clays. Canadian Geotech. J., 13:430-441. (Also NGI Publ. 116.)

Lunne, T. and Christoffersen, H.P. (1983) Interpretation of cone penetration data for offshore sands. Proceedings, 15th Annual Offshore Technology Conf., Houston, Texas, 1:181-192. (Also NGI Publ. 156.)

Lunne, T., Christoffersen, H.P. and Tjelta, T.I. (1985a). Engineering use of piezocone data in North Sea clays. Proceedings, 11th Int. Conf. Soil Mechs. and Found. Eng., San Francisco, 2:907-912.

Lunne, T., Lacasse, S., Aas, G., and Madshus, C. (1985b). Design parameters for offshore sands: Use of in situ tests. Norwegian Geotech. Inst., Int. Rep. 52108-17 (65 pp).

Mitchell, J.K. and Gardner, W.S. (1975). In situ measurement of volume change characteristics. Proceedings, ASCE Spec. Conf. on In Situ Measurement of Soil Properties, Rayleigh, N.C., 2:279-345.

Mitchell, J.K. and Lunne, T. (1978). Cone resistance as a measure of sand strength. Proceedings, ASCE, 104:GT 7, pp. 995-1012. (Also in NGI Publ. 123.)

NGI (1976). A summary of work completed in the CRB (now RCA) calibration chamber (by N.C. Last). Norwegian Geotech. Inst., Int. Rep. 52108-1 (133 pp).

Parkin, A.K. (1977). The friction cone penetrometer: Laboratory calibration for the prediction of sand properties. Norwegian Geotech. Inst., Int. Rep. 52108-5.

Parkin, A.K. (1986a). Southampton seminar on CPT testing: Evaluation of calibration chamber data. Norwegian Geotech. Inst., Int. Rep. 52108-18 (31 pp).

Parkin, A.K. (1986b). Calibration chamber testing: Visit to ENEL and ISMES laboratories, Italy. Norwegian Geotech. Inst., Int. Rep. 52108-19 (25 pp).

Parkin, A.K. and Lunne, T. (1982). Boundary effects in the laboratory calibration of a cone penetrometer for sand. Proceedings, 2nd European Symp. Penetration Testing, Amsterdam, 2:761-768.

Rad, N.S. and Lunne, T. (1988). Direct correlations between piezocone results and undrained shear strength of clay. 1st Int. Conf. Penetration Testing, Orlando, Fl.

Robertson, P.K., and Campanella, R. G., (1983). Interpretation of cone penetration results, part I : sand. Canadian Geotech. J. 20:4 pp 718-783.

Schaap, L.H.J. and Zuidberg, H.M. (1982). Mechanical and electrical aspects of the electric cone penetrometer tip. Proceedings 2nd European Symp. Penetration Testing, Amsterdam, 2:841-851.

Schmertmann, J.H. (1974). Penetration pore pressure effects in quasi-static cone bearing q_c. European Symp. Penetration Testing, Stockholm, 2.2 pp. 345-351.

Schmertmann, J.H. (1975). Measurement of in situ shear strength. SOA Report. Proceedings, ASCE Spec. Conf. on In Situ Measurement of Soil Properties, Rayleigh, N.C., 2:57-138.

Schmertmann, J.H., (1976). An updated correlation between relative density, D_r, and Fugro-type electric cone bearing, q_c. Contract Report DACW 39-76-M 6646, US Army Waterways Exp. Station, Vicksburg, Miss.

Senneset, K., Janbu, N. and Svanφ, G. (1982). Strength and deformation parameters from cone penetration tests. Proceedings, 2nd European Symp. on Penetration Testing, Amsterdam, 2:863-870.

Senneset, K. and Janbu, N. (1984). Shear strength parameters obtained from static cone penetration tests. ASTM Symposium, San Diego.

Tomlinson, R. J. (1963). Foundation design and construction. Pitman, 785 pp.

University of Southampton (1984). Seminar on cone penetration testing in the laboratory. Part 1: Seminar papers (minuted by N.C. Last). Part 2: Southampton test equipment and results (by N.C. Last). Dept. of Civil Engineering, Nov. 1984.

University of Southampton (1985). An investigation of full scale penetrometers in a large triaxial calibration chamber. Progress Report to SERC and NGI (by N.C. Last), Dept. of Civil Engineering, Mar. 1985.

Veismanis, A. (1974). Laboratory investigation of electrical friction cone penetrometers in sands. Proceedings, European Symp. Penetration Testing, Stockholm, Vol. 2.2, pp. 407-419.

Vesic, A.S. (1970). Tests on instrumented piles : Ogeechee River site. Proceedings, ASCE, 96:SM2, pp. 561-584.

Vesic, A.S. (1972). Expansion of cavities in an infinite soil mass. Proceedings, ASCE, 98:SM3, pp. 265-290.

Villet, W.C.B and Mitchell, J.K. (1981). Cone resistance, relative density and friction angle. Cone penetration testing and experience. Proceedings, ASCE National Convention, St. Louis, Mo., pp. 178-208.

APPENDIX A

1 The electrical and mechanical calibrations required to obtain these outputs are treated in detail by Schaap and Zuidberg (1982).

2 A particular set of results, on dense NC Ticino Sand with B3 confinement and at a (nominal) vertical pressure of 100 kPa, is quoted by Bellotti et al. (and also by Bellotti, 1984) as evidence that q_c is not dependent on diameter ratio (D_c/d_c). This unexpected result conflicts with results elsewhere (on Hokksund Sand, Fig. 5), and needs further confirmation before being given credence, particularly as the associated sleeve friction values do not show the same independence from R_D (Parkin, 1986b). It is suggested here that, because this data set has a B3 boundary, the values of σ_h generated during test will not be the same for the three different cones. Accordingly, if q_c is correlated against σ_h, as recommended elsewhere in this paper, the relationship is unlikely to be independent of R_D (or cone size).

3 Some aspects of this problem have been studied by photoelastic analysis of penetration in crushed glass (of unspecified density), by Allersma (1982). This indicated that R_D 20 was inadequate to simulate a semi-infinite medium.

4 This represents a limited selection of data such as that tabulated by Bellotti et al. (1985) (and incompletely by Baldi et al., 1986).

5 Figure 3(b) shows a somewhat paradoxical situation wherein the B3 condition, despite being one of increased boundary stiffness, gives results that are everywhere lower (and appreciably so) than the B1 condition. No explanation for this is possible without data additional to that given by Bellotti (1984).

6 The indicated point on Fig. 3(a) would appear to be inconsistent with other B2 results. The possibility of a labelling error was discussed with V. Ghionna, but ruled out because there is no inconsistency in the relationship $\Delta\sigma_h$ v. R_D. No explanation is presently available.

7 Reproduced in log-log format by Parkin and Lunne (1982), wherein the $q_c A_c = F$ boundary is linearized. This condition is one where the thrust on the penetrometer is such that sample separation occurs at the top platen.

8 The attainment of OCR values greater than 8 is not easy in this equipment, and OCR 15, as in the Italian test programmes, is not likely to be exceeded. Note also that several results for the 5 cm^2 cone are included in Fig. 9 (R_D = 48), but as these were performed at σ_v = 100 kPa, the values have had to be corrected in accordance with the square root of stress.

9 Chapman and Donald (1981) found the q_c - σ_v relationship to be nearly linear for stresses above 75 kPa (NC sand), and Janbu and Senneset (1974) report results from the Florida test chamber being linear for σ_v up to 400 kPa (with an associated "attraction"). The latter results are, however, open to other interpretations.

10 Lunne and Christoffersen (1983) define M_o as the tangent modulus at in situ stress (σ'_{vo}), and M as the secant modulus over the interval to (σ'_{vo} + $\Delta\sigma$).

11 Baldi et al. note that if the stress and strain history of a deposit is not known, as is likely to be the case in practice, then correlations with OCR will be of dubious value. Accordingly, Fig. 14 shows only results for NC sand.

12 The mean of these points again represents about a 30% increase over the measurement ahead of tip. This, however, is not readily apparent because of difference in scale and origin.

13 Anisotropically consolidated to in situ stress, tested in undrained compression.

14 As these clays are all normally consolidated, an updated correction according to Aas et al. (1986) is probably not necessary.

15 Ambiguously referenced by Battaglio et al. (1986)

16 The method for deriving OCR is not given.

17 Rad and Lunne also show results to indicate that u/q_c is a function of OCR, as found by Lacasse and Lunne (1982).

18 Whilst this may indeed be adequate for clay, the writer would advise against extrapolating from results obtained on sand (such as Parkin and Lunne, 1982).

19 Some fairly basic chambers, up to 1 m by 1 m, have been in use at Oxford University (Houlsby - see Univ. of Southampton, 1984), but for testing model piles rather than the CPT. No details of sample preparation are given.

Penetration Testing 1988, ISOPT-1, De Ruiter (ed.)
© 1988 Balkema, Rotterdam, ISBN 90 6191 801 4

New developments in penetration tests and equipment

James K.Mitchell
University of California, Berkeley, USA

ABSTRACT: New developments in penetration tests and equipment since ESOPT II in 1982 are both substantial and significant. The standard penetration test has been improved by the introduction of automatic drop hammers and instrumentation for measurement of energy delivered to the rods. The static cone penetration test has been augmented through the addition of special measuring devices within and behind the cone tip and friction sleeve. These devices include, in addition to pore pressure sensors, lateral stress sensors, acoustic monitoring devices, electrical resistivity cells, nuclear density gages, and seismic wave detectors. Combination cone-pressuremeter systems are under development. The need for better information about in situ lateral stresses and a simple, practical means for obtaining it, is motivating many of the new developments. More research on the stress changes and deformations accompanying the insertion of penetrometers into the ground is needed so that pre-penetration conditions can be reliably deduced from quantities deduced during and after penetration.

1 INTRODUCTION

The purpose of this paper is to describe recent developments in penetration tests and penetration test equipment. It is based on a review of developments that have occurred since the Second European Symposium on Penetration Testing held in 1982. Literature that appeared prior to that conference was not included in this review.

Although emphasis is on new test methods and equipment, with minor refinements in existing penetration tests left for consideration in other sessions of this Symposium, it is perhaps significant that many of the new developments have been associated with the most commonly used tests; i.e., the standard penetration test (SPT), the cone penetration test (CPT), the pressuremeter test (PMT), the dilatometer test (DMT), and combinations of these tests. In addition to all the usual reference sources, additional information for this paper was solicited from the ISOPT I test reporters, reviewers and lecturers and from all members of the International Society for Soil Mechanics and Foundation Engineering Committee on Penetration Testing of Soils (TC16). The response to this solicitation was very gratifying, and I express my appreciation to all of those who sent data, papers, and illustrations.

The scope and organization of this paper are indicated by the following categories of testing systems, which are presented in the order listed.

- Dynamic penetration devices and systems
- Cone penetration devices and systems
- Pressuremeter systems
- Dilatometer test developments
- Special purpose devices

A review of new deep water seabed testing systems, which are proving very useful for penetration testing of seafloor sediments, is not included because of length limitations and because these systems have been well-described recently by Richards and Zuidberg (1985), van der Wal et al. (1986), and Power and Eastland (1986).

With any penetration test or other type of in situ test it is important to be clear about what you want to measure, what you think you measure, what you do measure, and what the results of the measurements mean relative to the actual soil behavior and properties of interest in the particular geotechnical problem to be solved. This paper concentrates on the first three of these issues. The last is more appropriately the subject of the other sessions of this Symposium.

2 DYNAMIC PENETRATION DEVICES AND COMBINATION DYNAMIC - STATIC SYSTEMS

2.1 Standard penetration test (SPT)

The SPT continues in wide usage throughout the world owing to its long history, the existence of many useful correlations, readily available equipment in most areas, and, of the widely used penetration tests, it is the only one that provides representative samples. Refinements continue to be made in order to improve its reliability and reproducibility. Two recent advances include the availability of automatic drop hammers and the approval of ASTM Standard D4633-86, "Stress Wave Energy Measurement for Dynamic Penetration Testing Systems." This new ASTM Standard describes the procedure for measuring that part of the drive weight (hammer) kinetic energy that enters the connector rod column of any penetrometer, cone, blade, sampler, or other device driven from the ground surface.

The force-time record for the first compressive wave produced following the hammer impact is measured using a load cell mounted on the penetrometer connector rods. A processing instrument integrates the force over the duration of the pulse to yield the stress wave energy. There is an approximately linear relationship between this energy and the incremental advance of the penetrometer, and, therefore, an inverse relation between this energy and the number of blows per unit length of penetration. Thus the results of the measurements make it possible to correct SPT values to a common standard as well as to evaluate different operator and equipment systems.

2.2 The PENELAT

A light weight dynamic penetrometer, called the PENELAT, which contains a static lateral penetrometer within its tip has been developed in France. The purpose of this instrument, which is shown schematically in Fig. 1, is to retain the advantages of dynamic penetration while permitting a horizontal static penetration test at any depth without the adverse effects of side friction. The lateral penetrator is advanced using nitrogen pressure at 6 MPa. The ratio of point area to lateral penetrator area is four. The rate of lateral penetration is 10 mm/min, and lateral penetration resistances from 2 to 24 MPa can be measured. This device is described by French Patent No. 80.01557, European Patent No. 0.032.648, and U.S. Patent No. 4,436,647.

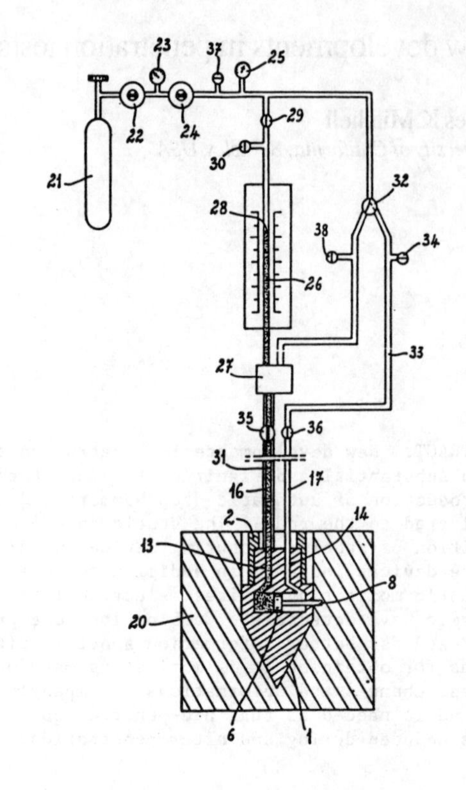

Fig. 1 Schematic diagram of the PENELAT, a dynamic penetrometer with static lateral penetration testing capability. The lateral penetrometer is indicated by 8. (Patent drawing provided by Guy Sanglerat)

2.3 "DINASTAR" penetrometer

Both dynamic and static penetration tests are possible using the Dinastar developed by Tecnotest. This system is shown in Fig. 2. The total weight of the system is about 250 kg, and it disassembles into four pieces for easy transport. It is operated hydraulically. For dynamic penetration, points having a diameter larger than the rods are used. The driving hammer weighs 30 kg and the fall distance is 200 mm. This corresponds to the German Standard DIN 4049 medium energy specifications. The hammer is lifted by a hydraulic motor at an adjustable speed from 0 to 80 rpm.

An extra-long friction sleeve is used to make the system self anchoring for static penetration tests using a standard Dutch mantle cone. This sleeve connects to the apparatus frame while a piston advances a static cone point. Anchor sleeves with

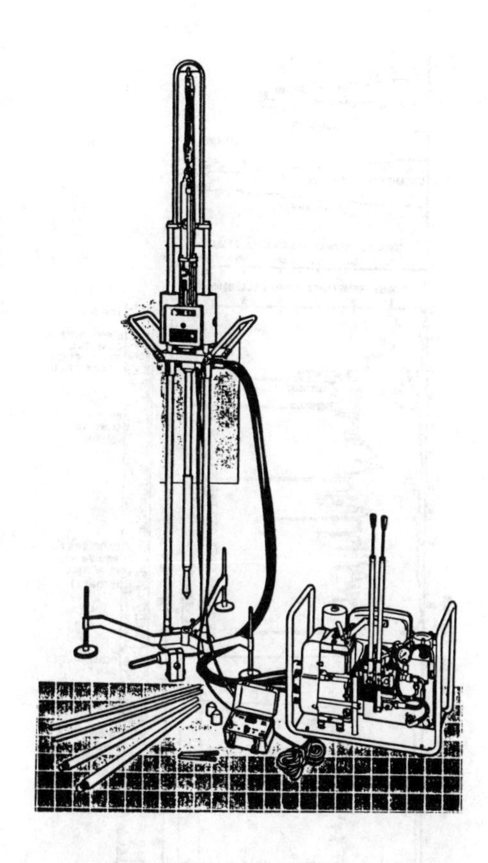

Fig. 2 DINASTAR light static and dynamic self-anchoring penetrometer (from Tecnotest brochure)

extensions are available to give total friction surfaces of 500, 700, and 900-sq cm. The system is reported to work well in soils having a friction ratio greater than about 1.1 percent. For lower values of friction ratio the length of friction sleeve required to provide the needed anchor resistance becomes excessive.

The DINASTAR penetration testing system is described more completely and test data are presented in the paper submitted to this Symposium by Triggs and Liang (1988).

3 CONE PENETRATION DEVICES AND SYSTEMS

The static cone penetration test has now perhaps become the most popular type of penetration test for the investigation of soils that do not contain gravel or other obstructions and which can be penetrated using reaction loads up to 15 to 20 tons. Reasons for this include (1) the test is simple and relatively economical, (2) continuous records with depth are obtained, (3) the results are interpretable on both empirical and analytical bases, (4) a variety of different sensors can be incorporated within the penetrometer tip and friction sleeve, and (5) a large experience base is now available. As a result there has been continuing research on the CPT from the standpoints of both the penetrometer system itself and methods for interpretation of the data.

The greatest interest since ESOPT II has been centered on the development of the piezocone test (CPTU). As the piezocone is the subject of another special paper for this Symposium, it is not dealt with in detail here. Before leaving the subject of the piezocone, however, it is emphasized that measured pore pressures are very sensitive to test procedure and apparatus configuration. The latter point is well illustrated by data reported by Zuidberg et al. (1987) obtained using a "triple piezocone." The triple element piezocone penetrometer is shown in Fig. 3, and pore pressures recorded during penetration through seabed soil are shown in Fig. 4. Clearly, the location of the pore pressure sensing filter element must be taken into account when making analytical or empirical interpretations of CPTU data.

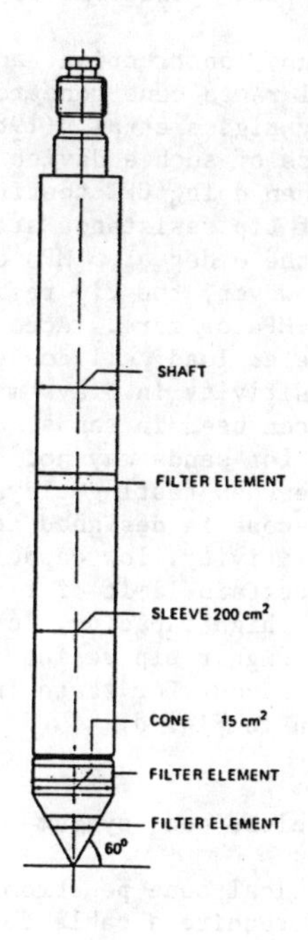

Fig. 3 Triple element piezocone penetrometer (from Zuidberg et al., 1987)

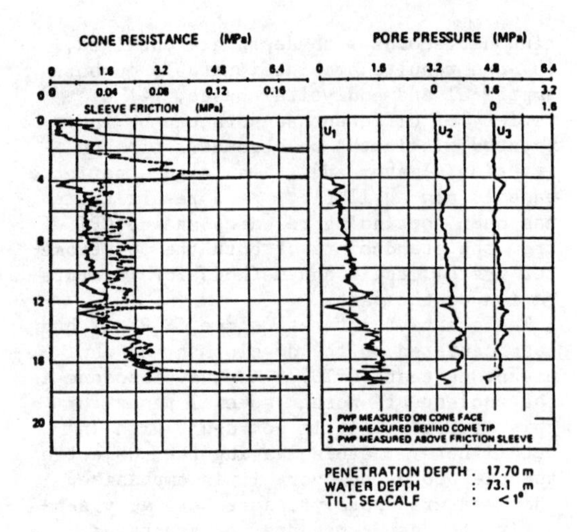

Fig. 4 Results of a triple piezocone test on the seafloor (from Zuidberg et al., 1987)

There have been a number of other significant recent developments in CPT apparatus and testing, and they are described in the following paragraphs.

3.1 Dual range penetrometer - Brecone

The design, construction, and application of a dual range cone penetrometer was described by Rigden et al. (1982), and the usefulness of such a device is noted again here. When doing CPT testing in clays the values of tip resistance are relatively low, of the order of 5 MPa or less. In sands, however, the tip resistance may be up to 40 MPa or more. Accordingly, a penetrometer load cell designed to give high sensitivity in clays may be overloaded when used in sands, and a load cell suitable for sands may not be sufficiently sensitive when testing clays. The dual range brecone is designed to measure on a high sensitivity, low capacity load cell up to a certain limit of tip resistance and then on a high capacity, lower sensitivity cell for higher tip resistances. Results using this cone for tests in a glacial till are shown in Fig. 5.

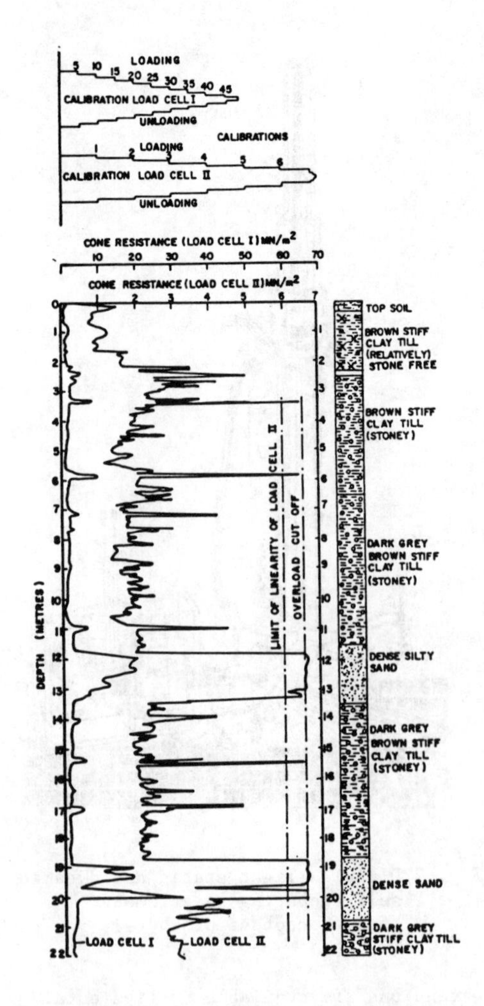

Fig. 5 Cone resistance profile in glacial till obtained using the dual range Brecone (from Rigden et al., 1982)

3.2 Cableless cone systems

An electrical cone penetrometer system that does not require a cable for data transmission to the ground surface during testing was described by Muromachi et al. (1982) at ESOPT II. In this "memory cone" the analog signals from the sensors in the cone are digitized and stored in a capsule located above the measuring unit. After the cone is recovered from the ground the memory capsule is detached, and the data are read into a micro-computer for processing. The lack of real-time information about the values being read by the sensors is a disadvantage of this system, because malfunctions, anomalies, etc. go undetected until the cone is withdrawn from the ground.

A different type of cableless CPT system, termed an ultra sonic cone penetrometer, has been developed by Geotech A/B in Sweden. Data from the penetrometer are transmitted by acoustic signals through the cone rods to the ground surface. The Geotech CPT apparatus also incorporates a patented (U.S. No. 4499954) mud injection system that reduces the required reaction weight for penetration by more than 50 percent through reduction of friction along the push rods. Drilling mud is pumped in low volumes down the cone rods as the cone is pushed and is ejected in the ground above the cone.

3.3 Acoustic cones

An acoustic cone in which an acoustic sensor is incorporated in a cone penetrometer tip was first described by Muromachi (1974) at ESOPT I. His phono-sounding apparatus was useful for distinguishing sand and gravel from clayey soils and detection of sand seams in cohesive soils. The usefulness of acoustic response as an indicator of soil type and profile stratigraphy was further illustrated at ESOPT II by Tringale and Mitchell (1982). For a given apparatus and constant rate of penetration, the amplitude of the acoustic signal is related to mean particle size of the soil as can be seen in Fig. 6, from Tringale (1983).

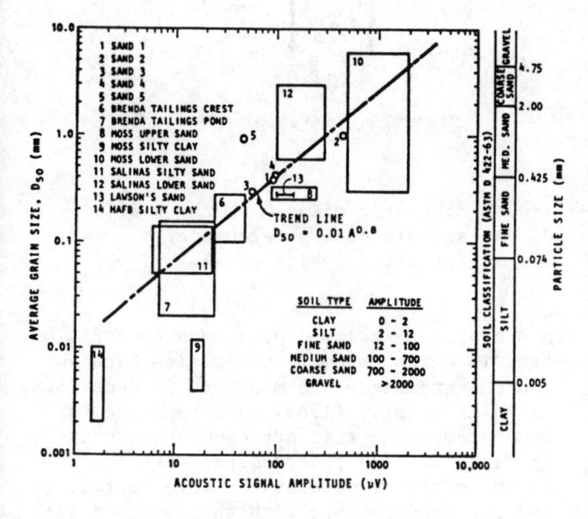

Fig. 6 Correlation between acoustic signal amplitude and average grain size for constant rate of cone penetration (from Tringale, 1983)

When measurements are made in the audible frequency range, as was the case for the studies noted above, the microphone also registers surrounding ground disturbances and noise generated by the mechanical testing system. Massarsch (1986) describes an acoustic penetrometer that measures in a higher frequency range to avoid these problems. Also, because high frequency signals attenuate more rapidly, only signals generated close to the penetrometer tip are recorded. This improves the possibility that even very thin layers will be detected.

Massarsch did tests both on sand in the laboratory and in a cohesive soil deposit near Stockholm. The acoustic emission rate was about 100 times greater in sand than in clay, which is quite consistent with the results shown in Fig. 6. It was also determined by visual inspection that sand and silt layers as thin as 1 mm could be identified. Such thin layers were not detectable using conventional penetration tests.

3.4 Lateral stress cone

The in situ horizontal stress is one of the most important parameters needed for many geotechnical analyses, yet reliable knowledge of its value is difficult to obtain. Recently, studies have been undertaken to measure the in situ lateral stress using the cone penetrometer (Huntsman et al., 1986). The friction sleeve of an electrical penetrometer of standard dimensions was instrumented so that the radial horizontal stress can be measured during penetration and while adding rods.

Owing to soil disturbance during advancement of the penetrometer, it cannot be expected that the measured lateral stress will equal its initial value. A series of calibration chamber tests on sand as well as some theoretical analyses were done to determine the relationship between the measured value of horizontal stress and the initial value as a function of the initial relative density (Huntsman, 1985). A field test was done in a caisson-retained island at the Amerk location in the Canadian Beaufort Sea. Results obtained using the lateral stress sensing cone penetrometer (LSSCP), a piezocone, and a self-jetting pressuremeter are compared in Fig. 7. The agreement between the tip resistances obtained using the two cone penetrometers and between the lateral stress determined using the LSSCP and the pressuremeter is very good.

Additional research is in progress at the University of California, Berkeley, aimed at the development of a more rugged and more sensitive LSSCP than used for the initial tests and on a more comprehensive analysis of the effects of ground displacement and disturbance on the measured lateral stress. A cone tip that measures total radial stress

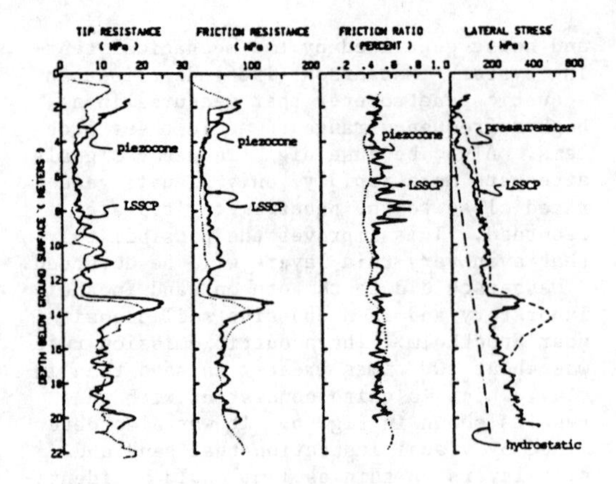

Fig. 7 Comparison of lateral stress values determined by a lateral stress sensing cone penetrometer and a self-jetting pressuremeter (from Huntsman et al., 1986)

is also under development by Jamiolkowski and Bruzzi (personal communication, 1987) in Italy.

3.5 Seismic cone

A new test, called the seismic cone penetration test (SCPT) involves the incorporation of small velocity seismometers and accelerometers into an electronic cone or piezocone (Campanella et al., 1986a, 1986b). The cones have either a 10 or 15 sq cm base area and a 60 degree apex angle. A cross section of the cone arrangement is shown in Fig. 8. Special features include the possibility to measure the pore pressure, the cone inclination, and the temperature. The layout for a standard downhole test is shown in Fig. 9. Hammers and explosive caps have been used as seismic sources. A Nicolet 4094 digital oscilloscope with floppy disk is used to detect and record the wave traces. The system provides very accurate timing and high signal resolution.

Tests are done at several depths during cone penetration. The shear wave velocity is readily calculated from the measured arrival times and the known geometry of the system. Agreement between shear wave velocities determined using the seismic cone and by usual crosshole tests is excellent. An example is shown in Fig. 10. The seismic cone offers the advantage, relative to the crosshole method for shear wave velocity measurement, that only one hole is needed, and it can be developed during the course of a regular cone penetration test.

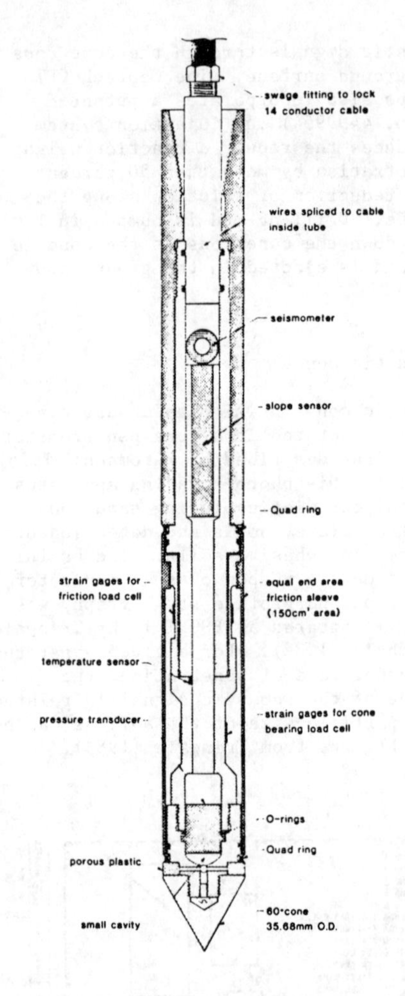

Fig. 8 Schematic diagram of the U.B.C. seismic cone penetrometer (from Campanella et al., 1986b)

A crosshole seismic piezocone penetration test in which the wave velocities between two penetrometers are measured is described by Baldi et al., (1988) in a paper to this Symposium. The test arrangement is shown in Fig. 11. The penetrometer with the 15 sq cm tip is used as the wave source, and the penetrometer with the 10 sq cm tip containing a biaxial geophone is the receiver. Continuous measurements of tip resistance q_c, pore pressure u, local shaft friction f_s, and inclination i are obtained during penetration. Seismic waves are generated by dropping a hammer on the top of the rods of the 15 sq cm cone.

The arrivals of the compression and shear waves are clearly distinguishable, allowing determination of the corresponding wave

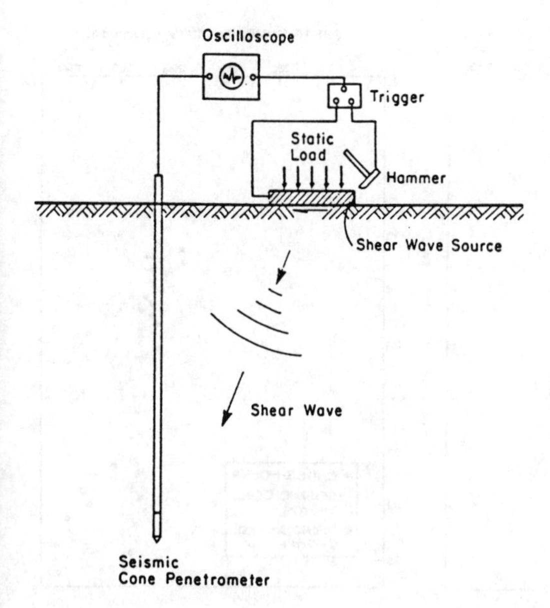

Fig. 9 Schematic layout of downhole seismic cone penetration
(from Campanella et al., 1986b)

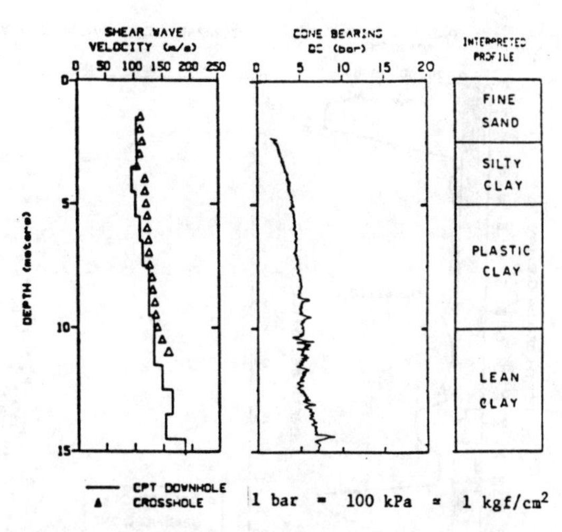

Fig. 10 Comparison of shear wave velocities determined using seismic CPT down-hole and crosshole tests for the Museumparken site, Drammen, Norway
(from Campanella et al., 1986b)

velocities. Profiles of shear and compression wave velocities as a function of depth in Po River sand are shown in Fig. 12 from Baldi et al. (1988). Values are shown for penetrometer spacings of 5 m and 10 m. Values of wave velocities determined by conventional cross hole tests are also shown for comparison. The reasons for the discrepancies in values for depths greater than 29 m are not known, but the authors indicated they may result from local soil variability or test errors. Nonetheless, the values compare very well through most of the profile, and for soil deposits that can be penetrated using the CPT, the seismic cone would seem a very convenient means for determination of shear wave velocities.

3.6 ISMES vibratory cone (CPTV)

An electric friction cone penetrometer having a 42 mm diameter cone and friction sleeve has been equipped with an electrical vibrator as shown in Fig. 13 (PWRI, 1985). The vibrator draws a power of 260 W and operates at a frequency of 200 Hz. During penetration of a saturated sand layer,

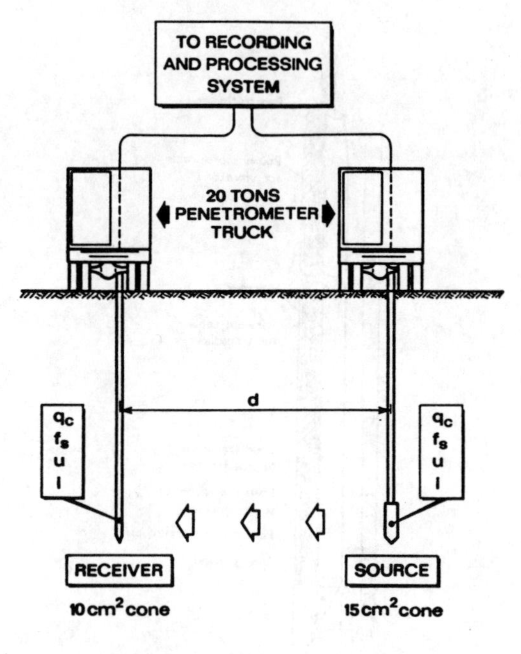

Fig. 11 Seismic crosshole test using two cone penetrometers
(from Baldi et al., 1988)

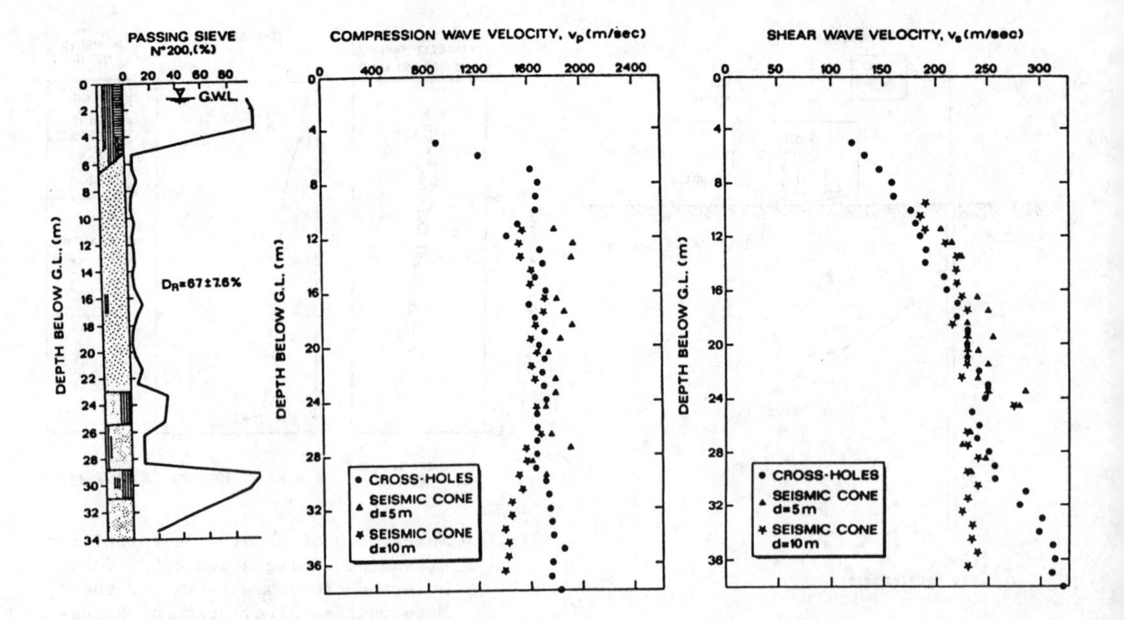

Fig. 12 Seismic wave velocities in Po River sand (from Baldi et al., 1988)

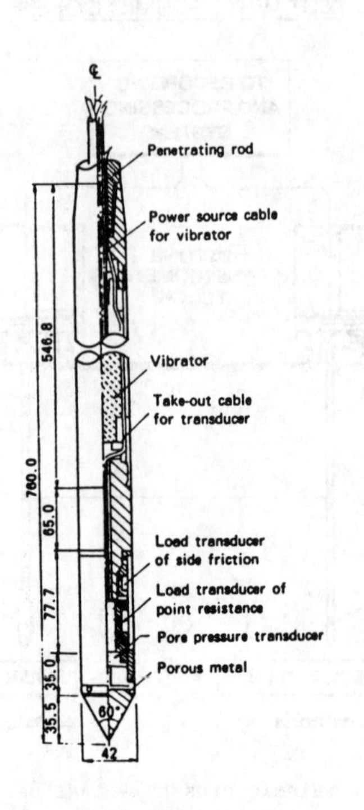

Fig. 13 ISMES vibratory cone (from PWRI, 1985)

vibrations can cause an increase in the pore water pressure in the soil near the cone, with a concurrent decrease in the penetration resistance. An example of penetration resistance profiles with and without vibration during penetration is shown in Fig. 14.

If the penetration resistance without vibration is defined as q_c^s, and the penetration resistance with vibration is defined as q_c^v, then a parameter D can be defined according to

$$D = \frac{q_c^s - q_c^v}{q_c^s} \qquad (1)$$

The larger the value of D the easier it is for pore pressures to build up, and the more readily the sand may liquefy. Thus the vibratory cone is intended as a rapid means for liquefaction potential assessment. At the present state of development the test is useful qualitatively; more studies are needed to establish quantitative correlations between D and liquefaction potential.

3.7 Miniature cone penetration test (MCPT)

A miniature CPT system for testing soils under conditions where full size CPT testing

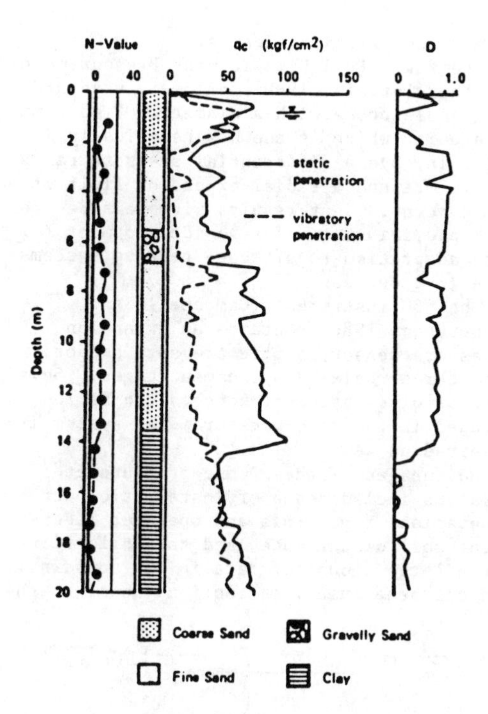

Fig. 14 Example of test results using the ISMES vibratory cone (from PWRI, 1985)

is not practical has been developed by the Earth Technology Corporation (personal communication, May 12, 1987). Conventional full size CPT technology is incorporated into a system that is built around a 2-ton four wheel drive pickup truck with an on-board computer data acquisition system. It is well suited for situations requiring rapid testing to shallow depths, but it can be used also to test to depths of 25 m in soft soils.

Two instrument sizes have been developed: 0.5 in diameter, giving a tip area of 0.20 sq in (1.29 sq cm) and a friction sleeve area of 3.69 sq in (23.81 sq cm); and 0.75 in diameter, giving a tip area of 0.44 sq in (2.84 sq cm) and a friction sleeve area of 6.7 sq in (43.22 sq cm). Tip pressures in excess of 400 tsf (40 MPa) can be sustained without damage to the instruments. Comparison tests using the miniature penetrometers and full size cones done in a 30 in (760 mm) diameter drum have indicated the same values of tip resistance, but friction measurements with the miniature cones were about 10 percent higher than with the full size penetrometer. Reasons for this difference in friction sleeve resistance values have not yet been determined.

3.8 Self cleaning, self grouting cone penetrometer

The possibility of vertical contaminant migration has restricted the use of penetration tests at hazardous waste sites. When holes are drilled or driven, then it is necessary that they be completely filled and sealed to close off any new pathways for chemical flow. If conventional CPT tests are used, then the penetrometer must be withdrawn and the hole repenetrated with grout pipes to try to seal the hole. Hole deformation and collapse, uncertainty about alignment, and possible chemical transport with the penetrometer and pipes all lead to uncertainty about the final result.

A self grouting CPT method has been developed by the Earth Technology Corporation (personal communication, May 12, 1987) that overcomes these problems. The system is shown schematically in Fig. 15. After an electric friction cone penetration test the system converts to a grouting tool. An expendable cone tip is left in the ground, and bentonite or cement-bentonite slurry is pumped out of the nose of the cone.

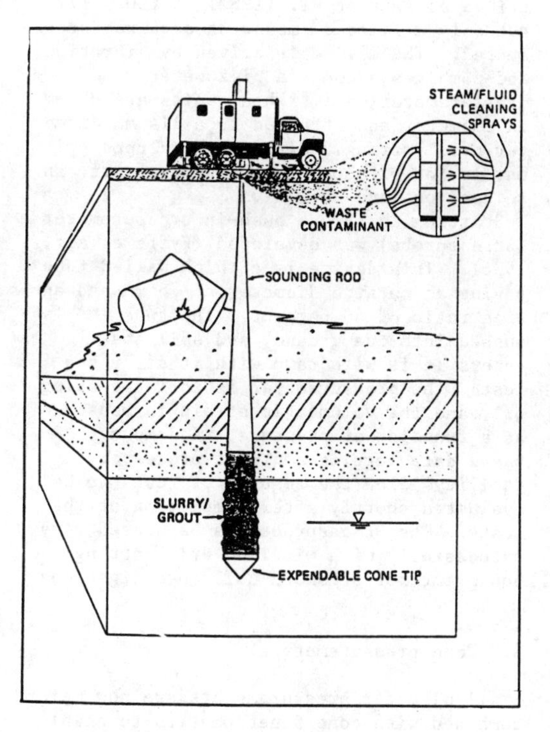

Fig. 15 Self-grouting self-cleaning penetrometer (courtesy of The Earth Technology Corporation)

The system also has a self-contained rod cleaning attachment in which a hot water jet is directed from a steam cleaner onto the rods. The sealed washer unit collects the waste fluids which are then pumped into drums for disposal.

4 PRESSUREMETER SYSTEMS

4.1 Background

The classical pressuremeter test (PMT) is not a penetration test; it is a cylindrical expansion test in a borehole. Thus it might be considered outside the scope of this Symposium. However, test devices that combine the CPT and the PMT have been developed and are in use. The data from each test type can supplement that from the other. In particular, the CPT provides continuous profiling, a means for soil identification from knowledge of the friction ratio, and a basis for strength and bearing capacity estimation. The PMT is useful for estimation of in situ lateral stress and deformation modulus values.

The LPC pressio-penetrometer was described by Amar et al. (1982) at ESOPT II, but only a limited amount of data was presented. The device is driven by vibration and combines a cone, a piezometer, and a pressuremeter cell. These units are 89 mm in diameter and attached to a 114 mm diameter drill string. Although developed primarily for offshore investigations, it can be used on land as well.

Starting in 1975 a push-in pressuremeter (Stressprobe) was developed (Fyffe et al., 1986). This device is a thick walled tube having an outside diameter of 78 mm and an area ratio of 40 percent. The tube is pushed into the ground, and soil that enters it is withdrawn with it after the test. The pressuremeter is inflated using oil, and the volumetric strain is measured as a function of pressure. A computer based data logging and post processor facility allow the results of tests to be evaluated shortly after completion of the tests. The Stressprobe has been used very successfully with wireline drill string equipment for offshore soil investigations.

4.2 Cone pressuremeters

Small diameter pressuremeters are now being combined with cone penetrometers to enable PMT testing as a part of the CPT operations. Both Withers et al. (1986) and Campanella and Robertson (1986) describe devices that follow directly behind 15 sq cm base area cones and friction sleeves.

The Fugro Full Displacement Pressuremeter (Withers et al., 1986) follows a standard Fugro piezocone with a diameter of 43.7 mm. The unit, which is shown schematically in Fig. 16, has a differential pressure range of 10 MPa and a radial expansion limit of 50 percent. Test results with this device are promising, and further development for its adaptation to offshore testing systems are in progress.

The UBC instrument (Campanella and Robertson, 1986) contains an expansion pressure sensor, a pore pressure sensor, and three radial displacement gages. Seismic or other types of sensors can be included in the cone-pressuremeter system if desired as well.

Advantages of these cone-pressuremeter systems include ease of installation and operation, repeatable and operator independent soil disturbance, and the collection of a large amount of data in one sounding. On the other hand, the soil adjacent to the

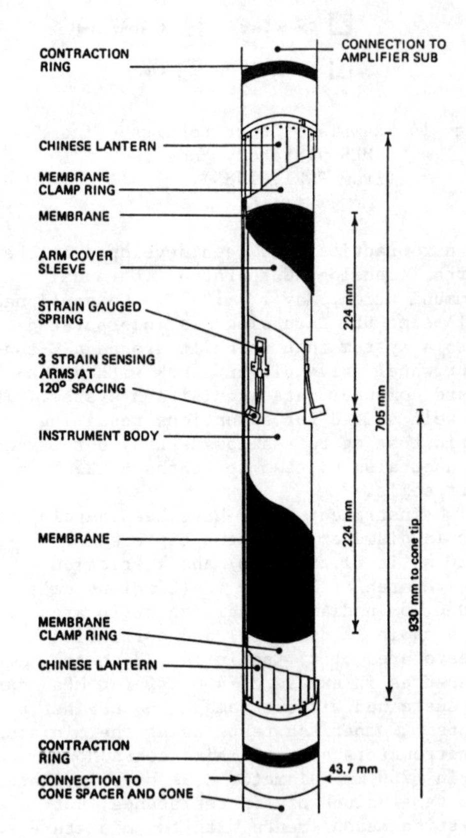

Fig. 16 Fugro full displacement pressuremeter for attachment behind a 43.7 mm diameter cone tip (from Withers et al., 1986)

pressuremeter will have been significantly disturbed because of the full displacement of the surrounding soil and passage of the cone tip and friction sleeve. This will mean, as is true also for the lateral stress sensing cone penetrometer systems described earlier, that the lateral pressures and moduli that are measured are not necessarily the same as existed prior to insertion of the device. Research is needed to develop suitable correlations.

5 DILATOMETER TEST DEVELOPMENTS

The Marchetti flat plate dilatometer is perhaps the newest of the penetration test devices that are currently used in routine practice. The dilatometer test and its interpretation are the subject of a separate session of this Symposium, so significant new developments in the apparatus and procedure are covered only briefly here.

The use of high strength steel blades has reduced problems of bending, breaking, scoring, and erosion. High strength membranes with better scoring and puncture resistance are now available, and there are improved control boxes giving better sensitivity in the low range.

A dilatometer for offshore use has been developed at the Norwegian Geotechnical Institute, as shown in Fig. 17. This device has reduced blade and diaphragm dimensions compared to the standard Marchetti device so that it will fit within the casings conventionally used for seafloor drilling and in situ testing. An electronic data acquisition system has been developed for use with this instrument so that it can be controlled from shipboard. A computerized data acquisition and data reduction system for the regular dilatometer has been developed by Hogentogler & Company.

An important new development in dilatometer test apparatus and procedure is the attachment of a low pressure gage and needle valve to the console that permits measurement of the pressure on the diaphragm if the system is deflated back to the A-reading position after the B-reading is made. This measurement, called the C-reading closure pressure and discussed in more detail by Lutenegger in another paper to this Symposium, may indicate the ambient pore water pressure in sands and the excess pore pressure generated in clays by the insertion of the dilatometer.

6 SPECIAL PURPOSE DEVICES

Special purpose devices that can be driven, pushed, or placed in the ground or incorporated within other penetrometers continue to be developed. Usually these instruments are designed to measure one specific property or condition. Such devices include, for example, the Iowa Stepped Blade for lateral stress measurement, heat flow probes, electrical conductivity probes, and nuclear density probes. Such devices are described briefly below.

6.1 K_o stepped blade and the tapered blade

In the continuing search for a rapid, reliable, practical, and economical means for determining the lateral stress state of soils in situ, Handy et al. (1982) report on the development and testing of the Iowa stepped blade. The concept is illustrated in Fig. 18. A series of pressure sensing membranes, each fixed to a blade of increasing thickness is pushed into the ground. The pressure recorded on each membrane when it is positioned at the test depth of interest is plotted versus the corresponding blade thickness. The plot is extrapolated to zero thickness to give an estimate of what the lateral stress would have been if the blade had not been inserted. In its present form the device has four steps instead of the three shown in Fig. 18.

The methods for interpretation of the data are described and test data are presented by Handy et al. (1982) and Lutenegger and Timian (1986). Reasonable values of K_o have been obtained for many soils; however, they appear unreasonably high in dense overconsolidated soils and may not be reliable in dense sands.

As the discontinuity between successive steps of the blade may cause significant remolding to occur, a continuously tapered blade that would act as a penetrating wedge was suggested by Lutenegger and Timian (1986). Such a device has now been developed, as shown in Fig. 19 (Lutenegger, personal communication, July 1987) and is undergoing evaluation. This instrument has a 3 degree taper and contains seven sensors along one face. Lutenegger has suggested that such a device may be useful for generation of "p-y" curves as well as determination of lateral pressure.

6.2 Thermal conductivity probe

In several instances thermocouples or thermisters have been incorporated into cone penetrometers for the purposes of determining soil temperature and the change in temperature caused by the penetration

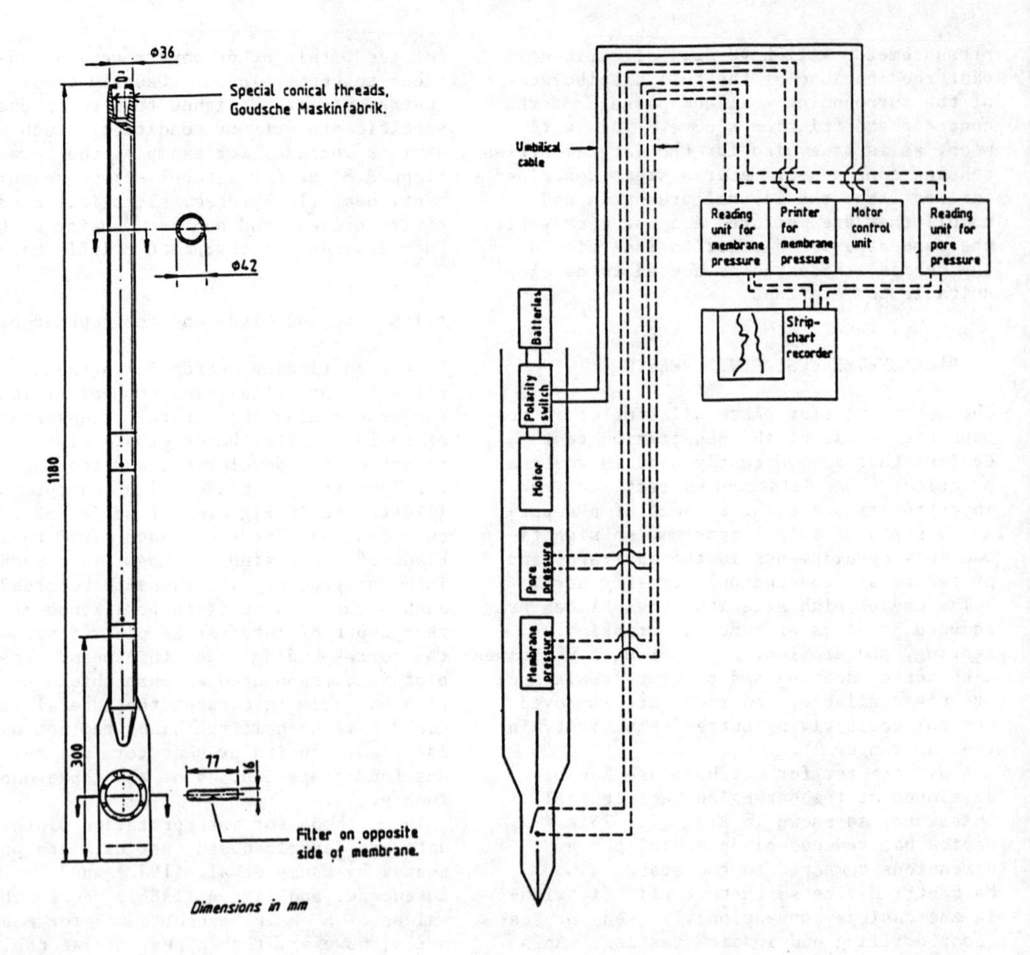

a. Dilatometer unit

b. Data acquisition system used in 1985

Fig. 17 Offshore dilatometer developed at the Norwegian Geotechnical Institute (furnished by S. Lacasse, Norwegian Geotechnical Institute)

process. Schaap and Hoogendoorn (1984) describe a thermal conductivity probe, shown schematically in Fig. 20, that can be used for determination of in situ thermal conductivity of a soil during penetration testing. By measuring the temperature increase as a function of time for a constant rate of heat input to the heating element, the thermal conductivity is readily computed (Mitchell and Kao, 1978).

The temperature difference between elements T_1 and T_2 is maintained as close to zero as possible to maintain a radial heat flow pattern around measuring sensors T_2, T_3, and T_4, thus conforming to the conditions assumed for the computation of thermal conductivity. The conductivity measurement can be done in less than 15 minutes. A probe for thermal conductivity measurements offshore is described by Zuidberg et al. (1987).

6.3 Electrical conductivity cone

A standard electric friction cone with electrically conducting rings or pins embedded into an insulating body behind the friction sleeve is described by Zuidberg et al. (1987) and shown schematically in Fig. 21. A hard ceramic material is used for insulation to provide greater strength and abrasion resistance. A low frequency alternating current is used rather than direct current to minimize electro-kinetic effects. Knowledge of

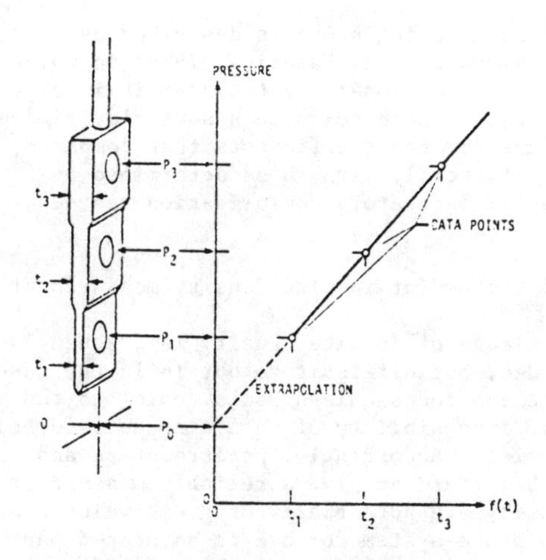

Fig. 18 Principle of the Iowa K_o stepped blade (from Handy et al., 1982)

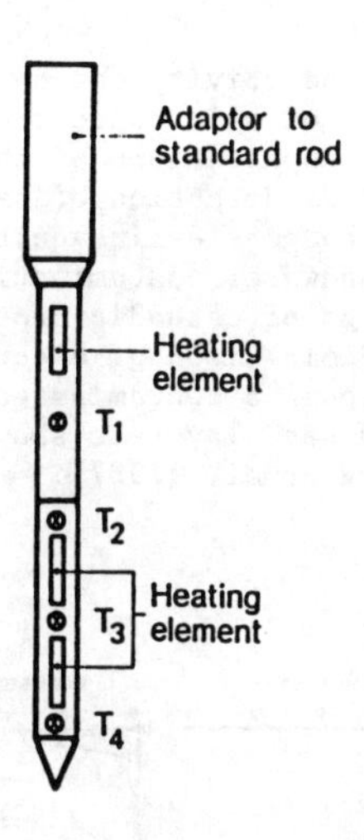

Fig. 20 Thermal conductivity probe (from Schaap and Hoogendoorn, 1984)

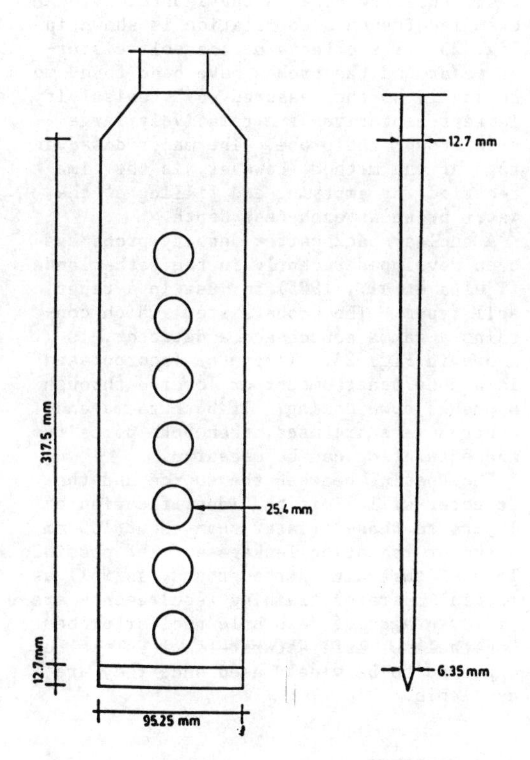

Fig. 19 Tapered blade with seven sensors for lateral pressure measurement (courtesy A. J. Lutenegger)

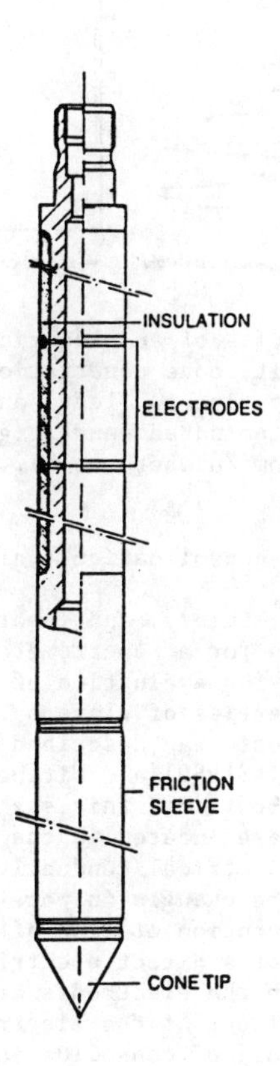

Fig. 21 Electrical conductivity cone penetrometer (from Zuidberg et al., 1987)

electrical conductivity can be used for determination of salt concentrations in the pore water. Applications of the test results include detection of salt water-fresh water boundaries in aquifers, location of groundwater contamination plumes, and the design of cathodic protection systems. A comparison of electrical conductivities through a contaminated and non-contaminated sand layer is shown in Fig. 22 from Zuidberg et al. (1987).

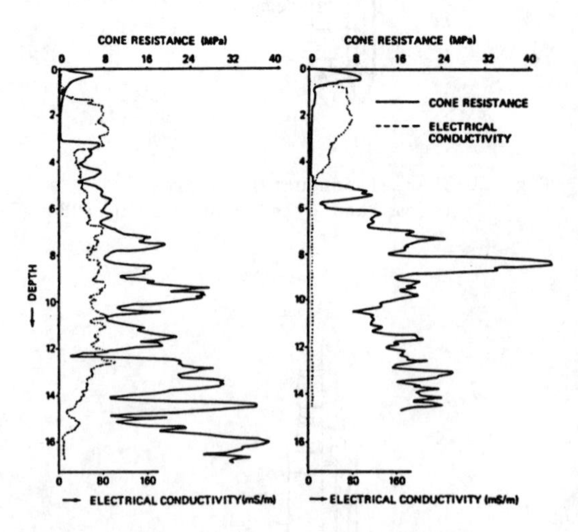

Fig. 22 Results of an electrical conductivity cone penetration test in contaminated (left) and non-contaminated sand (right) (from Zuidberg et al., 1987)

6.4 In-situ consolidation evaluation device

The concept, theory, and laboratory chamber test results for a penetrometer probe that can be used for evaluation of the consolidation properties of clays by means of electro-osmosis was described by Banerjee and Mitchell (1980) and Mitchell and Banerjee (1980). In this system ring electrodes are located in the same manner as in the electrical conductivity probe, Fig. 21. The changes in pore water pressure as a function of time after the initiation of a direct electrical current flow between the electrodes are measured. From a knowledge of the electrical field and the theory of consolidation by electro-osmosis it is possible to deduce the coefficient of volume change and the coefficient of consolidation.

A self boring probe is described by Vitaysupakorn and Banerjee (1986) in which the electro-osmosis test system is incorporated. Field tests in a soft clay yielded values for the coefficients that compared satisfactorily with those determined by regular laboratory consolidation tests.

6.5 Probes for in situ density measurements

Knowledge of in situ density is frequently needed, but difficult to obtain in the case of clean cohesionless soils, owing to the near impossibility of obtaining undisturbed samples. Accordingly, penetrometers and probes that can give a reliable measure of density are potentially of great value. A two probe system for use in saturated sands has been described by Vlasblom (1977), and is shown in Fig. 23. The electrical resistivity of the soil plus water is measured by the soil probe, and the resistivity of the water alone is measured by the water probe. The ratio of the porewater resistivity to that of the saturated soil is related to the porosity by calibration tests that are done in the laboratory. An example of such a correlation is shown in Fig. 23. The effects of the soil disturbance around the probes have been found to be small, so the measured soil resistivity is representative of a relatively large area around the probe. The major disadvantage of the method, however, is the time requried for emptying and filling of the water probe at each test depth.

A nuclear backscatter density probe has been developed recently in the Netherlands (Tjelta et al., 1985) for use in a range of soil types. The probe system, which contains a gamma source and a detector, is shown in Fig. 24. It can be incorporated in a cone penetrometer or lowered through a pushed down casing. If both gamma and neutron rays are used, then both porosity and saturation can be measured.

The spacing between the source and the detector will limit the identification of layers to those greater than about 100 mm. Risks of radiation leakage and the possible loss of the unit in the ground, as well as special operator training requirements are disadvantages of down hole nuclear probes. Nonetheless, safe accurate devices can be expected to be widely used once they are available.

7 CONCLUSIONS

New developments in penetration tests and equipment since ESOPT II in 1982 have been

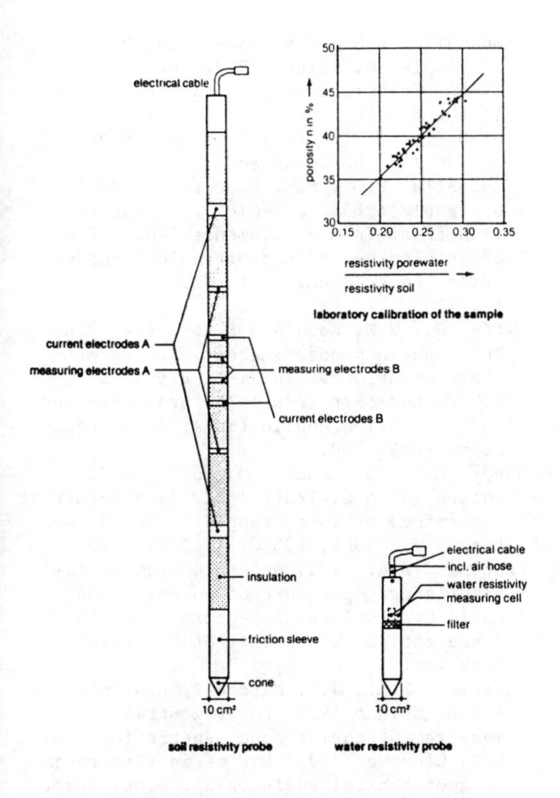

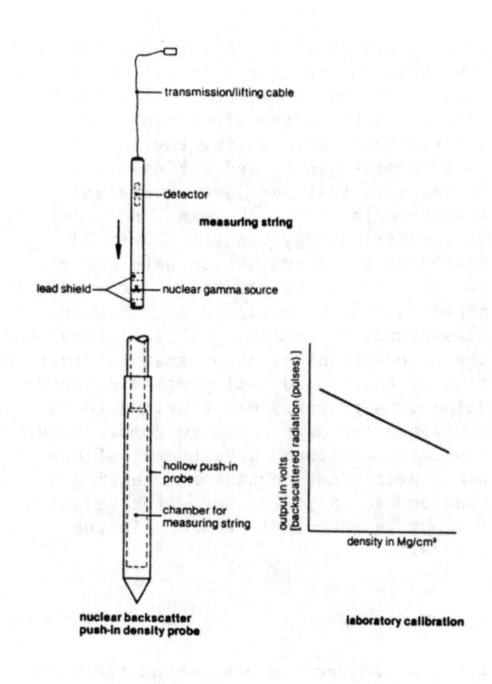

Fig. 24 Nuclear backscatter in-situ density measurement (from The Netherlands Commemorative Volume ICSMFE 1985, Chapt. 2, p. 34)

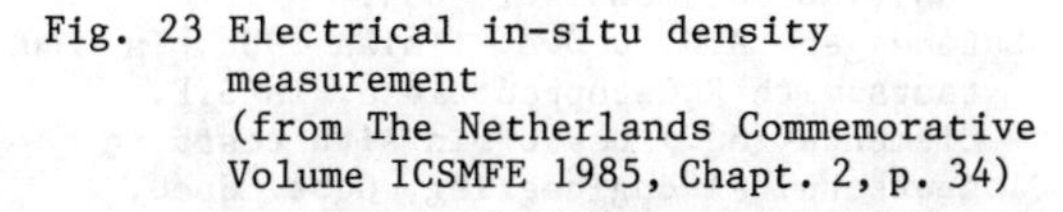

Fig. 23 Electrical in-situ density measurement (from The Netherlands Commemorative Volume ICSMFE 1985, Chapt. 2, p. 34)

both substantial and significant. Improvements in the standard penetration test have included the introduction of automatic drop hammers and the development of instrumentation for the measurement of delivered energy. Versatile lightweight combination dynamic and static penetration test systems are now available.

The static cone penetration test continues to increase in popularity, and a number of special measuring devices have been developed or are in the process of development for inclusion in multi-unit cone devices. These include piezometers at several locations, lateral stress measuring cells, seismic wave detectors, acoustic monitoring devices, electrical resistivity measuring devices, systems for electrokinetic measurements, nuclear density gages, temperature sensors, thermal resistivity testing units, and inclinometers. Larger diameter cones (15 sq cm base) are being used in some cases because of their greater strength and the ability to house more sensors, electronics, and wires in the friction sleeve and rods. A miniature cone system that operates from a pickup truck has been developed.

The concern about cross contamination and leakage of chemicals at hazardous waste sites has stimulated the development of a self cleaning, self grouting cone penetrometer. A vibratory cone system for use in assessment of the liquefaction potential of sands is under development. A cableless cone system that uses acoustic signals for transmission of data from the cone to a recorder at the ground surface is now available.

Presently, extensive research is in progress on the fabrication, testing and evaluation of combination cone penetrometer and pressuremeter devices. Such units would offer the potential to provide much information about the soil profile and many of the soil properties needed for analysis of most geotechnical problems. The effect of the extensive soil displacement during penetration on the results of the pressuremeter test using such systems will need detailed evaluation.

A number of the new developments in penetration test devices would appear motivated by the need for better information concerning the in situ lateral stress and a simple, practical means for obtaining it. The challenge is in the fact that the process of trying to measure it changes it, so more research on the stress changes and deformations accompanying the insertion of penetrometers is needed. Finally, it must be realized that new instruments and measurements, while exciting and promising at conception, may require continued development and evaluation before their reliability can be assured and their full potential can be utilized in practice. Measurement of something in the ground and the proper translation of that information into an envisioned soil profile or specific value of a property may continue to be a challenge for many years to come. Nonetheless, the continued development of new penetration test devices and testing procedures can only lead to better solutions for our geotechnical problems in the future.

ACKNOWLEDGMENTS

I thank many colleagues around the world who responded so generously to my request for information about new penetration tests and equipment. Without their help preparation of this paper would not have been possible.

Sai Wing Lai, graduate student in geotechnical engineering, assisted in the assembly, review, and organization of information.

Financial support for the preparation of this paper was provided by the National Science Foundation under Grant MSM-8612084 for research on "Horizontal Stresses In Situ by Cone Penetrometer and Related Studies."

REFERENCES

Amar, S., F. Baguelin & J.F. Jezequel 1982. Pressio-penetrometer for geotechnical surveys on land and offshore. Proc. 2nd European Symp. on Penetration Testing. Amsterdam. 2:419-423.

Baldi, G., D. Bruzzi, S. Superbo, M. Battaglio & M. Jamiolkowski 1988. Seismic cone in Po River sand. Proc. 1st Int. Conf. on Penetration Testing, Disney World, Florida.

Banerjee, S. & J.K. Mitchell 1980. In-situ volume change properties by electro-osmosis: theory. J. Geot. Eng. Div., ASCE. 106 (GT4):347-365.

Campanella, R.G. & P.K. Robertson 1986. Research and development of the UBC cone pressuremeter. Proc. 3rd Canadian Conf. on Marine Geot. Eng. Newfoundland.

Campanella, R.G., P.K. Robertson, D. Gillespie, N. Laing & P.J. Kurfurst 1986a. Seismic cone penetration testing in the Beaufort Sea. 3rd Canadian Conf. on Marine Geot. Eng., Memorial University, St. John's, Newfoundland.

Campanella, R.G., P.K. Robertson & D. Gillespie 1986b. Seismic cone penetration test. In S.P. Clemence (ed.), Use of in situ tests in geotechnical engineering, Geot. Spec. Publ. No. 6, ASCE, 116-130.

Fyffe, S., W.M. Reid & J.B. Summers 1986. The push-in pressuremeter: 5 years offshore experience. In J.-L. Briaud & J.M.E. Audibert (eds.) The pressuremeter and its marine applications, ASTM Spec. Tech. Publ. 950.

Handy, R.L., B. Remmes, S. Moldt, A.J. Lutenegger & G. Trott 1982. In situ stress determined by Iowa stepped blade, J. Geot. Eng. Div., ASCE, 108(GT11):1405-1422.

Huntsman, S.R. 1985. Determination of in-situ lateral pressure of cohesionless soils by static cone penetrometer. Ph.D. dissertation, University of California, Berkeley.

Huntsman, S.R., J.K. Mitchell, L.W. Klejbuk & S.B. Shinde 1986. Lateral stress measurement during cone penetration. In S.P. Clemence (ed.) Use of in situ tests in geotechnical engineering, Geot. Spec. Publ. NO. 6, ASCE, 616-634.

Lutenegger, A.J. & D.A. Timian 1986. In situ tests with K_o stepped blase. In S.P. Clemence (ed.) Use of in situ tests in geotechnical engineering, Geot. Spec. Publ. No. 6, ASCE, 730-751.

Massarsch, K.R. 1986. Acoustic penetration testing. Proc. 4th Int. Geot. Seminar, Field Instrumentation and In-Situ Measurements, Nanyang Tech. Inst., Singapore.

Mitchell, J.K. & S. Banerjee 1980. In-situ volume change properties by electro-osmosis: evaluation, G. Geot. Eng. Div., ASCE, 106(GT4):367-384.

Mitchell, J.K. & T.C. Kao 1987. Measurement of soil thermal resistivity, J. Geot. Eng. Div. ASCE, 104(GT10):1307-1320.

Muromachi, T. 1974. Phono-sounding apparatus --discrimination of soil type by sound. Proc. 1st European Symp. on Penetration Testing, 2:1:110-112, Stockholm.

Muromachi, T., H. Tsuchiya, Y. Sakai & K. Sahai 1982. Development of multi-sensor cone penetrometers. Proc. 2nd European Symp. on Penetration Testing, 2:727-732, Amsterdam.

Power, P.T. & D. Eastland 1986. "Seasprite" -a new seabed soil testing system. Underwater Systems Design. 8(2)32-35.

PWRI, 1985. Newsletter No. 19, January.

Richards, A.F. & H. Zuidberg 1985. In situ testing and sampling offshore in water depths exceeding 300 m. In Offshore Site Investigation, London: Graham & Trotman, 129-163.
Rigden, W.J., S. Thorburn, A. Marsland & A. Quartermain 1982. A dual range cone penetrometer. Proc. 2nd European Symp. on Penetration Testing, Amsterdam, 2:787-796.
Schaap, L.H.J. & H.G. Hoogedoorn 1984. A versatile measuring system for electric cone penetration testing, in Kovari, K. (ed.) Field Measurments in Geomechanics, Int. Symp. Proc. 1:313-324, Rotterdam, Balkema.
Tjelta, T.I., A.W.W. Tieges, F.P. Smits, J.M. Geise & T. Lunne 1985. In-situ density measurements by nuclear backscatter used in offshore soil investigation. Proc. Offshore Tech. Conf., Houston, paper 4917.
Triggs, J.F. Jr. & R.Y.K. Liang 1988. Correlation between static and dynamic cone penetration resistance from use of a light-weight, portable penetrometer. Submitted to First Intl. Symp. on Penetration Testing, Orlando Florida.
Tringale, P.T. 1983. Soil identification in-situ using an acoustic cone penetrometer. Ph.D. dissertation. University of California, Berkeley.
Tringale, P.T. & J.K. Mitchell 1982. An acoustic cone penetrometer for site investigations. Proc. 2nd European Symp. on Penetration Testing, 2:909-914, Amsterdam.
van der Wal, J., C. van Bergen Henegouw, H.G. Hoogendoorn & A.F. Richards 1986. Sealion: A Snellius-II Expedition automatic system for in situ geotechnical testing in water depths of 6000m. In Oceanology, p. 241-246. London: Graham & Trotman.
Vitayasupakorn, V. & S. Banerjee 1986. Initial results from an in-situ consolidometer, In S.P. Clemence (ed.) Use of In Situ Tests in Geotechnical Engineering, Geot. Spec. Publ. No. 6, ASCE, 1249-1262.
Vlasblom, A. 1977. Density measurements in-situ. LGM-Mededelingen, XVIII (4): 69-70.
Withers, N.J., L.H.J. Schaap, & J.C.P. Dalton 1986. The development of a full displacement pressuremeter. In J.L. Briaud & J.M.E. Audibert (Eds.) The Pressuremeter and Its Marine Applications, ASTM Spec. Tech. Publ. 950.
Zuidberg, H.M., ten Hope, J. & J.M. Geise Advances in in situ measurements. Symp. on Field Measurements in Geomechanics. Kobe, Japan.

Penetration Testing 1988, ISOPT-1, De Ruiter (ed.)
© 1988 Balkema, Rotterdam, ISBN 90 6191 801 4

New correlations of penetration tests for design practice

M.Jamiolkowski, V.N.Ghionna, R.Lancellotta & E.Pasqualini
Technological University, Torino, Italy

ABSTRACT: A critical review of the significant progress and innovation in interpretation of in-situ penetration testing is presented. Emphasis is placed on the Standard Penetration Test (SPT), Cone Penetration Test (CPT) and Flat Dilatometer Test (DMT). Use of these test methods in geotechnical practice to evaluate the basic design parameters of initial state variables and stress-strain-strength characteristics of cohesive and cohesionless soils is presented.

KEY WORDS: In-Situ Testing, Standard Penetration Test, Flat Dilatometer Test, Cone Penetration Test, Calibration Chambers, Indirect Approach, Correlations, Cohesionless Soil, Cohesive Soil.

1 INTRODUCTION

The present lecture attempts to summarize the existing knowledge of Penetration Testing in Geotechnical Design Practice. The last fifteen years have been characterized by significant developments in the area of in-situ testing. These developments have resulted in the invention of new tests, and in the innovation, improvement and standardization of the existing ones. However, the most relevant feature of this period, is a better understanding of the relationships between the results of in-situ tests and basic soil behaviour.

This last fact has contributed to a remarkable rationalization of the interpretation of different kinds of in-situ tests and of the use of their results in design.

More detailed information concerning the role, advantages and limitations of the in-situ testing techniques as applied in Geotechnical Practice can be found in works by Ladd et al. (1977), Mori (1981), Robertson and Campanella (1982), Wroth (1984), Jamiolkowski et al. (1985) and Robertson (1986).

To avoid duplications with other lectures in the programme of ISOPT-I, the following discussion is limited to the Standard Penetration Test (SPT), Static Cone Penetration Test (CPT), and Flat Dilatometer Test (DMT) (Marchetti, 1975; 1980).

2 INNOVATION AND PROGRESS IN PENETRATION TESTING

The penetration test is an old and well established method of in-situ experimental soil engineering. At present, recent advances in electronic sensors and data acquisition systems have largely improved the capabilities of Penetration Tests to contribute in a cost-effective way to the solution of important design problems.

The most relevant innovations are:

a. Development of the static electric cone penetrometer which permits continuous measurements of cone resistance q_c, and local shaft friction f_s as well as the monitoring of the deviation of the CPT tip from vertical (De Ruiter, 1971; 1981; 1982; Schaap and Zuidberg, 1982; Robertson and Campanella, 1984).

b. Development of large calibration chambers (CC) in which different penetration tools are tested under strictly controlled laboratory conditions (Holden, 1971; Veismanis, 1974; Schmertmann, 1976; Marcuson et Bieganousky, 1977, 1977a; Parkin et Lunne, 1982; Bellotti et al., 1982).

c. Recognition of the energy delivered to the rods during driving of the SPT sampler as an important parameter which must be considered when interpreting the results of this test (Schmertmann and Palacios, 1979; Kovacs and Salomone, 1982).

d. Use of the Becker Penetration Test, a large-scale dynamic penetration test, initially developed in Canada in the late fifties, for testing deposits of gravel and cobbles. Penetration resistance is recorded in the form of blows to penetrate each 30 cm using a double-walled closed-end casing. Recent work by Harder and Seed (1986) accounts for variation in energy delivered to the casing by a double-acting diesel hammer during driving.

e. Incorporation of the piezometer sensor into the standard electric CPT tip (Baligh et al., 1981; Campanella and Robertson, 1981; Muromachi, 1981; Tumay et al., 1981; De Ruiter, 1981; Smits, 1982), which allows the measurement of pore pressure, u, during the penetration process.

f. Development of a number of new specific penetration devices among which the following are of immediate practical interest to designers.

- Flat Dilatometer (Marchetti, 1975; 1980; Marchetti and Crapps, 1981; Lutenegger, 1988);
- Seismic Cone (Campanella and Robertson, 1984; Campanella et al., 1986; Baldi et al., 1988);
- Lateral Stress Cone (LS-CPT) and Piezo-Lateral-Stress Cell (PLSC), which allow measurement of the lateral stress on the cone shaft (Baligh et al., 1985; Morisson, 1984; Huntsman, 1985; Huntsman et al., 1986; Jefferies and Jonsson, 1986; Bruzzi, 1987).
- Vibratory Cone (V-CPT), which creates the hope for evaluating the susceptibility of cohesionless deposits to liquefaction (Sasaki et al., 1984; Bruzzi, 1987).
- Pressio-Cone, also called Full-Displacement Pressuremeter, which combines the features of the CPT with that of the pressuremeter test (Jezequel et al., 1982; Hughes and Robertson, 1985; Whiters et al., 1986).

3 PENETRATION TESTING AND GEOTECHNICAL DESIGN

The use of penetration testing results in geotechnical design may be split into the following two distinct approaches.

a. Direct Approach, which gives the opportunity to pass directly from in-situ measurements to the performance of foundations without the need to evaluate any intermediate soil parameters.
This approach is frequently used in the evaluation of the settlement of shallow foundations in cohesionless deposits and to assess the ultimate and service limit states of piles subjected to both axial and horizontal loadings.
The direct approach leads to empirical methods in which quality is strictly linked to the number and quality of the case records upon which the approach has been established. Valuable examples of this approach are the works by Burland and Burbridge (1984), Bustamante and Gianiselli (1982), Reese and Wright, (1979), and Reese and O'Neil (1987).

b. Indirect Approach, which leads to interpretation methods that allow evaluation of the parameters describing the stress-strain-strength and consolidation behaviour of soils. This appproach, although basically more sound and rational than the direct approach, suffers from the fact that it requires the solutions of very complex boundary value problems that, in the case of the penetration tests, are rarely feasible (Baligh, 1985, 1986). In this respect it is worth considering the following three categories:

- The solution of a more or less complex boundary value problem can lead to the determination of stress-strain and strength characteristics. All soil elements strained during the test follow very similar effective stress paths. Therefore, with appropriate assumptions about the drainage conditions during the test and the stress-strain relationship of the tested soil, it is possible to evaluate deformation and strength characteristics.

This category of tests includes the pressuremeter test, especially the SBPT, and seismic tests.

- The strained soil elements follow different effective stress paths depending on the geometry of the problem and on the magnitude of the applied load. In this case a rational interpretation of the test is very difficult. Even with appropriate assumptions concerning the drainage conditions and soil model, the solution of a complex boundary value problem leads to something like "average" soil characteristics. Comparisons between these average values and the behaviour of a typical soil element tested in the laboratory or their use in the specific design calculation are far from straigthforward.
  Typical examples of in situ tests in this category are the plate load test as well as the CPT and CPTU when interpreted for evaluating soil strength.
- The in situ tests results are empirically correlated to selected soil properties. Typical examples are the widely used correlations between deformation moduli and penetration resistances measured in the CPT and SPT. Because of the purely empirical nature of these correlations, they are subjected to many limitations which are not always fully recognized by potential users. In addition, it should be recognized that these correlations are formulated for either fully undrained or fully drained conditions.

In view of what is stated above and keeping in mind the extremely broad spectrum of the topic covered by this lecture, the following presentation is restricted mainly to the use of the indirect approach to assess the design parameters of cohesionless and cohesive deposits from SPT's, CPT's and DMT's.

4 INITIAL STATE PARAMETERS

All geotechnical analyses and especially those which, more or less rigorously, overcome the limitations of linear isotropic elasticity and rigid-plastic behaviour require knowledge of the <u>initial state</u> of the soil deposit. This term incorporates the information concerning:

- The initial total vertical (σ_{vo}) and horizontal (σ_{ho}) geostatic stresses.
- The initial pore pressure (u_o), not necessarily hydrostatic.
- The macro-and micro-fabric of the deposit.
- The initial void ratio (e_o) and/or relative density (D_R) in the case of cohesionless deposits.
- The preconsolidation pressure (σ'_p) and overconsolidation ratio (OCR). A more complete definition of the stress history of a deposit should include an estimate of its cyclic strain history (originated by low intensity earthquakes, ocean waves, etc.) which is especially important in cohesionless deposits but, at present, is almost impossible to assess.

Within the topic of the present lecture, it is appropriate to discuss the evaluation of D_R, σ_{ho} and OCR from the results of penetration tests.

Since the late forties (Terzaghi and Peck, 1948), tentatives have been made to correlate qualitatively the SPT resistance, N_{SPT}, with the in-situ state of densification of the cohesionless deposits.

Furthermore, quantitatively empirical correlations of the type $D_R = (N_{SPT}, \sigma'_{vo})$ have been developed by Gibbs and Holtz (1957), Bazaraa (1967), Peck and Bazaraa (1969), and Marcuson and Bieganousky (1977, 1977a). Regarding these correlations the following comments apply:

a. The Gibbs and Holtz (1957) correlation (GH), still widely applied in practice, may be approximated by the following formula (Meyerhof, 1957):

$$D_R = \left(\frac{N_{SPT}}{23\,\sigma'_{vo} + 16} \right)^{0.5} \qquad (1)$$

where:

N_{SPT} = SPT resistance in blows/30 cm, and

σ'_{vo} = effective overburden stress acting at the depth of the SPT test, expressed in bars (1bar=98.1 kPa).

The GH correlation has been obtained for clean predominantly silica sands. By analogy with what has been ascertained for the CPT performed in sands (Schmertmann, 1976; Baldi et al.,1985; 1986; Jamiolkowski et al., 1985), because this correlation is referred to σ'_{vo}, its application should be restricted to normally consolidated (NC) sands. The use of this correlation in overconsolidated (OC) sands leads to an overestimate of the insitu D_R, unless a correction similar to the one suggested by Skempton (1986) is adopted. Since the rod energy achieved during the Gibbs and Holtz (1957) CC tests is unknown, it is impossible to account for the influence of the specific driving procedure used during the SPT (Seed et al.,1984; Seed and De Alba, 1986; Skempton, 1986). This represents an additional uncertainty when evaluating D_R.

b. The Peck and Bazaraa (1969) correlation corresponds to the upper limit of $D_R = f(N_{SPT}, \sigma'_{vo})$ for dense coarse quaternary sands deposits. Otherwise, all other comments already mentioned in the case of the GH correlation apply.

c. Marcuson and Bieganousky's (1977, 1977a) correlation (MB) obtained in fine and coarse sands is the only one that attempts to take into account the influence of OCR. In this case the level of the rod energy is known, leading to an energy ratio ER ≈ 83% (ER=actual rod energy/theoretical energy).

d. The GH and MB correlations have been established on the basis of CC tests performed on samples reconstituted in the laboratory. Recent re-analysis by Skempton (1986) of the available SPT's performed in NC natural and man-made sand deposits, where ER, D_R and age of the deposits are known, suggests that the empirical relations as the one given by eqn.(1) may be influenced by aging. This phenomenon is reflected in the increase of the ratio:

$$\frac{(N_1)_{60}}{D_R^2} = a + b \qquad (1a)$$

with increasing the age of the NC deposit, see Fig.1. In this case:

$(N_1)_{60}$ = SPT blow/count for ER=60%, normalized with respect to σ'_{vo} = 1 bar.

On the basis of these findings, one can argue that the use of the existing D_R vs N_{SPT} correlations established on the basis of CC tests can lead to an overestimate of in-situ D_R in all sand deposits, except recently man-made fills.

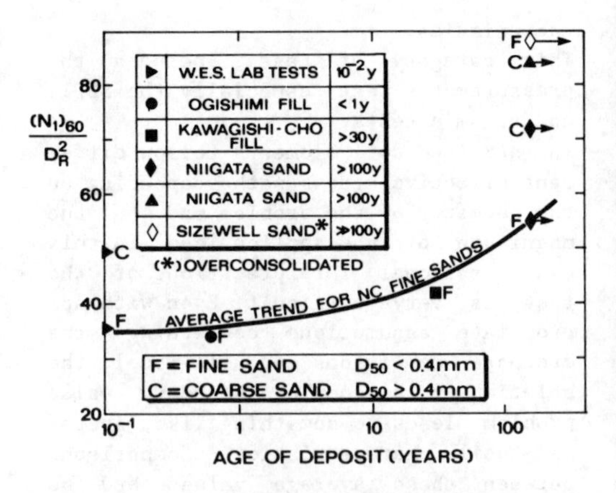

Fig.1 Influence of aging on Standard Penetration Resistance of NC sands (Adapted from Skempton, 1986)

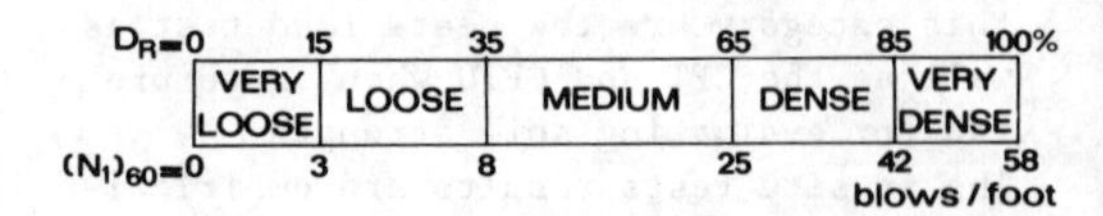

• FOR $D_R \geq 35\%$ ⟶ $\frac{(N_1)_{60}}{D_R^2} \simeq 60$

• FOR COARSE SANDS N_{SPT} SHOULD BE REDUCED IN THE RATIO $\frac{55}{60}$

• FOR FINE SANDS N_{SPT} SHOULD BE INCREASED IN THE RATIO $\frac{65}{60}$

Fig.2 Revised Terzaghi-Peck classification (1948) for NC sands (Adapted from Skempton, 1986)

e. All the available D_R vs N_{SPT} correlations have been established for predominantly silica sands. Their use in more crushable and compressible sands, like calcareous sands or even silica sands containing a non-negligible amount of fines, may lead to an underestimate of D_R (Tatsuoka et al., 1978).

f. In Fig.2 Terzaghi and Peck's (1948) classification for NC silica sands as revised by Skempton (1986) is given so that one can refer to the normalized SPT blow/count, $(N_1)_{60}$.

In the last fifteen years, comprehensive series of CC tests have been performed on numerous uncrushable and moderately crushable silica sands with the aim to validate and improve the existing correlations between q_c and engineering parameters of sands.

This effort yielded a series of D_R vs q_c correlations (Schmertmann, 1976; Baldi et al., 1983; 1985; 1986; Lancellotta, 1983) obtained on pluvially deposited, unaged and uncemented sands. Regarding these correlations the following comments apply:

a. These correlations have been worked out under the assumption that for a given sand the q_c is mainly controlled by the level of the consolidation stress tensor and by the relative density. Other factors like degree of saturation, stress and strain history *(with the exception of increase of σ'_{ho} as result of the mechanical overconsolidation, which concurs to the value of the relevant stress tensor)*, and environmental factors (cementation, aging, etc.) are assumed to play a secondary role (see Harman, 1976). However on the basis of the conclusions reached by Skempton (1986) regarding the SPT, the above postulations must be critically reconsidered; in fact, analogously with what is observed in the case of the SPT, one may suppose the D_R vs q_c correlation is also influenced by aging.
If this is true, the D_R vs q_c correlations obtained on freshly deposited clean sands will lead to an overestimation of D_R when applied to natural sand deposits.

b. On the other hand, it must be pointed out that the same correlations will cause an underestimation of D_R if applied to more crushable and compressible sands than those used in CC research, or to sands containing more than 5 to 10% fines.

c. Keeping in mind the above considerations, it is possible to summarize as follows the D_R vs q_c correlations resulting from the Italian CC research. Fig.3 gives the D_R vs q_c through σ'_{vo} correlation as obtained in moderately crushable silica Ticino Sand (TS). The same figure shows a similar correlation worked out by Schmertmann (1976) on the basis of CC tests performed on 6 different sands.
The D_R = f (q_c, σ'_{vo}) for TS is given by

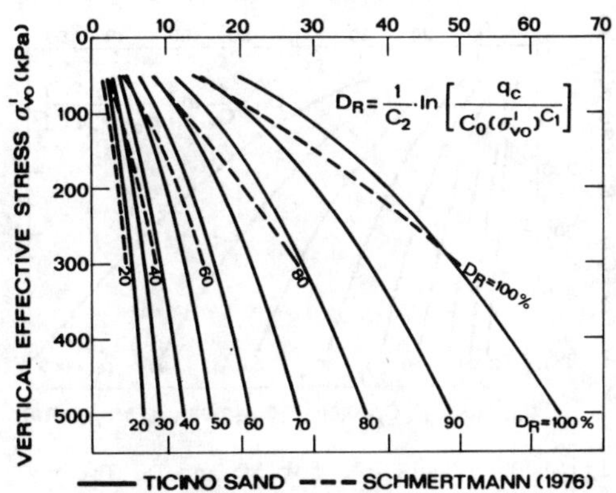

Fig.3 D_R versus q_c for NC sands

the following equation which fitted, very satisfactorily, 124 CC tests performed on NC TS:

$$q_c = C_o \cdot (\sigma'_{vo})^{C_1} \cdot \exp(C_2 \, D_R) \text{ (kPa)} \qquad (2)$$

where: C_o = 172; C_1 = 0.51 and C_2 = 2.73 are empirical constant, and
σ'_{vo} = effective overburden stress in situ, or the axial consolidation stress at the midheight of the CC sample (kPa).
The same equation, applied to 41 CC tests performed on samples of Hokksund sand (HS), yielded: C_o = 88, C_1 = 0.55 and C_2 = 3.57. Since the above mentioned correlations are referenced to σ'_{vo}, they apply to NC sands only.

d. The use of eqn.(2) in OC deposits leads to an overestimate of D_R, with the magnitude increasing with increasing OCR. To make such a correlation applicable to both NC and OC sands, it is necessary to replace σ'_{vo} in eqn.(2) with the mean effective stress σ'_m. By applying this form of fitting equation to the 228 CC tests performed on both NC and OC samples of TS, one finds:
C_o = 205, C_1 = 0.51, C_2 = 2.93. The resulting curves relating D_R to q_c through σ'_m are shown in Fig.4.
Analogously, on the basis of 67 CC tests performed on both NC and OC samples of HS, one obtains:

C_o = 149, C_1 = 0.53, C_2 = 3.33.

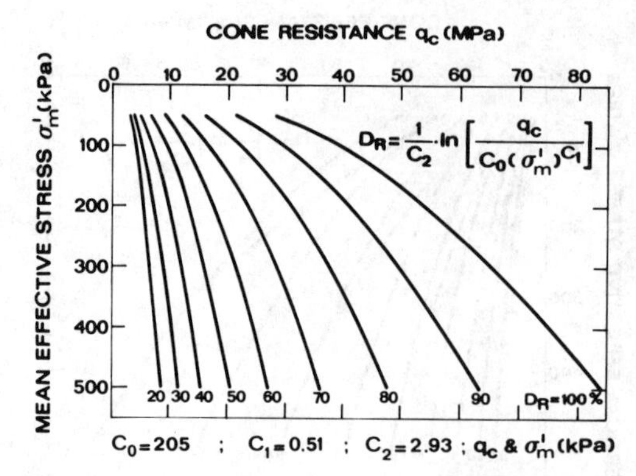

Fig.4 D_R versus q_c, for NC and OC Ticino sand

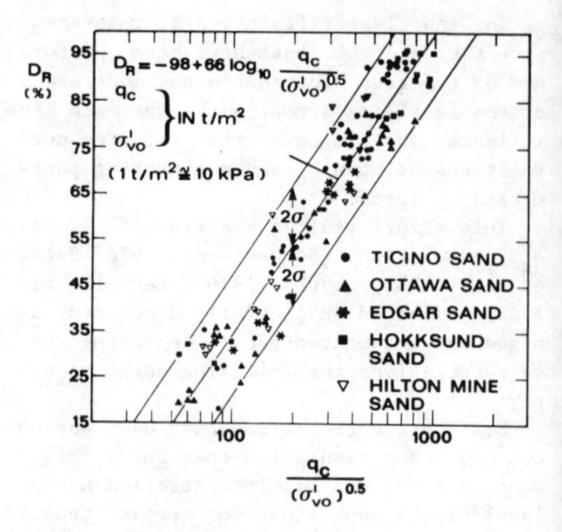

Fig.5 Correlation between D_R and q_c through σ'_{vo} (Lancellotta, 1983) for NC sands

All above mentioned D_R vs q_c correlations have been developed after q_c measured in the CC was corrected for the chamber size effect (Parkin and Lunne, 1982; Jamiolkowski et al., 1985; Eid, 1987).
Lancellotta (1983) *[for more details see Jamiolkowski et al. (1985)]*, using results of 144 CC tests performed on five different NC sands, worked out the correlation between D_R and q_c through σ'_{vo} shown in Fig.5. This correlation allows one to estimate the uncertainties involved in the evaluation of D_R from q_c measurements.

To compare values of D_R from different correlations between D_R and penetration test results, a reference study was made of the quaternary deposit of the Po River silica sand. For this geotechnically well investigated deposit, reliable results of SPT's with rod energy measurements and of CPTU's are available (Bellotti et al., 1986; Bruzzi et al., 1986). Fig.6 gives the values of D_R for Po River sand as obtained by means of the different N_{SPT} and q_c vs D_R correlations from which the following information of practical interest can be inferred:

- The GH correlation yields D_R values about 8 to 10% higher than those resulting from the formulae suggested by Skempton (1986) for NC natural sand deposits. In the case that the GH calibration chamber tests have been carried out with ER < 60%, as one would expect, the comparison between GH and Skempton (1986) correlation would lead to an even larger difference.
- At least in principle, the D_R vs q_c correlations obtained with laboratory sand specimens should lead to the value of D_R in agreement with GH values also obtained in the freshly deposited sands. This is indeed true in the first 17 m of depth. Below that depth the values of $D_R = f(q_c)$ match well with those yielded by Skempton's (1986) correlations which are valid for aged natural NC sands.
  The reasons for such an apparent discrepancy are not known. The empirical nature of the correlations used and the local soil variability might be responsible for this.

In conclusion, the D_R vs penetration resistance correlations suffer at present from some uncertainties linked with the fact that all of them have been established on freshly deposited sands and due to the fact that by referencing them to σ'_{vo}, their applications are, strictly speaking, correct only in NC unaged sand deposits.

The tentatives to infer σ_{ho} (σ'_{ho}) and/or OCR from the results of the penetration tests are quite recent and are still at a preliminary stage of validation.

The insertion of the penetration tool into the soil changes drastically the geostatic stress conditions. In case of undrained penetration in saturated cohesive deposits, this change is mainly reflected in a large increase in excess pore pressu-

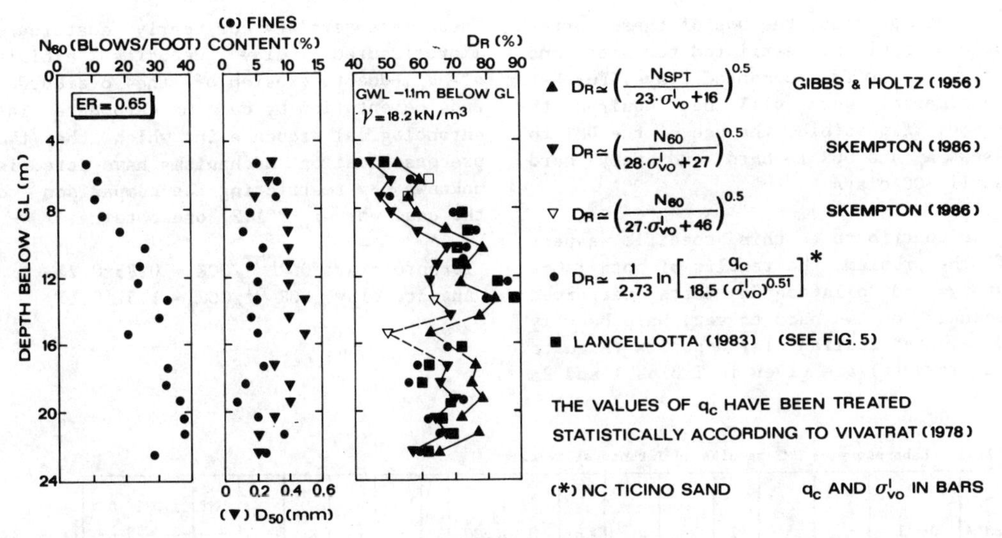

Fig.6 Relative density of Po River sand from SPT's and CPT's

re (Δu), and only modest changes of the horizontal effective stress are noticed (Campanella et al., 1985; Baligh et al., 1985; Morisson, 1984).

After the complete dissipation of the induced pore pressure, Δu, the effective horizontal stress σ'_h acting on the in-situ device can be equal to or higher than σ'_{ho}, depending on the stress-strain-strength characteristics, sensitivity and stress history of the soil.

After drained penetration (Δu=0) into cohesionless soil, the effective horizontal stress surrounding the device increases to a value $\sigma'_{hp} > \sigma'_{ho}$. For a given soil, the magnitude of $(\sigma'_{hp}-\sigma'_{ho})$ depends on D_R and effective confining stresses. In dense soil the σ'_{hp} may tend to decrease with time, due to the relaxation phenomenon.

The above stated soil-penetrometer interaction illustrates the difficulties faced when trying to infer the parameters describing the initial state (K_o, σ_{ho}, OCR) of soil deposits from any kind of penetration test.

As far as cohesive deposits are concerned, only the DMT (Marchetti, 1980; Marchetti and Crapps, 1981) offers an empirical approach to evaluate both K_o and OCR as function of Horizontal Stress Index, K_D, and Material Index, I_D.

The proposed correlations are:

$$K_o = \left(\frac{K_D}{1.5}\right)^{0.47} - 0.6 \qquad (4)$$

$$OCR = (0.5\ K_D)^{1.56} \qquad (5)$$

with the latter being valid for soils having $I_D \leq 1.2$.

Eqn.(5) results from eqn.(4) based on the assumption that normalized behaviour "a la SHANSEP" (Ladd and Foott, 1974) applies. The equations have been validated against the results of laboratory tests in soft and medium to stiff uncemented clays (Marchetti, 1980; Lacasse and Lunne, 1988). This validation, however, is hampered by the fact that the real in-situ value of K_o is unknown and the values of the preconsolidation stress σ'_p obtained from oedometer tests, which concur with the reference values of OCR, are affected to variable degrees by sample disturbance.

The experience gained seems to indicate that:

a. Eqns.(4) and (5) tend to overestimate both K_o and OCR in clays ($I_D \leq 1.2$) having $K_D > 8$.

Lacasse and Lunne (1988) in the paper presented at this symposium suggest some modifications to the correlations under discussion.

b. At present, the use of these correlations should be restricted to soft and medium to stiff uncemented clays. Further experimental work will be required to extend, if possible, the use of the DMT to assess K_o and OCR in hard and very hard heavily OC clays.

To contribute to this specific aspect of the problem, the results of both laboratory and dilatometer tests performed recently on two hard to very hard heavily OC cemented Italian clay deposits (Augusta and Taranto) are given in Tables 1 and 2. These late tertiary or early quaternary microfissured clay deposits exhibit $\sigma'_p \gg \sigma'_{vo}$ due to erosion of the overburden and cementation by calcium carbonate. The chronological sequence in which the two preconsolidation mechanisms have acted is unknown. By restricting the comparison to the case when $I_D \leq 1.2$, one obtains:

Taranto clay: $OCR^{DMT}/OCR = 0.85 \pm 0.22$
Augusta clay: $OCR^{DMT}/OCR = 1.32 \pm 0.17$

Table 1. Laboratory vs DMT results in Taranto silty-clay.

BH/S	Depth		σ'_{vo}	γ	PI	C_u UU	σ'_p	OCR	M	$CaCO_3$	CF	p_o	p_1	K_D	I_D	C_u^{DMT}/C_u	OCR^{DMT}/OCR	M^{DMT}/M
No.	from m	to m	kPa	kN/m^3	%	kPa	MPa	-	MPa	%	%	MPa	MPa	-	-	-	-	-
1/1	2.5	3.1	54.9	19.8	39.9	783	-	-	-	29.0	44	1.14	2.77	20.8	1.43	N.A.	-	-
1/2	5.5	6.1	113.8	19.9	29.7	823	-	-	-	26.5	33	1.21	2.93	10.6	1.42	N.A.	-	-
1/3	7.9	8.5	158.9	19.6	42.7	886	-	-	-	26.0	38	2.49	3.91	15.7	0.57	0.52	-	-
1/4	10.9	11.5	188.4	18.7	29.7	747	-	-	-	25.5	34	2.49	4.26	13.1	0.72	0.58	-	-
3/2	5.5	6.1	106.9	20.1	30.7	305.1	1.77	16.5	35.3	29.5	34	1.01	1.63	9.4	0.62	0.53	0.68	1.48
3/3	7.0	7.6	121.6	19.4	28.7	252.1	1.49	12.3	29.5	26.5	34	1.21	1.91	9.8	0.59	0.77	0.97	2.04
3/4	8.5	9.1	136.4	20.0	30.8	302.1	2.28	16.7	39.2	27.0	32	1.45	2.24	10.3	0.56	0.77	0.77	1.77
3/5	10.0	10.6	151.1	21.0	27.1	483.0	≈3.13	20.7	59.8	27.5	37	1.75	2.79	11.2	0.61	0.59	0.71	1.49
3/6	11.5	12.1	165.8	20.8	22.3	485.0	≈3.27	19.7	88.3	27.0	39	2.21	3.44	12.9	0.57	0.77	0.93	1.33
3/11	22.0	22.6	268.8	20.4	25.2	676.4	3.12	11.6	54.0	24.0	30	2.92	4.17	10.2	0.45	0.67	1.10	2.03
3/11*	22.0	22.6	268.8	20.3	25.2	676.4	2.75	10.2	48.6	24.0	30	2.92	4.17	10.2	0.45	0.67	1.25	2.25
3/12*	26.5	27.1	312.9	20.5	-	-	≈4.44	15.8	108.9	-	-	3.25	4.67	9.7	0.47	-	0.72	1.13
3/14	35.5	36.1	401.2	20.6	26.4	845.6	4.91	12.3	81.4	49.5	35	4.09	5.62	9.4	0.41	0.72	0.91	1.60
5/4	9.0	9.6	113.8	19.9	27.0	200.0	1.62	14.2	27.7	25.0	35	0.37	1.51	3.2	3.78	N.A.	0.30	2.16
5/6	13.6	14.2	158.9	20.0	25.2	475.8	2.35	14.7	38.5	26.5	36	0.69	1.69	3.6	1.74	N.A.	0.28	1.38
5/8	17.5	18.1	197.2	20.3	26.6	-	2.97	15.0	59.8	28.5	32	1.60	2.54	7.3	0.65	-	0.50	1.21

NOTES: GWL ≈ 0.00 ABOVE M.S.L.; GL ≈ +8.0, +5.1, 2.3 RESPECTIVELY BH-1, BH-3 and BH-5;
M = TANGENT CONSTRAINED MODULUS AT ≈ 0.9 • σ'_p;
* = CRS-OEDOMETER TEST, ALL OTHER IL TESTS

Table 2. Laboratory vs DMT results in Augusta clay.

BH/S	Depth		σ'_{vo}	γ	PI	C_u UU	C_u CK_oU	σ'_p	OCR	M	$CaCO_3$	CF	p_o	p_1	K_D	I_D	C_u^{DMT}/C_u	OCR^{DMT}/OCR	M^{DMT}/M
No.	from m	to m	kPa	kN/m^3	%	kPa	kPa	kPa	-	MPa	%	%	kPa	kPa	-	-	-	-	-
1/1	5.3	5.9	81.4	18.2	51.1	130.5	126.1	784.8	9.6	11.3	14.0	51	777.4	1191.9	9.3	0.55	0.95	1.15	2.55
1/1*	5.3	5.9	81.4	18.2	51.1	130.5	126.1	833.9	10.2	13.7	21.0	51	777.4	1191.9	9.3	0.55	-	1.08	2.10
1/2	7.0	7.6	97.1	18.0	51.9	142.2	-	932.0	9.6	19.6	21.0	54	1044.8	1554.9	10.4	0.51	1.18	1.36	2.29
1/4	12.0	12.6	141.3	18.5	48.3	234.5	-	1275.3	9.0	21.6	16.5	51	1648.1	2261.2	11.1	0.39	1.13	1.61	2.57
1/4**	12.0	12.6	141.3	18.5	48.3	234.5	-	1422.5	10.0	24.0	16.5	51	1648.1	2261.2	11.1	0.39	-	1.45	2.31
1/6	17.0	17.6	185.4	18.3	50.5	181.5	-	1471.5	7.9	20.6	21.0	58	1819.8	2469.7	9.1	0.39	1.49	1.35	2.63
1/6*	17.0	17.6	185.4	18.3	50.5	181.5	-	1569.6	8.5	21.6	21.0	58	1819.8	2469.7	9.1	0.39	-	1.25	2.51
1/8	25.0	25.6	256.0	18.1	49.7	-	197.7	1569.6	6.1	21.1	16.5	60	2138.6	2952.8	7.5	0.40	1.49	1.29	2.95

NOTES: GWL AT 3.6 M BELOW GL; M = TANGENT CONSTRAINED MODULUS AT ≈0.9 σ'_p; * CRS OEDOMETER TESTS;
** INSTRUMENTED OEDOMETER TESTS ALLOWING MEASUREMENT OF RADIAL STRESS, ALL OTHER IL TESTS

where:

OCR^{DMT} = overconsolidation ratio estimated from DMT using eqn.(5), and

OCR = overconsolidation ratio estimated from the end-of-primary consolidation oedometer curve using Casagrande's (1936) procedure.

For these clays having a similar age, stress history, cementation and fabric and referring to tests on samples having the same degree of disturbance, the DMT on average underestimates OCR by 15% in Taranto clay and overestimates OCR by 30% in Augusta clay. Considering the inherent difficulties and uncertainties connected with the assessment of OCR, the results yielded by DMT's in Taranto and Augusta clays look promising, and further validation of the device in similar soil deposits is encouraged (see example shown in Fig.7).

Despite the present uncertainties linked with the assessment of the K_o and OCR values from DMT results, this test gives an useful qualitative information on the stress history of the deposit through the K_D vs depth profile which exhibits a trend similar to that of OCR.

Another possibility of evaluating the OCR of fully saturated cohesive deposits originates from an indication by Baligh et al. (1980) that the pore pressure measured during undrained cone penetration may reflect in some way the stress history of a deposit. This specific topic is covered by a lecture presented at this symposium by Campanella and Robertson (1988). For further details regarding the potential and the limitations of this approach see also Wroth (1984), Baligh (1985a), Battaglio et al. (1986) and Robertson (1986).

Other newly developed penetration tools like PLSC (Baligh et al., 1985; Morisson, 1984) and the Iowa Stepped Blade (Handy et al., 1982), have the potential to evaluate σ'_{ho} and K_o in soft to medium clays, but their discussion is beyond the scope of this lecture.

In cohesionless soils, attempts have been recently made to correlate the σ'_{hp} acting on the penetration tools after penetration with σ'_{ho}. This implies at least a qualitative understanding of the factors upon which the difference (σ'_{hp} - σ'_{ho}) depends. These factors can be grouped as follows:

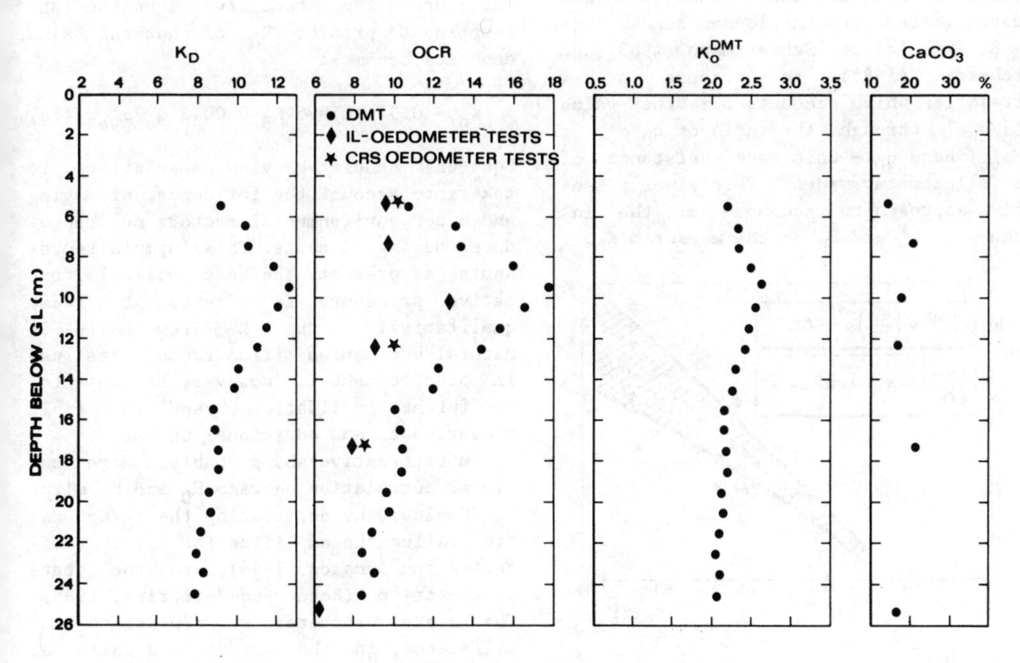

Fig.7 Stress history of cemented OC Augusta clay from DMT

- Material dependent, i.e. grading, mineralogy, shape of grains, etc.
- Stress level dependent, i.e. mean consolidation stress, shear strain level, etc.
- Fabric dependent, i.e. anisotropy, cementation, etc.
- Geometry dependent, i.e. physical dimensions of the penetration tool, position where σ_{hp} is measured, etc.

Considering the large number of parameters implicated and keeping in mind the complexity of the deep penetration process, particularly in cohesionless soils, the difficulties involved in the numerical or analytical evaluation of $(\sigma'_{hp}-\sigma'_{ho})$ are evident. Under these conditions, all interpretation procedures available at present to assess σ_{ho} or K_o using penetration devices (Marchetti, 1985; Huntsman, 1985; Robertson, 1986a) are purely empirical.

Marchetti (1980), on the basis of limited experience, postulated that eqn.(4) might be also applicable to sands.

Further CC tests and field experience led to the conclusions that this empirical formula largely overestimates the value of K_o in dense and very dense sands and can underestimate it in loose sands, see Fig.8. This induced Schmertmann (1983) and Marchetti (1985) to propose a new correlation which accounts for the value $(\sigma'_{hp}-\sigma'_{ho})$ through the ratio of q_D/σ'_{vo} or q_c/σ'_{vo} (where q_D = unit base resistance of the dilatometer wedge). This gives a tentative approach to account for the influence of σ'_m and D_R on the measured K_D.

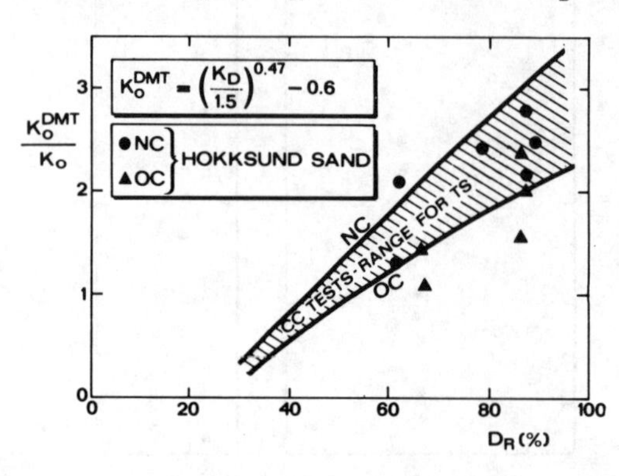

Fig.8 K_o^{DMT} vs measured K_o of Ticino sand using original Marchetti correlation (1980)

Based on the results of DMT's performed in CC on pluvially deposited specimens of TS and HS (Baldi et al., 1986a), it was possible to establish the following tentative correlation between K_o and K_D:

$$K_o = D_1 + D_2\ K_D + D_3\ q_c/\sigma'_{vo} \tag{6}$$

where:

D_1, D_2, D_3 = empirical constants, and

q_c = cone resistance at the depth where K_D has been measured.

For pluvially deposited sands tested in the CC, the empirical constants have the following values:

$$D_1 = 0.376;\ D_2 = 0.095 \text{ and } D_3 = -0.0017.$$

Use of eqn.(6) in the Po River sand, where σ'_{ho} can be estimated from the SBPT's and on the basis of the geological information (Baldi et al. 1986a; Bruzzi et al. 1986), still overpredicts K_o.

To eliminate this discrepancy (Baldi et al. 1986a), it was arbitrarily decided to search by trial and error for a multiplier of D_3 so that the $K_o = f(K_D)$ from eqn.(6) would coincide with "a best estimate" of K_o in-situ. This led to D_3=-0.0046. Therefore the tentative equation attempting to predict K_o of natural sand deposits becomes:

$$K_o = 0.376+0.095\ K_D-0.0046\ q_c/\sigma'_{vo} \tag{6a}$$

Eqn.(6a) should be view as an attempt to take into account the influence of aging and other environmental factors not reproduced during CC tests. This formula represents, at present, the best available tentative procedure to assess, at least qualitatively, the K_o from DMT's in natural uncemented silica sands. Its use in practice should, however, be subjected to further validation based on field measurements and additional CC tests.

An alternative and probably a more rational correlation between K_D and K_o might be developed by correlating the K_D/K_o ratio, called the amplification factor (Jefferies and Jonsson, 1986), to the state parameter ψ (Been and Jefferies, 1985). The state parameter ψ represents the difference in the current void ratio, e, of the material at a given σ'_m and the void ratio on the Steady State Line, e_{ss}, at the same mean effective stress. The meaning of this parameter is illustrated in Fig.9.

The parameter ψ reflects the combined influence of both e, or D_R, and σ'_m on the behaviour of the cohesionless soil.

The value of ψ allows discrimination between contractive ($\psi > 0$) and dilatative ($\psi < 0$) behaviour of soil and generally correlates well with the behaviour of cohesionless soils at or close to failure (ϕ', q_c; p_o, p_o-p_1, K_D, etc.); see Been et al. (1986), Konrad (1987), Jefferies and Jonsson (1986). To correlate ψ and other engineering parameters for different sands, it was suggested to normalize ψ with respect to (e_{max}-e_{min}) or to some limiting negative or positive value of the state parameter (Hird and Hassana, 1986; Been and Jefferies, 1986; Konrad, 1987), with:

e_{max}= maximum void ratio } of the test
e_{min}= minimum void ratio } sand

Fig.10 presents the plot of the dilatometer amplification factor K_D/K_o vs ψ as obtained for TS and HS from 56 CC tests. In this figure K_o represents the coefficient of earth pressure at rest measured during the one-dimensional straining of the CC specimen.
From this figure it appears that K_D/K_o correlates quite well with ψ and that the relationship between these two parameters can be successfully fit by means of the following formula:

$$K_D/K_o = a \exp (m \psi) \quad (7)$$

where:
a, m = empirical coefficients.

In principle, eqn.(7) offers the possibility of evaluating K_o = f (K_D,ψ), but use in practice is hampered by the fact that calculation of ψ requires assessment of σ'_m which in turn implies knowledge of K_o.

To proceed further along this line, it becomes necessary to set-up a second relationship between ψ and the result of other in-situ tests. An attempt in this direction was tried by Jefferies and Jonsson who used as the second equation the correlation between q_c and ψ reported by Been et al. (1986).

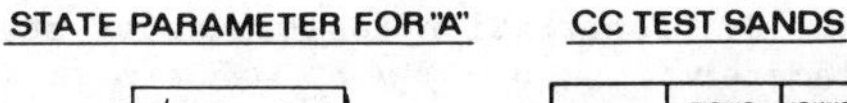

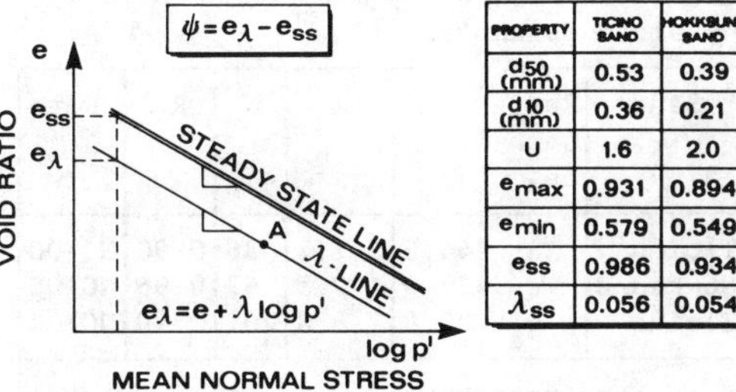

PROPERTY	TICINO SAND	HOKKSUND SAND
d_{50} (mm)	0.53	0.39
d_{10} (mm)	0.36	0.21
U	1.6	2.0
e_{max}	0.931	0.894
e_{min}	0.579	0.549
e_{ss}	0.986	0.934
λ_{ss}	0.056	0.054

Fig.9 State parameter for sand (Been & Jefferies, 1985)

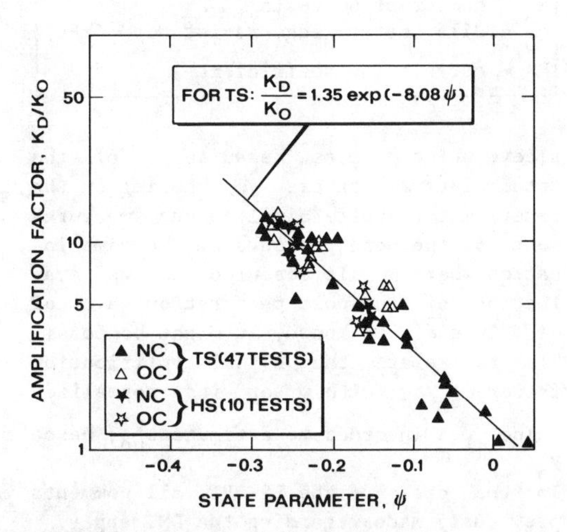

Fig.10 Dilatometer amplification factor from CC tests

Table 3 gives the correlation between q_c and ψ and q_D and ψ as obtained for TS and HS. With the help of these correlations, one can attempt to solve a system of two equations in two unknowns for ψ and K_o (Jefferies and Jonsson, 1986).

At present, the above mentioned possibilities to obtain K_o = f (K_D, ψ) represent nothing else than a tentative idea to rationalize the evaluation of K_o on the basis of DMT results in sands. The basic idea behind this approach is that the value of ψ should be linked with the magnitude of (σ'_{hp}-σ'_{ho}).

The recently developed LS-CPT (Huntsman, 1985; Huntsman et al., 1986; Jefferies and Jonsson, 1986) is an electrical CPT tip which incorporates a transducer or load cell on or just behind the friction

Table 3. Penetration resistance vs state parameter: $(q-\sigma_m) / \sigma'_m = a \exp (m \psi)$. ψ = state parameter

Sand	Resistance	a	m	N	R	Note
Ticino	q_D	44.5	-7.7	36	0.90	NC+OC
Hokksund	q_c	37.0	-9.5	67	0.95	NC+OC
Ticino	q_c	30.8	-8.8	201	0.90	NC+OC

q_c = cone resistance (kPa)
σ_m = total mean stress (kPa)
σ'_m = effective mean stress (kPa)
N = number of CC tests
q_D = dilatometer wedge resistance (kPa)
R = correlation coefficient

sleeve which permits measurement of the total lateral stress σ_{hp} acting on the penetrometer shaft. Simultaneous measurement of the pore pressure at the same location where σ_{hp} is measured allows evaluation of the post-penetration value of σ'_{hp}. Once σ'_{hp} is known, it might be possible to relate the LS-CPT amplification factor $\sigma'_{hp}/\sigma'_{ho}$ with ψ, or its normalized value $\overline{\psi}$, in order to estimate σ'_{ho}, hence K_o.
In the case of the LS-CPT, all comments previously made regarding the DMT apply.

In both cases the use of the amplification factors vs correlations in order to assess the σ'_{ho} and K_o of the sand deposits in-situ requires further research and validation.

Additional uncertainties when attempting to assess σ'_{ho} and K_o from the results of penetration tests arise from the fact that the distribution of σ_{hp} (LS-CPT) and p_o (DMT) along the shaft of the penetration device seems to be highly non homogeneous (Campanella and Robertson, 1981; Hughes and Robertson, 1985). This topic requires additional research with the aim of establishing the optimum geometry of penetration tools used to investigate horizontal stress existing in the ground prior to penetration.

5 DEFORMATION CHARACTERISTICS

Since the appearance of the penetration tests, the engineers have been attempting to assess the deformation characteristics and/or settlement of structures from their results (Terzaghi and Peck, 1948; De Beer, 1948; Meyerhof, 1956). This approach is of great practical interest in cohesionless and other soil deposits where undisturbed sampling is still impossible, unreliable or not cost effective. However, as mentioned before, interpretation of in-situ testing suffers from many limitations which make the assessed deformation characteristics very difficult to be linked to the relevant drainage conditions and stress or strain level of the specific project. This renders correlations between penetration resistances and deformation characteristics of soils purely empirical, falling into the third category in Section 3b.

The deformation characteristics of soils are generally defined using the laws of continuum mechanics, usually under the assumption that the material behaves as a linear elastic isotropic material. More recently, by virtue of vast experimental evidence, geotechnical engineers are less reluctant to refer to the more realistic linear elastic cross anisotropic model (Lekhnitzkii, 1977; Wroth and Houlsby, 1985).
Its use in every day practice is however hampered by the difficulties to determine experimentally the relevant five independent elastic constants.

Even accepting the simplifications involved with the use of elastic models for determining the deformation characteristics of soils, this should implicity imply the followed effective stress path remains inside the stress space outlined by the current yield surface (Schofield and Wroth, 1968; Roscoe and Burland, 1968; Atkinson and Bransby, 1978).

Actually the deformation characteristics of a <u>given soil</u> depend on:
- Stress and strain history of the deposit intended in the broadest sense of the term (Jamiolkowski et al., 1985).
- Current level of the mean effective stress.
- Induced level of shear strain.

- Followed effective stress path reflecting both soil anisotropy and plasticity.
- Time factor which, by phenomena like viscous hardening (aging) and creep in shear, influences the soil stress-strain characteristics.

Therefore, the correct and safe use of correlations between penetration resistances and soil moduli presented below are subordinated, at least qualitatively, to the skill of the engineer to account for all the factors mentioned above.

The use of penetrometers to assess the deformation moduli of cohesive deposits is practically limited to the correlation between dilatometer modulus E_D and the constrained modulus M, proposed by Marchetti (1980). The drained soil modulus (M) is correlated to the undrained dilatometer expansion which occurs in soil already strained by the blade penetration, indicating the highly empirical nature of this correlation. Despite this problem, available experience (Marchetti, 1980; Schmertmann, 1986; Lacasse and Lunne, 1988) suggests that the DMT allows the prediction, with an acceptable degree of precision, of the tangent constrained modulus at the the vertical stress equal to σ'_{vo}. This appears to be especially true in soft and medium to stiff deposits having M < 20 MPa.

The existing M vs E_D correlation might be less reliable for highly OC very stiff to hard cemented clays, as it emerges from the data concerning the Augusta and Taranto clays shown in Tables 1 and 2. In these Tables the reference oedometer modulus corresponds to M evaluated far beyond σ'_{vo} at 0.9 σ'_p [the influence of the sample disturbance has been reduced by an unloading-reloading loop performed starting at (2 to 3) σ'_{vo})]. Despite this assumption, results from Marchetti's (1980) procedure lead to M^{DMT} values higher than those inferred from oedometer tests.

Much wider and more relevant to the engineering practice is the use of penetration resistance vs soil deformation modulus correlations in cohesionless deposits (D'Appolonia et al. 1968; D'Appolonia and D'Appolonia, 1970; Schmertmann, 1970, 1978; Parry, 1971, 1977, 1978; Mitchell and Gardner, 1975). Only in the last decade, however, has a better theoretical understanding of the stress-strain behaviour of sands been combined with a large number of accumulated experimental evidence to provide a rational outlook for the reliability and limitations of such correlations. These findings, of great practical interest, may be summarized as follows:

a. The influence of overconsolidation on a cohesionless soil can be divided into two factors. The first one is the strain hardening of the material produced by the accumulated plastic strain. The second one corresponds to an increase in σ'_{ho} for a given level of σ'_{vo} and is responsible for the well known fact that $K_o^{OC} > K_o^{NC}$, with; $K_o^{OC} = f$ (OCR) (see Ladd et al., 1977 and Jamiolkowski et al., 1985). This last factor is conventionally linked with only mechanical overconsolidation only *[Even if a recent work by Mesri and Castro (1987) shows that in clays K_o increases with aging]* while the plastic hardening of the soil appears as a consequence of all types of preconsolidation mechanisms, i.e. aging, cementation, dessication, low strain cyclic stress history induced by earthquakes and wind loadings, etc.

b. Small and large scale (CC) laboratory tests have shown that while all kinds of deformation moduli are strongly influenced by both plastic hardening and σ'_{ho}, penetration resistances are influenced by the current level of σ'_{ho} and remain almost insensitive to the effect of the accumulated plastic strain. This indicates that large strains caused by the penetration of the devices like SPT, CPT, DMT, etc., mostly obliterate the effects of stress and strain history in the soil surrounding the penetrometer. This phenomenon has been observed in the special CC tests for: Dynamic Cone Penetration Test *[According to the writers all findings regarding DCPT applies qualitatively also to SPT]* by Hababa (1984), Clayton et al. (1985), (1986), CPT by Lambrechts and Leonards (1978), Jamiolkowski et al. (1985), Bellotti et al. (1986), Baldi et al. (1985), (1986) and for DMT by Baldi et al. (1986a). These important facts can be inferred from results of the special CC tests summarized in Fig.11 and 12.

c. The results of CC tests show that all kinds of penetration resistances are more sensitive to σ'_{ho} than to the σ'_{vo}. Clayton et al. (1985) report qualitatively

• "IN-SITU" TESTS IN CC, PERFORMED ALWAYS AT Ⓐ
• CYCLIC LOADING BETWEEN Ⓐ AND Ⓑ

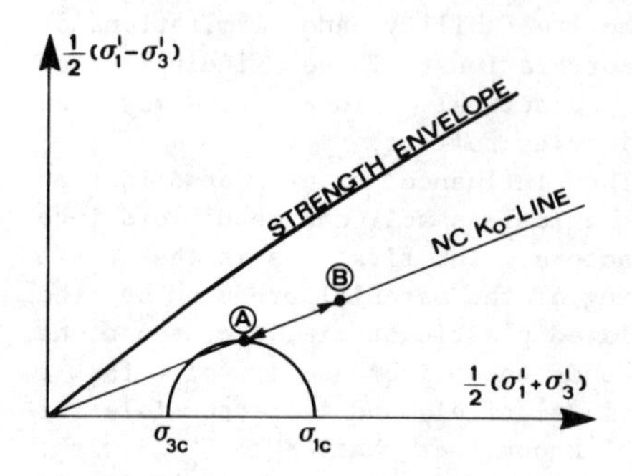

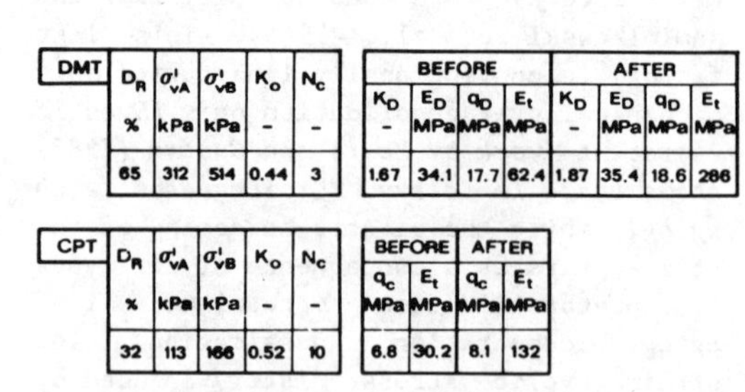

DMT	D_R	σ'_{vA}	σ'_{vB}	K_o	N_c
	%	kPa	kPa	–	–
	65	312	514	0.44	3

BEFORE				AFTER			
K_D	E_D	q_D	E_t	K_D	E_D	q_D	E_t
–	MPa	MPa	MPa	–	MPa	MPa	MPa
1.67	34.1	17.7	62.4	1.87	35.4	18.6	286

CPT	D_R	σ'_{vA}	σ'_{vB}	K_o	N_c
	%	kPa	kPa	–	–
	32	113	166	0.52	10

BEFORE		AFTER	
q_c	E_t	q_c	E_t
MPa	MPa	MPa	MPa
6.8	30.2	8.1	132

E_t = TANGENT YOUNG'S MODULUS AT A
N_c = NUMBER OF UNLOADING - RELOADING CYCLES

Fig.11 Effect of prestraining on results of CPT and DMT in Ticino sand

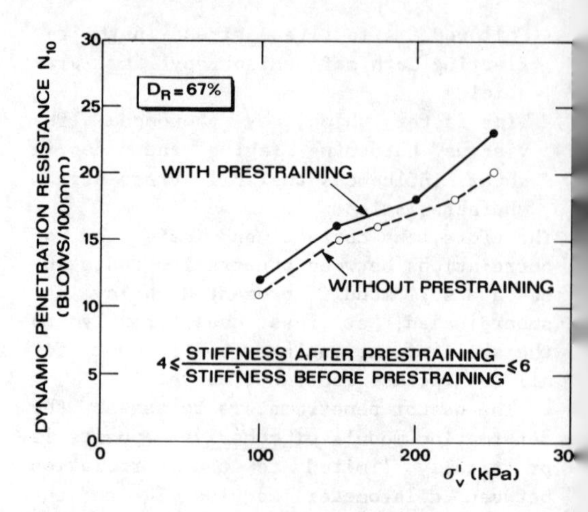

Fig.12 Effect of prestraining on DCPT results in Leighton Buzzard sand (Adapted from: Clayton et al., 1985; Leonards et al., 1986)

that the blow/count of DCPT is two times more sensitive to σ'_{ho} than to σ'_{vo}. Baldi et al. (1986), by fitting their CC test data for TS and HS, have obtained the following empirical equation which allows separation of the influences of σ'_{ho} and σ'_{vo} on the measured value of q_c:

$$q_c = C_o p_a \left(\frac{\sigma'_{vo}}{p_a}\right)^{C_1} \left(\frac{\sigma'_{ho}}{p_a}\right)^{C_2} \exp\left(C_3\ D_3\right) \quad (8)$$

with the empirical coefficients C_o thorough C_3 (for p_a = 1 kPa) given in Table 4.

In view of what is stated above, it appears obvious that, for the same sand, a unique correlation between penetration resistance and non-linear deformation modulus cannot exist.
This is confirmed by the results of the CC tests summarized in Figs.13 through 15. They show M vs q_c and E vs q_c correlations as obtained for TS and HS. In Fig.13, M

Table 4. Empirical coefficients C_o to C_3 in the formula ...(8)*.

Sand	C_o	C_1	C_2	C_3	No.	R^2
Ticino	220	0.065	0.440	2.93	228	0.93
Hokksund	149	0.140	0.400	3.38	64	0.96

*Obtained from CC tests with
$0.30 < D_R < 0.98$;
$50 \le \sigma'_v \le 800$ kPa; $1 \le OCR \le 8$;
using q_c values corrected for CC size effect

corresponds to the tangent constrained modulus measured at σ'_v at which the penetration test in the CC specimen has been performed.

The E_{15}, E_{25} and E_{50} in Figs.14 and 15 correspond to the drained Young's moduli obtained from the triaxial CK_oD compression tests performed on the pluvially deposited specimens of the test sands. They have been evaluated at deviatoric stress levels corresponding, respectively, to 15%, 25% and 50% of the failure stress. The average values of the axial strain ϵ_a at which these moduli have been determined are given in Table 5.

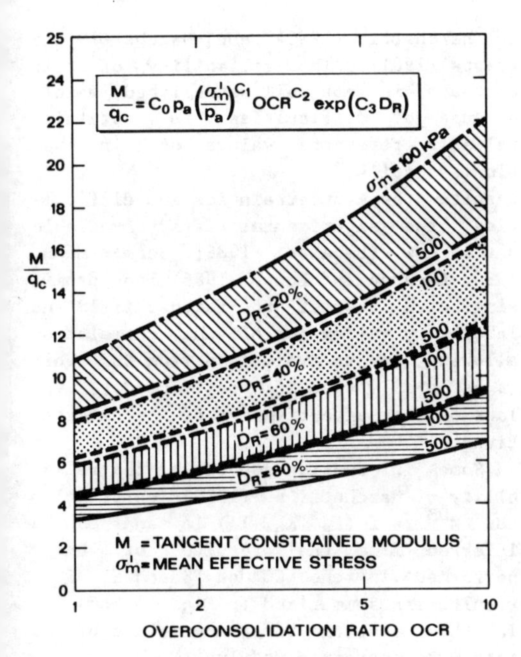

Fig.13 M versus q_c correlation for Ticino sand (Jamiolkowski, 1986)

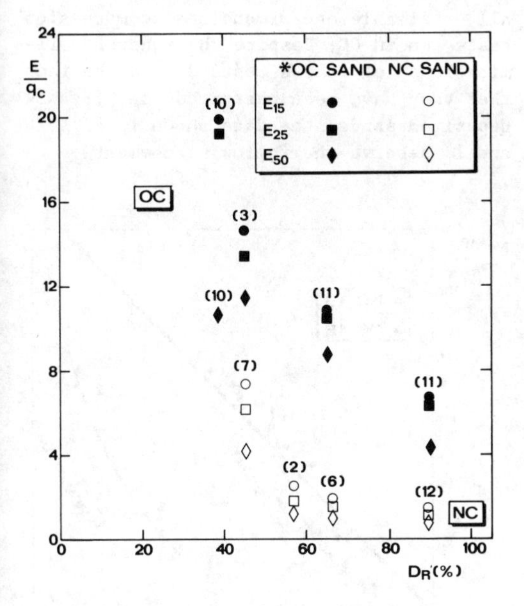

Fig.14 Young modulus vs cone resistance in Hokksund sand

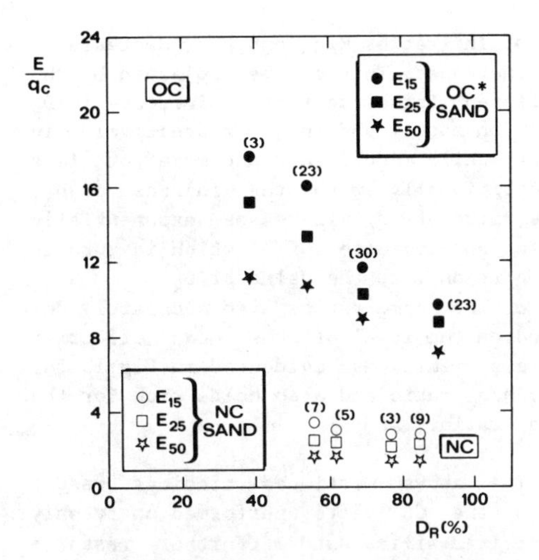

Fig.15 Young modulus vs cone resistance in Ticino sand

Table 5. Average Axial Strain corresponding to Young's Moduli in Figs.14 and 15.

Sand	OCR	$(\epsilon_a)_{15}$ %	$(\epsilon_a)_{25}$ %	$(\epsilon_a)_{50}$ %
Ticino	1	0.113	0.232	0.694
Ticino	2 to 8	0.039	0.069	0.159
Hokksund	1	0.152	0.316	0.939
Hokksund	2 to 8	0.036	0.060	0.215

Analysis of the ϵ_a values displayed in Table 5 suggests that E_{15} and E_{25} of OC test sands correspond to the "elastic" stiffness evaluated inside the current yield surface, while the E_{50} of OC sands and obviously all the NC values of E reflect the elastic-plastic stiffness at the yield surface.

The results shown in Figs.13 through 15 indicate the following important trends in terms of the discussed correlations:

a. The ratios M/q_c and E/q_c are much higher for the mechanically OC sands than for the NC ones. This means that, without an "a priori" knowledge or assumption of the stress history of the deposit, it is impossible to select a reliable value of the design modulus vs q_c ratio.

b. The ratios M/q_c and E/q_c decrease as D_R increases. This can be explained by the different influence that an increase in D_R has on moduli and on q_c, respectively. In fact, while moduli increase more or less proportionally with the increase of D_R, the value of q_c increases exponentially with an increase of ϕ' which in turn is proportional to the value of D_R.

c. The same ratios also moderately depend on the level of the mean effective stress. This is evidenced in Fig.13 for the M/q_c ratio and also holds true for the E/q_c ratio.

The above mentioned findings result from the CC tests performed on freshly deposited silica sands. Further research is necessary to find out to what extent the findings are applicable to natural and not necessarily silica sand deposits. The writers believe, however, that at least qualitatively similar overall trends should also be expected in natural aged sands.
This is especially true as far as the influence of the stress and strain history is concerned. The confirmation of this can be inferred from the methods which allow computation of the settlements of shallow foundations in sands from the SPT, CPT and DMT results based on observed settlements of full-scale structures. All these methods make a clear distinction between NC and OC sands as far as the relationship existing between the penetration resistance and settlement is concerned.

Regarding the correlations between deformation moduli and SPT resistance, there is nothing substantially new with respect to the works by D'Appolonia et al. (1968), D'Appolonia and D'Appolonia (1970), Parry (1971, 1977, 1978) except the valuable direct approach recently proposed by Burland and Burbridge (1984). This area probably requires a new research effort and a critical revision of the available experimental data, especially as far as the impact of rod energy on the correlations is concerned.

The DMT allows evaluation of M as a function of the dilatometer modulus E_D following Marchetti's (1980) correlation. According to such a correlation, the $M^{DMT}/E_D = R_M = f(K_D, I_D)$, where the formulae allowing the evaluation of R_M can be found in Marchetti (1980) and Marchetti and Crapps (1981). The reliability of this approach is not well established, mainly because of difficulties in obtaining reliable reference values of M in cohesionless soils.
Despite these uncertainties and difficulties, the experimental data available until now (Lacasse, 1986; Schmertmann, 1986; Lacasse and Lunne, 1986) look promising with respect to further field and laboratory validation of the correlation allowing one to evaluate M from DMT. This is especially true if one considers the lack of cost-effectiveness of the alternatives.

Some information concerning the reliability of Marchetti's (1980) correlation for $M^{DMT} = f(E_D, K_D, I_D)$ in sands can be inferred from the results of DMT's performed in the CC on TS and HS. Such results are summarized in Figs.16 and 17. In this case the reference values of the tangent constrained modulus M have been computed for the specific conditions of every CC test (D_R, σ'_h, σ'_v) using the equation given by Baldi et al. (1985): $M = f(D_R, \sigma'_v, OCR)$ which fits results of all available one-dimensional compression tests run in CC. Despite the inherent limitations of these tests due to the fact that they have been performed on freshly deposited sands, the data shown in Figs.16 and 17 deserve the following comments:

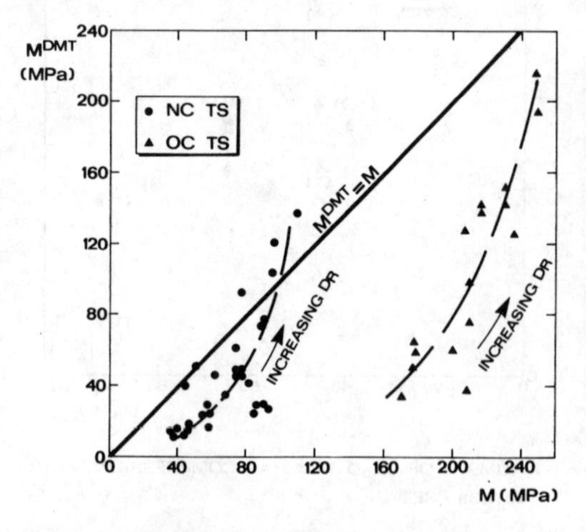

Fig.16 Constrained modulus measured in CC vs DMT modulus

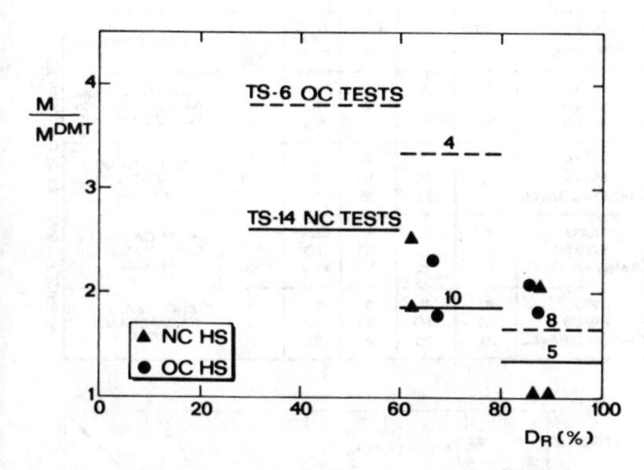

Fig.17 Measured M vs calculated M^{DMT} from CC tests for Ticino and Hokksund sands

a. The ratio of M/M^{DMT} is higher in OC sands than in NC sands. The trend is similar to that observed for the E vs q_c correlations presented in Figs.14 and 15. Yet, the difference between $(M/M^{DMT})_{OC}$ and $(M/M^{DMT})_{NC}$ is smaller than the difference between $(E/q_c)_{OC}$ and $(E/q_c)_{NC}$. This indicates an advantage of the DMT with respect to the CPT by virtue of the fact that consideration in the correlation of an additional parameter (K_D) compensates for, at least partially, the obliterating effect that the dilatometer penetration has on the stress and strain history of the sand.

b. The M/M^{DMT} ratio decreases as D_R decreases, suggesting that for a given sand the existing $M^{DMT} = f(E_D, K_D, I_D)$ correlation is more conservative at low than at high relative densities. The results of the CC tests suggest a need for further improvement of the M^{DMT} vs E_D correlation.

In the meantime, more research is advocated towards the development of correlations between E_D and E in the vertical direction because this latter parameter more intimately relates to the evaluation of settlements of foundations in sands. In this regard the recent paper by Leonards and Frost (1987) presents an interesting attempt to link E_D with E_{25} suggesting that the ratio $E_{25}/E_D = R_{25}$ is equal to 0.7 and 3.5, respectively, in NC and OC soils.

Further improvements in correlations between non-linear deformation moduli of cohesionless soils must take into account the physical significance of the measured dilatometer modulus which emerges from the results shown in Fig.18. This figure shows the results of a test performed using the research dilatometer (Baldi, 1987) which allows measurement of both deflection and internal pressure during expansion of the dilatometer membrane. The obtained experimental data are similar to those shown by Campanella et al. (1985) and allow the following comments.

a. The shape of the pressure versus deflection curve resembles a pressuremeter expansion curve.

b. The slope between A and B proportional to E_D is one order of magnitude smaller than the slopes of the small unloading-reloading loops performed during the dilatometer expansion.

This latter experimental evidence tends to suggest that, even in an OC deposit, the disturbance caused by penetration of the dilatometer blade determines the situation in which the measured E_D corresponds to an "elastic-plastic" behaviour of soil at the current yield locus, while the slope of the unloading-reloading loops reflects the "elastic" stiffness of the material.

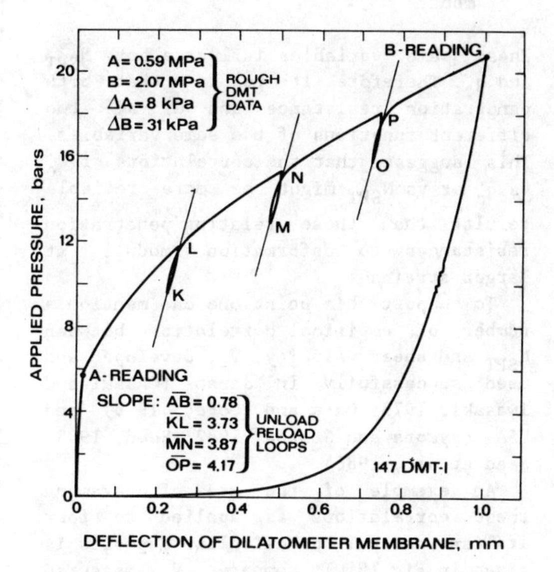

Fig.18 Typical pressure-displacement diagramme from DMT in dry Ticino sand (Baldi, 1987)

The above exposed facts and the related comments lead to a quite negative attitude as far as the reliability of the existing correlations between penetration resistance and non-linear deformation moduli are concerned. A notable exception in this respect is represented by the "maximum" dynamic shear modulus G_o measurable [resonant column test (RCT) and in-situ tests using seismic methods] at a shear strain level $\gamma < 10^{-5}$, see Ohta and Goto (1976, 1978), Sykora and Stokoe (1982), Robertson and Campanella (1982, 1984), Bellotti et al. (1986), Rix (1984) and Lo Presti (1987). The large number of available experimental data show that G_o of cohesionless soils is influenced very little by the stress and strain history of the sand. For a given cohesionless soil this modulus is mostly a function of the following variables [see Yu and Richart (1984), Lee and Stokoe (1986) and Ni and Stokoe (1987)],

$$G_o = f\,(D_R,\ \sigma'_a,\ \sigma'_b) \tag{9}$$

where:

σ'_a = effective stress acting in the direction of seismic wave propagation, and

σ'_b = effective stress acting in the direction of soil particle displacement.

These same variables influence both N_{SPT} and q_c. Therefore it results that both penetration resistance and G_o are two different functions of the same variables. This suggests that the correlations of G_o vs q_c or vs N_{SPT} might be more reliable results than those relating penetration resistances to deformation moduli at larger strains.

To support this point one can mention a number of empirical correlations between N_{SPT} and shear velocity, V_s, developed and used successfully in Japan (Ohsaki and Iwasaki, 1973; Ohta and Goto, 1978) and USA (Sykora and Stokoe, 1982; Seed, 1983; Seed et al., 1986).

An example of the use of one among these correlations as applied to four Italian natural cohesionless deposits is given in Fig.19. It compares V_s measured using the crosshole method against V_s evaluated from N_{SPT}, using the Ohta and Goto (1978) empirical formula adapted by Seed et al. (1986):

SITE BOREHOLE	DEPTH m	FINES %	GC %	D_{50} mm	D_{60}/D_{10}	SOIL TYPES
PO RIVER SAND	5 TO 39	3 TO 28	0	0,1 TO 0,5	2 TO 7	FINE TO MEDIUM SAND WITH THIN LENSES OF SILT
GIOIA TAURO SAND+GRAVEL	5 TO 39	1 TO 17	0 TO 79	0,1 TO 7	2 TO 90	FROM SAND TO SAND AND GRAVEL
DORA RIVER SAND+GRAVEL	8 TO 48	8 TO 22	10 TO 71	0.34 TO 10.5	13 TO 320	FROM GRAVELLY SAND TO SANDY GRAVEL
PO RIVER SAND+GRAVEL	6 TO 30	7 TO 25	0 TO 75	0.2 TO 15	6 TO 100	FROM SAND TO SAND AND GRAVEL

← INCREASING AGE OF DEPOSIT

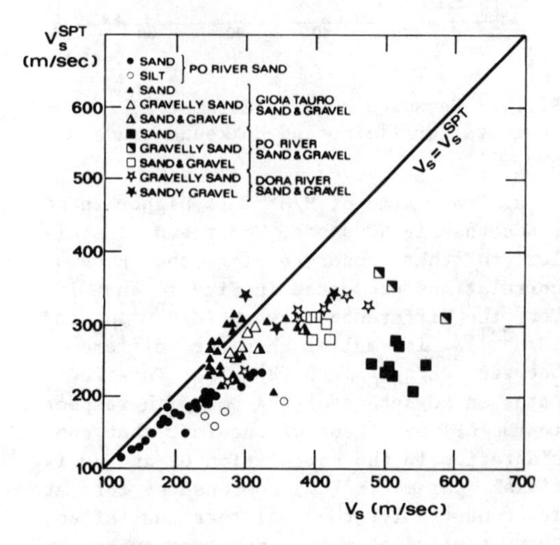

Fig.19 Shear wave velocity from N_{SPT} (Ohta & Goto, 1978)

$$V_s^{SPT} = C\,(N_{60})^{0.17}\,Z^{0.193}\,f_A f_G \text{ (m/sec)} \tag{10}$$

where:

C = empirical constant=53.5,

Z = depth in meters,

f_A = factor depending on the age of deposit, see Table 6, and

f_G = factor depending on the soil grading, see Table 6.

The formula has the advantage of considering at least approximately both the age and the grading of the deposit; it leads, for the four considered deposits, to the average ratios of estimated V_s^{SPT} over measured V_s^{CH} given in Table 7.

The V_s^{SPT}/V_s^{CH} ratio resulting for the four Italian natural cohesionless deposits seems to indicate that the reliability of the empirical correlation expressed by eqn.(10) decreases with increasing age of the deposit and/or with increasing value of N_{60}.

Table 6. Age and Grading Factors in Eqn.(10) - Otha and Goto (1978).

Soil Grading: f_G	Clay	Fine Sand	Medium Sand	Coarse Sand	Sand and Gravel	Gravel
	1	1.09	1.07	1.14	1.15	1.45
Deposit's Age: f_A		Holocene 1.0		Pleistocene 1.30		

Table 7. V_s-Cross-hole vs V_s^{SPT}.

Site	Grading	$\frac{V_s}{V_s^{SPT}}$ + 1SD	Estimated Age of Deposit
River Po V-4017	Fine to medium	0.84±0.06	Holocene
Gioia Tauro	Sand and gravel	0.92±0.11	Holocene
River Dora	Gravel and sand	0.84±0.14	Lower Holocene
River Po, TV	Gravel and sand	0.60±0.14	Holocene-Pleistocene

The writers are not aware to what extent this is due to the range of the SPT resistances for which the correlation has been established or to the inadequate account via f_A for the age and stress history of the deposit.

Analogously, Rix (1984), Baldi et al. (1986) and Bellotti et al. (1986) have shown that q_c can be correlated reliably with G_o. An example of such a correlation as obtained by Baldi et al. (1986) comparing the results of CC tests and RCT's both performed on pluvially deposited specimens of the TS is reported in Fig.20. The same figure shows also the values of G_o/q_c resulting for the Po River sand taking into account G_o obtained from the V_s measured using ISMES's seismic cone (Bruzzi, 1987).

The G_o/q_c ratios of the Po River sand (having the grading and mineralogy similar to those of TS) shown in Fig.20, are for sand containing less than 10% fines, in which CPTU penetration occurred under completely drained conditions.

Preliminary experimental data available indicate that E_D from the DMT can also be correlated against G_o.

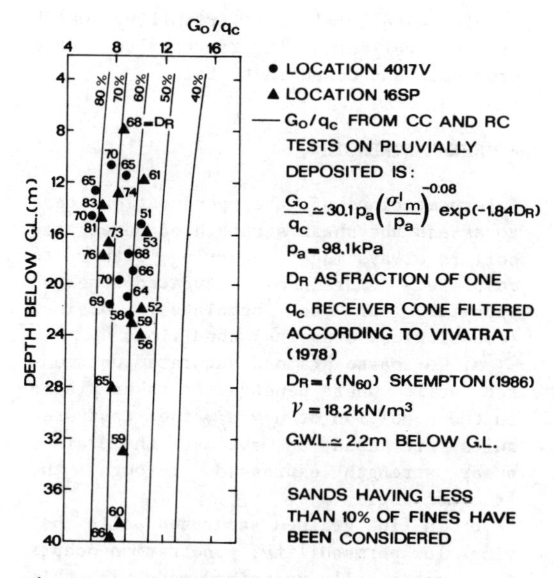

Fig.20 Correlation of G_o vs q_c from seismic cone in Po river sand

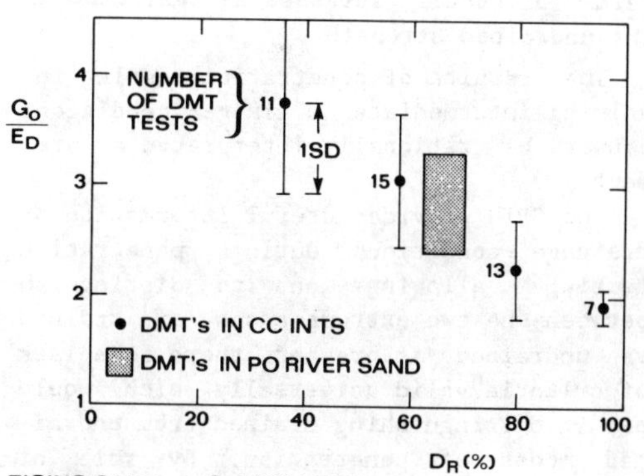

Fig.21 G_o vs E_D for Ticino and Po river sands

In fact, in this case one might take advantage of the additional parameters measured, trying to establish a correlation similar to eqn.(10) in which $f_A = f(K_D)$ and $f_G = f(I_D)$.
As an example, Fig.21 shows the G_o/E_D values obtained in Po River sand. The same figure shows the G_o/E_D ratio for TS as inferred from CC tests and RCT's. Experience gained so far in correlating the G_o-field obtained from seismic methods vs results of the penetration tests indicates that these kinds of relationships might suffer from the fact the penetrometer is very

sensitive to local soil variability which is not reflected in the results of the crosshole and downhole tests.

6 SHEAR STRENGTH

Interpretation of the penetration test to assess the shear strength of penetrated soil is always made referring either to completely undrained (saturated cohesive deposits) or to completely drained (cohesionless deposits) conditions.

a. In coarse grained saturated uncemented soils where penetration takes places in the condition of $u \approx u_o$, the test results are used to evaluate the drained shear strength expressed through the friction angle, ϕ'.

b. In fine grained saturated soils having low permeability, penetration occurs in an essentially undrained mode. In this case the results of penetration tests are used for evaluation of shear strength in terms of total stresses as reflected in the undrained strength c_u.

The results of penetration testing involving intermediate drainage conditions cannot be rationally interpreted at present.

The CPTU provides useful information on drainage conditions during penetration testing, allowing one to distinguish between the two extreme cases of drained vs undrained. At present, there is a lack of criteria valid universally which would permit distinguishing drained from undrained modes of penetration. By rule of thumb, Baligh and Levadoux (1980) and Baligh (1985a) indicate that if 50% of the pore pressure excess, Δu, caused by CPTU penetration requires more than 2 to 5 minutes to dissipate, one can argue that the test was run in virtually undrained conditions. This refers to the 10 cm^2 electric Fugro-type cone penetrating at a rate of 2 cm/sec.
Moreover the comparison of Δu and q_c, obtained during tests run at different penetration rates can give useful indications on drainage conditions at specific sites.

Procedures to evaluate c_u in saturated cohesive deposits from the results of CPTU's and CPT's have been the subject of a large number of theoretical and experimental studies, a review of which is beyond the scope of this paper. Generally speaking, all available approaches can be grouped as either theoretical or empirical. The former refers to bearing capacity theories, based on different failure mechanisms and constitutive soil models. As typical examples one can mention Berezantzev (1967), Vesic (1972) and Baligh (1986) approaches which refer, respectively, to rigid-plastic; linearly-elastic perfectly-plastic and non-linearly elastic perfectly plastic soil behaviour. Among them, the simple pile (SP) solution obtained by Baligh (1985, 1985a, 1986), using the strain path method and assuming non-linear elastic perfectly plastic soil behaviour, takes into account the most factors affecting the undrained penetration resistance and (it) represents the most rational solution available at present. According to this theory, the approximate value of c_u can be evaluated from the following formula:

$$c_u \approx \frac{q_{SP} - \sigma_m}{N_{SP}} \tag{11}$$

where:

- c_u = undrained shear strength of an isotropic clay, corresponding in the first approximation to the undrained shear strength of an anisotropic (=natural) clay obtained from undrained direct simple shear tests;
- q_{SP} = unit simple pile penetration resistance $\approx q_{CT}$;
- σ_m = initial mean total stress at the depth where q_{CT} has been measured;
- N_{SP} = simple pile penetration resistance factor; and
- q_{CT} = total cone resistance corrected for unequal area effect (see Jamiolkowski et al., 1985)

Fig.22 shows the values of N_{SP} as given by Baligh (1985); N_{SP} is a function of the deviatoric yield strain, E_y, and the ratio of G_{max}/G_{min}, whose physical meaning is similar to Vesic's rigidity index I_r *[within the frame of the constitutive model used by Baligh (1986) E_y coincides with E_f at failure]*. In principle, the values of E_y, G_{max} and G_{min} should be established through DSS tests (see Fig.22), but as a first approximation they might also be assessed via CK_oU triaxial compression (TC) tests by means of the following relationships:

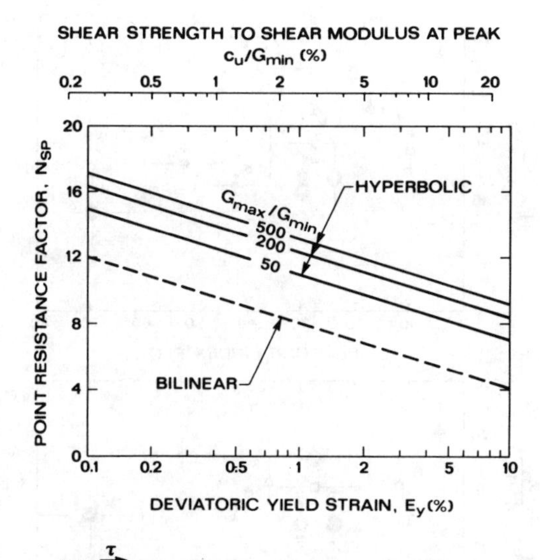

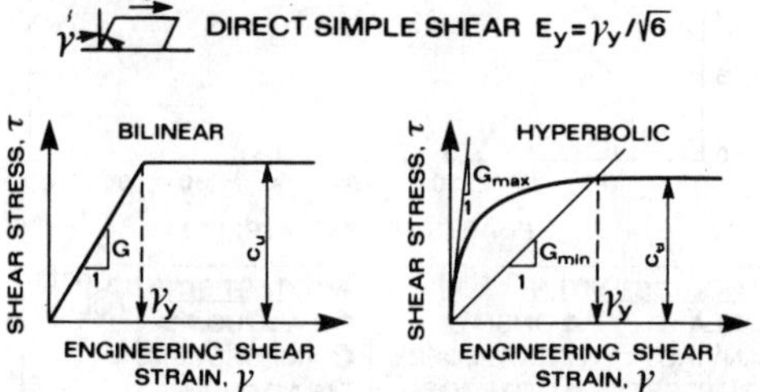

Fig.22 Predicted point resistance of simple pile in clays (Baligh, 1985)

$$E_y = \frac{1}{\sqrt{2}} \epsilon_f \text{ (TC)}; \quad G = \frac{E_u}{3}$$

where:

ϵ_f = axial strain at failure, and

E_u = undrained Young modulus at a given strain level.

Despite its remarkable refinements, the SP solution must still be considered as approximate, because it does not incorporate the factors like anisotropy, strain softening, and strain-rate dependency exhibited by real soils and because it refers to a simplified geometry of the cone tip.

Preliminary validation of this formula is given by Baligh (1985). Further experimental research in this direction is required to check the reliability of eqn.(11) for large variety of natural soils.

The theories of expanding cavities in linear-elastic perfectly plastic soil can be used to predict c_u on the basis of the penetration pore pressure measured during the CPTU (see the lecture by Campanella and Robertson (1988) presented at this symposium).

The complexity of the undrained penetration process and the difficulties in a realistic modelling of soil behaviour lead to the use of empirical approaches. In this case undrained strength is estimated from the following formula:

$$c_u = \frac{q_c - \sigma_{vo}}{N_c} \qquad (12)$$

where:

N_c = non-dimensional empirical cone factor.

Establishment of reliable empirical correlations between q_c and c_u and their correct use in practice is strictly related to adequate consideration of the following aspects.

a. The undrained strength is not a uniquely defined soil property but reflects rather an overall complex soil behaviour (Ladd et al., 1977; Wroth, 1984). This makes it necessary to refer the q_c vs c_u correlations to a well defined and clearly declared strength test. The most reliable and controlled correlations can be obtained basing them on c_u resulting from the field vane (FV) tests, from undrained triaxial compression tests performed on the samples reconsolidated under in-situ effective geostatic stresses (TC-CK_oU) or from DSS-CK_oU tests.

b. The empirical correlations presently available are based on measured q_c. With the wider use of the CPTU, it will be more rational to refer them to q_{CT}. This might be especially important in soft and medium clays.

c. The use of q_c to obtain c_u should be limited to non-fissured clays without highly developed macrofabric. Ignoring this limitation might easily lead to an overestimate of the field undrained strength (Marsland and Powell, 1979). At present, the more reliable correlations have been validated for soft to stiff intact relatively homogeneous clays (Lunne et al., 1976; Lacasse et al., 1978; Kjekstad et al., 1978; Baligh et al., 1980; Lunne and Kleven, 1981; Jamiolkowski et al., 1982; Keaveny, 1985; Aas et al., 1986; Keaveny and Mitchell, 1986).

d. The data for soft and medium to stiff clays have been mainly collected at MIT and NGI using the electric Fugro-type cone and referring to c_u (FV). This leads to the values of N_c summarized in the upper part of Fig.23. The lower part of Fig.23 shows N_c values obtained after correction of the c_u (FV) for anisotropy and strain rate effects as recommended by Bjerrum (1973), Ladd et al. (1977). This leads to the average value of $\bar{N}_c = 14 \pm 5$ which according to Lacasse et al. (1978) allows to assess the <u>operational</u> undrained strength.

e. Assuming that "a la SHANSEP" (Ladd and Foott, 1974) normalized behaviour holds, Marchetti (1980) suggested the procedure for evaluating the ratio of c_u/σ'_{vo} as function of OCR^{DMT} as:

$$\left(\frac{c_u}{\sigma'_{vo}}\right)_{OC} = \left(\frac{c_u}{\sigma'_{vo}}\right)_{NC} \cdot (OCR)^{0.8} \approx$$

$$\approx (0.23 \pm 0.04)(0.5K_D)^{1.25} \qquad (13)$$

This formula implies that the average operational c_u along a potential failure plane of a NC clay is $(0.23 \pm 0.004)\sigma'_{vo}$ (see Mesri, 1975; Larsson, 1980; Jamiolkowski et al., 1985). The reliability of this formula is directly related to the reliability of the assessed value of the $K_o = f(K_D)$ and the postulated links between K_o and c_u/σ'_{vo} through OCR.

f. The existing experience with the use of eqn.(13) in soft to stiff uncemented nonfissured clays is quite positive. The paper by Lacasse and Lunne presented to this Symposium reports an interesting set of comparisons between c_u^{DMT} and the results of the FV, DSS-CK_oU and TC-CK_oU tests as obtained for seven different clay deposits. One can deduce that the values of c_u^{DMT} corresponding to the lower limit of eqn.(13) matches satisfactorily with the TC and the corrected FV strengths, while it overpredicts the c_u resulting from the DSS-CK_oU. At present, the use of eqn.(13) in highly OC cemented and/or fissured clays is more uncertain. The experimental data shown in Tables 1 and 2 indicate that for these types of clays further calibration of the DMT correlation is required.

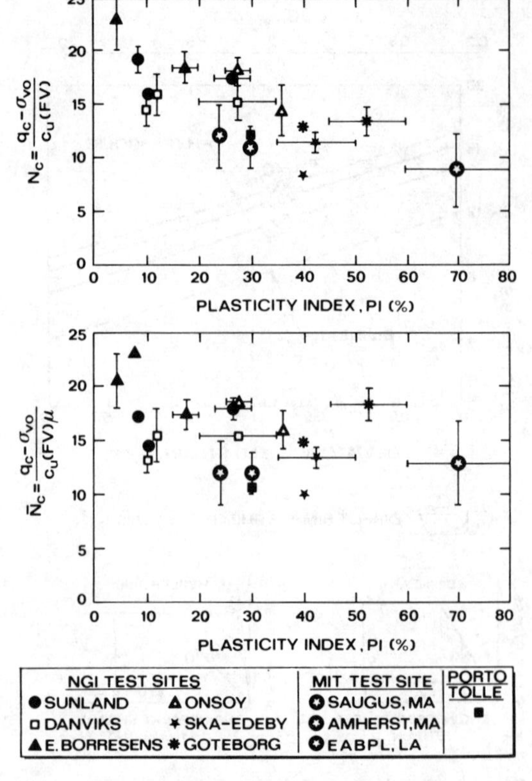

Fig.23 Empirical cone factors (Adopted from Baligh et al., 1980)

The shear strength of uncemented <u>cohesionless</u> soils is usually related to the mobilized angle of friction, ϕ', expressed in terms of effective stress. One of the most relevant aspects of the behaviour of granular materials at failure is their curvilinear strength envelope (De Beer, 1965; Vesic and Clough, 1968; Yaroshenko, 1967; Berezantzev, 1967; 1970). This well documented experimental fact induced Baligh (1975) to formulate the following strength criterion:

$$\tau_{ff} = \sigma'_{ff}\left[\tan\phi'_o + \tan\alpha\left(\frac{1}{2.3} - \log_{10}\frac{\sigma'_{ff}}{p_a}\right)\right] \qquad (14)$$

where:

τ_{ff} = shear stress on the failure plane at failure;

σ'_{ff} = effective normal stress on the failure plane at failure;

p_a = reference stress, assumed equal to 1 bar = 98.1 kPa;

Table 8. ϕ' from q_c - Available Approaches.

Examples of available approaches	Stress-strain Relationship	Curvilinear Strength Envelope	Compressibility	Progressive Failure	Relevant Stress Tensor	Other Data Required
Schmertmann (1978)	No Assumption-Empirical	no	no	no	none	D_R, GSD
Been et al. (1986)	No Assumption-Empirical	$\phi'=\phi'_s$	no	no	σ'_m	$e'_o, e_{ss}, \lambda_{ss}$ K_o
Durgunoglu & Mitchell (1973)	Rigid-Plastic	$\phi'=\phi'_s$	no	no	σ'_{vo}	K_o
Bolton (1984, 1987)	Stress-Dilatancy	$\phi'=\phi'_s$	no	no	p'_f	ϕ'_{cv}, K_o
Vesic (1972)	Elastic-P.Plastic	$\phi'=\phi'_s$	yes	no	σ'_m	K_o, G, ϵ_v
Baligh (1975)	Elastic-P.Plastic	yes	yes	no	σ'_{ho}	K_o, G, ϵ_v

Note: none of the approaches take into account the anisotropy of ϕ', see Tatsuoka (1987)

ϕ'_o = secant angle of friction at σ'_{ff}=2.72 p_a; and

α = angle which describes the curvature of the failure envelope.

As shown by Baldi et al. (1986) in silica sands, α increases with increasing D_R. In the first approximation, the variation of α with D_R can be matched by means of the following empirical relation:

$$\alpha \approx [(D_R-0.2)/0.8] \cdot 10° \geq 0° \quad (15)$$

Because of the non-linearity of the strength envelope, the angle ϕ' of a given sand is not uniquely defined but it depends on the magnitude of σ'_{ff}. Therefore any value of ϕ' inferred on the basis of the SPT, CPT and DMT results corresponds to a secant angle of friction, whose magnitude is controlled by the average value of σ'_{ff} acting on the failure plane around the penetration device. The estimate of this value of σ'_{ff} is very difficult. At present, it can only be determined in a very approximate manner following the indications reported by De Beer, 1965; Schmertmann, 1982 and Bellotti et al., 1983). Knowledge of σ'_{ff} is however essential to link the ϕ' value inferred from the penetration tests with ϕ' value of a specific design problem (Baldi et al., 1986). The existing experience in assessing ϕ' from the results of penetration tests can be summarized as follows (see also Table 8).

a. A number of empirical correlations exist like:

- ϕ' vs N_{SPT} correlation (De Mello, 1971);
- ϕ' vs D_R relationship taking into account soil-grading (Schmertmann, 1978), where D_R can be assessed both from q_c and N_{SPT} values;
- ϕ' vs ψ correlation, suggested recently by Been et al. (1986).

b. There are procedures based on the bearing capacity theories of a rigid plastic body (Durgunoglu and Mitchell, 1973; Berezantzev, 1967; 1970). These methods, through measured q_c and q_D values, allow determination of the secant ϕ' corresponding to triaxial and plane strain conditions, respectively. To validate these methods against the results of laboratory tests, it is necessary to estimate the average value of σ'_{ff} around the penetrometer, which usually involves the assumption of $\sigma'_{ff} \approx (1+\sin\phi')\sigma'_{vo}$ (Schmertmann, 1982; Mitchell, 1984) and which therefore involves an iterative process when computing ϕ'. With the above mentioned assumption, the experience so far gained in using the Durgunoglu and Mitchell (1973) theory (Schmertmann, 1982; Bellotti et al., 1983; Keaveny, 1985; Mitchell and Keaveny, 1986; Baldi et al., 1986) indicates that this method leads for silica sands to ϕ' values which are, on average, 1° to 2° lower than peak ϕ' resulting from TC tests. Fig.24 shows an example of the evaluation of ϕ' on the basis of q_D from DMT's performed in the CC in TS using the Durgunoglu and Mitchell (1973) theory. In this case the push force F_D was measured by means of a load cell located just above the dilatometer blade. To infer the net value of the dilatometer wedge resistance q_D, it was necessary to subtract from F_D the force absorbed by the friction on the blade. This was attempted following two distinct methods indicated in Fig.24, both leading to almost the same value of q_D.

The plane strain values of $\phi'_{PS} = f\ (q_D)$ are compared with the ϕ'_{PS} of TS obtained from laboratory tests; this latter was obtained by transforming triaxial ϕ'_{TX} to ϕ'_{PS} on the basis of Lee and Lade's (1976) empirical relation.

$$\left.\begin{array}{l}\bullet\ \tau_s = f_s^{CPT} \\ *\ \tau_s = p_o^{DMT}\tan\delta\end{array}\right\}\ q_D = \frac{F_D - \tau_s\left(300 + \frac{40}{3}\right)}{12.8}$$

φ'_{PS} EVALUATED FROM TC TESTS
$\varphi'_{PS} \simeq 1.5\cdot\varphi'_{TX} - 17°$ FOR $\varphi'_{TX} \geq 34°$

F_D = PUSH FORCE MEASURED JUST ABOVE DILATOMETER BLADE

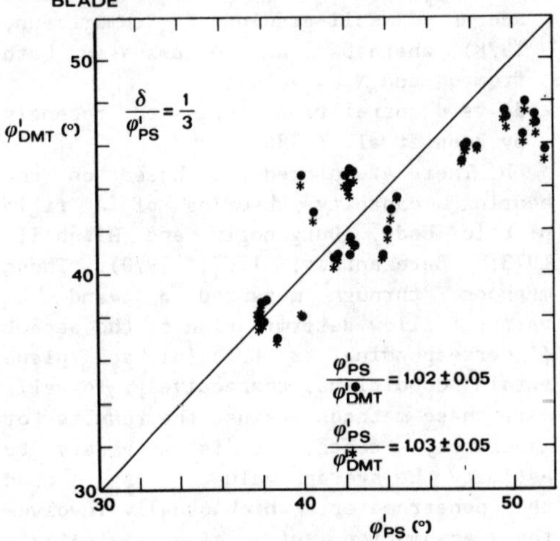

Fig.24 ϕ'_{PS} of Ticino sand from DMT using Durgunoglu & Mitchell's theory

The results show that the Durgunoglu and Mitchell (1973) theory used in conjunction with q_D sligthly underpredicts the value of ϕ'_{PS}. The difference $[\phi'_{PS}(lab)-\phi'_{PS}(q_D)]$ increases with increasing sand density. This probably testifies to the increasing importance of crushing and progressive failure phenomena around the penetrating blade.

c. The rigid-plastic bearing capacity theories are unable to account for the influence of the soil deformability on the ultimate collapse load. Therefore, they fail to predict q_c and q_D of sands with the exception of the uncrushable or moderately crushable silica sands. To predict more reliably q_c and q_D in more compressible materials, it is necessary to refer to the bearing capacity formulae based on the theory of expanding cavities in a linear elastic perfectly plastic medium (Vesic, 1972; Baligh, 1975). This approach however requires, in addition to the penetration resistance, at least an approximate knowledge of additional soil parameters like K_o, volumetric strain and G all concurring to the assessment of the reduced rigidity index I_{rr} (Vesic, 1972; Keaveny, 1985) which renders the use of these theories in practice difficult.

Recent comprehensive reviews of the procedures used in the evaluation of ϕ' from q_c (Keaveny, 1975; Mitchell and Keaveny, 1986) show the following.

- In case of the compressible and crushable sands, only the theories of expanding cavities can correctly predict ϕ' from q_c.
- To obtain a good agreement between $\phi'(q_c)$ and ϕ'(Lab), the Authors suggest the use of spherical cavity theory when $I_{rr} < 250$ and cylindrical cavity theory when $I_{rr} \geq 250$.
- Based on the laboratory results (Miura & Toki, 1984, 1984a), Keavany (1985) postulated that the volumetric strain measured during the triaxial extension test is more pertinent to the problems of expanding cavities than volumetric strain from the triaxial compression test. Therefore, the results of extension tests should concur with the assessment of I_{rr}.

Fig.25 reports the values of ϕ' for five

NC and OC sands having different compressibilities as predicted from q_c using Vesic's (1972) theory.

The agreement is excellent, and especially remarkable is the capability of the modified Vesic (1972) theory to predict ϕ' of the highly compressible Chattahoochee River sand containing 10% mica.

d. The shear strength of cohesionless soils is related to the rate of dilation which in turn depends on their relative density, level of mean effective stress and compressibility.

The above factors are reflected in Rowe's (1962) stress-dilatancy theory which has recently received a simple but conceptually sound formulation by Bolton (1984), (1986). This formulation represents an excellent tool in attempting to assess ϕ' from q_c or q_D.

e. The main features of Bolton's (1984) stress-dilatancy theory may be summarized as follows.

During shearing at large strains, the sand reaches the critical state. At this stage further shearing deformations occur at constant volume and the shear strength of the material is controlled by the critical value of the angle of friction, ϕ'_{cv}. It varies roughly between 33° (quartz) and 40° (feldspar) and depends principally on the mineralogical composition of the grains. Factors like the grading, shape of the grains, mean stress level and direction of shearing are of minor importance.

Before reaching the critical state condition, the difference between the mobilized angle of friction ϕ' and ϕ'_{cv} is controlled by the rate of dilation $(-d\epsilon_v/d\epsilon_1)$, where ϵ_v and ϵ_1 are, respectively, the volumetric and major principal strains.

In particular, the peak value of friction angle, ϕ'_{max}, is related to the maximum dilation rate, and according to Bolton (1984) the difference $(\phi'_{max}-\phi'_{cv})$ in triaxial and plane strain tests can be approximated, respectively, by means of the formulae:

Triaxial test: $\phi'_{max} - \phi'_{cv} = 3\ DI$

Plane strain test: $\phi'_{max} - \phi'_{cv} = 5\ DI$

where: DI = dilatancy index.

The value of DI is empirically related to the D_R and mean effective stress at failure p'_f through the following relationship:

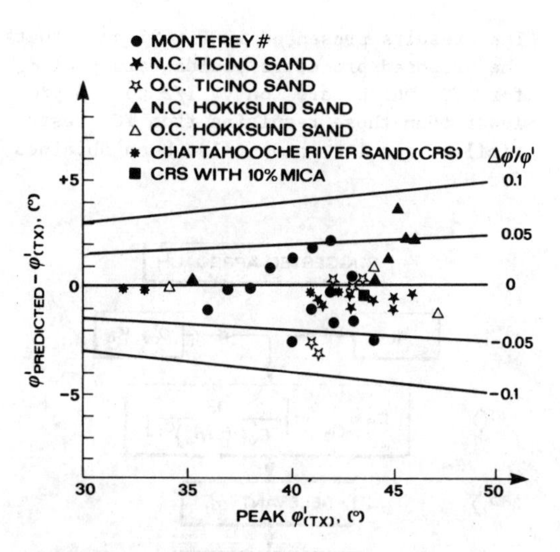

Fig.25 Difference between $\phi'_{(TX)}$ and ϕ' predicted using modified Vesic (1972) theory (Keaveny, 1985)

$$DI = D_R\ (Q - \ln p_f) - 1 \qquad (16)$$

where:

Q = factor depending on the mineralogy and shape of the soil particles.

According to Bolton (1984), Q decreases with increasing compressibility of the soil. For typical silica sands, a value of Q=10 is suggested.

f. In an attempt to assess ϕ'_{max} from q_c, the approach presented in Fig.26 is suggested. The use of this approach requires knowledge of K_o, ϕ'_{cv} and an assumption about the value of p'_f. Fig.27 reports the validation of the proposed approach for TS; the ϕ'_{max} from TC tests are compared with ϕ'_{max} obtained from q_c measured in CC tests. Because of the non-linearity of the strength envelope, the above mentioned validation has been referenced to σ'_{ff}=2.72 p_a(=267 kPa); this leads to the ϕ'_o as defined in Baligh's (1975) curvilinear strength envelope. The reciprocal relation between σ'_{ff} and p'_f being given by the following formula:

$$p'_f = \sigma'_{ff}\left[1 + \tan^2\phi' - \frac{\sin\phi'}{3\cos^2\phi'}\right] \qquad (17)$$

which for the range of ϕ'_{max} (34° to 46°) of practical interest leads to:

$$1.2 \le \frac{p'_f}{\sigma'_{ff}} \le 1.6$$

The results presented in Fig.27 show that the proposed procedure yields values of ϕ'_o for TS which are only 1/2 to 1 degree lower than those resulting from TC tests. Similar results have also been obtained for HS.

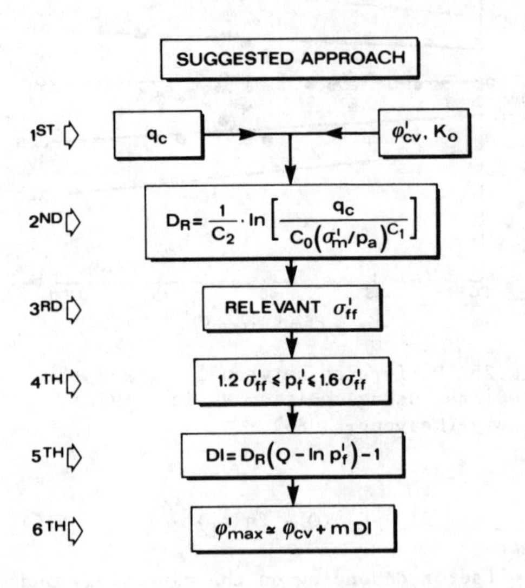

SUGGESTED APPROACH
1ST
2ND
3RD
4TH
5TH
6TH
q_c
φ'_{cv}, K_o
$D_R = \frac{1}{C_2} \cdot \ln\left[\frac{q_c}{C_0(\sigma'_m/p_a)^{C_1}}\right]$
RELEVANT σ'_{ff}
$1.2\,\sigma'_{ff} \leq p'_f \leq 1.6\,\sigma'_{ff}$
$DI = D_R(Q - \ln p'_f) - 1$
$\varphi'_{max} \cong \varphi_{cv} + m\,DI$
TRIAXIAL TEST: m=3; PLANE STRAIN TEST: m=5
p_a= REFERENCE STRESS = 98.1kPa

Fig.26 $\phi'_{max} = f(q_c)$ of sand from Bolton's (1986) stress dilatancy theory

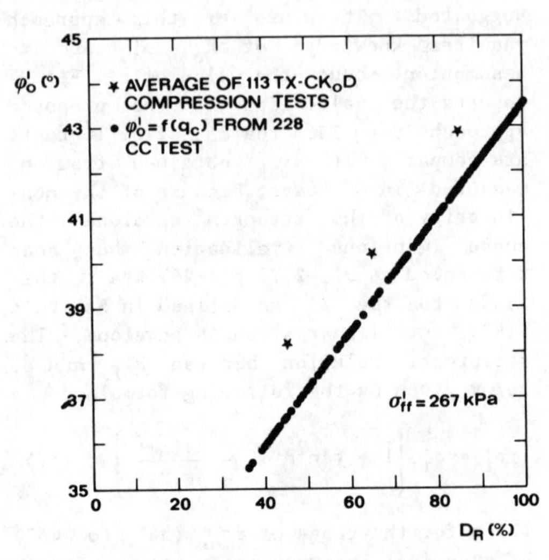

Fig.27 $\phi'_o = f(q_c)$ for NC and OC Ticino sand using Bolton's stress dilatancy theory (1986)

g. For use in practice, Fig.28 reports the relation between ϕ'_o and q_c/σ'_{vo} for values of K_o of interest. Fortunately, considering the difficulties in the estimate of K_o in cohesionless soils, Fig.28 indicates that the estimate of ϕ'_o is not too strongly influenced by the assumption concerning this parameter.

This figure has been obtained by adopting ϕ'_{cv}=34° which is that of TS, while the ϕ'_{cv} of HS is equal to 33°. The average value of ϕ'_{cv} of 23 different silica sands, including the two mentioned above, is equal to 33.6° ± 2.5° (see also Hanna and Youssef, 1987).

Once the procedure based on Bolton's (1984) stress-dilatancy theory outlined in Fig.26 is accepted and validated, it can be used to obtain the value of ϕ'_{max} at any value of p'_f as a function of σ'_{ff} which is of interest for design. An alternative approach might consist in the evaluation of ϕ'_o from Fig.28 combined with an assumption of $\alpha = f(D_R)$ appearing in Baligh's strength envelope, using thereafter eqn.(14) to obtain secant ϕ' at any desired level of σ'_{ff}.

Additional information and details concerning Bolton's (1984), (1986) stress-dilatancy theory can be found in Tatsouka (1987), Bolton (1987), Tatsuoka and Pradham (1988), Hanna and Youseff (1987) and Hanna et al. (1987).

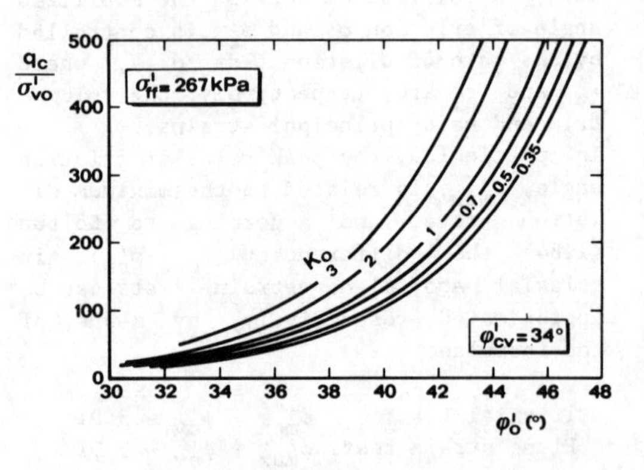

Fig.28 Angle of shearing resistance ϕ'_o using Bolton's stress dilatancy theory (1986)

7 CONCLUSIONS

This paper presents a critical review of new possibilities offered by penetration testing as an important aid in geotechnical design. The topic covered by the title of this lecture is extremely broad. The Authors have, however, chosen to limit the topic to addressing:

- The indirect approach, which permits, evaluation of the basic design parameters (initial state variables, stress-strain- strength and consolidation characteristics) from the results of penetration testing.
- Discussion of the three most widely used penetration devices in practice, namely the SPT, CPT and DMT.

The following conclusions can be drawn from this critical examination.

1. Initial State Parameters

a. The existing correlations between D_R and penetration resistances (N_{SPT}, q_c) are applicable only to NC unaged silica sands. Their use in other sands might cause either overestimation (in the case of aged and/or OC deposits) or underestimation (in the case of more compressible sands) of the in-situ density.

b. The use of Marchetti's Flat Dilatometer to infer the stress history of saturated cohesive deposits looks very promising. Existing experience indicates that K_D profiles in clays are similar in shape to the OCR profiles established through oedometer tests performed on high-quality undisturbed samples. Because this correlation holds in clay deposits which have been subjected to different σ'_p mechanisms, values of σ'_p resulting from the DMT should be considered as a vertical yield stress along the K_o stress path.

c. The assessment of σ'_{ho} and/or K_o in sands from penetration test results is in a very early stage of development. The recently postulated ideas that σ'_h or K_o might be inferred through correlations between amplification factor (K_D/K_o or more generally $\sigma'_{hp}/\sigma'_{ho}$) and state parameter ψ requires further experimental and theoretical validations.

2. Deformation Characteristics

a. The main area of interest in the use of penetration tests is to assess the deformation characteristics of cohesionless deposits. Existing experience is mainly limited to uncrushable and moderately crushable silica materials.

b. Because insertion of the penetrometer obliterates the stress and strain history of the deposit, no unique correlation between penetration resistance and nonlinear or elasto-plastic soil modulus can exist. To evaluate the magnitude of medium-to-large strain deformation moduli of cohesionless soils on the basis of penetration tests, one must estimate at least qualitatively the stress history of the deposit.

c. Experience gained so far indicates that the results of penetration tests can be correlated in a quite reliable manner to the linear-elastic maximum shear modulus G_o.

3. Shear Strength

a. The undrained shear strength of soft-to-stiff saturated intact cohesive deposits can be inferred in a quite reliable manner from CPT and DMT results. The use of the above mentioned correlations in hard clays with highly developed macro-fabric might be questionable.
When evaluating c_u from penetration tests, it is always necessary to keep in mind the reference strength to which the specific empirical or theoretical procedure is referred.

b. The cone resistance q_c offers a possibility to assess ϕ' of freely draining cohesionless deposits. In silica sands the use of the rigid plastic bearing capacity theory by Durgunoglu and Mitchell (1973) yields satisfactory results which are generally sligthly on the safe side. An estimate of ϕ' from q_c of more compressible (calcareous or micaceous) sands requires theories of expanding cavities to take into account the influence of plastic volumetric strain on the measured q_c.
The recently presented stress dilatancy theory by Bolton (1984, 1986) offers an excellent tool for assessing ϕ' from q_c.

Further validation of this theory, especially in more compressible sands, will be extremely welcome.

c. When assessing ϕ' from q_c using all of the above mentioned procedures, due attention should be paid to the fact that the strength envelope of most natural sands is curvilinear. This means that the inferred ϕ' corresponds to the secant angle of friction which depends on the average σ'_{ff} existing around the penetration device. To link with a specific design problem, one must take this stress (σ'_{ff}) dependency into account.

ACKNOWLEDGEMENTS

The writers acknowledge the contribution of Dr. G. Baldi (ISMES), Dr. R. Bellotti (ENEL CRIS) and Dr. D. Bruzzi (ISMES) whose fruitful discussions and informations on the results of their recent research have greatly contributed to the present work. Most of the experimental data presented in the paper represent the outcome of joint research among ENELCRIS of Milan, ISMES of Bergamo and the Technical University of Turin.

The writers wish to thank Dr.K. Stokoe (U. of Texas) for his critical review of the manuscript. Also, the careful typing of C. Ramella and drafting of V. Manerba and S. Bacci are sincerely appreciated.

REFERENCES

Aas, G., Lacasse, S., Lunne, T. and Hoeg, K. 1986. Use of In-Situ Tests for Foundation on Clay. In-Situ '86, Proc. Spec. Conf. GED ASCE, Virginia Tech., Blacksburg.

Atkinson, J.H. & Bransby, P.L. 1978. The Mechanics of Soils - An Introduction to Critical State Soil Mechanics. Mc Gra-Hill Inc. London.

Baldi G. 1987. Personal Communication to M. Jamiolkowski.

Baldi, G. et al. 1983. Prova Penetrometrica Statica e Densità Relativa della Sabbia. XV CIG Spoleto, Italy.

Baldi, G. et al. 1985. Laboratory Validation of In-Situ Tests. AGI Jubilee Volume, XI ICSMFE, San Francisco.

Baldi G. et al. 1986. Interpretation of CPT's and CPTU's. II Part: Drained Penetration on Sands. Proc. IV Int. Geotech. Seminar on Field Instrumentation and In Situ Measurements, Nanyang Tech. Inst., Singapore.

Baldi G. et al. 1986a. Flat Dilatometer Tests in Calibration Chambers. In-Situ '86. Proc. Spec. Conf. GED ASCE, Virginia Tech., Blacksburg.

Baldi G. et al. 1988. Seismic Cone in Po River Sand. Proc. ISOPT-I Orlando, Fla.

Baligh, M.M. 1975. Theory of Deep Site Static Cone Penetration Resistance. Res. Report R75-56, No.517, Dept. of Civil Eng., MIT, Cambridge, Mass.

Baligh, M.M. 1985. The Strain Path Method. JGED. ASCE GT9.

Baligh, M. 1985a. Interpretation of Piezocone Measurements during Penetration. MIT Special Summer Course: Recent Developments in Measurement and Modelling of Clay Behaviour for Foundation Design. Cambridge, Mass.

Baligh, M. 1986. Undrained Deep Penetration, Geotechnique, No.4.

Baligh, M.M. & Levadoux, J.N. 1980. Pore Pressure Dissipation After Cone Penetration. Research Report R.80-11, MIT, Cambridge, Mass.

Baligh, M.M., Vivatrat, V. & Ladd, C.C. 1980. Cone Penetration in Soil Profiling. JGED, ASCE, GT4.

Baligh, M.M. et al. 1981. The Piezocone Penetrometer. Proc. Symposium on Cone Penetration Testing and Experience, ASCE, National Convention, St. Louis, Missouri.

Baligh, M.M., Martin, T.R., Azzouz, A.S. & Morison, M.J. 1985. The Piezo-Lateral Stress Cell. Proc. XI ICSMFE, San Francisco.

Battaglio, M. et al. 1986. Interpretation of CPT's and CPTU's. Ist Part: Undrained Penetration of Clays. Proc. IV Int. Geotech. Seminar on Field Instrumentation and in Situ Measurements, Nanyang Tech. Inst., Singapore.

Bazaraa, A.R.S. 1967. Use of the Standard Penetration Test for Estimating Settlement of Shallow Foundations on Sand. Ph. D. Thesis, University of Illinois, Urbana, USA.

Been, K., & Jefferies, M.G. 1985. A State Parameter for Sands. Geotechnique, No.2.

Been, K., Crooks, J.H.A., Becker, D.A. & Jefferies, M.G. 1986. The Cone Penetration Test in Sands: Part I, State Para-

meter and Interpretation. Geotechnique, No.2.
Bellotti, R., Bizzi, G., Ghionna, V. 1982. Design, Construction and Use of a Calibration Chamber. Proc. ESOPT II, Amsterdam.
Bellotti, R. et al. 1983. Evaluation of Sand Strength from CPT. International Symposium on Soil and Rock Investigations by In-Situ Testing, Paris.
Bellotti, R. et al. 1986. Deformation Characteristics of Cohesionless Soils from In-Situ Tests. In-Situ '86. Proc. Spec. Conf. GED ASCE, Virginia Tech., Blacksburg.
Berezantzev, W.G. 1967. Certain Results of Investigation of the Shear Strength of Sands. Proc. of the Geotechnical Conf. on Shear Properties of Natural Soils and Rocks, Oslo.
Berezantzev, W.G. 1970. Calculation of the Construction Basis. Leningrad (in Russian).
Bjerrum, L. 1973. Problems of Soil Mechanics and Construction on Soft Clays. SOA Report, Session 4, Proc. VIII ICSMFE, Moscow.
Bolton, M.D. 1984. The Strength and Dilatancy of Sands. University of Cambridge, CUED/D-Soils/TR 152.
Bolton, M.C. 1986. The Strength and Dilatancy of Sands. Geotechnique No.1.
Bolton, M.C. 1987. The Strength and Dilatancy of Sands. Author's reply in Discussion. Geotechnique No.2.
Bruzzi, D. 1987. Personal Communication to M. Jamiolkowski.
Bruzzi D. et al. 1986. Self-Boring Pressuremeter Tests in Po River Sand. The Pressuremeter and its Marine Applications, II Int. Symp., ASTM STP 950, Texas A&M University.
Burland, J.B. & Burbidge, M.C. 1984. Settlement of Foundations on Sand and Gravel. Glasgow and West of Scotland Association Centenary Lecture.
Bustamante, M. & Gianeselli, L. 1982. Pile Bearing Capacity Prediction by Means of Static Penetrometer CPT. Proc. ESOPT II, Amsterdam.
Campanella, R.G. & Robertson, P.K. 1981. Applied Cone Research. Univ. of British Columbia, Vancouver, Canada. Soil Mechanics Series No.46.
Campanella, R.G. & Robertson, P.K. 1983. Flat Plate Dilatometer Testing: Research and Development. Univ. of British Columbia, Vancouver, Canada. Soil Mechanics Series No.68.
Campanella, R.G. & Robertson, P.K. 1984. A Seismic Cone Penetrometer to Measure Engineering Properties of Soil. LIV Annual International Meeting and Exposition of the Society of Exploration Geophysicists, Atlanta, Georgia.
Campanella, R.G., Robertson, P.K., Gillespie, D.G. & Grieg, J. 1985. Recent Developments in In-Situ Testing of Soils. Proc. XI ICSMFE, San Francisco.
Campanella, R.G. et al. 1986. Seismic Cone Penetration Test. In Situ '86. Proc. Spec. Conf. GED, ASCE, Virginia Tech., Blacksburg.
Campanella, R.G. & Robertson, P.K. 1988. Current Status of the Piezocone Test. ISOPT I, Orlando, Fla.
Casagrande, A. 1936. The Determination of the Pre-Consolidation Load and Its Practical Significance. Proc.I ICSMFE, Cambridge, Mass.
Clayton, C.R.I., Hababa, M.B. & Simons, N.E. 1985. Dynamic Penetration Resistance and the Prediction of the Compressibility of a Fine-Grained Sand-a Laboratory Study. Géotechnique No.1.
Clayton, C.R.I., Hababa, M.B. & Simons, N.E. 1986. Author's reply in Discussion. Geotecnique No.2.
D'Appolonia, D.J., D'Appolonia, E., Brissette, R.F. 1968. Settlement of Spread Footings on Sand. JSMFE Div.ASCE, SM3.
D'Appolonia, D.J., D'Appolonia, E. 1970. Use of the SPT to estimate settlement of footing on Sand. Proc. Symp. Foundations on Interbedded Sands, Perth, Australia.
D'Appolonia, D.J., D'Appolonia, E., Brissette, R.F. 1970. Closure to Settlement of Spread Footings on Sand. JSMFE Div.ASCE, SM4.
De Beer, E. 1948. Settlements Records of Bridges Founded on Sand. Proc.II ICSMFE, Rotterdam.
De Beer, E.E. 1965. Influence of the Mean Normal Stress on the Shear Strength of Sand. Proc. VI ICSMFE, Montreal.
De Mello, C.F.B. 1971. The Standard Penetration Test. SOA, Report IV Panamerican Conf. on SMFE, San Juan, Puerto Rico.
De Ruiter, J. 1971. Electric Penetrometer for Site Investigation. JSMFE Div.ASCE, SM2.
De Ruiter, J. 1981. Current Penetrometer Practice. Proc. Symposium on Cone Penetration Testing and Experience, ASCE

National Convention, St. Louis, Missouri.

De Ruiter, J. 1982. The Static Cone Penetration Test. SOA Report, Proc. ESOPT II, Amsterdam.

Durgunoglu, H.T. & Mitchell, J.K. 1973. Static Penetration Resistance of Soils. Research Report Prepared for NASA Headquarters, Washington, D.C., Univ. of California, Berkeley.

Eid, W.K. 1987. Scaling Effect in Cone Penetration Testing in Sands. Ph.D. Thesis in Civil Engineering, Virginia Polytecnic Institute and State University.

Gibbs, H.J. & Holtz, W.G. 1957. Research on Determining the Density of Sands by Spoon Penetration Testing. Proc. IV ICSMFE, London.

Handy et al. 1982. In Situ-Test Determination by Iowa Stepped Blade. JGE Div., ASCE GT11.

Hanna, A.M. & Youssef, F.H. 1987. Evaluation of Dilatancy Theories of Granular Materials. Proc. Conf. on Prediction and Performance in Geotechnical Engineering, Calgary.

Hanna, A.M., Massoud, N. & Youssef, H. 1987. Prediction of Plane-Strain Angles of Shear Resistance from Triaxial Test Results. Proc. Conf. on Prediction and Performance in Geotechnical Engineering, Calgary.

Harder, L.F. & Seed, H.B. 1986. Determination of Penetration Resistance for Coarse-Grained Soils Using the Becker Hammer Drill. UCB/EERC-86/06, Univ. California, Berkeley.

Harman, D.E. 1976. A Statistical Study of Static Cone Bearing Capacity, Vertical Effective Stress, and Relative Density of Dry and Saturated Fine Sands in a Large, Triaxial Test Chamber. M.Sc. Thesis, University of Florida.

Hird, C.C. & Hassana, F. 1986. Discussion on "Been K. & Jefferies M.G. 1985". Geotecnique No.1.

Holden, J. 1971. Research on the Performance of Soil Penetrometers. Churchill Fellowship 1970, Country Road Board of Victoria.

Hughes, J.M.O, & Robertson, P.K. 1985. Full Displacement Pressuremeters Testing in Sands. Canadian Geotechnical Journal, No.3.

Huntsman, S.R. 1985. Determination of In-Situ Lateral Pressure of Cohesionless Soils by Static Cone Penetrometer, Ph.D. Thesis, Univ. of California, Berkeley.

Huntsman, S.R., Mitchell, J.K., Klejbuk, L.W., Shinde, S.B. 1986. Lateral Stress Measurement During Cone Penetration. In Situ '86, Proc. Spec. Conf. GED ASCE, Virginia Tech., Blacksburg.

Jamiolkowski, M., Lancellotta, R., Tordella, L. & Battaglio, M. 1982. Undrained Strength from CPT. Proc. ESOPT II, Amsterdam.

Jamiolkowski, M., Ladd, C.C., Germaine, J.T., Lancellotta, R. 1985. New Developments in Field and Laboratory Testing of Soils. Theme Lecture, Proc.XI ICSMFE, San Francisco.

Jefferies, N.G. & Jönsson, L. 1986. The Cone Penetration Test in Sands: Part 3 Horizontal Geostatic Stress Measurements during Cone Penetration. Draft, submitted to Geotechnique for pubblication.

Jezequel, J.F., Lamy, J.L. & Perrier, M. 1982. The LPC-TLM Pressio-Penetrometer. Proc. of Symposium on the Pressuremeter and Its Marine Applications, Paris.

Keaveny, J.M. 1985. In-Situ Determination of Drained and Undrained Soil Strength Using the Cone Penetration Test. Ph.D. Thesis University of California, Berkeley.

Keaveny, J.M. & Mitchell, J.K. 1986. Strength of Fine-Grained Soils Using the Piezocone. In Situ '86, Proc. Spec. Conf.GED ASCE, Virginia Tech., Blacksburg.

Kjekstad, O., T., Lunne & C.J.F., Clausen 1978. Comparison between in Situ Cone Resistance and Laboratory Strength for Overconsolidated North Sea Clays. Marine Geotechnology, No.4. Also Norwegian Geotechnical Institute Publication, 124.

Konrad, J.M. 1987. The Interpretation of Flate Plate Dilatomer Tests in Sands in Terms of the State Parameters. Draft, Submitted to Geotechnique for Publication.

Kovacs, W.D., & Salomone, L.A. 1982. SPT Hammer Energy Measurements. ASCE, JGED, GT4.

Lacasse, S. 1986. Interpretation of Dilatometer Test. NGI Confidential Report 40019-28.

Lacasse, S., Ladd, C.C. & Baligh, M. 1978. Evaluation of Field Vane. Dutch Cone Penetrometer and Piezometer Testing Device. US Dept. of Transportation,

FHWA, Washington, DC.
Lacasse, S., Olsen, T.S. & Vage, T. 1985. Dilatometer Tests in Holmen Sand. NGI Internal Report.
Lacasse, S. & Lunne, T. 1986. Dilatometer Tests in Sands. In Situ '86. Proc. Spec. Conf. GED ASCE, Virginia Tech., Blacksburg.
Lacasse, S. and Lunne, T. 1988. Calibration of Dilatometer Correlations. Proc. ISOPT I, Orlando, Fla.
Ladd, C.C. & Foott, R. 1974. New Design Procedure for Stability of Soft Clays. JGED, ASCE, GT7.
Ladd, C.C., Foott, R., Ishihara, K., Schlosser, F. & Poulos, H.G. 1977. Stress-Deformation and Strength Characteristics. SOA Report, Proc. of IX ICSMFE, Tokyo.
Lade, P.V. & Lee, K.L. 1976. Engineering Properties of Soils. Report UCLA-ENG-7652. University of California at Los Angeles, p.145.
Lambrechts, J.R. & Leonards, G.A. 1978. Effects of Stress History on Deformation of Sand. JGED, ASCE, No.GT11.
Lancellotta, R. 1983. Analisi di Affidabilità in Ingegneria Geotecnica. Atti Istituto Scienza Costruzioni. No.625. Politecnico di Torino.
Larsson, R. 1980. Undrained Shear Strength in Stability Calculation of Embankments and Fountations on Soft Clays. Canadian Geotechnical Journal, No.4.
Lee, S.H.H. & Stokoe, K.H. 1986. Investigation of Low-Amplitude Shear Wave Velocity in Anisotropic Material. Geotechnical Engineering Report GR86-6 Civil Eng. Dept. The Univ. of Texas at Austin.
Lekhnitzkii, S.G. 1977. Theory of Elasticity of an Anisotropic Body. Edit. Science, Moscow (in Russian).
Leonards, G.A. et al. 1986. Discussion on "Clayton C.R.I., Hababa, M.B. & Simons N.E. 1985". Geotechnique No.2.
Leonards, G.A. & Frost, J.D. 1987. Settlement of Shallow Foundations on Granular Soils. Paper Submitted for Publication to JGED, ASCE.
Lo Presti, D. 1987. Behaviour of Ticino Sand During Resonant Column Test. Ph.D. Thesis, Technological University of Turin.
Lunne, T., Eide, O. and De Ruiter, J. 1976. Correlation between Cone Resistance and Vane Shear Strength in Some Scandinavian Soft to Medium Stiff Clays. Canadian Geotechnical Journal, No.4.
Lunne, T. & Kleven, A. 1981. Role of CPT in North Sea Foundation Engineering. Proc. ASCE National Convention at St.Louis Cone Penetration Testing and Experience.
Lutenegger, A. 1988. Current Status of the Marchetti Dilatometer Test. Proc.ISOPT I, Orlando, Fla.
Marchetti, S. 1975. A New In Situ Test for the Measurement of Horizontal Soil Deformability. Proc. ASCE Spec. Conf. on In Situ Measurement of Soil Properties, Raleigh, N.C.
Marchetti, S. 1980. In Situ Tests by Flat Dilatometer, JGED, ASCE, GT3.
Marchetti, S. & Crapps, D.K. 1981. Flat Dilatometer Manual. Schmertmann and Crapps Inc. Gainesville, Fla.
Marchetti, S. 1985. On the Field Determination of K_o in Sand. Panel Discussion to Session 2A, Proc.XI ICSMFE, San Francisco.
Marcuson, W.F., Bieganousky 1977. Laboratory Standard Penetration Test on Fine Sand. JGED, ASCE GT6.
Marcuson, W.F., Bieganousky 1977a. SPT and Relative Density in Coarse Sand. JGED, ASCE, GT11.
Marsland, A. & Powell, J.J.M. 1979. Evaluating the Large Scale Properties of Glacial Clays for Foundation Design. Proc. BOSS'79, London.
Mesri, G. 1975. Discussion on New Design Procedure for Stability of Soft Clays. JGED, ASCE, GT4.
Mesri, G. & Castro, A. 1987. C_α/C_c Concept and K_o during Secondary Compression. JGED, ASCE, GT3.
Meyerhoff, G.G. 1957. Discussion of Session 1. Proc. IV ICSMFE, London.
Mitchell, J.K. 1984. Personal Communication to M. Jamiolkowski.
Mitchell, J.K. & Gardner, W.S. 1975. In Situ Measurement of Volume Change Characteristics, SOA Report, Proc. ASCE Spec. Conf. on the In Situ Measurement of Soil Properties, Raleigh, N.C.
Mitchell, J.K. & Keaveny, J.M. 1986. Determining Sand Strength by Cone Penetrometer. In Situ '86, Proc. Spec. Conf.GED ASCE, Virginia Tech., Blacksburg.
Miura, S. & Toki, S. 1984. Elasto-plastic Stress Relationship for Loose Sands with Anisotropic Fabric under Three-Dimensional Stress Conditions. Soils and Founda-

tions. No.2.
Miura, S. & Toki, S. 1984a. Anisotropy in Mechanical Properties and its Simulation of Sands Sampled from Natural Deposits. Soil and Foundations No.3.
Mori, H. 1981. Soil Exploration and Sampling. General Report Session 7. X ICSMFE, Tokyo.
Morrison, M.J. 1984. In-Situ Measurements on A Model Pile in Clay. Ph.D. Thesis, MIT, Cambridge, Mass.
Muromachi, T. 1981. Cone Penetration Testing in Japan, Proc. of Symposium on Cone Penetration Testing and Experience. ASCE National Convention at St. Louis, Missouri.
Ni, S.H., Stokoe, K.H. 1987. Dynamic Properties of Sand Under True Triaxial Stress States from Resonant Column Torsional Shear Tests. Geotechnical Engineering Report GR87-8, Civil Eng. Dept. the Univ. of Texas at Austin.
Ohsaki, Y. & Iwasaky, R. 1973. On Dynamic Shear Moduli and Poisson's Ratio of Soil Deposits. Soils and Foundations No.4.
Ohta, Y. & Goto, N. 1976. Estimation of S-Wave Velocity in Terms of Characteristic Indices of Soil. Bitsuri-Tanko, 29(4), 34-41 (in Japanese).
Ohta, Y. & Goto, N. 1978. Empirical Shear Wave Velocity Equations in Terms of Characteristic Soil Indexes, Earthquake Engineering and Structural Dynamics, Vol.6.
Parkin, A.K. & Lunne, T. 1982. Boundary Effects in the Laboratory Calibration of a Cone Penetrometer for Sand. Proc.ESOPT II, Amsterdam.
Parry, R.H.G. 1971. A Simple Driven Piezometer. Geotechnique, No.2.
Parry, R.H.G. 1971. A Direct Method of Estimating Settlements in Sand from SPT Values, Proc. Symp. Interaction of Structures and Foundations, Midlands SMFE Soc. Birmingham, U.K.
Parry, R.H.G. 1977. Estimating Bearing Capacity in Sands from CPT Values. JGED ASCE, GT9.
Parry, R.H.G. 1978. Estimating Foundation Settlements in Sand From Plate Bearing Test. Geotechnique 28.
Peck, R.B. & Bazaraa, A.R.S. 1969. Discussion on Settlement of Spread Footings on Sand. JSMF Div., ASCE, SM6.
Reese, L.C. & Wright, S.J. 1977. Drilled Shaft Manual. Vol.1, U.S. Dept. of Trasportation, Offices of Research and Development, Implementation Div. HDV-2, Washington D.C.
Reese, L.C. & O'Neill, M.W. 1987. Drilled Shafts: Construction Procedures and Design Methods. Draft prepared for U.S. Department of Transportation Federal Highway Administration Office of Implementation Mc Lean, Virginia.
Rix, G.J. 1984. Correlation of Elastic Moduli and Cone Penetration Resistance. Master of Sc. Thesis, the University of Austin of Texas.
Robertson, P.K. & Campanella, R.G. 1983. Interpretation of Cone Penetration Tests. Canadian Geotechnical Journal Vol.20, 1983.
Robertson, P.K. & Campanella, R.G. 1984. Guidelines for Use and Interpretation of the Electronics Cone Penetration Test. Univ. of British Columbia, Vancouver, Canada, Soil Mechanics Series No.69.
Robertson, P.K. 1986. In-Situ Testing and Its Application to Foundation Engineering. Canadian Geotechnical Journal No.4.
Robertson, P.K. (1986a). In-Situ Stress Determination in Sands Using Penetration Devices. Univ. of British Columbia, Vancouver, Canada, Soil Mechanics Series No.99.
Roscoe, K.H. & Burland, J.B. 1968. On the Generalized Stress-Strain Behaviour of "Wet" Clay. In Engineering Plasticity, Cambridge Univ. Press, (ed. by J. Heyman).
Rowe, P.W. 1962. The Stress-Dilatancy Relation for Static Equilibrium of an Assembly of Particles in Contact. Proc. Royal Soc.
Rowe, P.W. 1975. Application of Centrifugal Models to Geotechnical Structures. Proc. of the Symp. on Recent Developments in the Analysis of Soil Behaviour and their Application to Geotechnical Structures; Univ. of New South Wales.
Rowe, P.W. 1975. Discussion on Granular Material, Session 1, Proc. Conf. on Settlement of Structures, Cambridge, U.K., Edit. Pentech Press, London.
Sasaki, Y. et al. 1984. US-JAPAN Cooperative Research on In-Situ Testing Procedures for Assessing Soil Liquefaction (No.1). XVI Joint Meeting. Washington, D.C.
Schaap, L.H.J., Zuidberg, H.M. 1982. Mechanical and Electrical Aspects of the Electric Cone Penetrometer Tip. Proc.

ESOPT II, Amsterdam.
Schmertmann, J.H. 1970. Static Cone to Compute Static Settlement Over Sand. JSMF Div., ASCE, SM3.
Schmertmann, J.H. 1976. An Updated Correlation between Relative Density, D_R, and Fugro-Type Electric Cone Bearing, q_c. Unpublished report to WES, Vicksburg, Miss.
Schmertmann, J.H. 1978. Guidelines for Cone Penetration Test Performance and Design. Report No.FHWA-TS-78-209, U.S. Department of Transportation, Federal Highway Administration, Washington, D.C.
Schmertmann, J.H. 1982. A Method for Determining the Friction Angle in Sands from the Marchetti Dilatometer Test. Proc. ESOPT II, Amsterdam.
Schmertmann, J.H. 1983. Revised Procedure for Calculating K_o and OCR from DMT with $I_D>1.2$ and which Incorporates the Penetration Force Measurements. DMT Workshop, Gainesville, Florida.
Schmertmann, J.H. 1986. Dilatometer to Compute Foundation Settlement. In Situ '86, Proc. Spec. Conf. GED, ASCE, Virginia Tech., Blacksburg.
Schmertmann, J.H. & Palacios, A. 1979. Energy Dynamics of SPT. JGED, ASCE, GT8.
Schofield, A. & Wroth, P. 1968. Critical State Soil Mechanics, McGraw-Hill Inc., New York.
Seed, H.B. 1983. Evaluation of the Dynamic Characteristics of Sand by In-Situ Testing Techniques. Int. Symp. on Soil and Rock Investigations by In-Situ Testing. Paris.
Seed, H.B., Tokimatsu, K., Harder, L.F. & Chung, R.M. 1984. The Influence of SPT Procedures in Soil Liquefaction Resistance Evaluations. Report No.UCB/EERL-84/15, Earthquake Engineering Research Center, Univ. of California, Berkeley.
Seed, H.B. & De Alba, P. 1986. Use of SPT and CPT tests for Evaluating the Liquefaction Resistance of Sands. In Situ '86, Proc. Spec. Conf. GED ASCE, Virginia Tech., Blacksburg.
Seed, H.B., Wong, R.T., Idriss I.M. & Tokimatsu, K. 1986. Moduli and Damping Factors for Dynamic Analyses of Cohesionles Soils. JGED, ASCE, GT11.
Skempton, A.W. 1986. Standard Penetration Test Procedures and the Effects in Sands of Overburden Pressure, Relative Density, Particle Size, Ageing and Overconsolidation. Geotechnique, No.3.
Smits, F.P. 1982. Penetration Pore Pressure Measured with Piezometer Cones. Proc. ESOPT II, Amsterdam.
Sykora, D.W. & Stokoe, K.H. 1983. Correlations of In-Situ Measurements in Sands of Shear Waves Velocity, Soil Characteristics and Site Conditions. Geotech. Eng. REP GR 83-33. Texas Univ., Austin.
Tatsuoka, F. 1987. The Strength and Dilatancy of Sands. Discussion, Geotechnique, No.2.
Tatsuoka, F. et al. 1978. A Method for Estimating Undrained Ciclic Strength of Sandy Soils Using Standard Penetration Resistances. Soil and Foundations. No.3.
Tatsuoka, F. & Pradham, T.B.S. 1988. Discussion on Direct Shear Tests on Reinforced Sand to appear in Geotechnique.
Terzaghi, K., Peck, R.B. 1948. Soil Mechanics in Engineering Practice. J. Wiley & Sons Inc., New York.
Tringale, P.T., Mitchell, J.K. 1982. An Acoustic Cone Penetrometer for Site Investigations. Proc. ESOPT II, Amsterdam.
Tumay, M.T., Boggers, R.L., Acar, Y. 1981. Subsurface Investigations with Piezocone Penetrometer. Proc. Symposium on Cone Penetration Testing and Experience. ASCE National Convention at St. Louis, Missouri.
Veismanis, A. 1974. Laboratory Investigation of Electrical Friction Cone Penetrometers in Sand. Proc. ESOPT I, Stockholm.
Vesic, A.S. 1972. Expansion of Cavities in Infinite Soil Mass. JSMF Div., ASCE, SM3.
Vesic, A.S. & Clough, G.W. 1968. Behaviour of Granular Materials under High Stresses. JSMF Div., ASCE, SM3.
Vivatrat, V. 1978. Cone Penetration in Clays. Ph.D. Thesis, MIT, Cambridge, Mass.
Whiters, N.J., Schaap, L.H.J., Kolk, K.J. & Dalton, J.C.P. 1986. The Development of the Full Displacement Pressuremeter, The Pressuremeter and Its Marine Applications, II Int. Symp., ASTM STP 950, Texas A&M University.
Wroth, C.P. 1984. The Interpretation of In-Situ Soil Tests. XXIV Rankine Lecture, Geotechnique No.4.
Wroth, C.P. & Houlsby, G.T. 1985. Soil Mechanics - Property Characterization and Analysis Procedures. Theme Lecture, Proc.XI ICSMFE, San Francisco.
Yaroshenko, V.A. 1964. Interpretation of

the Results of Static Penetration in Sands. Fundamenty Proekt, N.3.
Yu, P. & Richart, F.E. Jr. 1984. Stress Ratio Effects on Shear Modulus of Dry Sand. JGE Div., ASCE, GT3.

Penetration Testing 1988, ISOPT-1, De Ruiter (ed.)
© 1988 Balkema, Rotterdam, ISBN 90 6191 801 4

Cone penetration problems and solutions involving non-purpose-built deployment systems

J.M.O.Hughes
Hughes Insitu Engineering Ltd, Vancouver, BC, Canada

ABSTRACT: There is the potential to make almost any drill rig into a deployment system for a cone, thus making the cone very accessible to almost any project where the ground conditions are suitable. This paper discusses the problems with this approach, and some of the solutions which have been adopted to solve them. The approach suggested may seem inefficient, compared with that used for purpose-built vehicles. However, it is possible in suitable terrain to test up to 200 m per day, with no mobilization of heavy equipment. Further, there are few limits to the type of material that can be tested. The field experience which has provided the basis of this paper has been obtained in Canada or the United States, both on land and over shallow water.

INTRODUCTION

The difficulties encountered with cone testing are, for the most part, related to the methods of deployment of the cone. Deploying cones from purpose-built vehicles, enclosed and environmentally-controlled, is quite a different situation from operating with a drill rig from a floating barge, in a swamp or even on firm ground.

Although considerable effort has been devoted to develop standards for the external geometry of the cone, the internal details and some of the external details of cones manufactured by different companies vary. The impact of this on commercial testing is that cones, even in calibration, can yield different results. In most instances, this is not a problem. However, if different types of cones produced by the same or different manufacturers are used in the same formation, there are situations in which they will give different values for friction and porepressure.

The need for further understanding of the reasons for these anomolies, and the need to set standardisation guidelines, is well recognised. No doubt with time all these anomolies will be resolved. However, at present, the cone contractor and the design engineer must recognise these problems.

The object of this article is to concentrate on the use of cones, particularly from non-purpose-built deployment systems. Although it is recognised that the above problems exist, they are not addressed in this paper.

Most of the background for this paper has been based on experience in North America, where engineers, with a few exceptions, are not familiar with the capability of the cone. Purpose-built deployment vehicles are few and far between, but drill rigs of all shapes and sizes are readily available.

COMMON GEOTECHNICAL SITE INVESTIGATION TEST PROCEDURES IN NORTH AMERICA

In North America the cone, particularly the piezocone, is relatively speaking a recent innovation. Hence, site investigations are often not designed to make the optimum advantage of the cone. Ideally, it should be used to identify the strata in which high-quality samples should be obtained. Unfortunately, all too often, it is used to provide a check on zones which cannot be sampled or where there is uncertainty about the samples. If the project is large enough, the cone is sometimes used as an afterthought, to confirm the traditional approach of routine sampling and SPT testing.

Except in small geographical regions of North America where the geology is suitable and where equipment is readily available, such as in the South-eastern United States, the projects on which the cone is used are spread out over a considerable area. Often these projects are small and will not support large mobilization costs. This is changing as more engineers become familiar with the potential of the cone in their site evaluations. Often, use of the cone may start out as some minor afterthought, but as the engineers see the data, the involvement becomes more extensive.

MOBILE DEPLOYMENT SYSTEM

To take advantage of the opportunities that exist necessitates making use of deployment equipment that is either available or readily obtainable on site, thus mobilizing the minimum of heavy equipment. There is a considerable variety of drill rigs mounted on many types of carriers such as standard and four-wheel-drive trucks, all-terrain wheeled and tracked vehicles, as well as very light helicopter-transportable systems. These rigs are located all over the country. Hence, by suitable choice of rig, it is possible to operate in any soil condition. There is, however, a significant interfacing difficulty between the drilling contractor and the cone operator.

The cone operators are required to operate with relatively delicate equipment, particularly from the point of view of maintenance and data acquisition. The cables which have to be strung about the rig from the cone or at least the drill head to the data recording system, are always in the way of drillers. The drillers, with few exceptions, have no experience with such equipment. The problem compounds when the drilling company and the testing company have not worked together. The drill owners, and more particularly the operators, are unlikely to be mechanical engineers. They have learned their trade by experience, and are capable of doing their routine job well, but cone work is not the same as driving an SPT. The rods are smaller and the pushing forces are often higher than those to which the drillers are accustomed. It is not uncommon for the driller to provide a universal joint on his pushing head and attempt to push cone rods unsupported over 2 m. This method will certainly work if the rods are 100mm in diameter, but the chance of success is limited with 36mm rods.

The importance of having a rigid pushing head which will not rotate in a vertical plane cannot be overstressed. The next limitation on the load that can be placed on the rods is the length of rod unsupported laterally between the ground and the pushing head. This can usually be limited by setting a length of casing in the ground and supporting the top from lateral movement against the drill rig. With a rigid head and the minimum unsupported buckling length, the next limitation is the power of the hydraulic rams to move the head both downwards and upwards. Finally, the load is limited by the dead weight of the rig, or the amount that the rig can be anchored to the ground. In general, few drillers know the power of their rams, or the load that can be applied through the head. It is invariably over-estimated.

Having overcome all these minor difficulties, the next is the control of the pushing rate. Most modern rigs which are equipped with flow control regulators have little problem obtaining an even push. If their controls are more rudimentary, some skill on the part of the driller is required to maintain a constant pushing rate. Usually this is a problem only if the ground is very soft.

The pulling forces must not be neglected, particularly as hydraulic rams do not produce the same force in both directions. This is not a problem in sands. However, in some clays the pulling force can equal and even exceed the pushing forces.

As a step between the use of the drill rig alone to the purpose-built deployment system, some groups have attached a separate hydraulic ram used only for cone deployment. This is a relatively portable unit, often with its own power unit, that is designed to adapt to a particular drill rig. This approach uses the rig as dead load only. The hydraulic power units are not readily adapted to rigs other than those for which they are constructed, but they do have the advantage of providing a sound solution to all the problems outlined above, i.e. a rigid pushing head, minimum buckling length, and a known ram power. The inclusion of a portable ram to the drilling system, while solving many problems, does increase the mobilization costs and reduces the flexibility as to the choice of rig.

With a drill rig operation the pushing forces are limited. Usually a maximum of about 100 kN is all that is available - often much less. Hence, there are limits to the depth of penetration, particularly if the friction forces get too high or the material at depth is too hard. However, there is the immediate capability of transferring to the drilling mode and drilling out the hard layers and then proceeding. This drilling-out procedure also has its problems, and in certain situations can lead to the loss of the cone through buckling of the rods.

There are three approaches to dealing with a drill-out:

1. If the rig is equipped to auger, a hollow-stem auger can rapidly drill through the hard zone, provided, of course, the material is suitable (sands below the water table can present a problem). The cone can then be pushed through the hollow auger. If the internal diameter of the auger is much larger than the rods (i.e. greater than 50 mm), casing may have to be set inside the augers to prevent buckling.

2. The auger can be withdrawn and the hole backfilled with coarse sand or fine gravel. The cone can then be pushed from the surface, in the normal manner, through the loose filling. The loose filling will provide the lateral support to the cone rods, provided the hole is completely full and there are no voids. Care must be taken if the filling material is fine and there is water present. In this situation the filling material is liable to arch across the hole and prevent the hole from being properly backfilled. This situation is particularly dangerous if high pushing forces are required in the material which has yet to be probed.

3. Another alternative, and from personal experience the most satisfactory, is to rotary drill a 50mm diameter hole through the hard layers.

If the cone starts to move off vertical as the push starts, this may signal the onset of buckling. However, if the likely buckling section is further up the rods, the problem can sometimes be detected by the excessive spring-back of the rods after each push. This is difficult to observe unless the pushing system is rigidly anchored to the ground.

In either the first or third alternative approaches to predrilling, clamps are required to assist in lowering the cone and rods down the open hole. These clamps will usually have to be provided, as they are not likely to be part of the driller's equipment.

OVERWATER SITUATIONS

Cone testing overwater for offshore testing is routine, and many ingenious systems have been developed, especially for deep water. Several international groups have extensive expertise in this area. In general, their systems are probably too expensive to mobilize for small overwater projects such as for pier investigations on local bridges. Conventional site investigation for such work is usually conducted as an extension of the land-based approach. A drill is mounted on a barge anchored over the hole location. The conventional drilling investigation is conducted in the usual manner, allowing for the additional complication of tide or wave action. In sheltered water this is not a major difficulty. Occasionally the work may have to be halted if the sea movement is excessive. Using a cone in such conditions poses a few more problems. The major ones are those of buckling.

If interest is in the material properties at some depth below the seabed, then heavy casing can be set into the sea bed. The lateral support for the cone rods will then be provided by the casing. The maximum vertical force that can be applied to the cone will probably be somewhat less than on land, particularly if the barge is light. However, if the interest is in the determination of the characteristics of the seabed near the surface, the casing, which provides lateral support for the cone rods, must "rest" on the sea bed.

Usually this requires the base of the casing to be connected to a large plate or footing, to reduce the settlement of the casing and to provide some lateral support. In this situation, the pushing forces are limited, and the danger to the cone increases.

It hardly needs to be mentioned that the cone depth sensor on the barge must be referenced to the sea bed, not the barge.

Using a drill rig in the above manner requires a degree of manual control to maintain a constant rate of push. In general, this is not a problem unless the wave motion is large.

Specialised systems, developed for deep offshore work, are available. These fully compensate for wave action but are not readily available.

CARE OF CONES

When cones are supplied by the manufacturer they will almost certainly be in good working order. However they will often be used in a very harsh environment. There is no knowing what they will hit or at what angle they will hit an unseen object below the surface. By carefully monitoring the tip load, it is possible to limit the maximum load that can be applied to the cone, thus controlling the ram pressure either electrically or manually. However, unless the feedback system is very fast, some danger will always exist.

By far the greater concern is that caused by bending. In the limiting case, these bending forces will either bend the load cell permanently, or deflect the cone to such an extent that the rods will snap. If the cone is equipped with a slope sensor, continuous monitoring of the change in slope will help indicate a potential problem.

Less traumatic is the problem of minor damage to the cone as it hits gravel or small rocks. The damage this can cause is often only cosmetic; however with some designs, it is possible to damage the edge of the tip such that there is a direct mechanical link to the sleeve. Grit between the sleeve and the body of the cone can also be a problem. If it is not cleaned out, some materials will become cemented on drying, and thus restrict the tip and the friction sleeve from acting independently.

Finally, the surface of the tip and the sleeve will become worn to such an extent that the values measured in the same formation will be different from what they were before the wear occurred. The amount of wear that is acceptable is addressed in the standards. Hence, even though the cones may well start life in perfect condition, they will have to be carefully looked after to ensure they are operating as intended, within calibration limits.

Cone operators who manufacture their own cones have very few problems, whereas the operators with equipment that has been purchased often experience numerous difficulties. Probably one of the major reasons for this is the lack of technical support. It is difficult for a supplier to establish support centers close to every operator.

In view of the fact that the likelihood of damage with the above deployment system is relatively high, and the project can be well away from the base, the operators have to be very familiar with their equipment or carry an extensive array of spares.

CONCLUSION

In contrast to the purpose-built deployment system in which all the necessary components to operate and perform the tests are contained on one vehicle, the danger to damage or loss of equipment, or of just neglecting to mobilize one minor item, is very high. However, with this approach, the opportunity of performing cone tests and of expanding the interest in insitu testing is opened up extensively. To conduct cone testing successfully, the operators of the equipment must be familiar with all aspects of the system, as the lines of support, in North America at least, can be very long indeed.

ACKNOWLEDGEMENTS

The background for this paper has been obtained from extensive personal experience in using cone equipment over many regions of the United States and Canada. Almost all of this experience has been obtained by working with cone deployment systems adapted to drill rigs.

Many people and groups have helped build up this experience. The author would like to acknowledge the following organisations, for either their direct input into this paper, or for helpful assistance while working together:- Earth Technology Corp., U.S. Geological Survey, Fugro- McClelland Inc., Gulf Canada, Esso Resources Canada, Dome Petroleum Ltd., British Columbia Dept. of Highways, British Columbia Hydro, J. & W. A.B., Geotech. A.B., Shannon and Wilson Inc., Knight Piesold Inc., CH2MHill Inc., Law Eng. Inc., Syncrude Canada Ltd., Kring Drilling Inc. Kenner Drilling Inc., Mobile Auger Ltd., Associated Drilling, Foundex Exploration Ltd., Dominion Soils Ltd., Hart Crowser Inc., University of British Columbia, Thurber Consultants Ltd.

The author also wishes to acknowledge the assistance of Paul White and John Graham, who were primarily responsible for introducing commercial cone testing into Western Canada.

Penetration Testing 1988, ISOPT-1, De Ruiter (ed.)
© 1988 Balkema, Rotterdam, ISBN 90 6191 801 4

Penetration testing – A more rigorous approach to interpretation

C.P.Wroth
Department of Engineering Science, University of Oxford, UK

ABSTRACT: The paper is concerned with the problems of interpreting the results of penetration tests and correlating them with engineering properties of soils. The need is stressed for relationships to be based on physical insight, set against a theoretical background and be expressed in suitably dimensionless form. Accepted correlations are examined, and suggestions made for modifications to existing parameters and for new parameters to be adopted.

1 INTRODUCTION

The purpose of this paper is to consider some of the general principles which form the basis of a proper interpretation of in situ tests. Many in situ tests used for geotechnical purposes were originally devised before present-day understanding of soil behaviour was established; their early interpretation was based almost exclusively on empirical correlations,many of which in spite of their importance are now considered to be unsatisfactory. The reasons why some of these correlations should be abandoned and others should be modified are presented in the ensuing arguments.

2 RELATIONSHIPS BETWEEN SOIL PROPERTIES

The choice of properties that should be used in any attempted correlation is crucial. Any successful relationship that can be used with confidence outside the immediate context in which it was established should ideally be:

(a) based on a physical appreciation of why the properties can be expected to be related;

(b) set against a background of theory, however idealised this may be;

(c) expressed in terms of dimensionless variables so that advantage can be taken of the scaling laws of continuum mechanics.

An illustration of these points is provided by considering correlations of the undrained shear strength s_u of a clay. All soils are basically frictional materials with the strength being provided by the frictional resistance between soil particles governed by the effective stress to which they are subjected. Starting *ab initio*, the first relationship to be explored would be

$$\frac{s_u}{p_f'} = f(\phi) \qquad (1)$$

where p_f' is the mean principal effective stress at failure and ϕ is the angle of shearing resistance. In any real situation the value of p_f' will not be known, and it has to be replaced by some other stress variable. If the initial conditions are selected, and the mean principal effective stress p_o' is used, then its relationship with p_f' depends on the excess pore pressures generated during shearing to failure, which in turn depends on the overconsolidation ratio OCR of the clay. Hence a second approach would be to consider the relationship

$$\frac{s_u}{p_o'} = f(\phi, OCR) \qquad (2)$$

However, this relationship will in practice be subject to much uncertainty because the in situ mean principal effective stress p_o' is unlikely to be known or to have been estimated with any accuracy. The single stress variable that can be estimated with most reliability is the in situ vertical effective stress σ_{vo}'. Since this is related to p_o' as a function of OCR, its use in lieu of p_o' will not increase the number of variables in the relationship. Consequently a good engineering compromise is to adopt the expression

$$\frac{s_u}{\sigma_{vo}'} = f(\phi, OCR) \qquad (3)$$

and to define s_u/σ_{vo}' as the *undrained strength ratio.*

2.1 Undrained shear strength

In the preceding paragraphs it has been tacitly assumed that there is a unique property of a given clay termed the undrained shear strength s_u. However, it is now well recognised that this property does not have a single value when determined experimentally for two reasons:
(i) the strength is affected by the condition and size of the sample tested and
(ii) the strength depends on the type of test by which it is determined.

In comparing undrained strengths of a soil, it is necessary to distinguish carefully between specimens in the following conditions:

(a) reconstituted soil (as used in index tests, i.e. with the specimen having been dried, some of the large particles possibly removed and the remainder being reconstituted with new pore fluid),
(b) remoulded soil (all soil fabric destroyed, but with the same constituent parts, i.e. no change of pore fluid),
(c) 'undisturbed' soil tested in the laboratory, and
(d) 'undisturbed' soil tested in situ.

The first two of the above categories might be considered at first sight to be irrelevant to the interpretation of in situ tests of soil strength. However, they have played an important part not only in establishing relationships between index properties and engineering properties of soils (e.g. Wroth and Wood (1978) and Wroth (1979)), but also in providing a consistent and accurate set of data on which critical state soil mechanics (CSSM) - and other similar approaches to understanding soil behaviour - have been based.

For example, the expression proposed by Ladd et al. (1977) relating the undrained strength ratios for normally consolidated (nc) and overconsolidated (oc) clays with the OCR

$$\frac{(s_u/\sigma_{vo}')_{oc}}{(s_u/\sigma_{vo}')_{nc}} = (OCR)^m \qquad (4)$$

was originally derived empirically. However, it has been shown by Wroth and Houlsby (1985) that this expression can be obtained theoretically from concepts of CSSM and, moreover, that the constant exponent m is related to well-recognised physical properties of the clay in question being replaced by Λ, the plastic volumetric strain ratio.

The third and fourth types of undrained strength may differ markedly due to
(i) different degrees of disturbance of soil,
(ii) different sizes of specimen and effects of fabric (such as the presence of fissures), and
(iii) different types of effective stress path to which the specimen is subjected.

A further difficulty and major uncertainty arises from the basic, but deficient, definition of undrained shear strength

$$s_u = \tfrac{1}{2}(\sigma_1 - \sigma_3) \qquad (5)$$

i.e. half the difference between the major and minor principal stresses, or the radius of the largest Mohr circle. This is an unsatisfactory definition as it neither takes account of the intermediate principal stress σ_2 nor distinguishes between the different types of test which are well known to give different results for identical soil specimens.

It is essential to distinguish between different test results by an inelegant plethora of suffixes as follows:

in the laboratory

s_{utc}	triaxial compression test
s_{ute}	triaxial extension test
s_{upsa}	plane strain active test
s_{upsp}	plane strain passive test
s_{udss}	direct simple shear test

in the field

s_{uspt}	standard penetration test
s_{ucpt}	cone penetration test
s_{ufv}	field vane test
s_{upm}	pressuremeter test

An attempt has been made by the writer, Wroth (1984), to express these different strengths in terms of the angle of shearing resistance ϕ and to show that there is a hierarchy of such strengths for a given soil with a given geological history. For example, for a deposit of normally consolidated clay it is suggested that

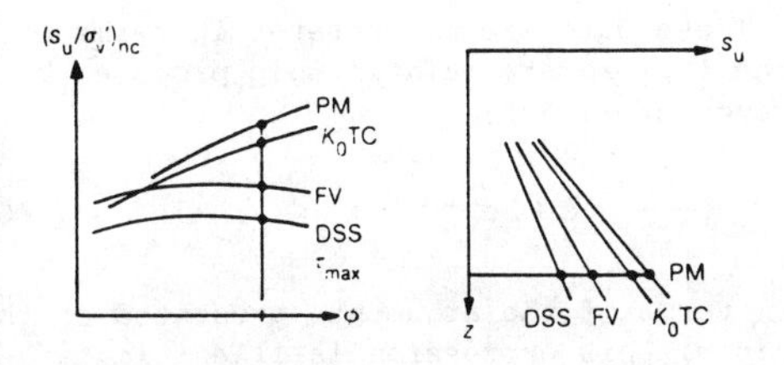

Fig.1 Possible profiles of the undrained shear strength in a soft clay deposit measured in different tests

typical profiles of strength would be as indicated in Fig.1.

Two examples of sites which support these ideas are shown in Fig.2 with profiles of strengths at Porto Tolle and Panigaglia, obtained by Ghionna et al. (1983) who were concerned in interpreting and comparing the results of pressuremeter tests with other tests, so that the individual points and scatter bars refer only to pressuremeter data and the straight lines to best-fit profiles of the other test results.

All the above comments about the determination of undrained shear strength mean that great care must be exercised in using any correlation for interpreting in situ soil tests to give values of strength. Many correlations have been based on data obtained relatively early in the history of soil mechanics, well before all these difficulties and distinctions were fully appreciated. Often incomplete information is given in the supporting papers, and the results embrace different types of test, thereby adding unnecessarily to the apparent scatter of the data. It is strongly recommended in future that

(i) publications supporting such correlations carefully record all the relevant information about the state of the specimen, type of test, etc. and

(ii) ideally one type of test - the undrained triaxial compression test (after appropriate reconsolidation of the specimen) - is used exclusively in all such correlations.

2.2 Hierarchy of correlations

In the Introduction it was stressed that any correlation should be based on a physical appreciation of why the various properties can be expected to be related. For example, the end bearing q_t measured in the cone penetration test (CPT) (appropriately corrected for pore-pressure effects) is some measure of the *strength* - but *not* the

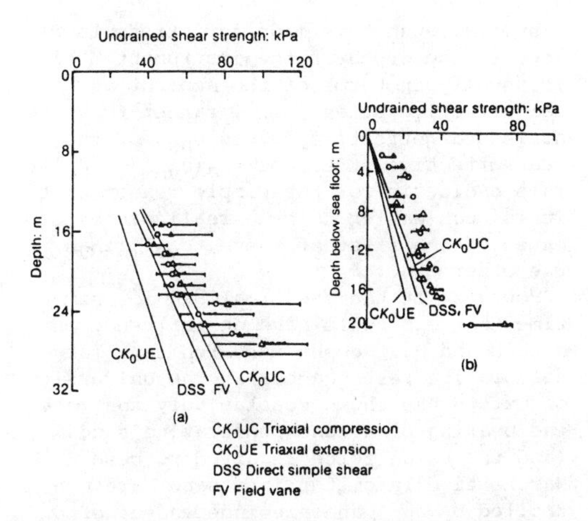

Fig.2 Profiles of the undrained shear strength measured in different tests at two soft clay sites in Italy (after Ghionna et al. 1983): (a) Porto Tolle; (b) Panigaglia

stiffness - of the soil being tested. Hence the *primary* correlation must be between q_t and s_u.

Moreover the single observed quantity q_t can only lead to *one* independent interpretation of soil properties; any additional interpretation, if truly independent, must depend on some other information. For example, any correlation that is suggested between q_t and the undrained Young's modulus of the soil E_u is a *secondary* relationship that is dependent on some other relationship between s_u and E_u. This implicit dependence on another correlation means that the secondary relationship is a weaker one, with more scatter in the data, and on which less reliance can be placed. Hence it is suggested that all correlations should be categorised as primary or secondary.

3 PENETRATION TESTS IN COHESIONLESS SOILS

The basic measurement in the Standard Penetration Test (SPT) and Cone Penetration Test (CPT) is the effort required to advance the instrument into the ground: the number of blows N (per 30 cm penetration) for the former, and the corrected end bearing stress q_t for the latter. Both of these quantities have their primary correlation with strength, which in the case of cohesionless materials will be in terms of effective stresses conventionally represented by stress level, relative density D_r and angle of shearing resistance.

When allowance is made for stress level (due to the depth of the instrument) this is usually in terms of the ambient vertical effective stress σ_{vo}', rather than the horizontal effective stress σ_{ho}' or the mean effective stress $p_o' = \frac{1}{3}(\sigma_{vo}' + 2\sigma_{ho}')$. This choice is for the simple reason that it is much easier to make realistic estimates of the first of these, σ_{vo}', than of the other two.

However, it has been realised for some time that σ_{ho}' will have an influence on both N and q_t. Recent research in a large calibration test chamber at the University of Oxford has shown conclusively that the end bearing of a cone penetrating sand (and the axial force required to push a Marchetti dilatometer into sand) are controlled by σ_{ho}' and are independent of σ_{vo}', as reported by Houlsby and Hitchman (1988). This finding means that it is preferable to use σ_{ho}' to p_o', and not to use σ_{vo}' when considering the level of stress relevant to a particular test.

3.1 Standard penetration tests in sands

Notable research work reported by Schmertmann (1979) and Schmertmann and Palacios (1979) has done much to enhance the doubtful reputation of the SPT by helping to standardise the test and by advancing its interpretation significantly. Subsequent work has been collated and reviewed by Skempton (1986). Results of tests conducted at the US Bureau of Reclamation and the Waterways Experiment Station are presented in Fig.3 where the influences of σ_v' and D_r on N are shown.

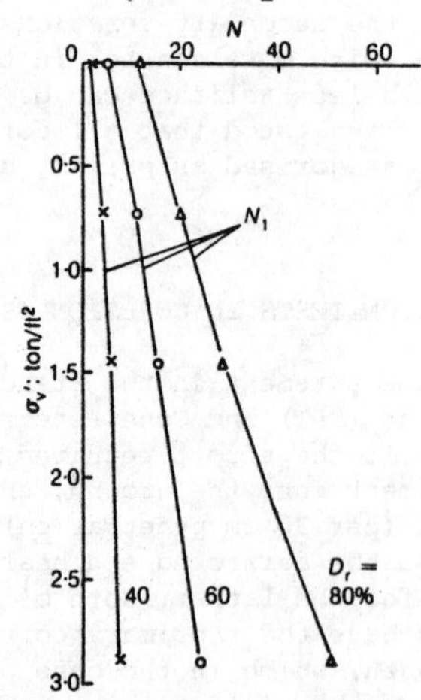

Fig.3 Results of laboratory standard penetration tests on sand of different densities and under different pressures (after Skempton, 1986)

These data are interpreted in terms of the approximate relationship proposed by Meyerhof (1957):

$$\frac{N}{D_r^2} = a + b\sigma_v' \qquad (6)$$

In terms of the arguments presented in this paper, this expression is illegitimate because it is dimensionally inconsistent, with the so-called constants a and b being dependent on the choice of units used to express σ_v'. Although N is quoted as a number and taken as dimensionless, strictly it has the dimension of 'work' (Schmertmann has shown a typical efficiency of 60% so that the energy imparted to the soil by each blow is about 285 joules). Consequently a has the same dimension as N, whereas b has the dimension of (length)³.

Hence a much more satisfactory formulation of the relationship would be

$$N = \pi r^2 \ell . D_r^2 (a'p_a' + b'\sigma_v') \qquad (7)$$

where r is the radius and ℓ is the length of the standard penetration tool, p_a' is some reference pressure (preferably atmospheric) and a' and b' would be properly dimensionless constants for a sand.

Further unsatisfactory features of this approach are that

(i) the effects of D_r and σ_v' are dissociated and

(ii) for $\sigma_v' = 0$, N has a finite value.

Both of these effects are physically untenable. These objections can be overcome by adopting the ideas proposed by Been et al. (1986 and 1987) for interpretation of CPT results in sand which are discussed briefly in the next section.

The above comments relate to the *primary* interpretation of SPT data in terms of soil strength. Much effort has also been expended in attempting to relate N to soil compressibility in terms of a constrained modulus or Young's modulus; such interpretation leads to a *secondary* correlation depending on highly questionable relationships between strength and stiffness of cohesionless soils. Any such correlation should be treated with considerable caution.

3.2 Cone penetration tests in sands

In two important papers Been et al. (1986 and 1987) have extended the understanding and interpretation of cone penetration tests in sand by using the concept of a state parameter ψ to describe the condition of a sand. This parameter (defined as the difference between the voids ratio e of the sand and its voids ratio e_{ss} at the 'steady state' at the *same* mean effective pressure p') is embodied in the concepts of CSSM; it was used in the early research work at the University of Cambridge by Wroth and Bassett (1965). The essential feature of the parameter is that it *combines* the influences of relative density and stress level in a coherent and satisfactory way.

Been et al. suggest that the mean effective pressure p_o' is the appropriate measure of stress and that CPT data should be represented by the relationship

$$\frac{q_t - p_o}{p_o'} = k\ \exp(-m\psi) \qquad (8)$$

where k and m are soil constants, both related to the value of λ_{ss} (the gradient of the steady-state line in a plot of e against $\ell n\ p'$). Expressions are given for k and m which apply reasonably well to a wide range of sands. For the interpretation of CPT data, equation (8) requires the independent measurement (or estimation) of (i) σ_{ho}', to arrive at a value of p_o', and (ii) λ_{ss}.

This promising development confirms the importance of the ambient value of the horizontal effective stress in the performance of penetration tests in sand. No longer is it adequate practice to ignore its influence and to normalise values of end bearing by use of σ_{vo}' alone.

4 PENETRATION TESTS IN COHESIVE SOILS

Significant developments have taken place in the last few years in the cone penetration test, both regarding the equipment and its instrumentation, and in the interpretation of results. This development is set to continue, particularly with the piezocone which is likely to become the dominant tool for the preliminary characterisation of clay deposits. It is vital therefore that the interpretation of CPT data is carried out as intelligently as possible; the remainder of this paper is concerned with this topic, to the exclusion of comments about the use of the SPT or the Marchetti dilatometer in clays.

4.1 Interpretation of end bearing

The primary measurement in the cone penetration test is that of the end bearing pressure q_c. This observation has to be corrected to allow for the pore-pressure acting on the back of the cone to give the corrected value q_t. In correlating this with undrained strength, allowance must be made for the ambient level of total pressure in the ground at the horizon of the test. Although recent work has suggested that this should be by subtraction of σ_{ho}, the standard approach has been to subtract the total overburden pressure σ_{vo} and interpret the results in the form

$$(q_t - \sigma_{vo}) = N_{kt} s_u \qquad (9)$$

where N_{kt} is an empirically derived cone factor, dependent on the type of undrained test used for the relationship. There has been much debate about the proper value to be chosen for N_{kt}, within a range as wide as 9 to 30. This wide variation is due both to the type of test used to determine s_u, and to the state of the clay itself.

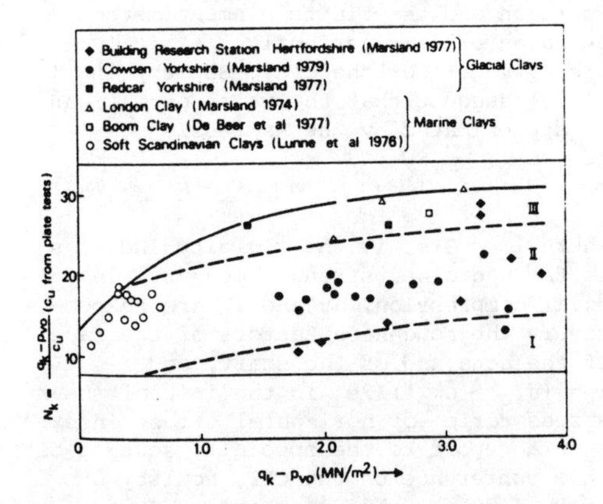

Fig.4 Values of cone factor using values of undrained shear strength back-analysed from large in situ plate tests (after Marsland and Quarterman, 1982)

For example, Fig.4 shows results reported by Marsland and Quarterman (1982) of variation of N_{kt}, back-analysed from plate bearing tests, with $(q_t - \sigma_{vo})$, i.e. with strength of the clay. In view of the desirability of using dimensionless variables it is suggested that it would be better to plot N_{kt} against $(q_t - \sigma_{vo})/\sigma_{vo}'$ which can be thought of as $N_{kt}(s_u/\sigma_{vo}')$ and which is related to OCR, as shown later in the paper.

The problem of deriving theoretical values for N_{kt} is a very complex one, as has been amply demonstrated in the literature. A major effort has been made at Oxford to model the penetration of a cone numerically, using both the strain path

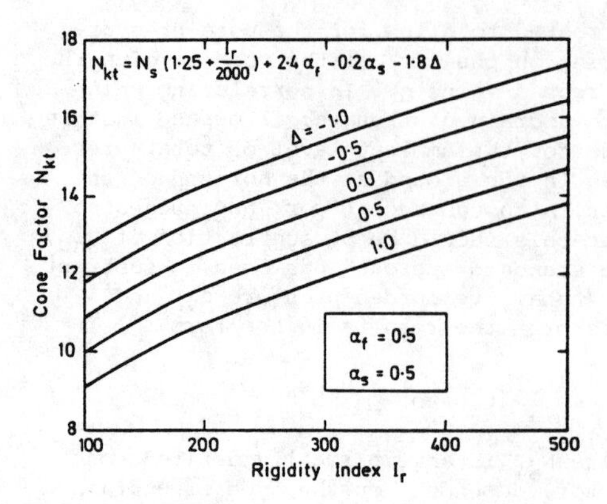

Fig.5 Variation of computed cone factor with rigidity index and initial shear stress ratio (after Houlsby and Wroth, 1988)

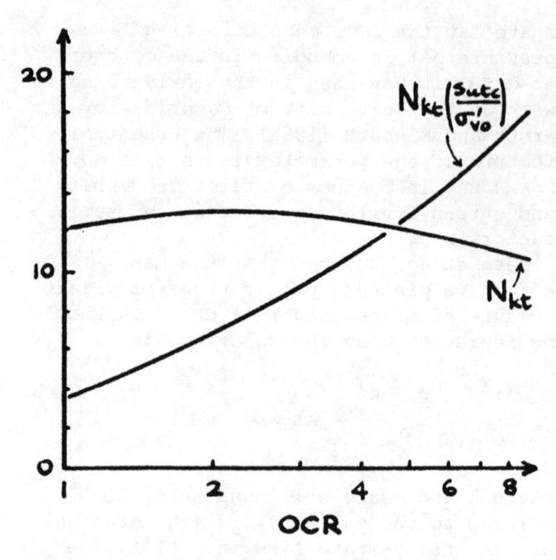

Fig.6 Variation of computed values of cone factor N_{kt} and $N_{kt} \cdot s_{utc}/\sigma_{vo}'$ with OCR for Boston Blue clay

approach and the finite element method. Using an elastic perfectly plastic (von Mises) model for the clay, Houlsby and Teh (1988) suggest that the value of N_{kt} can be approximated by the expression

$$N_{kt} = \frac{4}{3}(1 + \ln I_r)\ (1.25 + I_r/2000) + 2.4\alpha_f - 0.2\alpha_s - 1.8\Delta \quad (10)$$

where $I_r = G/s_u$ is the rigidity index, s_u is the undrained strength measured in triaxial compression, α_f and α_s are respectively the roughness factors of the face of the cone and of the shaft, and $\Delta = (\sigma_{vo}' - \sigma_{ho}')/2s_u$ is the initial shear stress ratio (or horizontal stress index).

In a report to the specialty session of this conference on the CPT, Houlsby and Wroth (1988) include a chart, reproduced as Fig.5, showing how the computed value of N_{kt} varies with I_r and Δ for the case when α_f and α_s are both taken as 0.5. This chart suggests the possibility of a relatively wide range of values of N_{kt}; however this turns out not to be the case for one given soil, because variations of I_r and Δ with OCR tend to cancel eath other out in their contributions to N_{kt}.

This can be illustrated by taking some typical values for Boston Blue clay. The relationship between Δ and OCR was established by Wroth and Houlsby (1985) in their equation (37) and Fig.16 (where the symbol f was used in place of Δ), and values are presented in their Table 1. Typical values for I_r as a function of OCR are taken from Fig.23 of the same publication for the case of a'factor of safety' of 2, i.e. using G_{50}, the secant value of G corresponding to 50% of the shear stress at failure. These values are presented in

Table 1. Theoretical values of the cone factor N_{kt} for Boston Blue clay for the case of full interface friction on the cone and shaft.

OCR	Δ	$(I_r)_{50}$	N_{kt}	$\frac{s_{utc}}{\sigma_{vo}'}$	$N_{kt} \cdot \frac{s_{utc}}{\sigma_{vo}'}$
1	0.8139	200	12.07	0.307	3.71
1.5	0.4644	210	12.83	0.420	5.39
2	0.2828	210	13.15	0.525	6.90
3	0.1093	170	12.92	0.719	9.29
4	0.0223	130	12.45	0.899	11.19
5	- 0.0122	100	11.94	1.069	12.76
6	- 0.0609	83	11.64	1.231	14.33
8	- 0.0989	55	10.91	1.539	16.79

Table 1, and values of N_{kt} plotted in Fig.6 for the case of full interface friction, $\alpha_f = \alpha_s = 1$.

Although these computations for N_{kt} for Boston Blue clay are based on several questionable assumptions, it is believed that they are of value qualitatively in indicating the likely trend with OCR. The curve reflects essentially the variation of I_r with OCR, which increases from the normally consolidated state to a maximum at about OCR of 2.5 and thereafter decreases with increasing OCR. For engineering purposes an average value of $N_{kt} \sim 12$ over the range $1 < OCR < 8$ would be appropriate in this case. Similar curves can be expected for other clays having different properties.

Reverting to Fig.4 it has already been suggested that it would be better to plot N_{kt} against $(q_t - \sigma_{vo})/\sigma_{vo}'$. Besides being dimensionless and accounting in a simple way for the depth of the test in the ground by means of the effective vertical stress, it is directly related to OCR, because

$$\frac{(q_t - \sigma_{vo})}{\sigma_{vo}'} = \frac{(q_t - \sigma_{vo})}{s_u} \cdot \frac{s_u}{\sigma_{vo}'} = N_{kt}\left(\frac{s_u}{\sigma_{vo}'}\right) \qquad (11)$$

The undrained strength ratio (s_u/σ_{vo}') varies with OCR in a well defined way - see equation (4) - and typical values for Boston Blue clay have been included in Table 1, and used to produce the plot of $N_{kt} \cdot s_u/\sigma_{vo}'$ against OCR in Fig.6.

There are few reported instances of sites where good quality CPT data can be correlated with reliable data of OCR. One of the best researched sites is that at Onsøy used by the Norwegian Geotechnical Institute. From the information reported by Lacasse and Lunne (1982) values of $(q_t - \sigma_{vo})/\sigma_{vo}'$ have been calculated for depths at intervals of 2 m up to 20 m. These have been plotted in Fig.7 against the relevant value of OCR taken from the profiles resulting from conventional oedometer tests.

This well-conditioned relationship suggests that values of $(q_t - \sigma_{vo})/\sigma_{vo}'$ are a good indicator of values of OCR, and further that they should be used in place of $(q_t - \sigma_{vo})$ in charts used for identification of soil type.

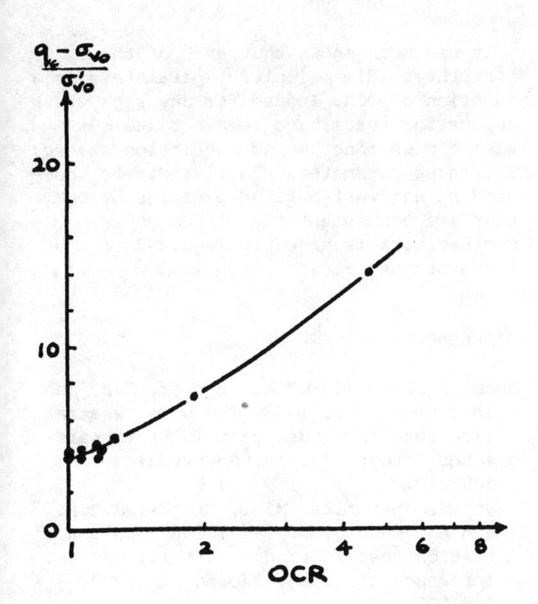

Fig.7 Variation in $(q_t - \sigma_{vo})/\sigma_{vo}'$ with OCR at Onsøy

4.2 Interpretation of friction ratio

The friction ratio is defined as the ratio of the unit sleeve friction f_s divided by the total end bearing pressure q_t. The value of f_s is directly related in the case of a clay to the undrained strength s_u. In a manner analogous to the assessment of the axial capacity of piles it can be simply expressed as

$$f_s = \alpha\, s_u \qquad (12)$$

where α is a factor dependent on the OCR of the clay.

The end bearing q_t is also a measure of the strength of the clay, but the total overburden pressure is deducted before a value of s_u is computed. Hence the writer has suggested, Wroth (1984), that it would be more satisfactory to divide the unit shaft friction by $(q_t - \sigma_{vo})$ instead of q_t, and to define a modified friction ratio

$$f_r^* = \frac{f_s}{(q_t - \sigma_{vo})} = \frac{f_s}{s_u} \cdot \frac{s_u}{(q_t - \sigma_{vo})} = \frac{\alpha}{N_{kt}} = f(\mathrm{OCR}) \qquad (13)$$

Both α and N_{kt} vary with OCR in a consistent manner so that the revised friction ratio f_r^* could be expected to be an indicator of OCR. For example, for a normally consolidated clay $\alpha \sim 1$, $N_{kt} \sim 12$ so that f_r^* would be about 8%, whereas for a heavily overconsolidated clay with $\alpha \sim 0.5$ f_r^* would be halved to about 4%.

4.3 Interpretation of pore pressure

The addition of a pore pressure transducer to a standard cone has added considerably to the value and importance of cone penetration tests, particularly regarding the identification of soil strata and the finer details of stratigraphy. Interpretation of the observed magnitudes of pore pressure to provide quantitative information about soil properties has been keenly debated, and is not satisfactory to date. Much of the difficulty has been due to the different locations of the porous element in various types of piezocone, see for example Campanella et al. (1982).

Various parameters have been suggested for interpretation of the pore pressures;

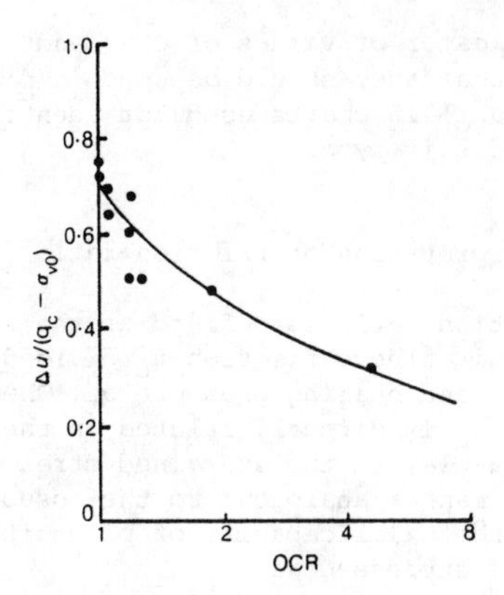

Fig.8 Variation of piezocone pore pressure ratio with OCR at Onsøy

it now seems to be generally agreed that the most appropriate is that denoted as

$$B_q = \frac{\Delta u}{(q_t - \sigma_{vo})} \qquad (14)$$

in which Δu is the excess pore pressure. This parameter is analogous to the pore pressure parameter A used for the interpretation of undrained triaxial tests and is directly related to the OCR of a clay. Results from the Onsøy site are given in Fig.8 where B_q is plotted against OCR derived from consolidation tests (as is the case for Fig.7).

Konrad and Law (1986) have attempted to deduce the preconsolidation pressure, and hence the OCR, from a combination of the end bearing pressure and the excess pore pressure measured by a piezocone. Their novel approach looks promising and warrants further investigation.

5 CONCLUSIONS

The present period in the steady development of soil mechanics and foundation engineering is seeing a rapid growth in the range and quality of in situ tests in general, including penetration tests. The purpose of this paper has been to consider the interpretations of the SPT and the CPT and to recommend the use of new and revised parameters for their improvement.

It has been emphasised that any correlation between results of such tests and soil properties should be

(a) based on physical insight

(b) set against a theoretical background and

(c) expressed in dimensionless form.

There is particular difficulty with correlations involving undrained shear strength because of the different results obtained on a soil in a given condition in different types of test.

For CPT tests in sand it is recommended that the approach proposed by Been et al. (1986, 1987) should be adopted with the parameter

$$\frac{(q_t - p_o)}{p_o'}$$

being related to the state parameter ψ.

For CPT tests in clays it is recommended that (i) instead of q_t or $(q_t - \sigma_{vo})$ correlations and charts should be in terms of

$$Q = \frac{(q_t - \sigma_{vo})}{\sigma_{vo}'}$$

(ii) instead of the friction ratio, a revised friction ratio should be used

$$F = \frac{f_s}{(q_t - \sigma_{vo})}$$

and (iii) the pore pressure parameter B_q is to be preferred, namely

$$B = \frac{\Delta u}{(q_t - \sigma_{vo})}$$

It has been shown that each of these last three (dimensionless) parameters is a function of OCR. Indeed for any site where successful tests have been carried out with a piezocone having a friction sleeve, all three parameters should indicate independent and well-defined profiles of OCR that are consistent one with another. Further work is urgently required to test these proposals.

REFERENCES

Been, K., Crooks, J.H.A., Becker, D.E. and Jefferies, M.G. 1986. The cone penetration test in sands: part I, state parameter interpretation. Géotechnique 36: 239-249.

Been, K., Jefferies, M.G., Crooks, J.H.A. and Rothenburg, L. 1987. The cone penetration test in sands: part II, general inference of state. Géotechnique 37: 285-299.

Campanella, R.G., Gillespie, D. and Robertson, P.K. 1982. Pore pressures during cone penetration testing. Proc.

2nd European Symp. on Penetration Testing Testing, Amsterdam 2:507-512
Ghionna, V.N., Jamiolkowski, M., Lacasse,S. Lancellota, R. and Lunne, T. 1983. Evaluation of self-boring pressuremeter. Proc. Int. Symp. on In Situ Testing, Paris 2:294-301.
Houlsby, G.T. and Hitchman, R. 1987. Calibration chamber tests of a cone penetrometer in sand. Oxford University Engineering Laboratory Research Report 1699/87.
Houlsby, G.T. and Teh, C.I. 1988. Analysis of the piezocone in clay. Proc. 1st Int. Symp. on Penetration Testing, Orlando (in press).
Houlsby, G.T. and Wroth, C.P. 1988. Research on the analysis of the piezocone at Oxford University. Report to 1st Int. Symp. on Penetration Testing, Orlando (in press).
Konrad, J-M and Law, K.T. 1986. Preconsolidation pressure from piezocone tests in marine clays. Proc. 39th Canadian Geotechnical Conf. Ottawa 219-227.
Ladd, C.C., Foott, R., Ishihara, K., Schlosser, F. and Poulos, H.G. 1977. Stress-deformation and strength characteristics: state of the art report. 1977. Proc. 9th Int. Conf. Soil Mech. and Found. Eng. Tokyo 2: 421-494.
Marsland, A. and Quarterman, R.S. 1982. Factors affecting the measurements and interpretation of quasi-static penetration tests in clay. Proc. 2nd European Symp. on Penetration Testing, Amsterdam 2:697-702.
Meyerhof, G.G. 1957. Discussion on research on determining the density of sands by spoon penetration testing. Proc. 4th Int. Conf. Soil Mech. and Found. Eng. London 3:110.
Schmertmann, J.H. 1979. Statics of SPT. J.Geotech.Eng.Div.ASCE 105:655-670.
Schmertmann, J.H. and Palacios, A. 1979. Energy dynamics of SPT. J.Geotech.Eng. Div.ASCE 105:909-926.
Skempton, A.W. 1986. Standard penetration test procedures and the effects in sands of overburden pressure, relative density, particle size, ageing and overconsolidation. Géotechnique 36:425-447.
Wroth, C.P. 1979. Correlations of some engineering properties of soils. Proc. 2nd Int. Conf. on Behaviour of Offshore Structures, London 1:121-132.
Wroth, C.P. 1984. The interpretation of in situ soil tests. 24th Rankine Lecture. Géotechnique 34:449-489.
Wroth, C.P. 1984. Site investigation practice - assessing BS 5930, Field testing I. Proc. 20th Regional Meeting, Engineering Group, Geol. Soc. Guildford 2:31-35.
Wroth, C.P. and Bassett, R.H. 1965. A stress-strain relationship for the shearing behaviour of a sand. Géotechnique 15:32-56.
Wroth, C.P. and Houlsby, G.T. 1985. Soil mechanics - property characterization and analysis procedures, Theme lecture No.1. Proc. 11th Int. Conf. Soil Mech. and Found. Eng. San Francisco 1:1-55.
Wroth, C.P. and Wood, D.M. 1978. The correlation of index properties with some basic engineering properties of soils. Canadian Geotech.J. 15:137-145.

Technical papers: Standard penetration test

Penetration Testing 1988, ISOPT-1, De Ruiter (ed.)
© 1988 Balkema, Rotterdam, ISBN 90 6191 801 4

Penetration testing in tropical lateritic and residual soils – Nigerian experience

Larry A. Ajayi
Foundation Engineering (Nigeria) Limited
L.A. Balogun
University of Lagos, Nigeria

ABSTRACT: The two most widely used penetration testing methods in tropical lateritic and residual soils of Nigeria are the dynamic Standard Penetration Test in boreholes and the static (Dutch) Cone Penetration Test using 2 to 20 tonne capacity machines with hydraulic measuring devices. Experience over thirty years of operation has revealed that some considerations need be given both in equipment design and results application to the peculiar nature of these deposits if their usage for foundation design purposes was to be further encouraged. Problems encountered in carrying out the two penetration tests with the available standard commercial equipment and results interpretation procedure based on european soil conditions are highlighted. An analysis of results from both methods has been made to correlate with laboratory results for strength and deformation characteristics. These results were obtained from routine soil investigation works at about fifty sites where the static penetration tests have been carried out at close proximity of boreholes in which standard penetration tests were performed and undisturbed samples obtained for laboratory testing.

1. INTRODUCTION

Penetration testing basically involves pushing or driving a steel tube or cone into the subsurface and monitoring the resistance to penetration mobilised in the soil. Although penetration testing has been extensively used in some parts of the world, particularly in Europe and America, it is only relatively recently that the technique has achieved widespread use in the wider english speaking world.

Penetration testing for foundation design was introduced into Nigeria in about 1955 and had since gained wide application as a supplement to borehole and soil sampling method of subsoil investigation. The equipment and method of application have been strictly restricted to the practice in Europe and in particular, in accordance with the relevant British Standard.

The two most widely used penetration test methods are:-

(i) the dynamic penetration test in which a falling hammer drives a string of rods connected to a split tube sampler or steel cone (Standard Penetration Test) and

(ii) the static (Dutch) cone penetration technique using the 2 to 20 tonne capacity machines.

In testing recently deposited sedimentary soils prevalent in the southern part of Nigeria, (see Figure 1) these methods of testing have presented no problems in execution and results application. However, experience in the greater part of the country underlain by lateritic, residual or redeposited soils of the Pleistocene to Pre-cambrain era has indicated that the adoption of these tests has not produced results which are as reliable as they are known for non-residual european or temperate climate soil conditions. To illustrate this phenomenon results from the two methods of testing

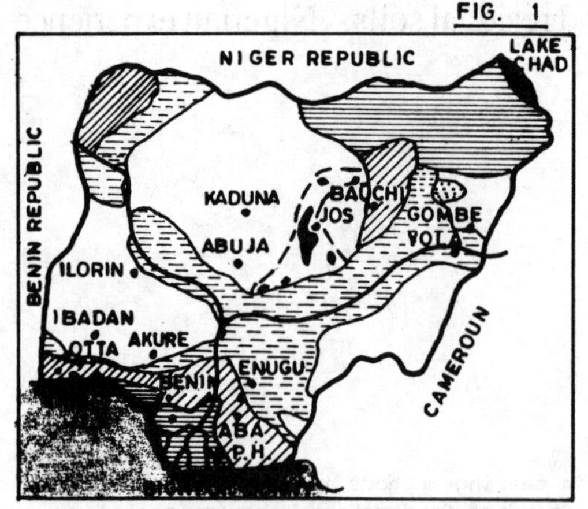

NIGERIA GEOLOGY

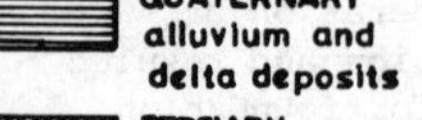

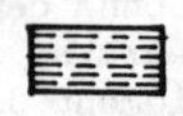

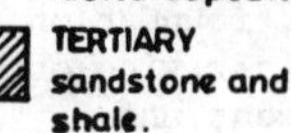

from routine subsoil investigation jobs have been collected and collated from sites where the static cone penetration tests were performed near boreholes in which the dynamic penetration tests were carried out. The aim of this paper is therefore to correlate the results of the SPT N-values with the CPT and laboratory shear strength for lateritic and residual soils of Nigeria.

2. TROPICAL SOILS

From the geological map of Nigeria, (Figure 1), it can be observed that major part of the country is underlain by basement complex rocks the weathering of which had produced lateritic materials spread over most of the area. Some of these materials have either been eroded by the abundance of rainfall and redeposited in the lower plains or remained in situ with or without some relict features of the parent rock. The more familiar typical lateritic and residual soil profiles are shown in Figure 2.

Unlike sedimentary soils and those of recent geological deposition along the southern coast line and in the various river flood plains, the soil deposits generally referred to here as tropical lateritic and residual soils have been known to possess complex macro and micro-fabic (Da Fontoura 1985), texture and structure (Melfi 1985), due to their morphological features. In Nigeria, they are normally encountered as firm to hard, unsaturated deposits and with obvious structural inconsistency both laterally and horizontally (Ajayi 1985) due to their mode of formation.

The sandy types of these deposits generally have high silt or clay contents often with some gravel size particles or cemented fragments. The types normally identified as clays also contain varying but significant sand admixture, with gravel or cemented fragments. A summary of the geotechnical properties of the samples used in this exercise is presented in Table 1.

TABLE 1. Summary of geotechnical properties

PARAMETER	RANGE
Moisture Content (w)	11 - 40%
Liquid Limit (W_L)	32 - 75%
Plastic Limit (W_P)	10 - 24%
Plasticity Index (I_P)	21 - 58%
Liquidity Index	- 0.28 to 0.43
Bulk Density (γ_b) kN/m³	19.2 - 21.8
Passing 2mm Sieve	57 - 100%
" 0.425mm Sieve	42 - 92%
" 0.063mm Sieve	21 - 81%
Clay Content $<$ 0.002mm	13 - 50%
Friction Ratio ($100 q_f/q_c$)	1.0 - 12.0

Because of their extreme susceptibility to disturbance during sampling, some form of reliable in situ testing technique for assessment of strength and deformation characteristics become imperative and the penetration testing appear most logical.

Fig. 2

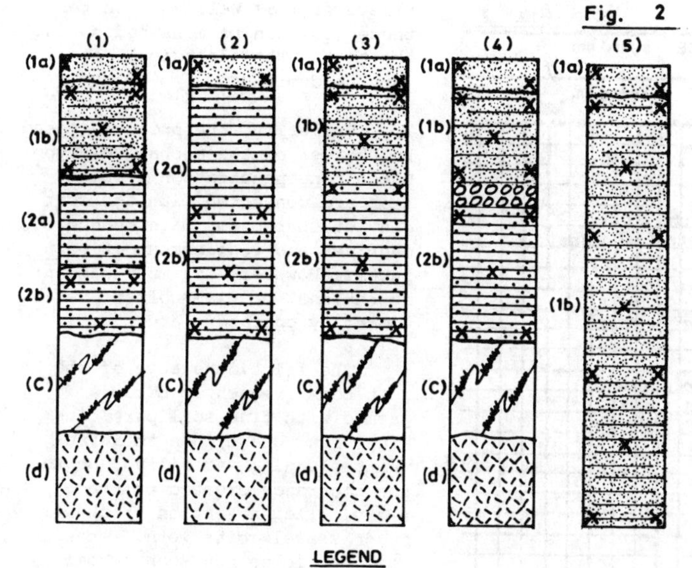

1a humous stained soil horizon
1b top soil
2a Laterite horizon
2b mottled zone
c pallid zone (weathered rock)
d sound parent rock

	PROFILE TYPE				
	1	2	3	4	5
Nature of the profiles	Residual matured	Residual fairly matured	Residual young	Residual non residual young	Non residual young
Local topographic location	Peneplain remnant, hill top, upper slope	Upper slope middle slope	Middle slope lower slope	Lower slope	Flat ground level

stone line divides residual and non-residual sections of profile (No4).

IDENTIFICATION OF LATERITE SOIL PROFILES.
(After Gidigasu 1972)

3. STANDARD PENETRATION TEST

Standard penetration tests are carried out at the base of boreholes as boring progresses. The method and equipment generally used are those described in B.S. 1377 : 1975. The test utilises a 51mm diameter split barrel sampler of thick wall construction which is fitted with a cutting shoe or steel cone tip. The assembly is connected by drill rods to the surface and driven under the impact of a 63.5kg hammer with a free fall of 760mm into the undisturbed soil at the base of boreholes. The sampler tube is subsequently extracted and dismantled to recover the sample.

Semi-automatic trip hammers are now gaining wide application but regular maintenance of the equipment is essential to guarantee a constant falling height. Some operators still use the rope-pull system which invariably yields unreliable results.

When the test is performed in gravelly soils or saprolites of the lower horizon of the residual or lateritic soil profiles where rock fragments are frequently encountered, the cutting shoe of the sampler is replaced with a 60° apex steel cone which has the advantage of eliminating the potential damage to the shoe but precludes the added advantage of recovering a sample of the soil penetrated.

Drive blows are preferably recorded for every 75mm penetration and the total for the third to sixth 75mm penetrations is normally quoted as the N-value.

4. STATIC CONE PENETRATION TEST

The static cone penetration testing technique and equipment commonly used in Nigeria are generally in accordance with the Dutch system and particularly, the 2, 10 and 20 tonne capacity machines with hydraulic measuring device. The test cones are of 60° apex with 10 sq.cm cross-sectional area, forced into the soil at a rate of 20mm per second. Two types generally used are the Mantle Cone and Friction Jacket Cone provided with a friction sleeve having a 150sq.cm surface area. With the latter type of test cone soil adhesion around the

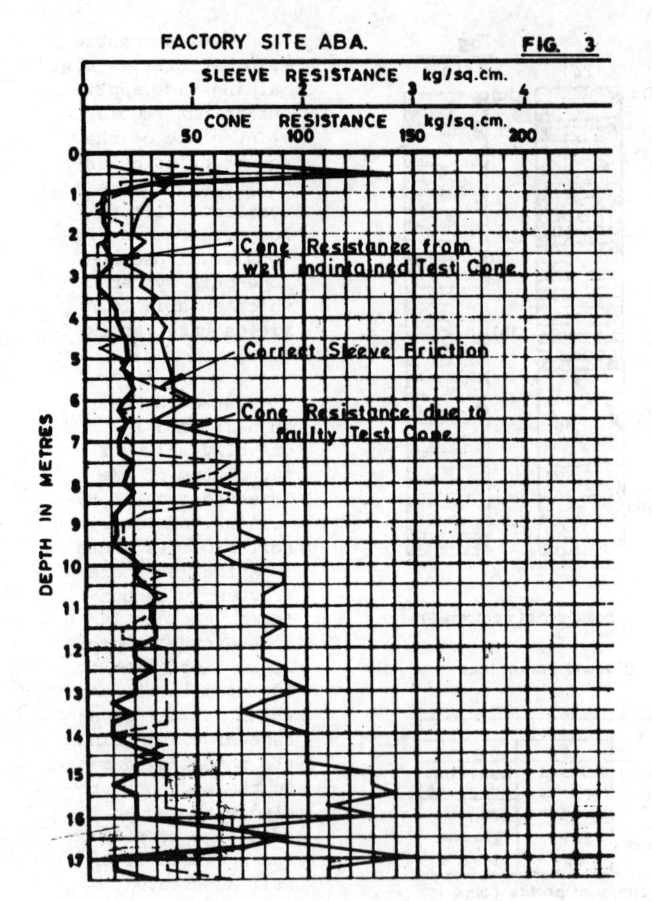

sleeve (fs) as well as cone resistance (q_c) can be measured during the test. The electrical cone is very seldomly used.

Primarily, the problems posed by the use of the mechanical type Dutch cone penetration testing gear are connected with the scarcity of spares for maintenance of machines and replacement of worn parts. However, the most common operational problems often overlooked by operators are:

The friction sleeve of the test cone frequently becomes jammed with fine soil particles after long usage, due to infrequent cleaning and oiling, or to the clearance created by the sleeve sliding section which progressively gets worn. These soil particles can soon become tightly packed and cause load transfers between the cone tip and sleeve. An example of result from such faulty cone compared with that from a good quality cone is show in Figure 3.

Adhesion (skin friction) developed along the penetrometer rods by tropical and residual soils can be so high that deep penetration is usually restricted.

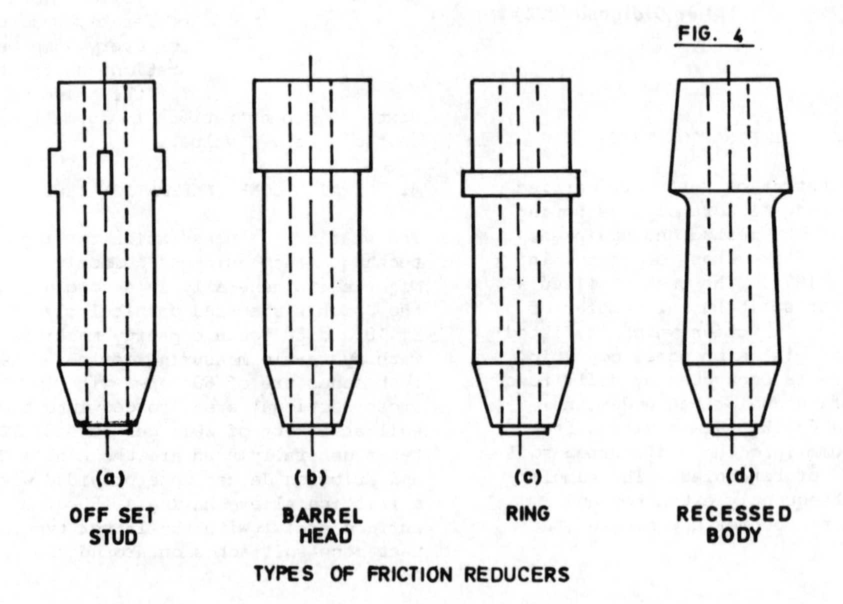

TYPES OF FRICTION REDUCERS

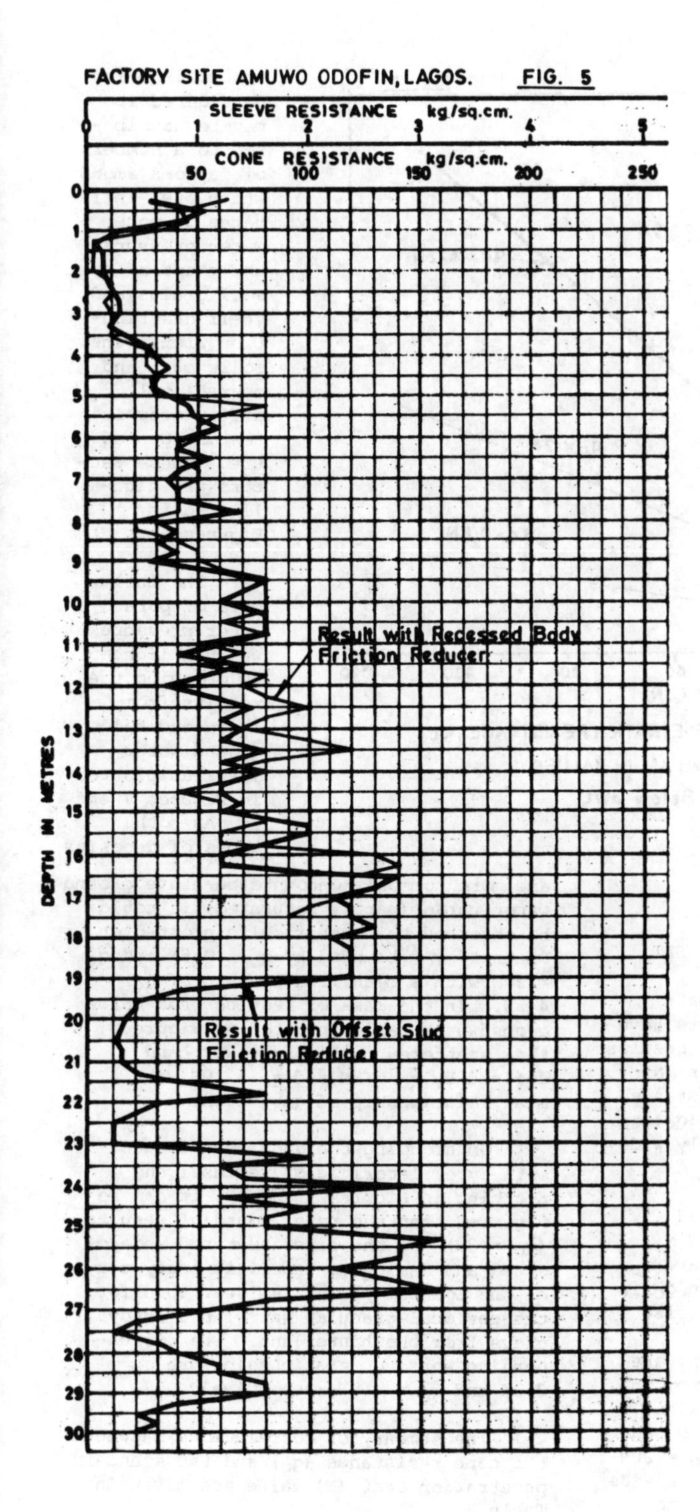

Hence, friction reducers are normally fitted above the cone assembly to minimise soil adhesion on the string of rods which pushes the cone to testing positions. The four most commonly used types of friction reducers are shown in Figure 4. Efficiency of usage of these friction reducers vary with soil types and improper test termination has very often occurred due to wrongful usage of the elements. The offset stud type of friction reducer (Figure 4a) allowed farthest penetration in the tropical stiff sandy clay soils. The next efficient type of friction reducer is the enlarged head type, (Figure 4b), then the ring type (Figure 4c), while the least effective in the residual and lateritic soils is the recessed body type (Figure 4d) which is most easily worn by intense usage. However, 4d has proved most efficient in sedimentary sandy deposits.

Besides the restricted penetration effect, misuse of these attachments also leads to erroneous penetration results. An example of such results is given in Figure 5 showing deeper penetration with stud type reducer than with recessed body type. Preboring and testing from bottom of boreholes can eleiminate some of these problems.

Infrequent recalibration of the pressure gauges used for resistance measurements leads to production of unreliable results.

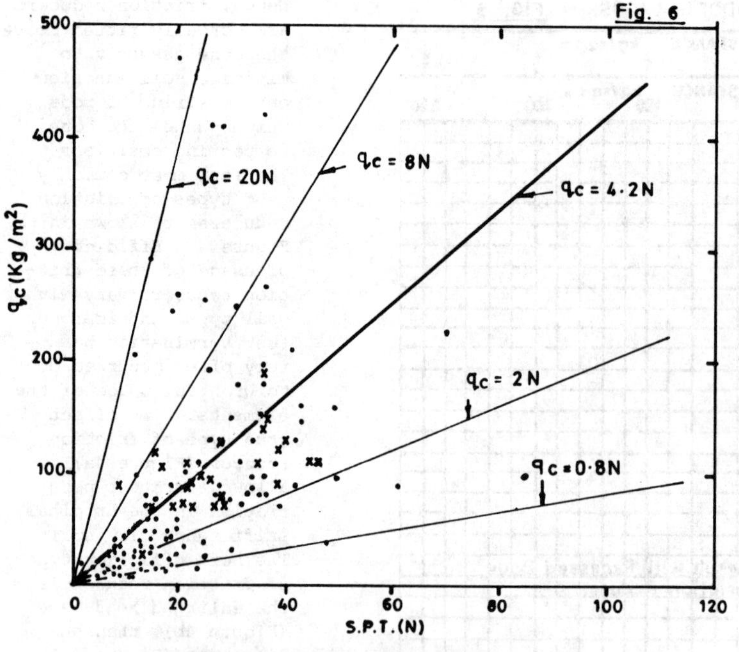

VARIATION OF CONE PENETRATION RESISTANCE (q_c) AGAINST SPT(N) ALL IN VALUES

• Above GWL x Below GWL

the soils being considered it may be attributed to a number of factors among which are: influence of pore water pressure, the shape of soil grains and their distribution within the soils mass and operational differences in testing procedure. These factors are highlighted in Figures 7 to 10.

The influence of pore water pressure was highlighted by grouping the results from above and below ground water level as plotted in Figures 7 and 8. As both types of deposits are more commonly encountered above ground water table (partly saturated) it could be observed that the results plotted on Figure 7 are bounded by q_c = 0.8N and q_c = 20N with a mean relationship of q_c = 4.5N. In the case of results from below groundwater table, plotted on Figure 8, they scattered within narrower band defined by q_c = 2N and q_c = 10N with a mean relationship of q_c = 4.0N.

On grouping the data according to soil type, lateritic or residual and plotting in Figures 9 and 10, it can be observed that the upper bound of results (q_c = 8N) for the finer textured redeposited and desicated lateritic sandy clays of the coastal plains is lower than for the residual granular saprolitic sandy clays from the hinterland. The corresponding working relationships are q_c = 3.2N and q_c = 4.5N respectively.

The summary of correlations between the cone resistance (q_c) and the standard penetration test (N) value are given in Table 2.

5. DISCUSSION OF RESULTS

5.1 Correlation between CPT and SPT

For almost as long as the two tests have been in practice, researchers have been attempting to correlate the results of the standard penetration tests (N) with the static penetration resistance measured in the Dutch cone test (q_c). Most reports indicate a relatively wide scatter of individual results.

Various authors have proposed different correlations for q_c and N, particularly, Sanglerat (1972) gave q_c varying between 2N and 18N, where q_c is measured in kg/cm². In this exercise average results of both tests at corresponding locations and depths are plotted as shown in Figure 6. From the results plotted a reasonable working relationship of q_c (kg/cm²) = 4.2N can be assumed, with q_c = 0.8N and q_c = 20N as the lower and upper bounds. The wide scatter of the results is similar to that reported by Kruizinga (1982) for sandy and clayey soils and in respect of

TABLE 2. Summary of N and q_c relationship.

Soil Type	q_c/N Range	Mean
All soil types and conditions	0.8 - 20.0	4.2
All types of soil above water table	0.8 - 20.0	4.5
All types of soil below water table	2.0 - 10.0	4.0
Lateritic Sandy Clay	0.8 - 8.0	3.2
Residual Sandy Clay	0.8 - 20.0	4.2

Schmertman (1970) proposed a more comprehensive correlation between N and q_c for different soil types as presented in Table 3.

Although no previous correlation in respect of tropical residual or lateritic soils has been reported, the correlation observed by the authors generally fall within the range given by Schmertman (1970) for other types of soil.

5.2 Correlation by means of regression analysis

The field readings were introduced into an APPLE 2 Micro-Computer programme to determine the "best fit" line by means of linear regression analysis. The degree of "fit" is expressed by the correlation co-efficient (r).

The function investigated for all types of soil was of the form $y = a_o + ax$ and the results are given in Table 4. The wide scatter of the results is relected in the low value of r in each case.

5.3 Effect of overburden pressure on penetration test results

The bulk density of the soils in this study range between 19 and 22kN/m³ and to investigate the effect of overburden pressure on the penetration results, Figure 11 shows the variation of cone resistance (q_c) with total overburden

TABLE 3. Relationship between Static Point resistance q_c (kg/cm²) and N-Values of SPT (after Schmertman, 1970).

Soil Type	q_c/N
Silts, sandy silts and slightly cohesive silt - sand mixture	2
Clean fine to medium sands and slightly silty sands	3 - 4
Coarse sands and sands with little gravels	5 - 6
Sandy gravels and gravel	8 -10

TABLE 4. Result of Regression Analysis of Data Available.

Soil Type	Regression Equation	Correlation Coefficient (r)
All soil types and conditions	q_c =4.29-4.6	0.630
All soil types above water table	q_c =4.49N-15.9	0.685
All soil types below water table	q_c =1.85N+52.6	0.538
Lateritic sandy clay	q_c =3.6N-5.4	0.575
Residual sandy clay	q_c =3.94N+48.9	0.410

pressure (P_o). It can be observed from the figure that the points are so scattered that no reliable deductions could be made. This is contrary to the experience of some researchers (Baldi et al 1982) with sands prepared at a given relative density. The indications from our exercise are that q_c is not governed by overburden pressure. This may be attributed to the material being normally encountered in partly saturated conditions and due to structural inconsistency and micro structures inherited

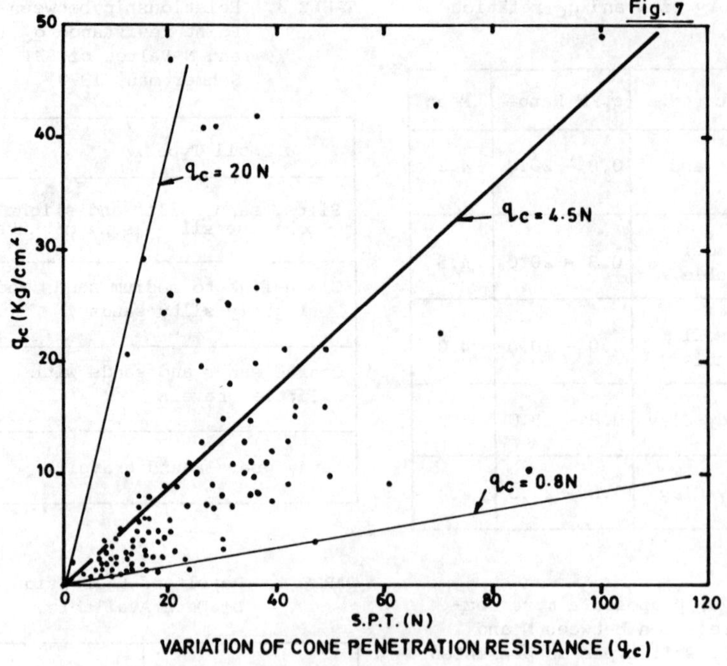

VARIATION OF CONE PENETRATION RESISTANCE (q_c) AGAINST SPT (N) FOR SOILS ABOVE WATER TABLE

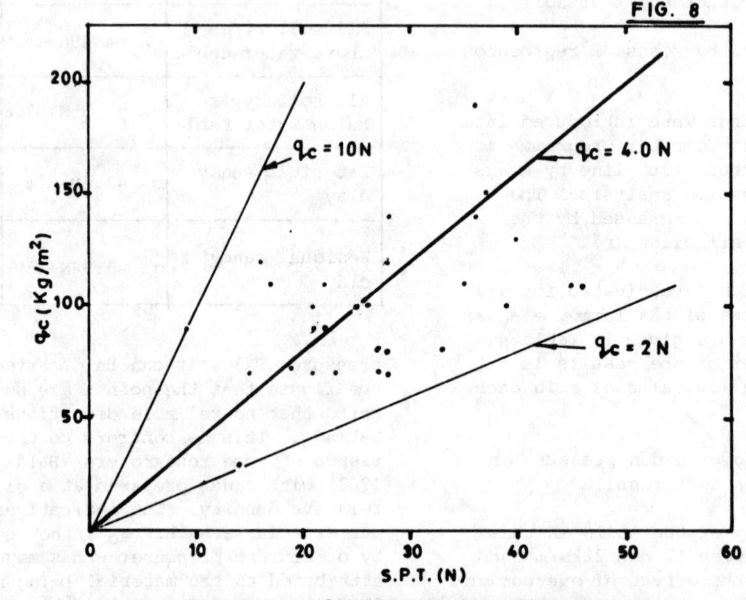

VARIATION OF CONE PENETRATION RESISTANCE (q_c) AGAINST SPT (N) FOR SOILS BELOW WATER TABLE

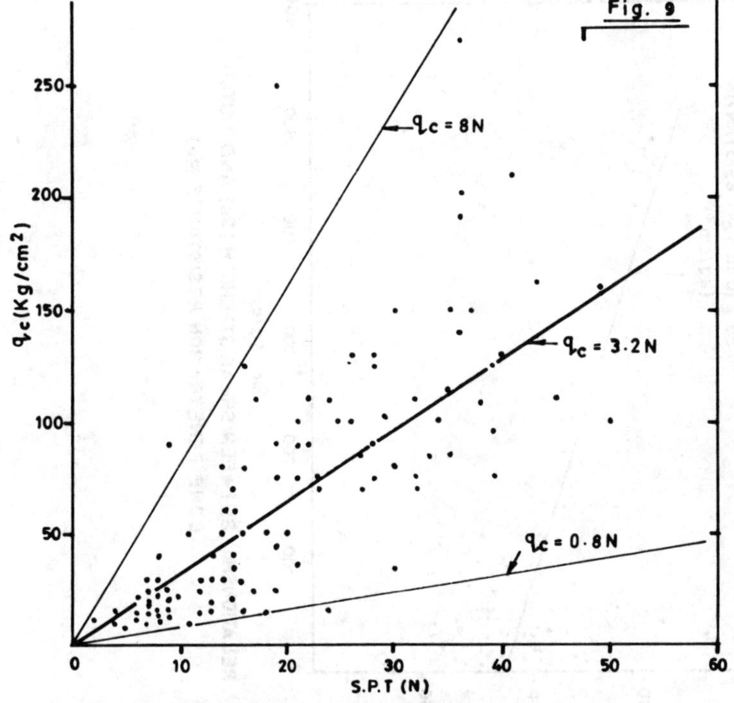

VARIATION OF (q_c) AGAINST SPT(N) FOR LATERITIC SANDY CLAYS

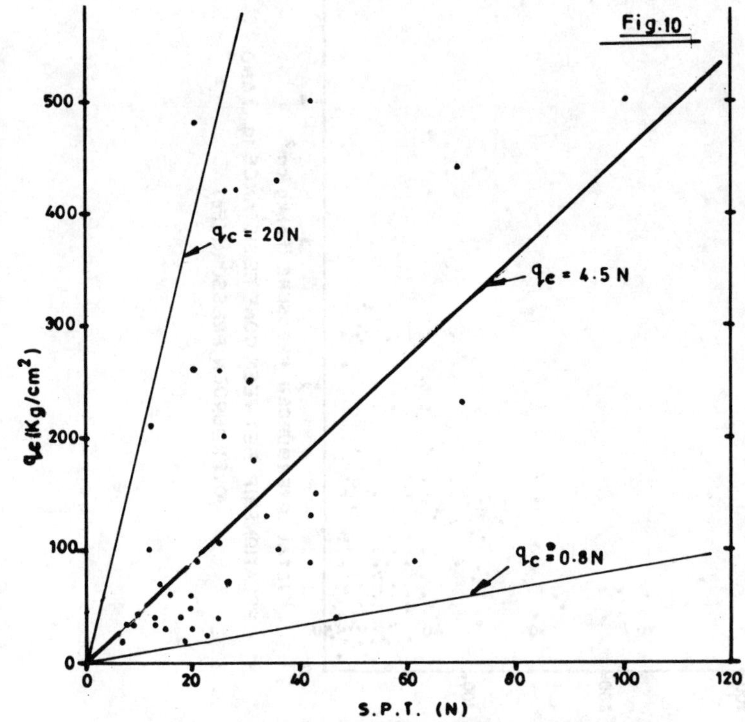

VARIATION OF (q_c) AGAINST SPT(N) FOR RESIDUAL SANDY CLAYS

from mode of formation.

5.4 Correlation of laboratory shear strength with CPT results.

In attempting to correlate the results of in situ tests with laboratory determination of shear strength the authors have collected data from those boreholes and cone penetrometer tests which were located within 5 metres of each other.

Various researchers have found that cone resistance varies directly with the undrained shear strength of clay (S_u) and the formula generally used to equate the two is:

$$S_u = \frac{q_c - P_o}{N_c} \quad .....(1)$$

Where

P_o = total overburden pressure

N_c = cone factor.

The shear strength (S_u) is usually defined as the value determined from the in situ vane test in fairly homogeneous fully saturated clays with ϕ_u assumed zero. However, the residual and lateritic soils were tested for this exercise in undrained unconsolidated triaxial compression by multistage technique using 100mm diameter specimens, same as obtained from site sampling and in

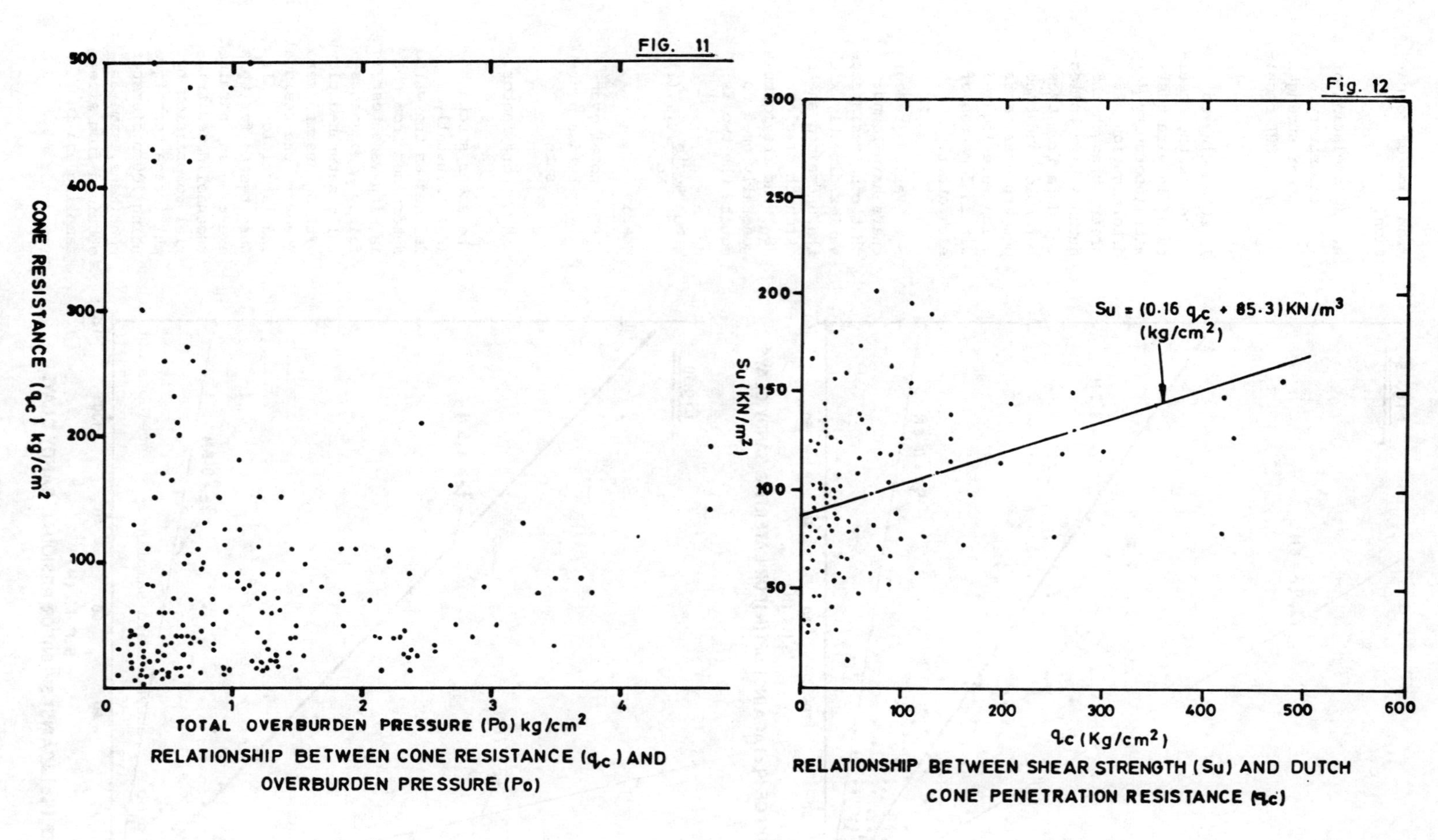

RELATIONSHIP BETWEEN CONE RESISTANCE (q_c) AND OVERBURDEN PRESSURE (Po)

RELATIONSHIP BETWEEN SHEAR STRENGTH (Su) AND DUTCH CONE PENETRATION RESISTANCE (q_c)

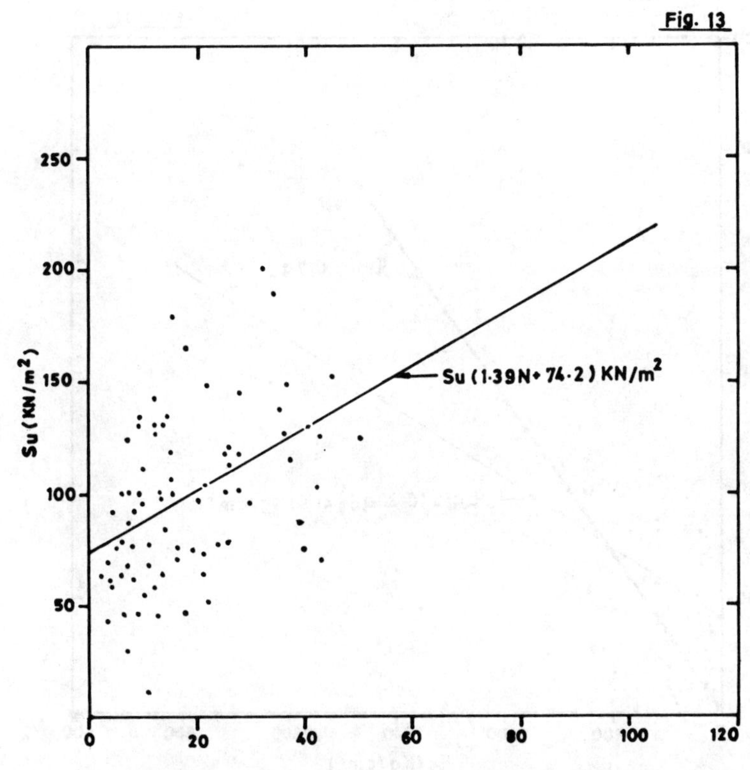

RELATIONSHIP BETWEEN SHEAR STRENGTH (Su) AND SPT (N) VALUES

consideration that they are generally partly saturated. They exhibit appreciable angle of shearing resistance, varying from 3° to 31°.

The results have been plotted against cone resistance at the appropriate depth in Figure 12. The shear strengths have been calculated using the general expression:

$$S_u = C_u + P_o \tan \phi_u \ldots\ldots\ldots\ldots (2)$$

Where C_u and ϕ_u are the apparent cohesion and angle of shearing resistance with respect to total stress

P_o is the total overburden pressure at the appropriate level.

Figure 12 shows a large scatter, much of which is probably due to field sampling disturbance. Analysis of the result by linear regression gave the relationship:

$$S_u = (0.16q_c + 85.3)\ \text{kN/m}^2 \ldots\ldots\ldots (3)$$

Where

S_u is the undrained shear strength of the soil in kN/m².

q_c is the cone penetration resistance in kg/cm³.

Equation (3) can be re-arranged as follows

$$S_u = \frac{q_c + (85.3 \times 6.25)}{6.25} = \frac{q_c + k}{N_k} \ldots\ldots (4)$$

Comparing equations (3) and (4), it is observed that the numerator has a positive rather than a negative sign for the constant (k). N_k is approximately equal to 6 which may be regarded as the cone factor. This appears to confirm that for the residual and lateritic soils in Nigeria, the cone penetration resistance does not depend on overburden only as demonstrated by Figure 11.

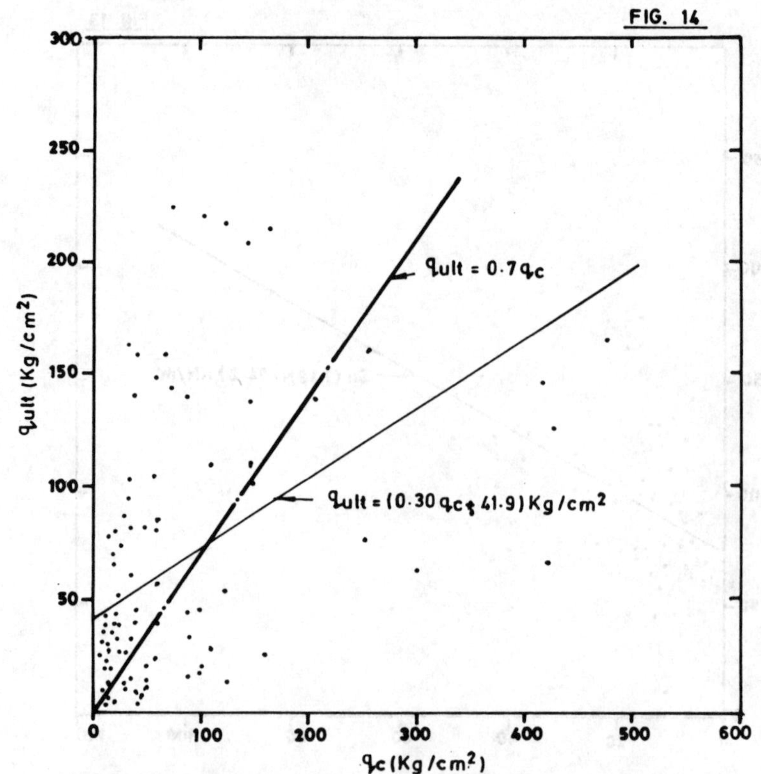

RELATIONSHIP BETWEET ULTIMATE BEARING CAPACITY (q_{ult}) AND DUTCH CONE PENETRATION RESISTANCE (q_c)

5.5 Correlacíon of laboratory shear strength with SPT

The laboratory shear strength measurements and the corresponding SPT N-values at the appropriate depths have been plotted in Figure 13. Again the very wide scatter observed in the figure is probably due to field sampling disturbance on results of both tests. Linear regression analysis of the results presented gave the "best fit" line as:

$$S_u = (1.39N + 74.2) \text{ kN/m}^2 \quad \ldots\ldots\ldots (5)$$

The co-efficient of correlation is 0.405 which reflects the wide scatter of the results.

Previous researchers had observed similar wide scatter of the results. The straight lines obtained represented the best fit for widely scattered observations. Their correlations took no account of the effect of overburden nor did they separately take into account the effects of over consolidation, as has been in this exercise.

5.6 Correlation of penetration resistance with ultimate bearing capacity

Both penetration tests have been considered as measurements of failure of the soil against a dynamic or quasi-static force applied on a circular surface. Since the test depth is always greater than five times the diameter of the loaded area, the test can also be regarded as a deep foundation failure with conditions similar to those implied in Meyerhof (1956) analysis - that ultimate failure occurs with significant downward movement before the full shearing resistance of the soil is mobilised. Consequently, the ultimate bearing capacity at various levels was determined from the formula:

$$q_{ult} = cN_c + P_o N_q \quad \ldots\ldots\ldots (6)$$

using Meyerhof's N_c and N_q values for

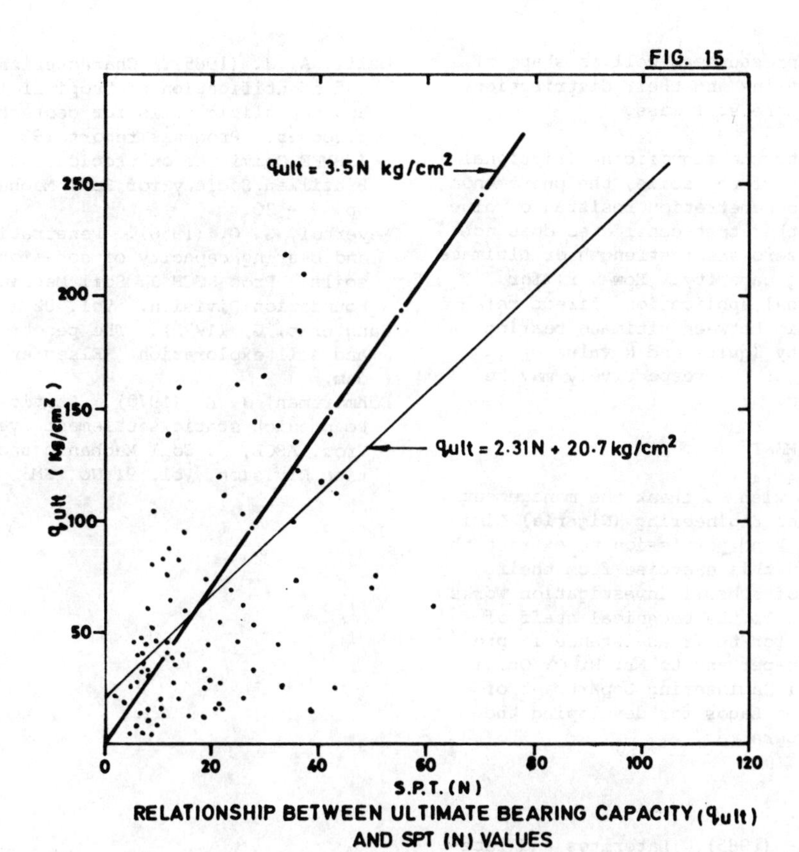

RELATIONSHIP BETWEEN ULTIMATE BEARING CAPACITY (q_{ult}) AND SPT (N) VALUES

deep foundation.

Ultimate bearing capacity has been plotted against q_c in Figure 14 and against N in Figure 15. As previously discussed, by linear regression analysis, the "best fit" line for these relationships have been defined by:

$$qult = (0.30q_c + 41.9)\ kg/cm^2 \ldots\ldots (7)$$
$$and (2.31N + 20.7)\ kg/cm^2 \ldots\ldots (8)$$

For practical applications, the mean relationship for the plotted results was established as:

$$qult = 0.70q_c\ (kg/cm^2) \ldots\ldots\ldots\ldots (9)$$
$$or\ 3.5N\ (kg/cm^2) \ldots\ldots\ldots\ldots (10)$$

6. CONCLUSIONS

From the results of this study the following conclusions can be made:-

1. The behaviour of dynamic standard penetration (SPT) and static Dutch cone penetration (DCP) tests in tropical residual and lateritic soils of Nigeria is not entirely similar to that experienced in temperate climate sedimentary soil deposits. Consequently, modification of results application must be developed to ensure efficient practice.

2. The best friction reducer to use in cone penetration testing in tropical residual and lateritic soils is the 'offset stud' type.

3. Correlation between the Dutch cone penetration resistance (q_c) and SPT (N) values is rather poor but a general and realistic relationship may be taken as $q_c = 4.2N$, where q_c is in kg/cm².

4. Penetration test results seem not to be affected by overburden pressures in these deposits but they appear to be significantly affected by pore

water pressure as well as shape of soil grains and their distribution within the soil mass.

5. Owing to the significant frictional element of the soils, the phenomenon of zero penetration resistance value for either test considered does not imply zero shear strength or ultimate bearing capacity. However, for practical application, direct relationship between ultimate bearing capacity (qult) and N value or q_c of 3.5 or 0.7 respectively may be adopted.

ACKNOWLEDGEMENT

The authors wish to thank the management of Foundation Engineering (Nigeria) Limited for the kind permission to extract the data used in this exercise from their many years of subsoil investigation works in Nigeria. To the technical staff of the company for their assistance in preparing the paper and to Mr. N. O. Obaji of the Civil Engineering Department of University of Lagos for developing the computer programmes.

REFERENCES

Ajayi, L. A. (1985). Laterites and lateritic soils of Nigeria. ISSMFE Golden Jubillee Commemorative Volume - Nigeria, NGA Lagos. pp 7 - 15.

Baldi, G. Bellotti, R. Ghionna, V. Jamiolkowski, M. and Pasqualini, E. (1982). Design parameters for sand from CPT. Proc. second european symposium on penetration testing, Amsterdam, Vol.1i pp. 425 - 432.

B.S. 1377 (1971). Methods of test for soils for civil engineering purposes. British Standard Institution, London.

Da Fontoura S.A. (1985). Shear strength of undisturbed tropical lateritic soils. Progress report 1982 - 1985 ISSMFE Committee on tropical soils. Brazilian Society for Soil Mechanics pp. 47 - 66.

Gidigasu, M. D. (1972). Mode of formation and geotechnical characteristics of laterite materials of Ghana in relation to soil forming factors. Engineering Geology, Amsterdam, Vol. 6 pp 75 - 150.

Kruizinga, J. (1982). SPT-CPT correlations. Proc. second european symposium on penetration testing. Amsterdam, Vol. 1 pp. 91 - 94.

Melfi, A. J. (1985). Characterization and identification of tropical lateritic and saprolitic soils for geotechnical purposes. Progress report 1982 - 1985 ISSMFE Committee on tropical soils. Brazilian Society for Soil Mechanics pp. 9 - 20.

Meyerhof, G. G. (1956). Penetration tests and bearing capacity of cohesionless soils. Proc ASCE J. Soil Mechanics and Foundation Division. Vol. 82 (1) P.866.

Sanglerat, G. (1972). The penetrometer and soil exploration. Elsevier Amsterdam.

Schmertman, J. K. (1970). Static cone to predict static settlement over sand. Proc. ASCE, J. Soil Mechanic and Foundation Division, Vol. 91 No. SM3.

Penetration Testing 1988, ISOPT-1, De Ruiter (ed.)
© 1988 Balkema, Rotterdam, ISBN 90 6191 801 4

Preliminary geotechnical characterization of pyroclastic rocks in Rionero in Vulture (Basilicata, Italy)

G.Baldassarre & B.Radina
Department of Geology and Geophysics, University of Bari, Italy
C.Cherubini
Institute of Engineering Geology and Geotechnics, University of Bari, Italy

ABSTRACT: The soils studied are pyroclastic rocks present in the volcanic area near Rionero in Vulture (Basilicata, Italy). These pyroclastic rocks are also differentiated in the field by means of boreholes sunk into three units which are designated from bottom to top: pyroclastic rocks in thick layers and layers, thinly layered pyroclastic rock and pyroclastic rocks mainly in thick layers, with blocks. The three units, by the point of view of geotechnical characterization, can hardly be differentiated with regard to particle size. These soils can be classified as medium dense to dense with rather high friction angles (32°-45°) measured directly in consolidated drained direct shear tests and determined also by means of some correlations with number of blows of Standard Penetration Test.

1 INTRODUCTION

In Basilicata the areas of outcroppings of volcanic deposits (the Monte Vulture region - Hieke Merlin, O., 1967) are situated in numerous cases where there are residential and industrial zones. However these deposits have only been analysed from a geologico-technical point of view after the earthquake of 23 November 1980 affecting the areas of Campania and Irpinia (Baldassarre G., Francescangeli R., Radina B., 1984). By studying in greater depth one of the above-mentioned residential areas (Rionero in Vulture) it has been possible to extend our analysis of the subsoil and to determine geotechnical properties, also in situ, by means of Standard Penetration Tests. It has been considered worthwhile to make our findings known as, for the time being, they represent the first geotechnical characterization of the volcanic deposits of the Monte Vulture.

Moreover, these results may also constitute a preliminary orientation for those concerned with the restoration and/or reconstruction of residential areas and their related infrastructures, as well as being useful for purposes of comparison with other similar types of deposits in Italy (Penta F., Croce A. and Esu F., 1961; Croce A. and Pellegrino A., 1967).

2 THE SOILS ANALYSED

The volcanic deposits of the residential zone of Rionero in Vulture and the surrounding area have already been classified as belonging to three distinct lithostratigraphic units on the basis of studies carried out (La Volpe L. and Piccarreta G., 1971): for pratical, geotechnical purposes they may be distinguished as follows (see Figure 1):

1 pyroclastic rocks, mainly in thick layers, with blocks (PB);
2 thinly layered pyroclastic rocks (PS);
3 pyroclastic rocks in layers and thick layers (Psb).

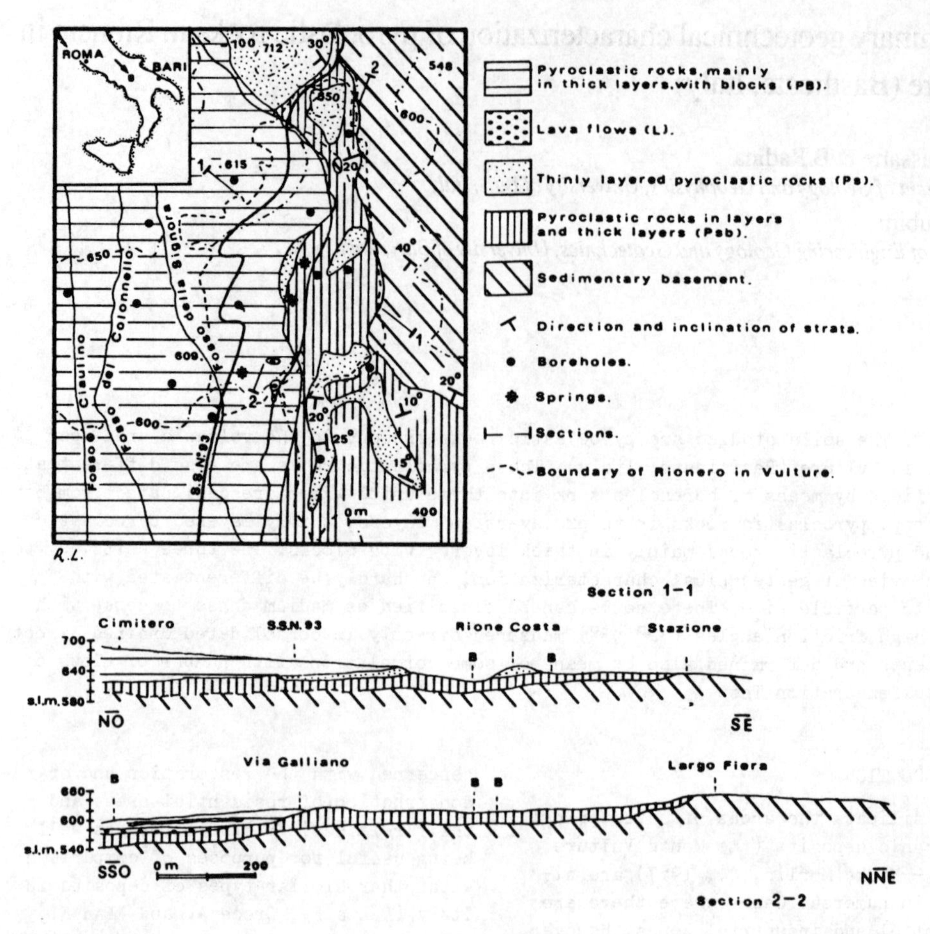

Fig. 1 Geolithological map of Rionero in Vulture

It has been ascertained by means of 19 boreholes sunk up to a depth of 30 m below ground level that these units occur from bottom to top in the order given above. By correlating the stratigraphic columns of the boreholes, moreover, it has been noted that these units are present in the subsoil with varying thicknesses (the highest values are those of the PB units), while the Ps unit is the least present both in the subsoil and on the ground surface. It is also worth mentioning that within the PB unit and between the PB unit and the underlying Psb unit there are lava flows of several meters in thickness. The above-mentioned conditions of the subsoil are interpreted in the two sections of Figure 1.

The areas of outcroppings of the volcanic deposits, which are pratically impermeable, are of a gently undulating nature with culminations generally in a N-S direction.

3 GEOTECHNICAL CHARACTERISTICS

It may be worthwhile making a few observations on the results of certain laboratory and field tests carried out on three types of the pyroclastic rocks that had been previously identified.

It should be pointed out straight away that this geotechnical characterization in only of a preliminary nature, especially in view of the low number of tests on which it has been possible to make observations and hypotheses.

Geotechnical analyses carried out in

the laboratory have enabled us to acquire data in the following parameters:

1 Grain size distribution
2 Total unit weight
3 Effective shear strenght parameters measured in direct shear tests

The field analyses, on the other hand, consisted exclusively of Standard Penetration Tests.

As regards grain size, it should be noted that the three types of rock have substantially the same ditribution; in other words, they are all constituted predominantly of (fine and large-grain) sands with percentages of not more than 20% of gravel and/or silt (Figure 2).

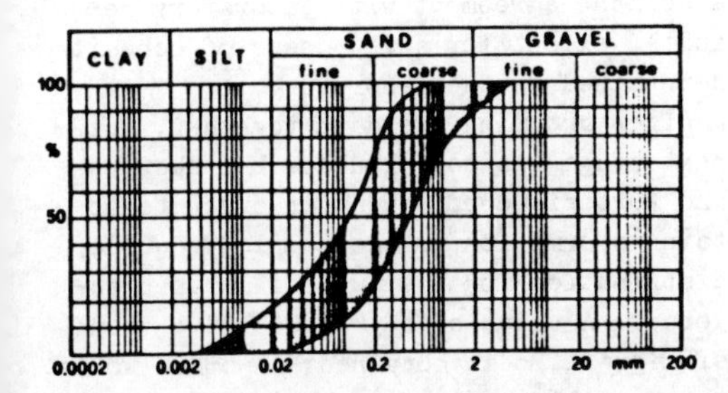

Fig. 2 Grain size distribution of soil studied

Data relating to total unit weight are very variable, ranging from 1.29 to 1.97 gr/cm^3 with values most frequently around 1.65 gr/cm^3. It has been observed that PS soils usually have the lowest values.

The shear tests carried out have provided the results that are summarised in Figure 3. In the ordinates the effective friction angle φ' is reported while in the abscissae the effective cohesion, whenever it occurs, is reported.

Thus a possible "field of strenght parameters" can be identified between the two continuous-lined curves marked in the same figure.

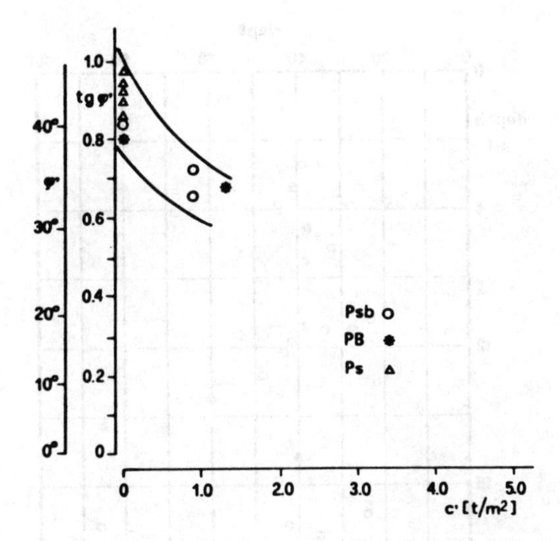

Fig. 3 Effective cohesion and friction angle values determined on shear box apparatus

The data are not particularly numerous but they do highlight the fact that the above-mentioned soils generally possess a high shear strenght due largely to the high friction angles (from 32° to 45°) and zero or rather low cohesion values, probably linked to limited phenomena of diagenesis. First laboratory results substantially confirm some results of Croce A. and Pellegrino A. (1967).

Given the larger number of determinations, we may draw useful conclusions from the results of the Standard Penetration Test (Pasqualini E., 1983). These, however, only concern the Psb and PB units.

In the diagram in Figure 4 are reported the values of the number of blows in Standard Penetration Test (N_{SPT}) as a function of the depth of the two units.

Dispersions of considerable entity have been noted as regards the Psb soils (from a minimum of 15 to a maximum of 86 blows) while the results of the tests carried out on the PB soils show a more limited range of variability (from 28 to 60 blows).

The relative density that may be hypothesized according to Gibbs H.J. and Holtz W.G. (1957) and Bazaraa A.R.S.S. (1967) is mainly situated between 60 and 75%.

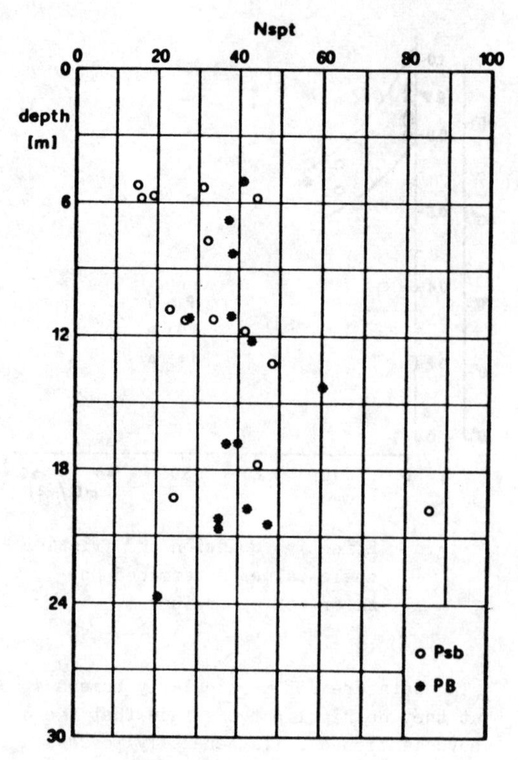

Fig. 4 Standard Penetration Test determinations as a function of depth

After having modified the rough results so as to take into account the effect of depth, it is possible to make an albeit approximate calculation of the friction angle.

For this purpose there are two laws available (Shioi Y. and Fukui J., 1982):

$$\varphi = \sqrt{15\ N_{SPT}} + 15 \qquad (1)$$

$$\varphi = 0.3\ N_{SPT} + 27 \qquad (2)$$

It has thus been possible to calculate a series of values for φ' for the PB and Psb soils by using law (1). The results obtained by using law (2) do not vary noticeably from the former. Some of the most significant statistical indices are reported below:

Psb soils
Maximum value φ' = 45.00°
Minimum value φ' = 33.17°
Mean value φ' = 38.86°
Standard deviation 3.59°

The corresponding coefficient of variation is just above 9%.

PB soils
Maximum value φ' = 43.98°
Minimum value φ' = 34.75°
Mean value φ' = 38.27°
Standard deviation 3.79°

Also in this case the coefficient of variation is just above 9%.

It can be observed immediately that, without resorting to particular statistical tests, there are no substantial differences between the two soils, at least as regards the data for friction angles deducible from Standard Penetration Tests with the relationship provided in the literature.

Moreover, if we observe the variation ranges for φ' we note that they are in excellent agreement with laboratory results, except for a few cases of cohesion determined in the laboratory.

After grouping the data together, the next stage was to calculate the skewness and kurtosis coefficient in order to obtain information on the sample frequency distribution for the two soils taken together. A value of β_1 = 0.06 and a value of β_2 = 2,35 is obtained; in other words, the overall sample ditribution is somewhere between the normal and the uniform distributions.

Further information may be deduced by processing the friction angles values, obtained in an indirect mannen, as a function of depth.

By meansof two randomness tests such as "runs test" and "mean squared successive difference test" (Wonnacott T.H. and Wonnacott R.J., 1977) it has been possible to ascertain that even major variation that may be observed for the friction angles in each single unit does not depend on depth. Thus no significant trends or groupings have been found and thus the variation in the φ' parameter is, at least for the depths analysed (0-24 m), completely random.

4 CONCLUSIONS

The soils of the volcanic complex of

Monte Vulture (Basilicata, Italy) have never been characterized from the geotechnical point of view. The first results from Standard Penetration Tests carried out on some of these volcanic soils (pyroclastic rocks) are reported and compared with the corresponding data deriving from laboratory tests. Results show a close agreement between the different methodologies used and also in relation to the data available for similar soils in other parts of Italy.

REFERENCES

Baldassarre G., Francescangeli R. and Radina B. 1984. Le cavità del centro storico di Rionero in Vulture (Basilicata) in relazione a problemi tecnici. Geol. Appl. e Idrogeol., vol. XIX, Bari

Bazaraa A.R.S.S. 1967. Use of the Standard Penetration Test for Estimating Settlement of Shallow Foundations on Sand. Ph. D. Thesis. University of Illinois Urbana, U.S.A.

Croce A. and Pellegrino A. 1967. Il sottosuolo della città di Napoli. Caratterizzazione geotecnica del territorio urbano. A.G.I., VIII Convegno, Cagliari.

Gibbs H.J. and Holtz W.G. 1957. Research on Determining the Density of Sands by Spoon Penetration Testing. Proc. IV Int. Conf. on Soil Mech and Found. Eng. London.

Hieke Merlin O. 1967. I prodotti vulcanici del Monte Vulture (Lucania). Mem. Ist. Geol. Min. Università di Padova, 26.

La Volpe L. and Piccarreta G. 1971. Le piroclastiti del Monte Vulture (Lucania). Le pozzolane di Rionero e Barile. Rend.Soc.It.Min. e Petr., 27, Pavia.

Pasqualini E. 1983. Standard Penetration Test. XI Ciclo di conferenze di Geotecnica di Torino.

Penta F., Croce A. and Esu F. 1961. Caratteristiche geotecniche dei terreni vulcanici. Geotecnica, vol. VIII, n.2, Roma.

Shioi Y. and Fukui J. 1982. Application of N-value to Design of Foundations in Japan. Proceedings of the II European Symp. on Penet. Testing. Amsterdam.

Wonnacott T.H. and Wonnacott R.J. 1977. Introductory Statistics. Wiley. New York.

Penetration Testing 1988, ISOPT-1, De Ruiter (ed.)
© 1988 Balkema, Rotterdam, ISBN 90 6191 801 4

SPT-CPT correlations for granular soils

Chung-Tien Chin, Shaw-Wei Duann & Tsung-Chung Kao
Moh and Associates, Inc. Taipei, Taiwan

ABSTRACT: With the increasing use of the Cone Penetration Test (CPT), it would be of significant value to establish a reliable correlation between the cone tip resistance, q_c, and the Standard Penetration Test (SPT) blow count, N-value. Based on recent data obtained from sand deposits, a historical review on SPT-CPT correlations is presented. For sands, the q_c/N ratio decreases significantly with increasing fine content. This paper suggests that the q_c/N ratio can be better correlated with fine content instead of the mean grain size for granular soils. It is important for geotechnical engineers to be aware of the scatter of the q_c/N ratio caused by the inherent variability of the penetration tests. The SPT N-value used to establish the local correlation should be corrected by the energy level.

1 INTRODUCTION

Up to the present, the Standard Penetration Test (SPT) is still one of the most commonly used in-situ tests for site investigation. Many empirical relations have been established between the SPT blow count, N-value, and other engineering properties of soils. Although geotechnical engineers use these correlations in foundation design, continued effort has been made recently for the standardization of the SPT. It is believed that the application of a measured energy correction factor will lead to more repeatability and reliability of the SPT N-value in the future (Campanella and Robertson, 1982; Kovacs and Salomone, 1982).

The Cone Penetration Test (CPT) is becoming increasingly popular for its unequalled ability to delineate soil stratigraphy and measure soil properties rapidly and continuously. Similar to the use of the SPT results, many correlations need to be established for the direct application of CPT results. Before these relations can be set up, it is very valuable to correlate the cone tip resistance, q_c, to SPT N-value so that the available data base of the field performances and property correlations with the N-value could be effectively utilized.

The purpose of this paper is to review some of the previous researches on the q_c/N ratio and to present a correlation for granular soil between the q_c/N ratio and its fine content.

2 HISTORICAL REVIEW

Compiling a number of studies, Robertson et al. (1983) presented the q_c/N ratio as a function of mean grain size, D_{50} (Fig. 1). This presentation provides a very useful guideline to convert the cone tip resistance to the equivalent N-value for soils with D_{50} varying between 0.001 mm to 1 mm, namely, from clay to gravelly sand. It should be noted that each data point represents the result of one site. Energy correction of the N-value has not been applied to most of their data. Figure 1 indicates that the q_c/N ratio increases with increasing mean grain size. They have also pointed out that the scatter of the q_c/N ratio also increases with increasing mean grain size.

In the last few years, many research efforts have been made to establish more reliable local correlations between SPT and CPT. For example, Moh (1985) has collected many research results on the correlations between q_c and N-value. In his report, the q_c/N ratio for silty sand can vary between 3 and 6 as proposed by different researchers.

It is interesting to note the results presented by Kasim et al. (1986) on naturally and hydraulically filled sand in Alameda, California, USA. In Fig. 1, each data point of their results is based on one energy corrected SPT blow count (N_{55}) and its corresponding q_c and D_{50}. The Alameda data show that the curve suggested by Robertson et al. (1983) is a good average, but the scatter around this average is significant. For the mean grain size varying between 0.14 mm and 0.28 mm, the q_c/N ratio ranges from 2 to 7. It is also noted that the standard deviation of D_{50} is relatively small, yet the fine content of these sandy materials varies significantly.

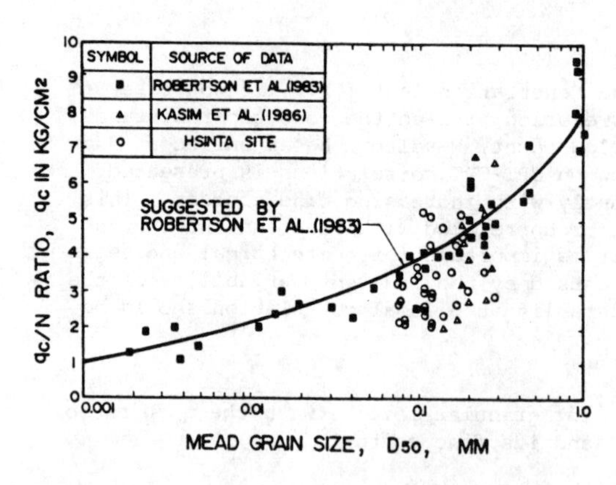

Fig. 1 Variation of q_c/N Ratio with Mean Grain Size

Generally speaking, these available data imply that the correlation suggested by Robertson et al. (1983) provides a good framework to start with, but the direct application of the average curve in engineering practice may lead to significant deviation. This paper, therefore, attempts to establish a more stringent correlationship between q_c and N-value.

3 SITE INVESTIGATION

In this research, the site studied is the Hsinta Power Plant in Taiwan. In order to prepare the preliminary design for two generator units, seven boreholes were drilled and eighteen cone penetration tests were conducted. Within the area and depth where the correlation tests were conducted, the subsoil conditions can be subdivided into three layers. The top layer consists of hydraulic sand fill and natural sand with a total thickness of approximately 7 m. Underlying this sand layer is a 13 m clay layer. Immediately below this cohesive deposit is a sand layer which can be as thick as 35 m. The groundwater table is typically at 2.5 m below ground surface. Grain size analyses were carried out on split spoon samples following ASTM Standard D422-63 (1972). According to the Unified Soil Classification System, both the hydraulically filled sand and natural sand are generally classified as SM.

Standard Penetration Tests were conducted by using a rope and cathead assembly to raise and drop the donut type hammer. Many variables which may affect the N-value have been carefully taken into account. The kinetic energy computed from the impact velocity is compared to the theoretical free fall energy. The energy correction factor was then applied to the measured N-value to calculate N_{55} which corresponds to 55% of the standard energy.

Cone penetration soundings were made using a Hogentogler type electronic cone. Tip resistance, skin friction and cone inclination were continuously recorded during penetration. Pore water pressures measured in a few tests indicated that the difference between the uncorrected tip resistance, q_c, and the corrected tip resistance, q_t (Jamiolkowski et al., 1985) is within 5%.

A total of 35 data points of sand deposits were selected from this investigation. A summary of these field measurements and laboratory test results are tabulated in Table 1.

4 CORRELATION ANALYSIS

Results obtained from Hsinta site were superimposed on Fig. 1. Similar to the Alameda data, these results also show that the curve presented by Robertson et al. (1983) can be best served as a reasonable average, while the direct utilization of this curve may end up with significant deviations. It should be mentioned that data plotted on Fig. 1 are based on SPT N-values corrected by the energy level. If this correction would not have been made, the scatter of the q_c/N ratio is much greater. It is noticed that the D_{50} of Hsinta data has a mean value of 0.13 mm with a standard deviation of 0.05 mm. However, the fine content varies from 13% to almost 50%. This suggests that the large variation of q_c/N for granular soils (except gravelly sands) may be better reflected by the fine content rather than mean grain size. Therefore, Fig. 2 was developed to evaluate the relationship between the q_c/N ratio and the fine content. It clearly illustrates that the

Table 1 Penetration and Grain Size Data of Hsinta Site

Depth m	N_{55}	q_c kg/cm^2	FR*	FC**	D_{50} mm
2.0	12.85	48.14	0.26	35	0.100
4.0	5.00	13.46	0.43	25	0.170
38.5	39.96	134.84	0.78	25	0.095
44.5	77.07	170.65	1.27	44	0.083
21.0	30.68	106.59	0.91	23	0.100
31.0	29.29	62.32	0.30	37	0.084
33.0	19.52	63.65	0.28	46	0.078
37.0	23.71	76.19	0.40	45	0.081
1.0	14.97	64.87	0.73	28	0.120
2.0	11.23	49.06	0.49	25	0.170
3.0	8.73	42.02	0.48	18	0.140
4.0	14.97	46.21	0.56	21	0.200
21.0	41.16	99.45	0.82	18	0.110
31.0	24.95	73.24	0.45	37	0.120
37.0	47.40	131.07	1.03	46	0.080
39.0	51.14	153.51	1.25	34	0.120
41.0	52.39	184.93	1.82	33	0.130
5.0	16.73	48.55	0.04	13	0.290
37.0	51.19	114.04	0.63	24	0.170
5.0	11.75	54.16	0.02	17	0.180
20.5	41.17	88.03	0.81	18	0.120
28.5	26.11	94.55	0.10	24	0.250
34.5	45.69	108.43	0.29	25	0.140
36.5	18.28	93.53	0.54	34	0.120
40.5	56.13	150.25	0.89	48	0.077
42.5	67.88	169.93	1.45	32	0.110
48.5	48.30	104.86	1.17	46	0.080
2.0	8.56	44.68	0.41	17	0.110
3.0	7.14	36.92	0.29	15	0.180
5.0	17.12	61.71	0.29	23	0.260
21.5	27.12	70.99	0.93	36	0.110
33.5	29.97	61.40	0.36	31	0.120
37.5	41.39	129.44	1.22	28	0.110
39.5	68.51	208.08	1.22	21	0.110
45.5	67.01	183.80	1.78	47	0.080

* FR: Friction Ratio, %.
** FC: Fine Content, %.

effect of fine content on the q_c/N ratio is signicant and that the ratio increases with decreasing fine content. A simple relationship is established based on the available date:

$$q_c/N_{55} = 4.70 - 0.05 \times \text{Fine Content (\%)}$$

It still should be noted that the data scatter caused by the variability of penetration tests cannot be eliminated. A more sophisticated statistical approach is under investigation so that this effect can be identified and minimized.

From the practical point of view, the use of fine content as the index in the correlation is more convenient. The fine content can be easily estimated by the percentage of materials retained on #200 sieve, but D_{50} can only be obtained by a more tedious sieve analysis.

The disadvantage of using either Fig. 1 or Fig. 2 is that the equivalent N-value cannot be directly calculated from CPT results. A laboratory test is required to estimate the gradation of the soil sample. Hence, many classification charts have been constructed in order to directly convert the q_c to N-value by the use of other measured parameters during cone penetration, for example, skin friction or pore pressure. Based on the friction ratio and corrected tip resistance, Robertson (1986) proposed a simplified soil classification chart with suggested q_c/N ratio for each soil type, as shown in Fig. 3. In this plot, cone bearing, q_t, is the measured cone resistance, q_c, calibrated by the pore pressure and the net area ratio. Measured data from Hsinta site were plotted on Fig. 3 by using different symbols. Each symbol represents a different q_c/N ratio. Since pore pressure was not measured in Hsinta, uncorrected tip resistance, q_c, was used instead of q_t. It is believed that the difference between q_c and q_t is not very significant since the excess pore pressure during cone penetration in granular deposits is relatively small. All the Hsinta data plotted on Fig. 3 fall within Zone 8 and Zone 9 of the chart which are described as sand to silty sand and sand, respectively. Notice that solid symbol on Fig. 3 means that the measured q_c/N ratio coincides exactly with suggested value. Only 7 data points (solid symbols in Fig. 3) show the exact match. However, there also has 7 data points with their measured q_c/N ratios equal to only half of the suggested values. This comparison indicates that this general classification

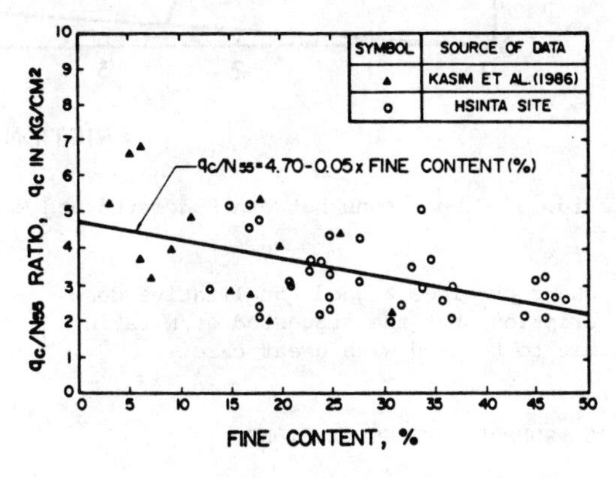

Fig. 2 Variation of q_c/N Ratio with Fine Content of Granular Soils

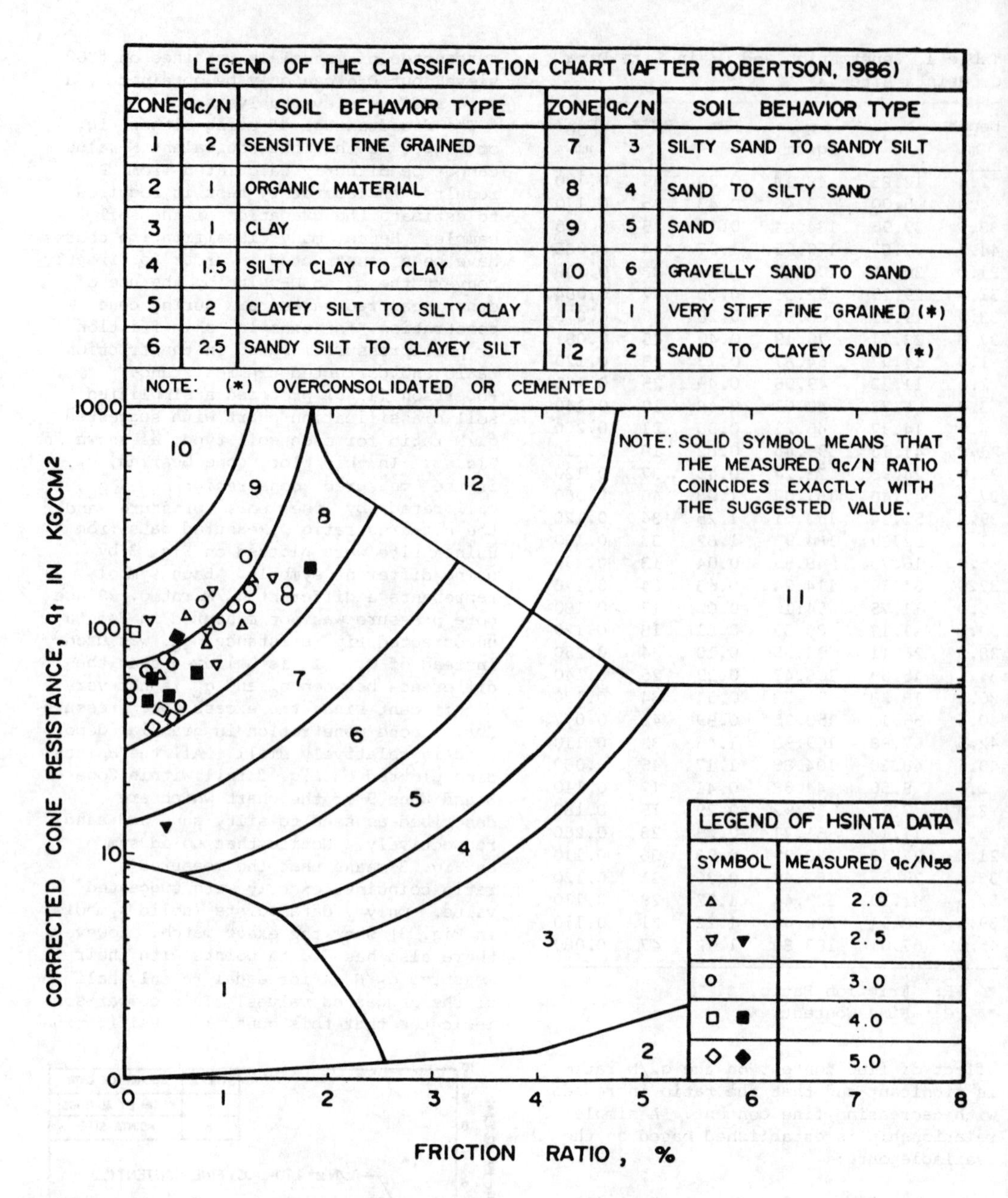

Fig. 3 Comparisons between Suggested and Measured q_c/N Ratio

chart provides a good qualitative description, but the suggested q_c/N ratio has to be used with great care.

5 SUMMARY AND CONCLUSIONS

With the increasing use of CPT, it is useful to correlate the cone tip resistance to the SPT blow count so that the available abundant experiences on SPT can be utilized. The focus of this paper is on the SPT-CPT correlations for granular soils.

It is very important for geotechnical engineers to appreciate the variability of these two penetration tests. Therefore, it is suggested that the SPT N-value should be carefully corrected for the energy level. It is also recommended that the cone resistance be corrected for its unequal area and excess pore pressure effects whenever it is possible.

This paper presents the findings of the SPT-CPT correlations for SP and SM materials based upon the data obtained from well-documented and carefully conducted site investigations. It suggests that the relationship between the q_c/N ratio and D_{50} proposed by Robertson et al. (1983) can be regarded as a reasonable average. However, the scatter around this average is very significant even within a small range of D_{50}. On the other hand, this study clearly illustrates that the q_c/N ratio is much smaller for sands with higher fine contents than for clean sands. It is believed that the correlation between the q_c/N ratio and fine content for granular soils is more meaningful and more convenient to use. A simple relationship is thus proposed based on the available data. More well conducted site investigations are needed to confirm or improve the correlation presented in this paper.

ACKNOWLEDGEMENT

The authors gratefully acknowledge the assistance of all the MAA colleagues involved in this project. Special thanks are extended to Dr. Z.C. Moh, Dr. S.M. Woo and Mr. Joseph Sun for their discussions and suggestions. The cooperation of the Taiwan Power Company is greatly appreciated.

REFERENCES

Campanella, R.G., & Robertson,P.K. 1982. State of the art in in-situ testing of soils: developments since 1978. Engineering Foundation Conference on Updating Subsurface Sampling of Soils and Rocks and Their In-situ Testing, Santa Barbara, CA, USA.

Jamiolkowski, M., Ladd, C.C., Germaine, J.T. & Lancellotta, R. 1985. New developments in field and laboratory testing of soils. 11th International Conference on Soil Mechanics and Foundation Engineering, San Francisco, CA, USA.

Kasim, A.G., Chu, M.Y. & Jensen C.N. 1986. Field correlation of cone and standard penetration tests. Journal of Geotechnical Engineering, American Society of Civil Engineers, Vol. 112, No. 3:368-372

Kovacs, W.D., & Salomone, L.A. 1982. SPT hammer energy measurements. Journal of Geotechnical Engineering, American Society of Civil Engineers, Vol. 108, No. 4:599-620

Moh, Z.C. 1985. Site investigation and in-situ testing. Commemorative Volume in Celebration of the 50th Aniversary of the International Society of Soil Mechanics and Foundation Engineering, Southeast Asian Geotechnical Society.

Robertson, P.K. 1986. In-situ testing and its application to foundation engineering. Canadian Geotechnical Journal, Vol. 23, No. 4:573-594

Robertson, P.K., Campanella, R.G. & Wightman, A. 1983. SPT-CPT correlations. Journal of Geotechnical Engineering, American Society of Civil Engineers, Vol. 109, No.11:1449-1459

Penetration Testing 1988, ISOPT-1, De Ruiter (ed.)
© 1988 Balkema, Rotterdam, ISBN 90 6191 801 4

Dynamic cone, wave equation and microcomputers: The Mexican experience

R.Abraham Ellstein
Laboratorios Tlalli, S.A., Mexico City

ABSTRACT: Treating the rod assembly of the SPT as a micropile and analyzing results with the Wave Equation may be a novel approach to interpreting field results of dynamic penetrometers. The combination of SPT with WE has been used since 1985 for in-situ measurements of the bearing capacity of the hard layer in Mexico City, with results that seem reasonable when compared with direct measurements. Hard to sample soils have also been investigated and when these are homogeneous enough to show a linear increase of the bearing capacity with depth, it is possible to use a computer solver to find their shear strength parameters. Limited experience shows good agreement between field and lab results not only for the peculiar case of Mexico City, but elsewhere too.

KEYWORDS: Bearing capacity, Dynamic penetrometer, in-situ measurements, Microcomputer, Pile tests, Standard Penetration Test, Strength, Wave equation.

INTRODUCTION

The Valley of Mexico is an endorreic lacustrine basin formed by at least six different lakes, not all of them still in existence since Mexico City occupies the whole of the former Lake of Mexico and parts of Lake Texcoco and Lake Xochimilco. It has been argued whether the lakes grew during hot or cold periods (Ellstein 1972), because there is plenty of evidence about successive variations in the volume of lakes. Sometimes they swelled becoming one big water body, others they shriveled and fragmented. Recently Mooser (1986) settled the question showing the existence of large morrainic deposits all over the basin that prove that although the City is at 19° latitude, glaciers developed during glaciations. Through correlation with formations in the hills around the former lake, Mooser has determined that there were several "lakes" at different times where the City sits now. The first lake corresponds to the most recent glaciation, the Wisconsinian, and the second to the Illinoian. Between them there was the Sangamon interglaciation during which the "hard layer" formed, a stratum used extensively to sustain piled foundations.

To the West of the City there are volcanic formations, mostly tuffs and lahera that may contain veins or layers of pumice. These materials constitute the basal rock of the recent deposits, dipping below them to depths that are right now being determined.

Since Marsal (1959) the Mexico City clays have been extensively studied and their properties are common knowledge, but the hard layer is incomparably less well known due to the difficulties found in its sampling.

BEARING CAPACITY OF THE HARD LAYER

Being the hard layer the closest competent material where to attain sufficient resistance for end bearing piles, it was used for that purpose long before Soil Mechanics came of age, and today is still playing the same role. Several authors, among them Resendiz (1964a,b) and Tamez (1964), endeavored to determine the bearing capacity of the layer, measuring it at its top by means of prototype tests with instrumented piles. Resendiz took advantage of large housing projects to perform arrays of tests with several piles of different types at each site and Tamez studied several sites with one pile apiece.

Table 1. Results of load tests.

Reference	Bearing capacity kN/m2	R e m a r k s
Resendiz, 1964a	13440	Mean of eight piles at Jardín Balbuena
Resendiz, 1964b	12753	One pile at Nonoalco-Tlatelolco
Tamez, 1964	13047-15794	Range of eight tests throughout the City

Not long ago it became practical the application of the Wave Equation to routine problems due to spreading efforts of authors like Coyle et al. (1977) and the use of software written for the microcomputer (e.g. Wiseman and Zeitlen 1981). On the other hand it is well known that the Standard Penetration Test was developed as a means to test in-situ the driveability of piles, and that in its inception the tool was a micropile formed by a steel pipe driven by a drop hammer. If the split spoon sampler now affixed to the end of the sounding rod assembly is substituted by a conical point, the micropile status is recovered, becoming amenable to analysis by the Wave Equation as a dynamic penetrometer.

Seed (1985) has studied the SPT and has concluded that the procedure employed to lift and release the weight providing the energy to drive the assembly is crucial, and that when a cathead with a rope are the means, the blow efficiency is around 45%.

Assuming that 90% of the static resistance of the micropile comes from the tip and that the balance goes for friction losses with the borehole walls, lateral vibration, etc. it is possible to determine the driving curve for a given assembly, be it the ASTM norm (D-1586) or any variation thereof. Directly from the Wave Equation it is possible to obtain the function RUP vs. UP, where RUP is the ultimate static bearing resistance of the soil under the conical point and UP is the average unit penetration attained after a given number of blows has been discharged to the assembly. Fig. 1 shows one such driving curve obtained for the particular rig used by the author. From experience it is known that the function is quite robust against variations in the equipment's specifications and also against relatively large variations in the total length of rods used. Nonetheless, important changes occur when different blow energies are put in.

These physical and analytical tools have been used to measure the bearing capacity of the hard layer at five different sites over the City and in the three zones in which it has been divided: the hills, where basal rock either outcrops or is at no more than 10 m depth with no compressible materials above it; the transition zone where the layer is typically found near a depth of 15 m; and the lake zone where depths of 30+ meters are the usual.

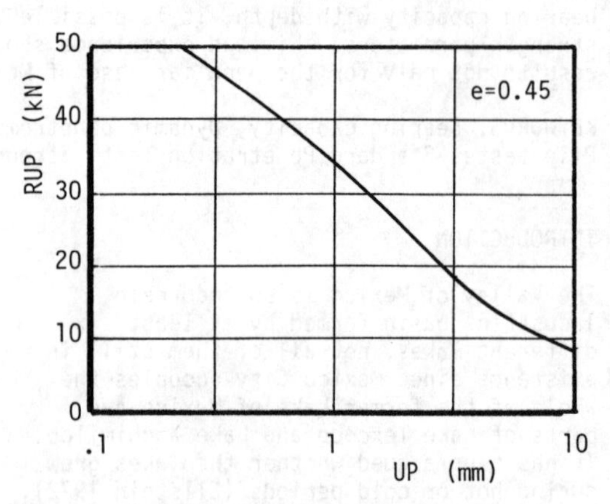

Fig. 1 Driving function for AW rods

One immediate advantage of this approach is that bearing capacities can be measured both at the top of the layer and at discrete depths within it; something hard to attain with test piles.

Site 1 is in the hills zone where no hard layer as such exists, but the basal rock where piles or piers usually rest has been customarily called the hard layer. Sites 2 and 3 belong to the transition zone and sites 4 and 5 to the lake zone.

If the lake values shown in Table 2 are compared with those appearing in Table 1 it is seen that they are in the same league. Nonetheless, one feature readily apparent in the table is that in the lake zone the hard layer has a sandwichlike structure, something that went undetected with the instrumented pile tests and on the unsafe side. Tests in site 1 show clearly a sudden increase in the bearing values due to the fact that the upper two values belong to recent deposits whilst the lower ones were taken from the basal rock. Both cases tend to show the intimate knowledge that can be gained at a given site by using the dynamic cone.

Table 2. Results at five sites throughout the City.

Site No.	Boring No.	Depth m	Bearing capacity kN/m^2
1	1	8.8	9123
1	1	10.7	9123
1	1	12.1	20601
1	1	13.3	24623
2	1	16.6	11968
2	1	18.1	16677
3	1	14.2	12851
3	1	15.4	13342
3	1	20.5	24230
4	1	30.0	18149
4	1	31.0	12851
4	1	33.0	19326
5	1	32.0	12851
5	1	33.0	7357
5	1	34.3	6796
5	1	36.4	9614
5	2	31.0	12851
5	2	32.4	9418
5	2	33.7	12655

BEARING CAPACITY OF HARD TO SAMPLE SOILS

Pumitic soils combine a pair of features that make their sampling quite difficult; on the one hand they are fairly hard, so rotation is called for, but on the other they erode easily and normal rock procedures cannot be employed. Use of special samplers is required and that tends to discourage undisturbed sampling.

To solve a slope stability problem at a site called Santa Fe, created by the existence of a fault scarp, 13 undisturbed samples were carved from the scarp's face and tested at the lab by means of unconfined undrained triaxials (UU). One particular feature of the formation at this site is its red color, which identifies it so uniquely that geologists use it for a marker. Even if there are other red and yellow soils in the hills around the City, this stratum is the reddest and easily identifiable. At another site called Tlalnepantla, distant several kilometers from the first, a couple of standard penetration borings were taken and in the process six dynamic cone tests performed at different depths. When split spoon disturbed samples were classified it was recognized that the six field tests had been performed in the red marker soil.

Table 3. Results of 13 UU triaxial tests

Test No.	Avg. moisture %	fi deg.	c kN/m^2
1	20.2	31.5	61.9
2	13.7	35.9	59.7
3	13.6	34.1	70.8
4	14.6	37.8	74.0
5	13.1	36.2	73.1
6	13.3	36.3	62.4
7	20.4	39.8	32.6
8	12.9	37.2	29.0
9	11.5	35.1	44.9
10	10.7	35.9	59.4
11	18.4	32.9	52.6
12	15.9	37.8	20.6
13	20.4	35.1	29.4

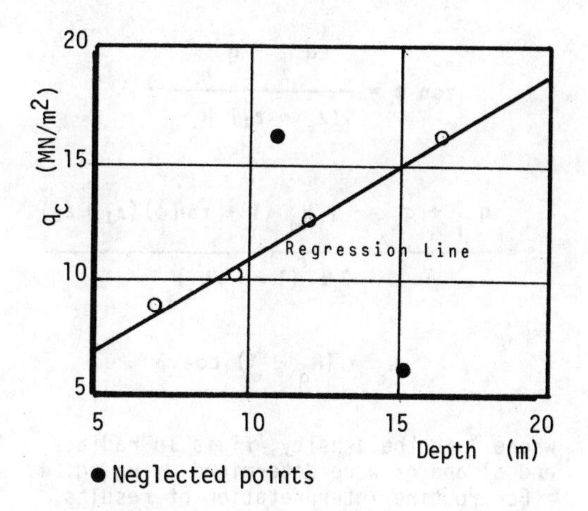

Fig. 2 Plot of measured bearing capacities in Tlalnepantla

The measured bearing capacities are shown in Table 4, and it can be seen in Fig. 2 that four of them are practically aligned, so if points 3 and 5 are neglected, a regression line is obtained which reads as follows

$$q = 792Z + 3138 \qquad (1)$$

where Z is the depth in meters.

Table 4. Dynamic cone results at Tlalnepantla.

Test No.	Boring No.	Depth m	Moisture %	Bearing capacity kN/m^2
1	1	7.2	43.6	9220
2	1	9.6	23.7	10300
3	1	11.2	31.6	16380
4	2	12.3	27.4	12753
5	2	15.2	25.5	6279
6	2	16.5	28.5	16383

The most widely used procedures to determine c and fi from field tests are based on the knowledge of two bearing capacities q1 and q2 at two depths Z1 and Z2; like those by De Beer (in Sanglerat, 1972) and Mitchell-Durgunoglu (1973). Santoyo and Olivares have proposed a semi-graphic procedure to expedite the tedious iterative process for solving the four simultaneous equations

$$N_q = \frac{q_2 - q_1}{\gamma(1 + \tan\phi)(z_2 - z_1)}$$

$$\tan\phi = \frac{q_2 - q_1}{\gamma(z_2 - z_1) N_q} - 1$$

$$c = \frac{q_1 + q_2 - \gamma N_q (1 + \tan\phi)(z_1 + z_2)}{2 N_c (1 + N_q/N_c)}$$

$$N_c = (N_q - 1)\cot\phi$$

where γ is the density, fi is in radians and q1 and q2 were determined using eq. 1.

For routine interpretation of results, instead of the abovementioned methods, use of a computer solver (Software Arts, 1983) in the iterative mode and an estimate of fi to start the process, made it possible to solve the equations without any manual labor. In the Tlalnepantla case fi=34.9° and c=40.02 kN/m2.

Averages of the 13 triaxial series shown in Table 3 are of 35.8° and 51.6 kN/m2, that when compared with the field values show differences of 2.6% in fi and 28.9% in c. An explanation of the relatively large difference in the c's might be found in the fact that the average moistures for the materials in Tlalnepantla and Santa Fe are respectively 30.1% and 15.3%.

These results were good enough to encourage additional investigations to find how accurately the dynamic cone can measure shear strength parameters. Cape St. Luke (Cabo San Lucas) is at the southmost tip of the Baja California peninsula. There a dune formed by homogeneous sands was probed with 25 dynamic cone tests and results compared with five series of quick direct shear tests on samples taken near the surface. The value of fi measured in the field was 32.8° and the mean of the five lab series was 34.3°, a difference of 4.5%.

CONCLUSIONS

Using a couple of well established Soil Mechanics tools; the Standard Penetration Test rig and the Wave Equation, it is possible to perform a simple and straightforward test that is very much like the standard test. It only takes the indent--ation of the soil at the bottom of a clean borehole by means of a conical point driven by a few hammer blows. A penetration of 1 to 2 inches is usually enough to obtain a reliable measure of the penetration per blow (UP), which is then employed as argument in Fig.1. Use of this apparatus and interpretation of the tests with the aid of the Wave Equation and a microcomputer has resulted in easy measurements of the bearing capacity of the hard layer in Mexico City. Accuracy seems to be adequate when comparing the recently measured values with results obtained by independent researchers working with instrumented piles. Moreover, the new procedure is suited for performing several tests at different depths and it can discover details that give the engineer an intimate feel of the investigated materials.

Another use for this procedure is in-situ testing of difficult to sample materials and so far results have been encouraging enough to merit further investigations, not only in and around Mexico City but also elsewhere.

This should not be taken as encouragement to embellish excessively the Standard Penetration Test, since an inordinate proliferation of penetration techniques is certainly unwanted. What is being proposed is basically a different way to interpret results, with practically no penalty in standardization.

REFERENCES

Coyle, H.M. et al. 1977. Wave equation analysis of piling behavior. Numerical Methods in Geotechnical Engineering, p. 272-296. NY:McGraw-Hill.

Ellstein, A. 1972. La carga de preconsolidación y las lluvias glaciales. Boletín SMMS No. 22.

Marsal, R. and Mazari, M. 1959. El subsuelo de la Ciudad de México. México. UNAM.

Mitchell, J.K. and Durgunoglu, H.T. 1973. In-situ strength by static cone penetration test. Procs. VIII ICSMFE. Toronto.

Mooser, F. et al. 1986. Características geológicas y geotécnicas del Valle de México. COVITUR.

Reséndiz, D. 1964a. On a type of point bearing pile through sinking subsoil. Procs. Conf. on Deep Foundations. p. 385-403. México.

Reséndiz, D. 1964b. Estudio de campo sobre pilotes de concreto reforzado. Revista Ingeniería Vol. 34 No. 1:101-110.

Sanglerat, G. 1972. The penetrometer and soil exploration. USA:Elsevier.

Santoyo, E. and Olivares, A. 1981. Penetrómetro estático para suelos blandos y sueltos. Series del Instituto de Ingeniería No. 435. UNAM.

Seed, H.B. et al. 1985. Influence of SPT procedures in soil liquefaction resistance evaluations. J. of the Geotech. Engrg. Div. ASCE. Vol. 111 No. 12:1425-1445.

Software Arts. 1983. TK!Solver v. TK-1(2J). USA.

Tamez, E. 1964. Pilotes electrometálicos en las arcillas del Valle de México. Procs. Conf. on Deep Foundations. p. 279-291. México.

Wiseman, G. and Zeitlen, J.G. 1981. Wave equation analysis of pile driving using personal computers and programmable calculators. Technion, Israel Institute of Technology. Haifa.

Penetration Testing 1988, ISOPT-1, De Ruiter (ed.)
© 1988 Balkema, Rotterdam, ISBN 90 6191 801 4

Penetration resistance and shear strength of cohesive soils

Ervin Hegedus & Jon H.Peterson
Applied Construction Technologies, Inc., Cleveland, Ohio, USA

ABSTRACT: The use of penetration testing as an insitu test method to predict the undrained shear strength of cohesive soils are discussed. The results of the nearly 470 standard penetration, unconfined compression and other soil properties tests are included. The data for the soil type tested was statistically analyzed. Empirical relationships were developed to forecast shear strength from penetration resistance. Best correlations were obtained utilizing logarithmic and power type equations to fit the experimental data. The empirical correlations were compared with previously published data.

1 INTRODUCTION

Drive sampling operations have been commonly utilized during subsurface exploration to obtain data concerning the state of occurrence of subsoils and coincidently to retain samples of the formations penetrated (Hvorslev, 1949). The sampling procedure was standardized by the American Society for Testing & Materials (ASTM) in 1967. To date, it remains a standard sampling procedure. It involves driving the 2" (51 mm) outside diameter split spoon sampler attached to drill rod(s), with a 140 pounds (63.5 kg) drop hammer freefalling through a distance of 30" (76.2 cm). When subsurface conditions permit, the sampler is driven 18" (45.7 cm) and the number of hammer blows required to advance the sampler is recorded for each 6" increments. The sum of the blowcounts associated with the last 12" penetration interval is termed the Standard Penetration Resistance (N).

The method is considered fairly reliable to estimate the relative density and shear strength parameters of granular soils. The Standard Penetration Test (SPT) is generally accepted for classification of cohesive soils as to their consistencies. It is not considered reliable for forecasting their shear strength characteristics (ASCE Task Committee 1972). There are published correlations between unconfined compressive strength and penetration resistance (Terzaghi & Peck, 1948; Sowers, 1954; Stroud, 1947). The published data suggest a linear variation between unconfined strength and penetration resistance satisfying the boundary condition that when the penetration resistance is zero, the unconfined strength is also zero. Different linear correlations were also developed for clays of low or high plasticity characteristics. It is suspected that different empirical relationships would also apply to soils of different geologic origins. In granular soils, the penetration resistance is considered dependent on the overburden pressure, density and lateral pressure, water table, etc. (Fletcher, 1965). For clays, the factors affecting penetration resistance are less understood.

The present paper deals with the correlation of penetration resistance with the undrained shear strength and other pertinent properties of silty clays encountered primarily along the southern shores of Lake Erie of the Great Lakes of North America. The soil types discussed include glacial lake deposits and glacial till formed during the several glacial periods affecting the area. (Hough, 1958). The results were statistically analyzed using least square techniques and curve fitting procedures to establish an empirical relationship between unconfined strengths and penetration resistance. Polynominal, power, exponential and logarithmic relationships were considered in fitting the data points with the appropriate curve. Confidence limits were established based on standard deviation.

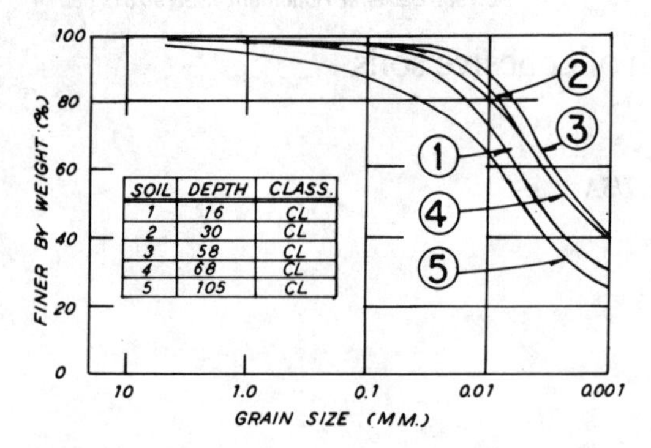

Fig. 1 Typical Grainsize Curves for glacial lake deposits

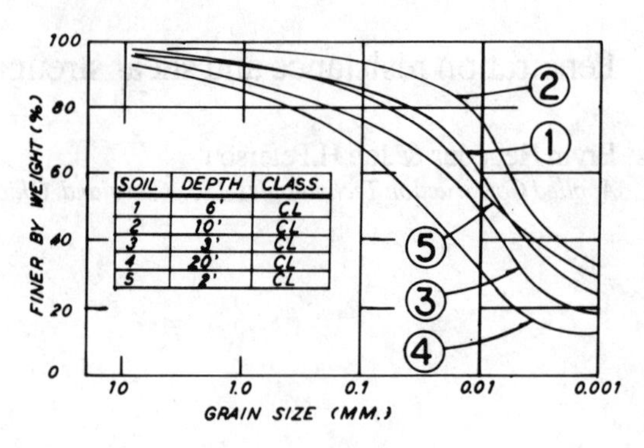

Fig. 2 Typical grain size curves for glacial till soils

2 DESCRIPTION OF SOILS

The glacial lake deposits discussed in this paper typically occur at and near the south shores of Lake Erie, either near the ground surface or under substantial manmade fill deposits. The fill overlay thickness along the Cleveland lakefront is approximately 30' (9.1 m) deep. Bedrock is encountered at the sampling locations at between approximately 40' (12.2 m) and 115' (35 m) below the top of ground. The bedrock typically consists of shale except in the western Lake Erie region where it is limestone. Representative subsoils in this group include between approximately 50% to 75% of claysize particles and between 2% and 15% of combined sand and gravel fraction (Fig. 1).

The lake deposits include relatively uniform soft to medium clays and varved clays with generally stiff clays of varying thicknesses overlying the shale bedrock. These soils exhibit low to moderate plasticity characteristics (Fig. 3).

In accordance with the Unified Classification System (ASTM D-2487, 1985), the glacial lake deposits can be represented by group symbol CL. Generally, the lake bottom sediments exhibited water contents near the liquid limit. The degrees of saturation varied between 0.9 to 1.0.

The glacial till deposits tested were relatively thin and rarely exceeded 40' (12.2 m) in depth over shale or sandstone bedrock. Most test data was obtained within sites with less than 15' (4.6 m) of overburden on top of rock. The till deposits also fell into the generalized soil group of CL as per the Unified Soil Classification System. However, these soils were more heterogeneous than the lake deposits and included between roughly 7% and 37% of sand and gravel and between 20% and 55% of clay (Fig. 2 and 3).

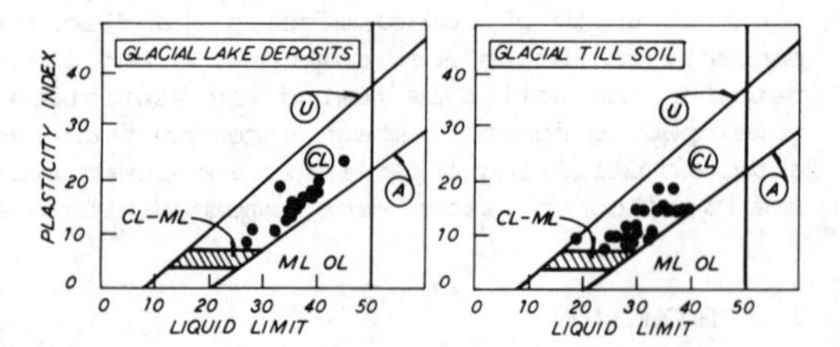

Fig. 3 Plasticity characteristics of soils tested

Generally, the till deposits were highly overconsolidated, however, contained occasional soft clays. The degrees of saturations of the formations tested varied between less than 0.7 and 1.0.

3 SAMPLING AND TESTING PROGRAM

The data included in this paper was obtained in foundation test borings. Sampling techniques included obtaining split spoon, i.e. drive samples and undisturbed samples by means of thin walled Shelby tube samplers at intervals. Drive samples were taken by means of the two inch (51 mm) outside diameter split spoon sampling device. Undisturbed samples were taken by either the 2" (51 mm) or the 3" (76 mm) diameter thin walled Shelby tube samplers pressed into the subsoils by steady static force. The drive samples were classified as a minimum. The undisturbed samples were subjected to unconfined compression tests or occasionally to unconsolidated-undrained triaxial shear tests. The natural moisture contents and dry densities of the samples were also determined. In most cases, undisturbed sampling was routinely followed by drive sample acquisitions. In all other cases, the results of SPT conducted nearest to the undisturbed

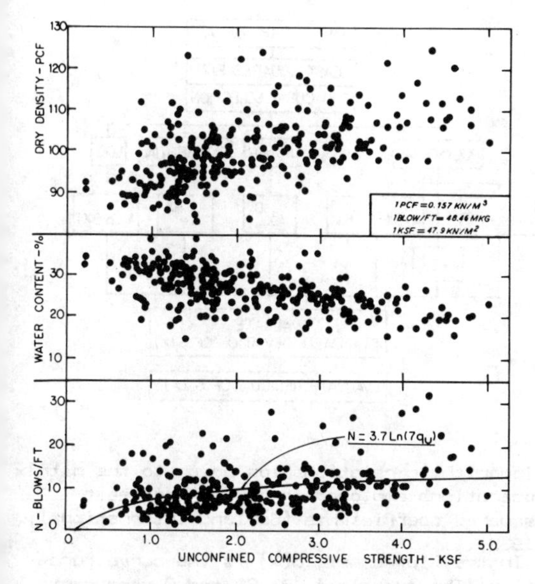

Fig. 4 Test data for glacial lake deposits

sample elevations were utilized to establish correlation between the results of drive sampling and the undrained shear strength. Maximum sampling intervals did not exceed 3.5' (1.1 m).

The soil samples were obtained predominantly by two drilling companies using either rotary drive hollow or solid stem flight augers followed when required by wash boring procedures. For the purposes of this paper, it has been assumed that the drilling and sampling procedures were conducted in accordance with standardized drilling and sampling techniques. Inaccuracies in sampling and insitu testing, which may be due to equipment, personnel, inconsistencies in subsurface conditions, which tended to make the test data non-reproducible, was assumed negligible.

A total of 299 test results were available for analysis from the glacial lake deposits and 168 data points from the glacial till soils.

4 INVESTIGATION RESULTS

The results are presented in Figures 4 through 7. In Figures 4, 5 and 6, the unconfined compressive strength of the two major soil types discussed above were plotted against:

1. The standard penetration resistance obtained at or near the undisturbed sample location, and

2. The moisture content and the dry density of the sample involved in the uniaxial compression testing.

The degree of saturation (S_r) of the glacial lake deposits ranged between 0.9 and

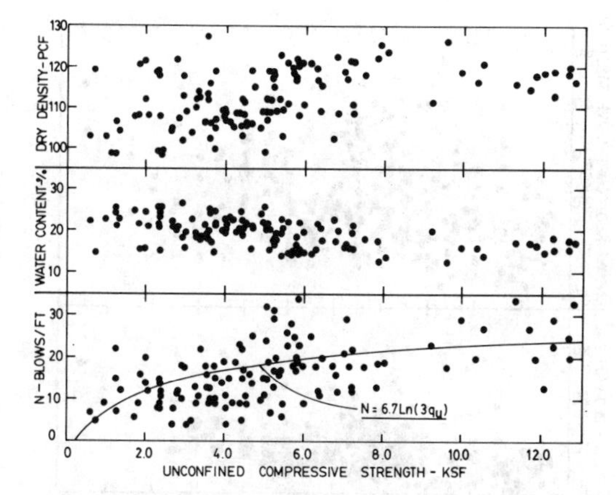

Fig. 5 Test data for glacial till soils ($S_r \geq 0.9$)

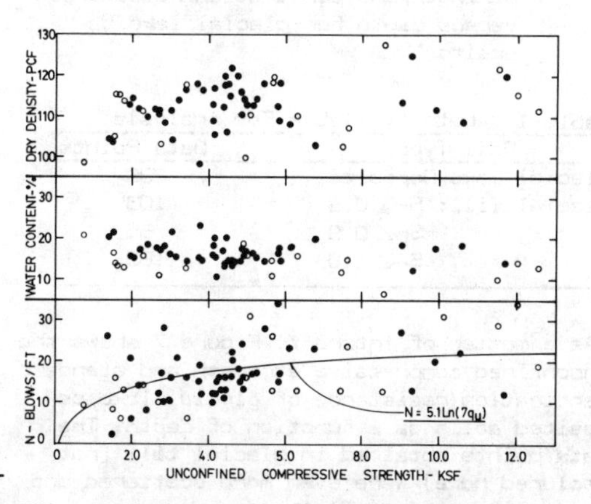

Fig. 6 Test data for glacial till soils ($S_r < 0.9$)

1.0. Most samples were fully saturated.

The test results for the glacial till soils were arbitrarily sorted according to degrees of saturation. Fig. 5 shows the data for samples with a degree of saturation between 0.9 and 1.0. In Fig. 6, the dark circles represent degrees of saturation between 0.8 and 0.9. The empty circles indicate saturation less less than 0.8

This was done to investigate the effect the degree of saturation may have on the empirical correlation to be established between the undrained strength and the penetration resistance.

The general trend of data indicates an increase in compressive strength with increasing densities and penetration resistances and decreasing moisture contents. This trend is more pronounced for the more uniform lake deposits. It further appears that the data is more consistent for soils having high degrees of saturations.

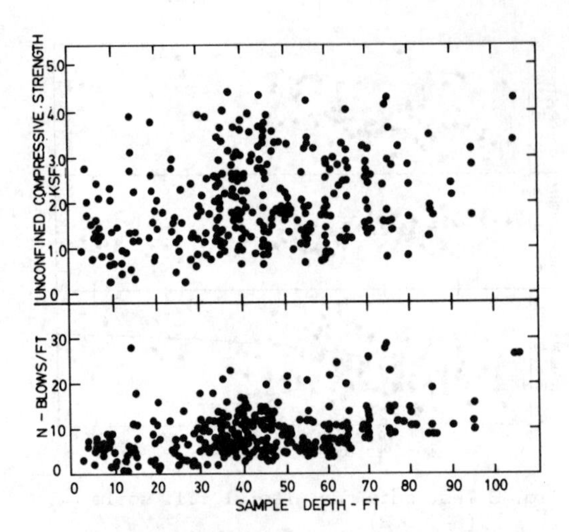

Fig. 7 Strength and penetration resistance versus depth for glacial lake deposits

Table I. Data Available For Analysis

Soil Type	Data Points
Glacial Lake Deposits	299
Glacial Till: $S_r \geq 0.9$	103
$S_r \angle 0.9$	65
$S_r \angle 1.0$	168

As a matter of interest, Figure 7 shows the unconfined compressive strength and standard penetration resistance of glacial lake deposited soils as a function of depth. The data points obtained in glacial till (not included here) were even more scattered and do not suggest a definite trend.

5 STATISTICAL ANALYSIS

The data concerning unconfined compressive strength and penetration resistance of soils were sorted with respect to the categories shown in Table I.

The Least Squares Technique was utilized to analyze this data. This procedure employs the following stipulations in developing an analytical relationship:

$$\sum (Y - Y_O) = 0 \qquad (1)$$

$$\sum (Y - Y_O)^2 = \text{minimum} \qquad (2)$$

Where Y_O represents the original observations and Y reflects the computed values. The analysis rational, together with the analytical structural forms used to review the data in terms straight line, second order, power, expotential and logarithmic type relationships are shown in the flow chart below. Where necessary, these relationships were transformed to a linear format via

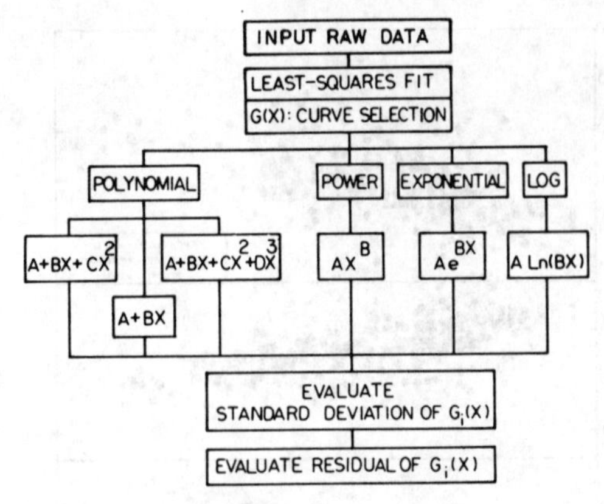

logarithmic relationships prior to the matrix operations employed to obtain the least squares coefficients (Borland International, 1986).

In the flow chart, G(X) is the curve function. The symbols A, B, C, and D represent curve fitting coefficients. X and Y are variables normally indicated on the abscissa and the ordinate, respectively, of a cartesian coordinate system. Ln = natural logarithm.

Table 2 summarizes the least square coefficients and standard deviations obtained for each soil category considered. It should be noted that the analysis of the residual data indicated negligible improvements of the analytical representations.

A realistic boundary condition requires the candidate analytical function to essentially go through the origin (X=0, Y=0). Thus all polynomial functions can be eliminated from final consideration. The exponential format also fails to meet the boundary requirement. The logarithmic format has a somewhat smaller standard deviation and thus is selected as the most appropriate empirical relationship.

The resulting final analytical forms, after interpreting the variable Y as N - the SPR and X as q_u - the unconfined compressive strengths in kips per square foot are shown below.

Glacial Lake Deposits:

$$N = 3.7 \text{ Ln } (6.9\ q_u) \qquad (3)$$

Glacial Till:

$$S_r \geq 0.9 \quad N = 6.7 \text{ Ln } (3.3\ q_u) \qquad (4)$$

$$S_r \angle 0.9 \quad N = 5.1 \text{ Ln } (6.9\ q_u) \qquad (5)$$

$$S_r \angle 1.0 \quad N = 6.2 \text{ Ln } (3.1\ q_u) \qquad (6)$$

Curves of the above equations for each soil type are shown superimposed in the previously referenced Fig. 4, 5, and 6. A comparison of the above diagrams, with previously published data, is included in Fig. 8.

TABLE 2. LEAST SQUARES FIT OF SOILS DATA

MEDIUM / LEAST SQUARE CURVE	LEAST SQUARES COEFFICIENTS A	B	C	D	STANDARD DEVIATION
(1) Glacial Lake Deposits					
Straight Line	5.08	2.05	-	-	4.75
2nd Order	5.92	1.19	0.18	-	4.75
3rd Order	3.58	5.05	-1.59	0.23	4.75
Power	0.23	0.47	-	-	4.88
Expotential	4.83	0.24	-	-	4.86
Logarithmic	3.69	6.93	-	-	4.80
(2) Glacial Till					
(a) Degree of Saturation $S_r \angle 0.9$					
Straight Line	12.24	1.03	-	-	7.57
2nd Order	10.16	1.86	-0.06	-	7.60
3rd Order	4.88	5.74	-8.11	0.03	7.60
Power	0.66	0.37	-	-	7.67
Expotential	11.08	0.07	-	-	7.77
Logarithmic	5.13	6.95	-	-	7.53
(b) Degree of Saturation $S_r \geq 0.9$					
Straight Line	7.22	1.60	-	-	5.57
2nd Order	7.66	1.42	0.14	-	5.60
3rd Order	7.88	1.26	0.05	-0.002	5.63
Power	0.25	0.47	-	-	5.81
Expotential	7.94	0.11	-	-	5.74
Logarithmic	6.69	3.27	-	-	5.90
(c) All data Points $S_r \leq 1.0$					
Straight Line	8.88	1.43	-	-	6.49
2nd Order	8.31	1.67	-0.02	-	6.51
3rd Order	6.21	2.49	-0.33	0.02	6.51
Power	0.33	0.45	-	-	6.63
Expotential	8.84	0.10	-	-	6.69
Logarithmic	6.21	3.16	-	-	6.61

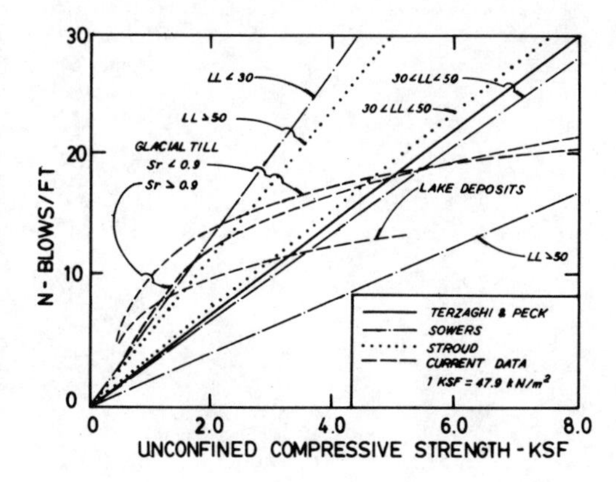

Fig. 8 Comparison of published and current data

6 CONCLUDING REMARKS

The logarithmic type empirical formula arrived at by means of the statistical data analysis appears to be the best fit for data points for both the glacial lake deposits and the glacial till soils tested. A higher confidence level in estimating the unconfined strength from penetration test data is indicated for the glacial lake deposits than for the glacial till. This circumstance is probably due to the relatively uniform nature of the lake deposits which permitted the acquisition of better quality undisturbed samples and more consistent penetration testing than it is normally feasible in the heterogeneous glacial till. The presence of rock fragments, the friable and brittle character and fissures which are often encountered in the till severely limits the feasibility of obtaining good quality samples or reliable insitu or laboratory test data.

It is probable that cohesive soils with different geological histories would require the development of an empirical relationship for each soil type in order to permit the estimation of undrained shear strength from standard penetration test data. Even then, an over or under estimation of the strength by several fold could well occur.

The degree of saturation for highly overconsolidated glacial till appear to have very little influence on the empirical formula developed for the test data in that soil.

The general data trend, for both types of soil tested, indicate that with increased penetration resistances, an increase in the undrained shear strength and the dry densities can be anticipated. Coincidentally, the soil moisture content tends to vary inversely with these parameters.

The penetration resistance in the deep test borings conducted in the glacial lake deposits appeared to increase with depth within soils of similar consistencies and strength characteristics. This is probably due to the loss in energy during driving and the wall friction or adhesion of the borehole. (At 100' depth, the drill rods weigh about 400 pounds.)

REFERENCES

ASCE Task Committee 1972. Subsurface investigation for the design and construction of foundations of buildings. ASCE, J.SMFD Vol. 98.SM6.

ASTM 1985. Penetration test and split-barrel sampling of soils (D-1586). ASTM Standards, Volume 04.08, Philadelphia.

ASTM 1985. Classification of soils for engineering purposes. (D-2487). ASTM Standards, Vol. 04.08, Philadelphia.

Borland International, Inc. 1986. Turbo pascal numerical methods tool box. Scotts Valley: Borland International.

Fletcher, G.F.A. 1965. Standard penetration test: its uses and abuses. ASCE, J.SMFD Vol. 91.SM4.

Hough, J.L. 1958. Geology of the great lakes. Urbana: Univ. Illinois Press.

Hvorslev, M.J. 1949. Subsurface exploration and sampling of soils for engineering purposes. Vicksburg: Waterways Experiment Station.

Sowers, G.F. 1954. Modern Procedures for underground investigations. ASCE, J.SMFD Vol. 80,SM6.

Stroud, M.A. 1974. The standard penetration test in insensitive clays and soft rocks. ESOPT I, Stockholm.

Terzaghi, K & Peck, R.B.V. 1948. Soil mechanics in engineering practice. New York: John Wiley & Sons.

Penetration Testing 1988, ISOPT-1, De Ruiter (ed.)
© 1988 Balkema, Rotterdam, ISBN 90 6191 801 4

Comparison between relative density, static and dynamic strength estimations based on SPT data with laboratory results on very fine sands

C.Olalla, V.Cuellar, F.Navarro & J.C.de Cea
Laboratorio de Geotecnia, Madrid, Spain

ABSTRACT: Waxed and frozen soil samples are obtained and used as undisturbed material in order to get laboratory results from identification tests and static and dynamic triaxial apparatus. SPT results are used together with empirical formulae to predict the soil behaviour under dead loads and natural hazards. As a result of the comparison it is shown how traditional laws relating D_r with N values underestimate values of Relative Density in these very fine and uniform silty-sands. By the same way static friction angles are also underestimated. Dynamic strengh estimations, taking into account fines content produce different values with great dispersion. However intermediate values are in good agreement with triaxial dynamic test results.

1 INTRODUCCION

The paper deals with the advantages and limitations found when using SPT results to predict basic geotechnical properties such as relative density (D_r), friction angle (ϕ') and dynamic shear strength of non cohesive Quarternary soils.

The materials investigated belong to the seabed delta deposits of the Motril harbour, which is located in a seismic hazardous zone in the South of Spain. The mantle at the site is mainly composed of fine sand silty sand and sandy silt (especially below 20 m), up to a depth of 50 m below the sea bottom.

Because of the economical and technical importance of the installations, together with the possibility to extend the conclusions reached to other points in the South coast of Spain an estensive campaign of in situ and laboratory tests was undertaken with the aim of comparing results obtained through different techniques.

2 IN SITU WORK

A total of 6 boreholes with an initial diameter of 190 mm have been made using rotary drilling and avoiding the circulation of water under pressure. The maximum depth reached has been 53 m below the sea bottom.

To recover 33 undisturbed samples three different types of sampling tubes have been used:

a) A thin-wall sampler with ball valve with an inner diameter of 76.5 mm and a length of 1 m.

b) A piston sampler of the Osterberg type manufactured in Spain with an inner diameter of 71 mm and a length of about 0.60 m.

c) A piston sampler similar to the one previously described but imported from the USA and having an inner diameter of 71 mm and a length of 0,92 m.

Bentonite slurry was not used either in drilling or sampling.

Some of the samples were sealed with wax immediately after having been recovered, while others were kept in a vertical position for about 2 or 3 days to drain out excess water existing in the pores of the saturated materials. Then they were frozen at the sampling site with the aid of liquefied nitrogen.

Also 67 penetration tests have been run according to the recommended standard for the SPT test elaborated by the Subcommittee on Standardisation of Penetration Testing in Europe who belongs to the International Committee on Soil Mechanics and Foundation Engineering (ICSMFE).

All this work has been implemented with continuous seismic proliling of the seabed sediments using an Uniboom system.

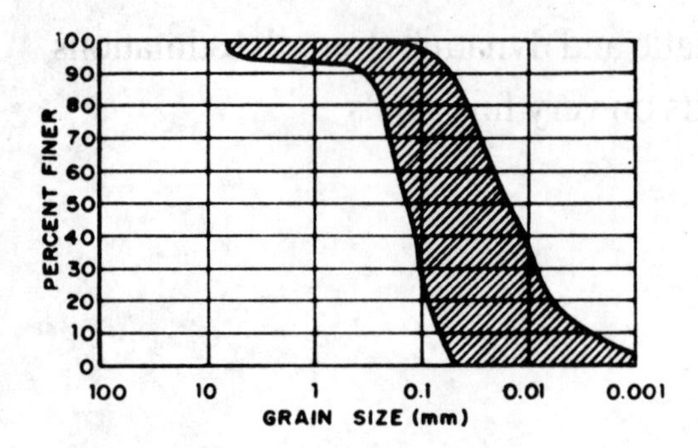

Fig. 1.- Limits of grain size distribution curves

3 BASIC GEOTECHNICAL PROPERTIES

Identification tests and shear strength tests have been carried out in the laboratory using disturbed and undisturbed samples collected at the site.

Up to a depth or 20 m, the results obtained indicate the existence of a soil profile with a smooth variation in its geotechnical properties. Below that level they remain practically constant.

Figure 1 shows the grain size distribution curves which embody all the results obtained in the mechanical analysis.

Figures 2, 3, and 4 illustrate respectively the variation with depth of the percent of fines passing the ASTM # 200 mesh, the mean particle diameter (D_{50}) and the uniformity coefficient (C_u).

From the consistency limits determined, all the material tested may be classified as being non-plastic.

Figure 5 gives the variation with depth of the dry unit weight. To come up with each of the values plotted in figure 5 a comparison was made among the values obtained by the following three ways:

1. As the result of rutinary determinations on soil specimens carried out when running laboratory tests.
2. As the ratio between the weight and volume of soil contained in thin-wall sampling tubes.
3. As the value computed from the water content and the unit weight of the solids.

The differences found among the three distinct procedures can not be considered as being significant. Neither have been the differences obtained when comparing results worked out from samples waxed and frozen at the site.

The unit weight of the solids varies between 27.6 kN/m^3 and 28.1 kN/m^3 with a mean value of 27.8 kN/m^3.

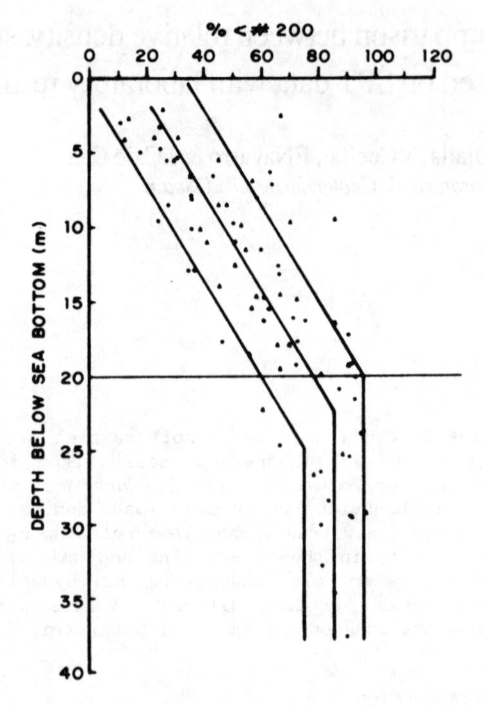

Fig. 2.- Percent finer than # 200 versus depth

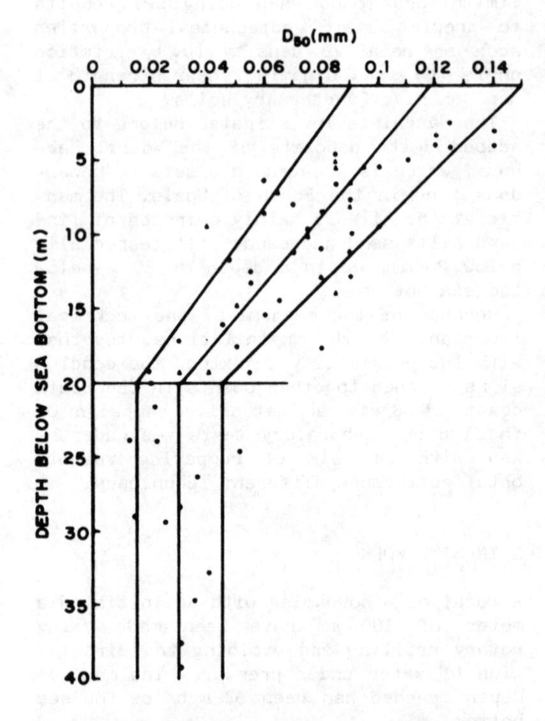

Fig. 3.- Mean particle diameter versus depth

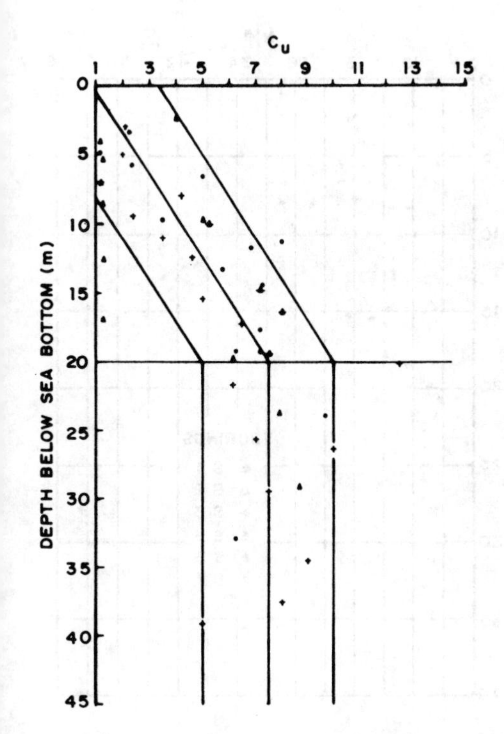

Fig. 4.- Uniformity coefficient versus depth

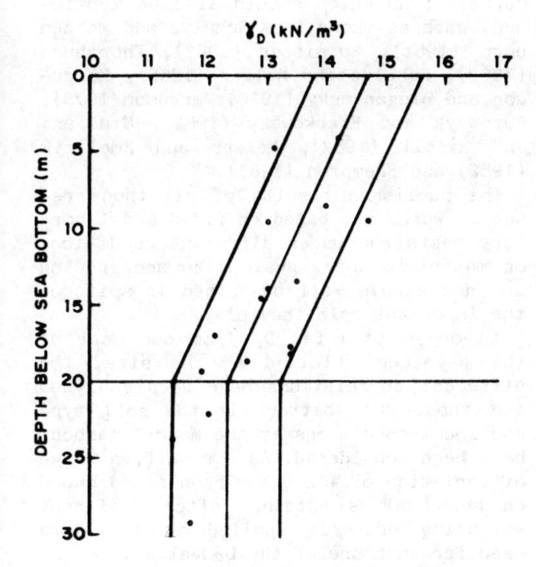

Fig. 5.- Dry unit weight versus depth

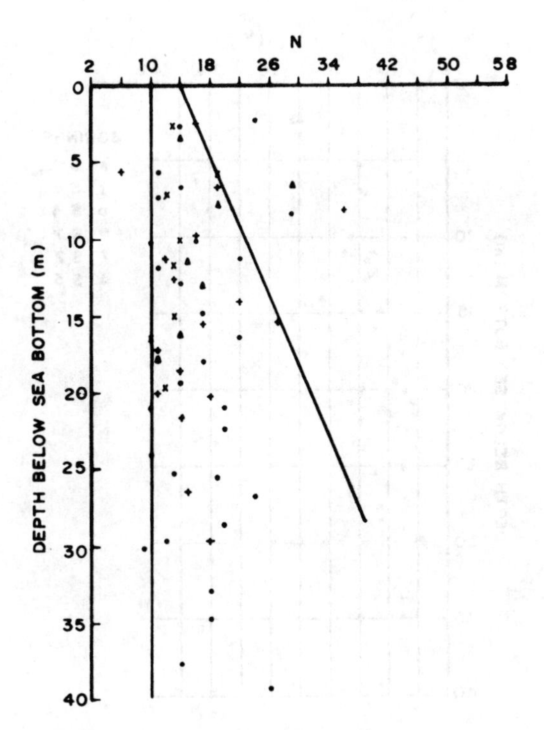

Fig. 6.- SPT blow counts versus depth

4. SPT RESULTS

In order to interpret the results obtained from the SPT tests, the following methodology has been adopted:

- First, N-values exceedingly high or low have been removed from the bulk of data collected (see figure 6).

- Secondly, the blow count values have been normalised to an effective overburden pressure of 100 kN/m^2 to account for the effect of increasing N-value with increasing confining stress. Thus the N-values have been multiplied by a depth factor, or normalizing function C_N, to obtain the corrected blow count values N^*. Based on the work by the Waterways Experiment Station, Peck et al. (1973), Seed (1976,1979) and Marcuson at al. (1977) proposed different expressions for C_N. More recently, Skempton (1986) examined the matter in detail and suggested new expressions for C_N. Figure 7 illustrates the variation with depth of the mean N* value obtained when applying to each value of N the different expressions for C_N referred to earlier.

- Thirdly, a correction has been made to account for excess pore-water pressures

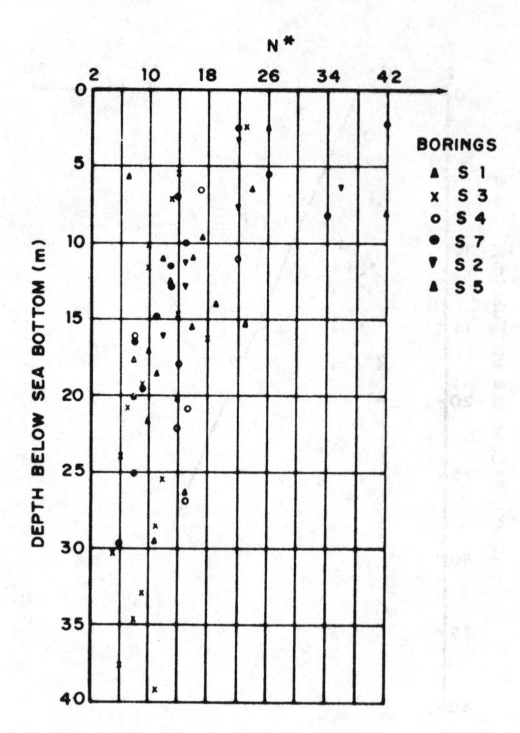

Fig. 7.- Corrected N values versus depth

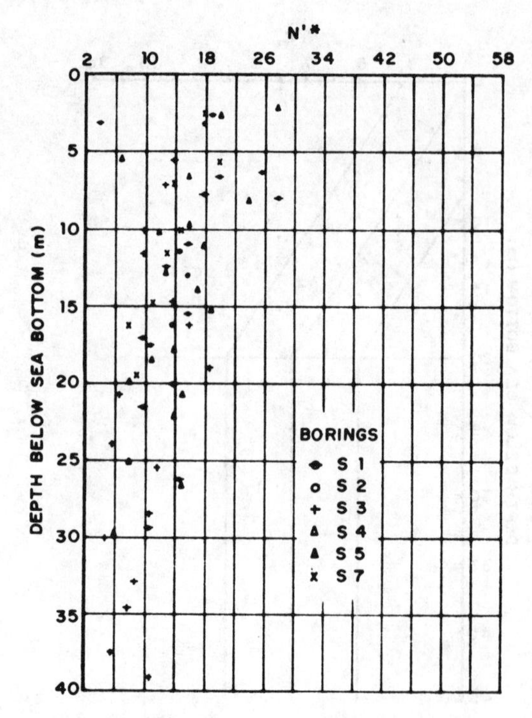

Fig. 8.- Corrected N values versus depth

set up during driving below the water table if the soil consists of very fine or silty sand. Terzaghi and Peck (1948), in their Soil Mechanics textbook, suggest the use of the following equation:

$$N' = 15 + \frac{1}{2}(N - 15) \quad (1)$$

being $N > 15$

Although this recommendation is not included in the second edition of the same book, Sanglerat (1972) incorporates it in his well Known text on the penetrometer.

Since almost 50% of the blow counts obtained in the silty sand of the Motril harbour are less than 15, and the remaining values are only slightly superior, the use of equation (1) means only minor modifications on the SPT values. Applying to each value of N' the different expressions for C_N, a mean value N' can be obtained (see figure 8).

Since Terzaghi and Peck (1948) first proposed an approximate relation between the number of blows N of the SPT and the relative density, D_r, many others studies have been carried out in different ways. Perhaps the results of the works done by Gibbs and Holtz (1957) and Bazaraa, (1967) are the best known, but there are other correlations which should also be mentioned, such as those from Schulze and Menzenbach (1961), Burmister (1962), Thornburn (1963), Schulze and Melzer (1965), Marcuson and Bieganousky (1976), Marcuson (1978), Borowzyk and Frankowsky (1981), Giuliani and Ricoll (1982), Melzer and Smoltzyk (1982) and Skempton (1986).

The published results of all those research works are based on field and laboratory tests run under different conditions on materials whose grain size and grading are not always well described in spite of the important role they play.

In order to infer D_r from each one of the N-values obtained at the site, the different correlations have been analysed and those that better fit the soil type and local conditions at the Motril harbour have been considered. As a result, a range of variation of $\pm$ 2.5 has been fixed based on mean values obtained after different weighting criteria applied to the data used for each one of the D_r values.

All the values obtained have been plotted against depth in figure 9 (derived from the N* profile) and figure 10 (corresponding to the N'* profile). The inspection

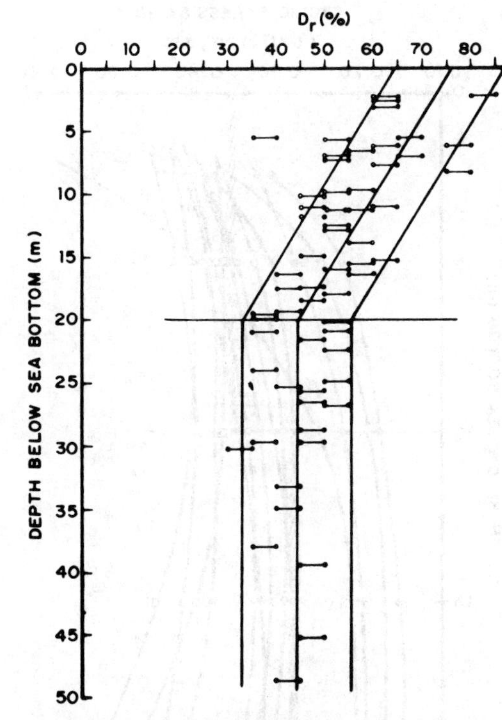

Fig. 9.- Relative density derived from N* values

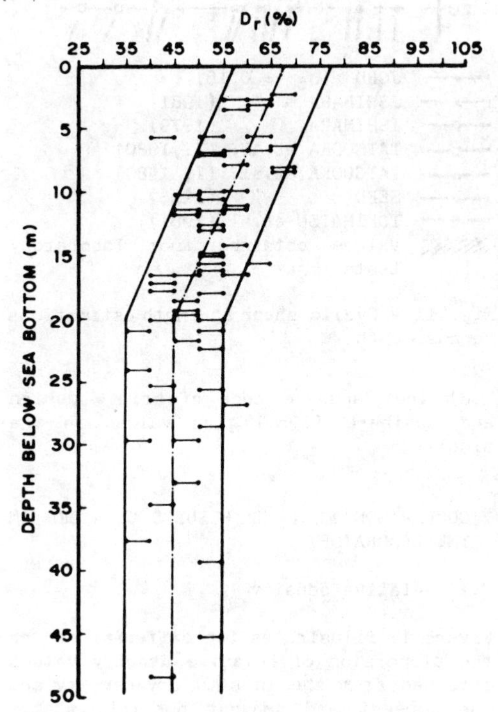

Fig. 10.- Relative density derived from N'* values

of these graphs leads to the following conclusions:

- Equation 1 has little influence in the determination of the relative density of the Motril harbour sediments.
- The variation of relative density with depth shows a tendency similar to the one exhibited by other geotechnical parameters.

5 ESTIMATING ANGLE OF SHEAR RESISTANCE

The data obtained with the SPT can also be used to estimate the friction angle of the materials. Although there is a smaller number of expressions in the literature correlating these two variables (even some of them link them through D_r) the works done by Meyerhof (1959), Bjerrum et al. (1961), Alpan (1964), D'Appolonia (1968), De Mello (1968), Saxena et al. (1982) and Giuliani et al. (1982) are relevant.

From these studies and with due consideration of the effect that the particular characteristics of the Motril soils may have on the different factors included in those expressions an estimation (as indicated in Table 1) of the most representative values of the angle ϕ' was made before running the laboratory tests.

TABLE 1

ESTIMATION OF FRICTION ANGLE

Depth Below Sea Bottom (m)	SPT Range (N*)		Friction Angle ϕ'	
	From	To	From	To
0	20	30	32°	34°
10	10	25	32°	34°
20	6	15	31°	33°
20	6	15	31°	33°

6 ESTIMATING DYNAMIC SHEAR STRENGTH

The possibility of being able to predict the dynamic behavior of soils by resorting to other index parameters has increased in the last few years although the amount of existing empirical formulas to do so is still limited.

By assembling a vast quantity of cyclic triaxial test data obtained in Japan, Tatsuoka et al. (1980) and Ishihara (1979) recommended the use of the following expressions:

$$(\tau/\sigma'_0) = 0{,}0676\sqrt{N^*} + 0{,}225\log_{10}\left(\frac{0{,}35}{D_{50}}\right) \quad (2)$$

(Tatsuouka et al. I 1980)

for 0.04 mm. D_{50} 0.6 mm

$$(\tau/\sigma'_0)_{20} = 0{,}0676\sqrt{N^*} + 0{,}0035\ C \quad (3)$$

(Tatsuoka et al. II 1980)

$$(\tau/\sigma'_0)_{20} = 0{,}009\ (N^* + 13 + 6{,}5\log_{10} C) \quad (4)$$

(Ishihara, 1979, 1985)

$$(\tau/\sigma'_0)_{20} = 0{,}0676\sqrt{N^*} + 0{,}085\log_{10}\left(\frac{0{,}50}{D_{50}}\right) \quad (5)$$

(Ishihara, 1981, 1985)

where (τ/σ'_0) denotes the cyclic stress ratio required to cause a 5% double amplitude strain in 20 cycles of shear stress application and C is the content of fines in percent passing the sieve 200 mesh.

There are also the recommendations included in the Seismic Code for Bridges in Japan (JCBD), although they are best suited for soils coarser than those found in the Motril harbour with D_{50} ranging between 0.15 mm and 0.35 mm.

Lately, Tokimatsu and Yoshimi (1981), for granular soils with a mean particle diameter (D_{50}) less than o.15 mm and Seed (1983, 1984), for soils with an amount of fines of about 35%, produced new empirical laws to estimate the value of τ/σ'_0.

Figure 11 shows the results obtained when applying the aforementioned expressions keeping in mind the peculiar characteristics of the Motril soils. To come up with the curves drawn in that picture use has been made of the mean laws given in figures 7 and 8 for N* and N'*. It has been assumed that there is risk of liquefaction below a depth of 20 m.

From an overall view of the curves included in figure 11 the following conclusions can be drawn:

a) There is a significant spread on the cyclic strees ratio predicted by the different empirical rules.

b) Equations (2) and (3) recommended by Tatsuoka yield the higher values.

c) When using the curves suggested by Tokimatsu and Yoshimi (1983) and Seed (1983) the cyclic strength shows the lower values.

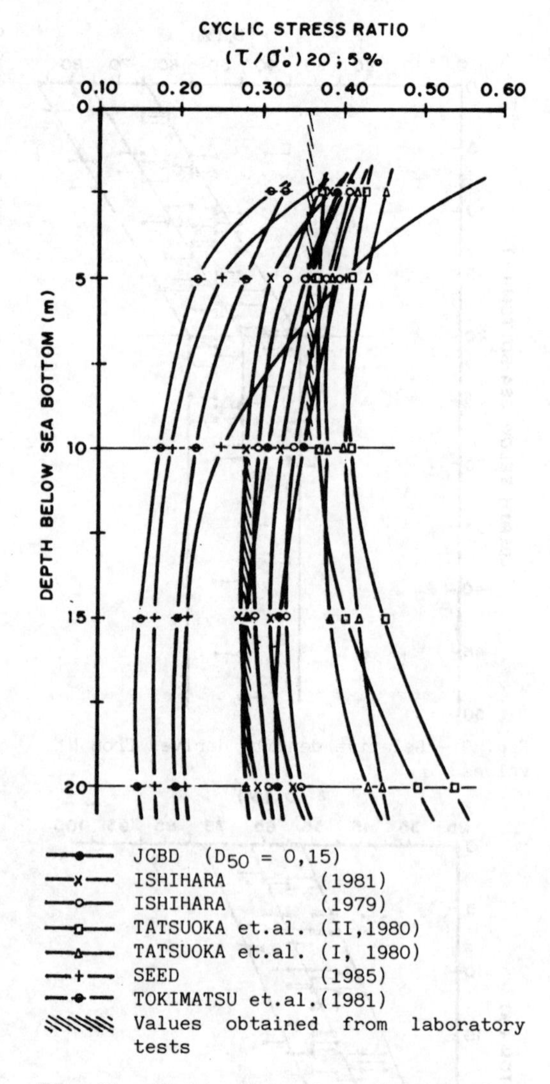

Fig. 11.- Cyclic shear strength estimations versus depth

d) The Japanese code of bridge design and Ishihara (1981) give values in the middle.

7 COMPARISON WITH THE RESULTS OBTAINED IN THE LABORATORY

7.1 Relative density

Figure 12 illustrates for different depths the dispersion of relative density values obtained from the in situ dry density and the loosest and densest possible states reached in the laboratory. The dispersion

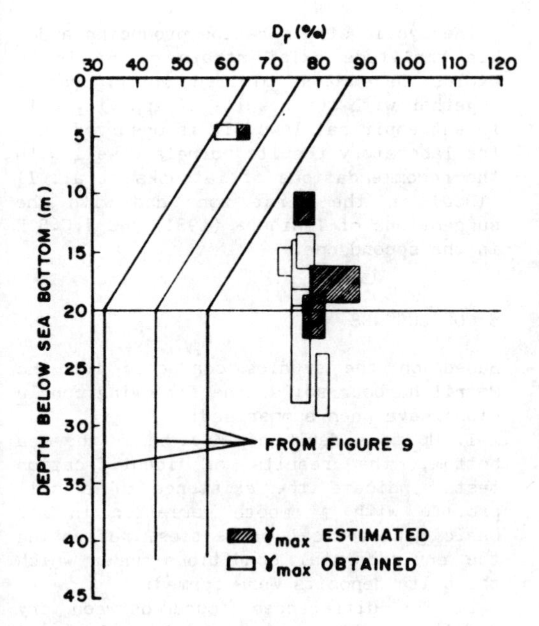

Fig. 12.- Relative density obtained from in situ and laboratory works

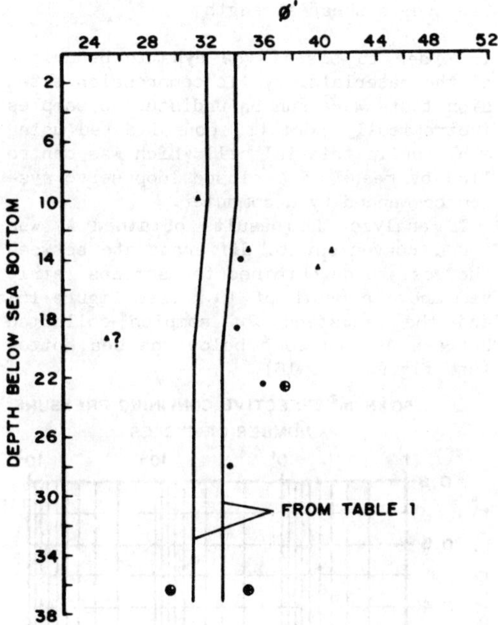

Fig. 14.- Friction angle versus depth

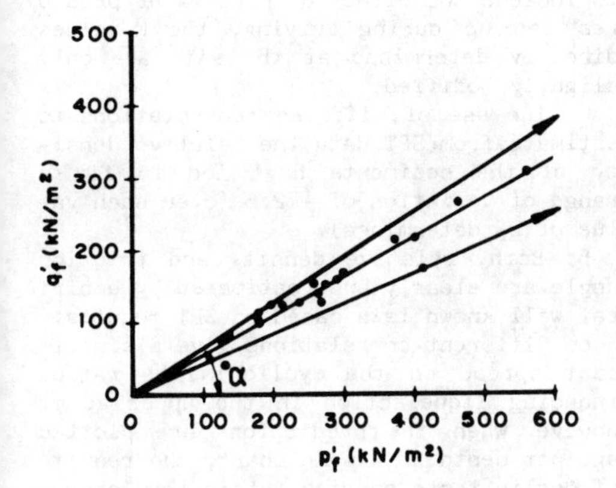

Fig. 13.- Friction angles obtained

inherent to the different procedures used to estimate the density in the field and the need to mix in some cases material of the same characteristics, but recovered at different levels in the borings, have also favoured the display of data by means of the rectangles drawn in that picture. There have also been included in it the limits and mean law estimated from N* values in figure 9.

It may be seen that the use of empirical laws based on SPT blow counts clearly underestimates the degree of compactness for the silty sand and silt of Motril harbour. This, may be in part due to the fact that the grain size distributions used by the authors of the various correlations applied belong to sands somehow coarser than those found at the site.

7.2 Friction angle

Figure 13 shows in a p-q diagram the bounds and the best fitting straight line determi̲ned in the regression analysis carried out with the results of the triaxial strength tests run on the undisturbed samples. Both waxed and frozen samples were tested without any significant difference in the results obtained.

The friction angles obtained are plotted against depth in figure 14 together with the limits of variation of Ø' estimated from the N-values of the SPT. It can be seen that, in general, the correlations used to predict the friction angle by means of the SPT results also underestimate the shear strength of the materials.

7.3 Cyclic shear strength

In order to assess the dynamic behaviour of the materials, cyclic compression extension tests were run on undisturbed samples isotropically normally consolidated using a hydraulic triaxial cell which was controlled by means of a closed loop servo system commanded by a computer.

To analyse the results obtained it was found convenient to differenciate between the resistance obtained for samples retrieved above a depth of 10 m (see figure 15) and the resistance of samples collected between 10 and 20 m below the sea bottom (see figures 11, 16).

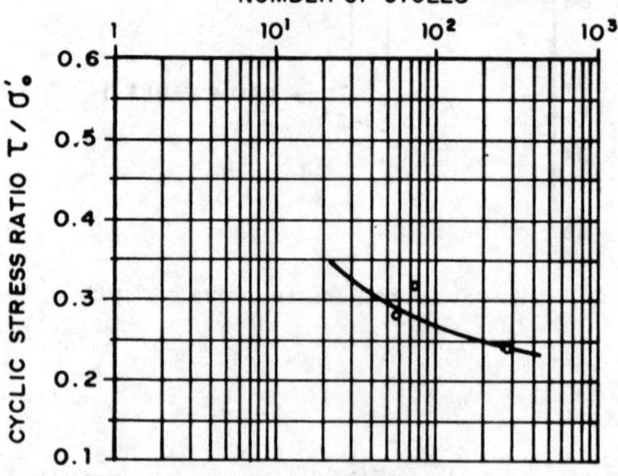

Fig. 15.- Cyclic shear strength of fine sand from above a depth of 10 m.

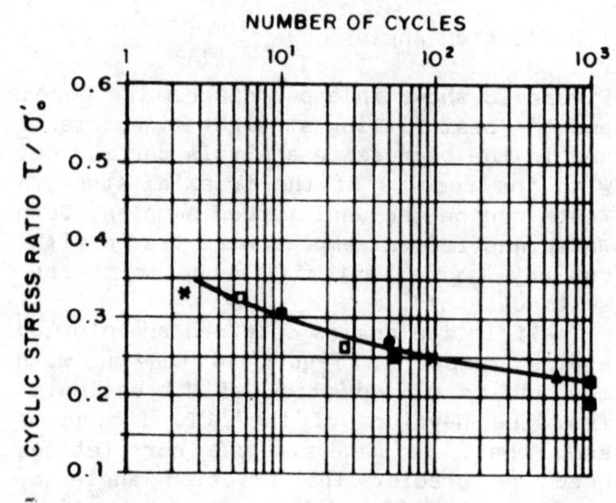

Fig. 16.- Cyclic shear strength of silty sand from 10-20 m.

The cyclic stress ratios producing a double amplitude axial strain of 5% in 20 cycles have been plotted in figure 11 together with the results of applying different empirical laws. It is observed that the laboratory results correlate well with the recommendations of Tatsuoka et al. II (1980) in the first zone and with the suggestions of Ishihara (1981) and J.C.B.D. in the second one.

8 CONCLUSIONS

Based on the studies conducted for the Motril harbour soils, the following conclusions have been summarised:

1. Up to a depth of 20 m below the sea bottom, the results of identification tests indicate the existence of a soil profile with a smooth variation in its basic geotechnical properties, reflecting the environmental conditions under which the delta deposits were formed.

2. The differences found between dry densities determined in waxed samples and frozen samples recovered at the site by means of piston samplers of the Osterberg type are not significant.

3. When SPT blow counts are corrected to include the effect of pore-water pressures set up during driving, the N-values directly determined at the site are only slightly modified.

4. The use of different correlations to estimate from SPT data the relative density of the sediments have led to fix a range of variation of ± 2.5 (over each value of D_r determined).

5. Both, relative density and friction angle are clearly underestimated by empirical well known laws based on SPT results.

6. Different correlations give a significant spread on the cyclic stress ratios inducing liquefaction in the upper 20 m. However when the predictions are plotted against depth in a same chart, the results of cyclic tests carried out in the laboratory correlate well with those empirical laws located in the middle of the bunch of curves.

REFERENCES

Alpan, I. 1964. Estimating settlements of foundations on sands", Civil Engineering and Public Works Rev., Vol. 59, p. 1415.

Bazaraa, A.R.S.S. 1967. Use of the Standard Penetration Test for estimating settlement of shallow foundations on sand, PH.D. Thesis, University of Illinois, Urbana.

Bjerrum, L., Kringstad, S. & Kummeneje, O, 1961. The shear strength of fine sand, 5th ICSMFE, París, Vol. I, p. 29.

Borowzyk, M. & Frankowski, Z. 1981.Dynamic and static sounding results interpretation. X ICSMFE Stockholm, 2: pp. 451-454.

Burminster, D.M. 1962. Field Testing of Soils, ASTM SPT 322, American Society for Testing and Materials, pp. 67-97.

D'Appolonia, D.J. et al., 1968. Settlement of Spread Footing on Sand, ASCE, Vol. 94, SM 3, p. 735.

De Mello, J. 1971. Standard Penetration Test. 4th Panamerican Congress on Soil Mechanics and Foundation Engineering. Puerto Rico, June, Vol. I.

Gibbs, H.J. & Holtz, W.G. 1957. Research on determining the density of sands by spoon penetration testing. Proc. 4th Int. Conf. Soil Mech. Fdn. Engng. London 1, pp. 35-39.

Giuliani, F. & Nicoll, F.L. 1982. New analytical correlations between SPT overburden pressure and relative density. Proceedings of the Second European Symposium on Penetration Testing. Amsterdam, pp. 47-55.

Ishihara, K. 1985. Stability of Natural Deposits During Earthquakes. Proc. of the Int. Conf. of Soil Mechanics and Foundation Engineering. XI. ICSMFE. San Francisco. USA.

Marcuson, W.F., III. & Bieganousky, W.A. 1977. Journal Geotechnical Engineering Division. American Society of Civil Engineers. Vol. 103, No. GT6, pp. 565-587.

Marcuson, W.F. III. & Bieganousky, W.A. 1977. Journal of the Geotechnical Engineering Division. American Socity of Civil Engineers. Vol. 103, No GT11, pp. 1295-1309.

Marcuson, W.F. 1978. Determination of in situ density of sands. Dynamic Geotechnical Testing. ASTM, SPT 654, American Society for Testing and Materilas, pp. 318-340.

Melzer, K. & Smoltczyk, U. 1982. Dynamic Penetration Testing. State of the art report. Proceedings of the Second European Symposium on Penetration Testing. Amsterdam, pp. 191-202.

Meyerhof, G.G. 1969. Penetration Tests and bearing capacity of cohesionless soils. Proc. American Society Civil Engineering Jour. Soil. Mechanics and Foundation Div. 82, paper 866.

Peck, R.B.; Hanson, W.E. & Thornburn, T.H. 1973. Foundation Engineering 2nd. ed. John Wiley and Sons.

Sanglerat, G. 1972. Penetrometer and Soil Exploration. Elsevier Publishing Company. Amsterdam.

Saxena, S.R. & Srivasulu, G.T. 1982. Prediction of engineering behaviour of sands from standard penetration test. Proc. 2nd. Symp. Penetration Testing. Amsterdam. Holland.

Schulze, E. & Melzer, K.J. 1965. The Determination of the density and the modulus of compressibility of non-cohesive soils by soundings. Proc. 6th Int. Conf. Soil Mech. Montreal I, pp. 354-358.

Seed, H.B.; Tokimatsu, K.; Harder, L.F. & Chug, R.M. 1984. The influence of SPT peocedures in soil liquefaction resistance evaluations. Report No. UCB EERC-84/15, Berkeley.

Skempton, A.W. 1986. Standard penetration test procedures and the effects in sands of overburden pressure, relative density, particle size, aging and overconsolidation. Geotechnique 36, nº 3, pp. 425-447.

Taksuoka, F.; Iwasaki, T.; Tokida, R; Yasuda, S.; Hirosi, M.; Imai, T. & Kon-no, M. 1980. Standard penetration tests and soil liquefaction potential evaluation, Soils and Foundations, Vol. 20, No. 4, pp.95-111.

Terzaghi, K. & Peck, R.B. 1948. Soil mechanics in engineering practice. New York. John Wiley and Sons.

Thornburn, S. 1963. Tentative correction chart from the standard penetration test in non-cohesive soils. Civ. Eng. R.S. Wks. Rev. 58:6:pp. 752-753.

Penetration Testing 1988, ISOPT-1, De Ruiter (ed.)
© 1988 Balkema, Rotterdam, ISBN 90 6191 801 4

Effect of overburden pressure on SPT N-values in cohesive soils of north Iran

A.M.Oskorouchi
Sharif University of Technology, Tehran, Iran
A.Mehdibeigi
Geotechnical Consulting Engineer, Tehran, Iran

ABSTRACT: The results of more than 100 standard penetration tests on fine grained soils of north Iran performed at different depths and locations have been chosen to be analyzed in this paper, undrained triaxial compression and occasionally Unconfined compression tests have been carried out on Undisturbed samples of the same soils. Correlation between "N" Value and cohesion of these soils have shown that the over-burden pressure has a great influence on the N-Values of medium to stiff clays and little effects on soft clays. Suggestions have been made for the corrections of "N" due to overburden pressure on these soils. Correlations between "N"(Corrected N' due to overburden) and the cohesion for Soils in this region have been presented.
Keywords : SPT "N" Values , cohesion , Overburden pressure , fine grained Soils.

Introduction

Gilan and Mazandaran are two northern provinces of Iran which together comprize a region with profound similarities as far as climate geology and Soil type are concerned. The region with a total area of about 62000 Km^2 is located to the north of the Alborz mountains and south of Caspian sea and geologically is the interconnection between alluvial plains of northern Alborz and southern lake deposit of the Caspian sea. Surfacial soils of the region are composed of fine sands, silts, and medium to highly plastic clays underlain by marls and conglomerates. The area is tightly covered with forests and has long wet seasons of heavy rains and shallow water tables. The reason why SPT results performed in this areas were chosen to be analysed was to minimize the factors having influence on these correlation for similar tayp of soils.

Field and laboratory data

Field and laboratory tests from more than 10 subsurface soils investigations performed for different projects in different locations of Gilan and Mazandaran have been considered in this parer. As a routine in these investigation programs SPT's have been run and undisturbed soil samples have been secured during drilling operations at different depth.
UU triaxial compression tests and occasionally unconfined compression test have been performed on undisturbed samples in the laboratory. From these test results about a 100 have been used. Of course files of these soil investigations have been studied carefully and unreasonably high or low N-Values have been omitted form the analysis. Data used for this analysis is tabulated in table 1 and 2. These data have been plotted in Figures 1 through 4 and using linear regression analysis correlations between 'C' and'N' have been estabilished as best fitted straight lines. Although the soils from Gilan and Mazandaran provinces are generally similar. They have been grouped and analysed seperately to minimize the differences , while analysing the C-N correlations for these soils. It was found that if they were further grouped to soft and to medium to stiff according to their C or N-Values, better correlations could be obtained. So Figs 1 & 2 are for Mazandaran Soils (soft and medium to stiff) , and Figs 3 & 4 are for Gilan soils(soft and medium to stiff). Further it was found that if medium to stiff soils are seperated between two depths, 0-5 m and 5-15 m, better

correlations are obtained. Soft soils studied, were under ground water level-

Effect of overburden pressure on N' Values

As it can be seen from Figs 1 & 3 the overburden pressure has a considerable influence on the N-Values obtained on soils with cohesion more than 0.6 kg/cm^2 (58.8 kpa). In other words for soils with the same cohesion N-Values increase with depth , provided cohesion is greater than 0.6 kg/cm^2. On the other hand the effect of overburden pressure is very little on soft clays with an N- Value of less than 12. If cohesion of about 0.6 kg/cm^2 (58.8 kpa) or N-Value of about 12 is taken as the border line between soft and medium to stiff clays, for medium to stiff soils N-Values should be corrected for overburden pressure while for soft soil the effect of overburden is diminished and there is no need for the correction. Figs 2 & 4 show the data for soft soils and the way the effect of overburden pressure diminishes, such that no seperation between two depth is possible. The following equations are suggested to be used to correct the effect of overburden pressure (p):

For $N' \leq 12$ No correction needed

For $N' > 12$ $N = 12 + \frac{N'-12}{6}$

For Gilan Soils:

For $N' \leq 10$ No corrections needed

For $N' > 10$ $N = 10 + \frac{N'-10}{2.8}$

N' = Actual field SPT blow count
N = Corrected N' for overburden pressure

Correlation between N,C and natural unit weight

Tables 3 and 4 show the correlation between N, the corrected SPT blow count for overburden; C, cohesion from laboratory tests (UU triaxial or Unconfined tests); γ_n, natural unit weight for Mazandaran and Gilan soils respectively.

Conclusions

Standard penetration test results on fine-grained soils from two northern provinces of Iran were analysed in order to find correlation between C and N and effect of overburden pressure. The conclusions obtained can be summerized as follows;

1- For medium to stiff clays the overburden pressure has a considerable effect on N for soils with the same cohesion, Fige 1 & 3.

2- Linear regression analysis was carried out between C - N for deep and surface soils and the resulted relationship lines crossed each other at C = 0.6 kg/cm^2 (58.8 kpa), which means for soft clays the effect of overburden pressure is diminished.

3- The above conclusion was supported by data obtained on soft clays in the same region as plotted in Figs 2 & 4.

4- Formulas for correcting N' for the effect of overburden was suggested for the areas studied.

References

Subsurface investigation reports of darupakhsh building, telecommunication center of Sarry, Bandar Torkaman, Bandar Gas, Mahmoodabad in Mazandaran and telecommunication center of Hashtrud, Rezvan shahr, police station Langrud; Aggricultural college, Hospital and Fertilizer factory of Rasht; Aidoghmish Bridge in Hashtpar in Gilan. These reports were studied in Technical Library of Soil and Technical Laboratory of Highway & Transportation Department of Islamic Republic of Iran.

Table 1- Test results for Gilan soils

USC	Depth (m)	N	C kg/cm^2	Test type
CH	0-5	10	0.13	UU
CH	,,	5	0.6	UU
CH	5-15	19	1.0	UU
CH	,,	16	0.9	UU
CH	,,	18	0.38	UU
CH	,,	17	0.50	UU
CL	5-15	18	0.5	UU
CL	5-15	17	0.55	UU
GW	,,	18	0.26	UU
CL	,,	21	0.5	UU
CL	,,	28	0.55	UU
CL	0-5	17	0.65	UN
CL	,,	12	1.19	UN
CL	,,	28	1.42	UN
CL	,,	38	1.32	UN
CL	,,	30	2.69	UN
CL	,,	25	2.21	UN
CL	,,	17	0.62	UU
CL	,,	18	0.84	UN
CL	,,	20	0.90	UU
CL	,,	20	1.02	UN
CL	,,	22	1.13	UN
CL	,,	22	1.63	UU
CL	,,	21	2.07	UN
CL	5-15	20	0.85	UN
CL	,,	20	1.14	UN
CL	,,	20	1.30	UU
CL	,,	22	1.12	UN
CL	,,	21	1.2	UU
CL	,,	23	0.84	UN
CL	,,	23	1.30	UU
CL	0-5	9	0.3	UU
CL	,,	14	0.35	UU
ML	,,	17	0.20	UU
CL	5-15	16	0.3	UU
CL	,,	29	0.5	UU
CL	,,	32	0.45	UU
CL	0-5	6	0.4	UU
CL	,,	15	0.9	UU
CL	,,	9	1.0	UU
CL	,,	10	1.0	UU
CL	,,	10	1.0	UU
CL	,,	10	0.35	UU
SC	5-15	13	0.5	UU
CL	,,	12	0.6	UU

UU unconsolidated, undrained triaxial test
UN unconfined compression test

Table 2- Test results for Mazandran soils

USC	Depth (m)	N	C kg/cm^2	Test type
CL	0-5	15	1.3	UU
CL	,,	15	0.7	UU
CL	,,	15	0.75	UU
CL	,,	13	1.30	UU
CL	,,	18	1.0	UU
CL	5-15	20	1.1	UU
CL	,,	24	0.95	UU
CL	,,	11	1.15	UU
,,	,,	20	0.5	UU
,,	,,	35	0.8	UU
,,	,,	55	0.5	UU
,,	,,	11	0.37	UU
,,	,,	12	0.38	UU
,,	,,	10	0.25	UU
,,	,,	11	0.48	UU
ML	0-5	8	0.35	UU
CL	,,	7	1.70	UU
CL-ML	,,	7	0.5	UU
CL-ML	,,	7	0.52	UU
CL	5-15	5	0.25	UU
CL	5-15	6	0.25	UU
CL	,,	7	0.7	UU
CL	,,	11	0.35	UU
CL	,,	10	0.25	UU
CL	,,	8	0.85	UU
CL	,,	3	0.13	UU
CL	,,	5	0.5	UU
CL	,,	10	0.16	UU
CL	,,	17	0.50	UU
CL	,,	5	1.2	UU
CL	0-5	4	1.5	UU
CL-ML	,,	2	0.15	UU
ML	,,	2	0.3	UU
CL-ML	5-15	5	0.4	UU
CL	,,	7	0.4	UU
CL	,,	18	0.3	UU
CL	,,	3	0.9	UU
CL	,,	3	0.2	UU
CL	,,	4	0.9	UU
ML	0-5	5	0.3	UU
ML	,,	8	0.5	UU
CL-ML	5-15	5	0.25	UU
MH	,,	11	0.65	UU
ML	,,	8	0.25	UU
CH	0-5	8	0.2	UU
ML	,,	6	0.4	UU
Sc-ML	5-15	11	0.35	UU
CL	,,	11	0.2	UU

N versus C- Mazandran soils

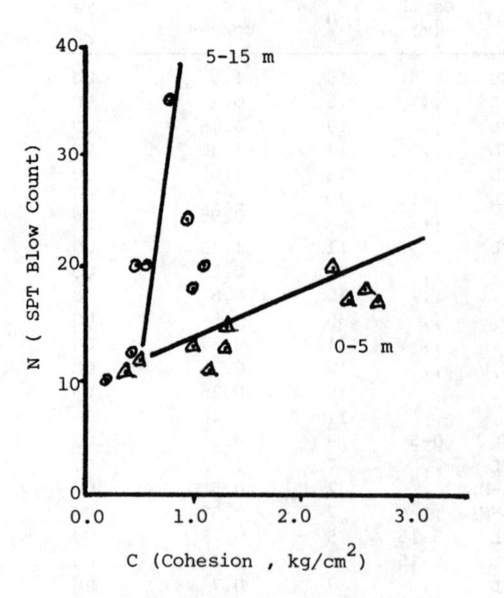

Fig.1 : Medium to stiff clays

△ - Depth 0-5 m
⊙ - Dept 5-15m

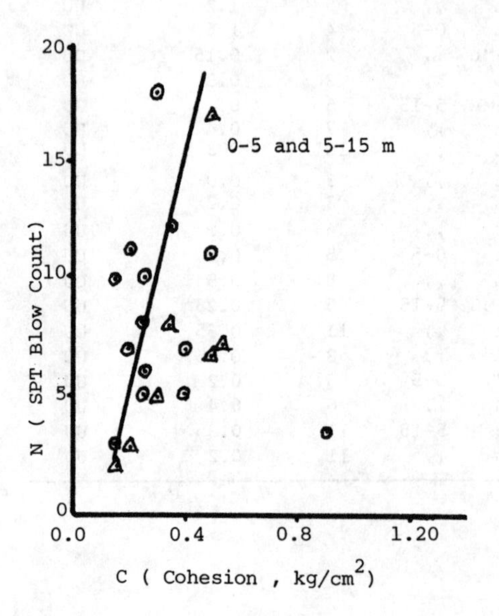

Fig.2 : Soft clays

N versus C - Gilan soils

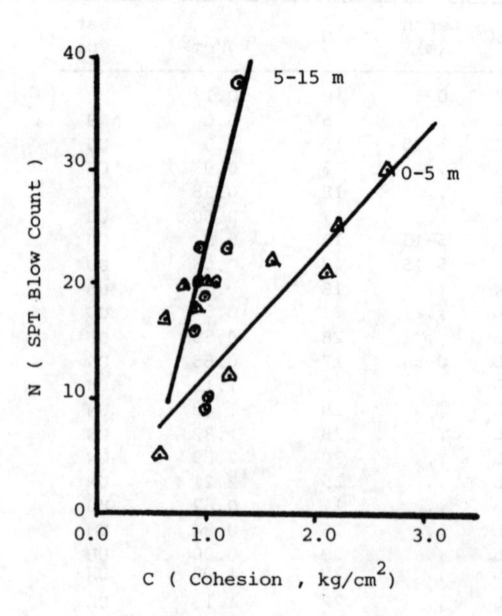

Fig.3 : Medium to stiff clays

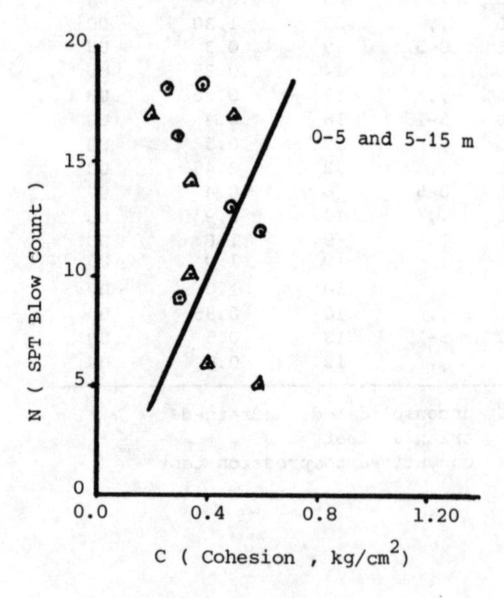

Fig. 4 : Soft Clays

Table 3- Mazandaran soils

Soil consistency	N	C(kg/cm^2)	γ_n (g/cm^3)
Soft	2–4	0.1-0.4	1.80 to 1.99
	4–9	0.2-0.55	
	9–12	0.3-0.65	
medium to stiff	12–16	0.6-1.3	1.78 to 2.10
	16–20	1.3-2.4	
	20–24	2.4-3.0	

Table 4- Gilan soils

Soil consistency	N	C(kg/cm^2)	γ_n (g/cm^3)
Soft	2–6	0.1-0.3	1.77 to 2.05
	6–10	0.15-0.6	
medium to stiff	10–16	0.6-1.0	1.80 to 2.10
	16–20	1.0-1.3	
	20–29	1.3-1.6	

1.0 kg/cm^2 = 98.1 kpa
1.0 g/cm^2 = 9.80 kN/m^3

Penetration Testing 1988, ISOPT-1, De Ruiter (ed.)
© 1988 Balkema, Rotterdam, ISBN 90 6191 801 4

Static pile capacity based on penetrometer tests in cohesionless soils

Hari D.Sharma
Fluor Canada Ltd, Calgary, Alberta

ABSTRACT: The applicability of the commonly used penetrometer tests, such as: SPT, DCPT and CPT, for pile capacity estimations has been reviewed. The estimated pile point (base) capacity for the expanded base compacted piles calculated from the available correlations and formulas with penetrometer values, has been compared with the values obtained from pile load tests. The results are in good agreement. The measured skin friction values on two drilled pile shafts, in cohesionless soil, were found to be slightly higher than the values estimated from penetrometer correlations. In general, penetrometer correlations provided good estimates for pile capacities.

KEY WORDS: Penetrometer, load tests, drilled piles, Franki piles, pile point capacity, shaft resistance, skin friction.

1 INTRODUCTION

This paper first provides a brief review of the empirical relationships between pile capacities and the common types of Penetrometer test values. The three Penetrometer tests that have been discussed here are the Standard Penetration Test (SPT), the Dynamic Cone Penetration Test (DCPT) and the Static Cone Penetration Test (CPT) methods. Following this, the expanded base compacted (Franki type) pile point (base) capacities obtained from field tests have been compared with the corresponding values estimated from SPT and CPT correlations and a dynamic driving formula. Some new empirical relationships have also been proposed for the dynamic driving formula. Data indicate good agreement between the test results and the values calculated from formulas. Finally, the measured shaft friction for two drilled piles in a cohesionless soil, has been compared with the values estimated from empirically related skin friction with SPT and CPT values. Results indicate that the measured shaft friction is slightly higher than the shaft friction values estimated from penetrometer data. The results, in general, indicate that the penetrometer test values can be used to estimate both the pile point and the shaft resistance values for the cases presented here.

2 PENETROMETER TEST ESTIMATES: A REVIEW

2.1 Penetrometer test types:

The three common types of penetrometer tests for estimating soil parameters that are considered here for pile design, are the Standard Penetration Test (SPT), the Dynamic Cone Penetration Test (DCPT) and the Static Cone Penetration Test (CPT). Their advantages and disadvantages are discussed by Kovacs and Solomone (1972). Robertson (1986), and Schmertmann (1977).

Many factors, such as, equipment, procedures and operator characteristics affect the reproducibility of the SPT. However, there is a considerable wealth of experience and design confidence based on SPT correlations. Therefore, engineers like to obtain equivalent SPT (N) values for DCPT (q_{cd}) and CPT (q_c) values. One such correlation, originally presented by Robertson et al (1983), is given in Figure 1. This relationship can be used to obtain equivalent N from field measured q_c values. Interpretation using DCPT is generally based on locally established correlations. One such relationship is provided in Figure 2 for a cohesionless soil in which an expanded base compacted (Franki type) pile was installed and load tested. This load test will be discussed in the following section (3.1).

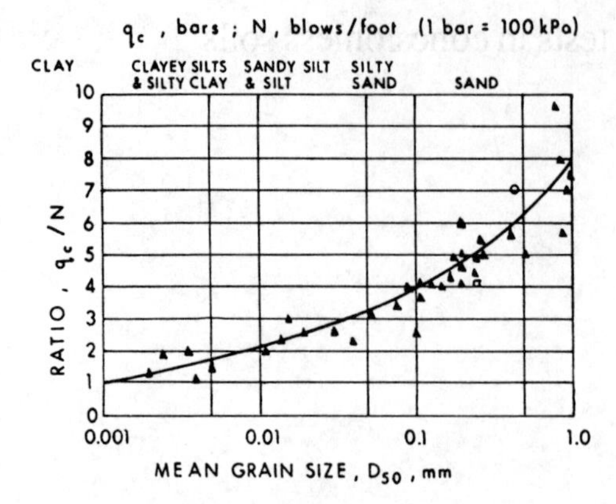

Fig. 1 Variation of q_c/N ratio with mean grain size (Robertson et al, 1983 and Robertson, 1986).

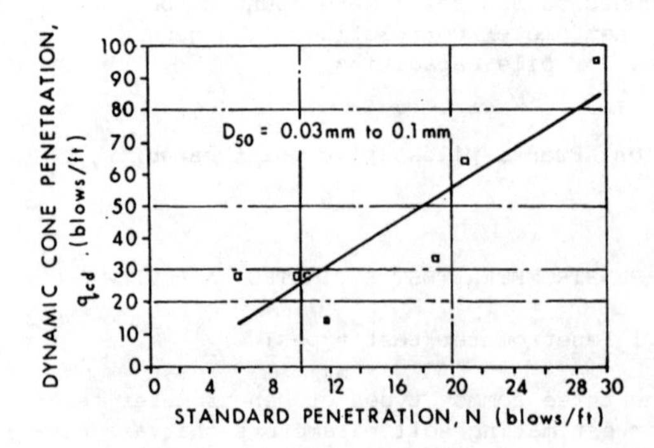

Fig. 2 Relationship between N and q_{cd} for a cohesionless soil.

2.2 Pile capacity estimates based on penetrometer values:

The unit resistance q_{ult}, of a pile is made up of unit point resistance, q_p, and the unit friction resistance, q_f, and is given by the following relationship:

$$q_{ult} = q_p + q_f \qquad (1)$$

q_p and q_f values in kPa can be obtained from Table 1.

Relationships (2), (4) and (5) are applicable for driven piles. For Franki piles about twice the values given by equations (2) and (5) can be used (Meyerhof, 1976). Robertson et al (1985) showed that equation (4) yields an excellent assessment of ultimate pile capacity for driven steel pile.

Table 1. Unit point and friction resistance values for pile capacities.

Relationship	Equation #	Reference
$q_p = q_c \frac{D_b}{10B} \leq q_1$	(2)	Meyerhof (1976)
$q_f = f_s = f_c$	(3)	Meyerhof (1976)
$q_p = \frac{q_{c1}+q_{c2}}{4} + \frac{q_{c3}}{2}$	(4)	De Ruiter and Beringen (1979)
$q_p = \frac{40N\ D_b}{B} \leq 4N$	(5)	Meyerhof (1976)
$q_f = f_s = 2\bar{N}$	(6)	Meyerhof (1976)
Dynamic formula (for Franki piles): $L_w = \frac{WHB'(V)^{2/3}}{K}$	(7)	Nordlund (1982)

Footnotes:

(a) q_c = static cone resistance, f_c = unit resistance of local sleeve of cone penetrometer. Generally $f_c < 1\%$ of q_c and $q_c > 3000$ kPa for sand.

(b) N = average SPT near pile tip and $\bar{N}$ = average SPT within the embedment length.

(c) L_w = allowable working load on Franki pile base which was driven with WH impact energy requiring B' number of blows to ram a unit volume of bulk concrete into the base. V is the bulk volume of concrete in the base. K is an empirical factor and can be related with N.

This method will not be used here. Readers should refer to the original paper.

3 CAPACITY OF EXPANDED BASE COMPACTED PILES BASED ON PENETROMETER VALUES

3.1 Test pile and load test:

An expanded base compacted (Franki type) pile was load tested as per ASTM D 1143-81 (1986). Figure 3 shows the load settlement curve and surrounding soil conditions for this pile. Installed shaft diameter for this pile was about 500 mm. The failure load interpreted by Butler and Hoy (1977), Davisson (1972) and Fuller and Hoy (1970) methods were 2359 kN (530 kips), 2403 kN (540 kips) and 2403 kN (540 kips), respectively.

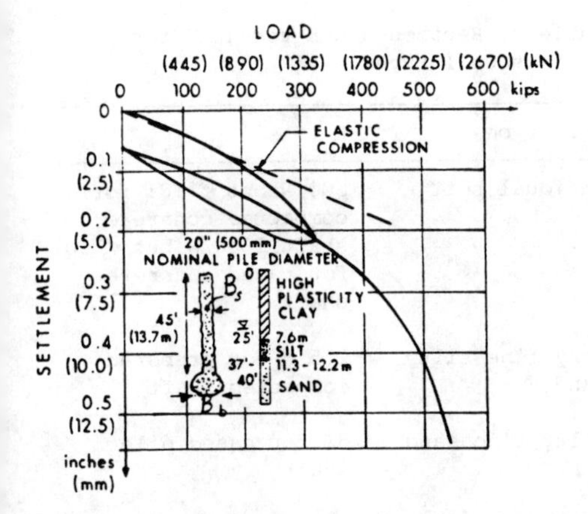

Fig. 3 Load-settlement curve for an expanded base compacted pile.

The pile installation method consisted of preboring through the top 6 m of clay and then driving the casing. Pile base formation was started at 13.7 m depth with 160 kN-m energy. Three concrete buckets, each with 0.14 cu m (5 cu ft) concrete, were used in the base. The first bucket of concrete required 20 blows, the second required 23 blows and the third required 30 blows of 160 kN-m impact energy for concrete expulsion into the base. Laboratory and field testing established that the clay had an undrained strength of 65 kPa and the undrained strength for silt was 38 kPa. The skin friction between pile and the clay was determined by using a reduction factor of 0.5 on the undrained strength. The length of the pile through the clay deposit was 7.6 m and through the silt was 4.6 m. Based on this, the ultimate skin friction, Q_f, for the pile was estimated at 672 kN (151 kips). The ultimate load carried by pile point will then be 2403 - 672 = 1731 kN (388 kips).

3.2 Estimated pile capacity based on SPT, N, Values:

Using equation (5), q_p = 986 kPa (21 ksf) when N = 13.4, D_b = 0.92 m and B = 0.50 m. The value of D_b is calculated based on the assumption that the total volume of concrete in the base is within an equivalent sphere of diameter D_b. The ultimate pile point load = 986 x 0.66 = 651 kN (146 kips) for a 0.66 m² base area. The ratio of ultimate pile point load estimated from load test to that calculated from SPT value above, is then = 1731/651 = 2.6.

Similarly an analysis of the installation and pile load test data provided by Nordlund (1982) on many expanded base compacted piles, indicates that the ratio of the ultimate pile point load estimated from load test to the calculated load from SPT by using equation (5) varies between 2.4 and 3.

Above analysis indicates that, equation (5) can be modified for expanded base compacted piles by multiplying it with a factor of at least 2.

3.3 Estimated pile capacity based on CPT, q_c values:

Grain size analysis of the cohesionless soil into which the expanded base compacted pile was installed indicated D_{50} ranging between 0.03 mm and 0.1 mm. From Figure 1, for this range of D_{50} value, $q_c/N \simeq 3.5$, where N = 13.4. Then using equation (2) will yield q_p = 865 kPa and Q_p = 571 kN.

Again, the ratio of ultimate pile point load estimated from load test to the value calculated from CPT, q_c, is equal to 1731/571 = 3.0. Therefore, as for SPT, the bearing capacity value could be obtained from equation (2), provided it is multiplied by a factor of at least 2. Further load test values are, however, required and analysis must be performed in terms of q_c to confirm this relationship.

3.4 Pile capacity based on dynamic formula:

For the load-settlement data of Figure 3, the pile was installed with WH = 160 kN-m, B' = 30/0.14 blows per cu m, V = 0.57 cu m and L_w = 1731/2.5 = 692 kN (156 kips) for a FS = 2.5. Substituting these values in equation (7), we get K = 34. Since average N = 13.4, the K = 2.5 N for this case.

Nordlund (1982) provides the B', V, N and L_w values for various soil types based on load test data on expanded base compacted piles. K values can then be obtained from these data based on equation (7). Table 2 provides a summary of this analysis. This table indicates that the constant K can be related with N values for various soils. Thus with known K, the allowable pile point load, L_w, can be estimated with reasonable accuracy from equation (7). Based on the data presented in Table 2, K values can be estimated for various soil types and pile installation conditions and are recommended in Table 3. In this table K values for residual soil are as proposed by Nordlund (1982).

Table 2. K Values for various soil types and shaft conditions.

Test	Soil-pile data	K/N	Shaft condition
1	Very fine silty sand, N=13.4, L_w=692 kN, K=34	2.5	Prebored compacted
2	Fine sandy silt (residual), N=25, L_w=570 kN, K=79	3.2	Cased
3	Fine sandy silt (residual), N=22, L_w=1086 kN, K=27	1.3	Compacted
4	Silt with sand (residual), N=13, L_w=854 kN, K=46	3.5	Compacted
5	Fine sandy silt (residual), N=14, L_w=721 kN, K=43	3.1	Compacted
6	Silty fine sand N=10, L_w=347 kN, K=29	2.9	Cased
7	Silty fine sand N=10, L_w=445 kN, K=30	3.0	Cased
8	Silty fine sand N=10, L_w=222 kN, K=30	3.0	Cased
9	Silty fine sand N=10, L_w=365 kN, K=30	3.0	Cased
10	Coarse to medium sand, N=4, L_w=676 kN, K=14	3.5	Cased

Footnotes:
(a) Test 1 is for load-settlement plot of Figure 3.
(b) Test details for test numbers 2 through 10 are reported by Nordlund (1982).
(c) Pile for test no. 1 was predrilled in top 6m length.
(d) Piles for test no. 2, 6, 7, 8, 9 and 10 had cased piles above the base.
(e) Piles for test no. 1, 3, 4 and 5 had compacted concrete shaft above the base.

Table 3. Recommended K versus N for various soil types.

Soil type	K
Residual soil	600/N but ≮ 18: for compacted concrete shaft. 1800/N but ≮ 50: for cased concrete shaft
Very fine silty sand	2.5 N for prebored compacted shaft
Silty fine sand	3N for cased pile shaft
Coarse to med. sand	3.5N for cased pile shaft

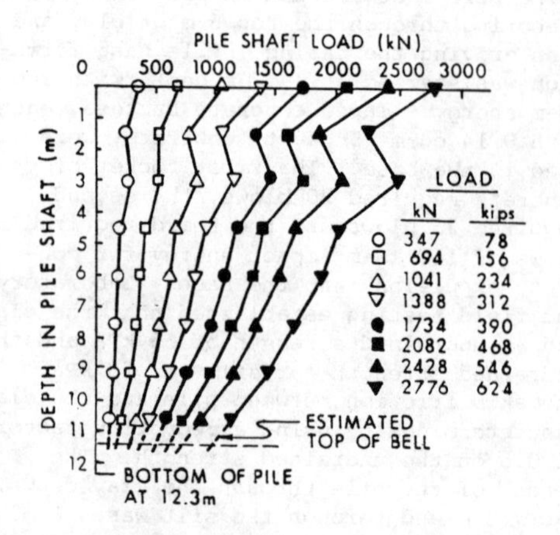

Fig. 4 Pile shaft load with depth from strain gauge readings for TP-2 (sharma et al 1986).

4 SHAFT FRICTION FOR PILES DRILLED THROUGH A COHESIONLESS SAND TILL

4.1 Test piles and load tests:

Two drilled and belled cast-in-place concrete piles (TP-1 & TP-2) were load tested as per ASTM D 1143-81 (1986). Vibrating wire strain gauges were installed at 1.5 m intervals down the pile shaft to measure load distribution along the shaft during compression. Generalized soil conditions at the site consisted of 0.5 m to 4.9 m thick sand fill over about 9.0 m of sand till overlying oil sand. Two piles that were load tested were of 0.66 m shaft diameter, 0.914 m bell diameter and were about 12.5 m long. Pile shafts were installed entirely in the sand till. Figure 4 shows the typical pile shaft loads measured during load testing on the pile TP-2. Sharma et al (1986) provide further details of the measured soil properties and pile load test data.

The average skin friction, f_s, along pile shaft was obtained by dividing (Q_{top} - Q_{base}) by $\pi B_s L$. Where Q_{top} and Q_{base} are measured loads at pile top and pile base, respectively. B_s is pile shaft diameter and L is pile length. These average mobilized skin friction values are plotted against load at the top of the pile in Figure 5. This figure indicates that the skin friction appears to be reaching an ultimate value of about 85 kPs.

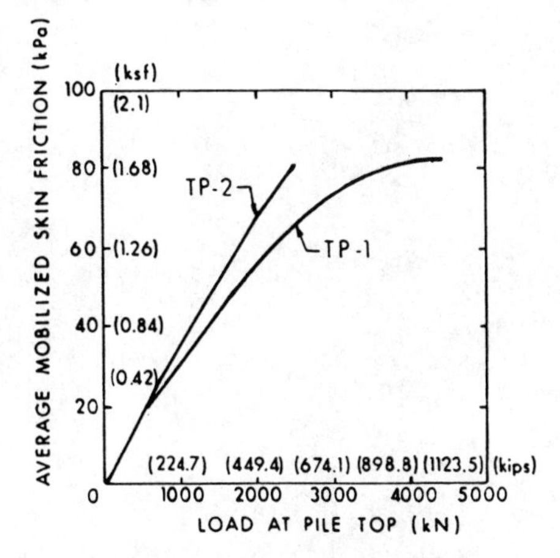

Fig. 5 Average mobilized skin friction versus load at pile top (Sharma and Joshi, 1986).

4.2 Estimated skin friction based on SPT values:

The soil stratigraphy at the pile location consisted of 1.8 m thick sand-gravel over 9.1 m thick deposit of sand till. The average SPT, N value for sand-gravel was 15 and for sand till was 39. Using equation (6), the weighted average estimated skin friction, f_s, was 70 kPa. When compared to the maximum mobilized field measured value of 85 kPa, the calculated value is about 17% lower than the average measured value.

4.3 Estimated skin friction based on CPT interpretation:

Field CPT values for the soils at the site are not available. However, laboratory grain size analysis for the cohesionless material indicated that D_{50} value for sand-gravel fill is about 0.4 mm and for sand till is about 0.1 mm. Using Figure 1, the weighted average q_c was estimated to be 14,386 kPa. When using f_s = 0.5% q_c, we get f_s = 72 kPa which is closer to the value obtained above. This confirms earlier observations that $f_s < 1\%\ q_c$ and $q_c >$ 3000 kPa for sands (Robertson, 1986).
When compared to the maximum mobilized field measured value of 85 kPa, the average calculated skin friction value is about 15% lower than the average measured value. Average skin (shaft) friction values estimated from SPT and q_c relationships are, however, in agreement if f_s/q_c is taken as 0.5%. The value of 0.5% for f_s/q_c may be a reasonable estimate since, as indicated above, the soil has an equivalent high cone bearing ($q_c >$ 3000 kPa).

5 CONCLUSIONS

For the soil-pile test data analyzed here, the following conclusions can be made:

(1) For expanded base compacted piles, the ultimate pile point (base) capacity before actual pile installation can be estimated with reasonable accuracy by using either equation (2) or equation (5) and multiplying them by a factor 2.

(2) The dynamic pile driving formula, equation (7), can be used to estimate the allowable pile point (base) capacity during an expanded base compacted pile installation when energy level WH, blows B' and concrete volume in the base, V, are known. The constant K, in this equation can be empirically related with the soil type, pile installation method and SPT values as recommended in Table 3.

(3) The measured ultimate shaft friction for two drilled pile shafts through a dense cohesionless material was about 16 percent higher than the weighted avarage values estimated from SPT and CPT correlations. Thus, for preliminary estimates, the ultimate shaft friction can be conservatively calculated either from equation (6) or from equation (3) by using f_s = 0.005 q_c for dense cohesionless material.
The above conclusions do not replace the requirements for pile load tests to obtain the accurate pile capacity for a specific site. Further analysis of the existing and new load test data should be carried out to confirm the applicability of above conclusions for other cases.

REFERENCES

ASTM Designation 1986. D 1143-81 Standard method of testing piles under static axial compressive load. Annual book of ASTM standards, Vol 04-08, pages 239-254.
Butler, H.D. & Hoy, H.E. 1977. Users manual for the Texas quick-load method for foundation load testing. Federal highway administration, office of

development, Washington, 59 pp.
Davisson, M.T., 1972. High capacity piles. Proceedings, lecture series, innovations in foundation construction, ASCE, Illinois section, 52 pp.
De Ruiter, J., and Beringen, F.L., 1979. Pile foundations for large north sea structures, marine geotechnology, 3 (3) pp 267-314.
Fuller, F.M. & Hoy, H.E. 1970. Pile load tests including quick-load test method, conventional methods and interpretations. HRB 333, pp 78-86.
Kovacs, W.D., and Salomone, L.A., 1982. SPT Hammer energy measurements. ASCE journal of the geotechnical engineering division, 108 (GT4), pages 599-620.
Meyerhof, G.G., 1976. Bearing capacity and settlement of pile foundations. ASCE journal of the geotechnical engineering division, 102 (GT3), pages 197-228.
Nordlund, R.L. 1982. Dynamic formula for pressure injected footings. ASCE geotechnical engineering division, Vol. 108, no. GT3, pp. 419-437.
Robertson, P.K., Campanella, R.G. and Brown, P.T. 1985. Design of axially and laterally loaded piles using in-situ tests: a case history. Canadian geotechnical journal, 22 (4), pages 518-527.
Robertson, P.K. 1986. In-situ testing and its application to foundation engineering, Canadian geotechnical journal, 23 (4), pages 573-594.
Robertson, P.K., Campanella, R.G., and Wightman, A., 1983. SPT - CPT correlations. ASCE journal of the geotechnical division, 109 (11), pages 1449-1459.
Schmertmann, J.H. 1977. Use the SPT to measure dynamic soil properties? yes, but ______ ! in dynamic geotechnical testing. American society for testing and materials, special technical publication 654, pages 341-355.
Sharma, H.D. and Joshi, R.C. 1986. Comparison of in-situ and laboratory soil parameters for pile design in granular deposits, 39th Canadian geotechnical conference, Ottawa, pages 131-138.
Sharma, H.D., Harris, M.C., Scott, J.D., and McAllister, K.W., 1986. Bearing capacity of bored cast-in-place concrete piles in oil sand. ASCE journal of geotechnical engineering, 112 (12), pages 1101-1116.

Penetration Testing 1988, ISOPT-1, De Ruiter (ed.)
© 1988 Balkema, Rotterdam, ISBN 90 6191 801 4

Bearing capacity of footings on Guabirotuba Clay based on SPT N-values

A.X.Tavares
Universidade Federal do Paraná, Curitiba, Brazil

ABSTRACT: This paper presents the relationships between the bearing capacity of footings on Guabirotuba Clay and the N values of the standard penetration test. The Guabirotuba Formation is from the Quaternary period and is located in the region of the city of Curitiba in the South of Brazil. The Guabirotuba Clay is preconsolidated. To consider the shape and the depth of the footing, shape and depth factors are presented. Comparison between the estimated and observed values of the ultimate bearing capacity of shallow footings on Guabirotuba Clay is shown in the text.

1 INTRODUCTION

The standard penetration test is the most common dynamic penetration test carried out during the drilling of any exploratory standard wash boring in Brazil. It is normalized and carried out every meter of depth. It is a kind of test which is very familiar to the Brazilian engineer. Because of its widespread use any attempt to get a better understanding of its results and limitations is worthful.

In the following paragraphs the results of an investigation undertaken with the purpose of estimating the bearing capacity of footings on the preconsolidated Guabirotuba Clay based on the SPT N values are presented.

2 GEOLOGY

The geology of the region includes the Crystaline base formed by Pre-Cambrian metamorphic rocks which are mostly composed of migmatites. Disposing over the Pre-Cambrian rocks are the Pleistocenic sedimentes of Guabirotuba Formation. It is composed by sediments consolidated lightly and formed mainly by clays containing sporadic thin layers of feldspatic sands.

Overlaying the Guabirotuba Formation, are the Holocenic deposits of plain flooding areas of the rivers of the Curitiba region.

3 GUABIROTUBA CLAY

The Guabirotuba Clay is a preconsolidated quaternary silty clay with fine sand. It is gray to brown-gray with black spots. The typical properties of the clay are shown in Table 1.

The short-term and long-term coefficient of settlements indicated in Table 1 are compressibility parameters for Guabirotuba Clay which behaves like an elastic material when submitted to pressure below the preconsolidation pressure (Tavares 1985, 1986).

Fig. 1 shows a soil profile at Portão district in Curitiba, together with the variation with depth of the N value, water content, consistency limits and undrained shear strength of Guabirotuba Clay.

4 BEARING CAPACITY

The short-term ultimate bearing capacity of a footing on clay is given by the formula

$$q_u = c_u N_c + \gamma D_f \quad (1)$$

where c_u is the undrained shear strength of the soil, N_c is the bearing capacity factor and γ is the unit weight of the soil within the footing depth D_f.

Bearing capacity factors were determined by plate-loading tests and model tests carried out on Guabirotuba Clay (Tavares 1978). The mean value of N_c for a circular footing at the surface of the clay was 6,9. For deep circular footings the N_c mean value almost coincides with the Skempton's value of 9 (Skempton 1951). The values of N_c for circular footings are shown in Table 2.

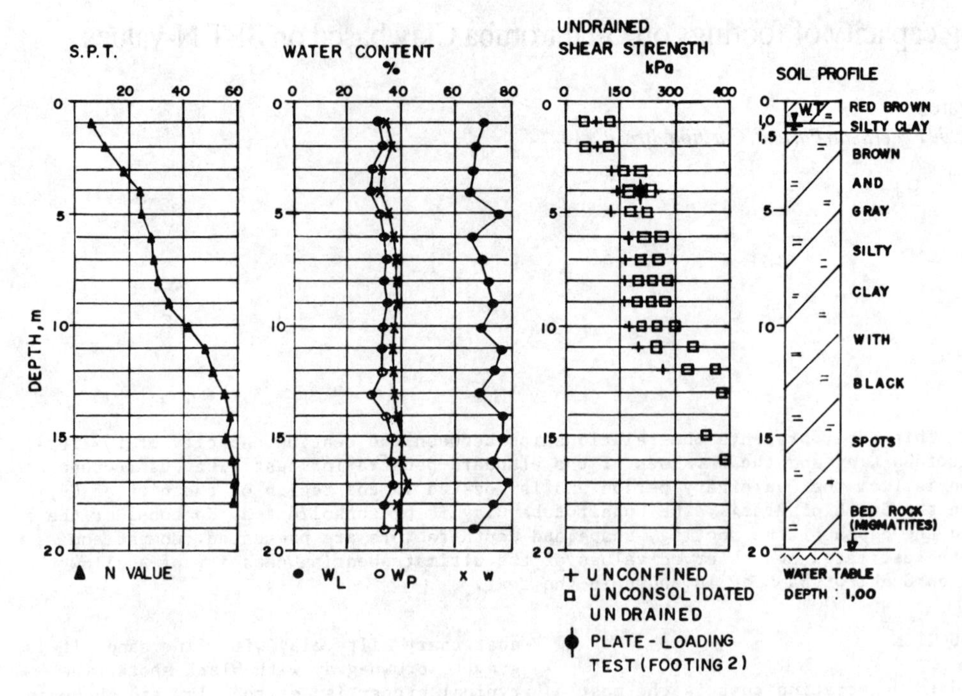

Fig. 1 Soil profile at Portão district. Variation of soil parameters

Table 1. Typical properties of brown-gray Guabirotuba Clay

Index property	Value
Water content	18 to 40
Liquid limit	61 to 87
Plastic limit	23 to 45
Plasticity index	38 to 42
Clay fraction	50 per cent
Unit weight	18,7 kN/m^3
Preconsolidation pressure	300 to 1000 kPa
Undrained shear strength	70 to 400 kPa
Coefficient of consolidation	10 to $20x10^{-4}cm^2/s$
Poisson's ratio drained	0,18 to 0,25*
Compression index	0,160 to 0,280
Swelling index	0,027 to 0,035
Overconsolidation ratio, O.C.R.	3.7 to 8.2
Short-term coefficient of settlement	0,015 to 0,025*
Long-term coefficient of settlement	0,018 to 0,035*

* Tavares (1985, 1986)

Table 2. Bearing capacity factors of circular footings on Guabirotuba Clay

Depth/breadth radio, D/B	N_c
0	6,9
1	7,7
2	8,3
4	9,0

4.1 Shape and depth factors

Based on experimental data it is suggested that the ultimate bearing capacity of a rectangular footing at depth D on Guabirotuba Clay can be given by the insertion of shape and depth factors in formula (1) (Meyerhof 1963), and it becomes

$$q_u = c_u \quad N_c \quad \lambda_c \quad d_c \tag{2}$$

in which q_u is the net ultimate bearing capacity of the footing, λ_c is the shape factor and d_c is the depth factor given for Guabirotuba Clay by the expressions

$$\lambda_c = 1 + 0,2 \frac{B}{L} \quad (3)$$

and

$$d_c = 1 + 0,1 \frac{D}{B} \quad (4)$$

where B is the footing width and L is the footing length.

Formulas (3) and (4) are valid for $\frac{D}{B} \leqslant 2$.

N_c in equation (2) is the Terzaghi's bearing capacity factor for a continuous footing which is 5,7 (Terzaghi 1943). As requirement for stability a factor of safety against bearing capacity failure of 3 should be used to obtain the net allowable soil pressure.

5 CORRELATION BETWEEN UNDRAINED SHEAR STRENGTH AND N VALUE

The undrained shear srength of Guabirotuba Clay was determined through the conduction of undrained triaxial tests, unconfined tests and plate-loading tests.

The triaxial tests have been carried out on undisturbed clay samples which were obtained by thin walled sampling or carved from the walls of test pits or piers. The samples were tested unconsolidated undrained.

The plate-loading tests were of the constant rate of penetration type. The undrained shear strength for these tests was obtained using equation (1) in which N_c is the bearing capacity factor given in Table 2. The undrained shear strength obtained from the plate-loading test referred as footing 2 in Table 4 is plotted in Fig.1. The test was performed at the base of a test pit 1,20m deep and of the same diameter as the plate. The test pit was made at the bottom of a large excavation for a building basement which was at a depth of 2,80m. For Guabirotuba Clay it was found that the in-situ undrained shear strength determined by plate-loading tests is near the mean of the laboratory results (Tavares 1978).

Variation with depth of the N value and the undrained shear strength of Guabirotuba Clay for a typical soil profile is shown in Fig.1.

To obtain the relationship between the undrained shear strength of Guabirotuba Clay and the N value of the standard penetration test, the penetration resistance

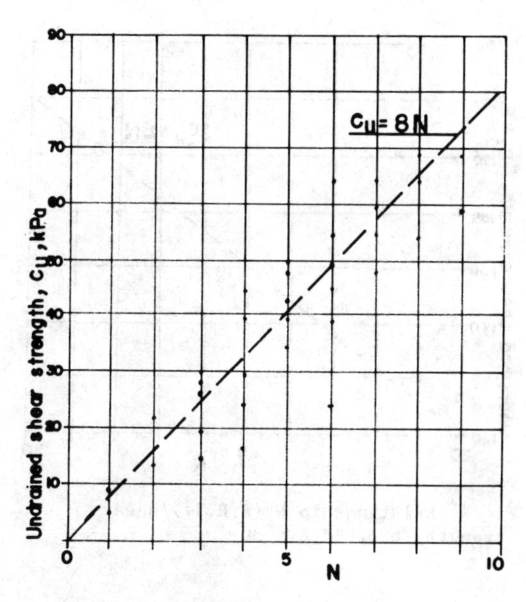

Fig. 2 Relationship N (S.P.T.)/shear strength for N < 10

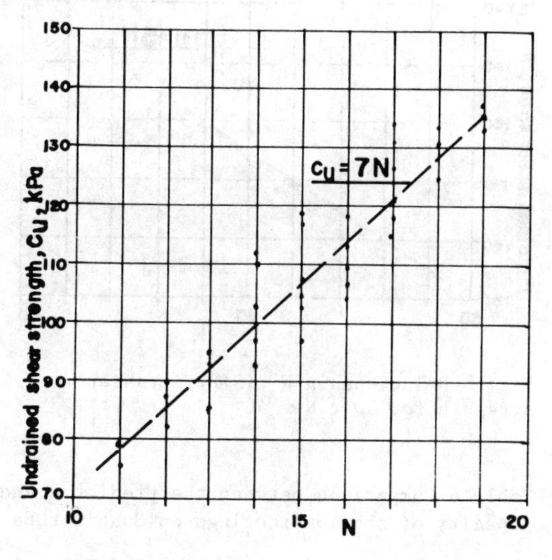

Fig. 3 Relationship N (S.P.T.)/shear strength for 10 < N < 20

was divided into four groups which are indicated below:

Group 1 - clay soil with N value less than 10
$N \leqslant 10$

Group 2 - clay soil with N value more than 10 and not exceeding 20
$10 < N \leqslant 20$

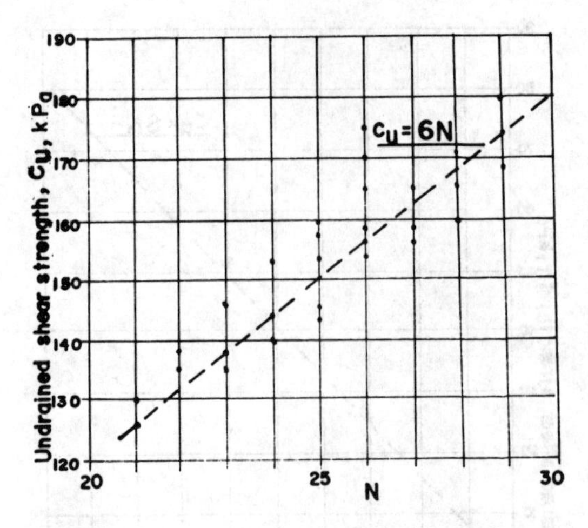

Fig. 4 Relationship N (S.P.T.)/shear strength for 20 < N < 30

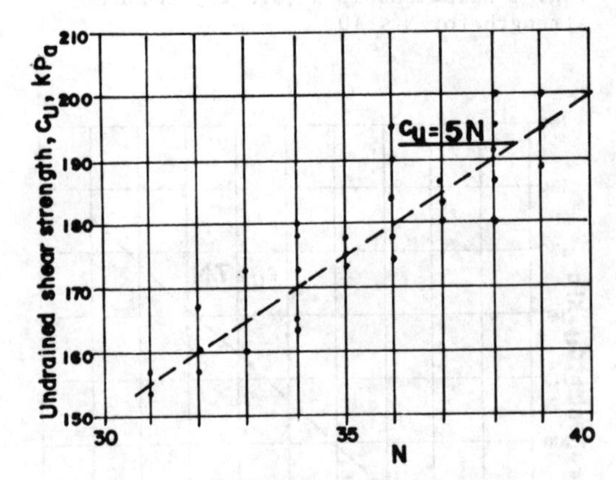

Fig. 5 Relationship N (S.P.T.)/shear strength for 30 < N < 40

Group 3 - clay soil with N value more than 20 and not exceeding 30 $20 < N \leqslant 30$

Group 4 - clay soil with N value more than 30 and not exceeding 40 $30 < N \leqslant 40$

The undrained shear strength and N value relationships for the four groups of penetration resistance are shown in Figs. 2 to 5.

In Table 3 the correlation N value and undrained strength is indicated.

6 RELATIONSHIP BETWEEN THE ULTIMATE BEARING CAPACITY AND N VALUE

Inserting the values of cu given in Table 3 in equation (2), the corresponding values of the net ultimate bearing capacity of footings on Guabirotuba Clay were obtained which are given by the following expressions

$$q_u = 45,60 \quad N \quad \lambda_c \, d_c \qquad (9)$$

for $N \leqslant 10$

Table 3. Undrained shear strength of Guabirotuba Clay on the basis of the N values of the S.P.T.

Range of N	Undrained shear strength in kPa	
$N \leqslant 10$	Cu = 8N	(5)
$10 < N \leqslant 20$	Cu = 7N	(6)
$20 < N \leqslant 30$	Cu = 6N	(7)
$30 < N \leqslant 40$	Cu = 5N	(8)

Table 4 Comparison between the predicted and measured values of the net ultimate bearing capacity of shallow footings on Guabirotuba Clay

Footing Number	Footing Shape	B (m)	L (m)	B/L	D/B	N S.P.T.	Net ultimate bearing capacity in kPa: Estimated value	Measured value	Ratio measured/estimated
1	circular	0,30	–	1	2	39	1600	1795	1,12
2	circular	0,60	–	1	2	25	1231	1550	1,26
3	circular	0,30	–	1	1	12	632	1390	2,20
4	square	0,30	0,30	1	2	18	1034	1370	1,32
5	square	1,00	1,00	1	1	14	737	1098	1,49
6	rectangular	0,60	1,20	0,5	2	35	1317	1510	1,15

$$q_u = 39{,}90 \; N \lambda_c d_c \tag{10}$$
for $11 < N \leq 20$

$$q_u = 34{,}20 \; N \lambda_c d_c \tag{11}$$
for $23 < N \leq 30$

$$q_u = 28{,}50 \; N \lambda_c d_c \tag{12}$$
for $32 < N \leq 40$

q_u units: kPa

7 ALLOWABLE SOIL PRESSURES

The allowable soil pressure for the footing is obtained by adopting a factor of safety of 3 into expressions (9) to (12) to get

$$q_a = 15{,}20 \; N \lambda_c d_c \tag{13}$$
for $N \leq 10$

$$q_a = 13{,}30 \; N \lambda_c d_c \tag{14}$$
for $11 < N \leq 20$

$$q_a = 11{,}40 \; N \lambda_c d_c \tag{15}$$
for $23 < N \leq 30$

$$q_a = 9{,}50 \; N \lambda_c d_c \tag{16}$$
for $32 < N \leq 40$

where q_a is the net allowable soil pressure in kPa and λ_c, d_c are valid for $\frac{D}{B} \leq 2$.

The lower limits of the interval of N variation were adjusted to eliminate ambiquity of values. For an N value which falls between the upper and the lower limits of two contiguous classes of N variation, q_u and q_a given by equations (9) to (16) are obtained by linear interpolation between the extreme values of the two classes.

In the above equations N is the average value within the depth B below the base of the footing.

8 COMPARISON BETWEEN THE PREDICTED AND MEASURED VALUES OF THE ULTIMATE BEARING CAPACITY

Table 4 shows a comparison between the predicted and measured values of the ultimate bearing capacity of some footings on Guabirotuba Clay. The predicted net values were obtained through the use of equations (9) to (12). By observation of the values indicated in Table 4 it is seen that the computed and the measured values of the net ultimate bearing capacity of the footings are reasonably in close agreement to each other for very stiff clay. When the consistency of the clay decreases the difference between the two values becomes quite appreciable.

9 CONCLUSIONS

The method described above for estimating the bearing capacity of footings on Guabirotuba Clay through the N values of the standard penetration test takes into account the shape and the depth of the footing.

Comparison between the predicted and measured values of the ultimate bearing capacity of footings on Guabirotuba preconsolidated clay shows that for very stiff clay the predicted values are less than the measured ones within 30%. For N values less than 15 the differences between the measured and predicted values increase and can achieve more than 100%. This emphasizes the aproximate nature and limitations of the standard penetration test when applied to clay soils.

Despite the differences pointed out above, for very stiff clays of Guabirotuba Formation the method gives reasonable results.

REFERENCES

Meyerhof, G.G. 1963. Some recent research on the bearing capacity of foundations. Can. Geotech. J. Vol.1. No 1.

Skempton, A.W. 1951. The bearing capacity of clays. Bld. Res. Congress. Div.1.Pt3: 180-189.

Tavares, A.X. 1978. Investigação da capacidade de carga de estacas em argilas. Titular Professor thesis. Universidade Federal do Paraná. Brazil.

Tavares, A.X. 1982. Capacidade de carga de fundação direta em argilas obtida através de ensaios de penetração constante. Relatório de pesquisa nº820003980. Universidade Federal do Paraná. Brazil.

Tavares, A.X. 1985. Settlement of foundation on clay by C.R.P. plate-loading tests. Proceedings of the Eleventh International Conference on Soil Mechanics and Foundation Engineering. Vol.2. San Francisco: 945-946.

Tavares, A.X. 1986. Recalque de Fundação em argila pré-adensada. Anais do VIII Congresso Brasileiro de Mecânica dos Solos e Engenharia de Fundações. Vol. VI. Fundações. Porto Alegre. Brazil: 213-222.

Terzaghi, K. 1943. Theoretical soil mechanics. New York: John Wiley & Sons, Inc.

Penetration Testing 1988, ISOPT-1, De Ruiter (ed.)
© 1988 Balkema, Rotterdam, ISBN 90 6191 801 4

Empirical formulas of SPT blow-counts for gravelly soils

Yasuo Yoshida & Motonori Ikemi
Research Laboratory, CRIEPI, Japan
Takeji Kokusho
Soil Mechanics Section, CRIEPI, Japan

ABSTRACT: In order to expand the applicability of the Standard Penetration Test to gravelly soils, a series of penetration tests have been carried out in a large scale container in the laboratory. Empirical formulas based on these tests on gravelly soils with different grain size distributions have been proposed in which the soil density and the shear wave velocity are interconnected with the SPT blow counts (the N-value) for a wide range of overburden stress as a parameter. The Large Penetration Test (LPT) has also been conducted to have a correlation with SPT. Field soil investigations at two gravelly soil sites revealed satisfactory agreements in the soil density and the shear wave velocity between the field and the empirical formulas.

INTRODUCTION

The Standard Penetration Test is one of the most popular soil investigation methods in Japan. Various design parameters like bearing capacities and settlements of foundations, liquefaction potentials, etc. are often determined based on empirical formulas incorporating the N-value proposed for cohesionless soils.

However, the application of SPT to gravelly soils is so far rather limited because natural gravelly soils are in most cases stiff enough as a bed layer supporting normal structures, and SPT is often stopped when it reaches gravelly layers. Furthermore, it is generally believed SPT is sometimes less reliable for gravels with large grains because the N-value is apt to vary from one point to another due to individual grain particles. Breakage of the probe at the cutting shoe by hitting hard grains may also lead to erroneous results.

Recently as important structures are increasingly constructed on hard gravelly soil layers of the pleistocene epoch, field investigation methods for gravelly soils are much more needed than before. Thus, it is quite meaningful to expand the applicability of SPT to gravelly soils by establishing empirical relationships between the SPT blow-counts and key design parameters like the soil density and the shear wave velocity.

For this purpose, laboratory tests were carried out in a large soil container in which the dynamic penetration test was systematically conducted for gravelly soils with different grain size distributions under different soil densities and different overburden pressures.

TEST METHOD

A large cylindrical steel container of 2.0 m in diameter and 1.5 m in depth was developed to make the penetration test in the laboratory. As shown in Fig. 1 a/b rigid cap of the container has five openings through which penetration tests can be done. Moist gravelly soils were compacted into the container to make the tested layer of about 100 cm in thickness with different densities and saturated by raising the water table. The soil layer was then vertically loaded by a hydraulically pressurized rubber bag which was located right under the cap but still allowed the penetration test through built-in holes. Underneath the tested layer was a dummy sand layer in which eight pressure cells to measure the vertical stress and a steel bar for shear wave generation were installed besides three pressure cells for the horizontal strell attached on the wall.

Two kinds of penetration tests were carried out; SPT and LPT, the specifications of which are listed in Table-1. In both tests, the hammer was dropped by the trip

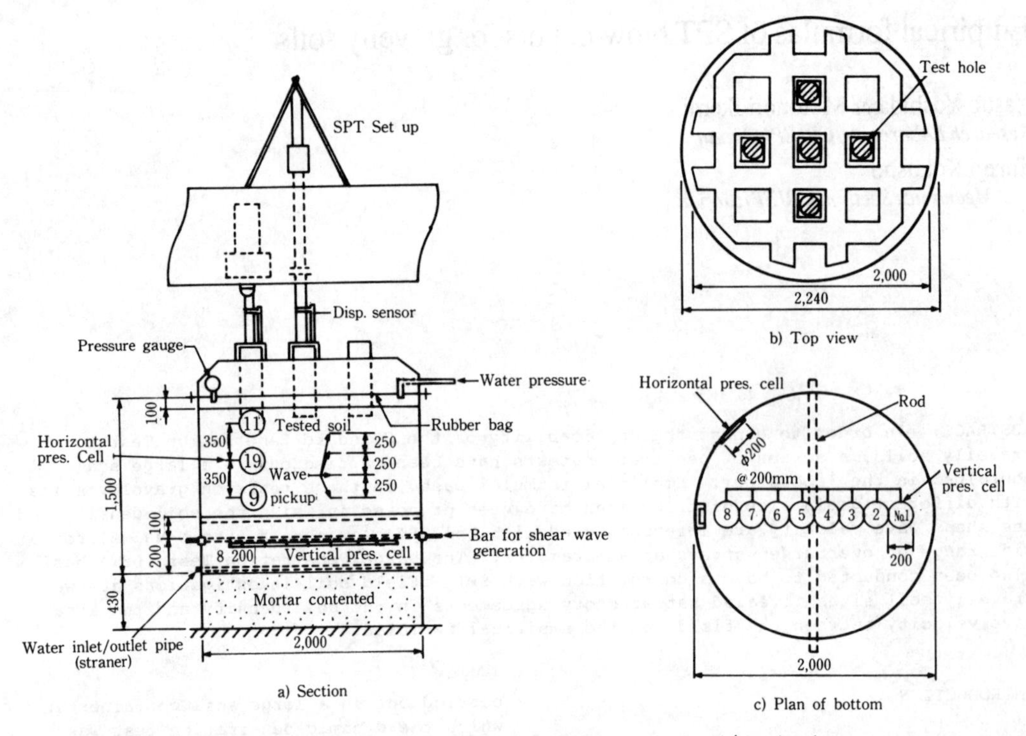

Fig. 1. Sketch of large soil container for Laboratory penetration tests

Table 1. Details of penetration probes and conditions of penetration test

No.	(1)	(2)
Details of penetration probe	175, 560, 75, 35, 51	560, 70, 50, 73
Test name	Standard Penetration Test	Large Penetration Test
Symbol	S.P.T.	L.P.T.
Drive method	Fall weight	Fall weight
Weight	622.3N	980N
Drop height	75 cm	150 cm
Drive length	30 cm	30 cm

monkey method. The same practice as in the field was followed including initial 15 cm driving followed by blow-counting by further 30 cm driving. The metal of the cutting shoe of the probe was hardened to preclude errors due to its breakage during penetration as far as possible, which actually was found very effective.

Two kinds of gravel and one kind of sand were employed in the test, the physical properties and the particle gradations of which are presented in Table-2 and Fig. 2. The density of the soil layer in the container was regulated by adjusting the time of compaction with a mechanical tamper. The density of the soil was determined by direct measurements of the thickness of the compacted soil layer in the container, the soil weights and the moisture content. The maximum dry densities of soil materials listed in Table-2 were determined by applying a electric vibrator attached to a steel disc to the top of dry soil in a mould of 30 cm in inner diameter and 35 cm in depth. The minimum dry density was decided by gently pouring dry soils into the same mold.

Due to the skin friction of the container wall, the stresses in the soil layer varied both vertically and horizontally violating uniform stress conditions which should have

Table 2. Physical properties of soils

	Fine sand	Gravel fraction 25 %	Gravel fraction 50 %
Max. dry density (g/cm^3)	1.705	2.004	2.151
Min. dry density (g/cm^3)	1.374	1.706	1.867
Uniformity coefficient	1.95	5.65	11.25
Mean particle size (mm)	0.340	1.13	2.28
Maximum void ratio	0.966	0.567	0.429
Minimum void ratio	0.584	0.334	0.24
Specific gravity	2.701	2.674	2.668

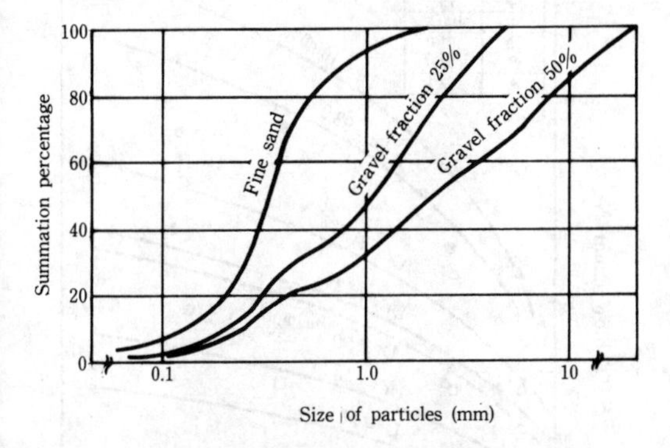

Fig. 2. Particle size distribution curve

been desirable. To overcome this, the stress conditions in the soil layer were monitored by eleven pressure cells both in vertical and horizontal direction. Fig. 3 shows an example of the measured stress distributions for different applied rubber bag pressures of 50 to 700 kPa. Therefore reference stresses, p_{vm}, interpolated from these measurements for tested points in the layer were employed for overburden stresses referred in the subsequent analyses. The stress ratio between vertical and horizontal stresses (σ_v' and σ_h' respectively) was on the average σ_h'/σ_v' = 0.45 for loose soils and decreased down to σ_h'/σ_v' = 0.25 with increasing density.

The shear wave velocity of the soil was measured before each penetration test by buried wave pickups which detected the horizontal shear wave vertically propagating from the wave source of the steel bar.

Five penetration tests were conducted in the same saturated soil layer through the

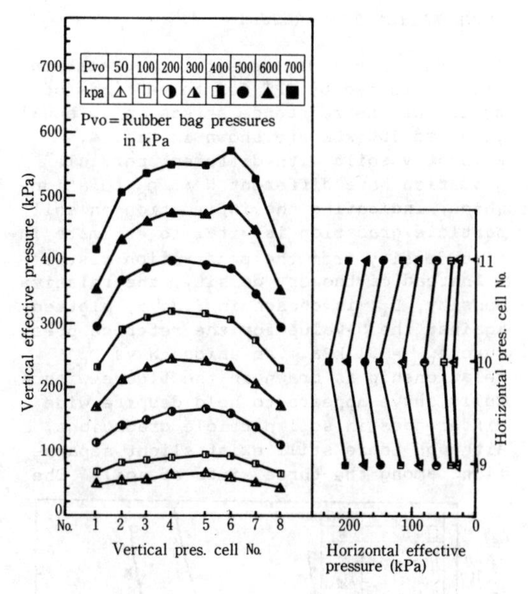

Fig. 3. Earth pressure distribution in the soil container

five holes in the cap either with different overburden pressures or with different kinds of test; SPT, LPT, etc. The distance between the holes was 50 cm, while in the similar study by Gibbs and Holtz (1957) the corresponding distance was presumably around 25 cm. A preliminary study to examine interactions between neighboring holes revealed that previous neighboring penetration tests will not appreciably affect the later tests.

A short rod of 3.0 meter was employed throughout the test series. It is sometimes recommended by previous investigators e.q. Seed et al. (1983) to correct the driving energy for SPT using a rod shorter than 3.0 meter.

It seems to the present authors, however, that the first hammer impact was solely taken into account in calculating the energy in previous investigations despite the fact that the second or third inpact has some additional driving energy contribution for SPT using a short rod as demonstrated in basic studies by Uto et al. (1973) and also by the present research. Therefore, the correction factor for the rod length which may seem to exist between the previously proposed factor and no correction has not been defined in this research.

PENETRATION TEST RESULTS

N-value vs. soil dry density, ρ_d, relationships obtained by SPT for three kinds of soils for the reference stress, p_{vm}, equal to 50 to 600 kPa are shown in Fig. 4. Evidently soils with different particle gradation have different N vs. ρ_d relationships, indicating the information on soil particle gradation is vital to estimate insitu density from the penetration test.

Instead of the dry density, the relative density, D_r, is chosen in Fig. 5, plotted against the N-value for the refernce pressure p_{vm} = 81 kPa. An unique N vs. D_r relationship as drawn in the figure with a solid curve appears to hold despite wide differences in soil particle gradations, although there still exist slight separations among the three kinds of soil. The

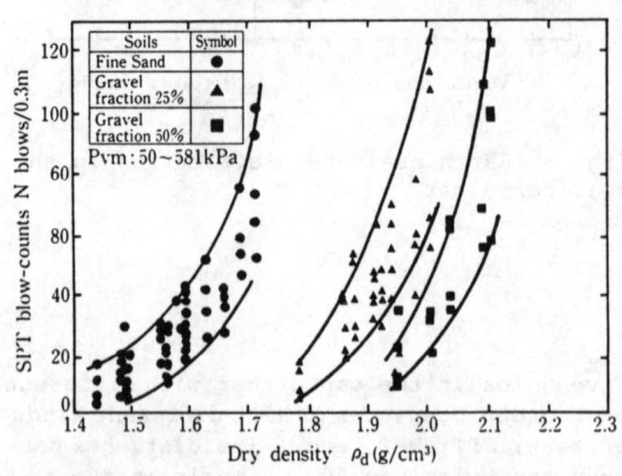

Fig. 4. SPT blow-counts (N) vs. dry density (ρ_d) relationship

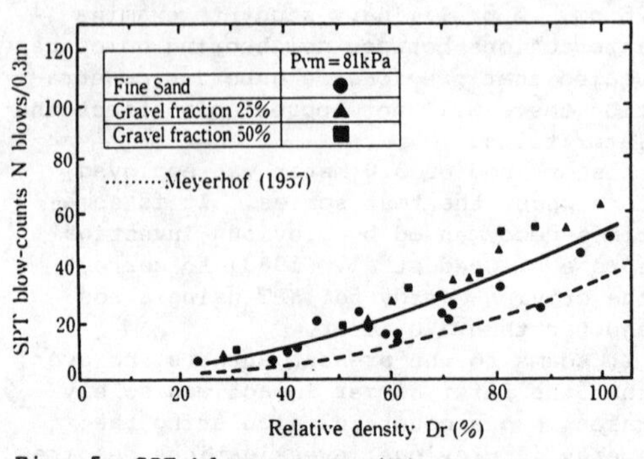

Fig. 5. SPT blow-counts (N) vs. relative density (ρ_d) relationship of the gravelly soils (P_{vm} = 81 kPa)

dashed curve shown in the same figure represents the Meyerhof's equation (Meyerhof 1957), which appreciably differs from the proposed curve.

N-value vs. reference vertical pressure, p_{vm}, relationships shown in Fig. 6 for the gravelly soils with the gravel fraction of 25 % indicate that the N-value is not linear with the overburden pressure. In a relatively narrow pressure range, a linear relashionship might be assumed as is employed in normal practice, while in a wide range of pressure increments of the N-value obviously decrease with increasing pressures indicating that this nonlinearity cannot be ignored for reliable estimate of design parameters.

Based on these systematic test data, empirical equations to interconnecting the N-value with the relative density, D_r, and the overburden stress, p_{vm}, have been derived by means of statistical methods. For the soil with the gravel fraction of 25 %,

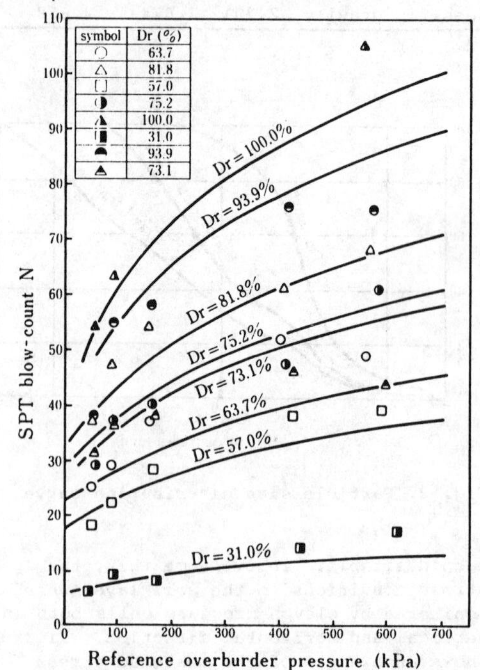

Fig. 6. Effect of over burden pressures on SPT blow-counts (N) (gravel fraction 25 %)

$$D_r = 18 * N^{0.57} * P_{vm}^{-0.14} \quad (1)$$

where P_{vm} and D_r are in kPa and percent respectively. Eq. (1) is compared with the experimental results in Figs. 6 and 7, indicating that the proposed equation can estimate the relative density from the N-value and the confining stress with satisfactory accuracy. Individual equations derived for the three soils as well as an unified equation for all soils are presented in Table 3.

Shear wave velocities, V_s, measured just before individual penetration tests are plotted against the corresponding N-values for different confining pressures for the soil

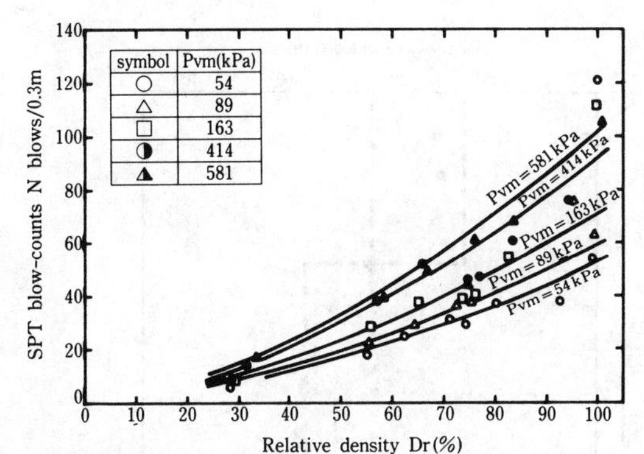

Fig.7. SPT blow-counts (N) vs. relative density (D_r) relationship of the gravel fraction 25 %

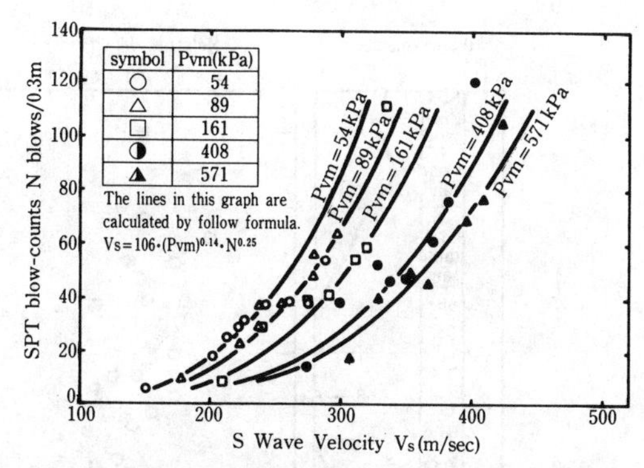

Fig.8. SPT blow-counts (N) vs. wave velocity (v_s) relationship of the gravel fraction 25 %

Table 3. Empirical equations of N-value vs. D_r relationship for gravelly soils

	Empilical formula
Fine sand	$D_r=22 \cdot N^{0.57} \cdot P_{vm}^{-0.14}$
Gravel fraction 25%	$D_r=18 \cdot N^{0.57} \cdot P_{vm}^{-0.14}$
Gravel fraction 50%	$D_r=25 \cdot N^{0.44} \cdot P_{vm}^{-0.13}$
All soils	$D_r=25 \cdot N^{0.46} \cdot P_{vm}^{-0.12}$

with the gravel fraction of 25% in Fig. 8. The solid curves drawn in the figure represent the following equation statistically derived

$$V_s = 56 * (P_{vm})^{0.14} * N^{0.25} \quad (2)$$

where V_s is in m/sec. Obviously the equation can satisfactorily estimate the shear wave velocity from the N-value and the effective overburden stress despite minor data scatters. In Table 4, equations derived in the same way for the three soils are listed.

In addition to SPT, Large Penetration Test (LPT) was conducted concurrently in

Table 4. Empirical equations of N-value vs. S wave velocity (V_s) relationship for gravelly soils

	Empilical formula
Fine sand	$V_s=49 \cdot N^{0.25} \cdot P_{vm}^{0.14}$
Gravel fraction 25%	$V_s=56 \cdot N^{0.25} \cdot P_{vm}^{0.14}$
Gravel fraction 50%	$V_s=60 \cdot N^{0.25} \cdot P_{vm}^{0.14}$
All soils	$V_s=55 \cdot N^{0.25} \cdot P_{vm}^{0.14}$

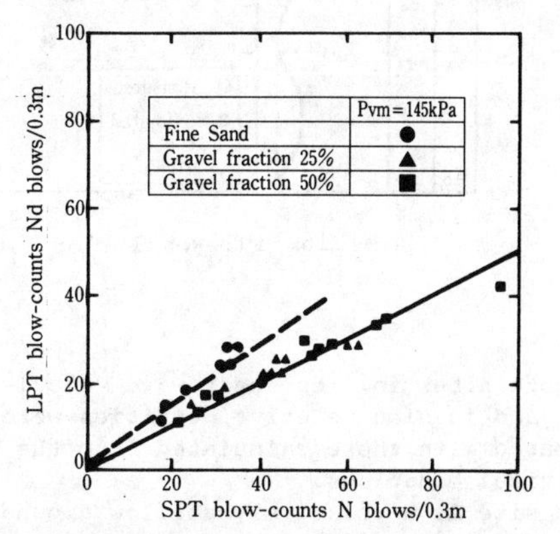

Fig.9. SPT vs. LPT relationship

the same test series. Fig.9 shows the blow-counts for LPT (the N_d-value) for the three soils against the corresponding SPT blow-counts (N-value). A linear approximation by the equation proposed by

$$N_d = N / 2 \quad (3)$$

will also be valid for gravelly soils, while a slightly different relationship as

$$N_d = N / 1.5 \quad (4)$$

will be applied to sand. As LPT seems to have greater reliability for larger gravels because the penetration resistance is less dependent on individual grains, the simple correlation between LPT and SPT will be conveniently used.

IN SITU TEST RESULTS

In order to study the applicability of the proposed equations, field soil investigations were conducted at two sites; A and B.

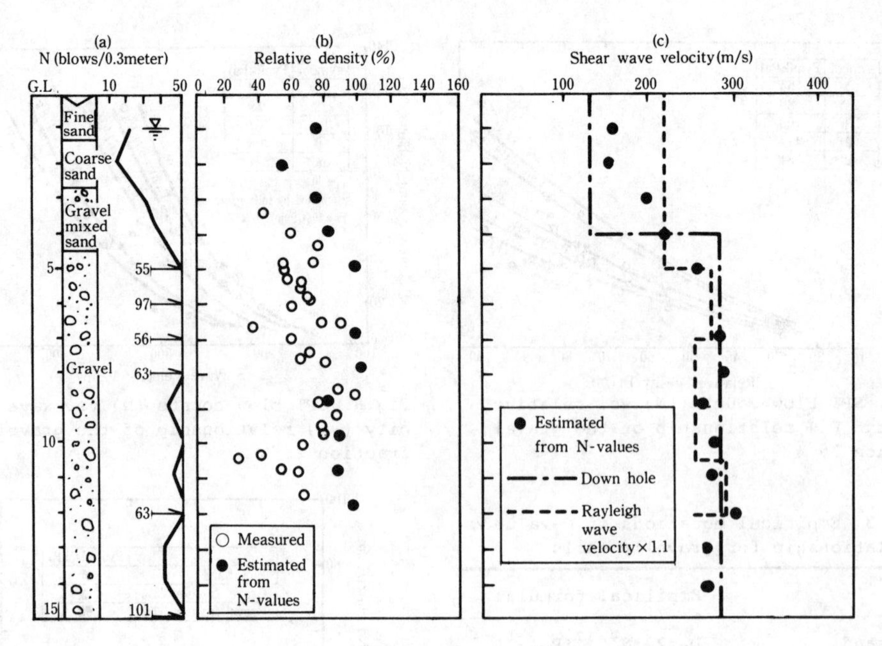

Fig. 10. Results of field investigations at site A

In both sites in-situ shear wave velocities and in-situ relative densities were compared with those calculated from the empirical equations.

In site A, a relatively shallow ground (down to 15 m) of Pleistocene epoch shown in Fig. 10 (a) was investigated. The ground consisted of a sandy and gravelly upper layer with the N-value of 10 to 40 and a gravelly lower layer with N-value over 40 to 50. The grain size distribution of the gravelly layer was similar to that of 50 % gravel fraction used in the laboratory study. The densities calculated from the equation for all soils given in Table 3 somewhat overestimate the in-situ densities which were determined from intact soil specimens sampled by in-situ freezing as shown in Fig. 10 (b). On the other hand as shown in Fig. 10 (c), in-situ shear wave velocities measured by the downhole method as well as the Rayleigh wave method are in good agreement with those calculated from the N-values by the equation for all soils given in Table 4.

The ground of site B was also of the Pleistocene epoch consisting of prevailing sand layers and thin sandy clay and gravelly layers with rather deep water table as indicated in Fig. 11 (a). In-situ relative densities determined from specimens recovered by the Modified Denison sampler are in fair agreement with relative densities calculated by the equation for all soils in Table 3 as indicated in Fig. 11 (b). A satisfactory estimate of in-situ velocity by the empirical equation for all soils given in Table 4 is also evidenced in Fig. 11 (c), in which fine measurements of in-situ velocities were made by means of a suspension-type wave logging survey.

CONCLUSIONS

Based on systematic penetration tests in the soil container in the laboratory and field soil investigations at two sites, the following conclusions have been reached.

1) The Standard Penetration Test can be applied to gravelly soils with more reliability if the metal of the cutting shoe of the probe is hardened so as to prevent breakage by hitting hard gravel particles.

2) In contrast to conventional assumption that the N-value is linear with the effective overburden stress, it has been found that the increment of the N-value will decrease with increasing pressure in the wide pressure range, and this nonlinear effect has been taken account in establishing the empirical formulas.

3) Based on the laboratory data, empirical equations to estimate the relative density

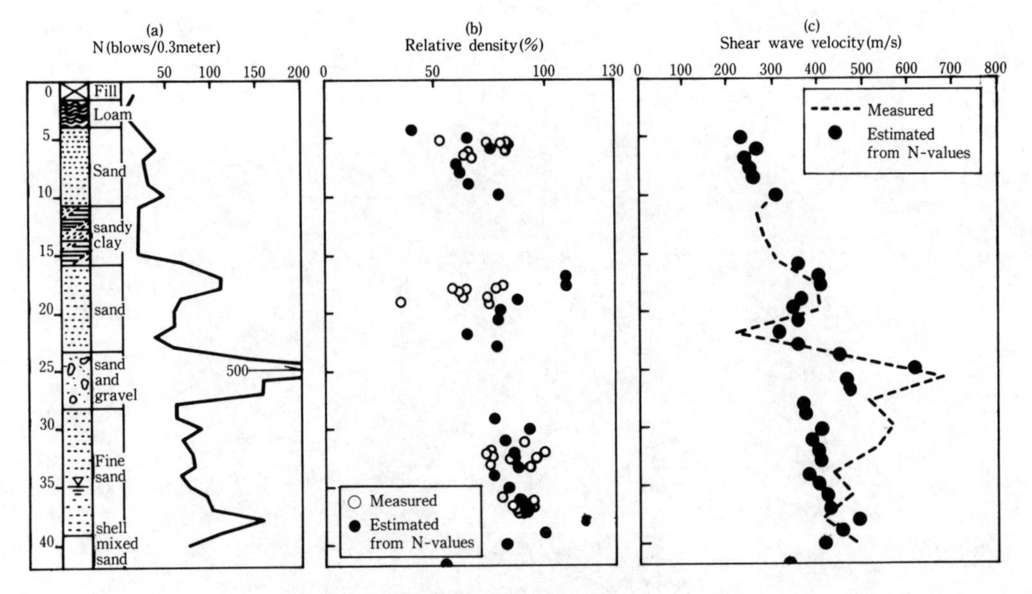

Fig. 11. Results of field investigations at site B

and the shear wave velocity from the N-value and the overburden pressure have been proposed for gravelly soils.

4) The correlation between the N-value of SPT and the N_d-value of the Large Penetration Test which seems to have greater applicability than SPT has been established based on the laboratory test for the gravelly soils.

5) In-situ density and in-situ shear wave velocity at two sites have been compared with those calculated from N-values of the same sites by the proposed equations. Despite the difference in geological background between the field and the laboratory, the equations can satisfactorily estimate the in-situ values, demonstrating great capability of the empirical formulas based on the penetration test for investigating gravelly grounds.

REFERENCES

Meyerhof, G. G. (1957), Discussion, Proc. 4th International Conference on SMFE, Vol III, pp. 110.

Gibbs, H. J. and Holtz, W. G. (1957), "Research on determining the density of sand by spoon penetration test," Proc. 4th International Conference on SMFE, Vol. I, pp. 35-39.

Seed, H. B., Idriss, I. M., and Arango, I (1983), "Evaluation of Liquefaction Potential Using Field Performance Data", Jour. of Geotechnical Engineering, Vol. 109, No. 3.

Morita, S., Uto, K., Fuyuki, M., Kondo, H., and Morihara, H., (1973), "Experimental Study on the Mechanism of Dynamical Penetration of Rod in the Standard Penetration Test (1)", Proc. of the Faculty of Engineering, Tokai University, 1973-No. 1.

Penetration Testing 1988, ISOPT-1, De Ruiter (ed.)
© 1988 Balkema, Rotterdam, ISBN 90 6191 801 4

Penetration testing application in China

Jian-Xin Yuan
Institute of Rock and Soil Mechanics, Wuhan, People's Republic of China

Abstract: The state-of the-art for the application of penetration testing in China is outlined, which mainly includes standard penetration tests, static penetration tests. The static penetration test and its applications, such as classification of soils, estimations of strength and deformation indexes of subsoils; pile capacity; collapsibility of loess and the possibility of liquefaction of sandy soils are discussed in detail. Finally, the trend of penetration testing and its application and some noticeable problems are reviewed in the paper.

Key word: Penetration test

1. Introduction

Penetration tests in this paper refer to standard penetration test (SPT) and static penetration test. By virture of simplicity, SPT has been widely used in China for various engineering exploration since 1953. And in 1974, it was adopted as a formal exploration method in the foundation design specification for industrial and civil buildings and the engineering geological exploration standard. The drop hammer used in SPT weights 63.5kg and the free falling height is 76cm. The maximum depth in tests is usually less than 30m. It was in 1956 that the static penetration test was first used in China to do some research on loesses. In the same year, the first static penetrometer, based on wire strain gauges, was developed, and a statistic formulation which showed the relationship between penetration resistances and bearing capacities was established for natural footings, after some a hundred comparing tests. At present, hundreds of organizations, ranging from building, railway, mechanics, traffic, hydro power, coal mining and other industrial divisions are engaged in research, and performing static penetration tests.

A variety of penetrometers have been manufactured. Among these, static penetrometers are extensively employed for the determination of bearing capacity of footings, compression modulus of subsoil, pile capacity, undrained shear strength of saturated clay and soil classifications. It is also used in evaluating compactions of filled soils and inspecting the qualities of foundation treatments like rolling, heavy tamping and sand pile compaction, as well as in estimating densities and possibilities of liquefaction of sandy soils. When the indexes obtained from swelling tests are available, the technique is suited for identifications of expansive soils and for evaluations of swelling potentials of the soils and deformations of footings, based on the correlation between cone resistances and the potentials. Tests carried out in Beijing area showed that the performance of penetration is related to the forming conditions of soils. Under the condition of same penetration resistances, the modulus obtained at laboratory of "Young" soils is remarkably lower than that of "old" soils.

Both SPT and static penetration tests, which are still in progress, should be used with caution about the applicability in special conditions. Generally speaking, SPT is adoptable for sandy soils and cohesive soils, but not for soils with cement agents, rock chips or cobbles. It is unsatisfactory for SPT to be used in soft clays, because of poor accuracy in these cases. Effects of some factors on blow counts in SPT are to be investigated. As to cobble strata or soil layers mixed with stones, due to the difficulty in penetrating and distortions occuring in penetration resistances, the static Penetrometer is not applicable. problems in penetration tests, such as mechanism of static penetration, affecting factors and their effects on penetration resistances, results interpretation and data processing, should be further studied.

2. Classification of soils

According to the average static point resistance q_c and friction ratio R_f(R_f=f_s/q_c), where f_s is side wall friction resistances, soil strata can be classified into (I) sand; (II) cohesive sandy soils, and (III) soft soils, Sandy clays and clays, suggested by Ministry of Railway as shown in Fig.1. Alternative classification is put forward by Hubei Electric exploration and Design Institute, as shown in Table 1.

3. Estimation of the strength and deformation parameters of subsoils

3.1 Undrained shear strength

The undrained shear strength Cu (kg/cm²) of saturated cohesive soils is usually estimated from correlation analysis of data obtained in static penetration tests. Table 2 shows some examples.

Where unit(A) to (D) refers to Wuhan combined group; the first Bureau of Navigation in Tianjin; the soft soil group of the Fourth Railway Exploration and Design Institute and Sichuan Institute of Building Research, in turn.

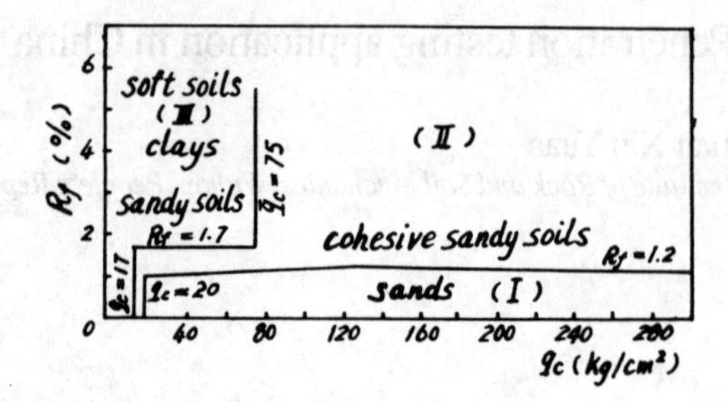

Fig.1 Classification of Soil

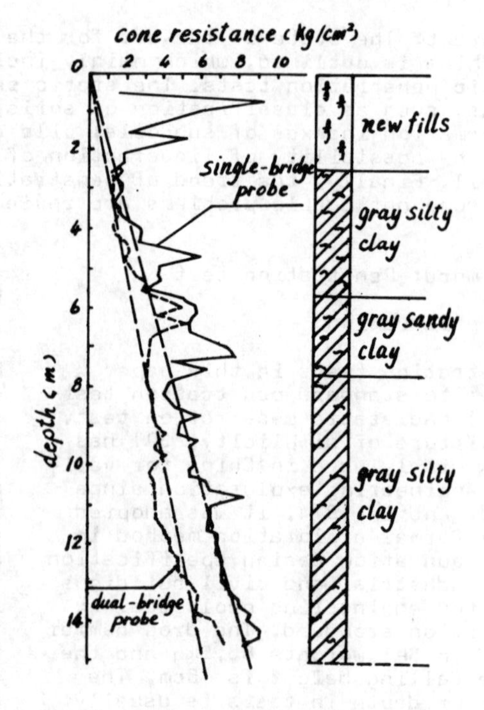

Fig.2 Resistant curves of single bridge probes and dual-bridge probes

Table 2 is based on data acquired in tests with single bridge probes. According the penetration tests carried out at Tianjin Xingang in northern China by Academy of Railway Sciences and Department of Beijing Railway, a comparison was made as shown in Fig. 2 for the results obtained with single bridge probes and dual-bridge probes. It

Table 1. Classification of soils

Type of soil	$R_s = \frac{f_s}{q_c} \times 100\%$	Shape of q_c-depth curve
CLAYS	>5	smooth
Sady clay	2-5	smooth
light sandy clay	<5	with fluctuations
silty clays	<1	smooth
fine sands	<1	with strong fluctuation

Table 2. Relations between undrained shear strength and penetration resistances

Unit	Empirical formula	Correlation coefficient (r)	Applicable range of p_s	Testing location
(A)	$Cu=0.0696Ps-0.027$	0.874	$3<p_s<12$	Wuhan, Ningbo
(B)	$Cu=0.038\ Ps+0.04$	0.90	$1<p_s<15$	Tianjin Xingang
(C)	$Cu=0.0519Ps+0.013$	0.779	$2.5<p_s<10$	Eastern China
(D)	$Cu=0.0543Ps+0.048$	0.84	$1<p_s<8$	Shanghai, Guangzhou

can be known from Fig.2 that the cone resistances for both kind of probes are approximately equal at the depth from 2 to 4 meters, while at the depth larger than 4 meters, the cone resistance obtained by dual-bridge probes is obviously lower than that by single probes. The results from more than 300 penetration tests performed in 1980 at different sites of various regions in eastern China by Single bridge probes can expressed by equation (1)

$$Cu=0.05q_c+0.016 \doteq \frac{1}{20}q_c \text{ (for saturated clays, } q_c<15) \quad (1)$$

As a comparison, the tests results obtained with dual-bridge probes at Zhenhai city in eastern China by Tong-ji University can be expressed by equation ....(1a)

$$Cu=0.0719q_c+0.0128 \doteq \frac{1}{14}q_c \quad (1a)$$

It has been proved that the relation between Cu and q_c varies with the changes of the shape of the probe, forming conditions of strata and the stress history.

3.2 Compression modulus of soils

The empirical formulations showing the relation between Compression modulus Es and total penetration resistance Ps including cone resistance q_c and side wall resistance f_s for different soils are listed in Table 3

Where units(A) and (B) are the same as in Table 2;
unit (C)-Beijing Exploration Division;
unit (D)-Ministry of Railway;
unit (E)-Hubei Exploration Institute.

3.3 Determination of bearing capacity of piles

Recently, the progress of estimations of pile capacity on the basis of static penetration tests has been greatly advanced. Several empirical formulas have been raised by units with experiences of their own. Only a brief introduction is presented in this paper.

When a dual-bridge probe (base area 20cm², surface area of the friction sleeve 300cm², and penetration rate 2cm/s) is employed in tests, the bearing capacity of concrete piles can be estimated by Eq.(2), suggested by the Ministry of Railway in the technique standard 1980.

$$Qu=\alpha q_c \cdot Ap+U \sum_{i=1}^{n} \beta_i L_i f_{si} \quad (2)$$

Table 3. Relations between Es and Ps

Unit	Empirical formula $Es(kg/cm^2)$	Applicable range of Ps	Applicable type of soil
(A)	Es=3.72Ps+12.63	3<Ps<50	silts, silty clays, clays
(B)	Es=3.63Ps+11.98	Ps<50	silts, silty clays, clays
(C)	Es=1.74Ps+49	10<Ps<90	"old" clays
	Es=1.71Ps+29	7<Ps<40	"new" sedimental clays
(D)	Es=2.14Ps+21.74	13<Ps<80	clays, sandy soils
(E)	Es=4.695Ps-6.27	10<Ps<35	clays

Where

α-general correction factor of the static point resistance

U-circumference of the pile

n-the number of penetrated strata

Li-thickness of ith stratum

βi-general correction factor of side wall friction

Ap-section area of the pile toe

Under conditions $q_c > 20 kg/cm^2$ and $\frac{f_s}{q_c} > 0.014$

$$\alpha = 1.257(\bar{q}_c)^{-0.25}$$

$$\beta = 1.798(f_s)^{-0.45} \qquad (3)$$

While for soils with $q_c < 20 kg/cm^2$ and $\frac{f_s}{q_c} < 0.014$

$$\alpha = 2.407(\bar{q}_c)^{-0.35}$$

$$\beta_i = 2.831(\bar{f}_s)^{-0.55} \qquad (4)$$

$\bar{q}_c$ is the mean of the average value of q_c over 4D below the pile toe and average value of q_c over 4D above the toe. If the latter is larger than the former, $\bar{q}_c$ takes the value of the former.

Shanghai pile Foundation Group proposed following formula in 1977.

$$Qu = \alpha \bar{q}_c Ap + U \sum_{i}^{n} \frac{\bar{q}_{ci}}{c_i} L_i$$

Where α is coefficient, α=1 for a long pile; $\bar{q}_{ci}$ is the average penetration resistance of the ith stratum. Values $\bar{q}ci$ and c_i refer to Tables 4 and 5

A-medium dense sands or clays with low water content

B-silt clays or clays with high water content

C-a soft stratum thicker than 2D within range of 3.75D below the tip

D-the tip penetrating into a soft stratum through a hard stratum

The calculation methods mentioned in the table are as follows:

Method I-Calculating the average penetration resistances above and below the pile tip respectively, and then take the mean of them as $\bar{q}_c$

Method II-taking the mean of the average resistance above the tip and the minimum below the tip as $\bar{q}_c$

Method III-Taking weighted mean as $\bar{q}_c$

Comparisons of pile capacities estimated from static penetration tests with those obatined from bearing tests, indicate that scatters generally being less than 30%. Table 6 shows the comparisons. It is clear that piles with scatter less than 20% are more than 80% of the total, which verifies the adoptability of static penetration tests.

4. Estimation of collapsibility of loesses

Based on the relation between the penetration resistance p_s and the Compression modulus Es, index δ_s, which indicates the collapsibility of loesses, can be expressed as

$$\delta_s = \frac{e_2 - e_2'}{1 + e_0} \qquad (6)$$

Where e_2 and e_2' denote void ratios before and after compaction of soils under a pressure of $2kg/cm^2$; e_0 refers to the void ratio of intact soils.

Table 4. Value range of $\bar{q}_c$

Bearing stratum of the pile tip	Value range: above the tip	Value range: below the tip	Method of calculation
A	8D	3.75D	I
B	3.75D	1D	I
C	8D	3.75D	II
D	1D	1D	III

D—Diameter of the pile

Table 5. Friction Conversion coefficient c_i in Shanghai area

soil stratum	Condition	c_i
gray silt clays, gray clays		20
dark green clays	$100^t/m^2 < q_c < 300^t/m^2$	20-30
	$q_{ci} = 100^t/m^2$	20
	$q_{ci} = 300^t/m^2$	$(q_{ci}/c_i) = 100^t/m^2$
light sandy clay, sandy clay, fine sands	$\bar{q}_{ci} < 500^t/m^2$	50
	$\bar{q}_{ci} > 500^t/m^2$	$(q_{ci}/c_i) = 100^t/m^2$
clays with high water content to soft plastic sandy clay	with fluctuations in penetration resistance curves	50
	smooth curves	$\frac{\bar{q}_{ci}}{c_i} = 100^t/m^2$
Shallow soils (6-8m beneath ground)		$20, \frac{\bar{q}_{ci}}{c_i} < 2^t/m^2$

Table 6. Comparisons between results from static penetration tests and bearing tests

Unit	Total Number of piles	Scatter	±(0-10%)	±(10-20%)	±(20-30%)	>±30%
Railway system	61	number of piles	33	17	8	3
		%	54	28	13	5
Shanghai pile base group	39	number of piles	25	6	2	6
		%	64.10	15.39	5.12	15.39

Eq (6) can be modified into

$$\delta s = 2\left(\frac{1}{E's} - \frac{1}{Es}\right)$$

or $\delta s = f(Ps, P_s')$ (7)

This means that index δs is a function of penetration resistances Ps and Ps', both before and after the soil being sunk into water. As a result, by means of injecting water int boreholes or the probe, Ps, Ps' as well as δs are acquisitive.

5. Appraisal of the possibility of liquefaction for sandy soils

A. By SPT method
According to the Chinese Earthquake Design Standard of Industrial and Civil Buildings, the liquefaction of sandy soils can be appraised by Eq. (8)

$$N_c' = \bar{N}_c'\left[1+0.125(d_s-3)-0.05(d_w-2)\right] \quad .....(8)$$

Where N_c' is the critical liquefaction blow count under assumptions that the depth of the saturated sandy soil is ds and the underground water level is located at d_w beneath the ground, $\bar{N}_c'$ is the critical liquefaction blow counts when d_s=3m and d_w=2m. If buildings with 7, 8 and 9 degrees of earthquake intensity scale are required, $\bar{N}_c'$ takes values of 6, 10 and 16 respectively. For saturated sandy soils deposited deeper than 15m, liquefaction is possible to occur, if the standard blow count is less than N_c' calculated from Eq(8). Here the standard blow count is specified as follows. First, let a hammer of 63.5kg weight freely drops 76cm onto a standard penetrometer and make it penetrate 15cm, and the blow count recorded for further 30m penetration accounts for the standard blow count. The method was tested during earthquakes occured in Hai Cheng and tangshan, and results were in a good agreement with practical happenings. Generally speaking, too many factors being able to exert influences on blow counts, the values of the blow counts obtained in tests are, usually, very scattered. Therefore, cautions should be taken during data processing.

B. By static penetration tests, Eq (9) has been set by Institute of Railway Sciences as a criteria to discriminate liquefaction of sandy soils.

$$P_s' = Pso\left[1+0.065(H_w-2)\right] \times \left[1-0.05(H_o-2)\right] \quad ....(9)$$

Where H_w is the distance between the ground and the underground water level; H_o denotes the thickness of the overburden layer,

Table 7. Values of P_{so}

Earthquake intensity	P_{so}
7	60-70
8	120-135
9	180-200
10	220-250

which is composed of non-liquefaction cohesive soils and overlaps on the saturated sandy soils; Pso refers to the critical penetration resistance when $H_o=H_w=2m$. The value of Pso refers to Table 7.

6. Concluding Remarks

Penetration tests including SPT and static and dynamic penetration tests have been playing important roles in China for various purposes, and have been stipulated as formal methods in corresponding standards. Relevant theoretic studies, like the study on mechanism of penetrations, have also been advanced recently.

SPT is a relatively rough in-situ test, which is able to reveal the capability of natrual sandy soils to resist the penetration under certain conditions, thus to appraise engineering properties of soils indirectly. When it is used to estimate bearing capacities of natural footing and piles, or evaluate liquefaction property of sandy soils by N values, attentions must be paid to the applicability of the method in different conditions. It should be pointed out that in the case of cohesive soils, SPT has no advantages over the others.

Static penetration tests have been well developed in China. Apparatus with penetration abilities of 20-30 tons and light apparatus of 10 kg have been manufactured, which can meet the needs in most projects. Great efforts have been made to diversify probes. It is possible at present to measure a variety of parameters such as the density, void pressure and inclination of boreholes, while in the past, only resistances and frictions could be measured. Investigations into relations between the variation of borehole pressure and cone point resistance or side frictions have come into notice. Studies are also aimed to improve the structure, form, sensitivity and setting of borehole pressure transducers. Automatic data acquisition and processing are in progress. Standardization of penetrometers is commonly concerned now and the sizes of transducers and friction sleeves have been specified in corresponding standards. Static penetration tests on seafloors has also started. A static penetrometer, developed by Institute of Ocean Sciences, Academia Sinica, is characterized by being able to work in the bottom of seas. Another trial was made by Institute of Shipping Design in Guandong province in 1975. Penetration tests were completed in shallow seas and river areas with a static penetrometer mounted on a drilling ship.

In order to put SPT and static penetration tests on a firm base, it is essential now to perfect measuring techniques, combined with theoretical analyses and Scale-down tests indoors. For important projects, vane tests, pressuremeter tests and other tests in-situ are also worth considering so as to provide reliable estimations.

REFERENCES

Ministry Hydro Power, China, 1980, Soil Testing Specifications (Second volume) SDS 01-79, Hydraulic Publishing Board,

Wang Chia-chun, 1984 Several Problems on Data Interpretation and Results Application of Static Penetration Testing in China, Engineering Geology Division of Tong-ji University

Tang Xian-qiang, Yeh Chi-ming 1983, Static Penetration, Chinese Railway Publishing Bureau

Wang Zhong-qi, 1986 In-Situ Testing Techniques for Soils, Chapter 6, Geotechnical measurement Techniques, Chinese Building Industry Publishing Bureau.

Technical papers: Dynamic probing / Weight sounding test

Penetration Testing 1988, ISOPT-1, De Ruiter (ed.)

Soil characteristics from penetration test results: A comparison between various investigation methods in non-cohesive soils

Ulf Bergdahl & Elvin Ottosson
Swedish Geotechnical Institute, Linköping

ABSTRACT: The results from comparative investigations at 14 test sites have been analyzed. Penetration test results as well as results from pressuremeter tests and laboratory investigations have been evaluated and compiled. The results have been processed in a computer. Comparisons between various parameters and penetration test results thus could be freely chosen and presented in diagrams. From these diagrams, as well as from existing relationships in the literature, a comparative table with penetration resistances and the soil characteristics, angle of internal friction and settlement modulus, have been proposed.

Keywords: Penetration testing, cone penetration test, weight sounding test, dynamic probing, Standard Penetration Test, pressuremeter tests, angle of internal friction, settlement modulus, comparisons, non-cohesive soils.

1 INTRODUCTION

In Sweden the bearing capacity in non cohesive soils (allowable bearing capacity) has for many years been determined from weight sounding tests. The soil is classified as dense or loose according to the Swedish Building Code (1980) or as loose, medium or dense according to the Swedish Bridge Building Code (1976). However, these classifications do not agree with international praxis, (Bergdahl & Eriksson, 1983).

The Swedish Geotechnical Institute (SGI) has, in co-operation with the Swedish Road Administration and the National Board of Physical Planning and Building, accomplished a comprehensive comparative work on results from different penetration testing methods as a basis for a more detailed classification on the stiffness of non cohesive soils in Sweden, adapted to international praxis. These comparisons can also be used to transfer foreign experiences from i.e. relationships between penetration resistance and soil characteristics (ϕ, E) or for direct calculations of the bearing capacity and settlement of slabs, (Bergdahl and Ottosson, 1982). Besides the investigations performed at SGI a large number of results presented by e.g. (Dahlberg 1975) and (Denver 1980) have been included in the comparison.

2 COMPILATION AND PROCESSING OF THE TEST RESULTS

The results from comparative investigations at 14 test sites have been compiled and processed. The penetration test results together with results from pressuremeter tests and laboratory investigations have been evaluated and compiled for every 1.0 m of the soil layers. Mean values have been evaluated for these layers from the penetration test results. However, apparent peak values caused by, for example stones, have not been considered. Values from depths less than 1.5 m below ground surface and from 0.5 m above and 0.5 m below the ground water table have not been included as these depths often cause a change in the test results, which cannot merely be attributed to the soil characteristics.

In total 19 variables have been compiled for each test site. The variables are: site (No.), depth (m), density (t/m3), distance from ground water table ($\pm$ m), soil, d_{60} (mm), d_{10} (mm), effective stress, σ_0' (kPa), friction angle ϕ' from triaxial tests (°), settlement modulus E from oedometer tests (MPa), water content (%), penetration resistance from cone penetration tests (CPT), static sounding tests (Trt), weight sounding tests (WST), dynamic probing (DP HfA), SPT tests and net limit pressure, p_l^*, (MPa), modulus, E_{pm}, (MPa) from pressuremeter tests.

The settlement modulus E from oedometer tests has been evaluated as a secant modulus in the stress interval between the particular effective stress (σ_o') and σ_o' + 300 kPa. If possible, undisturbed samples have been used. Otherwise the soil has been recompacted into the oedometer to the natural density.

The following testing methods have been used:

CPT (electrical cone penetration test according to the European standard) cone resistance - q_c

WST (weight sounding test according to the European standard) penetration resistance - N_{WST} (halfturns/0.2 m)

DP HfA (dynamic probing according to the Swedish geotechnical standard) N_{20} (blows/0.2 m). The HfA type of dynamic penetrometer is somewhat different from the recommended European DPB standard. Thus the height of free fall for the HfA test is 0.50 m and the cross-sectional area of the point is 16 cm^2. The HfA test results can be converted into DPB test values, since the resistance values according to the recommended European standard, are almost identical. It has also been found that the resistance $N_{20} = N_{30}$ from SPT-tests in cohesionless soils. This relationship is also valid for the present results. The skin friction in HfA has been measured with a slip coupling close to the point (Bergdahl and Möller, 1980). In these comparisons only the net point resistance has been used.

SPT (Standard Penetration Test according to the European standard) N_{30} (blows/0.3 m) using a free falling hammer.

The Menard pressuremeter type GC has been used for pressuremeter tests. Evaluation of the modulus E_{pm} and net limit pressure has been performed according to Menard (1975).

The evaluated results have been stored in a computer. By means of computer processing, comparisons between various parameters and penetration test results have been obtained and plotted in diagrams.

In the figures the compiled investigation material has been marked with triangular marks (▲) for sand to gravelly sand and plus signs (+) for silt and silty sand while the relationships from other authors are shown with lines or curves. Furthermore, limitations of the relationships proposed by SGI, are shown in dashed rectangles.

3 COMPARISONS BETWEEN RESULTS FROM VARIOUS INVESTIGATION METHODS

To be able to make a comparison with foreign classifications, the stiffness and characteristics of non cohesive soil from penetration test results as well as results from foreign investigation methods (cone penetration tests, SPT-tests and pressuremeter tests) have to be compared with typical Swedish methods (e.g. weight sounding tests and dynamic probing). Similar comparisons between various investigation methods have previously been presented by (Helenelund, 1966), (Bergdahl, 1974), (Tammerinne, 1974), (Dahlberg, 1975) and (Bergdahl & Eriksson, 1983).

Fig. 1 shows the weight sounding and the cone penetration test results from this compilation as well as results from (Bergdahl, 1974), (Helenelund, 1966) and (Muromachi and Kobayashi, 1982). The scatter is very large. A number of weight sounding tests in silt and silty sand show considerably larger penetration resistance than the average relationship for sand, which can be expressed by the equation:

$$q_c \text{ (MPa)} = 1.0 + 0.25\, N_{WST} \text{ (ht/0.2 m)} \quad (1)$$

In many publications comparisons between results from cone penetration tests and SPT-tests can be find, Fig. 2. However,

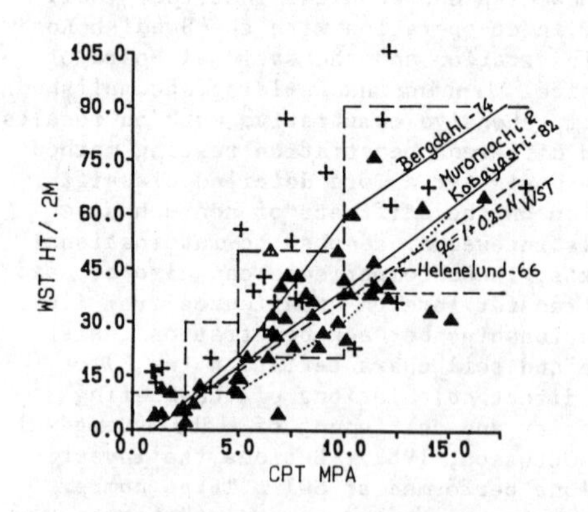

Fig. 1 Weight sounding test (WST) versus cone penetration test (CPT).

in this investigation no unique relationships like these previously presented have been found. For this reason, mainly previous compilations have to be trusted. This compilation show a large scatter in the results due to differences in the impact energy transmitted to the penetrometer rod. In the Swedish tests a freefalling hammer has been used while the old relationships often have been based on tests with the slip rope system. The tests in silt and silty sands are mainly followed by the often used relationship $q_c = 0.4 \cdot N_{30}$ while the results in sand show a larger scatter: q_c = 0.7 à 2.0 · N_{30}.

Fig. 3 shows the comparison between cone penetration resistance and dynamic probing results (net resistance, the skin friction resistance measured by the slip coupling has been subtracted). The scatter is fairly large due to the variations in the type of soil penetrated, (Rodin et al, 1974). Most of the points for the net resistance follow the relationship:

$$q_c \text{ (MPa)} = 0.8 \cdot N_{20} \text{ (blows/0.2 m)} \quad (2)$$

This relationship is close to that previously presented by, (Helenlund, 1966). The reason why this relationship differs from that shown by, (Bergdahl,1974), is primarily that the previous relationship was based on comparisons with SPT-tests performed without a free-falling hammer. Furthermore, the dynamic probing resistance has been reduced by the skin friction on the rods and the cone resistance at static penetration test has increased by using an electrical cone penetrometer with no area reduction at the cone base.

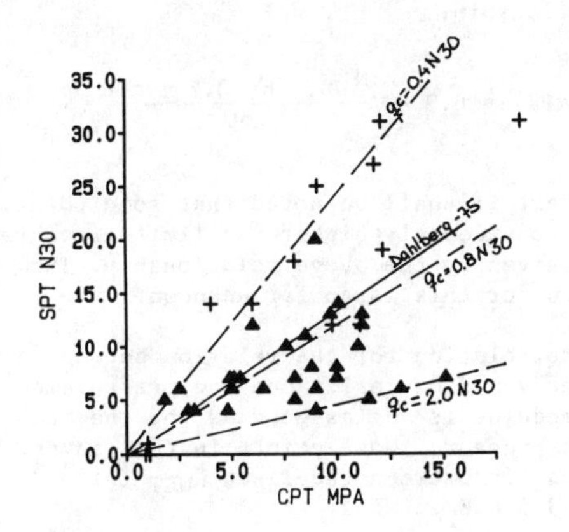

Fig. 2 SPT-test (SPT) versus cone penetration test (CPT).

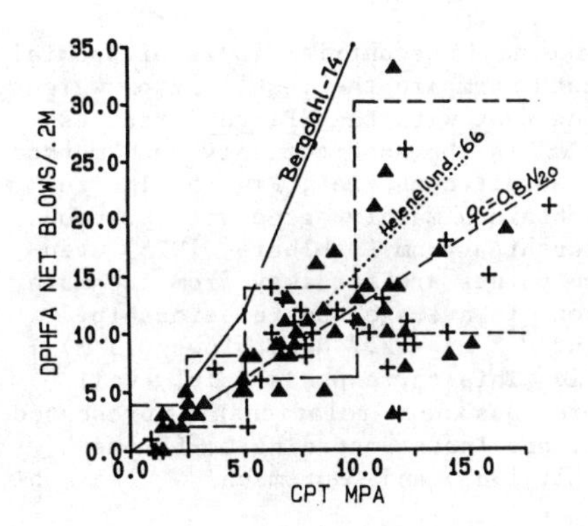

Fig. 3 Dynamic probing HfA point resistance (DP HFA NET) versus cone penetration test (CPT).

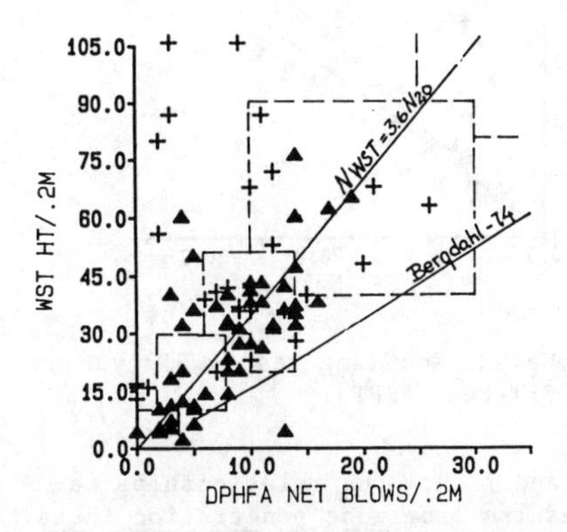

Fig. 4 Weight sounding test (WST) versus dynamic probing HfA point resistance (DP HFA NET).

A comparison between weight sounding and dynamic probing results, (net resistance) is shown in Fig. 4. The scatter is larger compared to the relation between cone penetration tests and dynamic probing, which indicates that the uncertainties from the weight sounding and dynamic probing are combined. It appears that, when large resistances have been measured with the weight penetrometer in silt and silty sand, the points are outside the normal relations. This can be due to the fact that the slip coupling in the dynamic tests maybe not always functions in silt. Therefore the skin friction reduction becomes too large. The regression line for sandy soils follows the relationship:

$$N_{WST} \text{ (ht/0.2 m)} = 3.6 \cdot N_{20} \text{ (blows/0.2 m)} \quad (3)$$

For Scandinavian countries it is of special interest to compare the results from weight sounding test with the SPT-test results as the WST is the most commonly used method in the Nordic countries, Fig. 5. The results obtained mainly agree with previous relationships from (Dahlberg, 1975) even if some points are far away from the main relation. In average the relationship: N_{WST} (ht/0.2 m) = 2.7 N_{30} (blows/0.3 m) is valid. This corresponds fairly well with previous known relationships presented by Miki and Inada according to (Broms & Bergdahl, 1982) and (Muromachi & Kobayashi, 1982).

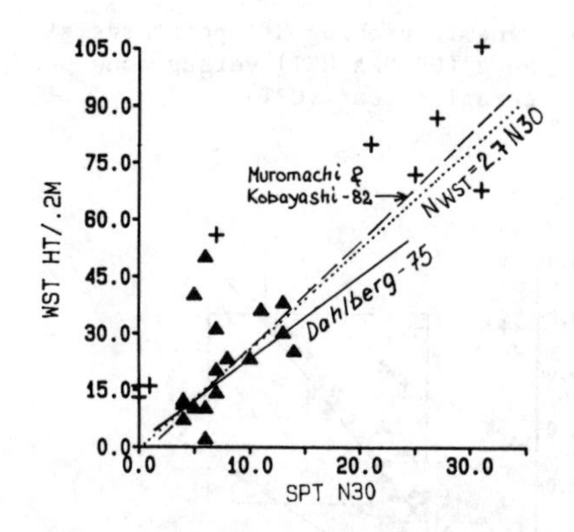

Fig. 5 Weight sounding test (WST) versus SPT-test (SPT).

Figs. 6 and 7 show the relationships between results from cone penetration tests, q_c, and pressuremeter tests, net limit pressure (p_I^*) and pressuremeter modulus (E_{pm}) respectively. Fig. 6 shows that the ratio between the cone penetration test and the net limit pressure, q_c/p_I^* = 5 to 15.

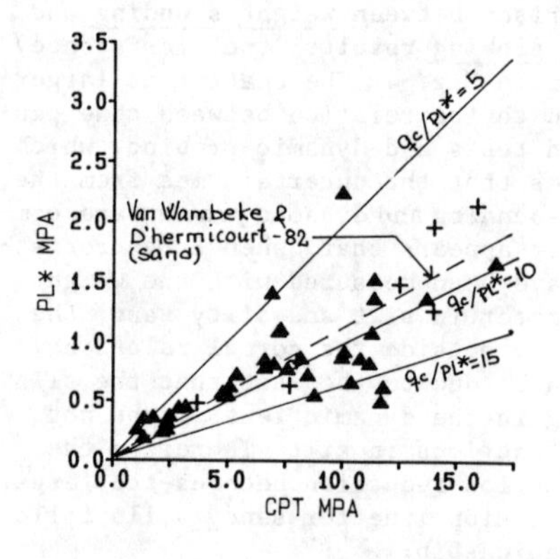

Fig. 6 Net limit pressure (PL*) versus cone penetration test (CPT).

The average test results obtained can be expressed by the relationship, q_c/p_I^* = 10, which closely corresponds to that presented by (van Wambeke & D'Hemricourt, 1982) for sand. A corresponding comparison with the pressuremeter modulus E_{pm}, Fig. 7, indicate that the point resistance q_c is on the average equal to the pressuremeter modulus, i.e.: q_c (MPa) = E_{pm} (MPa).

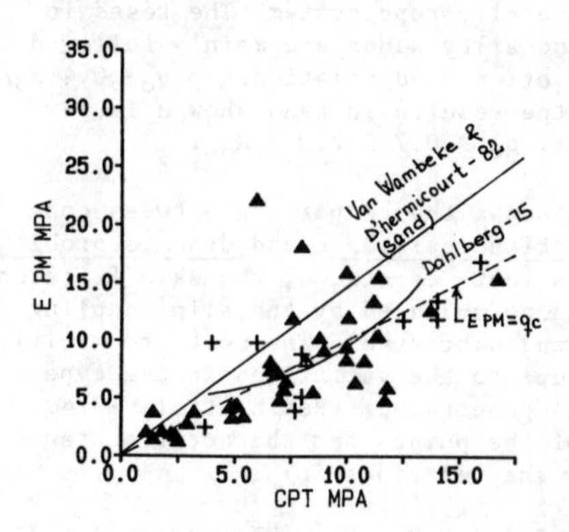

Fig. 7 Pressuremeter modulus (E PM) versus cone penetration test (CPT).

According to Figs. 8 and 9 a rather good correlation is obtained between results from the weight sounding tests and the net limit pressure, and the pressuremeter modulus. This may seem remarkable regarding their different way of working in the soil. The good correlation may depend on the fact that the weight sounding tip demands a large soil displacement at penetration. The mean relation between the weight sounding resistance, N_{WST}, and the net limit pressure, p_I^*, appears to follow the relationship:

$$p_I^* \text{ (MPa)} = 0.3 + \frac{N_{WST} \text{ (ht/0.2 m)}}{50} \quad (4)$$

However, it shall be noted that some points show considerably higher net limit pressures than given by the above relationship. The reason for this is so far unknown.

The correlation for the relation between weight sounding resistance and pressuremeter modulus is not as good as for the net limit pressure. Most points in this investigation lie between the lines N_{WST} (ht/0.2 m) = 3 à 6 E_{pm} (MPa).

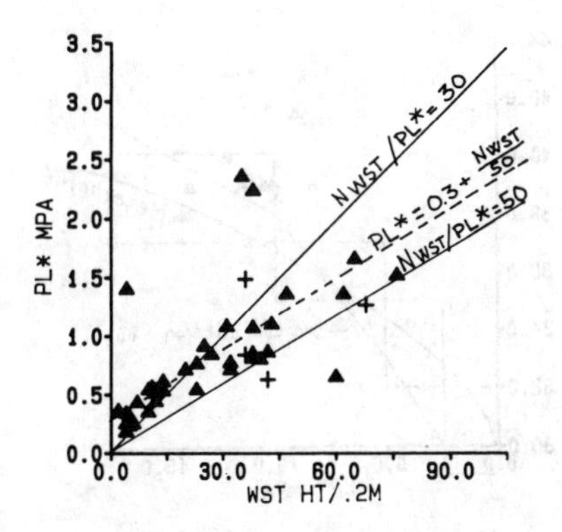

Fig. 8 Net limit pressure (PL*) versus weight sounding test (WST).

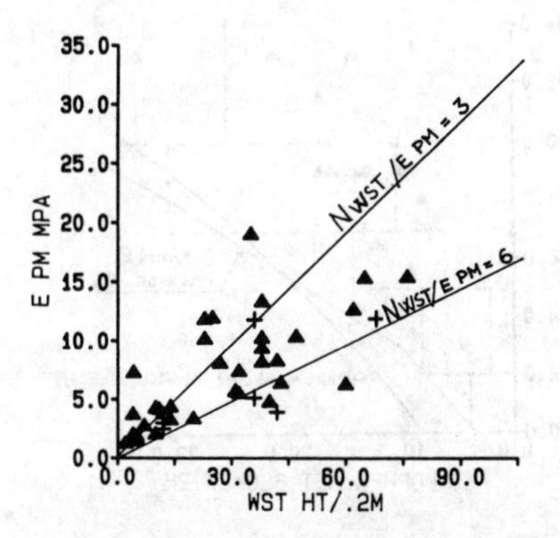

Fig. 9 Pressuremeter modulus (E PM) versus weight sounding test (WST).

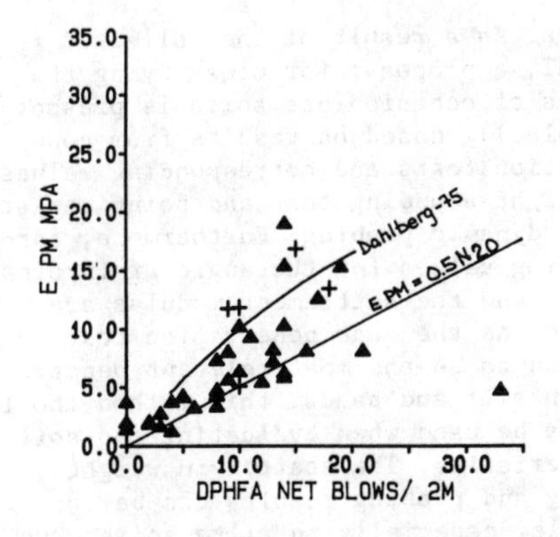

Fig. 10 Pressuremeter modulus (E PM) versus dynamic probing HfA point resistance (DP HFA NET).

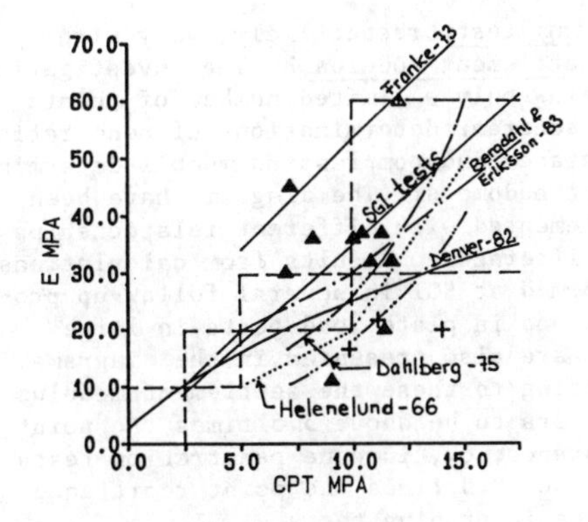

Fig. 11 Settlement modulus (E) versus cone penetration test (CPT).

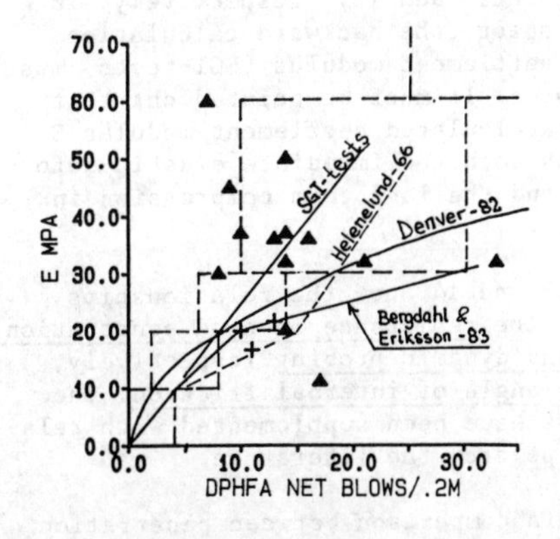

Fig. 12 Settlement modulus (E) versus dynamic probing HfA point resistance (DP HFA NET).

Comparisons between the point resistance in dynamic probing and the pressuremeter modulus is shown in Fig. 10. Good correlations are obtained between those two methods, however the scatter is larger at high penetration resistances.

A certain difference in correlation can be noted between the results for silt and for sand. The results from silty soils follow those shown by (Dahlberg, 1975). In average the pressuremeter modulus in MPa seems to be half the dynamic probing resistance in sand. Thus:

$$E_{pm}\ (\text{MPa}) = \tfrac{1}{2} \cdot N_{20}\ (\text{blows}/0.2\ \text{m}) \quad (5)$$

Figs. 11-12 show how the penetration resistance in cone penetration tests and dynamic probing tests respectively, vary with the settlement modulus E. The investigation contains only a limited number of points with adherent determinations of penetration resistance and compression moduli determined in the oedometer. The diagrams have been supplemented with different relationships from literature. Results from calculations performed at SGI in several follow-up projects and in plate load tests in large scale are also presented in the diagrams. According to these the settlement modulus E appears to be about 3.5 times the point resistance, q_c, in cone penetration tests and about 2.5 times the point resistance in dynamic probing tests.

The reason why the scatter is so large for a few investigation points may be due to the difficulty in taking undisturbed samples and to recompact the samples to an "adequate" density, respectively. In this respect, the backward calculation of the settlement modulus (SGI-tests) must be trusted. It must be pointed out that the backcalculated settlement modulus E contains both the immediate elastic deformation and the long term compression in the soil.

Figs. 13 and 14 show the relationships between the resistance in cone penetration tests and dynamic probing respectively, and the angle of internal friction. The diagrams have been supplemented with relationships from the literature.

Also this comparison between penetration resistance and the angle of internal friction includes a limited number of adherent determinations. The friction angles have been determined in drained triaxial tests. No adjustment in respect to the volume changes in the tests has been performed.

The relationships presented by (Bergdahl & Eriksson, 1983) are derived from a comprehensive literature survey. The dynamic probing resistance, N_{20}, has then been assumed to correspond to the resistance, N_{30}, in the SPT-tests.

4 CONCLUSIONS

The above compilation of various investigation results show that it is possible to obtain relationships between various penetration resistances, between various penetration resistances and results from pressuremeter tests and between penetration resistances and the soil characteristics as settlement modulus and angle of internal friction. As a result of the collected material, a proposal for classifying the stiffnes of cohesionless soils is presented (Table 1), based on results from cone penetration tests and corresponding values from weight sounding test and point resistance in dynamic probing. Furthermore, corresponding values for the angle of internal friction and the settlement modulus are proposed. As the cone penetration test has shown to be the most relevant penetrometer in silt and sands, this method should normally be used when evaluating the soil characteristics. The scatter in weight sounding and probing results can be considerable, especially in silty soils. Due to this, the lower values must be chosen when only weight sounding or dynamic probing results are available.

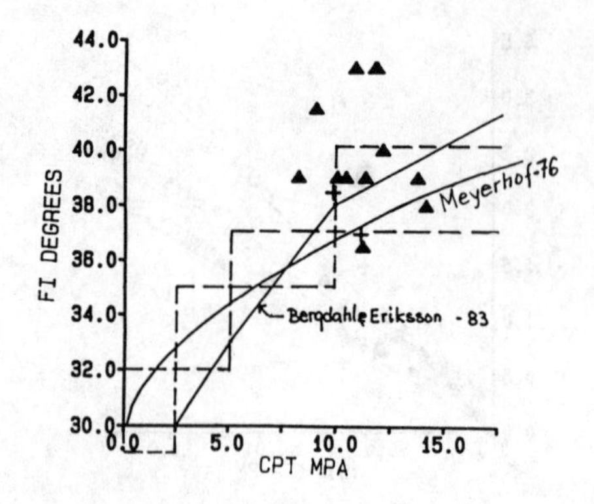

Fig. 13 Angle of internal friction (FI) versus cone penetration test (CPT).

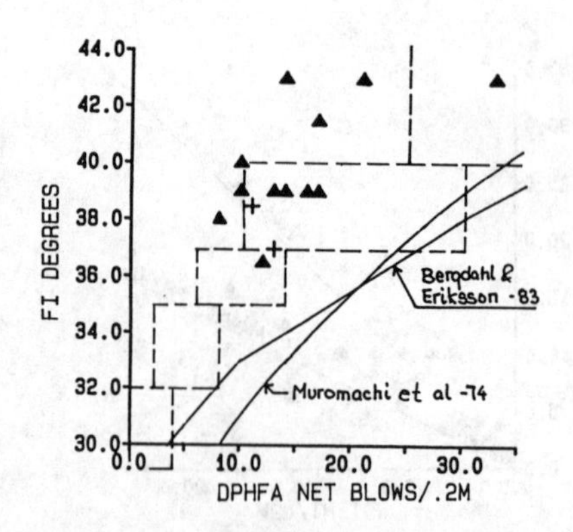

Fig. 14 Angle of internal friction (FI) versus dynamic probing HfA point resistance (DP HFA NET).

Table 1.

Rel. stiff-ness	CPT q_c (MPa)	WST[1)2)] halft/ 0.2 m	DP HfA[1)] blows/ 0.2 m	ϕ [3)] (°)	E (MPa)
very loose	0-2.5	0-10	0-4	29-32	<10
loose	2.5-5	10-30	2-8	32-35	10-20
medium dense	5-10	20-50	6-14	35-37	20-30
dense	10-20	40-90	10-30	37-40	30-60
very dense	>20	>80	>25	>40	>60

1) q_c-values to the left are normally correspondent to these values for WST and DP HfA

2) in silt and silty sands the weight sounding resistance should be reduced by a factor of 1.3 before classification.

3) stated values are valid for sand, for silty soils, reduce with 3° and for gravel, add 2°.

For the time being, the proposed relationships in Table 1 could be used. At design of footings and at settlement calculations, more direct methods may be used. These are based either on the penetration test results, (Bergdahl & Eriksson, 1983) or on pressuremeter tests, (Menard, 1975).

REFERENCES

Baguelin, F., Jézéquel, J-F., Shields, D.H., 1978. The pressuremeter and foundation engineering, Trans.Tech. Publication.

Bergdahl, U., 1974. Penetration tests in cohesionless soils in Sweden. Borros AB. Stockholm.

Bergdahl, U., 1979. Development of the dynamic probing test method. Proc. VIIth ECSMFE, Vol 2, Brighton.

Bergdahl, U., Möller, B., 1981. The Static-dynamic penetrometer. Proc. Xth ICSMFE, Vol 2, Stockholm.

Bergdahl, U., Ottosson, E., 1982. Calculation of settlements on sands from field test results. ESOPT II, Vol 1, Amsterdam.

Bergdahl, U., Eriksson, U., 1983. Bestämning av jordegenskaper med sondering - en litteraturstudie. (Estimation of soil characteristics from penetration test results. (In Swedish). Swedish Geotechnical Institute, Report No 22, Linköping.

Bergdahl, U., Hult, G., Ottosson, E., 1985. Calculation of settlement of footings in sands. XIth ICSMFE Vol 4, San Francisco.

Broms, B.B., Bergdahl, U., 1982. The weight sounding test (WST). State-of-the-art report. ESOPT II, Vol 1, Amsterdam.

Dahlberg, R., 1975. Settlement characteristics of preconsolidated natural sand. BFR document D1:1975.

Denver, H., 1980. Saetningsberegningar for fundamenter på sandlag. (Calculation of settlement for slab foundations in sand.) (in Danish). Laboratoriet for fundering. Danmarks Tekniske Höjskole.

Francke, E., 1973. Ermittlung der Festigkeitseigenschaften von nicht-bindigen Baugrund durch Sondierungen.

Helenelund, K.V., 1966. On the bearing capacity of frictional soils (in Finnish). VTT Sarja III Rohennus 97.

Menard, L., 1975. The interpretation of pressuremeter test results. Sols-Soils No 26.

Meyerhof, G.G., 1976. Bearing capacity and settlement of pile foundation. ASCE Geotech. Eng. Div. Vol 102, No GT 4, March.

Muromachi, T., Oguro, I., Miyashita, T., 1974. Penetration testing in Japan. ESOPT I, Vol 1, Stockholm.

Muromachi, T., Kobayashi, S., 1982. Comparative study of static and dynamic penetration tests currently in use in Japan. ESOPT II, Vol 1, Amsterdam.

Rodin, S. et al, 1974. Penetration testing in United Kingdom. ESOPT I, Vol 1, Stockholm.

SBN 1980. Swedish Building Code (in Swedish). National Board of Physical Planning and Building. Liber förlag, Stockholm.

Swedish Bridge Building Code, 1976 (in Swedish). Swedish Road Board. TB 103. Östervåla 1976.

Van Wambeke, A., d'Hemricourt, J., 1982. Correlation between the results of static or dynamic probings and pressuremeter tests. ESOPT II, Vol 2, Amsterdam.

Penetration Testing 1988, ISOPT-1, De Ruiter (ed.)
© 1988 Balkema, Rotterdam, ISBN 90 6191 801 4

Determination of CBR and elastic modulus of soils using a portable pavement dynamic cone penetrometer

Koon Meng Chua
Texas Transportation Institute, Texas A&M University System, USA

ABSTRACT: A theoretical approach towards modeling and interpreting results obtained using a portable Pavement Dynamic Cone Penetrometer [PDCP] is presented here. Testing using the PDCP basically involved dropping an 8 kg (17.6 lbs) sliding hammer over a height of 575 mm (22.6 inches) in order to drive a steel rod into the ground. The depth penetrated by the 60 degree cone of tempered steel located at the lower end of the steel rod as a result of a single blow from the hammer is an indication of the material properties of the medium. A one-dimensional model for penetration analysis of a rigid projectile into a ideally locking material was used to back-calculate the elastic modulus of the target medium. The model assumes the soil medium penetrated in one blow to be a horizontal disc and upon penetration, the projectile displaces the soil and a radial plastic shock wave propagates in the disc and plastic deformation takes place. The calculated elastic moduli are then compared with the California Bearing Ratio [CBR] values. Relationships of the elastic modulus and the CBR-value for various soils were developed using the analytical solution. The analytical solution was shown to be able to account for the variations in the CBR-elastic modulus relationships which are currently used. The confining pressure, and as a result the deviator stress at which the triaxial tests were conducted was found to be the main reason for the variation. An example problem is also presented. The solution presented is applicable for dynamic cone penetrometer of different configurations.

1 INTRODUCTION

The Pavement Dynamic Cone Penetrometer [PDCP] used in this paper was based on a design used in South Africa and extensively studied by Kleyn et al. (1982a; 1982b) and more recently by Harison (1986) and Livneh and Ishai (1987). To date, the PDCP was studied mainly in relation to applications with pavement structures and was primarily correlated with California Bearing Ratio [CBR] values. In view of the increasing importance of obtaining the in situ elastic modulus of soils for the structural evaluation of pavements and other shallow foundation applications, this study was initiated.

1.1 The Dynamic Cone Penetrometer

The PDCP is both portable and inexpensive. The instrument can be used to test in situ soils, thin pavements, as well as prepared roadbed or base course prior

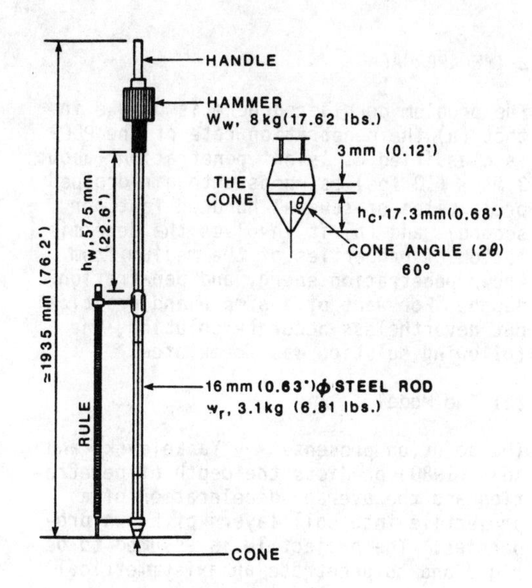

Fig. 1 The Pavement Dynamic Cone Penetrometer

to application of the finish/surface course.

Fig. 1 shows the dimensions of the PDCP used in this study. The PDCP measurements are reported as the penetration index which is the penetration depth (usually mm or inch) per blow resulting from a single drop of the hammer.

1.2 Dynamic Cone Penetration Analysis: the State-of-the-Art

Previous research of this type (dynamic/-impact penetration) can be broadly divided into three categories:

(a) experimental investigations to develop a qualitative understanding of the influence of various factors such as penetrometer shape and size, dynamics, soil properties, on the penetration phenomena (Dunlap, 1972),

(b) laboratory and field studies to develop empirical relationships for engineering applications (Young, 1969) and,

(c) theoretical/analytical approaches to modeling the penetration phenomena (Yankelevsky, 1980; Murff, 1972).

In dynamic penetration testing, perhaps the most comprehensive set of data may be found in Young (1969) who presented regression equations for predicting the penetration depth that can be achieved by a projectile in soils of known properties which is the more common trend in penetration analysis.

2 THE APPROACH

The problem considered here is unique in that (a) the penetration rate of the PDCP is classified as 'slow' penetration (about 3 ms^{-1} (10 fps) as opposed to air-dropped projectiles of several hundred feet per second), and (b) it involves the determination of properties of the medium from known penetration energy and penetration depth. For want of a simple and practical but nevertheless accurate solution, the following solution was formulated.

2.1 The Model

The solution presented by Yankelevsky and Adin (1980) predicts the depth of penetration and the average deceleration of a projectile into soil layers of known properties. The projectile is assumed to be rigid and to penetrate an axisymmetrical soil disc perpendicularly. Since the mechanism governing penetration is one of high volume changes, only plastic strains are considered. A propagating plastic shock wave defines the front boundary of a compressed plastic zone, and the elastic zone ahead is considered to have negligible contribution and is therefore disregarded.

Fig. 2 shows a sector of the soil disc behind the plastic shock front. The Mohr-Coulomb failure criterion was adopted (instead of the Tresca yield criterion used in the above mentioned paper) and the volumetric compressibility was idealized by an ideally locking material model, which is defined by its locking volumetric strain. This hydrostatic model is illustrated in Fig. 3.

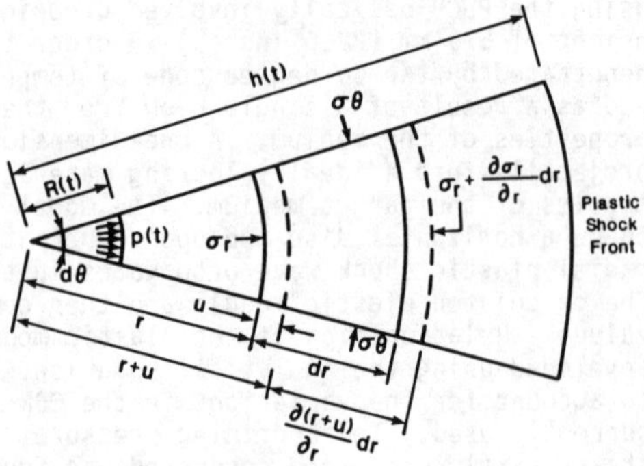

Fig.2 A sector of a soil-disc

The thickness of the soil disc is assumed to be the height of the cone nose which will be the only part of the device in contact with the soil. This solution requires the nose shape to be slender, that is, the cone apex angle needs to be an acute angle.

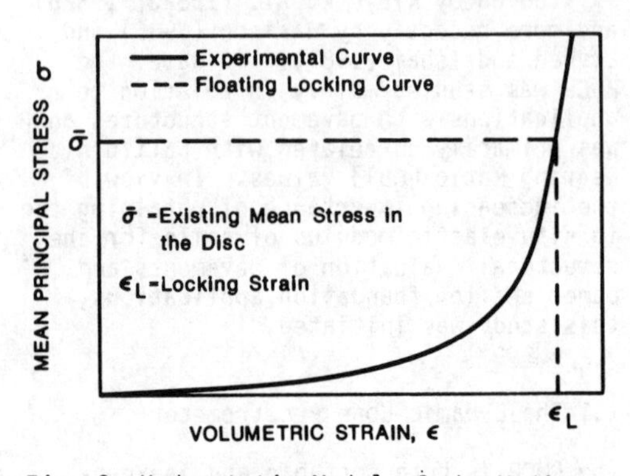

Fig. 3 Hydrostatic Model of the Medium

2.2 The Analytical Solution

The derivation of the interaction equation is found in Yankelevsky and Adin (1980) and will not be reproduced here. However, the equations which are of interest to this problem will be presented. The expression for the average mean stress, which is the mean of the radial stress σ_r and the tangential stress τ_θ, was given as,

$$\overline{\sigma} = 1/2.\ \tau_o.\ [(\overline{\varepsilon}_L - 1).\log_e\overline{\varepsilon}_L - \overline{\varepsilon}_L + 1/3] + 1/2.\ \rho_o.\ \dot{R}^2\ .[7/3 - \log_e\overline{\varepsilon}_L] + 1/2.\ \rho_o.\ R.\ \ddot{R}\ .(1 - \log_e\overline{\varepsilon}_L) \quad (1)$$

where τ_o is the principal stress difference at failure, ε_L is the mean volumetric locking strain, R is the radial displacement, $\dot{R}$ is the velocity and $\ddot{R}$ is the acceleration of the internal boundary; and ρ_o is the initial mass density.

The location of the shock wave front is defined by its instantaneous radius h, and

$$h = R/\varepsilon_L{}^{1/2} \quad (2)$$

where ε_L is the volumetric locking strain.

The radial stress behind the plastic shock front is given by,

$$h = \rho_o.\ \dot{R}^2 \quad (3)$$

2.3 Procedure Adapted to the PDCP

For the purpose of solving for the pressures and soil properties from PDCP readings, the values for the variables required in the above equations can be estimated from the following steps.

Step 1 Calculate the Dynamic Properties

The radial displacement, velocity and acceleration are determined from the vertical components at the cone tip.

The radial displacement is given by,

$$R = \frac{h_c}{2}.\ \tan(\theta) \quad (4)$$

where h_c is the height of the cone, and θ is half the cone angle.

The average radial velocity is,

$$\dot{R} = \frac{W_w.\ \sqrt{(2.g.h_w.)}}{(W_w + W_r).}\ .\ \tan(\theta) \quad (5)$$

where h_w, W_w and W_r are as defined in Fig. 1, and g is the gravitational acceleration.

The radial acceleration is obtained from the downward force calculated from the deceleration at the cone tip, and is given by

$$\ddot{R} = -\ \frac{g.h_w.\ W_w{}^2}{(W_w + W_r)^2.D}\ .\ \tan(\theta) \quad (6)$$

where D is the penetration index.

Step 2 Determine the Mean Stress over the Soil-Disc

The radial stress at the internal boundary of the disc is approximately given by,

$$p = \frac{h_w.W_w{}^2}{(W_w + W_r).\ D.\ h_c{}^2.\ \tan^2(\theta)} \quad (7)$$

The mean average stress over the plastic soil disc is approximately given by,

$$\overline{\sigma} = (p + \sigma_h)/2 - \tau_o/2 \quad (8)$$

and σ_h is obtained from Equation (3). (Note: the radial stress distribution unlike the static case, is not purely logarithmic, but include body forces due to particle acceleration).

Step 3 Determine the Principal Stress Difference at Failure

The principal stress difference at failure given by τ_o or $(\sigma_1 - \sigma_3)_f$ is given by,

$$(\sigma_1 - \sigma_3)_f = \frac{2\ c\ \cos\emptyset + 2\ \sigma_3\ \sin\emptyset}{1 - \sin\emptyset} \quad (9)$$

where σ_1 and σ_3 is the major and the minor principal stress, respectively, c is cohesion and $\emptyset$ is the friction angle.

Typical mid-range values of principal stress difference at failure for various soil types were obtained from the triaxial test data soils compiled by Duncan et al. (1980). Fig. 4 and Fig. 5 show the range and the mean obtained from the test data. The soils were grouped according to the Unified Soil Classification System. Table 1 summarizes the τ_o values which are to be used in Equation (1). The τ_o value for asphaltic concrete was estimated from results of unconfined compression tests reported by Carpenter (1986). This estimate is low and will be more suited to the surface-treatment course applied to thin pavements.

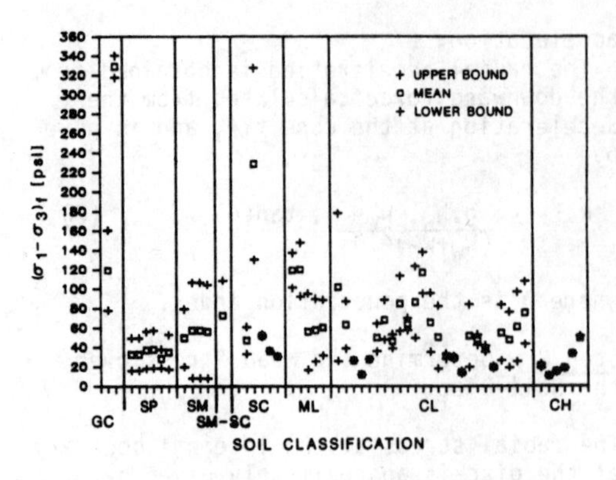

Fig. 4 Mean principal stress difference at failure for the undrained condition

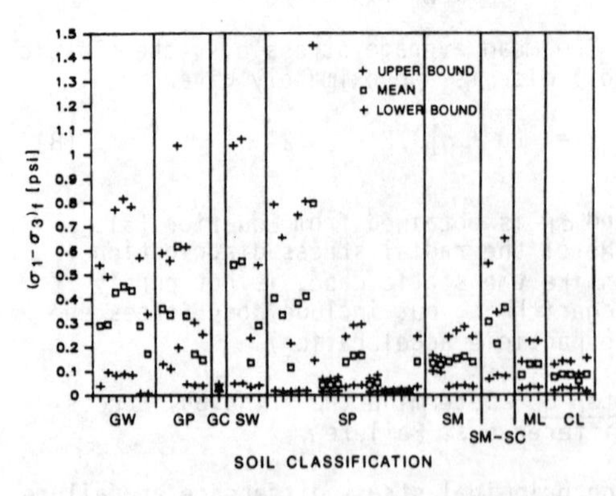

Fig. 5 Mean principal stress difference at failure for the drained condition

Table 1. Mid-range τ_o-values

Material Type	Principal Stress Difference at Failure, $_o$ [psi]
Gravel/Crushed Stones	300
Sandy Soil	100
Silty Soil	75
Clayey Soil (Low Plasticity)	50
(High Plasticity)	25
Asphalt Concrete	800

Step 4 Determine the mean volumetric locking strain

The mean volumetric locking strain is obtained by the trial-and-error approach using Equation (1) by adjusting e_L to obtain the value of the mean radial stress calculated using Equation (8).

Step 5 Determine the Elastic Modulus

The solution presented by Yankelevsky and Adin (1980) uses a known soil property, namely, the mean stress versus volumetric strain curve to predict soil penetration. However, by fitting a hyperbolic stress--strain curve (which is commonly used for soils) through the experimental data (of vertical stress versus vertical strain) presented in their paper as well as that from Yarrington (1978), a family of curves describing unique points (at a particular stress and strain) can be developed. The equation for describing the curves is as follows,

$$\sigma = \frac{\varepsilon_l}{\frac{1}{E_i} - \frac{1}{(a.E_i{}^b)} \cdot \varepsilon_l} \tag{10}$$

where ε_l is the linear strain (which is approximately $\varepsilon_L/3$), E_i is the initial tangent modulus, and a and b are constants. By fitting the data points shown in Fig. 6, the values of a and b are found to be 35.7 and 0.7, respectively. (Note:- the data points shown are for 'soil' and shale as well as for a median medium according to the references).

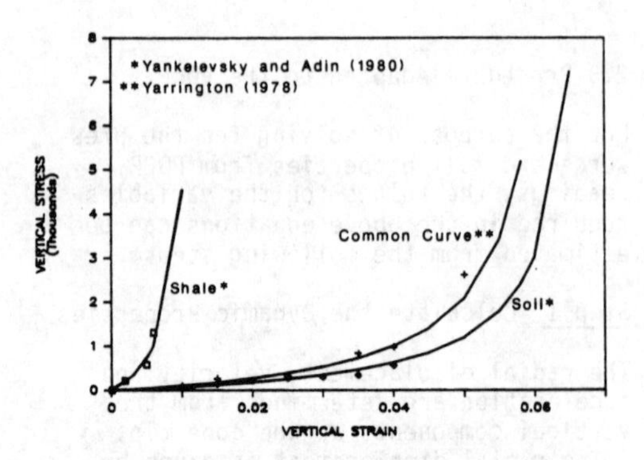

Fig. 6 Hyperbolic Stress-Strain Model for the Medium

The linear strain value (which is about one-third of the volumetric strain), is used in Equation (9) to solve for the initial tangent elastic modulus. The initial elastic modulus (E_i) can be found by the trial-and-error method, by adjusting the linear strain value to obtain a mean stress value equal to that determined in Step 2.

3 RELATING PDCP PENETRATION INDEX TO ELASTIC MODULUS AND CBR-VALUE

The following sections show how elastic moduli and CBR-values can be determined from PDCP readings.

3.1 Elastic Modulus and Penetration Index for Different Materials

Fig. 7 and Fig. 8 show the relationships between the elastic modulus and the penetration index for different materials and principal stress differences at failure shown in parenthesis. The mediums (soils) are broadly divided into clay of high and of low plasticity, silty soils, sandy soils, gravel and/or crushed stones and asphaltic concrete. Since only mid-range values were used for the different mediums, further refinement can be had by obtaining a band for each material showing the upper and the lower bound for the predictions. At this point, it should be realized that a different set of curves will be needed for a different impact-type cone penetrometer.

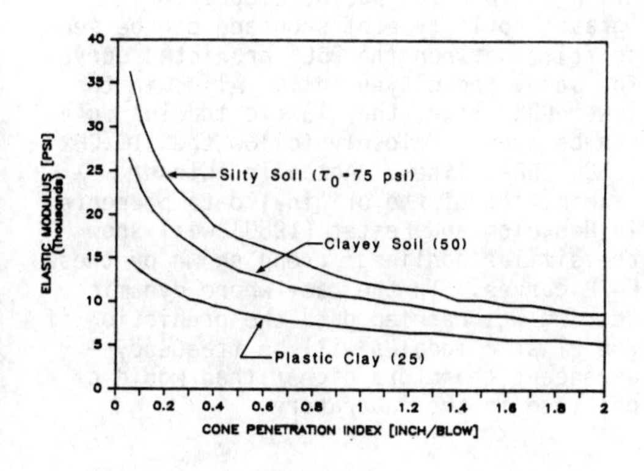

Fig. 7 Calculated elastic modulus vs Penetration Index fine-grained soils.

3.2 Relating the Penetration Index to CBR

Some of the best-fit regression equations relating the CBR-values to the penetration index (for the same PDCP) are shown in Fig. 9. Harison (1986) performed the PDCP tests on soil samples (clay, sand, gravel, soaked and unsoaked samples) in the standard CBR mold. Livneh and Ishai (1987) developed the relationship from data obtained in the laboratory and in the field. Both soaked and unsoaked fine-grained soil samples were tested in the

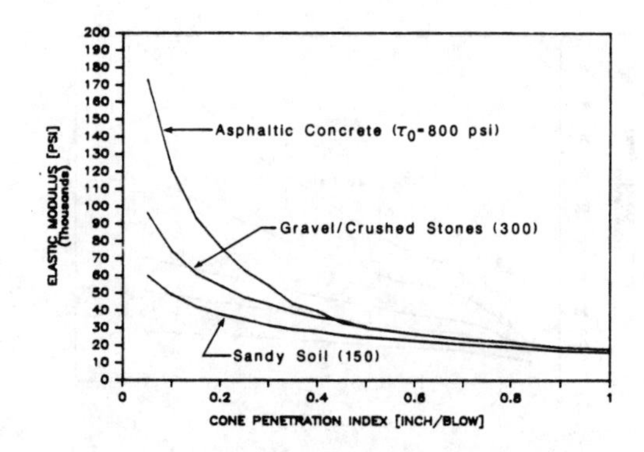

Fig. 8 Calculated elastic modulus vs Penetration Index for coarse-grained mediums

laboratory. Compacted granular soil samples were tested in flexible molds with variable controlled lateral pressure. The curve used by the NITRR [National Institute for Transport and Road Research, South Africa] is also shown in Fig. 9.

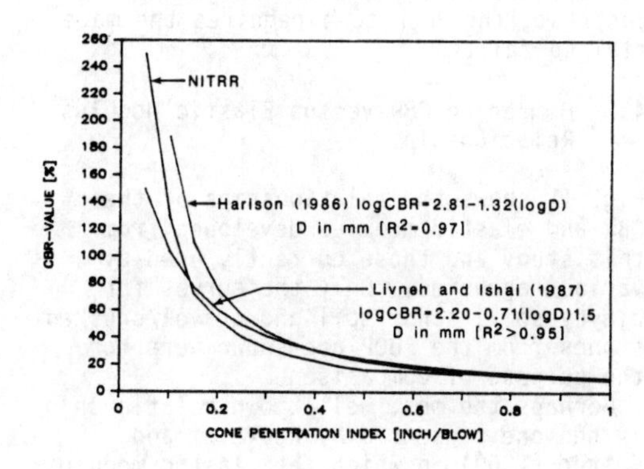

Fig. 9 CBR-values vs Penetration Index

4 RELATING THE ELASTIC MODULUS AND THE CBR-VALUE

The following section discusses the relationship between the CBR and the elastic modulus developed from PDCP results.

4.1 CBR versus Elastic Modulus

The elastic modulus can then be related to the CBR through the penetration index using both the equation presented by Livneh and Ishai (1987) and the relationship developed here. Fig. 10 shows the relationships for material with different

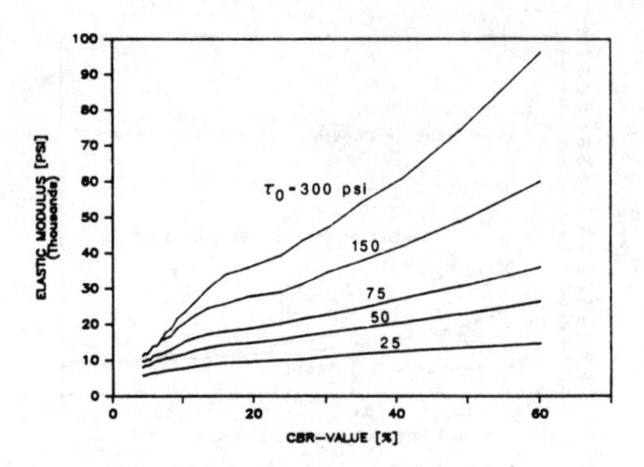

Fig. 10 Calculated Elastic Modulus vs CBR

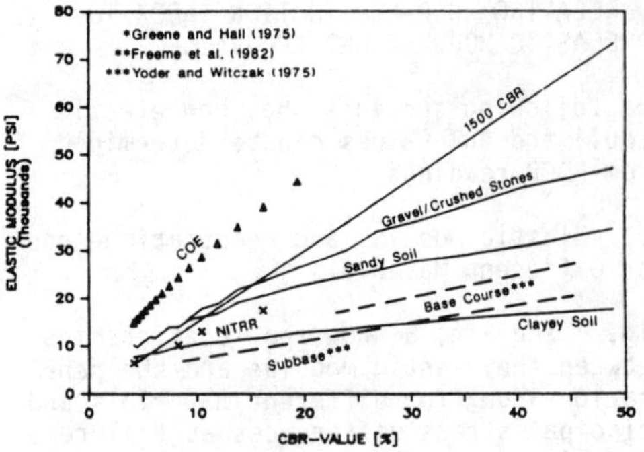

Fig. 11 Comparing elastic modulus vs CBR relationships

values of τ_0. As can be seen, the relationship is not unique in that no one general curve can describe all of the material tested. One reason may be that the failure characteristics of the materials tested do not show up in the CBR test results since the CBR is a measure of performance in the elastic range. In contrast, the PDCP test requires the material to fail.

4.2 Comparing CBR versus Elastic Modulus Relationships

Fig. 11 shows the relationships of the CBR and elastic modulus developed from this study and those currently used by various agencies. Only the curves for clayey soil, sandy soil and gravel/crushed stones from the PDCP are shown here for the purpose of comparison.

Perhaps the most well-known relationship is the one presented by Heukelom and Foster (1960) in which the elastic modulus is given as 110 CBR in kg/cm^2 (or 1500 CBR in psi). The relationship was developed mainly from data obtained from dynamic wave propagation testing for granular and fine-grained soils. The original data included materials with a dynamic elastic modulus up to 19720 kg/cm^2 (280000 psi). To include all of the data points, factors between 55 (750) and 220 (3000) are required. The Corps of Engineers [COE] (Greene and Hall, 1975) relationship was developed from vibratory testing, which may explain the higher modulus values.

The NITRR relationship (Freeme et al., 1982) applies only to subgrade and 'select' subgrade materials and was based on soaked CBR-values for 'gravel soil'. The relationships for base and subbase materials were inferred from the AASHO layer coefficient nomographs shown in Yoder and Witczak (1975).

In summary, the relationship of elastic modulus and CBR developed using the PDCP is more consistent and is shown to vary with material types and confining pressures. Referring again to Fig. 11, the NITRR curve which was developed for 'gravel soil' type of subgrade can be seen to falls between the PDCP predicted curves for sandy and clayey soil. Also, in the lower CBR range, the elastic modulus curve can be seen to closely follow the 110 CBR- (1500 CBR-) line . Actually, closer examination of the original data presented in Heukelom and Foster (1960) will show the similar nonlinear trend shown by these PDCP curves. In the case where dynamic testing was carried out, the prediction of the elastic modulus will be frequency--dependent therefore higher than would be obtained in the laboratory.

5 INTERPRETING FIELD DATA: AN EXAMPLE

One point to note in interpreting the PDCP reading is, the penetration depth obtained by the first blow (starting at the medium surface) will always be higher than subsequent readings. This is because resistance is offered by less than a soil-disc considered in the solution presented above. To rectify the situation, one can either ignore readings taken in the first half inch of penetration or average the first five blows. On the other hand, if a more accurate prediction is desired, one can develop a new set of curves for a range of soil-disc thicknesses.

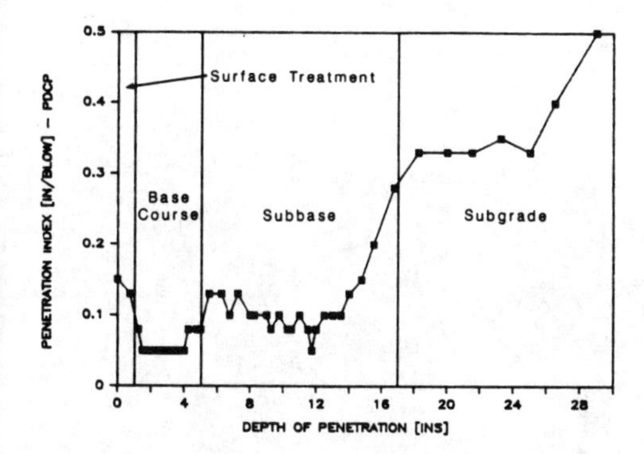

Fig. 12 Penetration Index for Test Section 11

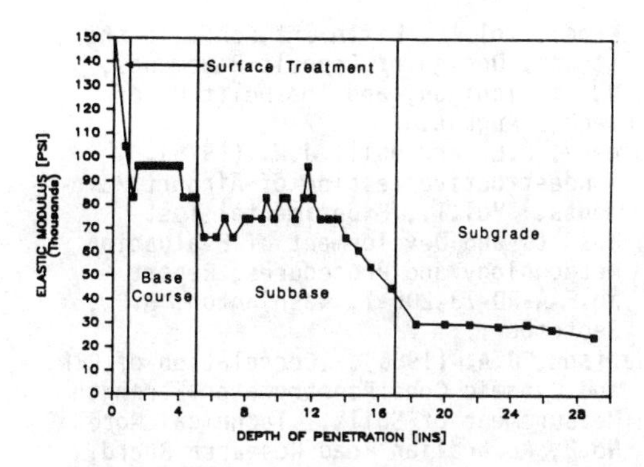

Fig. 13 Predicted elastic modulus profile for Test Section 11

5.1 Descriptions of the Test Section

The test section is located at the TTI (Texas Transportation Institute) Annex which has several different types of pavement structures constructed primarily for pavement research. Section No.11, considered here, is consisted of a 25mm (1") thick surface-treatment course, and a 100mm (4") thick unbounded granular course, over another 305mm (12") of the same type of material. The subgrade is a sand-gravel.

5.2 PDCP Readings

Depth of penetration were taken every five blows. The penetration index at each depth is shown in Fig. 12. As can be seen, it is possible to estimate pavement layer thicknesses from this type of plot. From the figure, the four different material zones can be seen. The penetration index for the four zones (downwards) is about 4, 13, 3 and 9 mm/blow (0.15, 0.5, 0.1 and 0.35 inch/blow), respectively.

5.3 Predicted Elastic Modulus Profile

The elastic modulus profile shown in Fig. 13 was developed using the relationships shown in Fig. 8. The predictions obtained are consistent and reasonable for the type of pavement considered.

6 CONCLUDING REMARKS

A new method of interpreting readings obtained from the PDCP is presented here. It had been shown that it is possible to obtain the in situ elastic modulus using a PDCP. Interpretative curves relating the elastic modulus of various media (clay of high and of low plasticity, silty soil, sandy soil, gravel/crushed stones and asphaltic concrete) to the penetration index as well as the CBR-value were presented. It was also shown that the relationship between the elastic modulus and the CBR-value is greatly influenced by the type of tests performed, the confining pressure applied and especially the principal stress difference at failure (shear strength) of the medium/soil.

The new approach towards interpreting the PDCP readings is shown to be reasonable and consistent and is a viable solution for engineers who desire an accurate and yet inexpensive method of in situ testing using a portable device.

7 REFERENCES

Carpenter, S.H. (1986). New Bituminous Mixture Analysis for Overlays. Present ed at the FHWA and Nat.Hwy.Inst. Training Course, Denver, Colorado, March 31 - April 4.

Duncan, J.M.; Byrne, P.; Wong, K.S. and Mabry, P. (1980). Strength, Stress-Strain and Bulk Modulus Parameters for Finite Element Analyses of Stresses and Movements in Soil Masses. Dept.Civ.-Engrg., U.Calif., Berkeley, Report No.UCB/GT/80-01, August.

Dunlap, W. A. (1972). Influence of Soil Properties on Penetration Resistance. Proc., Conference on Rapid Penetration of Terrestial Materials, Texas A&M University, February.

Freeme, C.R.; Maree, J.H. and Viljoen, A.W. (1982). Mechanistic Design of Asphalt Pavements and Verification Using the Heavy Vehicle Simulator.

Proc., Vol.I., Fifth Int.Conf. in the Struct. Design of Asphalt Pavements, U. of Michigan, and The Delft U. of Tech., August.
Greene, J.L. and Hall, J.W. (1975). Nondestructive Testing of Airport Pavements. Vol.I., Experimental Test Results and Development of Evaluation Methodology and Procedures, Report No.FAA-RD-73-205-I, Washington, D.C., September.
Harison, J.A. (1986). Correlation of CBR and Dynamic Cone Penetrometer Strength Measurement of Soils. Technical Note No.2, Australian Road Research Board, Vol.4.
Heukelom, W. and Foster, C.R. (1960). Dynamic Testing of Pavements. J. Struct.Div., ASCE, Vol.86, No.SM1, February.
Kleyn, E.; Maree, J.H. and Savage, P.F. (1982a). The Application of the Pavement DCP to Determine the In Situ Bearing Properties of Road Pavement Layers and Subgrades in South Africa. Proc.2nd European Symp. on Penetrometer Testing, Amsterdam, May.
Kleyn, E. and Savage, P.F. (1982b). The Application of the Pavement DCP to Determine the Bearing Properties and Performance of Road Pavement. Proc, Intl.Symp. on Bearing Capacity of Roads and Airfields, Trondheim, Norway, June.
Livneh, M. and Ishai, I. (1987). Pavement and Material Evaluation by a Dynamic Cone Penetrometer. Proc., 6th Intl. Conf. on Struct. Design. of Asphalt Pavements, University of Michigan, Ann Arbor, July.
Murff, J.D. (1972). An Analysis of Low Velocity Penetration of Clay Soils. A Ph.D. Dissertation submitted to the Texas A&M University, August.
Yankelevsky, D.Z. and Adin, M.A. (1980). A Simplified Analytical Method for Soil Penetration Analysis. Intl.J. for Num. and Analyt. Methods in Geomecha nics, Vol.4, 233-254.
Yarrington, P. (1978). A Comparison of Calculations with High Velocity Soil Penetration Data. SAND 78-0311, Sandia Labs., Albuquerque, New Mexico.
Yoder, E.J. and Witzcak, M.W. (1975). Principles of Pavement Design. 2nd Edition, John-Wiley & Sons, Inc., N.Y., 514-518.
Young, C.W. (1969). Depth Penetration for Earth-Penetrating Projectiles. J. Soil Mech. and Found.Div., ASCE, Vol.95, No.SM3, May.

Penetration Testing 1988, ISOPT-1, De Ruiter (ed.)
© 1988 Balkema, Rotterdam, ISBN 90 6191 801 4

Research on dynamic penetration testing of sands

C.R.I.Clayton, N.E.Simons & S.J.Instone
University of Surrey, Guildford, UK

ABSTRACT: This paper presents some of the results of research on dynamic penetration testing carried out over a period of more than 10 years at the University of Surrey, UK. The work has been divided into two parts: studies of the behaviour of dynamic penetrometers driven under controlled stress conditions in sand in a laboratory test chamber, and work on assessing the accuracy with which SPT related calculations can predict the settlement of spread footings on granular soils.

1 CHAMBER TESTS

The object of the chamber tests carried out at the University of Surrey has been to examine the influences of various factors on dynamic penetration resistance, rather than to provide data directly applicable t the SPT or any other form of dynamic penetrometer. Tests involved the driving of a 25 mm diameter, 60 degree cone-ended 200 mm high penetrometer by repeated blows of a 10 kg weight falling 430 mm, for a distance of 100 mm. The chamber, which has previously been described by Clayton and Dikran (1982), is relatively small, containing a specimen of 426 mm diameter and 630 mm height. A chamber of this siz has been shown by Clayton, Hababa and Simons (1985) to be adequate for the 25 mm diameter penetrometer used in these tests. The much higher ratios reported by Parkin and Lunne (1982) to be necessary for quasi-static penetration testing are not apparently required under the boundary conditions used (ie stress control at the bottom and side boundaries) for dynamic penetration testing. In a chamber such as ours, where water is used to control the lateral stress level and its drainage from the cell is relatively restricted, the effect of decreasing the diameter ratio of the chamber to the penetrometer below abou 15, however, is to produce a momentary increase in penetration resistance as a result of the increase in lateral (cell) pressure as the penetrometer is driven.

Two sands have been used in these tests:

1. Leighton Buzzard sand (D_{50}= 0.11mm, C_u= 1.6)
2. Woolwich Green sand (D_{50}= 1.0mm, C_u= 7.0)

Specimens were prepared at various initial densities, by raining through water, and were then subjected to a considerable variety of stress paths before the penetrometer was driven from the top of the chamber. Initially, testing was carried out to examine the uniformity of the specimens and the influence of the frictional top and bottom specimen boundaries. During the subsequent programme of penetration testing, measurements were made to determine not only penetration resistance but also the dynamic pore pressure response and the time during which the penetrometer was in motion.

2 INFLUENCE OF STRESS LEVELS AND STRESS HISTORY ON DYNAMIC PENETRATION RESISTANCE

The influence of vertical effective stress on dynamic penetration resistance has been the subject of much interest over a period of at least 3 decades (Gibbs and Holtz (1957), Marcuson and Bieganousky (1977 a,b)). More recently we have pointed out the important influence of horizontal stress (Clayton, Hababa and Simons (1985)).

Figure 1 shows the influences of both horizontal and vertical effective stress on the dynamic penetration resistance of normally consolidated Leighton Buzzard sand. It can be seen that vertical effective stress has a much smaller influence on dynamic penetration resistance than does horizontal effective stress level. This led to the hypothesis that the mean effective stress on the deposit immediately prior to testing might be a major factor controlling dynamic penetration resistance. Figure 2 shows

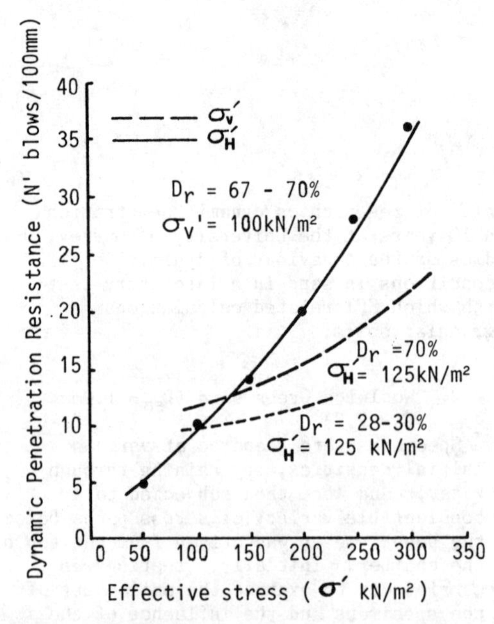

Fig. 1. Influence of horizontal and vertical stress on dynamic penetration resistance. (Hababa (1984))

that, for a given sand at a given density, dynamic penetration resistance can be correlated reasonably well with mean effective stress level, regardless of the stress history of the sand or of the current effective stress ratio applied to it. In fact, it has been repeatedly observed during our testing that dynamic penetration resistance is almost totally unaffected by stress history, simple prestressing, or overconsolidation along a zero lateral strain (Ko) stress path. How much the dependence of penetration resistance on horizontal effective stress is a matter of penetrometer geometry (eg length/diameter ratio) and how much is due to the dependence of end resistance on mean effective stress level remains to be determined. Intuitively it is expected that longer penetrometers would have a greater dependence on horizontal stress. The length/diameter ratio of our penetrometer is similar to that of the SPT, although a cone ended split spoon is used in only a few countries (eg in the UK, in gravels and the Chalk). In other countries much shorter cone-ended penetrometers of similar diameters to our penetrometer are in use (see German Standard DIN 4094).

One of the noticeable features of Figure 2, bearing in mind the considerable differences in the grading and density of the materials, is the relatively constant relationship between dynamic penetration resistance (N', blows per 100 mm penetration) and mean effective stress level (p'), from approximately N'= 0.05p' to N'= 0.15p' for this penetrometer. Thus at a given effective stress level, the

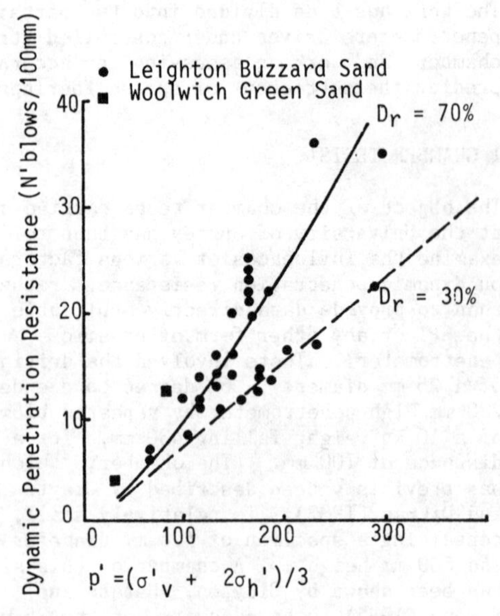

Figure 2. Influence of mean effective stress on dynamic penetration resistance (Hababa (1984))

combined effects of density, effective angle of friction, and particle angularity appear to be limited.

Hababa (1984) used the data from these tests to construct curves for the 'correction' of dynamic penetration resistance to the traditional vertical stress level of 100 kPa (1 kg/sq. cm. or 1 ton/sq. foot) - for example see Peck, Hanson and Thornburn (1974). The effect of horizontal stress increase due to Ko overconsolidation could also be included (Figure 3) to show that different OCR's would produce different depth correction curves. Hababa's correction factors appear to be considerably more dependent on vertical effective stress level than those proposed by Peck, Hanson and Thornburn (1974) and those suggested by Skempton (1986) on the basis of his interpretation of previous chamber and field SPT data.

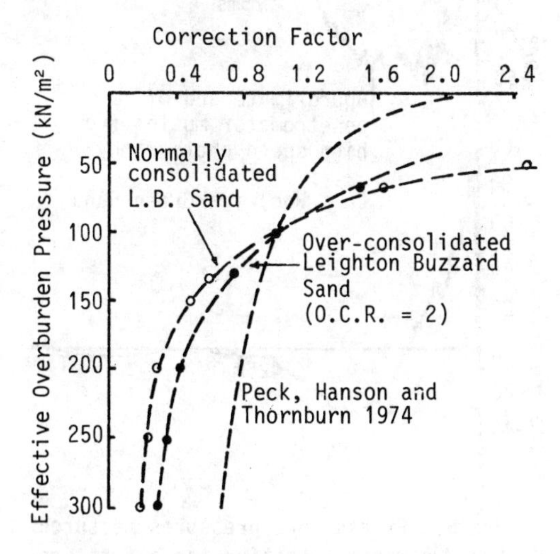

Figure 3. Influence of overburden pressure on penetration resistance. (Hababa (1984)).

Data of the sort shown in Figure 3 are of interest only because they lead to an understanding of a controlling factor of penetration resistance. They would seem to be of little use in settlement computations because

1. In general, the compressibility of sand is not well related to dynamic penetration resistance (Figure 4). Dynamic penetration resistance depends solely upon current effective stress level, but compressibility (however measured) is strongly related to stress history and can be significantly affected by even minor changes in stress history (Clayton, Hababa and Simons (1985)).
2. Even for normally consolidated granular soils, it is unlikely that an unique 'overburden correction curve' is valid, independent of soil type; overconsolidation alters the correction curve even for the idealised case of Ko overconsolidation. It is impossible, for a given site, to determine whether an increase in penetration resistance results from overconsolidation induced horizontal stress increase or from an increase in density, or from some other factor. These factors could reasonably be expected to have differing effects upon compressibility.
3. The relationships between vertical effective stress, penetration resistance and compressibility take different forms, so that soils with the same corrected penetration resistance cannot be expected to have the same compressibility characteristics. For example, for normally consolidated sands, compressibility is sometimes found to be linearly related to the square root of vertical effective stress, while penetration resistance is directly related to vertical stress. Thus, for example, the normally consolidated sands in Figure 4 show little variation of tangent modulus, but a very large variation in penetration resistance.

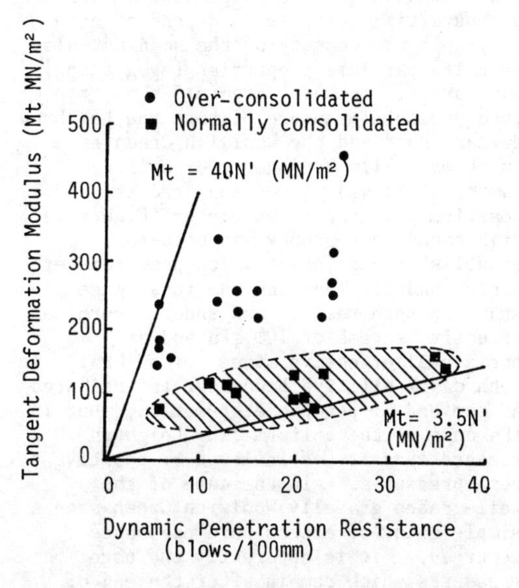

Figure 4. Relationship between Tangent Modulus and Dynamic Penetration Resistance

3 OBSERVATIONS OF PORE PRESSURES DURING DYNAMIC PENETRATION RESISTANCE

Initial experiments on Leighton Buzzard sand at relative densities of less than about 30 % produced liquefaction around the penetrometer in controlled chamber tests. These results, reported initially by Clayton and Dikran (1982), showed that liquefaction occurred when the measured positive excess pore pressure at the penetrometer tip exceeded the lowest of the horizontal or vertical normal effective stress applied to the sand immediately before penetration testing, provided that the duration of positive pore pressure was sufficiently large. The peak pore pressures recorded during these tests approximated to the maximum effective principal stress (in these cases in the vertical direction). These results are encouraging for the use of dynamic penetration resistance as an empirical guide to field liquefaction assessment.

Subsequent work (Hababa(1984)) has demonstrated that both the magnitude of the peak excess pore pressures generated during penetration, and the shape of the pore pressure/time response is dependent not only upon the density of the sand but also upon its particle properties (eg grading, angularity, etc.). Figure 5 shows the pore pressure response of both the Leighton Buzzard sand and the Woolwich Green sand (mathematically averaged over a large number of blows), when measured at position 1 on the penetrometer (Figure 6). High speed photography was used to establish the point at which penetrometer motion ended. The sand was in a dense state in both tests, and under a vertical effective stress of 100 kPa and a horizontal effective stress of 50 kPa. In both cases dilative behaviour is indicated at the end of penetrometer motion, but in the case of the uniform fine Leighton Buzzard sand it was followed by positive pore pressures. In the case of the well-graded gravelly Woolwich Green sand a simple decay of excess pore pressures occurred. It is apparently the pore pressures which remain after the end of normal penetrometer motion which produce the negligible penetration resistances associated with local liquefaction. Figure 6 shows that the maximum negative excess pore pressure measured during penetration is a function of the position of the sensor, as would be expected from work previously carried out on piezocones.

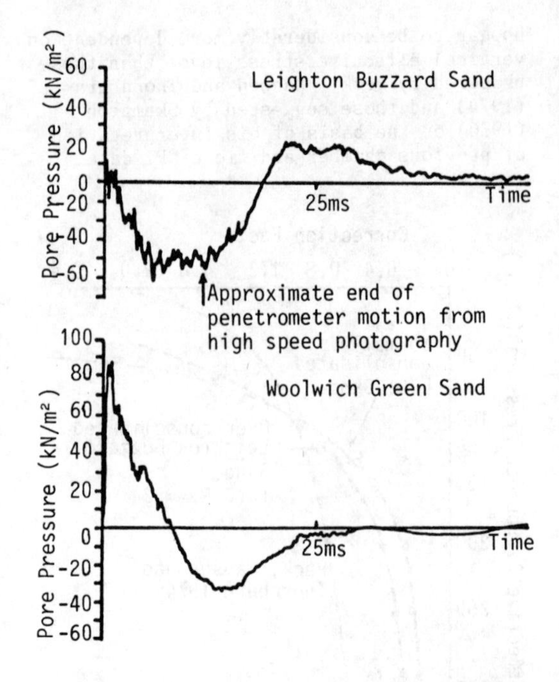

Figure 5. Excess pore pressures measured during dynamic penetration, as a function of time

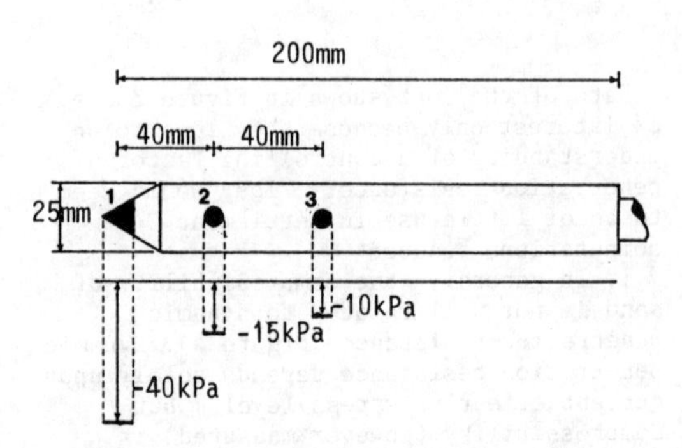

Negative pore water pressure developed during dynamic penetration on dense Leighton Buzzard Sand.

Figure 6. Excess pore pressures as a function of sensor position on the dynamic penetrometer

4 SPT BASED PREDICTIONS OF THE SETTLEMENT OF SPREAD FOOTINGS ON SANDS

The accuracy of the earliest methods of determining allowable bearing pressure or settlement for foundations on granular soil (for example, Terzaghi (1947)), which were presumably semi-intuitive, have been frequently shown to be very low.

Since 1947 the number of methods of settlement prediction has increased greatly, so that at present more than 20 SPT methods are available to the design engineer. The need to assess the relative accuracy of these methods has therefore become a matter of importance, and in the past ten years or so a considerable number of studies, generally using a fairly restricted database, has been made (for example Bratchell, Leggatt and Simons (1975), Simons and Menzies (1977), Talbot (1982), Milititsky et al. (1982), Jeyapalan and Boehm (1986)). Despite the introduction of new methods of analysis, the accuracy of available settlement predictions remains poor, because of

1. Borehole disturbance.
2. Lack of standardisation of the SPT, despite the publication of national and international standards.
3. The lack of dependence of dynamic penetration resistance on soil compressibility.

Logically, it would seem desirable to base prediction methods on the past performance of structures on granular soils, as suggested by Burland, Broms and de Mello (1977). Two methods based upon case record data currently exist (Schultze and Sherif (1973), Burland and Burbidge (1985)), and this section of the paper examines their performance.

In 1982 Burbidge produced an extremely valuable catalogue of 100 case records to be found in the literature for the settlement of structures on granular soil. Recently, this data has been used to examine the predictive capability of the two case record based methods. Burbidge's collection of case records contains some material which is not suitable for our purpose; therefore case records without direct SPT data (or indeed with non-standard split-spoons), where insufficient soil or settlement data was presented, or where compressible (ie non-sand) layers occur within the depth of influence of the foundation have all been ignored. The original 100 case records contain details of some 220 foundations or plate load tests, but only 54 % have associated SPT data. The other factors described above further reduced the data so that only about 90 foundations, involving about 200 settlement observations, could be used in this study.

Figures 7 and 8 show the results of a comparative study. In Figure 7 the results are plotted in the form of predicted settlement as a function of observed settlement, as has been done in a number of previous studies (recently, for example, by Jeyapalan and Boehm (1986)). The interesting features of this figure are that:

1. Observed settlements are most commonly smaller than 25 mm.
2. The two prediction methods only rarely produce settlements greater than 40 mm.
3. Settlements of up to 130 mm have been recorded.

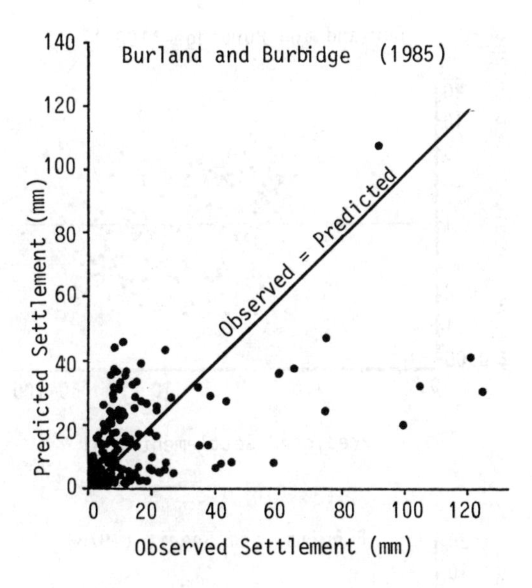

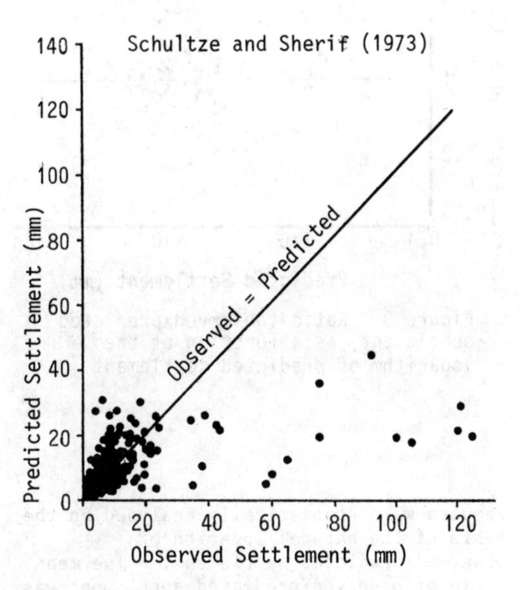

Figure 7. Predicted settlement as a function of observed settlement

Figure 8 presents the same data in a form suitable for assessing the likely range of actual settlements that might arise for a given value of predicted settlement. The

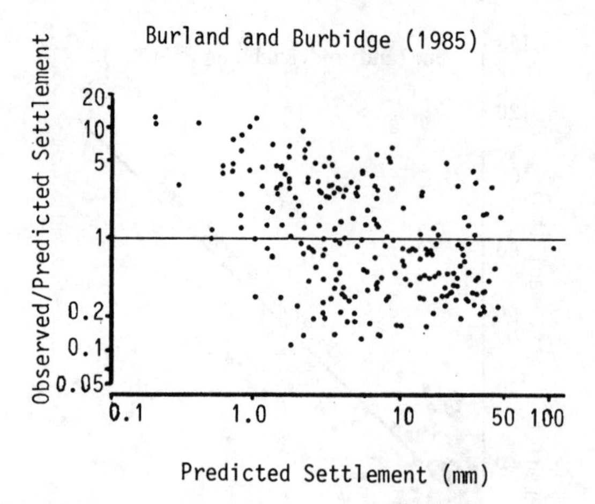

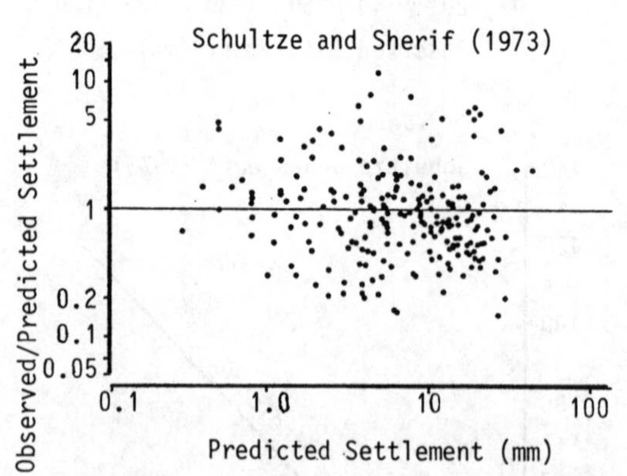

Figure 8. Ratio (observed/predicted settlement) as a function of the logarithm of predicted settlement

results were statistically analysed on the basis of the natural logarith of observed/predicted settlement. The mean ratio of observed/predicted settlement was then derived for each 5 mm band of predicted settlement (ie 0 - 5 mm, 5 - 10 mm, etc.), and is shown in Table 1 for each settlement prediction method. The average value of ln(observed/predicted settlement) was 0.045 for Burland and Burbidge's method and -0.085 for Schultze and Sherif's method, equivalent to observed/predicted settlement ratios of 1.05 and 0.92 respectively. The standard deviation for observed/predicted settlement was found to be 2.87 in the case of Burland and Burbidge's method and 2.23 in the case of Schultze and Sherif's method.

Table 1. Ratio of ln(observed/predicted) settlement for 5mm bands of predicted settlement.

Range of predicted settlement (mm)	Average of ln(observed/predicted) Burland & Burbidge	Schultze & Sherif
0 - 5	0.4347	0.0018
5 - 10	0.0568	-0.0545
10 - 15	-0.3843	-0.2326
15 - 20	-0.6338	-0.0907
20 - 25	-0.6104	-0.1392
25 - 30	-0.5232	-0.3845
30 - 35	-0.4582	-1.0471
35 - 40	-0.7026	0.7395
40 - 45	-0.3872	0.7219
45 - 50	-0.4715	
105 - 110	-0.1582	

If the results conformed to a normal Gauss distribution, 95% of the observed settlements would lie within 2 standard deviations of the mean, ie between ratios of observed/predicted settlements of 0.13-8.62 for Burland and Burbidge's method and 0.18-4.59 for Schultze and Sherif's method. The limits seen on Figure 8 show that Burland and Burbidge's method does rather better than the 95% confidence limits would suggest, and that a range of observed to predicted settlements from about 0.2 to 5 can be expected from both methods for predicted settlements greater than about 10 mm. When settlements of less than 10 mm are predicted the methods are less accurate. These levels of accuracy are unacceptable in engineering calculation of even the most routine kind.

For comparison, if the simple arithmetic mean of the observed settlements (12.5 mm) is used as the 'predicted' value of settlement, the average value of ln(observed/predicted settlements) is found to be -0.74 (equivalent to an observed/predicted settlement ratio of 0.48), and the standard deviation is equivalent to a ratio of observed/predicted settlement of 3.46. The 95% confidence limits are then equivalent to ratios of observed/predicted settlement of 0.04 to 5.7, and it can be seen that both the methods described above are not appreciably better than this.

If these methods represent the most accurate available for the prediction of settlements on sands using SPT data (and it is our belief that they are two of the best methods), there appears to be little value in using this test for this purpose. Rather than develop further settlement prediction methods it will be more practical to identify those factors which lead to exceptionally large settlements, for typically the settlement of a foundation on granular soil will be less than 25 mm. Taking the population of results used in the analyses above, and excluding narrow footings (ie 1 m or less wide, where bearing capacity failure may be developed), the average settlement is 13.9 mm.

Only 10% (a total of just 14) of the settlement records for foundations wider than 1 m had observed settlements of greater than 25 mm, and of these 70% (ie 10) were for foundations wider than 10 m. Of the remaining cases three involved narrow (1.2 m wide) foundations loaded in excess of 750 kPa, and the remaining record had an N value of 13 and a width of 6 m.

At this time, the use of the SPT to predict the settlement of foundations on granular soils cannot be justified. A more sensible engineering approach is to

1. Assume that 90% of settlements are likely to be less than 25 mm, and attempt to identify those limited structures which will be unable to tolerate the associated differential settlements.

2. Devote more resources to identifying those situations which have been observed to produce high settlements, such as
- narrow, heavily stressed foundations
- wide foundations
- non-quartz sands (eg shelly sands)
- foundations on metastable ('collapsing') soils, such as weathered dune sands and lightly cemented sands (eg sabkha)
- sands containing small amounts, or thin layers of organic or cohesive material.

3. Use long term static loading tests (eg plate or tank tests) to identify the compressibility characteristics of the ground, when fine estimates of settlement are required, or when high settlements are possible.

5 CONCLUSIONS

1. Laboratory research on dynamic penetration testing in sand has shown that
- low values of dynamic penetration resistance can occur as a result of liquefaction around the penetrometer.
- dynamic penetration resistance is strongly influenced by effective horizontal stress, and is unaffected by stress history if no change in effective stress results.
- correlations between dynamic penetration resistance and the compressibility of sand have been found to be poor.

2. Comparisons between observed and predicted settlements, using two case-history based methods, show that SPT predictions of the settlement of spread footings on granular soils continue to be unacceptibly inaccurate, despite a much improved approach in their derivation. More research is required to identify those factors which lead to the relatively small number of cases where significantly larger settlements are observed.

REFERENCES

Bratchell, G.E., Leggatt, A.J. and Simons, N.E. 1975. The performance of two large oil tanks founded on compacted gravel at Fawley, Southampton, Hampshire. Proc. Conf. on Settlement of Structures, Cambridge, 3-9.

Burbidge, M.C. 1982. A case review of settlement on granular soil. M.Sc. Dissertation, Imperial College, University of London.

Burland, J.B., Broms, B.B. and de Mello, V.F.B. 1977. Behaviour of foundations and structures - state of the art review. Proc. 9th Int. Conf. Soil Mech. Fdn Engng, 3, 395-546.

Burland, J.B. and Burbidge, M.C. 1985. Settlements of foundations on sands and gravels. Proc. Inst. Civ. Engnrs, 78, 1, 1325-1381.

Clayton, C.R.I. and Dikran, S.S. 1982. Pore water pressures generated during dynamic penetration testing. Proc. 2nd Eur. Symp. on Penetration Testing, Amsterdam, 245-250.

Clayton, C.R.I., Hababa, M.B. and Simons, N.E. 1985. Dynamic penetration resistance and the prediction of the compressibility of a fine-grained sand - a laboratory study. Geotechnique 35, 1, 15-31.

Dikran, S.S. 1983. Some factors affecting the dynamic penetration resistance of a saturated fine sand. Ph.D. Thesis, University of Surrey.

Gibbs, H.J. and Holtz, W.G. 1957. Research on determining the density of sands by spoon penetration testing. Proc. 4th Int. Conf. Soil Mech. Fdn Engng, 1, 321-376.

Hababa, M.B. 1984. The dynamic penetration resistance and compressibility of sand. Ph.D. Thesis, University of Surrey.

Jeyapalan, J.K. and Boehm, R. 1986. Procedures for predicting settlements in sands. Proc. ASCE Conf., Seattle, Spec. pub. no. 5, 1-23.

Marcuson, W.F. and Bieganousky, W.A. 1977a. Laboratory standard penetration tests on fine sands. Proc. ASCE, J. Geot. Engng Div., 103, GT6, 565-587.

Marcuson, W.F. and Bieganousky, W.A. 1977b. SPT and relative density in coarse sands. Proc. ASCE, J. Geot. Engng Div., 103, GT11, 1295-1309.

Milititsky, J., Clayton, C.R.I., Talbot, J.C.S. and Dikran, S.S. 1982. Previsao de recalques em solos granulares utilizando resultados de SPT: revisao critica. Proc. 7th Brazilian Conf. Soil Mech. Fdn Engng, Recife.

Parkin, A.K. and Lunne, T. 1982. Boundary effects in the laboratory calibration of a cone penetrometer for sand. Proc. 2nd Eur. Symp. on Penetration Testing, Amsterdam, 761-768.

Peck, R.B., Hanson, W.E. and Thornburn, T.H. 1974. Foundation Engineering. 2nd Edition. Wiley, New York.

Schultze, E. and Sherif, G. 1973. Prediction of settlements from evaluated settlement observations for sands. Proc. 8th Int. Conf. Soil Mech. Fdn Engng, 1.3, 225-230.

Simons, N.E. and Menzies, B.K. 1977. A short course in foundation engineering. 3rd Edition. Butterworths, London.

Skempton, A.W. 1986. Standard penetration test procedures and the effects in sand of overburden pressure, relative density, particle size, ageing and overconsolidation. Geotechnique, 36, 3, 425-447.

Talbot, J.C.S. 1982. The prediction of settlements using in-situ penetration test data. M.Sc. Dissertation, University of Surrey.

Terzaghi, K. 1947. Recent trends in subsoil exploration. Proc. 7th Texas Conf. Soil Mech. Fdn Engng, 1-15.

Penetration Testing 1988, ISOPT-1, De Ruiter (ed.)
© 1988 Balkema, Rotterdam, ISBN 90 6191 801 4

Shear strength parameters of lime sludge based on statistical analysis of dynamic soundings using probes with vane tips

E.Dembicki, I.Lewandowska & F.Loska
Department of Hydrotechnics, Gdańsk Technical University, Poland

ABSTRACT: The paper summarizes the experience gained by the authors in the use of dynamic soundings and vane tests in the lime sludge material. The proposed interpretation method of experimental data is oriented towards the possibility to obtain the relevant shear strength parameters for the stability calculations of storage yards. The coefficients of variation in underained shear strength have also been studied. A method which takes into consideration the great uncertainty caused by the non-homogeneity of the material has been proposed. The statistical method used in the analysis of the results from site investigation which is based on the assumption of Coulomb´s criterion, proved to be useful in the evaluation of shear strength parameters.

1 INTRODUCTION

Dynamic soundings and vane tests are widely used when investigating the location of layers in geological profile and evaluating the geotechnical parameters. The reliability of the results of those tests as well as the applicability of a given investigation method depends every time on a series of factors which generally influence the obtained values in various ways, namely:

1. natural heterogeneity of soil,
2. correlation of the measured quantities and the values of geotechnical parameters,
3. limited number of the test results (number of samples in a statistical analysis),
4. measurement errors.

This paper presents the method of adopting the dynamic soundings with vane test to determine the shear strength parameters ϕ_u and c_u for lime sludge which is a waste material of soda-production. The problem of heterogeneity of the tested material found in both laboratory and field tests has been analysed. The correlations assumed in the study of experimental data have been analysed from the point of view of evaluation of relevant values of shear strength parameters for stability calculations.

2 GENERAL CHARACTERISTICS OF THE PROBLEM INVESTIGATED

Waste materials obtained in soda-production process by Solvay´s method are stored in special artificial storage yards. The technology of their disposal is as follows: solid waste materials, except ash, are delivered by narrow-gauge railway to the sedimentary yards, where they are used to form superstructure of their outer embankments. Liquid waste materials are carried by means of hydro-transport to the yards in which the sedimentation of solid particles takes place. The clarified liquid is carried gravitationally by means of a drainage system, pipelines and trenches to the pump-station.

The lime sludge material stored within the embankments is characterized by a great variation of its physical and mechanical parameters resulting from the nature of the sedimentation process, the kind of delivered waste materials and time of storing and drying up. The sedimentation process of grains and particles which form sludge consists in settling of course-grained material at a little distance from the outlet of the pipeline used for hydro-transport; smaller particles settle at a greater distance while the finest particles settle at the greatest distance from the outlet of the pipe, close to the drain pit for clarified water.

The variability of properties of lime sludge in storage yards in vertical and horizontal direction has been confirmed by the results of the performed tests. It has been stated that there were two kinds of sediments: one of the water content 100 - 200%, and another-consolidated of the water content 60 - 80%. Earlier laboratory and field tests of sediments, carried out occasionally, showed the following values of physical and mechanical parameters of lime sludge in the embankment zone:

unit weight of solid particles	γ_s = 27-29 kN/m^3
unit weight of soil	γ = 12-15 kN/m^3
water content	ω = 60-600%
cohesion	c = 8-80 kPa
angle of internal friction	ϕ = 0 - 8°
coefficient of permeability	k = 0.78-1.33x10^{-6} m/s

The test carried out by the authors of this paper, consisted in performing a great number of measurements first of all of sediments soundings in ponds including their shearing in vane testing. The results of measurements formed a basis for a statistical interpretation method.

3 SITE INVESTIGATION

The site investigations were carried out in 1985 directly in the storage yards at Janikowo Soda Works (northern part of Poland). The average height of ponds was 17,0 m; their area about 112 ha. The experiments consisted in sounding of lime sludge by means of dynamic probes with vane tips which enabled the shear strength investigation. Two kinds of probes made in Poland were used: the I.T.B. probe and the S.L. probe. They differ from each other in weight of hammer and height of drop. The parameters of both probes are given

Table 1

Kind of probe	Hammer weight (kg)	Height of drop (m)	Rod diameter (mm)
I.T.B.	22.0	0.25	22
S.L.	10.0	0.50	22

in tab. 1.

In both cases the number of blows for each 10 cm of penetration of the probe as well as the shear strength at every 1 m of sinking of the probe have been recorded. All together 18 soundings with the light S.L. probe and 18 soundings with the I.T.B. probe to an average depth of 9.0 m below the sediment surface have been performed.
The soundings comprised the lime sludge material stored within the outer embankments of ponds in the zone about 40.0 m wide (Fig.1).

The obtained results showed the random character of the sediment variation at cross-section and in plane of the storage yards. Sudden increases of shear strength has been noted. It proved that shearing took place in hardened interbedding. As the distinction of harder and softer layers and their location turned out to be practically impossible, the results of measurements have been worked out in a statistical way.

4 METHOD OF ANALYSIS

Statistical method was used in the analysis of the site investigation data. Average numbers "i_m" of hammer blows per 10 cm of probe sinking for every 1 m of depth and their standard deviation $\sigma(i_m)$ (for each of used probes respectively) have been calculated. Then, the relations between the shear strength "τ" and the number of hammer blows "i" (regression lines) have been found. After that the computational regression lines for the assumed confidence level of 90% i.e. probability of non-appearance of values which are smaller than the determined ones have been determined From this lines the values of shear strength for average numbers of blows

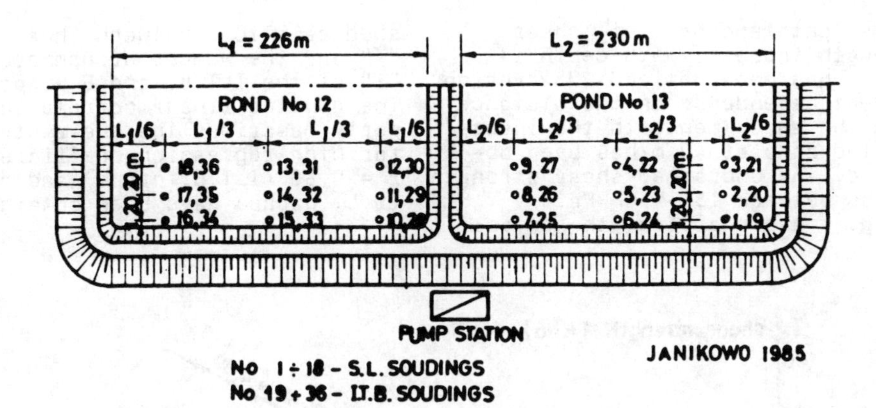

Fig. 1 Tests location

"i_m" in 1m layers have been obtained. Subsequently, these values have been compared in tables and the average values of shear strength and their standard deviation for successive depths have been calculated. For each layer of 1 m a coefficient of variation, which in this case is a measure of heterogeneity of the material, has been calculated:

$$V = \frac{\sigma(x)}{x_m} \cdot 100\,\%$$

where:
$\sigma(x)$ - standard deviation
x_m - mean parameter value

In view of variation of mechanical parameters (sudden increases of resistance when sounding in hard interbedding) there has been used a criterion of homogeneity consisting in rejection of values τ_i which do not belong to the intervals:

$$\tau_m - 2\cdot\sigma \leq \tau_i \leq \tau_m + 2\cdot\sigma$$

The great values of standard deviation were responsible for the fact that the values exceeding the mean of shear strength by $2\cdot\sigma$ have been neglected. As the next step a linear correlation between the shear strength τ and initial stresses at given depth has been established. The obtained line has been interpreted as Coulomb's line which determines the shear strength parameters of lime sludge in undrained conditions i.e. ϕ_u and c_u. It should be emphasized that the above mentioned values have been generalized for the sediments of each storage yard.

5 RESULTS

In Fig. 2 has been presented an example of the variation of shear strength obtained from I.T.B. tests with vane tip. The diagram concerns the investigations carried out at the bottom of the slope from the inner side of the banking of the pond (N°28), at the distance of 20 m (N°29) and 40 m (N°30), Fig.1.

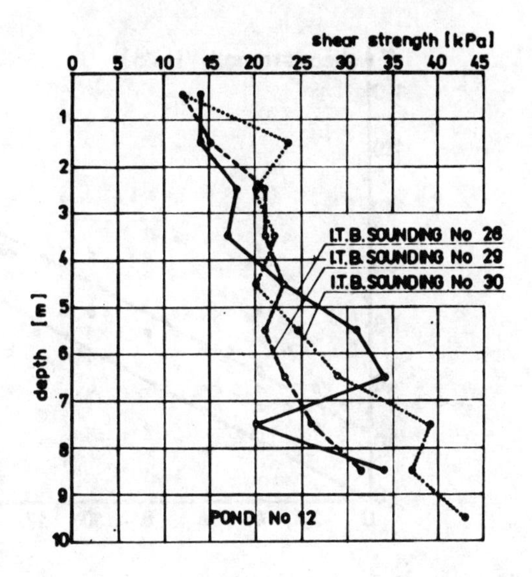

Fig. 2 Variation of shear strength

A slight tendency of shear strength increase with depth increase has been noticed. However no evident dependence on the distance from the embankments in the investigated zone of 40 m has been observed. The undrained shear strength varies between 15 - 30 kPa.

Fig. 3 and 4 present the established relations between shear strength "τ" and the number of hammer blows "i" of the I.T.B. and S.L. probes. The applied linear correlation properly describes the shear strength for high regression coefficients: r = 0.68 (I.T.B. probe) and r = 0.81 (S.L. probe) have been obtained.

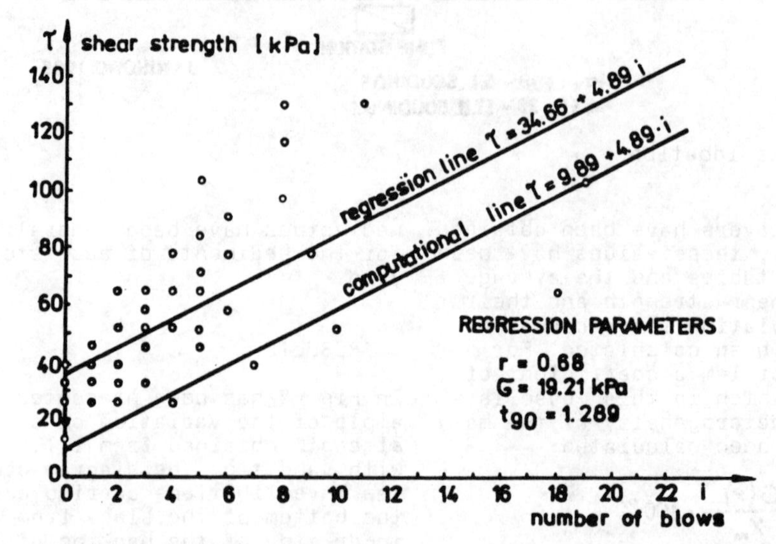

Fig. 3 Relation between shear strength "τ" and number of hammer blows "i" for the I.T.B. probe (storage yard N°12)

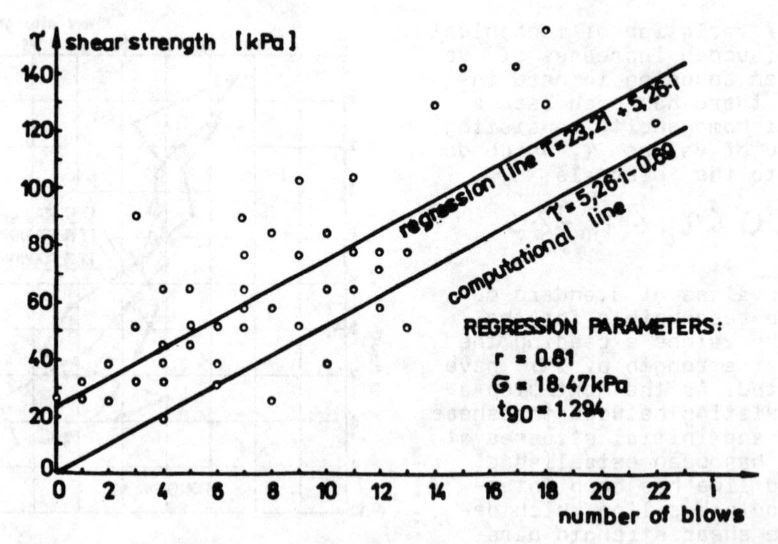

Fig. 4 Relation between shear strength "τ" and number of hammer blows "i" for the S.L. probe (storage yard N°12)

Fig. 5 presents an exemplary course with depth of the variation coefficient of shear strength determined on the basis of the interpretation of the I.T.B. tests results. The values of the variation coefficient concern values of the shear strength determined for the average numbers of hammer blows in consecutive 1 m layers. After introducing a homogeneity criterion for the analyzed sets i.e. after rejecting of the values which are much higher than the mean ones, the appropriate diagram of the variation coefficient has been marked, Fig. 5.

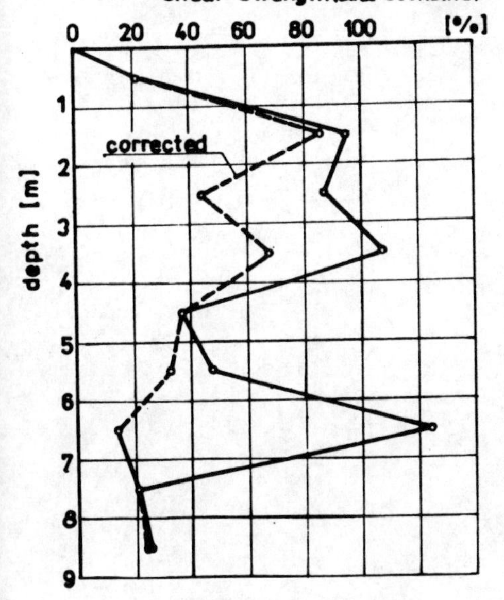

Fig. 5 Variation coefficient of shear strength determined from the I.T.B. tests

A considerable decrease of the variation coefficient is particularly evident for the layer of 6-7 m for which the results of 7 measurements showed values τ approximately equal to 30 kPa except one measurement for which τ = 183.5 kPa.

In Table 2 the results of the analysis of soundings in lime sludge stored in the storage yard N°12 have been summarized.

Taking into account a higher value of regression coefficient as well as the application of the results for the analysis of storage yard stability (construction safety) the values obtained from the I.T.B. tests have been accepted.

Table 2

Number of storage yard	Kind of probe	Regression parameters	
		Regression coefficient r	Mean standard deviation σ(kPa)
12	I.T.B	0.78	4.9
	S.L.	0.64	10.8

Number of storage yard	Kind of probe	Angle of internal friction ϕ_u (°)	Cohesion c_u (kPa)
12	I.T.B	9.3	18.70
	S.L.	13.4	26.30

6 CONCLUSIONS

1. The analysis has proved the applicability of the investigations carried put by means of dynamic sounding using probes with vane tips for the evaluation of shear strength parameters of lime sludge.
2. The interpretation of the results of soundings in lime sludge has showed a considerable scattering. In this case only the statistical analysis of the results enables the elimination of errors.
3. The application of statistically justified criteria of homogeneity of the data sets allows to reject from the analysis the values of the results that differ considerably from the mean ones. The rejection of the values that are much higher than the expected ones, eliminates a dangerous possibility to overestimate the final results which later are used for the stability calculations.
4. The assumption that shear strength of sediment is a function of angle of internal friction, cohesion and initial stress at a given depth i.e. the assumption of

the validity of Coulomb´s criterion has been confirmed by relatively high coefficients of correlation.

5. The authors are planning to continue the investigations of lime sludges by the application of continuous penetration devices such as the electrical static cone (CPT) or the piezocone (CPTU) and dynamic probing (DP).

Penetration Testing 1988, ISOPT-1, De Ruiter (ed.)
© 1988 Balkema, Rotterdam, ISBN 90 6191 801 4

Statistical evaluation of dynamic cone penetration test data for design of shallow foundations in cohesionless soils

A.Ş.Kayalar
Dokuz Eylül University, Izmir, Turkey

ABSTRACT: Random variations in soil properties requires statistical and probabilistic approach for the solution of geotechnical problems, when there is an adequate data base. Dynamic penetrometer may provide sufficient data for statistical analysis. A case study is presented. Dynamic penetration test data obtained from cohesionless soils are analysed for the design of shallow foundations in a light industry cooperation construction project. Reliability analysis is performed for the prediction of probability of failure, safety factor and foundation dimensions.

1 INTRODUCTION

The nonhomogeneous nature of soil deposits introduces unavoidable random variations in their physical and mechanical properties. Statistical and probabilistic modeling of this randomness can be performed by the methods given in the existing literature (e.g., Benjamin and Cornell (1970); Whitman (1984); Harr (1984)), when there is an adequate data base.

Dynamic penetrometer may provide sufficient data for reliability analysis.

The aim of this paper is to present a case study which contains statistical and probabilistic evaluation of dynamic cone penetration test data for design of shallow foundations on cohesionless soils.

Investigated area, soil profile and penetration test data are introduced. Relative frequency polygons and percentage cumulative frequency distributions of dynamic soil resistance to penetration for 8 different foundation widths have been plotted. Theoretical formulation of the reliability analysis is summarized. Variation of probability of failure with foundation width, for a central factor of safety of 4, has been determined. Also, foundation width-factor of safety curves for 5% and 10% probability of failures are obtained.

2 INVESTIGATION AREA, SOIL PROFILE AND FOUNDATION PROPERTIES

The investigated area is the construction site of a light industry cooperation and located on the old deltaic deposits of Gediz River, in the north-west of İzmir. Surface area is about 50×10^4 m^2 and nearly horizontal. Mainly, normally consolidated recent sediments are met down to about 25 m depth from the ground surface. Below this depth, overconsolidated soil formations take place. (Kayalar and Ülküdaş (1985)).

Soil profile in the investigated area has been determined with the soil samples taken by scraper bucket type samplers.

Soil profile consists of 0.30 m vegetative cover on top, stiff, brown, fissured, inorganic clay of medium to high plasticity between 0.30 m and 2.0 m depths and very loose, loose and medium dense nonplastic silty sand and silt-sand mixtures down to the investigated 10 m depths. Groundwater table is at a depth of 2 m from the ground surface and rises during wet seasons.

Because of presence of top clay layer and level of groundwater table, shallow foundation levels are planned at 2 m depth. Considering the industrial structures to be built and the heavy equipment to be installed, range of footing widths to be taken into account are 1-8 meters.

3 DYNAMIC CONE PENETRATION TESTS AND DYNAMIC SOIL RESISTANCES

Dynamic cone penetration tests have been performed at 12 different locations, about 200 m apart from each other, down to 10 m depth. Penetration locations are shown in Fig. 1.

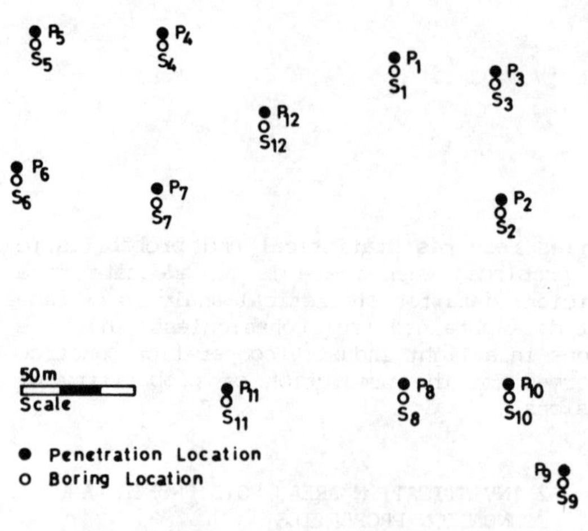

Fig. 1 Penetration and boring locations

Dynamic penetrometer used is of German Heavy Type (German standards DIN 4094, product of Gebr. LINDENMEYER GMBH & CO.), with hammer weight 0.50 kN, height of fall 0.5 m, penetrometer cone diameter 4.27×10^{-2} m, (section: 1.5×10^{-3} m^2), apex angle 90°, driving rod diameter 3.2×10^{-2} m, and driving rate 30 blows/minute. Dynamic penetrations are recorded as the number of blows required to advance 0.20 m.

Allowable bearing pressures for shallow foundations on cohesionless soils are estimated using the dynamic resistance R_d by the following, Dutch formula (Sanglerat (1972)),

$$R_d = M^2H/[A e (M + P)] \qquad (1)$$

Where, R_d = dynamic resistance, in kN/m^2; e = penetration per blow, in m; M = weight of the hammer, in kN; H = height of fall of the hammer, in m; P = weight of the pile, in kN, A = cross-sectional area of the pile, in m^2.

Dynamic soil resistances are calculated by the above equation, using the dynamic penetration numbers obtained from field tests. Dynamic soil resistances are given in Table 1. It is seen from Table 1 that the minimum, maximum values and the range of dynamic soil resistances are 600 kN/m^2, 16000 kN/m^2 and 15400 kN/m^2 respectively. These values are a good indication of relatively large variation in resistances.

Table 1. Dynamic soil resistances, R_d ($\times 10^2$ kN/m^2)

Depth (m)	Penetration location											
	P1	P2	P3	P4	P5	P6	P7	P8	P9	P10	P11	P12
2.0-2.2	13	13	19	13	25	32	13	19	13	13	25	19
2.2-2.4	32	13	19	19	25	25	19	45	19	19	25	25
2.4-2.6	57	32	19	19	19	6	32	45	19	19	19	57
2.6-2.8	83	38	19	19	13	6	32	19	19	19	38	32
2.8-3.0	102	45	19	38	6	19	25	70	19	77	45	64
3.0-3.2	95	53	30	83	6	18	41	71	24	83	18	77
3.2-3.4	124	47	47	119	12	18	41	65	30	77	53	59
3.4-3.6	142	77	41	119	30	24	41	65	30	65	47	47
3.6-3.8	124	89	36	119	59	36	36	89	36	30	59	47
3.8-4.0	89	71	36	89	24	71	36	83	24	18	59	36
4.0-4.2	50	55	44	67	39	72	55	83	33	72	67	55
4.2-4.4	72	83	50	50	39	94	67	55	22	89	33	72
4.4-4.6	83	83	72	100	50	105	78	28	28	89	55	78
4.6-4.8	94	100	83	128	55	67	83	94	39	100	50	78
4.8-5.0	100	83	105	139	44	33	78	28	67	89	61	67
5.0-5.2	115	89	99	125	31	26	78	31	63	104	83	89
5.2-5.4	109	83	104	125	26	21	57	31	52	57	83	63
5.4-5.6	99	94	94	125	26	31	78	57	47	68	63	63
5.6-5.8	125	99	115	120	16	42	94	57	52	78	73	68
5.8-6.0	125	89	115	125	36	26	78	31	52	16	63	68
6.0-6.2	133	93	123	108	69	118	54	29	84	25	69	64
6.2-6.4	123	113	113	108	54	128	88	29	59	39	84	79
6.4-6.6	128	108	84	113	59	64	108	34	49	34	88	88
6.6-6.8	88	128	79	133	54	34	39	34	59	29	98	93
6.8-7.0	20	103	69	147	49	34	79	54	49	29	108	93
7.0-7.2	56	74	74	144	65	42	70	33	60	37	112	112
7.2-7.4	93	84	107	116	56	51	112	33	70	51	98	79
7.4-7.6	60	79	70	112	56	88	102	70	93	70	107	42
7.6-7.8	33	65	56	102	60	84	88	107	79	74	107	33
7.8-8.0	70	88	60	93	56	88	74	107	74	70	107	23
8.0-8.2	110	106	106	93	57	75	71	115	71	75	106	35
8.2-8.4	102	115	128	115	75	79	75	146	66	79	115	44
8.4-8.6	79	110	110	115	66	79	79	160	53	110	93	53
8.6-8.8	57	115	106	110	62	57	71	160	53	106	57	79
8.8-9.0	53	115	71	110	75	57	62	160	79	128	71	88
9.0-9.2	59	105	55	115	71	63	42	160	67	97	59	67
9.2-9.4	67	109	63	101	67	63	42	160	55	76	55	59
9.4-9.6	71	84	71	80	67	59	34	130	63	63	55	42
9.6-9.8	84	84	71	67	71	63	42	84	71	76	59	46
9.8-10.0	101	80	88	55	71	59	42	84	88	63	63	29

4 STATISTICAL AND PROBABILISTIC EVALUATIONS

For the bearing capacity determination, Terzaghi and Peck (1967) and Meyerhof (1948) suggest that soil properties between foundation level and a depth of approximately equal to foundation width (B) below this level should be considered.

In the following analysis, dynamic resistances of the soil layer between foundation level and (B) below this level is considered for each foundation widths.

Frequency polygons and "or more" percentage cumulative distributions of dynamic soil resistance are plotted in Fig. 2 and Fig. 3 respectively, for different foundation widths (Hence, for different soil thicknesses below foundation level).

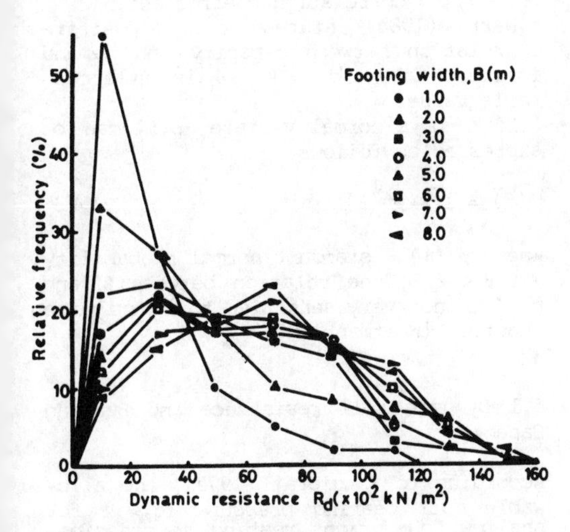

Fig. 2 Frequency polygons of dynamic resistance (R_d)

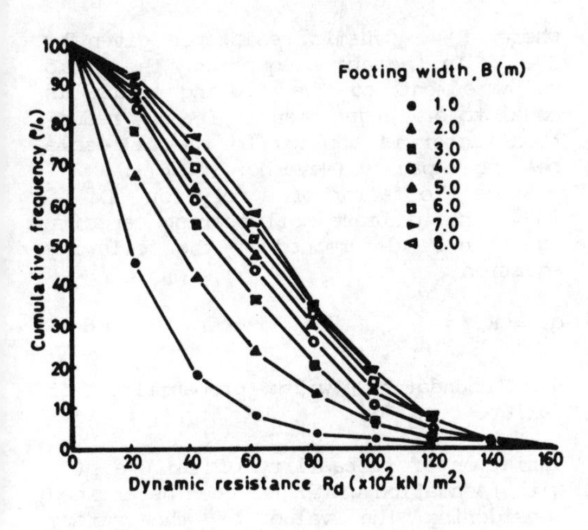

Fig. 3 Or more percentage cumulative distribution of dynamic resistance (R_d)

Distribution parameters are calculated and given in Table 2.

Table 2. Distribution parameters of dynamic soil resistance (R_d) for the depth intervals equal to footing widths.

Footing width, B(m)	1.0	2.0	3.0	4.0	5.0	6.0	7.0	8.0
Depth interval (m)	2-3	2-4	2-5	2-6	2-7	2-8	2-9	2-10
Ave. dyn. res. R_d (kN/m^2)	2852	4164	5072	5617	6042	6306	6714	6779
Standart dev. $\sigma(R_d)$ (kN/m^2)	1927	2912	3066	3259	3399	3325	3395	3305
Coef. of variation $V(R_d)$	0.68	0.70	0.60	0.58	0.56	0.53	0.51	0.59
Coef. of skewness β (1)	1.80	1.30	0.70	0.52	0.43	0.31	0.26	0.31
Coef. of kurtosis β (2)	6.18	4.26	2.81	2.36	2.21	2.17	2.30	2.54
Type of prob. distribution	I(J)	I(J)	I(Ω)	I(Ω)	I(Ω)	I(Ω)	I(Ω)	I(Ω)

Probability distribution curves are selected, using space of probability distributions given by Pearson and Hartley (1972), and presented in Table 2. First two distributions, for 1 m and 2 m foundation widths, are of reverse J shaped Type I (J), the remaining are of bounded from bottom and top and skewed to the right shaped Type I (Ω).

5 SAFETY FACTOR AND RELIABILITY

In the design of shallow foundations, the factor of safety (F) with respect to shear failure is defined as follows,

$$F = q_f / q_{ad}$$

where, q_f = ultimate bearing capacity, q_{ad} = allowable foundation pressure.

In general, bearing capacity and foundation pressure have uncertainties due to variability of soil properties, testing errors, variability in loading, analytical models and their assumptions. to account for the uncertainties in the components of the factor of safety, reliability theory is used. Basic elements of this theory for geotechnical engineering applications are given in Whitman (1984) and Harr (1984).

In civil engineering applications, because of its generality, capacity-demand concept is used in comparing the estimated resistance of the system to that of the imposed loading. The bearing capacity of the soil is the capacity and the column loads or the foundation pressure is the demand (Harr (1984)).

In this writing, ultimate soil bearing capacity q_f is used as capacity C and allowable foundation pressure q_{ad} is used as demand D. That is,

$$q_f = C \quad \text{and} \quad q_{ad} = D$$

Because of randomness in their nature, ultimate bearing capacity (capacity, C) and allowable foundation pressure (demand, D) possess probability distribution functions (Fig. 4).

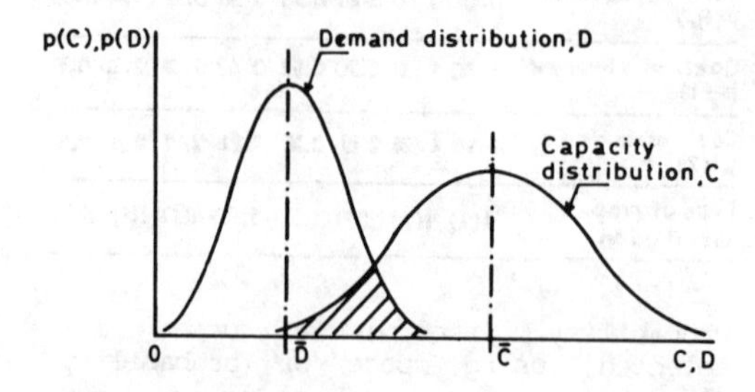

Fig. 4 Capacity-demand model

When probability distribution functions of capacity and demand overlap as shown in Fig. 4, a safety margin function (S), which is also a random variable, can be introduced as follows (Fig. 5),

$$S = C - D \qquad (2)$$

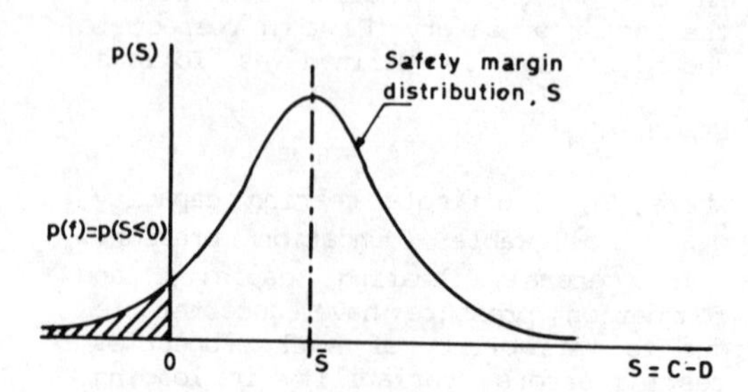

Fig. 5 Safety margin

The probability of safety margin function being less than or equal to zero is the probability of failure p(f),

$$p(f) = P[S \leq 0]$$

Reciprocal of the coefficient of variation of the safety margin is called reliability index,

$$\beta = 1/V(S) = \bar{S}/\sigma(S)$$

where, $\bar{S}$ = mean value, σ (S) = standard devitaion and V(S) = coefficient of variation of safety margin.

Reliability index can also be written in the following form (Harr (1984)),

$$\beta = (\bar{C}-\bar{D})/[\sigma(C)^2 + \sigma(D)^2 - 2\rho\sigma(C)\,\sigma(D)]^{1/2} \qquad (3)$$

where $\bar{C}$, $\bar{D}$ = mean, σ(C), σ(D) = standard deviation of capacity and demand respectively; ρ = correlation coefficient.

Harr (1984) states that, positive correlation between capacity and demand is reasonable and ρ =+ 3/4 is an acceptable value.

If β is a normal variate, p(f) can be expressed as follows,

$$p(f) = 1 - \emptyset(\beta) \qquad (4)$$

where ∅ (β) = standart normal probability. For β < 2.5, the relation between β and p(f) is not very sensitive to distribution function (Whitman (1984)).

5.1 Dynamic soil resistance and Bearing Capacity

According to Sanglerat (1972), the allowable soil bearing pressure (q_{ad}) for shallow foundations resting on cohesionless soils, with a factor of safety of 4, can be obtained by the following relation,

$$q_{ad} = R_d/20 \qquad (5)$$

where, R_d = dynamic resistance given by Eq. 1. In the above equation, the ratio of embedment to the footing width is equal to or larger than 1. If not, reduction factor is applied to the allowable bearing capacity (Meyerhof (1956)).

Since the factor of safety for Eq. 5 is 4, the ultimate soil bearing capacity (q_f) can be determined by the following equation,

$$q_f = R_d/5 \qquad (6)$$

5.2 Foundation width-probability of failure

Variation of probability of failure p(f) with foundation width has been determined, considering the value of the safety factor as 4. Results of computations are given in Table 3.

Table 3. Distribution parameters of dynamic resistance, capacity,demandand safety margin and probabilities of failure.

Footing width B(m)	1.0	2.0	3.0	4.0	5.0	6.0	7.0	8.0
Dynamic resistance								
$\bar{R}_d$ kN/m^2	2852	4164	5072	5617	6042	6306	6714	6779
$\sigma(R_d)$ kN/m^2	1927	2912	3066	3259	3399	3325	3395	3305
$V(R_d)$	0.68	0.70	0.60	0.58	0.56	0.53	0.51	0.49
Capacity								
$\bar{C}$ kN/m^2	570	833	1014	1123	1208	1261	1343	1356
$\sigma(C)$ kN/m^2	385	582	613	652	680	665	679	661
$V(C)$	0.68	0.70	0.60	0.58	0.56	0.53	0.51	0.49
Demand								
$\bar{D}$ kN/m^2	143	208	254	281	302	315	269	339
$\sigma(D)$ kN/m^2	72	104	127	141	151	158	135	170
$V(D)$	0.50	0.50	0.50	0.50	0.50	0.50	0.50	0.50
Safety margin								
$\bar{S}$ kN/m^2	427	625	760	842	906	946	1074	1017
$\sigma(S)$ kN/m^2	334	509	525	554	575	556	585	545
$V(S)$	0.78	0.81	0.69	0.66	0.64	0.59	0.54	0.54
Normal distribution								
β	1.278	1.228	1.448	1.520	1.576	1.701	1.836	1.866
$p(f)$	0.101	0.110	0.074	0.064	0.058	0.045	0.033	0.031
Beta distribution								
S_{min} kN/m^2	-575	-902	-815	-820	-819	-722	-681	-618
S_{max} kN/m^2	1429	2152	2335	2504	2631	2614	2829	2652
α	2.96	2.96	2.96	2.96	2.96	2.96	2.96	2.96
β	2.96	2.96	2.96	2.96	2.96	2.96	2.96	2.96
$p(f)$	0.10	0.11	0.06	0.05	0.045	0.033	0.026	0.024

In this table, distribution parameters of dynamic soil resistance (R_d) are taken from Table 2. According to Eq.6, C and $\sigma(C)$ are one fifth of R and $\sigma(R_d)$ respectively. V(C) is the same as $V(R_d)$. Since a central factor of safety of 4 is adopted between capacity and demand, $\bar{D}$ is one fourth of $\bar{C}$. V(D) does not have to be the same as V(C), because they have different probability distribution functions (Fig. 4). Here, V(D) is assumed to be 0.50. $\bar{S}$ is determined using Eq. 2. β and p(f) are found by Eq. 3 and Eq. 4 respectively. Denominator of Eq. 3 is $\sigma(S)$. considering β as a beta variate, p(f) values are determined using percentage point graphs in Harr (1984) and are given in Table 3. Minimum and maximum values of safety margin (S_{min}, S_{max}) are assumed as follows,

$$S_{min} = S - 3\sigma(S) \quad \text{and} \quad S_{max} = S + 3\sigma(S).$$

Probabilities of failure obtained by assuming β as a beta variate, do not differ very much from the ones obtained by assuming β as normal variate. Variation of probability of failure p(f) with foundation width (B) is given in Fig. 6 Probability of failure increases with decreasing foundation width.

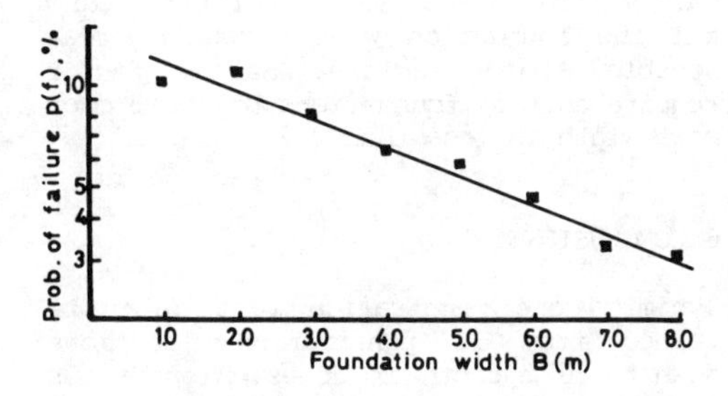

Fig. 6 Foundation width vs probability of failure

5.3 Foundation width-factor of safety-probability of failure

Eq. 3 can be arranged as follows,

$$[1 - \beta^2 V(D)^2]\,\bar{D}^2 + 2[\beta^2 \rho\,\sigma(C) V(D) - \bar{C}]\,\bar{D} + [C^2 - \beta^2 \sigma(C)^2] = 0$$

Assuming V(D) = 50 % and ρ = + 0.75, for different β values, corresponding D values can be obtained. β is assumed as normal variate and for two different values of probability of failure, namely, p(f) = 5 % and p(f) = 10 %, $\bar{D}$ values and hence corresponding factor of safeties are calculated. Results are shown in Fig. 7.

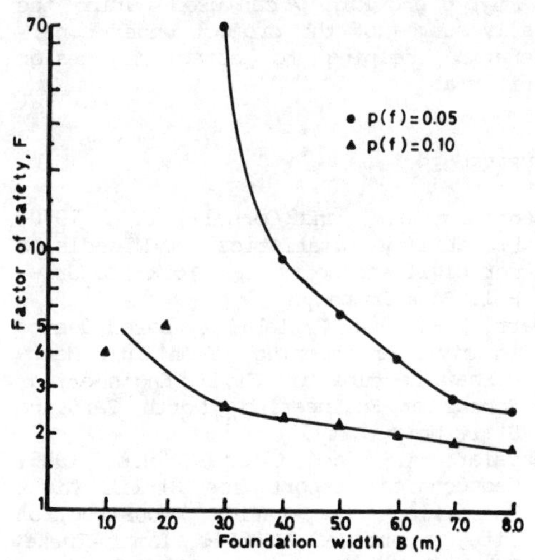

Fig. 7 Foundation width vs.factor of safety

For p(f) = 5 %, required factor of safety increases with decreasing foundation width very rapidly and foundation widths less than 4 meters are not practically applicable because of very high safety factor requirements. For p(f) = 10 %, all the foundation widths require reasonable safety factors. Safety factor requirements of foundations having 3 m or more width is less than 2.5.

6 CONCLUSIONS

Dynamic cone penetration test data obtained from 12 penetration locations down to 10 m depth in cohesionless soils has been analysed statistically. Reliability methods are applied to obtain variations of probability of failure with factor of safety and foundation width. For a safety factor of 4, which has been adopted as a design value in usual practice with dynamic penetrometers failure probabilities have been found to be in the range of 11.5 % - 3.0 % for 1 - 8 m foundation widths. Probability of failure decreases with increasing footing width.

Considering a probability of failure of 0.05 which is not a relatively small value, unusually large factor of safety requirements are met for foundations having a width of less than 4 m.

For a probability of failure of 0.10, which is relatively large probability, factor of safety requirements for all foundations have been found reasonable, being less than 5. For the foundations having 3 m or less width, safety factor requirement is less than 2.5.

Fig. 6 and Fig. 7 can used during the early stages of the project under consideration helping to establish design criteria.

REFERENCES

Benjamin, J.R. and Cornell, C.A. 1970. Probability, statistics, and decision for civil engineers. New York: Mc Graw-Hill Book Company.

Harr, M.E. 1984. Reliability-based desing in civil engineering, Twentieth Henry M.Shaw Lecture in Civil Engineering, School of Engineering, North Carolina State University.

Kayalar, A.Ş. and Ülküdaş, M.E. 1985. Geotechnical report for Birlik Light Industrial Cooperation construction site, Report No. GEO85P10, İzmir-Turkey Dokuz Eylül Univ., Faculty of Engrg. and Arch., Civil Engineering Department (In Turkish).

Meyerhof, G.G. 1948. An investigation of the bearing capacity of shallow footings on dry sand. Proceedings of the Second International Conference on Soil Mechanics and Foundation Engineering, Rotterdam, Vol. 1: 237.

Meyerhof, G.G. 1956. Penetration tests and bearing capacity of cohesionless soils. Soil Mech. Found. Div. Am. Soc. Civil Eng., 82 (1): 866.

Pearson, E.S. and Hartley, H.O. 1972. Biometrika tables for statisticians, Vol. II, Cambridge University Press.

Sanglerat, G. 1972. The penetrometer and soil exploration. Amsterdam: Elsevier Publishing Company.

Terzaghı, K. and Peck, R.B. 1967. Soil mechanics in engineering practice. New York: John Wiley and Sons.

Whitman, R.V. 1984. Evaluating calculated risk in geotechnical engineering. Journal of Geotechnical Engineering Division, ASCE, 110,2: 145-188.

Penetration Testing 1988, ISOPT-1, De Ruiter (ed.)
© 1988 Balkema, Rotterdam, ISBN 90 6191 801 4

Application of the dynamic cone penetrometer (DCP) to light pavement design

E.G.Kleyn & G.D.Van Zyl
Roads Branch, Transvaal Provincial Administration, Pretoria, RSA

ABSTRACT: The introduction of the portable Dynamic Cone Penetrometer (DCP) to road pavement evaluation, reported at ESOPT II in 1982, initiated the translation of sophisticated road pavement behaviour philosophy into more practical roadbuilding terms. The concept of Pavement Strength-Balance evolved from the DCP Layer-Strength Diagrams and concerns the natural law regarding the progression of the in situ strength profile which governs the behaviour of a structure such as a road pavement. The ability to evaluate the strength-balance of a pavement substantially enhances the potential of the traditional structure number design principle. Thus, correlation of the Pavement DCP Structure Number with the bearing capacity and performance of road pavements in practice as well as under Heavy Vehicle Simulator (HVS) testing, resulted in a very potent yet practical pavement design methodology. The paper discusses and illustrates the mentioned concepts as well as the methodology of evaluating in situ conditions and designing light road pavement structures to suit.

1 INTRODUCTION

Recent years have seen a pronounced increase in the understanding of pavement behaviour - assisted mainly by the introduction of full scale testing apparatus such as the HVS (Freeme, 1984), but certainly also because of improved portable soil surveying equipment like the Dynamic Cone Penetrometer (DCP) shown in Figure 1. (Kindermans 1977, Kleyn 1975 and 1982, Van Vuuren 1969)

Previously light pavements were generally not considered important enough to warrant 'sophisticated' design procedures and were relegated to 'rule of thumb designs'. However, the hypotheses regarding traffic moulding, pavement strength-balance and associated load sensitivity may now be applied with greater ease when designing relatively light pavement structures through implimentation of the HVS verified DCP Structure Number design model.

2 SUMMARY OF RESEARCH FINDINGS

The relevant research findings discussed in the references, will only be summarized in this paper.

2.1 Pavement strength-balance

It is a well known phenomenon that a pavement may at some stage during it's life cycle (commonly right at the onset thereof) start deforming noticeably under the traffic loading, only to stop doing so after a while and perform quite satisfactorily for the rest of it's life, albeit at a lower riding quality. Investigating this phenomenon by direct in situ shear strength measurements, using the DCP, led to the formulation of the hypothesis regarding traffic moulding and the pavement strength-balance curves, (Figure 2), which were subsequently substantiated by the HVS research program (Freeme 1984 and Kleyn 1985).

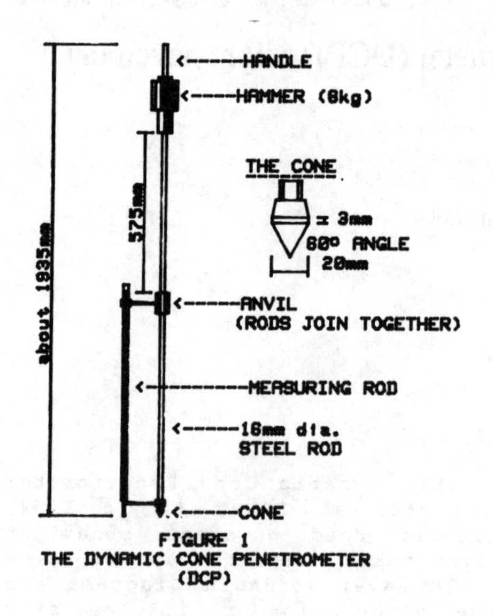

FIGURE 1
THE DYNAMIC CONE PENETROMETER
(DCP)

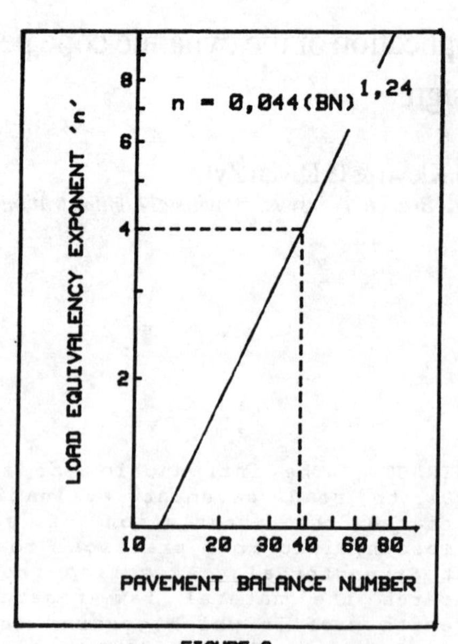

FIGURE 3
RELATION BETWEEN 'n' AND
PAVEMENT BALANCE NUMBER
(BN)

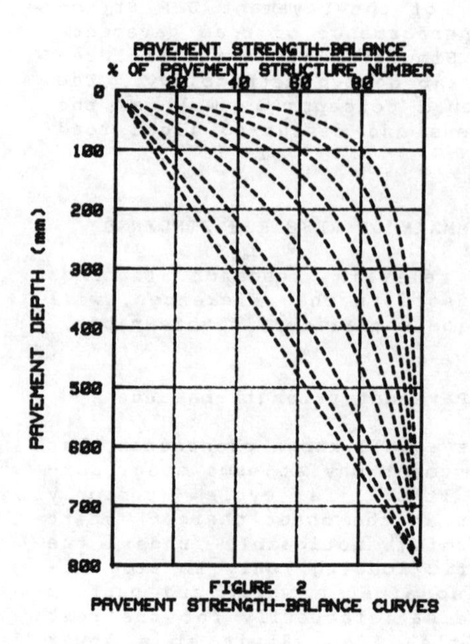

FIGURE 2
PAVEMENT STRENGTH-BALANCE CURVES

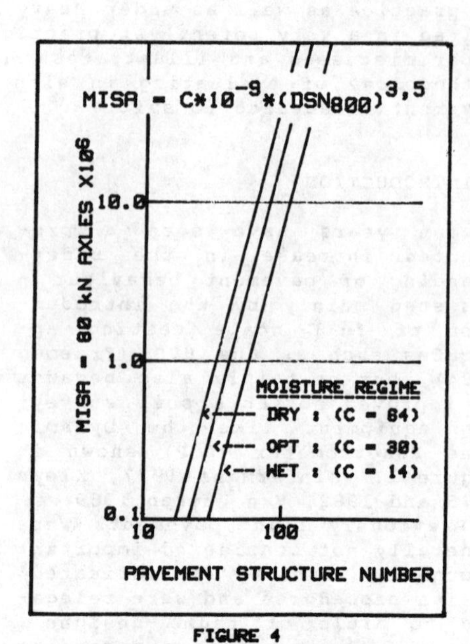

FIGURE 4
RELATION BETWEEN NUMBER OF
80 kN AXLES AND PAVEMENT
STRUCTURE NUMBER (DSN_{800})

What essentially came to light was that an overstressed pavement endeavours to achieve a state of stress equilibrium throughout the structure. In it's endeavour to achieve this 'balanced' state, the pavement, which normally is constructed in distinct layers, tends to be moulded by the load into a strength-balanced structure. The curves in Figure 2 illustrate the abovementioned concept of controlled reduction in the structural capacity of a pavement with depth, for various pavement compositions ranging from 'deep pavements' to 'shallow pavements'.

A deep pavement may be defined as a structure in which the strength diminishes relatively slowly with increasing depth whilst a shallow pavement may be defined as a structure in which the strength diminishes fairly rapidly with increasing depth. The concept of pavement strength-balance thus confirms what has all along been believed by pavement engineers - that it is possible to obtain the necessary bearing capacity by a virtually unlimited number of pavement compositions.

Note that, whereas the shear strength of a material measured by DCP is expressed in terms of mm/blow, called the DCP Number (DN), the structural value of a distinct layer of material is expressed in terms of hammer blow counts to penetrate the layer, and called the DCP Structure Number (DSN) for the particular layer. Thus the total structural value of a pavement is the summation of all the discrete layer DSN's (hammer blows to penetrate the entire pavement) and called the Pavement DCP Structure Number (DSNp). Since it has locally become customary to consider a material depth of 800mm for general purposes when evaluating a pavement, this Pavement DCP Structure Number(DSN_{800}) has been correlated with the bearing capacity of the pavement in terms of 80kN axle repetitions (Kleyn 1983 and 1984).

As a means of identifying the various balance curves and thus also the degree of 'shallowness' or 'deepness' of the pavement, a parameter called the Balance Number (BN) was defined as being the percentage of the pavement strength (DSNp) within the top 12,5% of the pavement. Thus a pavement with a BN = 40 means that 40% of the total pavement strength is within the top 12,5% of the pavement (or top 100mm for a material depth of 800mm). In the Transvaal it is found that pavements generally tend to have Balance Numbers within the range of 35 to 45 and pavements with a BN below that range are considered relatively deeper, whilst pavements with a BN above that range are considered relatively shallower.

The DCP Structure Number concept in combination with the pavement strengh-balance hypothesis forms a formidable evaluation or design tool and largely reduces the possibility of unjudicious pavement composition.

2.2 Load sensitivity

Experience has shown that apart from the behaviour of variously composed pavements differing under normal conditions, their reaction to overloading may differ startlingly and in many cases seem to bear no resemblance to the 'strength' of the pavement - leading to the understanding that some pavement compositions are just more load sensitive than others.

Viewing the pavement strength-balance concept, the hypothesis was put forward (Kleyn 1982 and 1984) that a shallower pavement should be more load sensitive than a deeper pavement - if likened to simply supported beam.In other words an overload may 'snap the back' of a shallow pavement much easier than it would a deeper pavement of similar strength or pavement structure number (DSNp).

In theory this would imply that the exponent 'n' in the well known load equivalency formula $F=(P/SA)^n$ (where P represents the specific axle load considered and SA represents the standard legal maximum axle load - 80kN in Southern Africa) is proportionate to the

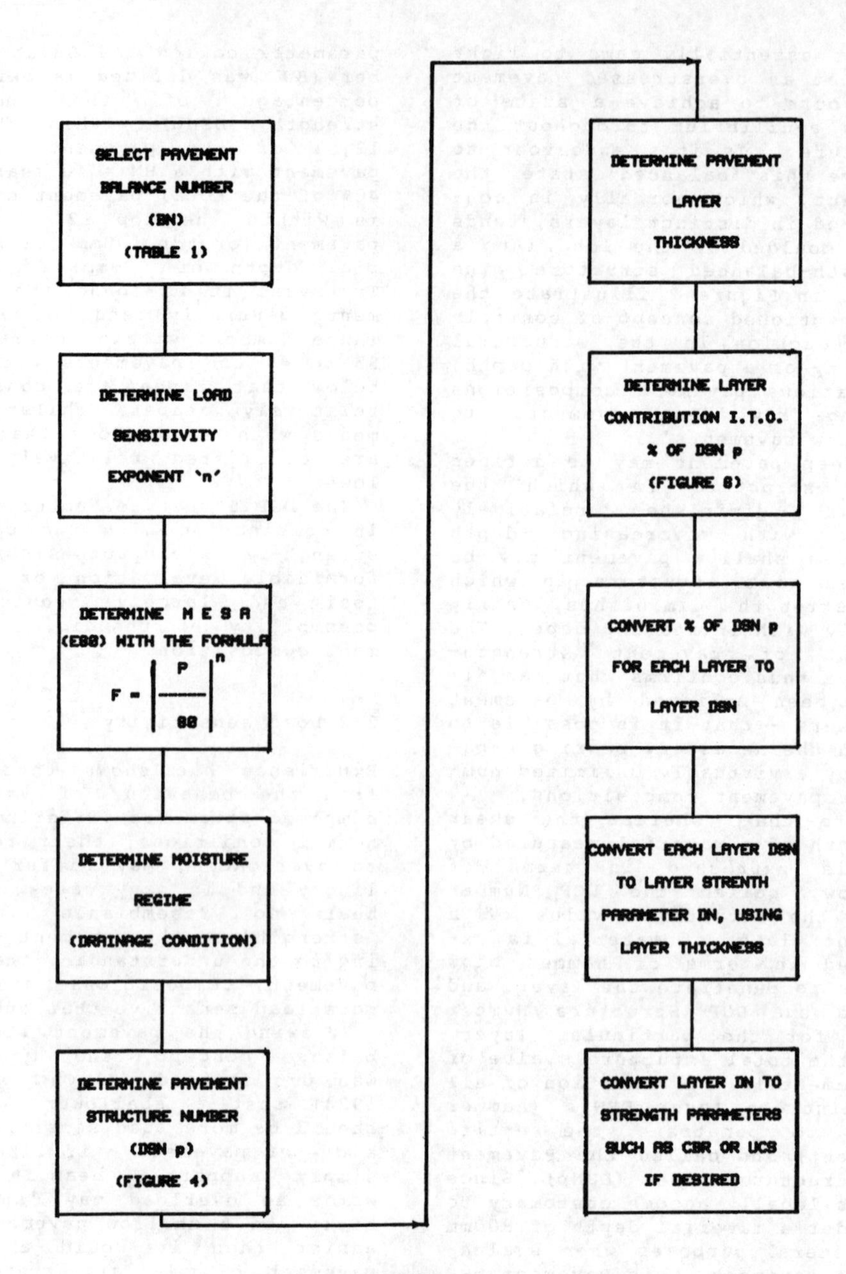

FIGURE 5

FLOW DIAGRAM OF DCP PAVEMENT DESIGN MODEL

'depth' or BN of the particular pavement. This hypothesis was once again evaluated and validated by the HVS research program and, since this forms part of the current research program, an interim relationship between BN and exponent 'n' was drawn up (Figure 3).

2.3 Pavement bearing capacity

Traditionally the bearing capacity of a pavement is given in terms of equivalent standard axle repetitions (ESA) and thus expressed as E80's.However, with the aforegoing hypothesis regarding load sensitivity in mind it is advantageous to express the bearing capacity of a pavement in terms of standard axle repetitions (SA) which by definition is not affected by the magnitude of the exponent 'n' or load sensitivity of the pavement. The traffic load on the other hand is converted to the appropriate equivalent standard axle repetitions (ESA) to correspond with the applicable n-value.

Thus, when talking in terms of millions of repetitions, as is normally the case, the bearing capacity of a pavement may be expressed in terms of MISA and the traffic load in terms of MESA. Figure 4 illustrates the correlation obtained between Pavement DCP Structure Number (DSNp) and pavement bearing capacity (MISA) as verified by HVS tests for various prevailing operational moisture conditions and total pavement depth of 800mm (Kleyn 1983 and 1984). The terminal condition around which this correlation was developed was taken as 20mm permanent deformation.

Observe that a partricular traffic spectrum may constitute totally different traffic loads to different pavements and may change throughout the life cycle of any one pavement due to environmental conditions and/or traffic moulding having taking place. This suggests that it would behove to dray up a traffic load sensitivity diagram assist in deciding which pavement composition to utilize.

3 APPLICATION OF RESEARCH FINDINGS

In accordance with the aim to utilize the in situ material as best possible, the design of light pavements may be divided into two main actions, namely:

a) Determine the design strength profile needed.

b) Integrate the design strength profile optimally with the in situ strength profile.

It goes without saying, that the other considerations normally associated with pavement design such as availability of funds, plant, personnel, etc., still enter into the design considerations.

3.1 Pavement design

Assuming that a balanced pavement composition as previously discussed is desired , the DCP Structure Number design model (Kleyn 1983 and 1984) may be used to obtain the required composition following the flow diagram in Figure 5.

Since the strength-balance of a pavement affects its load sensitivity and thus the traffic load (MESA), the Balance Number (BN) for the pavement to be utilized should first be decided upon. Selection of the most appropriate type of Balance Number in general depends on a combination of factors (par.2.2) as summarized in Table 1 and depending on experience, may warrant a traffic load sensitivity analysis as per paragraph 2.3. It is recommended that a pavement structure of average depth (BN=40) be selected to start off with and that this composition be amended if necessary to suit in situ conditions.

Having decided on the type of pavement Balance Number to be employed the design E80's constituted by the traffic may be calculated for the chosen design period, utilizing Figure 3 and applying the standard load equivalency formula (paragraph 2.2) to each axle load (or read off the load sensitivity analysis). The design E80's may then be entered into the graph

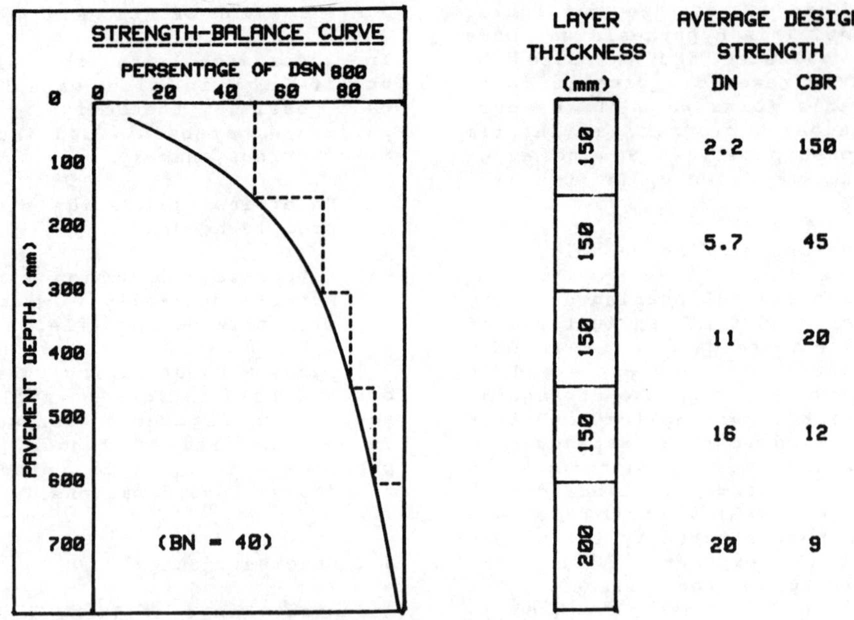

STRENGTH-BALANCE CURVE
PERSENTAGE OF DSN800
0 20 40 60 80
0 100 200 300 400 500 600 700
PAVEMENT DEPTH (mm)
(BN = 40)
LAYER THICKNESS (mm)
AVERAGE DESIGN STRENGTH
DN CBR
150 2.2 150
150 5.7 45
150 11 20
150 16 12
200 20 9

FIGURE 6
PAVEMENT DESIGN FOR 0,55 X 10^6 80kN AXLES (DSN_{800} = 120)

1000
300
100
30
10
3
1
.1 .3 1.0 3.0 10 30 100
CBR (%) and UCS (kPa x 100)
UCS
CBR
DN (mm/blow)

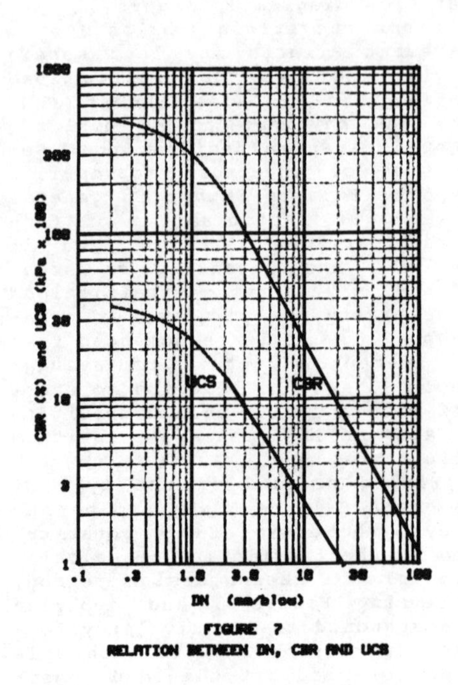

FIGURE 7
RELATION BETWEEN DN, CBR AND UCS

LAYER STRENGTH DIAGRAM
DN (mm/blow)
1 3 10 30
0 200 400 600 800
PENETRATION DEPTH (mm)
MIN.
AVE.
MAX.
CBR %
300 100 30 10 3

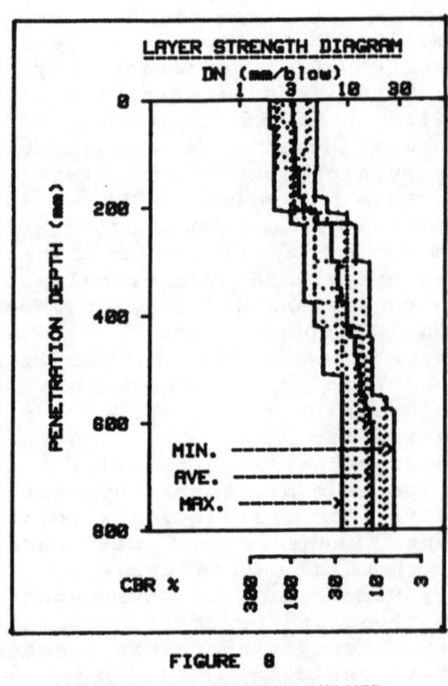

FIGURE 8
AVERAGE AND EXTREME VALUES

in Figure 4 to obtain the appropriate bearing capacity in terms of the Pavement DCP Structure Number (DSNp) for the pavement, depending on the moisture regime (drainage conditions) envisaged. Obviously this method results in an average value for the pavement load sensitivity and bearing capacity. However, if so desired, seasonal variability may be allowed for, depending on experience of the material to be utilized.

Table 1. Selection of BN

Aspect	Pavement Balance Type	
	Deeper (BN→12½)	Shallower (BN→80)
Traffic	Many overloads exp.	Few overloads exp.
Pavement support	Rel. good	Rel. poor
Pavement thickness	Unrestricted space	Restricted space
Material cost	Expensive base rel. to support layers	Expensive support layers rel. to base

The average 'in situ' strength (not 'soaked' value) of the various pavement layers may be obtained through use of the appropriate balance curve in Figure 2 after having decided on the layer thicknesses - which also depend on economy and practical limitations. (Conversely, the layer thickness corresponding to a specific material strength may be obtained in similar manner). This may be achieved in a number of ways, one of which is to move a vertical line, the length of which corresponds to the thickness of the layer in question, across Figure 2 at the appropriate depth until the bottom of the simulated layer intersects the chosen balance curve (see Figure 6). The percentage of the Pavement DCP Structure Number (DSNp) corresponding to this position minus that for the previous layer constitutes the structural contribution the layer in question makes to the total structural value of the pavement expressed as a percentage.

This figure may then be converted to the DCP Structure Number (DSN) for the layer and, through use of the chosen layer thickness, to average layer strength DN and % CBR (or UCS for cemented material) as desired through application of Figure 7. Note that the strength of the base course should not be allowed to drop below the tyre contact pressure of the heavy vehicles in the traffic spectrum as a precaution against shear failure in the base course.

Figure 6 shows an example of a pavement design for a bearing capacity of $0,55 \times 10^6$ 80 kN axles based on a BN value of 40 and DSNp = 120. Table 2 shows the strenght profiles for light pavements applied by the Transvaal Roads Branch (TRB 1984), compiled in similar manner. (The model has also been used successfully to design heavier pavements).

Table 2. Strength profiles for light pavement structures.

		Pavement bearing capacity		
$E80 \times 10^6$:		0,003-0,010	-0,030	-0,100
Class:		A	B	C
Base (150mm)	DN:	< 8	< 5	< 4
	CBR:	> 30	> 50	> 70
Subbase (150mm)	DN:	< 19	< 14	< 9
	CBR:	> 10	> 15	> 25
U/Select (150mm)	DN:	< 33	< 25	< 19
	CBR:	> 5	> 7	> 10
L/Select (150mm)	DN:	< 48	< 33	< 25
	CBR:	> 3	> 5	> 7

The DN and CBR values in Table 2 are in situ values for an optimum pavement moisture regime and BN=40.

Thus a design strength profile is obtained for the pavement structure which will satisfy the design traffic requirements and

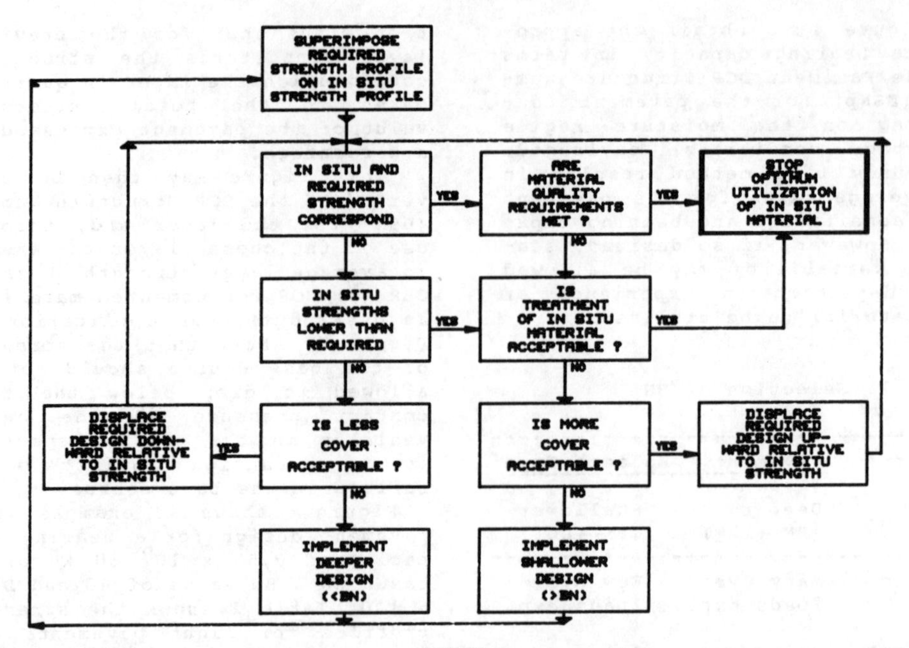

FIGURE 9
OPTIMIZING UTILIZATION OF IN SITU MATERIAL

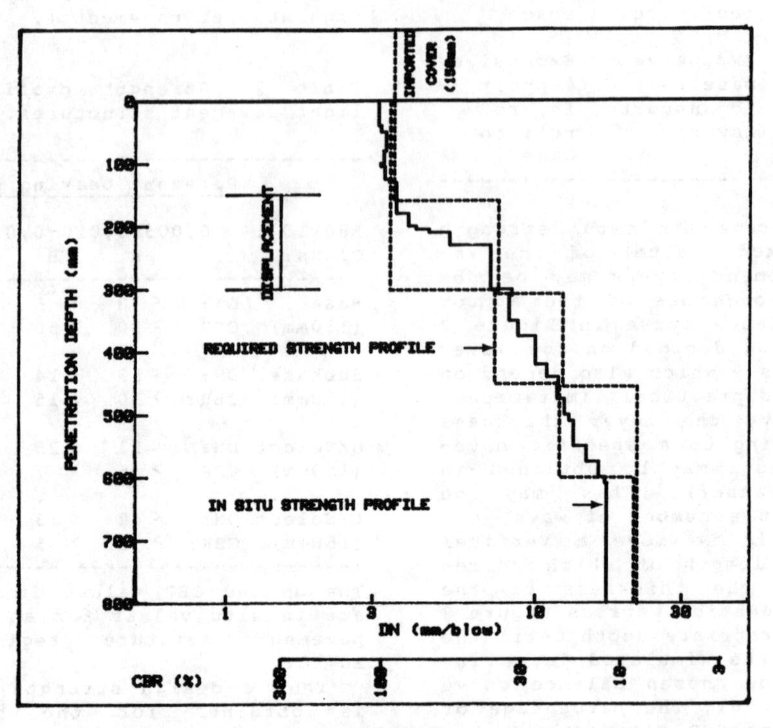

FIGURE 10
REQUIRED STRENGTH PROFILE SUPERIMPOSED ON IN SITU STRENGTH PROFILE

which may be superimposed on the in situ strength profile to evaluate utilization of the in situ material as described below.

3.2 Utilizing in situ material

In order that the design strength profile of the pavement, as determined in paragraph 3.1, may be integrated optimally with the in situ strength profile, it is necessary to conduct a DCP survey in addition to the survey normally done to obtain the inherent quality of the in situ material. This may be done concurrently.

The DCP information of the in situ material is collectively drawn on the Layer Strength Diagram (Kleyn 1982 and 1983) to obtain an in situ strength profile as shown in Figure 8. The minimum and maximum (or any other percentile) strength profile boundaries and hence the average or most probable in situ strength profile may be determined for each set of measurements as a further aid.

Keep in mind that the in situ strength profile which is to be evaluated against the design strength profile is sensitive to the ratio between the DCP 'survey' moisture regime and the anticipated 'service' moisture regime of the pavement - the Moisture Regime Ratio (MRR). Naturally it is preferable to do the DCP survey when the moisture regime of the existing material coincides with the anticipated in service moisture regime of the pavement. However when this is not possible reference may be made to Table 3 which suggests the range of the minimum in situ strength profile to be used, depending, inter alia, on the moisture sensitivity of the in situ materials.

Note that sections having widely divergent in situ readings should not be grouped together, since the in situ strength profile may become unrealistically biased (either too low or too high) for that section. This, of course, is modified by practical aspects such as available contruction equipment, techniques and costs.

The in situ strength profile thus obtained may be used in conjunction with the design strength profile as determined in paragraph 3.1. to optimize the utilization of the in situ material. The flow diagram illustrating the optimization process is given in Figure 9. Bear in mind that this process aims at optimizing only the utilization of the in situ material strength and quality and that other practical and economical implications also enter into the final decision making process.

Table 3: Percentile of minimum in situ strength profile to be used

MRR= Survey MR / Service MR	Percentile of minimum strength profile	
	Low moist. sensitive material	High moist. sensitive material
<1	50 - 75	> 75
≈1	25 - 50	50 - 75
>1	< 25	25 - 50

Figure 10 illustrates a typical situation where the design strength profile is superimposed on the in situ strength profile. In this case it is apparent that the in situ strengths at most depths are much less than required and that if, for instance, no treatment of the in situ material is allowed but additional cover on top of the existing material is allowed the design strength profile should be displaced upwards relative to the in situ strength profile. A displacement of one layer thickness (150mm in this case) results in all strength requirements being met by the in situ strength profile and if the material quality requirements are met this option may be accepted, resulting in 150 mm of material to be important on top of the existing material. Other options may be evaluated similarly.

Thus a method is obtained whereby the problem of determining what part of the in situ material may be utilized in the design pavement may now be approached with more confidence using the DCP.

4 CONCLUSIONS

It has been demonstrated that, contrary to the general drive for 'bigger and better', the gain in understanding of pavement behaviour over the last years may be applied equally cost effectively to the field of light pavements. In this respect the Dynamic Cone Penetrometer (DCP) proved very useful in translating the relatively sophisticated hypotheses regarding pavement behaviour, resulting from the Heavy Vehicle Simulator (HVS) research programme, into practical road building terms.

A guideline for the composition of light pavements was also given in the form of a catalogue, which was drawn up using the hypotheses regarding traffic moulding and pavement strength-balance. In this regard the use of the DCP Structure Number design model was also demonstrated as well as the procedure for the utilization of the in situ material.

5 ACKNOWLEDGEMENT

The authors wish to acknowledge the Executive Director, Roads for permission to publish this paper.

REFERENCES

FREEME, C.R. 1984. Heavy Vehicle Simulator Testing And Pavement Management Performance Predictive Models. ATC '84, Pretoria, South Africa.

KINDERMANS, J.M. 1977. De Lichte Slagsonde In De Wegenbouw- Informatiestudie En Proefnemingen. Researchverslag RV5/76. Opzoekingscentrum voor de Wegenbouw, Brussels.

KLEYN, E.G. 1975. The Use Of The Dynamic Cone Penetrometer (DCP). Report L2/74, Transvaal Roads Branch, Pretoria, South Africa.

KLEYN, E.G, MAREE J.H. van SAVAGE, P.F. 1982. Application Of A Portable Pavement Dynamic Cone Penetrometer To Determine In Situ Bearing Properties Of Road Pavement Layers And Subgrades In South Africa. Esopt II, Amsterdam.

KLEYN, E.G. and SAVAGE, P.F. 1982. The Application Of The Pavement DCP To Determine The Bearing Properties And Performance Of Road Pavements. International Symposium On Bearing Capacity Of Roads And Airfields. Trondheim, Norway.

KLEYN, E.G. and VAN HEERDEN, M.J.J. 1983. Using DCP Soundings To Optimise Pavement Rehabilitation. ATC '83, Johannesburg, South Africa.

KLEYN, E.G. 1984. Aspekte Van Plaveiselevaluering En -Ontwerp Soos Bepaal Met Behulp Van Die Dinamiese Kegelpenetrometer (Aspects Of Pavement Evaluation And Design As Determined With The Aid Of The Dynamic Cone Penetrometer) (MEng thesis). Report L6/84, TPA Roads Branch, Pretoria,South Africa.

KLEYN, E.G. FREEME, C. R. and TERBLANCHE, L.J. 1985. The Impact Of Heavy Vehicle Simulator Testing In The Transvaal. ATC '85, Pretoria, South Africa.

KLEYN, E.G. MAREE, J H and TERBLANCHE, L.J. 1986, Towards A Strategy For The Cost-Effective Allocation Of Available Road Pavement Funds. ATC 1986, Pretoria, South Africa.

SMITH, R.B. and PRATT D.N. 1983. A Field Study Of In Situ California Bearing Ratio And Dynamic Cone Penetrometer Testing For Road Subgrade Investigations. Australian Road Research 13(4), December 1983.

TRB. 1984. Beleid En Riglyne Ten Opsigte Van Die Betering Van Ligte Plaveisels. (Policy And Directives Regarding The Surfacing Of Light Pavements). Manual L02/84, Transvaal Roads Branch, Pretoria, South Africa.

VAN VUUREN, D.J. 1969.Rapid Determination Of CBR With The Portable Dynamic Cone Penetrometer, The Rodesian Engineer.

Penetration Testing 1988, ISOPT-1, De Ruiter (ed.)
© 1988 Balkema, Rotterdam, ISBN 90 6191 801 4

The relationship between in-situ CBR test and various penetration tests

Moshe Livneh & Ilan Ishai
Transportation Research Institute, Technion, Haifa, Israel

ABSTRACT: This paper describes four methods of penetration testing for the evaluation of CBR values of subgrades and flexible pavement structures. They are the Dynamic Cone Penetrometer (DCP), the American Airfield Penetrometer, the Standard Penetration Test (SPT) and the Dynamic Probing Type A Test (DPA). The correlations of each test with the CBR test are given. In addition to the discrete strength values obtained by these tests, it is possible also to evaluate the strength distribution with depth, namely the thicknesses of the different layers in the existing pavement. The DPA and the DCP methods do not require the opening of test pits on boring but only the extrusion of 100 mm diameter asphalt cones which is quite fast and cheap. On the other hand, the SPT requires the full drilling; however, this action is also much easier than the actions involved in opening and closing of test pits on drilling, especially when thick airport pavements (up to 1500 mm) are involved. Finally, it should be pointed out that three out of the four methods discussed (the DCP, SPT and DPA) have recently been successfully used for an actual pavement evaluation work in the main runway of the major Israeli international airport - Ben Gurion.

1 INTRODUCTION

The accelerated development which occurred in different western countries after World War II has created the fact that at the present time there is a significant decrease in pavement construction of transportation infrastructure systems (both highways and airports). Therefore, the technical center of gravity has recently been shifted from the development of new systems into the maintenance and rehabilitation of existing ones. This shift was quite essential also due to the fact that the rate of deterioration in performance of the existing transportation system was more rapid than the flow of national resources to maintain these systems at an acceptable level of service. This is true even in the most developed western countries.

These two basic facts have led the different responsible officials to be quite interested in the area of pavement maintenance and rehabilitation. The unequivocal conclusion reached after extensive studies of the subject was that the efficiency of the different maintenance and rehabilitation actions is primarily related to the existence of a Pavement Management System (PMS).

As is well known, one of the most important parameters in PMS is both the functional and structural capacity of the network. For this, the modern pavement evaluation methods usually involve the measurement of pavement response by a non-destructive testing (NDT) loading device (FWD, Road Rater, Dynaflect, etc.), and the computation of pavement structure moduli by a proper mechanistic model. Due to the variability of existing loading methods and mechanistic models, the variability of the structure moduli of a given pavement tested by different NDT methods and models is often very high and sometimes beyond possibility of comparative accuracy. Therefore, in a reliable pavement evaluation process, the NDT prediction is usually correlated and calibrated with physical measurement of individual material characteristics in pavement and subgrade layers. This has usually been done by a supplementary destructive testing of pavement and subgrade materials, using test pits and running in-situ CBR tests. Back calculations are carried out in order to evaluate the tested pavement strength.

Fig. 1: Performing the DCP test in a major runway in Israel.

Fig. 2: Performing the American Airfield Penetrometer test.

Fig. 3: Performing an SPT test in a major runway in Israel.

Fig. 4: The shape of the DPA penetrometer.

It is common knowledge that the execution of destructive testing is quite a cumbersome, slow and expensive process that cannot satisfy the PMS. Therefore, in many parts of the world, the paving technologists were searching for an alternative in different penetration testing. The attraction of these testings stems from their capacity to determine the soil's strength profile or the layer strength profile of existing roads, and this at a low testing cost. Moreover, the advantage of these tests is in their capacity to determine the strength and thickness of layers in existing roads and runways without boring or digging into them, except for core drilling in the asphalt layers only. In parallel, the advantage of these instruments is in determining the strength profile of natural soils also without boring or digging.

Therefore, the purpose of this paper is to present the use of several penetration tests for the evaluation of existing pavements and subgrades through the prediction and determination of in-situ CBR values.

2 THE VARIOUS PENETRATION TESTS

The various penetration tests used to determine the in-situ CBR values of pavement structures and subgrades are: the Dynamic Cone Penetrometer (DCP), the American Penetrometer, the Standard Penetration Test (SPT), and the Dynamic Probing Type A Test (DPA).

The DCP test (see Figure 1) has been widely described by Livneh (1987a, 1987b); therefore, only the summary of the findings is presented in this paper.

The American Penetrometer (see Figure 2) is used by military civil engineers to evaluate the bearing capacity of unsurfaced runways for aircraft operations. Therefore, for this test, a correlation between the measurement values (Airfield Index) and the CBR, was developed. In this paper, the Israeli experience in this correlation is presented.

The SPT (see Figure 3) has recently been used in Israel for bearing capacity evaluation of pavement structures which possess high thickness and strength values and at which the DCP is not suitable. The SPT at this new application is described by Livneh and Ishai (1987), but since the writing of that paper, new information has been collected which requires the specific quantitative correlation given by that paper. The corrected correlative equation is also given in this paper.

Finally, the DPA test has recently been used in Israel for evaluating the strength of different types of subgrades. Its application in pavement evaluation will also be discussed.

3 THE DYNAMIC CONE PENETROMETER (DCP) TEST

The Dynamic Cone Penetrometer (DCP) is currently used by several agencies in the world for determining the CBR values of existing pavements and subgrades. In Israel, its use for routine site investigation and site evaluation projects commenced about five years ago, following a comprehensive correlative study between the DCP values and the in-situ CBR values. This correlative study, which involved both laboratory and in-situ testing, is fully described by Livneh (1987a, 1987b).

The attained correlation for both cohesive and non-cohesive granular soils is:

$$\log \mathrm{CBR} = 2.20 - 0.71\,(\log \mathrm{DCP})^{1.5} \pm 0.075 \qquad R^2 = 0.96 \qquad N = 74 \qquad (1)$$

where:

the DCP is the ratio between the penetration (in mm) and the number of blows.

An international comparison involving 21 correlative equations has been made by Livneh (1987a, 1987b). It is concluded there that the use of the Israeli correlation is preferable for routine works. The other main conclusion from the abovementioned study is that the coefficient of variation of a CBR for any particular material is considerably higher than that of the cone penetrometer. Thus, it may be concluded that conducting dynamic penetrometer tests is preferable to the in-situ CBR. This remark is of great significance, as higher reliable values are gained with a simple tool which is much easier to operate on site, as it does not require any contra-weight.

Finally, the experience gained from routine projects has indicated the powerful capacity inherent in the cone penetrometer for detecting the strengths of the various pavement layers together with their subgrade. Thus, this tool can be

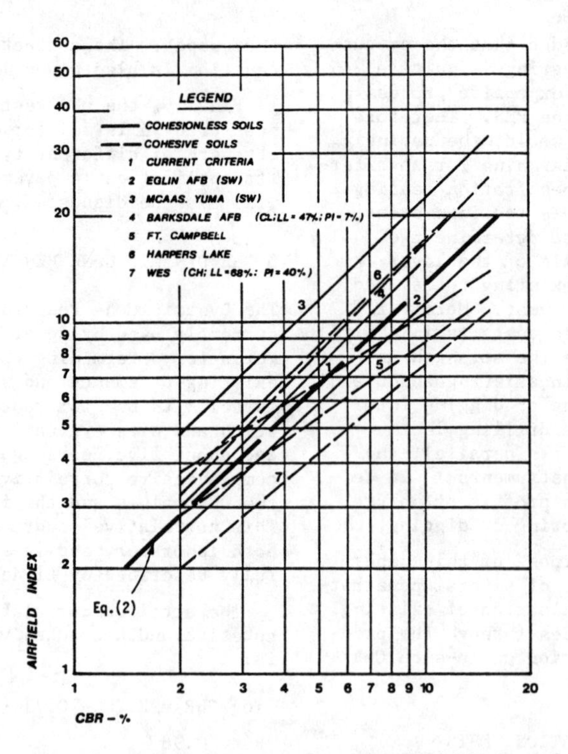

Fig. 5: CBR vs the American Airfield Index of different soils, after Turnage and Brown (1974).

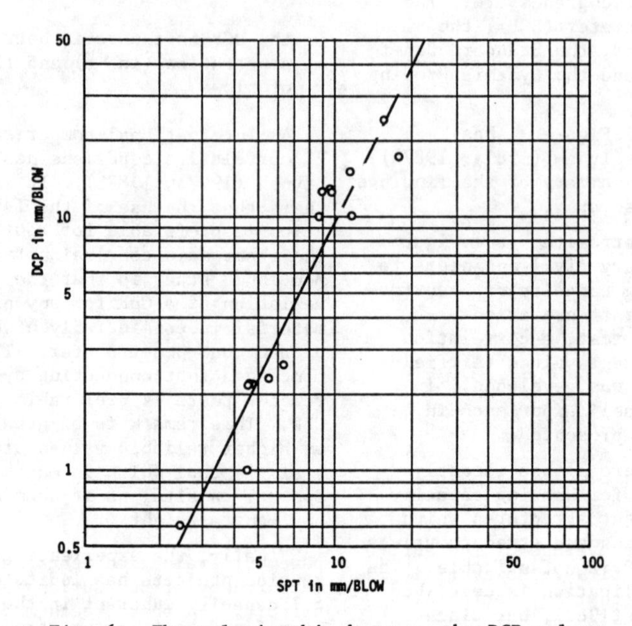

Fig. 6: The relationship between the DCP values and the SPT values.

used successfully for determining the various layer thicknesses and strength without digging any test pits in the pavement structure. The evaluation method based on the DCP apparatus is described by Livneh and Ishai (1987).

4 THE AMERICAN PENETROMETER TEST

The American penetrometer test, or in its other name - the "Airfield Cone Penetrometer," is a testing device which enables the determination of the soil strength index for both military and civil purposes. This device was developed by the American Corps of Engineers (Fenwick 1965), and it can be a supporting device to determine the bearing capacity of landing strips, trafficability of traffic paths, as well as for quality control for earth work and embankment construction. The device is compact and easy to use, also by people who do not have earlier experience in the determination of soil strength.

The device itself (see Figure 2) is constructed with a cone having an angle of 30° and a bore with a diameter of ½ inch (an area of 0.196 sq.in.). The cone is mounted at the end of a scaled rod where, at the other end, there is a spring, a force meter, and a handle. The reading in the force meter, which is also called the Airfield Index (AI), represents the resistance of the soil to penetration. Each reading is equivalent to a 10 lb. resistance. The maximum reading possible in this device is 15, i.e., a soil resistance of 150 lbs.

The research work on the correlation between the AI and the CBR is reported by L.C.I. (1985). According to this work, which includes tests on both cohensive and granular soils, the following relationship exists:

$$\log CBR = -0.22 + 1.10 \log AI + 0.13$$
$$R^2 = 0.83 \qquad (2)$$
$$N = 22$$

This relationship is graphically expressed in Figure 5. The figure also presents other correlations developed elsewhere. It can be seen that the Israeli relationship is close to the average of the entire range of the correlations. However, it is important to stress that the variability of the results in the above relationship is higher than that of the DCP (compare the average standard error of Equations (1) and (2)). This reduces, of course, the reliability in the determination of CBR values by the American penetrometer. Moreover, this device is limited to AI values up to 15, which is equivalent only to CBR values not higher than 12.

5 THE STANDARD PENETRATION TEST (SPT)

The well-known SPT is used here to determine the relation between the resistance to penetration and the CBR value of the soil medium. The resistance to penetration is expressed by the SPT parameter, which is the ratio between the penetration depth (in mm) and the number of blows, usually the following:

$$SPT = \frac{305}{N_s} \qquad (3)$$

where N_s is the number of standard blows to a penetration depth of 305 mm.

A preliminary correlation between the SPT and CBR is given by Livneh and Ishai (1987). This correlation is based on tests carried out on soil with only intermediate strength values (up to a CBR of 30%). A more recent work, which also includes stronger materials (a CBR of 100% and above), leads to the relationship given in Equation (4), (see Figure 6).

$$\log DCP = -1.05 + 2.03 \log SPT$$
$$R^2 = 0.93 \qquad (4)$$
$$N = 15$$

This equation is valid for $SPT < 10$ mm/blow. For $SPT > 10$ mm/blow, the old equation derived by Livneh and Ishai (1987) should be used.

The above correlation between the SPT and the DCP was used to express the CBR value with the aid of Equation (1), as follows:

$$\log CBR = 2.20 - 2.10 [\log(0.309\ SPT)]^{1.5} \qquad (5)$$

when $SPT < 10$ mm/blow.

The verification and control of the above correlation was recently made on pavement evaluation tests performed on the main runway of Ben Gurion international airport. In these tests made on granular base and subbase course materials, a comparison was made between CBR values obtained by the SPT tests (using Equation (5)) and CBR values obtained by direct

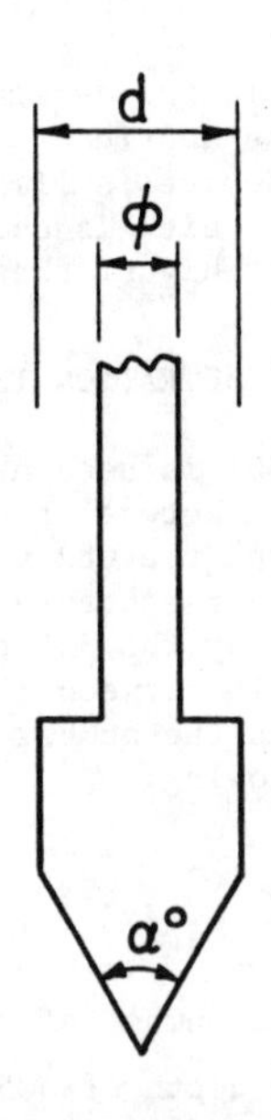

Fig. 7: The configuration of the cone.

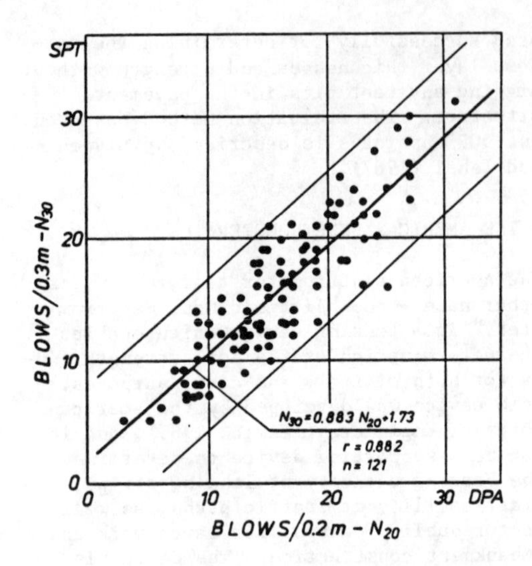

Fig. 8: Relationship between results of sounding by probe type SPT and probe type DPA, after Borowczyk and Frankowski (1981)

Fig. 9: Performing an in-situ CBR test in a test pit.

measurements in test pits performed several years earlier. The compliance of the results was quite satisfactory.

6 THE DYNAMIC PROBING TYPE A TEST (DPA)

The DPA test is a similar one to the DCP in its continuous recording of penetration vs. blows, in contrast, for example, to the SPT. In this test, the dimensions of the components and weight of the rummer are quite different. Table 1 presents the comparison of the two penetrometers (see also Figure 7).

Table 1. Comparison between the DCP and the DPA devices

Test	Falling Height (H),cm	Rummer Weight (W),kg	Angle of Cone Head (α),°	Cone Base Dia. (d),mm	Road Dia. (ϕ),mm
DCP	57.5	8.0	90	20	16
DPA	75.0	63.5	30	62	32

The detailed description of the DPA test is given by Bergdahl (1979) (see also Figure 4).

The resistance to penetration in this device is expressed by the DPA parameter, which is the ratio between the penetration depth (in mm) and the number of blows from that measured depth. Usually, it can be expressed as follows:

$$DPA = \frac{200}{N_s} \quad (6)$$

where N_s is the number of standard blows for a penetration depth of 200 mm.

The value of the relative soil resistance (R) in this device can be determined by the following expression, as presented by Durgunoglu and Mitchell (1974) and Kramer (1980):

$$R = \frac{WH}{d^{\beta} DPA \cdot \eta} \quad (7)$$

where:

η - a parameter depending on the cone angle (α) as given by Table 2;

β - a parameter varied between 1.40 and 2.30, depending on soil strength;

and all other parameters are defined in Table 1.

From Equation (7) and the values of Table 1, the following equation can relate the DCP and the DPA values:

$$\frac{DCP}{DPA} = \left(\frac{8.0}{63.5}\right)\left(\frac{57.5}{75.0}\right)\left(\frac{62}{70}\right)^{\beta} \cdot \left(\frac{27}{14}\right) = 0.186 \times 3.1^{\beta} \quad (8)$$

The preliminary correlation obtained in a comparative work between the DPA and DCP values was performed in a clayey site. The correlative relationship obtained is as follows:

Table 2. Values for different cone angles

α^o	180	90	60	30
α/2	90	45	30	15
η	70	27	20	14

Table 3. Comparison of DPA/SPT values according to Equations (13) and (14)

DPA	DPA/SPT Eq. (13)	DPA/SPT Eq. (14)
1	0.19	0.58
5	0.53	0.61
10	0.83	0.64
15	1.05	0.66

$$\log DCP = 0.40 + 0.74 \log DPA \quad (9)$$

or

$$\frac{DCP}{DPA} = \frac{2.51}{DPA^{0.26}} \quad (10)$$

hence

$$\beta = 2.30 - 0.47 \log DPA \quad (11)$$

Thus, when the soil is weak (i.e., DPA value is high), the value of β is in the lower part of the range (for example, if DPA = 80 mm/blow, then β = 1.40) and if the soil is strong (i.e., DPA = 1 mm/blow), the value of β is in the upper part of the range (β = 2.30).

It is quite clear that Equation (9) presents only a preliminary correlation. The combination of this equation with Equation (1) led to the following relationship:

$$\log CBR = 2.20 - 0.45\,[\log(3.47 \cdot DPA)]^{1.5} \quad (12)$$

The control of this relationship is made by comparing Equation (12) with Equation (5) as follows:

$$\frac{DPA}{SPT} = 0.19 \cdot DPA^{0.65} \quad (13)$$

This result can be compared with the findings of Borowczyk and Frankowski (1981) shown in Figure 8. The mathematical expression of their finding is as follows:

$$\frac{DPA}{SPT} = 0.58 + \frac{DPA}{176} \qquad (14)$$

According to Table 3, there is a reasonable agreement between the results of the two last equations only at low DPA values (a strong soil) up to a DPA value close to 10. Therefore, the use of Equation (12) should be limited only to DPA values smaller than 10.

7 SUMMARY AND CONCLUSIONS

This paper describes four methods of penetration testing evaluation of CBR values of subgrades and flexible pavement structures. They are the Dynamic Cone Penetrometer (DCP), the American Airfield Penetrometer, the Standard Penetration Test (SPT) and the Dynamic Probing Type A Test (DPA). In addition to the discrete strength values obtained by these tests, it is possible also to evaluate the strength distribution with depth, namely the thicknesses of the different layers in the existing pavement.

The DPA and the DCP methods do not require the opening of test pits on boring, but only the extrusion of 100 mm diameter asphalt cones, which is quite fast and cheap. On the other hand, the SPT requires the full drilling; however, this action is also much easier than the actions involved in opening and closing of test pits on drilling, especially when thick airport pavements (up to 1500 mm) are involved.

At this stage of the correlative work between the above penetration methods and the in-situ CBR values, the following can be summarized:

1. The DCP testing method is more reliable than the direct measurement of in-situ CBR. This provides additional advantage for the method;
2. The SPT method is applicable in cases where the pavement thickness is greater than 800 mm;
3. The AI testing method is not recommended due to its lack of reliability and its incapability to provide the distribution of strength with depth;
4. The DPA method seems to be promising, but its applicability to strong materials (DPA less than 2 mm/blow) should be further investigated in-situ;
5. The applicability of the relationships between DPA and SPT is limited at this stage to DPA values less than 10.

Finally, it should be pointed out that three out of the four methods discussed (the DCP, SPT and SPA) have recently been successfully used for an actual pavement evaluation work in the main runway of the major Israeli international airport - Ben Gurion, instead of performing an in-situ CBR test (see Figure 9).

REFERENCES

Bergdahl, U. 1979. Development of the dynamic probing test method. Design parameters in geotechnical engineering. VII ECSMFE, pp. 201-206, Brighton.

Borowczyk, M. & Frankowski, Z.B. 1981. Dynamic and static sounding results interpretation. X ICSMFE, 2:451-454, Stockholm.

Durgunoglu, H.T. & Mitchell, J.K. 1974. Influence of penetrometer characteristics on static penetration resistance. Proc., European Symp. on Penetration Testing, pp. 133-139, Stockholm.

Fenwick, W.B. 1965. Description and application of airfield cone penetrometer. U.S. Army Engineer Waterway Experimental Station, Corps of Engineers, Instruction Report No. 7.

L.C.I. Transportation Engineers Consultants 1985. Private communication (in Hebrew).

Livneh, M. 1987a. The use of dynamic cone penetrometer for determining the strength of existing pavements and subgrades. Proc., 9th Southeast Asian Geotechnical Conference, Bangkok.

Livneh, M. 1987b. The correlation between dynamic cone penetrometer (DCP) values and CBR values. Transportation Research Institute, Technion-Israel Institute of Technology, Publ. No. 87-065.

Livneh, M. & Ishai, I. 1987. The relationship between SPT and in-situ CBR values for subgrades and pavements in arid zones. Proc., 8th Asian Regional Conf. on Soil Mechanics and Foundation Engrg., Kyoto, Japan.

Livneh, M. & Ishai, I. 1987. Pavement & material evaluation by a dynamic cone penetrometer. Proc., 6th International Conference on Structural Design of Asphalt Pavements, University of Michigan, Ann Arbor.

Kramer, H.J. 1980. Gerateseitige einflussparameter bei ramm und drucksondierungen und ihre auswirkungen auf den eindringwiderstand. Dissertation for degree of Dr.Ing., Karlsruhe University.

Turnage, G.W. & Brown, E.H. 1974. Prediction of aircraft ground performance by evaluation of ground vehicle rut depth. U.S. Army Engineer Waterway Experiment Station, AD-775-744.

Penetration Testing 1988, ISOPT-1, De Ruiter (ed.)

Rotary-pressure sounding: 20 years of experience

Nils O.Rygg
Norwegian Road Research Laboratory (NRRL)
Arild Aa.Andresen
Norwegian Geotechnical Institute (NGI)

SYNOPSIS: The rotary-pressure sounding method was developed by the Norwegian Geotechnical Institute (NGI) and the Norwegian Public Road Administration (NPRA) in 1967. The aim was to design a rational and efficient sounding method, adjusted for Norwegian soil conditions, climate and terrain. The equipment is operated by a multipurpose drilling rig. The equipment consists of a bit extended by rods with flush couplings. It is forced into the ground at a constant rate of penetration (3 m/min.) and at a constant speed of rotation (25 RPM). The thrust necessary to maintain the constant speed of penetration is measured and plotted versus depth. After 20 yeras of experience with this method we may conclude that the aims to a great extent have been achieved. The major part of the Norwegian geotechnical institutions and firms employ rotary-pressure sounding in conventional site investigations.

The method dominates because it requires less work effort, reduces costs and provides more geotechnical information compared to traditional methods. The rotary-pressure sounding results give information about the overall ground conditions, and the method is widely used to locate quick clay deposits.

Rotary-pressure soundings cannot penetrate through coarser materials such as stone and boulders. Modifications towards a "total sounding" method which may penetrate all sorts of soils, are therefore presently underway.

Keywords: Site investigation, sounding, stratification, quick clay, identification, total sounding.

1 INTRODUCTION

In Scandinavia we generally encounter geologically young soils, deposited exclusively during the Quaternary age and covering very hard igneous rock. The soils have as rule not been preloaded.

In Norway the ground conditions change considerably from the higher regions where moraine, sand and gravel deposits are likely to be encountered, to the lower-lying regions with weak deposits of fine-grained soils, especially marine clays which often are quick. Soft clay deposits can be found up to the highest marine limit, more than 200 metres above the existing sea level. It is in these parts of the country that about 90 % of the population lives and that further development in communications, industries, towns, etc. is in great demand. The majority of soil investigations is therefore carried out in regions where the ground mostly consists of soft materials and also where the most challenging foundation conditions are found.

These general soils features in Scandinavia have led to a special "nordic" trend related to soil investigations on-shore. Already in 1922 the Swedish Geotechnical Committee (1) outlined the basis for current practice, first in Sweden and later in Finland and Norway.

An ordinary soil investigation starts with soundings to measure depths to bedrock and to obtain a picture of the relative variations in soil strength over the area. Based on these soundings sites are chosen where it is necessary to obtain further information, depending on the geotechnical problems involved. Soil sampling and vane borings are then performed to determine soil strength and other geotechnical properties. The ratio between number of soundings and sampling/vane borings will in general be of the order of 15:1 when performing investigations for a road project. Traditionally, soundings have been carried out by the Swedish weight sounding method, requiring mostly manual work.

The need to reduce the manual work involved and to provide automatic recording and handling of the sounding log, led to the rotary-pressure sounding method. This method was designed and developed by the Norwegian Geotechnical Insitute (NGI) and Norwegian Public Road Administration from 1967 (2).

The aim has been to design a rational and efficient sounding method, which gives a picture of soil conditions and depth to bedrock.

After 20 years of experience we may say that these aims to a great extent have been achieved.

2 EQUIPMENT AND METHOD

The equipment is operated by a multipurpose drilling rig (Fig. 1). The equipment consists of a bit extended by rods with flush couplings. It is forced into the ground at a constant rate of penetration (3 m/min.) and at a constant speed of rotation (25 RPM). The thrust necessary to maintain the constant speed of penetration is measured and plotted versus depth.

The maximum thrust plotted is 30 kN.

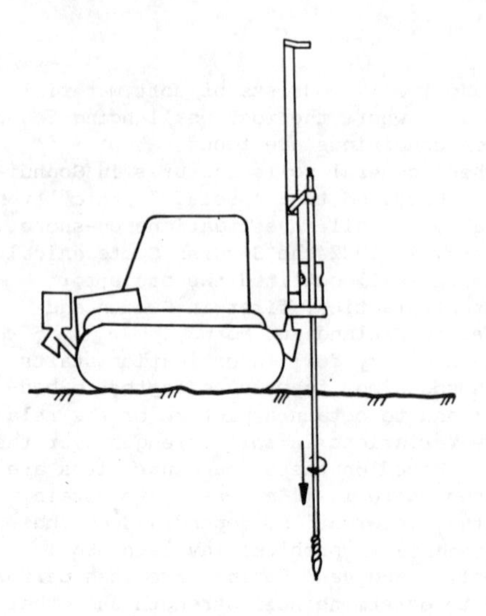

Fig. 1 Multipurpose drilling rig

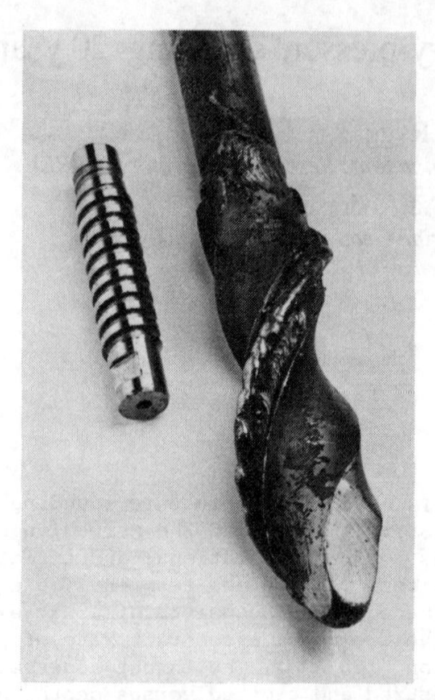

Fig. 2 Twisted flat bit with an eccentric hard welded seam

The original equipment had rods with a diameter of 33.5 mm, a cylindrical bit with a conical tip (120°), a diameter of 40 mm and a length of 150 mm with a "hard-welded" helical seam.

In 1980 the rod diameter was enlarged to 36 mm and the bit changed into a twisted flat steel bit with an eccentric hard welded seam (Fig. 2).

When it is not possible to penetrate at a rate of 3 m/min., the speed of rotation is increased and vertical pumping can be used.

The penetration ability and the sensitivity for recording and detecting soil layers depend on the shape and form of the bit.

The original round bit was more sensitive to soil stratification than the twisted bit. The latter penetrates, however, much better through frozen ground and gravelly soils.

If drilling through stones and boulders are required, rock drilling equipment is necessary. Penetration ability and the sensitivity for recording layers through frozen ground and gravelly soils are similar to those for the original bit.

3 EXPERIENCE

The rotary-pressure sounding method has an increasing range of use. Today nearly all the Norwegian geotechnical institutions and consulting firms use this sounding method. When performing investigations for road and bridge projects, about 70 % of all soundings are performed as rotary-pressure sounding, while the traditional methods, such as weight soundings and dynamic soundings are used less and less. Cone penetration testing (CPT) is used occasionally to evaluate special difficult foundation conditions.

The reasons why the rotary-pressure sounding method dominates are:

- adapted to light weight crawlers
- high production rate
- ability to identify stratification and type of soil

The rotary-pressure sounding method is operated by a hydraulically operated drill rig. For road investigations the rig is carried by a crawler tractor. The drill rig moves easily even on steep slopes and in snow to the location where soundings are to be performed.

Due to rational working operations and registrations of sounding logs, there are a minimum of working efforts necessary.

Rapid change of locations and fast drilling rates are essential to minimize time consumption and thus the costs. Overall cost by rotary-pressure sounding is about half the cost of weight-soundings.

Compared to other simple sounding methods, the rotary-pressure sounding results give more information about the overall ground conditions.

Figures 3 and 4 show examples of sounding logs from rotary-pressure sounding. Based on experience the sounding curves provide a basis to determine soil stratification, types of materials penetrated, and the sensitivity of clay deposits. Boundaries between various materials are easily located, especially when the tip penetrates from clay into granular soils and vice versa.

Rotary-pressure soundings penetrate ordinary Norwegian soil deposits to bedrock. Penetration will however stop in very hard layers and against stones and boulders. It is therefore necessary to modify the method to penetrate such layers as discussed below.

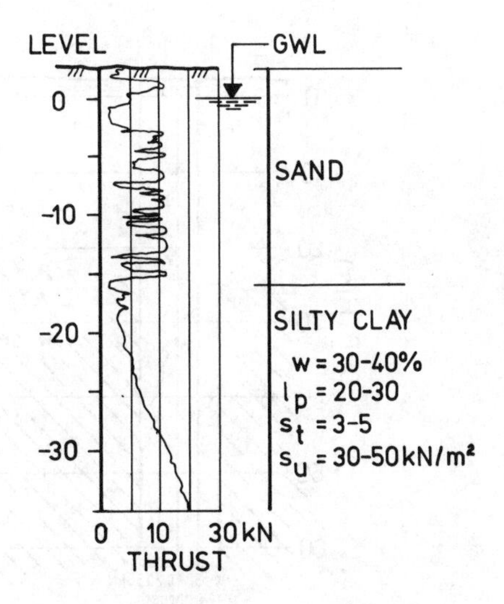

Fig. 3 Rotary-pressure sounding through sand into silty clay

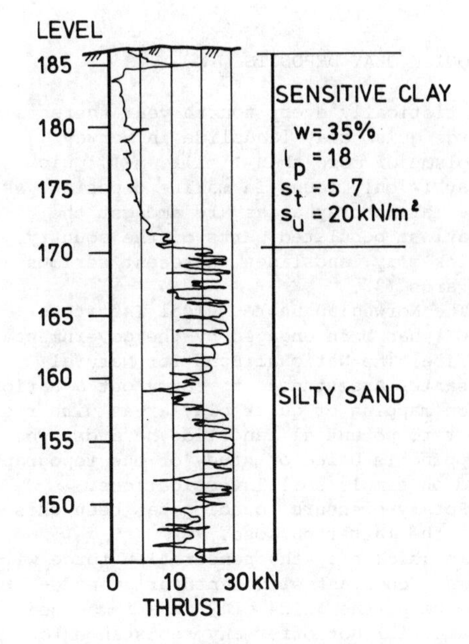

Fig. 4 Rotary-pressure sounding thorugh sensitiv clay into silty clay

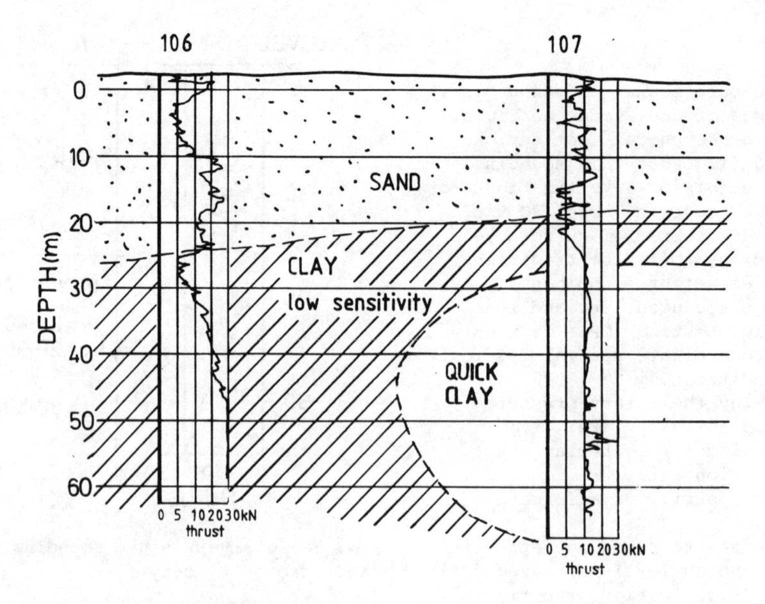

Fig. 5 Rotary-pressure sounding 107 indicates a layer of quick clay below a depth of 28 m

4 QUICK CLAY DEPOSITS

Statistically every fourth year there is a large quick clay landslide in Norway (volume of more than 1 mill. m^3). Quick clay is only found in marine deposits. As the marine clay areas are amongst the heaviest populated parts of the country, quick clay landslides represent serious hazards (3).

The Norwegian Geotechnical Institute (NGI) has been engaged by the governmental office "The National Fund for Natural Disaster Assistance" to carry out a nation-wide mapping of quick clay areas with respect to potential landslide hazards. The mapping is based on study of the topography and on simple soil investigations.

Rotary-pressure sounding has been chosen for the latter purpose.

In quick clay the penetration force will remain constant with depth or even decrease as the quick clay in a remoulded state will not offer any resistance to penetration. One may readily locate quick clay zones at depths under non-sensitive deposits, as illustrated in Fig. 5. In boring 106, the clay is not quick, while in boring 107 there is a quick clay zone from 28 m to a depth of more than 60 m.

The results of the soundings are easily interpreted and therefore the need for supplementary investigation have so far been quite small. Nine out of ten soundings profiles can be interpreted uniquely with respect to quick clay deposits. Penetration resistance through coarser materials overlying the quick clay has now and then been a practical problem. And this have added to other needs for the development of a "total sounding equipment".

5 TOTAL SOUNDING

The aim of "total sounding" (4) is to develop an equipment which will penetrate all sorts of soils, boulders and rock. The equipment now under testing at NGI is a hybrid of rotary-pressure sounding and percussion rock drilling equipment. The combination with rock drilling equipment rods at OD 44 mm is used, in connection with modified rock drill bit (Fig. 6). All these changes have been made to achieve better penetration. A number of drilling parameters are recorded, i.e. rate of penetration, rotation pr. min. and torque together with the thrust measured by rotary-pressure sounding, all versus drilled depth.

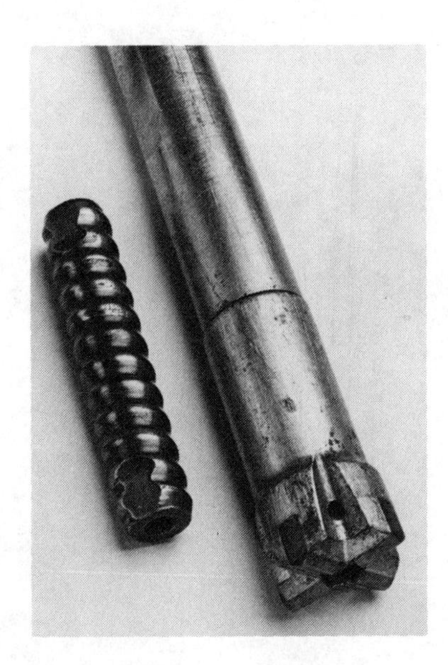

Fig. 6 Modified rock drill bit

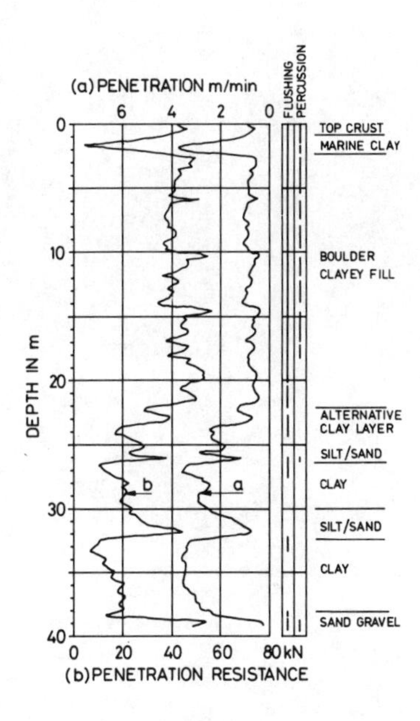

Fig. 7 shows the results of a total sounding at a site where it was impossible to penetrate by rotary-pressure sounding to more than 2-3 m

Percussion rock drilling requires flushing. Percussion on/off and flushing pressure are also recorded.

Drill log data is recorded on tape and the aim is to combine the drilling parameters in such a way that one or two curves will indicate type of soil and of soil resistance.

Figure 7 shows that the requirement to penetration capacity has been achieved through the modifications so far.

Flushing improves penetration but makes it difficult to obtain relevant figures of soil resistance. Flushing is therefor only used during percussion drilling. This includes the development of a patented check valve which will open when flushing is necessary.

REFERENCES

(1) Statens järnvägar, Sverige (1922). Statens järnvägars geotekniska kommission 1914-22; slutbetänkande avgivet till kungl. järnvägstyrelsen den 31. maj 1922. 180 p. Geotekniska meddelanden, 2.

(2) Andresen, A. and Rygg, N. (1975). Rotary-pressure sounding. European Symposium on Penetration Testing Sthm. 1974. Proceedings, Vol. 2:2, p. 15-18.

(3) Gregersen, O. and Løken, T. (1983). Mapping of quick clay landslide hazard in Norway. Criteria and experiences. Symposium on slopes on soft clays. Linköping 1982. Statens Geotekniska Institut, Linköping. Report, 17, pp. 161-174.

(4) Kolstad, P. and Eidsmoen, T. (1987). NGI report 57600-10. Totalsondering. Systembeskrivelse og resultatvurdering.

Penetration Testing 1988, ISOPT-1, De Ruiter (ed.)
© 1988 Balkema, Rotterdam, ISBN 90 6191 801 4

Cone penetrometer and in situ CBR testing of an active clay

R.B.Smith
Department of Main Roads, Parramatta, NSW, Australia

ABSTRACT: There has been a growing world-wide trend to move away from empirical methods of road and airport pavement design based on classification test data to a mechanistic design which requires measurement of the in situ properties of the material. This has meant the increased use of in situ California Bearing Ratio (CBR) testing and dynamic cone penetrometer (DCP) testing. This paper describes a field investigation into the relationship between the two. The results indicate that, for the data presented, there is a significant linear relationship between in situ CBR results and DCP results with the relationship being affected by moisture content and to a lesser extent by relative compaction. It is apparent, at least for this material, that any correlation developed for relating in situ CBR values to DCP results should also include a term for moisture content.

1 INTRODUCTION

There has been a growing world-wide trend to move away from empirical methods of road and airport pavement design based on classification test data to a mechanistic design which requires measurement of the in situ properties of the material. This has meant the increased use of in situ California Bearing Ratio (CBR) testing and dynamic cone penetrometer (DCP) testing.

Scala (1956) developed a dynamic cone penetrometer and carried out a correlation study between the results obtained using his DCP and in situ CBR testing. The dynamic cone penetrometer developed by Scala has a point angle of 30°, a cross-sectional area at its greatest width of $322mm^2$ (1/2 sq. in.) and a drop mass of 9.08kg (20lb) falling 508mm (20 in.). With metrication the various states in Australia adopted slightly different dimensions and tolerances.

A number of Australian authors have derived correlations between results from the Scala DCP and in situ CBR testing (eg Morris 1976; Morris, Potter and Armstrong 1977; Smith 1983; Smith and Pratt 1983). Scala (1956) noted that serious discrepancies were sometimes encountered between results of duplicate tests but this comment was not necessarily heeded by some of the authors.

Scala found that a log/log relationship showed the highest correlation. Smith (1983) found that his own data supported an inverse relationship but when the data were incorporated with those of Scala and reprocessed a log/log relationship was again indicated. For one of the materials investigated by Smith the log/log relationship only yielded a correlation coefficient of 0.28 whilst that for the inverse relationship yielded a correlation of 0.85.

Because the latter investigation was part of a pavement field trial periodic testing has continued since construction and one of the sections has been reconstructed. This has meant that quite an amount of additional information has become available since the papers of Smith (1983) and Smith and Pratt (1983). These new data have been incorporated with their previous data for the analyses provided in this paper.

2 THE MATERIAL

It was decided to restrict the analysis to the "imported" subgrade, a heavy clay, at the Rooty Hill field trial site. The clay component was 30% to 40% of the total sample. Of the clay component, 55% was kaolinite, 45% mixed layer expandable clay, and less than 5% was illite. Classification test data are given in Table 1. Laboratory testing showed that the unsoaked CBR value

Table 1. Properties of the material.

Property	Mean	Standard Deviation
Passing 19.0 mm sieve (%)	95.77	4.39
Passing 9.5 mm sieve (%)	93.40	1.77
Passing 4.75 mm sieve (%)	90.98	2.28
Passing 2.36 mm sieve (%)	88.46	2.43
Passing 425 µm sieve (%)	81.38	1.25
Passing 75 µm sieve (%)	45.47	2.15
Less than 13.5 µm (%)	32.64	2.29
Liquid Limit	38.38	3.59
Plastic Limit	13.82	1.55
Linear Shrinkage (%)	8.37	2.24
Max. Dry Density (t/cu.m)	1.90	0.03
Opt. Moisture Content (%)	13.03	0.96

decreased significantly as the moisture content at moulding increased (with standard compactive effort). Testing was performed 24h after compaction to allow pore pressures to dissipate. At 10.5% moisture content sample 102A had a CBR value of 50% whilst at 19% moisture content the CBR value was around 2%. Thirty samples were tested and each showed a similar trend. This indicates the susceptibility of the material to moisture.

3 THE INVESTIGATION

In the main investigation three in situ CBR tests were carried out at each selected site in a line at 150 mm spacing on the compacted surface of the layer. Moisture contents were taken from the layer 50 mm to 150 mm below the compacted surface at each penetration site.

Three DCP penetrations were made approximately 1 m apart to form a triangle with the in situ CBR test location as the centroid. The number of blows taken to drive the penetrometer 50 mm into the surface was taken as the DCP reading.

For experimental purposes each CBR value was recorded to the nearest 0.1 per cent for all values under 20. All moisture contents were calculated to the nearest 0.1 per cent. As is the custom, DCP readings were calculated to the nearest 0.1 blows/25 mm.

4 RESULTS

4.1 Relationship between CBR and DCP data

There were 43 sets of data available and these were processed using the STATPAK (Northwest Analytical 1984) suite of programs. Both CBR and DCP data were obtained in triplicate. Where one piece of data was missing the mean of the other two was substituted for it. The raw data are given in the Appendix.

A scattergram of the results (mean CBR value versus blows/25mm) is given in Figure 1. The data were initially analysed using simple regression with linear, exponential, logarithmic and power law (log/log) regression equations. The initial analysis was performed for the mean CBR value versus the mean DCP value. Correlations were then performed for mean CBR value versus mean DCP value plus one standard deviation; mean CBR value versus mean DCP value minus one standard deviation; mean CBR value minus one standard deviation versus mean DCP value minus one standard deviation; and mean CBR value minus one standard deviation. Correlation coefficients are presented in Table 2.

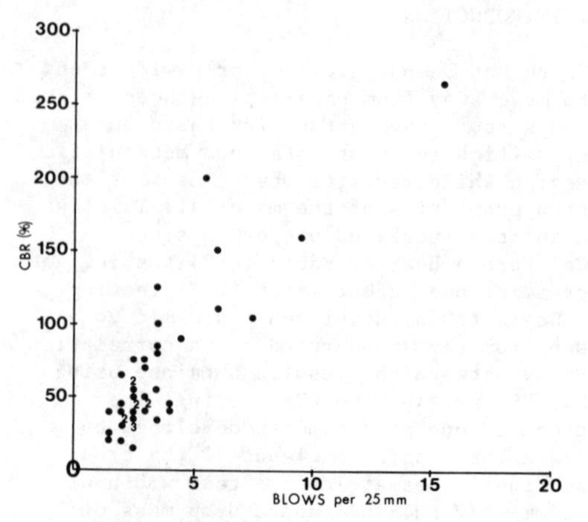

Fig.1 Mean CBR value versus DCP value (all data - n = 43)

As can be seen for all but one of the relationships the linear model yields the best correlation. Even in the case where the linear relationship (r = 0.85) was less than that for the logarithmic relationship (r = 0.88) the difference is only slight. For the data used in this study the appropriate relationship is considered, therefore, to be the linear one. The discussion which follows is restricted to the linear relationship (ie y = A + Bx) with CBR as

Table 2. Correlation coefficients for relationship between in situ CBR value and dynamic cone penetrometer results using various regression models.

Relationship	Correlation Coefficients			
	Linear Model	Exponential Model	Log Model	Power Law Model
Mean CBR v Mean DCP	0.86	0.73	0.84	0.80
Mean CBR v (Mean + sd)DCP	0.85	0.70	0.84	0.80
Mean CBR v (Mean - sd)DCP	0.76	0.68	0.68	0.65
(Mean + sd)CBR v (Mean + sd)DCP	0.85	0.70	0.88	0.82
(Mean - sd)CBR v (Mean - sd)DCP	0.62	0.52	0.53	0.46

the dependent variable and DCP as the independent variable.

It is now proposed to look at the statistics associated with each of the relationships for the linear model. Only the relationships using mean CBR are reported (Table 3).

Whilst the regression coefficients define the relationship between the results the correlation coefficient (r), the standard error of the estimate, and the coefficient of determination (or r^2) are the statistics of most interest in this discussion.

For 43 samples a correlation coefficient of at least 0.39 is required for the relationship to be significant at the p = 0.01 probability level. For each of the relationships reported in Table 2 the correlation coefficient is much greater than this indicating that the relationships are highly significant.

The statistic of particular importance for these data is the standard error of the estimate. Assuming a normal distribution of the results about the regression line the standard error of the estimate provides an estimate of the band within which approximately 68% of the results will lie.

The regression equations, to an appropriate order of accuracy, become:

CBR = [5 + 17.5 (mean DCP)] +/- 25

CBR = [17 + 11 ((mean + sd)DCP)] +/- 27

CBR = [-4 + 27 ((mean - sd)DCP)] +/- 32.5

Table 3. Statistics associated with linear regression relationship between mean in situ CBR and DCP.

Statistic	Relationship		
	Mean CBR v Mean DCP	Mean CBR v (Mean+sd) DCP	Mean CBR v (Mean - sd) DCP
"A" Regression Coefficient	5.01	17.02	-4.26
"B" Regression Coefficient	17.50	11.13	27.31
"A" Standard Error	6.67	6.25	10.47
"B" Standard Error	1.58	1.09	3.60
Standard Error of Estimate	25.19	26.75	32.48
Coefficient of Determination	0.75	0.71	0.58
Covariance	104.15	156.61	51.70
Correlation Coefficient	0.86	0.85	0.76

This means that the error of the estimate of CBR is quite large, and is to be expected if one observes the scatter of data in Figure 1.

The coefficient of determination, (r^2), is a measure of the proportion of variance in one variable "explained" by the other. As can be seen from Table 3 the proportion of variance in in situ CBR explained DCP value varies from 58% for the mean CBR versus mean DCP minus one standard deviation to 75% for the mean CBR versus mean DCP. As a result it was decided to add a further variable to the analysis to see whether the coefficient of determination could be improved.

Whilst ideally both in situ CBR value and DCP value should take complete account of moisture content and density it may not be the case. As the moisture content under the piston was available for all of the data stepwise multiple linear regression was performed on the data with DCP value as the first independent variable and moisture content under the CBR piston as the second independent variable. In this case the moisture content at the cone penetrometer site would have been more appropriate but it was not taken. Nevertheless the moisture

Table 4. Statistics for multiple linear regression of in situ CBR value with DCP value and moisture content.

Term	Coefficient	Standard Error	t-Statistic	Partial Correlation	Controlled R^2
A	110.502	24.928	4.43284	-	-
B	16.651	1.726	9.651	0.694	0.360
C	-9.015	2.103	-4.286	0.309	0.071

	Sum of Squares	Degrees of Freedom	Mean Square
Due to Regression	118636	2	59317.9
About Regression	22331.7	41	544.677
Total	140968	43	3278.32

R^2	0.842	Corrected R^2	0.834
F-Test	108.9	Standard Error of Regression	23.338

Table 5. Statistics for multiple linear regression of in situ CBR value with DCP value and moisture content and relative compaction.

Term	Coefficient	Standard Error	t-Statistic	Partial Correlation	Controlled R^2
A	219.788	271.25	0.81028	-	-
B	40.8056	19.791	2.062	0.515	0.503
C	-18.146	14.874	-1.220	0.271	0.176
D	-1.2465	2.284	-0.546	0.069	0.035

	Sum of Squares	Degrees of Freedom	Mean Square
Due to Regression	1251.81	3	417.269
About Regression	1123.32	4	280.829
Total	2375.12	7	339.303

R^2	0.527	Corrected R^2	0.172
F-Test	1.4858	Standard Error of Regression	16.758

content under the CBR piston is considered to be fairly representative of the moisture content at the test location.

When the CBR and DCP data were analysed using the linear model $z = A + Bx + Cy$, with DCP value and moisture content as the independent variables, the first step of the multiple linear regression yielded a coefficient of determination of 0.77 indicating that 77% of the variance in the CBR value was accounted for by the variance in DCP results. When the second variable (moisture content) was added the coefficient of determination increased to 0.84 indicating that 7% of the variance in CBR can be attributed to the variance in the moisture content. The statistics for the final analysis are given in Table 4.

The results of the multiple linear regression analysis indicate that the regression equation can be improved significantly by the addition of moisture content. The correlation coefficient has increased from 0.86 for the standard linear model to 0.92 for the multiple linear regression model with the addition of the moisture content. The coefficient of determination increased from 0.75 for the standard model to 0.84 for the multiple linear regression model (ie an additional 10% of the variation has been taken account of). The standard error of the estimate is still relatively high. The regression equation becomes:

CBR = [110 +17(mean DCP) - 9(mean mc] +/- 23

where
DCP is given in blows/25mm; and moisture content (mc) is given as a percentage.

The high standard error of the estimate indicates that, even with this refinement, there are dangers in relying on DCP values to determine the in situ CBR value. This is largely due to the inherent variability of the soil and probably due to the fact that it was in a compacted condition prior to testing.

4.2 Relationship between in situ CBR, DCP, moisture content and relative compaction

When one of the sections was reconstructed additional data became available and it was possible to relate the results from in situ CBR, rate of penetration, moisture content and relative compaction. The relative compaction was taken within 1m of the CBR test site and the moisture content refers to that in the top 50mm below the piston. None of the correlations was significant for the small sample size of 8. The linear relationship between CBR and DCP results again gave a higher correlation coefficient (r = 0.35) than the exponential (r = 0.52), logarithmic (0.56) or power law (r = 0.50) relationships. The scattergrams of DCP against moisture content and relative compaction are given in Figure 2 and Figure 3 respectively.

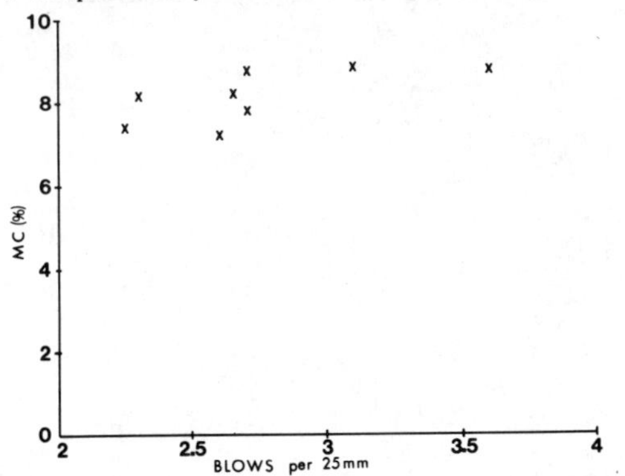

Fig.2 DCP values versus moisture content - new data (n = 8)

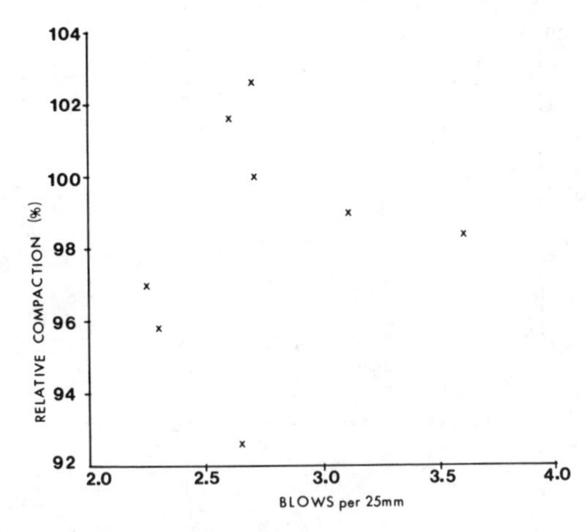

Fig.3 Relative compaction versus DCP value - new data (n = 8)

A stepwise forward multiple linear regression analysis was performed with in situ CBR as the dependent and penetration rate, moisture content and relative compaction as the independent variables (Table 5). In this case the model was of the form v = A + Bx + Cy + Dz. The regression analysis resulted in a significant relationship at a probability level of less than 0.01 level. The analysis indicated that 35% of the variance in the in situ CBR value was taken account of by the variance in rate of penetration, 14% by variance in moisture content and 3.5% by the variance in relative compaction. This left 47.5% of the variance unaccounted for. This would probably be higher with a larger sample size.

4.3 Relationship between penetration and moisture content

All of the above data were collected during the construction phase. Approximately 5 years after construction further penetration data were collected at 10 sites and moisture contents obtained for each 100mm of the 200mm depth of the material. The moisture content of the material had equilibrated over its service life and was generally higher than at construction. The data are presented in Figure 4. There was a highly significant correlation of -0.94 between the number of blows and the moisture. The results were significant at the p = 0.00001 level.

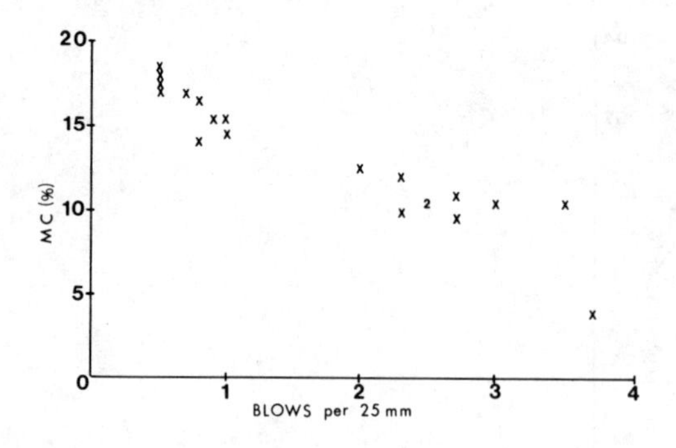

Fig.4 DCP value versus moisture content (n = 20)

5 DISCUSSION AND CONCLUSIONS

The results indicate, that for the data presented, there is a significant linear relationship between in situ CBR results and DCP results with the relationship being affected by moisture content and to a lesser extent by relative compaction. This is somewhat surprising as it was considered that moisture content and density would have been taken account in each of the tests. It may be that the results are influenced by the fact that the material was in a compacted condition when tested.

In each case there was a high standard error of the estimate indicating that any estimate of CBR from the DCP data could be unreliable particularly for relatively low CBRs or DCP results. This is consistent with the results found by Smith and Pratt (1983).

It is also clear that the moisture content is not affecting CBR and DCP results by the same relative amounts, the CBR being the more sensitive. The reason for this may be related to the difference in geometry of the two instruments, i.e. a narrow cone as opposed to a piston with relatively large diameter. A flatter cone may be more appropriate as the zone of generated pore pressure under a flatter cone would be closer to that generated under the CBR piston.

It is apparent, at least for this material, that any correlation developed for relating in situ CBR values to DCP results must also include a term for moisture content. This feature has not been part of the relationships developed by the authors of the earlier Australian studies. The development of equations based on multiple linear regression is a fairly straight forward task using appropriate statistical packages developed for micro-computers.

Until further studies are performed using multiple correlation techniques it considered that the published relationships between CBR value and DCP value should be used with caution.

6 ACKNOWLEDGEMENT

The author wishes to thank the Commissioner for Main Roads, Mr B G Fisk, for permission to publish this paper. The opinions expressed are those of the author and not necessarily those of the Department of Main Roads, New South Wales.

REFERENCES

Morris, P.O. 1976. Studies into the construction of low trafficked roads in the City of Salisbury, South Australia. Australian Road Research Board. Research Report, ARR No. 58.

Morris, P.O., Potter, D.W. and Armstrong, P. 1977. Assessment of subgrade moisture content and strength conditions in the Shire of Yallaroi. Australian Road Research Board. Internal report, AIR 230-3.

Northwest Analytical 1984. NWA STATPAK Version 3.1. Portland: Northwest Analytical Inc.

Scala, A.J. 1956. Simple methods of flexible pavement design using cone penetrometers. New Zealand Engineer 11(2): 34-44.

Smith, R.B. 1983. In situ CBR and dynamic cone penetrometer testing. Proceedings of International Symposium on Soil and Rock Investigations by In Situ Testing, Paris, Vol. 2: 149-154.

Smith, R.B. and Pratt, D.N. 1983. A field study of in situ California Bearing Ratio and dynamic cone penetrometer testing for road subgrade investigations. Australian Road Research 13(4): 285-294.

APPENDIX - CBR, Moisture Content and Penetration Results

CBR (%)		Moisture Content (%)		Penetration Rate (blows/25 mm)	
mean	sd	mean	sd	mean	sd
47.0	7.94	12.7	0.20	2.77	0.68
48.7	15.30	10.4	0.10	2.35	1.14
22.0	1.00	13.5	0.65	1.77	0.40
33.7	5.78	11.1	0.60	2.00	-
22.0	4.58	10.3	0.55	1.40	0.17
25.7	2.89	14.0	1.01	1.73	0.25
36.0	2.00	11.2	3.23	1.82	0.63
80.3	10.70	9.0	0.85	3.40	0.66
107.7	42.10	8.9	0.20	6.00	-
123.3	1.50	8.2	0.70	3.50	2.18
90.3	27.20	9.9	1.10	4.62	2.07
45.7	10.80	11.0	0.05	2.90	0.69
76.3	11.60	11.3	0.85	3.17	0.58
83.3	3.20	10.1	1.36	3.33	0.68
47.3	36.90	7.7	2.70	4.23	0.75
54.3	7.10	9.8	0.25	2.40	0.56
57.3	15.30	8.1	3.40	3.26	0.65
40.0	9.50	12.2	0.31	3.00	1.32
42.7	2.08	12.5	1.00	2.03	0.50
41.3	17.10	11.8	0.75	1.77	0.40
28.0	2.60	12.2	0.25	2.50	0.87
14.7	1.60	11.5	1.17	2.67	0.29
39.7	2.10	11.1	0.78	4.17	0.29
36.7	15.10	11.0	1.05	2.67	0.29
28.7	6.10	11.5	1.50	2.40	0.61
41.3	2.31	12.4	0.15	1.70	0.59
28.3	5.51	11.9	0.50	2.17	0.12
32.0	2.00	11.8	0.06	2.33	0.29
34.0	15.60	10.5	0.64	3.50	0.50
42.3	14.40	10.7	0.25	2.57	0.51
72.3	0.58	10.3	0.95	3.00	1.14
198.7	12.60	6.8	0.69	5.50	-
157.5	96.90	6.8	0.64	9.67	4.62
107.0	52.80	8.4	0.40	7.67	1.76
263.0	12.70	6.9	0.00	15.33	8.39
150.3	72.20	6.7	0.28	6.00	2.65
48.7	10.80	8.2*	0.20	3.10	-
45.7	10.06	8.7*	0.21	2.70	0.72
60.7	13.50	8.2*	0.26	2.63	0.72
66.3	14.90	7.4*	0.40	2.23	0.25
45.5	25.50	8.2*	0.70	2.30	0.30
99.8	13.20	8.7*	0.55	3.63	0.58
76.2	14.50	7.7*	0.17	2.67	0.29
58.0	5.30	7.1*	0.38	2.60	-

* moisture content 0-50 mm below the CBR piston, the remainder were taken 50-150 mm below the piston.

Penetration Testing 1988, ISOPT-1, De Ruiter (ed.)
© 1988 Balkema, Rotterdam, ISBN 90 6191 801 4

Development of and experiences from a light-weight, portable penetrometer able to combine dynamic and static cone tests

J.Fred Triggs, Jr.
EDP/TRIGGS Consultants, Inc., Willoughby Hills, Ohio, USA
Robert Y.K.Liang
Department of Civil Engineering, University of Akron, Ohio, USA

ABSTRACT: A Light-weight portable static-dynamic penetrometer has been developed to test soils, insitu, at locations where it would be difficult or impossible for conventional boring rigs or sounding trucks to reach or work. The features of light-weight and portability in a static cone penetrometer are realized by the use of a dynamically driven, extra-long, friction sleeve anchor. In soil conditions where this extra-long friction sleeve cannot provide sufficient anchorage for full reaction of the static cone penetration, or where dynamic cone penetration tests are more desirable, soils may be tested by dynamic cone penetration. The capability of this portable penetrometer for both static and dynamic cone penetration coupled with measurements of energy delivered to the rods has allowed the establishment of useful correlations between static and dynamic cone resistances. These results of correlation studies on loggings of 12 different sites are presented.

1 INTRODUCTION

In recent years, quasi-static cone penetration testing (CPT) has gained wide acceptance as an effective in-situ site characterization technique. However, despite the advances in CPT testing apparatus and interpretative methods, the standard penetration test (SPT) remains the workhorse of the practicing engineer in site explorations (Tavenas, 1986). In view of the advantages and limitations of SPT and CPT, Seed and De Alba (1986) suggested the desirability of a judicious combination of CPT and SPT tests, particularly for evaluating the liquefaction resistance of any given site.

The Dinastar penetrometer, a light-weight portable cone penetrometer which can be hand-carried into difficult access locations and can test subsoils to depths necessary for most foundation design, is being used commercially throughout Europe and North America. As the penetrometer is capable of providing both static and dynamic cone resistances with essentially the same driving apparatus, it provides an economical means to establish site-specific correlations between static and dynamic cone resistances. The major thrust of this paper is to present such a correlation study based on boring logs obtained from various test sites.

2 PURPOSE

The purpose of this paper is to provide its readers with:

1. A brief description of the design and operation of this portable, light-weight dynamic and static penetrometer.
2. A critical assessment of advantages and limitations of this portable penetrometer.
3. A presentation of field test results to demonstrate the wide range of potential applications.
4. Regression analyses of static and dynamic cone resistance data to provide useful correlations.
5. An illustration of dynamic instrumentation that allows measurements of energy delivered to the dynamic penetrometer.

3 DESIGN AND OPERATION

The most relevant design challenge for a light-weight, portable static cone penetrometer is the need to eliminate the traditional screw-anchor reaction system common to most existing light-weight static penetrometers of considerable capacity. Screw anchors consume too much human energy and time if installed by leveraged manpower. Hydraulic insertion of screw-anchors requires heavy servo-mechanisms contradicting the concept of a light-weight in-situ test-

DEPTH ft	m	BLOW COUNT	CONE RES. MPa	SKIN FRICT MPa	FRICT RATIO %	CONE RESISTANCE, MPa 0 5 10	TESTED CONSISTENCY
	0.1	-	-	-	-		
	0.2	-	-	-	-		
1	0.3	0	-	-	-		
	0.4	0	-	-	-		
	0.5	0	-	-	-		
2	0.6	0	-	-	-		
	0.7	0	-	0.382	-		
	0.8	0	-	0.120	-		
3	0.9	0	11.5	0.128	1.12	+	Compact SAND&GRAVEL
	1.0	129	6.4	0.092	1.45	*	Firm SAND&GRAVEL
	1.1	43	5.4	0.061	1.14	*	Firm SAND&GRAVEL
4	1.2	46	4.2	0.073	1.72	*	Firm SAND&GRAVEL
	1.3	33	6.3	0.061	0.98	*	Firm GRAVEL
	1.4	22	3.4	0.092	2.68	*	Firm SAND
5	1.5	26	2.3	0.120	5.32	*	Stiff CLAY
	1.6	22	2.2	0.137	6.34	*	Stiff CLAY
	1.7	33	2.5	0.148	6.03	*	Stiff CLAY
6	1.8	43	2.8	0.111	3.90	*	Stiff SAND/SILT/CLAY
	1.9	49	2.7	0.069	2.50	*	Loose SAND
	2.0	53	2.6	0.087	3.29	*	Stiff SAND/SILT/CLAY
7	2.1	42	2.5	0.092	3.77	*	Stiff SAND/SILT/CLAY
	2.2	26	3.0	0.111	3.65	*	Stiff SAND/SILT/CLAY
	2.3	33	3.1	0.106	3.37	*	Stiff SAND/SILT/CLAY
8	2.4	35	3.7	0.116	3.12	*	Medium SAND/SILT
	2.5	42	3.4	0.092	2.69	*	Firm SAND
	2.6	40	3.2	0.114	3.51	*	Stiff SAND/SILT/CLAY
9	2.7	44	3.2	0.106	3.26	*	Medium SAND/SILT
	2.8	35	3.2	0.090	2.79	*	Medium SAND/SILT
	2.9	43	2.7	0.098	3.56	*	Stiff SAND/SILT/CLAY
10	3.0	40	2.7	0.103	3.74	*	Stiff SAND/SILT/CLAY
	3.1	36	2.9	0.083	2.81	*	Medium SAND/SILT
	3.2	39	3.2	0.063	1.94	*	Firm SAND&GRAVEL
	3.3	41	2.6	0.068	2.56	*	Loose SAND
11	3.4	33	2.4	0.050	2.13	*	Loose SAND
	3.5	25	2.1	0.068	3.29	*	Stiff SAND/SILT/CLAY
	3.6	27	2.2	0.060	2.79	*	Loose SAND/SILT
12	3.7	20	2.1	0.045	2.19	*	Loose SAND
	3.8	27	1.9	0.036	1.92	*	Loose SAND&GRAVEL
	3.9	24	1.4	0.043	3.13	*	Medium SAND/SILT/CLAY
13	4.0	18	1.9	0.477	25.59	*	V.Hard CLAY
	4.1	15	1.3	-	-	*	
	4.2	18	3.0	-	-	*	

DEPTH ft	m	BLOWS per 10 cm	DYNAMIC CONE RESISTANCE, MPa	CONE RESISTANCE, MPa 0 5 10
	0.1	14	4.1	*
	0.2	28	8.3	*
1	0.3	27	8.0	*
	0.4	29	8.6	*
	0.5	22	6.5	*
2	0.6	20	5.9	*
	0.7	62	18.4	+
	0.8	300	88.8	+
3	0.9	250	74.0	+
	1.0	48	14.2	+
	1.1	38	10.6	+
4	1.2	17	4.7	*
	1.3	14	3.9	*
	1.4	16	4.5	*
5	1.5	19	5.3	*
	1.6	16	4.5	*
	1.7	14	3.9	*
6	1.8	17	4.7	*
	1.9	15	4.2	*
	2.0	13	3.6	*
7	2.1	14	3.7	*
	2.2	11	2.9	*
	2.3	12	3.2	*
8	2.4	14	3.7	*
	2.5	16	4.2	*
	2.6	14	3.7	*
9	2.7	14	3.7	*
	2.8	14	3.7	*
	2.9	16	4.2	*
10	3.0	7	1.8	*
	3.1	7	1.8	*
	3.2	7	1.8	*
	3.3	9	2.3	*
11	3.4	9	2.3	*
	3.5	9	2.3	*
	3.6	9	2.3	*
12	3.7	10	2.5	*
	3.8	13	3.3	*
	3.9	10	2.5	*
13	4.0	10	2.5	*
	4.1	9	2.1	*
	4.2	4	1.0	*

Figure 1 Static and dynamic cone penetration test logs at VPI

DEPTH ft	m	BLOW COUNT	CONE RES. MPa	SKIN FRICT MPa	FRICT RATIO %	CONE RESISTANCE, MPa 0 5 10	TESTED CONSISTENCY
	0.1	-	-	-	-		
	0.2	-	-	-	-		
1	0.3	0	-	-	-		
	0.4	0	-	-	-		
	0.5	0	-	-	-		
2	0.6	0	-	-	-		
	0.7	0	-	-	-		
	0.8	0	-	0.059	-		
3	0.9	0	-	0.033	-		
	1.0	0	1.4	0.073	5.29	*	Medium CLAY
	1.1	21	1.3	0.061	4.82	*	Medium CLAY
4	1.2	12	3.7	0.059	1.57	*	Firm SAND&GRAVEL
	1.3	26	3.9	0.064	1.64	*	Firm SAND&GRAVEL
	1.4	22	3.1	0.061	1.96	*	Firm SAND&GRAVEL
5	1.5	21	2.0	0.056	2.85	*	Loose SAND/SILT
	1.6	23	1.9	0.045	2.40	*	Loose SAND
	1.7	22	1.2	0.036	3.08	*	Medium SAND/SILT/CLAY
6	1.8	20	2.0	0.026	1.35	*	Loose SAND&GRAVEL
	1.9	16	0.9	0.026	2.99	*	Loose SAND/SILT
	2.0	13	0.8	0.021	2.69	*	Loose SAND/SILT
7	2.1	10	0.7	0.021	3.08	*	Soft SAND/SILT/CLAY
	2.2	10	0.5	0.024	4.85	*	V.Soft CLAY
	2.3	8	0.5	0.016	3.23	*	V.Soft SAND/SILT/CLAY
8	2.4	8	0.7	0.021	3.08	*	Soft SAND/SILT/CLAY
	2.5	9	0.7	0.021	3.08	*	Soft SAND/SILT/CLAY
	2.6	6	0.5	0.021	4.31	*	V.Soft CLAY
9	2.7	8	0.4	0.018	4.71	*	V.Soft CLAY
	2.8	8	0.4	0.015	3.83	*	V.Soft CLAY
	2.9	8	0.3	0.008	2.56	*	V.Loose SAND/SILT
10	3.0	7	0.3	0.008	2.56	*	V.Loose SAND/SILT
	3.1	6	0.2	0.010	5.11	*	V.Soft CLAY
	3.2	3	0.2	0.013	6.39	*	ORGANIC
	3.3	3	0.2	0.013	6.39	*	ORGANIC
11	3.4	4	0.2	0.015	7.67	*	ORGANIC
	3.5	5	0.2	0.018	8.94	*	ORGANIC
	3.6	5	0.2	0.018	8.94	*	ORGANIC
12	3.7	6	0.3	0.018	5.96	*	V.Soft CLAY
	3.8	7	0.3	0.014	4.86	*	V.Soft CLAY
	3.9	7	0.3	0.017	5.67	*	V.Soft CLAY
13	4.0	7	0.3	0.019	6.48	*	ORGANIC
	4.1	6	0.3	0.021	7.29	*	ORGANIC
	4.2	7	0.2	0.024	12.15	*	ORGANIC
14	4.3	8	0.3	0.024	8.10	*	ORGANIC
	4.4	9	0.3	0.024	8.10	*	ORGANIC
	4.5	10	0.5	0.029	5.83	*	V.Soft CLAY
15	4.6	10	0.7	0.029	4.17	*	Soft CLAY
	4.7	10	0.7	-	-	*	
	4.8	12	0.8	-	-	*	
16	4.9	12	0.9	-	-	*	
	5.0	0	0.8	-	-	*	
	5.1	0	-	-	-		

DEPTH ft	m	BLOWS per 10 cm	DYNAMIC CONE RESISTANCE, MPa	CONE RESISTANCE, MPa 0 5 10
	0.1	220	65.1	+
	0.2	31	9.2	*
1	0.3	23	6.8	*
	0.4	23	6.8	*
	0.5	24	7.1	*
2	0.6	20	5.9	*
	0.7	13	3.8	*
	0.8	11	3.3	*
3	0.9	15	4.4	*
	1.0	13	3.6	*
	1.1	23	6.4	*
4	1.2	18	5.0	*
	1.3	15	4.2	*
	1.4	10	2.8	*
5	1.5	7	2.0	*
	1.6	5	1.4	*
	1.7	4	1.1	*
6	1.8	1	0.3	*
	1.9	0	0.0	*
	2.0	0	0.0	*
7	2.1	0	0.0	*
	2.2	0	0.0	*
	2.3	0	0.0	*
8	2.4	0	0.0	*
	2.5	0	0.0	*
	2.6	0	0.0	*
9	2.7	0	0.0	*
	2.8	0	0.0	*
	2.9	1	0.3	*
10	3.0	2	0.5	*
	3.1	0	0.0	*
	3.2	0	0.0	*
	3.3	0	0.0	*
11	3.4	0	0.0	*
	3.5	0	0.0	*
	3.6	0	0.0	*
12	3.7	0	0.0	*
	3.8	2	0.5	*
	3.9	5	1.3	*
13	4.0	5	1.2	*
	4.1	4	1.0	*
	4.2	2	0.5	*
14	4.3	4	1.0	*
	4.4	8	1.9	*
	4.5	10	2.4	*
15	4.6	12	2.9	*
	4.7	13	3.1	*
	4.8	13	3.1	*
16	4.9	16	3.8	*
	5.0	17	3.9	*
	5.1	19	4.3	*

Figure 2 Static and dynamic test logs at Solon, Ohio

ing apparatus. The solution to this design challenge, as incorporated in the Dinastar penetrometer, was to anchor a reaction system using a dynamically advanced, extra-long, friction sleeve.

This self-anchoring system (or friction sleeve) is dynamically driven and needs to develop sufficient friction to resist the quasi-static cone penetration. For example, with a 10 cm^2 area cone, soils with a low sleeve friction to cone resistance ratio, such as 1.1%, would require a friction sleeve of 900 cm^2. Other smaller friction sleeves providing contact areas of 500 cm^2 or 700 cm^2 may also be used. As it turns out, the 500 cm^2 friction sleeve provides adequate anchorage in most soils except for sands and gravel.

Since the design of the anchoring system keeps reaction stresses within the rod string and out of the apparatus frame, a light-weight frame is successfully used.

This penetrometer's static cone has a 10 cm^2 cone area and 60 degree apex angle which conform to Dutch cone dimensions. The cone diameter to neck diameter ratio of the point is the same as that ratio for the classical Begemann point, a standard configuration for mechanical cones used in many of the familiar soil classification charts.

The dynamic driving system conforms with the German Standard DIN 4049 medium energy specifications. The 30 Kg hammer and 20 cm drop results in a theoretical energy of 0.6 Kg-m/cm^2.

When using the penetrometer in the purely dynamic cone penetration mode, either a fixed or a lost 10 cm^2 point is used. The fixed points are usually used in shallow tests where extraction at completion of the test is not difficult. The lost points facilitate extraction of the tubes after completion of deeper tests. During dynamic cone penetration, a viscous slurry is usually injected to fill the annulus between soil and tubes, thus eliminating the parasitic rod friction.

When using the penetrometer to determine the static cone resistance, the upper meter of soil is generally tested by the dynamically driven fixed-point cone. Once the friction sleeve can be completely embedded, the dynamic cone is pulled and the static cone is installed. During static cone penetration, the ultimate static cone resistance is read and recorded from either an electronic or hydrostatic load cell at 10 cm depth intervals. Following each static cone thrust, the friction sleeve is driven 10 cm, both to ready the cone for its next advance and to collect dynamic driving record for calculation of sleeve unit friction.

4 ADVANTAGES AND SHORTCOMINGS

To summarize the advantageous features of this light-weight portable penetrometer, the following observations can be made. First, this penetrometer's total weight is 250 Kg which can be disassembled into five parts, each weighing less than 56 Kg, thus allowing transportation by automobile and hand-carrying by two persons into difficult access test sites. Secondly, the total full working height of the penetrometer is 2.8 meters which allows operation in low head-room conditions. Thirdly, the penetrometer can provide both static and dynamic cone resistances, thus providing flexibility in selecting the most appropriate testing mode for various applications.

An obvious shortcoming of the long-sleeve penetrometer is associated with determination of the sleeve unit friction. The unit friction value is an average for the relatively long (40 cm, 55 cm, or 70 cm) friction sleeve. This average unit friction, with the calculated value assigned to the 10 cm long depth at mid-height of the sleeve, undoubtedly causes a reduction in the reliability of soil classifications. This shortcoming is more pronounced in thinly stratified soil profiles.

The other significant shortcoming of the portable penetrometer is related to insufficient anchorage of the friction sleeve, a situation which may occur in some sands, some gravels, some very stiff clays, and where a high strength stratum is encountered beneath a low strength stratum. Where these circumstances are prevalent, the penetrometer can be operated with a dynamic cone to obtain dynamic cone resistances.

5 SUMMARY OF FIELD APPLICATION EXPERIENCE

During the past three years, the authors have been personally involved in the use of the Dinastar portable penetrometer to help evaluate soil conditions at over 30 difficult access sites. The reasons for using this penetrometer on those sites are briefly summarized, followed by illustration of static and dynamic cone penetration test logs for two different sites.

Approximately half of these projects were considered to be difficult access because they involved working within existing buildings with narrow doorways and/or restrictive height ceilings. In several of these buildings, the soil exploration was for design of foundations for new machines. Some buildings were to be remodeled with an increase in existing column loadings. Other buildings had experienced slab and/or foundations settlement problems requireing idenification of the source of the settlement and a solution.

Site congestion was the second most prevalent reason for using a portable, lightweight penetrometer. This congestion took the form of closely spaced trees, narrow passageways, and steep topography, all factors why truck-mounted testing equipment could not reach the desired test locations.

Several sites were tested by the portable penetrometer because the site surface was too soft to support a testing vehicle. In one instance, dynamic cone testing from a styrofoam floating platform was used to measure quantity and consistency of sediment to be dredged from a silted-in lake.

At 12 previous test sites, both static and dynamic cone penetration tests were done in close proximity, primarily to gain correlations between static and dynamic cone resistances. With such correlations, dynamic cone resistances can be converted to equivalent static cone resistances.

At Virginia Polytechnical Institute, Blacksburg, Virginia, during the June, 1986 ASCE In-Situ Soil Testing Conference, this portable penetrometer was used in the demonstration project. Both static cone and dynamic cone penetration test logs at the same location are presented in Fig. 1. Based on the static cone and dynamic sleeve resistances profile, the site was characterized as consisting of a layer of sand and gravel, followed by a layer of stiff clay, and layers of mixtures of sand, silt, and clay of varying relative densities. The top one meter of the ground was tested by dynamic mode, therefore the static cone resistance was not shown. It may be seen that both static and dynamic cone resistances provide similar qualitative characterization of site soil consistencies.

Fig. 2 shows testing logs of an interesting site in Solon, Ohio, where a very soft peat deposit is overlain by firm, inorganic soils. Both the static and the dynamic cone logs clearly agree that the peat deposit has very low resistance. As expected, the relatively high friction ratio of a very soft peat deposit helps the penetrometer's self-anchoring system to work very well.

6 ENERGY FINGERPRINT OF THE PENETROMETER

To facilitate the correlation study, the energy loss characteristics of the dynamic hammer driving system were measured during the testing at the Warrensville Heights,

Figure 3 The instruments for measuring energy characteristics during DCP tests

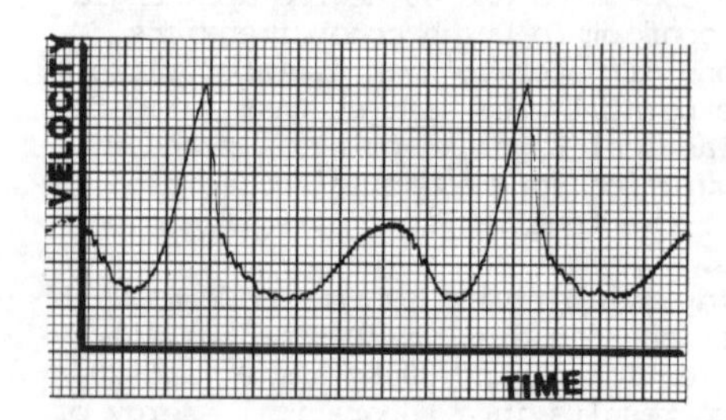

Figure 4 Velocity-time history of the driving hammer

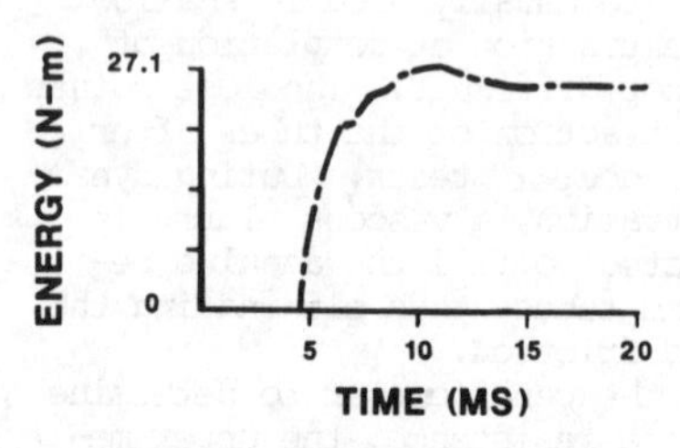

Figure 5 Energy delivered to the drill rods

Ohio site. Fig. 3 shows the layout of the instruments during testing. A radar gun was used to measure the hammer velocity while strain transducers and accelerators were used to measure the forces and accelerations in the sounding tube near the anvil. The data acquisition system used in this study was the PDI pile driving analyzer. Information from this detailed dynamic measurement is important because it provides an energy "fingerprint" of a particular penetrometer for comparison with other dynamic penetrometers, including SPT. It also provides information for converting dynamic cone resistance to equivalent static cone resistance, or to convert the dynamic driving record of friction sleeve into equivalent unit friction resistance.

Fig. 4 shows the velocity-time history of the hammer during dynamic cone penetration testing. Fig. 5 shows the PDI interpretation of the energy delivered to the rod strings during a typical hammer blow. As can be seen, the maximum energy delivered to the rod is about 27 Newton-Meter. Comparing to the theoretical potential energy of 58 Newton-Meter, an energy ratio of 46.5% is observed. Based on the velocity-time record of the hammer, it can be determined that the hammer velocity immediately before striking is about 1.54 meter/sec., the corresponding kinetic energy immediately before striking is about 35.5 Newton-Meter. Therefore, about 40% of the theoretical energy is lost prior to striking the anvil, and an additional 15% is lost during and after the strike. The total energy loss for the penetrometer was 55% of theoretical potential energy.

7 CORRELATION STUDIES

7.1 Previous work

Bolomey (1974) presented results of comparison between static and dynamic resistances in three different sites. The study involved the use of a mechanical type static cone with standard 10 cm^2 cone area and a dynamic penetrometer with theoretical input energy somewhat greater than the heavy type specified in the German DIN specification. The study indicated that where soils contained no particles larger than medium sand sizes, the dynamic cone resistance is about equal to the static cone resistance. However, for soils having gravel size particles, the dynamic cone resistance seemed to be smaller than the static resistance. Puech, et al (1974) arrived essentially at the same conclusion.

A study made by Waschkowski (19740 indicated that for normally consolidated clay and medium dense sand, the static and dynamic cone resistances are virtually equal. For dense sand, the dynamic cone resistance is less than the static cone resistance. This study indicated that the ratio between dynamic and static cone resistances lies within 1 to 2 for overconsolidated clays.

From these reviews, it may be concluded that previous correlation studies on static and dynamic cone resistances are limited both in quantities and their applicability. The main drawback of previous studies is the uncertain nature of actual energy input during dynamic testing. As a result, an empirical correlation factor has often been used to account for the energy loss in the system.

7.2 Theoretical consideration

In establishing the correlation between static and dynamic cone resistances, the major uncertainty is in calculating dynamic cone resistance. Determination of the dynamic cone resistance requires the use of an energy balance equation, which, according to Bolomey (1974), can be expressed as:

$$K(MH) + (M+P)e = R_d(e+c)A + (1-n^2)\frac{PMH}{M+P}$$

where M = mass of hammer
H = traveling distance of hammer
P = mass of driven parts
e = penetration resistance per blow
R_d = dynamic resistance of soil in the unit of pressure
K = correcting factor for energy loss through guiding installation
c = energy stored in the soil
n = correcting factor for partially elastic blow
A = contributing area

The Dutch formula is, in fact, a simplification of the above general equation by assuming n=0 (perfectly inelastic blow) and c=0 (no strain energy loss in rod and soil). The use of the Dutch formula is considered applicable to most soils except in firm ground where e is less than 1 cm (Bolomey, 1974). As discussed previously, the correcting factor for energy loss for this portable penetrometer can be taken as 0.5. Assuming that the second term on the left hand side of the equation is small and negligible and that n=0.7 for most soils, then the dynamic cone resistance and unit friction values can be calculated as:

$$R_d(\text{or } f) = 0.5\,\frac{M^2H}{(M+P)}\,\frac{1}{e}\,\frac{1}{A}$$

where f is the unit local friction resistance. For dynamic cone resistance calculation, A is taken as 10 cm^2; whereas, for unit skin friction calculation, A is the friction sleeve area.

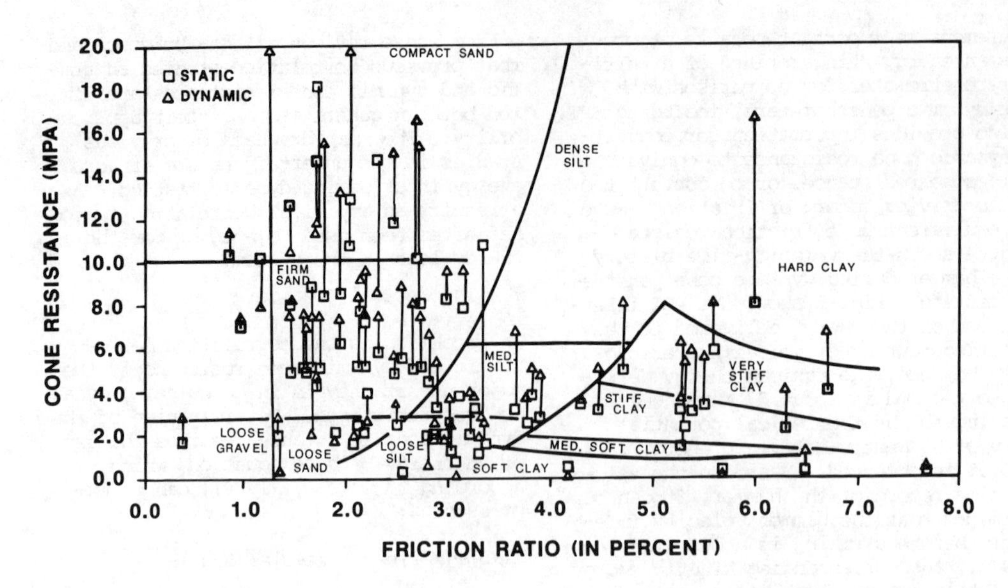

Figure 6 Cone resistances (static and dynamic) vs. friction ratios

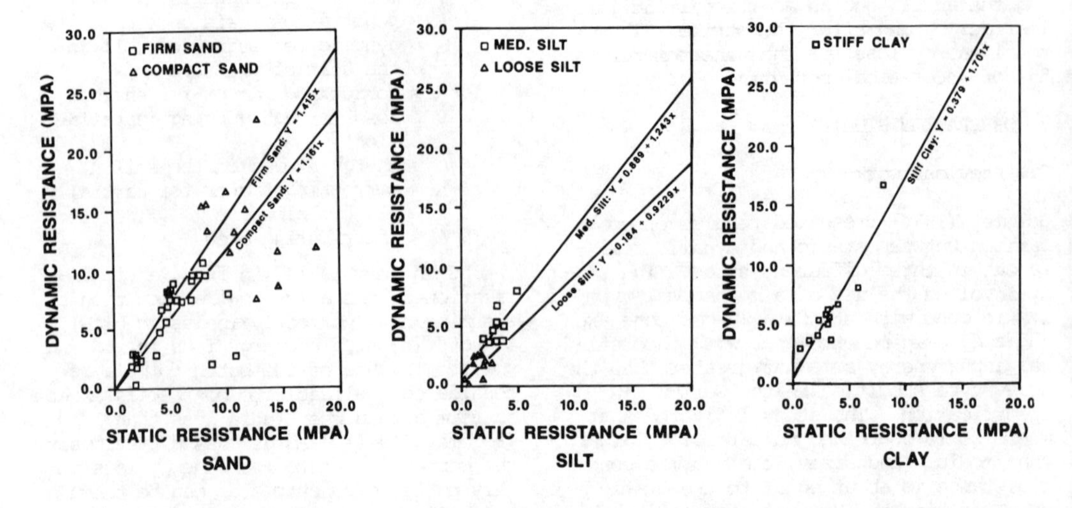

Figure 7 Regression analysis of relationship between static and dynamic cone resistance

In calculating long-sleeve unit friction values, the theoretically correct equation, which inclued a correction factor for energy loss, should be used. However, in calculating dynamic cone resistance, the commonly adopted procedure of using modified Dutch formula with the correction factor K taken as unity should be used. This inconsistency is reasonable considering that:

A. For geotechnical engineering interpretation of dynamic cone data, one normally converts the dynamic cone resistance into equivalent static cone resistance, after logging, using conversion factors that combine energy efficiency with soil type correlation coefficients.

B. Calculated sleeve unit frictions are used immediately and directly in logging soil types and relative densities from a static cone resistance/friction ratio chart, with the correction for sleeve driving energy loss made during logging.

7.3 Analysis of available field data

With the availability of both static and dynamic cone resistances data at 12 sites, correlations between these two quantities were studied. In selecting the data points for statistical analysis, the following guidelines were adopted:

A. There should be at least 3 nearly equal consecutive readings.

B. At a weak seam, the readings should be the lowest resistance.

C. Abrupt increases in readings should be disregarded.

Figure 6 shows the natural scale plot of cone resistances vs. friction ratio from 12 different sites, where the squares indicate the measured static cone resistance, and the circles indicate the calculated dynamic cone resistance. The boundary lines classifying different soil types and consistencies were drawn on the basis of the authors' CPT, SPT, and visual classification data bank, and are similar to those of Schmertmann's (1978).

A regression analysis of correlations between dynamic and static cone penetration resistances is shown in Fig. 7 with the best fit equations for different soil conditions indicated.

8 CONCLUSIONS

A. The self-anchoring mechanism provided by the extra-long friction sleeve has proven to work very well in a light-weight portable static-dynamic cone penetrometer.

B. With the detailed measurement of energy delivered by the dynamic driving system as described in this paper, the equivalent skin friction and dynamic cone resistance can be more accurately calculated from the dynamic cone penetration tests.

C. For this particular penetrometer, the relationship between static and dynamic cone resistances are summarized in Fig. 7. The dynamic cone resistance is calculated from Dutch formula which only considers the theoretical potential energy. The energy loss term is lumped into the correlation studies. Finally, it should be pointed out that because of the nature of dynamic operation, one cannot expect the same precision as one would from static operation. Consequently, a higher factor of safety in design of foundations is recommended when equivalent static cone resistance is used.

ACKNOWLEDGEMENTS

The second writer wishes to thank the support of the University of Akron through Faculty Research Grant RG.935. The assistance provided by GRL Associates, Inc. to measure the energy transfer during dynamic cone penetration test is acknowledged. Thanks are especially due to Mario Mambrini of Tecnotest s.n.c., for his technical assistance with the DINASTAR penetrometer.

REFERENCES

Bolomey, H. 1974. Dynamic penetration-resistance formulae. Proc. ESOPT1, 2:2: 39-46.

Puech, A., Biarez, J., Cassan, M., and Toutoungi, A. 1974. Contribution to the study of static and dynamic penetrometer. Proc. ESOPT1, 2:2:307-312.

Schmertmann, J.H. 1978. Guidelines for cone penetrometer test. FHWA-TS-78-209, Federal Highway Administration Report.

Seed, H.B. and De Alba, P. 1986. Use of SPT and CPT for evaluating the liquefaction resistance of sands. Insitu 86, Geotechnical Special Publication 6: ASCE 281-302.

Tavenas, F. 1986. In-situ testing: where are we? where should we go? Geotechnical News, 4:4.

Waschowski, E. 1982. Dynamic probing and practice. Proc. ESOPT2, 357-362.

Technical papers:
Dilatometer test / Pressuremeter test

Penetration Testing 1988, ISOPT-1, De Ruiter (ed.)
© 1988 Balkema, Rotterdam, ISBN 90 6191 801 4

Penetration testing of a desiccated clay crust

Gunther E. Bauer
Institute for Civil Engineering, Carleton University, Ottawa, Canada
Auro Tanaka
Department of Civil Engineering, University of Paraiba, Campina Grande, Brazil

ABSTRACT: This research study focusses on the determination of the in situ geotechnical properties of a fissured clay crust, in particular in relation to strength and deformation responses. Four in situ test devices were employed in this investigation, namely a monocellular pressuremeter, a Marchetti dilatometer, a driven screw plate, and a standard penetrometer.

The results from the four devices were compared and rated against one another. The results are presented as soil strength and modulus of deformation profiles with depth. In the evaluation of the strength and deformation parameters of the soil, the fissure pattern (density of fissures, direction, extent and spacings of the fissures) must be considered as well as the stress and strain paths imposed by the various in situ devices.

KEYWORDS: In Situ Test, Field Behaviour, Desiccated Crust, Deformation Modulus, Shear Strength, Pressuremeter, Screw Plate, Standard Penetration Test.

INTRODUCTION

The evaluation of the load-deformation response ('stiffness') and the determination of the shear strength characteristic of cohesive soil is an important aspect in the analysis, design and prediction of the foundation performance of civil engineering structures. Unfortunately, in many cases discrepancies exist between prediction and actual performance of structures founded in clays as documented in the geotechnical engineering literature. The evaluation of the geotechnical properties by conventional laboratory tests becomes an almost impossible task for desiccated clay crusts and therefore in situ testing is a logical alternative.

The regions of Eastern Canada, the St. Lawrence River and Ottawa River valleys are underlain by a post-glacial marine clay deposit which is highly sensitive and generally soft. Post-depositional changes, such as sub-aerial drying, weathering, consolidation, groundwater fluctuations and removal of overburden, have caused preconsolidation of the clay near the ground surface. This desiccated crust is highly fissured and its shear strength decreases with depth with a corresponding increase of natural water content. The overconsolidation ratio (OCR) also decreases with depth. Desiccation is not necessarily the only factor causing the formation of a thin stiff crust. Other possible causes are related to: erosional denudation, delayed consolidation under increased overburden, chemical processes in the sediments due to weathering and, syneresis due to spontaneous loss of water from a gel during aging. All these processes have caused discontinuities in the crust material leading to different fissure patterns with their own surface characteristics and geometries. In summary, it can be said that laboratory tests on small soil specimens of fissured clay, cannot produce representative soil parameters, in particular for strength and deformability, even if extreme care is exercised to avoid soil disturbance and stress relief. It appears then that the only two alternatives left to assess the geotechnical properties of these fissured crusts are in situ testing and back analysis of monitored or failed structures.

The range of in situ devices has been broadened over the past few years and the many testing techniques have been the object of many publications and conferences. This paper will report on the results obtained from four in situ devices; i.e. screw plate (SCT), monocellular pressuremeter (PMT), dilatometer (DMT) and standard penetrometer (SPT). An attempt is made to interpret the test results from each test device, in particular as these relate to shear and deformability characteristics of a stiff, fissured clay crust.

SCREW PLATE TEST (SCT)

The screw plate device consists, in general, of a single flight helical auger which is screwed into the ground to the desired test depth. In essence, it is a bearing plate located at a certain depth but where the load is applied at the ground surface. The sinkage of the plate under increments of load are recorded and the load-displacement curve for each test depth are interpreted to yield a deformation modulus of the soil. Other parameters, such as shear strength and consolidation coefficients can also be estimated.

Various techniques have been presented in the literature to evaluate the deformation and strength properties for cohesive soils (Janbu and Senneset, 1973; Schwab and Broms, 1977; and Selvadurai and Nicholas, 1979).

The results reported herein were interpreted using the analyses suggested by Selvadurai and Nicholas (1979) who have carried out a comprehensive analytical study of the screw plate test in cohesive soils. They considered the effects of soil disturbance caused during insertion, stiffness of the plate, and soil/plate interface conditions. Using their analysis the modulus of deformation can be estimated from the load-displacement response as follows:

$$E = \lambda(p/\omega)\,a \quad (1)$$

where

ω = settlement of plate
p = applied plate pressure
a = radius of plate
λ = coefficient that represents the combined effect of all factors introduced in the analysis

From theoretical and practical aspects the coefficient λ varies between 0.6 to 0.75.

Selvadurai et al. (1980) also found a solution for the undrained shear strength, which generally is given in the form of:

$$C_u = (p_{ult} - \sigma_v)/N_b \quad (2)$$

where

p_{ult} = applied failure pressure
σ_v = total overburden stress
N_b = bearing capacity factor

Selvadurai et al.'s solution yielded the following N_b boundary values for the ratio of applied failure stress to undrained shear strength:

$$9 \leq (p_{ult}/C_u) \leq 11.35 \quad (3)$$

The lower value was obtained for partial bonding and the upper limit corresponds to full bonding between plate and soil.

FLAT PLATE DILATOMETER TEST (DMT)

The device, test equipment and procedure are quite simple and complete details of the test and calibration procedure were given by Marchetti (1980), Marchetti and Crapps (1981), Schmertmann (1981,1982) and others. The pushing pressure of the dilatometer, p, the lift-off pressure of the flexible membrane, p_o, and the pressure p_1, to push the membrane 1 mm into the soil are recorded and corrected. Based on these corrected readings the following three parameters are obtained, the material index I_D, thee horizontal stress index K_D and the dilatometer modulus E_D.

$$I_D = \frac{p_1 - p_o}{p_o - u_o} \quad (4.a)$$

$$K_D = \frac{p_o - u_o}{\sigma'} \quad (4.b)$$

$$E_D = \frac{E}{(1-\nu^2)} = 38.2\,(p_1 - p_0) \quad (4.c)$$

where: p_1, p_0 = corrected pressure readings
u_0 = in situ hydrostatic water pressure
σ_v' = vertical effective stress
E_D = Young's modulus
ν = Poisson's ratio

From these three basic empirical soil index parameters, other correlations are derived, such as the overconsolidation ratio (OCR), the undrained shear strength, the K_0 values and the preconsolidation stress.

A computer program is used to interpret the DMT test data. To date there is very limited information available which allows comparison of test results from desiccated, fissured clay crusts.

PRESSUREMETER TEST (PMT)

The pressuremeter test is without doubt one of the most important developments in in situ testing. The derived soil parameters are well supported by theories. Again it is not within the scope of this paper to discuss the many different types and testing techniques, since this has been well documented by others (Baguelin et al. 1978, for example).

In this study on hand a monocellular probe was used (type Pencel) which had a diameter of 38 mm. Tests in the fissured clay crust were carried out in a predrilled hole in a downhole fashion. The hole was made by pushing a 39 mm diameter thin-walled Shelby tube into the ground. The undrained soil modulus can be estimated from the shear modulus which is derived from the linear part of the pressuremeter curve, as:

$$E_u = 2G\,(1+\nu) \qquad (5)$$

where: G = shear modulus
ν = Poisson's ratio

Since several load-unload cycles were performed, there are several possibilities to evaluate a shear modulus, from the initial expansion curve, from the unload-reload position and from the stress-strain curve of the pressure-deformation relationship. The merits and limitations of each method have been discussed by Tanaka (1986) with regard to the test results on hand.

The undrained shear strength can be obtained from the yield portion of the pressuremeter curve. The relationship proposed by Gibson and Anderson (1961) was adopted by this study for reasons listed by Tanaka (1986):

$$S_u = \frac{P_1 - P_0}{\beta} \qquad (6)$$

where

$$\beta = [1+\ln \frac{E}{2(1+\nu)\,Su}] = [1+\ln \frac{E}{S_u}]$$

P_0 = total in situ horizontal stress
P_1 limiting pressure

Some values for are reported for London clay and they seem to vary between 4.6 and 6.2 (Gibson and Anderson, 1961; Marsland and Randolph, 1977).

DESCRIPTION OF TEST SITE

The major part of the in situ test program was carried out in a desiccated crust in Ottawa, Canada, on the grounds of the National Research Council. This site was selected because geotechnical data from previous investigations were available (e.g., Eden and Crawford, 1975; Eden and Law, 1980). A borehole log of this site is shown in Figure 1. The top layer consisted of stiff, brown silty clay of 0.6 m thickness underlain by a very stiff, highly fissured, brown silty clay to a depth of 2m. This was followed by one metre of fissured, brown to gray clay. Below a depth of 3 m the clay changed gradually from a firm to a soft gray, very sensitive clay. The degree of fissuring also decreased with depth. The water table was encountered at a depth of 1 m. The natural water content increased from about 40% at the surface to about 80% at a depth of 4 m. From oedometer tests it was found that the overconsolidation ratio (OCR) averaged about 25 within the first 3 m and decreased rapidly with depth thereafter. The corresponding preconsolidation pressures averaged 500 kPa within the crust and decreased to about half that value at a depth of 4 m. Void ratios are generally in the order of 2 for this type of deposit. The fairly level test area was about 12 by 24 m in plan and the following in situ tests were carried out:

14 Dilatometer Tests (DMT)
5 Screw Plate Load Tests (SCT)
14 Pressuremeter Tests (PMT)
5 Standard Penetration Tests (SPT)

Several test holes were made for continuous sampling. These specimens were used for classification, oedometer- and triaxial compression testing, as will be discussed later.

TEST PROCEDURES AND RESULTS

Whenever available the standardized procedures for an in situ test was used, i.e., SPT and DMT. Due to limitation of space, only these test procedures will be briefly discussed for which no adopted standards exists.

Screw Plate Tests (SCT): Several attempts were made to utilize a 300 mm diameter plate, but due to the hard crust and the limitations of the trailer mounted drill rig (SIMCO-2400), smaller diameter plates had to be used. This drill rig was used for advancing the screw plate, the standard penetometer and for pushing Shelby tubes and sampling.

In view of the torque limitations of the drill, the screw plate diameter had to be reduced to 76 mm. The downward thrust and the rotational speed were adjusted such that the screwplate advanced one pitch per revolution. The intervals between test depth was kept at least 4 times the plate diameter or 0.3 m. After a test was completed the screw plate was rotated to the next depth. An independent reaction beam-hydraulic jack assembly was used for applying the bearing stress. The load was applied in equal increments of about 40 kPa and unloading was done in 20 kPa decrements. The load was monitored with a load cell and the settlement by a L.V.D.T. A particular load stage was considered complete when the settlement had ceased or after 20 minutes. The average time for one test sequence was about two and one-half hours.

A typical load-displacement curve is given in Figure 2. In this particular test (SCT-NRC-1), two unloading cycles were performed. This was done in order to observe the change of modulus of deformation with reloading. Figure 3 shows the variations of the modulus with depth. The modulus was determined according to Equation (1). As expected, the modulus increases with the number of load cycles. The estimated shear strength profile is shown in Figure 4. There are few tests only in the upper part of the crust and the strength values seem to be quite low compared to other results as will be discussed later.

Dilatometer Tests: The results of the dilatometer tests were quite consistent as given in Figure 5 which shows the variation of horizontal stress index with depth. Similarly the modulus profile is given in Figure 6. The shear strength values with depth are given in Figure 7 and show a clear decrease with depth and assume a constant value at about 3.5 m, that is, when the soft clay stratum was reached.

Other correlations with depth were obtained, such as the constraint modulus, M_D, the overconsolidation ratio (OCR) and the coefficient of earth pressure at rest K_0. Since these values are all related to the three basic indices given by Equations (4a, b, and c) they are not discussed here due to space limitations.

Pressuremeter Tests: The pressuremeter tests results were almost identical for different boreholes. Since for most tests four reloading cycles were performed, four different moduli values could be estimated. The boreholes were carefully trimmed, as mentioned earlier, in order to minimize soil disturbance effects. The maximum applied pressure varied from 1400 kPa at a depth of 1.5 m to values of about 400 kPa. All pressuremeter test curves resemble closely those which one generally associates with the "ideal" curve. This suggests that the insertion technique adopted in this study was adequate and that any scatter of results was in fact due to the non-homogeneity of the crust material. The derived modulus values with depth from the initial and first cycles are given in Figure 8. Additional load cycles produced slight increases in modulus values. On the average, values of 60, 70, 80 and 90 MPa were obtained at a depth of 1.5 m corresponding to the first, second, third and fourth cycle loading curves. At a depth 4.5 m the corresponding moduli were 15, 20, 25 and 30 MPa. Generally it can be said that the increase in modulus in the crust material from the initial to the first cycle was in thee order of 50 percent whereas subsequent load cycles caused an average increase of about 14 percent.

The shear strength was evaluated from the relationship proposed by Gibson and Anderson (1961) as given by Equation (6). The resulting strength profile is shown in Figure 9. These strength values are somewhat high when compared to those obtained from triaxial compression tests and to those reported by Eden and Law (1980) for that site.

DISCUSSION

A comparison of results for the modulus of deformation for the various test devices is given in Figure 10. It should be mentioned that each data point in this figure represents the mean of all the test points from the respective techniques. Also given are the mean values from the SPT profile converted to modulus values. An arbitrary conversion factor as suggested by Al-Khafaji et al.

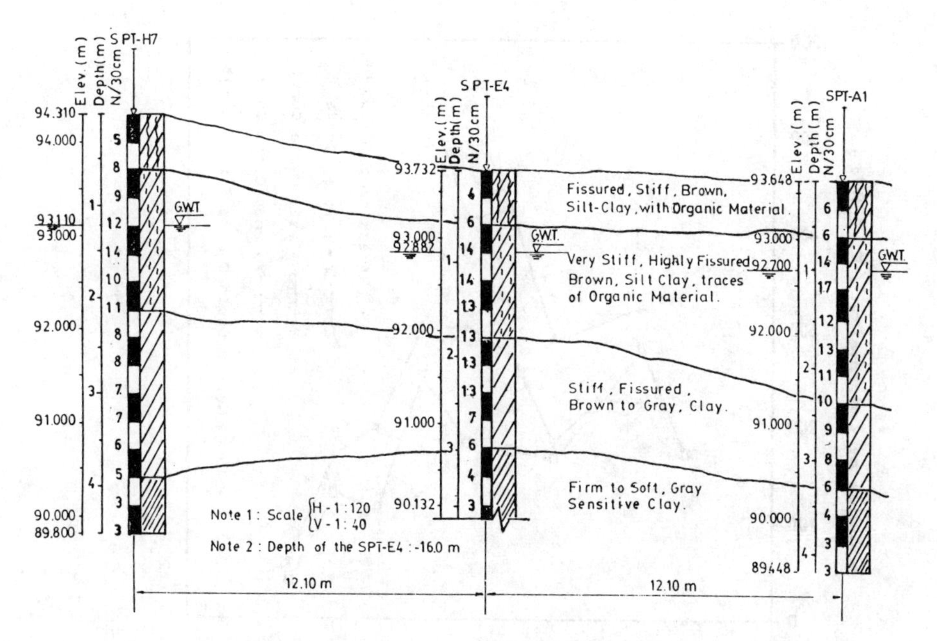

FIG. 1. SOIL PROFILE AT TEST SITE

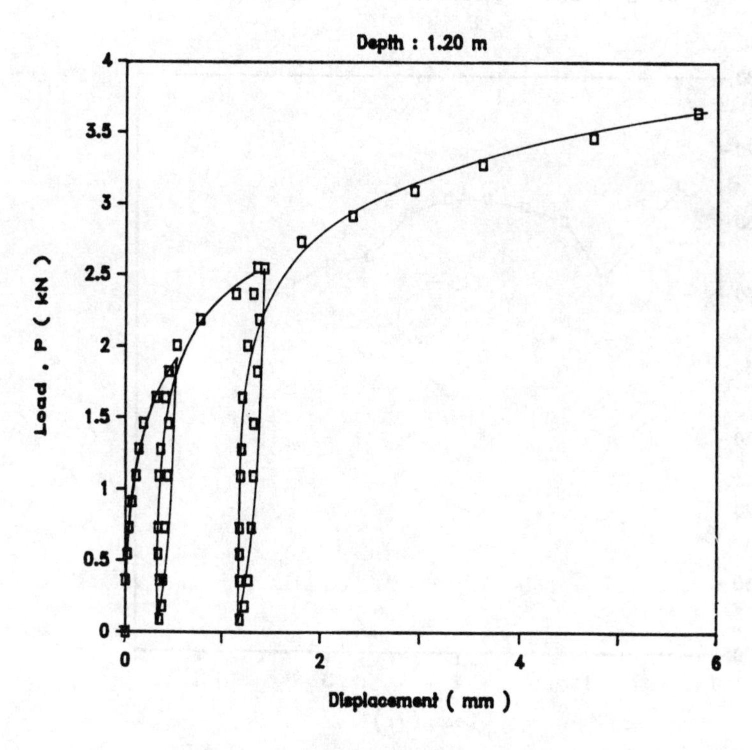

FIG. 2. SCREW PLATE TEST RESULT

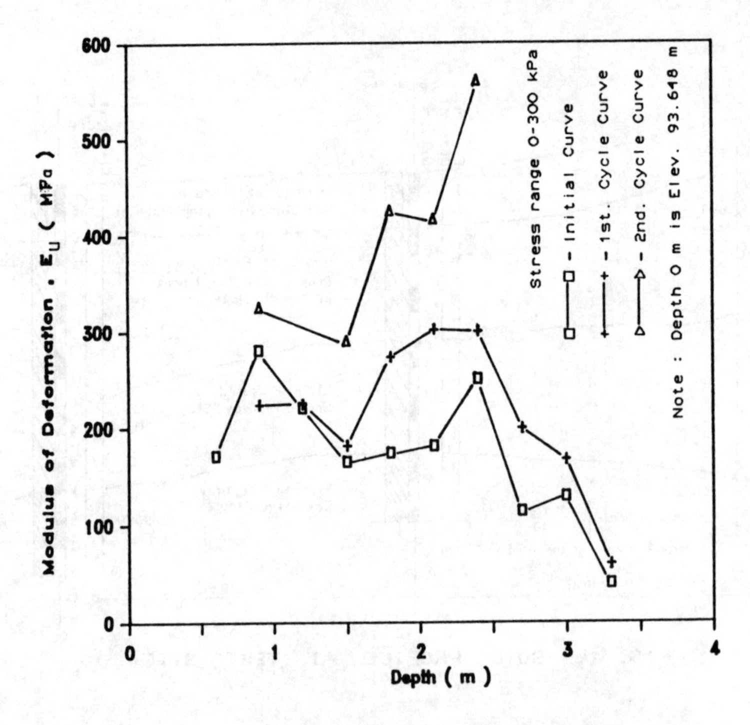

FIG. 3. SCT - MODULUS OF DEFORMATION

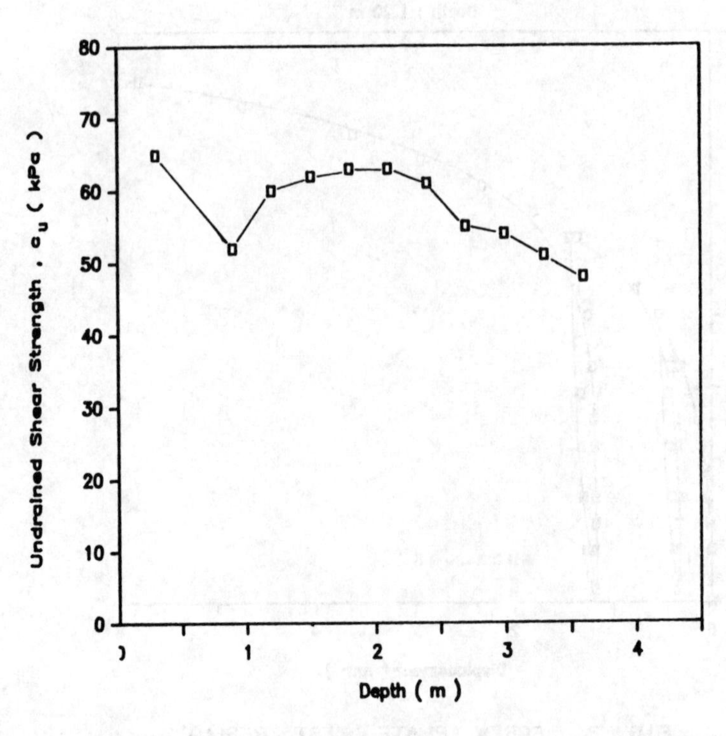

FIG. 4. SCT - UNDRAINED SHEAR STRENGTH

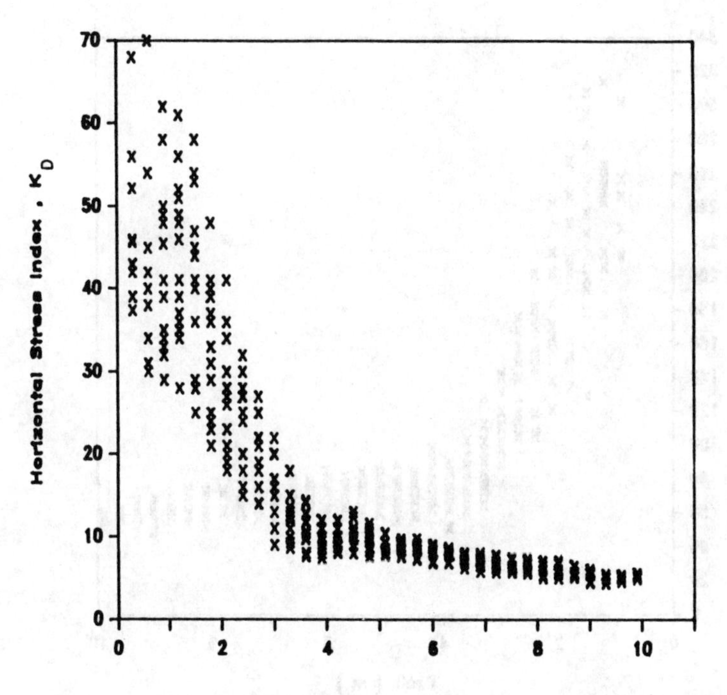

FIG. 5. VARIATION OF HORIZONTAL STRESS INDEX

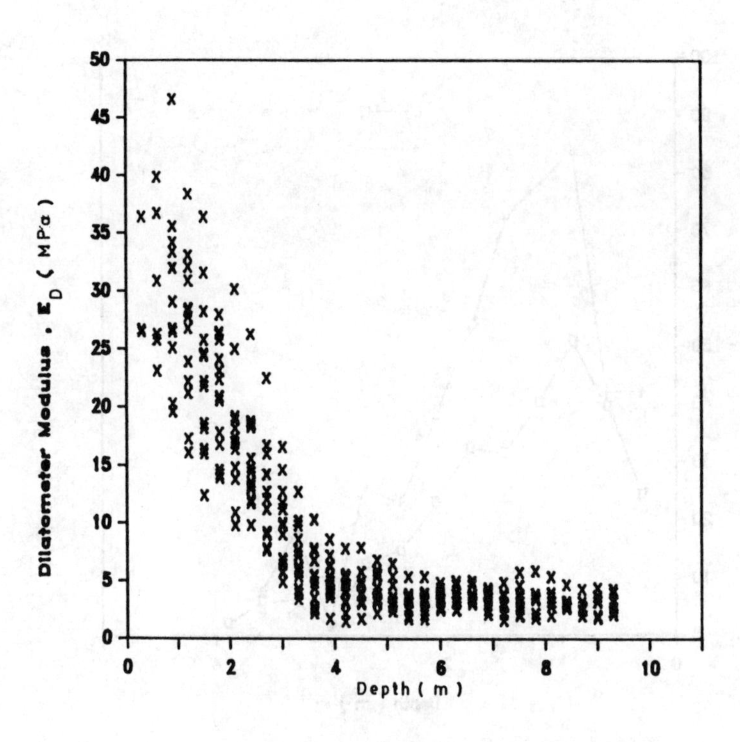

FIG. 6. DILATOMETER MODULUS

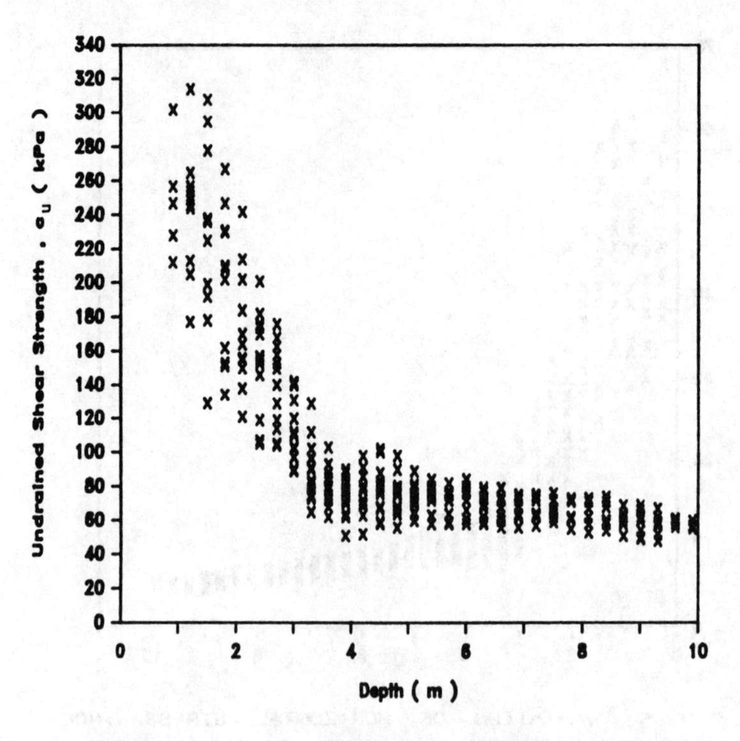

FIG. 7 . DMT - UNDRAINED SHEAR STRENGTH

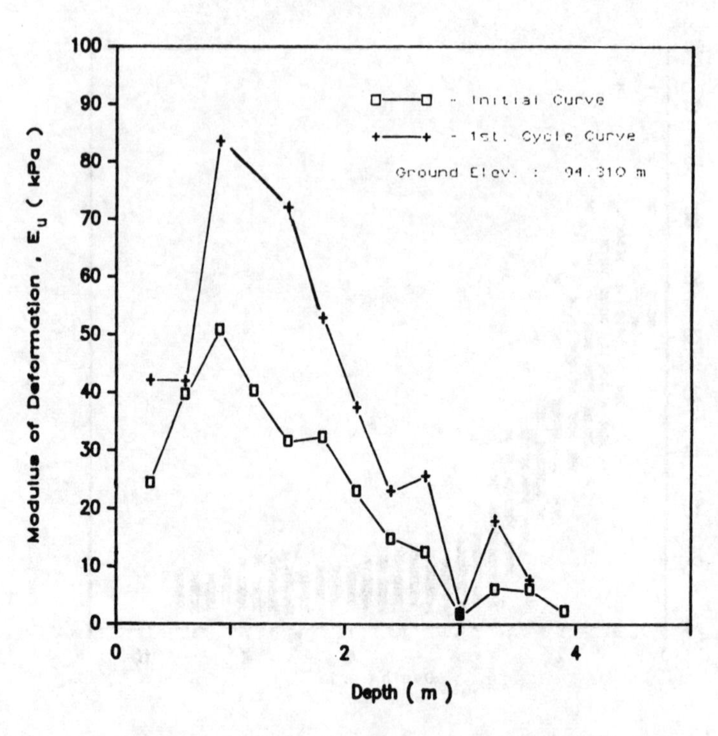

FIG. 8 . PMT - MODULUS OF DEFORMATION

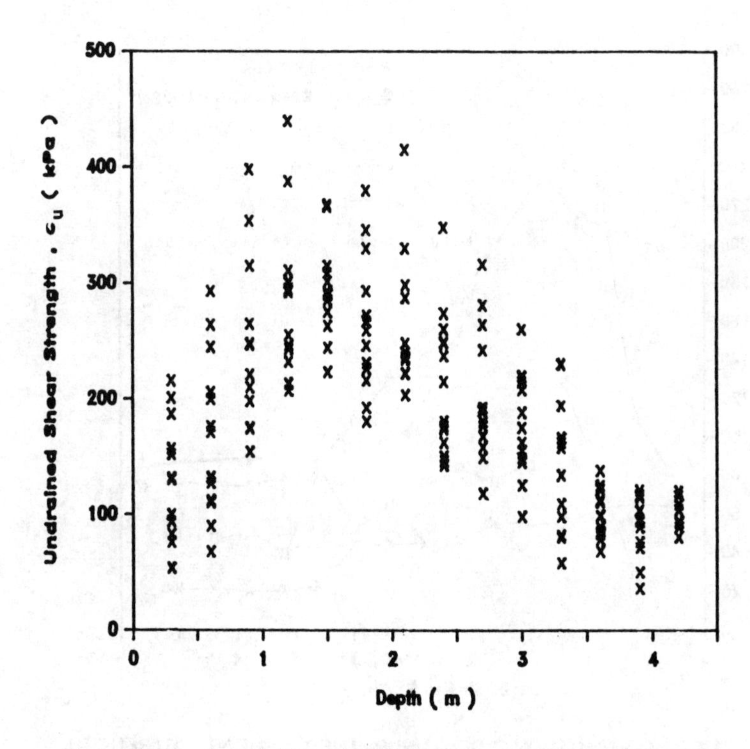

FIG. 9. PMT - UNDRAINED SHEAR STRENGTH

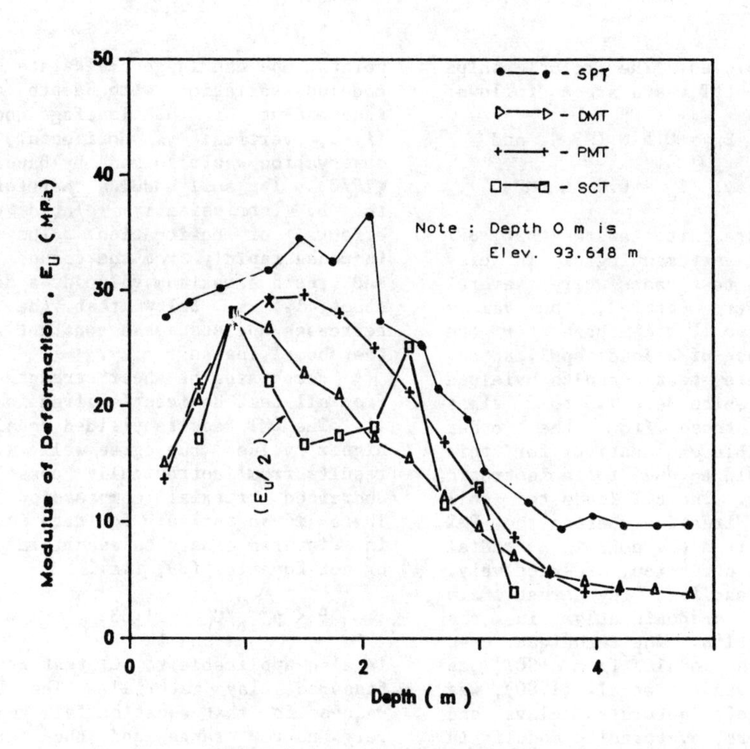

FIG. 10. COMPARISON OF MODULUS OF DEFORMATION

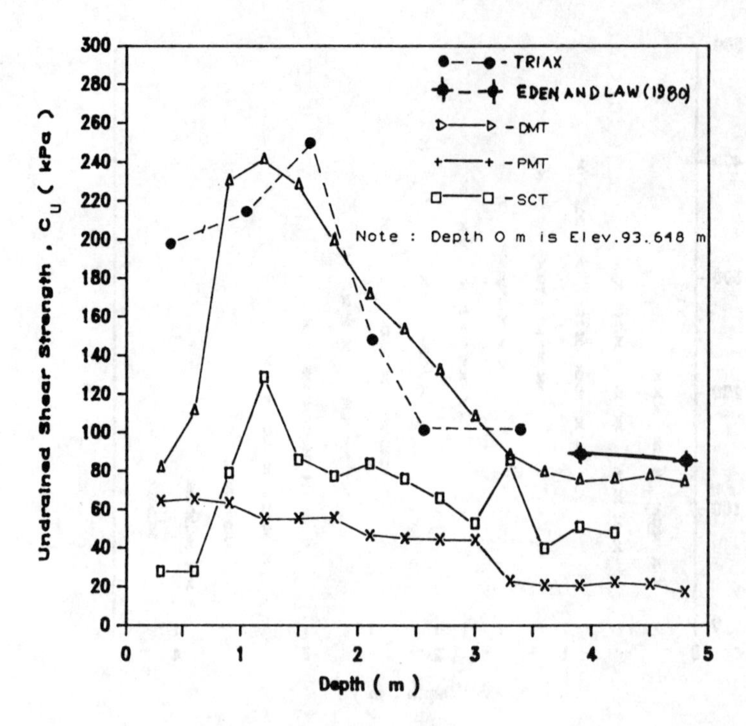

FIG. 11. COMPARISON OF UNDRAINED SHEAR STRENGTH

(1986) was employed. The relationships adopted for the SPT tests are as follows:

For the crust: E_u = 0.5 N (MPa), and

For the soft clay: E_u = 0.25 N (MPa)

The authors are quite aware that SPT values are not very meaningful in clay deposits. For test cases where several load cycles were carried, the values shown in Figure 10 are those for the initial response of a load application. The screw plate test results yielded modulus values which were 1.5 to 2 times greater than those from the other devices. Possible explanations for this phenomenon could be due to anisotropic soil behaviour. The SCT loads the soil in a vertical direction whereas the PMT and the DMT strain the soil in a radial and horizontal direction, respectively. As mentioned earlier, the crust main fissures are predominantly in the vertical direction. The technique used to interpret the modulus from a SCT, as proposed by Selvadurai et al. (1980), was intended for soft, saturated clays and might, therefore, overpredict moduli in fissured clays. Considering these points, one can in general state that the modulus variation with depth appeared independent of the loading conditions (i.e., vertical vs horizontal). This observation was also made by Bauer et al. (1973). The soil modulus therefore seems to be intrinsically related to the effects of desiccation. The values increase rapidly from the ground surface and reach a maximum value at a depth of about 1.5 m. Below that the modulus decreases and assumes a constant value in the underlying soft clay.

A comparison of shear strength values from all test devices is given in Figure 11. The DMT results yielded consistently higher values but agree well with the results from isotropically consolidated-undrained triaxial compression tests. There is so far very few data available in fissured crusts to ascertain whether or not Equation (3), i.e.,

$$9 \leq p_{ult}/C_u \leq 11.335$$

is also applicable to SCT test results in fissured clay material. The boundary values in that equation fall within a very narrow range and the graphical procedure to estimate the ultimate load,

p_{ult}, might not necessarily be indicative of the true strength of the desiccated crust. Values of undrained strengths for weathered Champlain Sea crust were reported by Lefebvre et al. (1987). Vane strength value were considerably higher than corresponding values from plate bearing and shear box tests. The latter two tests were carried out in a trench and the soil probably suffered from stress relief and opening up of fissures due to removal of overburden. Bauer et al. (1973) and Hanna (1976) made similar observations that strength values from plate bearing tests in trenches were considerably lower than those obtained from field vanes.

The strength values determined by Eden and Law (1980) from field vane and pressuremeter tests at at the same NRC site are also given in Figure 11 and agree well with those from DMT and PMT. Unfortunately, Eden and Law started their tests at a depth of 4 m below ground surface, i.e., below the stiff crust, and therefore a direct comparison with strength values within the crust is not possible.

CONCLUSIONS

From the present experimental investigation employing essentially three in situ devices and techniques (i.e., SCT, DMT and PMT) to estimate some of the geotechnical properties, the following conclusions can be drawn.

The three in situ devices yielded reasonable estimates of the modulus of deformation when compared to one another and to results published in the literature for this type of deposit.

The various techniques employed in this study provided qualitative and quantitative information for the intrinsic properties of the crust, despite subjecting the soil to different loading conditions, strain rates, etc. The results form each technique could give a profile of soil modulus and shear strength with depth.

The screw plate test simulates most accurately the load-deformation response of foundations, i.e., footings, rafts, etc., located on or within the clay crust. The modulus values derived from these tests were generally higher by a factor of two when compared to corresponding values from other techniques. Besides a SCT is quite time consuming, in particular when an independent load application system is needed.

The pressuremeter has the potential in providing a quick estimate of strength and deformation properties of the crust. The preboring technique, using thin-walled Shelby tubes, followed by downhole testing resulted in consistent and repeatable parameters of the crust. This technique proved to be unsatisfactory in the underlying soft clay due to soil yielding and in many cases loss of soil from the Shelby tubes. The self-boring technique would be more suitable in the softer deposits.

The dilatometer test (DMT) is quite simple and quick and the results were consistent and repeatable. The relationships to estimate the various geotechnical are quite empirical and the assumptions on which they are based are quite complex and not always convincing. For these reasons and until more data for comparison are available results from DMT should be considered for index and classification purposes only. These comments apply in particular to DMT results in desiccated, fissured crust material.

In summary, it can be concluded that in order to establish the soil properties of a stiff, desiccated crust, requires a larger number of tests than for homogeneous and uniform soil deposits. Field tests are more meaningful than laboratory testing due to reasons discussed before. In situ testing is in general quite simple and the results are obtained as soil-deformation responses or resistance values. The difficulty lies in obtaining meaningful interpretations of these basic results in order to obtain commonly accepted parameters, such as modulus and strength values.

ACKNOWLEDGMENTS

The authors wish to acknowledge the financial assistance of this study through a Natural Sciences and Engineering Research Council (NSERC) Grant held by the first author. The second author was given partial financial assistance through a grant provided by the Conselo Nacional de Pesquisa (CNPq) from Brazil.

REFERENCES

Baguelin, F., Jezequel, J.F., Shields, D.H. (1978). The Pressuremeter and Foundation Engineering. Trans. Tech. Publications, Clausthal, Germany.

Bauer, G.E.A., Scott, J.D. and Shields, D.H. (1973). The Deformation

Properties of a Clay Crust. Proc. of the 8th ICSMFE, Moscow, Vol.1.1, pp.31-38.

Eden, W.J. and Crawford, C.B. (1957). Geotechnical Properties of Leda Clay in Ottawa Area. Proc. of the 4th ICSMFE, London, Vol.1, pp.22-27.

Eden, W.J. and Law, K.T. (1980). Comparison of Undrained Shear Strength Results Obtained By Different Test Methods in Soft Clays. CGJ, Vol.17, pp.369-381.

Gibson, R.E. and Anderson, W.F. (1961). In Situ Measurements of Soil Properties With the Pressuremeter. Civil Engineering Public Works Review, London, pp.615-618.

Hanna, A.J. (1976). The Deformation Characteristics of Soil. M.A.Sc. Thesis, Department of Civil Engineering, University of Ottawa, Canada.

Janbu, N. and Senneset, K. (1973). Field Compressometer: Principles and Applications. Proc. of the 8th. ICSMFE, Moscow, Vol.1, pp.191-198.

Lefebvre, G., Pare, J.-J. and Dascal, O. (1987). Undrained Shear Strength in the Surficial Weathered Crust. Can. Geot. J., Vol.24, No.1, Feburary, pp.23-24.

Marchetti, S. (1980). In Situ Tests By Flat Dilatometer. JGED of the ASCE, Vol.106, GT.03, pp.229-321.

Marchetti, S. and Crapps, D.K. (1981). Flat Dilatometer Manual. Schmertmann and Crapps Inc., Gainesville, Fla.

Schmertmann, J.H. (1981). In Situ Tests by Flat Dilatometer. Discussion. JGED of the ASCE, Vol.107, GT.06, pp.831-832.

Schmertmann, J.H. (1982). A Method for Determining the Friction Angle in Sands from the Marchetti Dilatometer Test. Proc. ESOPT II, Amsterdam.

Schwab, E.F. and Broms, B.B. (1977). Pressure-Settlement-Time Relationship By Screw Plate Tests In Situ. Proc. of the 9th ICSMFE, Tokyo, Vol.1.

Selvadurai, A.P.S. and Nicholas, T.J. (1979). A Theoretical Assessment of the Screw Plate Test. Proc. of the 3rd. International Conference on Numerical Methods in Geomechanics, Aachen, Germany, Vol.3, pp.1245-1252.

Selvadurai, A.P.S., Bauer, G.E. and Nicholas, J.J. (1980). Screw Plate Testing of a Soft Clay. CGJ, Vol.17, No.4, pp.465-472.

Shields, D.H. and Bauer, G.E.A. (1975). Determination of the Modulus of Deformation of a Sensitive Clay Using Laboratory and In Situ Tests. Proc. of the ISMSP, ASCE, Raleigh, North Carolina, Vol.1, pp.395-421.

Tanaka, A. (1986). In Situ Measurements for Shear Strength and Deformation Characteristics of a Fissured Clay Crust. Ph.D. Thesis, Department of Civil Engineering, Carleton University, Canada.

Penetration Testing 1988, ISOPT-1, De Ruiter (ed.)
© 1988 Balkema, Rotterdam, ISBN 90 6191 801 4

Some experience with the dilatometer test in Singapore

M.F.Chang
Nanyang Technological Institute, Singapore

ABSTRACT: This paper describes four flat dilatometer tests (DMTs) carried out in the recent deposits of Singapore. Results of these tests were compared with those by other investigation methods including the piezocone, the field vane, the UU triaxial, and the oedometer tests. Special emphasis was placed on the stress history of the deposits as interpreted by various test methods. The validity of some existing empirical correlations for the interpretation of the DMT results is reviewed. The role of the DMT in site investigation as compared to the piezocone test is discussed.

1. INTRODUCTION

In the past 15 years, many new types of equipment and techniques for in-situ testing of soils have been developed. The newly invented flat dilatometer test (DMT) is a fine example. The test is gradually gaining acceptance of the geotechnical profession.

This paper presents the results of four dilatometer tests carried out in the Kallang Formation in Singapore. Results of these tests are interpreted with empirical correlations reported by Marchetti (1980). The DMT interpreted overconsolidation ratio, OCR, profile is compared with the pore pressure ratio, B_q, profile deduced from the piezocone test (CPTU) or the oedometer values. The DMT interpreted undrained shear strength and constrained modulus are also compared with the field vane strength and the oedometer result, respectively. The role of the DMT in site investigation as compared to the CPTU is discussed.

2. GEOLOGY AND SITE DESCRIPTION

The Kallang Formation of Singapore consists of soils with marine, alluvial, littoral and estuarine origins. The Singapore marine clay is by far the most important unit of the Formation. Peaty deposits, mostly clayey, are also common. The Singapore marine clay consists of the upper and the lower members, separated by an intermediate layer which is moderately overconsolidated due to weathering and desiccation. The lower member is lightly overconsolidated due to secondary compression and its overconsolidation ratio (OCR) is between 1.3 and 1.6 according to Hanzawa and Adachi (1983). The upper member is historically known to be normally consolidated, although earlier consolidation test results seem to suggest that the clay is lightly overconsolidated with its OCR ranging from 1.0 to 1.5 (Tan, 1983). Hanzawa and Adachi (1983) attributed the overconsolidation of the upper member to chemical bonding or cementation. The peaty deposits are marked by very large variations in moisture content and Atterberg limits. Typical engineering properties of the Singapore marine clay and the peaty deposits were reported by Tan (1983).

Test Sites A-1, A-2 and A-4 were underlain predominantly by the Singapore marine clay. Test Site A-2 was on a recently reclaimed land. Test Site A-3, located next to the Singapore river, involved primarily the peaty deposits and the lower marine clay. The index properties of the marine clay and the peaty clay present at various test sites are summarized in Table 1.

3. TEST EQUIPMENT AND PROCEDURES

A Marchetti dilatometer consisting of a 95 mm x 140 mm stainless steel blade

Table 1 : Summary of index properties of clays present at various test sites underlain by the Kallang Formation

Test Sites	A-1		A-2		A-3		A-4	
Soil Strata	Upper marine clay	Lower marine clay	Upper marine clay	Lower marine clay	Peaty clay	Lower marine clay	Upper marine clay	Lower marine clay
Liquid Limit LL, %	84	71	77	65	44-145	53	88	71
Plasticity Index, PI, %	59	47	46	38	26-106	37	50	38
Liquidity Index, LI	0.90	0.68	0.78	0.68	0.30-0.72	0.53	0.55	0.64
Sensitivity, S_t	3.1	2.9	4.2	3.9	3.7	-	5.3	3.7

with a 60 mm diameter expandable steel membrane mounted on one side of the blade was employed. Regulated nitrogen gas and a control unit equipped with a precision gauge were used. The dilatometer blade was jacked into the ground at a rate of 2 cm/sec using a cone penetrometer rig. At every 20 cm depth intervals the penetration was stopped. The pressure required to lift-off the membrane and that required to move the center of the membrane by 1.0 mm into the soil were recorded, each within 15 to 30 seconds. These two pressure readings yield P_0 and P_1, respectively, after corrections are made for membrane stiffness.

Apart from the dilatometer test, several well-established tests were also carried out either in-situ or in the laboratory for comparison purposes. These included the field vane test (FVT), the piezocone test (CPTU), the oedometer test and the UU triaxial tests. FVTs were carried out at all the test sites using the Geonor push-in type vane borer. CPTUs were carried out at Sites A-1 through A-3 using Delft's piezocones with a net area ratio of 0.8. The porous disc which acts as a filter for the piezocone is located at the base of the conical tip. A series of oedometer tests were carried out on 'undisturbed' samples recovered from Site A-4 using a load increment ratio of 1.0 and a load duration corresponding to the end of primary consolidation. Standard UU triaxial tests were also carried out at both sites A-1 and A-2 to determine the undrained shear strength.

4. TEST RESULTS

In the interpretation of DMT results, three dilatometer indices: the material index, I_D, the horizontal stress index, K_D, and the dilatometer modulus, E_D, expressed in terms of p_0 and p_1, are first calculated according to Marchetti (1980). Empirical correlations established by Marchetti (1980), which relate the DMT indices to various engineering parameters, are then adopted for the estimation of the OCR, the undrained shear strength, and the constrained modulus for the soils investigated.

Figure 1 shows the soil stratification and the CPTU and undrained strength profiles at Test Site A-1. The q_c-value seems to increase linearly with depth and indicate that the marine clay is normally to very lightly overconsolidated. A sudden drop in the q_c-value is noticeable in the organic silty clay immediately below the upper marine clay and the sandy to silty clay immediately below the lower marine clay. The undrained strength s_u profiles shown are based on FVTs and triaxial UU tests. The s_u-profiles appear to indicate that both marine members, especially the lower member, are overconsolidated to some extent.

The stress history and the degree of overconsolidation of the Singapore marine clay are still not fully understood due to a lack of high quality consolidation test data. To relate the CPTU result to the stress history of the soil deposit investigated, a pore pressure ratio, $B_q = (u - u_o)/(q_c - \sigma_{vo})$, where u is the total pore pressure, u_o is the in-situ water pressure, and σ_{vo} is the total overburden pressure, recommended by Senneset, Janbu and Svanø (1982), was calculated. The resulting B_q-profile is also shown in Figure 1. It is noted that the B_q-value varies from 0.5 to 0.75 for the upper marine clay. For the lower marine clay the B_q-value is between 0.65 and 0.85. Figure 2 shows a B_q versus OCR plot derived by Wroth (1984), on the basis of data reported by Lacasse and Lunne (1982), for Onsøy clay, which has a slightly higher sensitivity (S_t = 5 to 7) and a slightly lower plasticity index (PI = 23 to 36%) than the Singapore marine clay. Assuming that the plot is applicable to the marine members here, the interpreted OCR-values are between 1.0 and 1.7 for the upper member and 1.0 and 1.2 for the lower member. The lower marine clay is practically normally consolidated based on the CPTU data at Test Site A. This is different from what was understood historically and suggested by the s_u-profile. The degree of overconsolidation also appears to be slightly higher than that normally considered (OCR = 1 to 1.5) for the upper marine clay.

The interpreted OCR and s_u based on the DMT carried out at Site A-1 are presented in Figure 1 together with other soil data. It is noted that the OCR is around 1.9 to 2.0 for the upper marine clay and it decreases with depth from 2.2 to 1.0 for the lower marine clay, indicating that both marine members are lightly overconsolidated. These values seem higher than those normally quoted for the Singapore marine clay and interpreted from the CPTU. Cementation as quoted by Hanzawa and Adachi (1983) could be partially responsible for the high OCR-values obtained by the DMT at least for the upper marine clay. The DMT interpreted OCR is around 5.5 for the immediate layer, which is also higher than the range of 2.5 to 4.0 reported by Hanzawa and Adachi (1983). The B_q-profile, when plotted in a descending order as shown in the figure, however, matches reasonably well with the DMT interpreted OCR-profile. It is noted, that in the seem-meta-stable sandy to silty clay layer immediately below the lower marine clay, the DMT interpreted OCR is close to zero and the B_q-value is larger than 1.0.

The DMT interpreted s_u agrees reasonably well with the vane strength

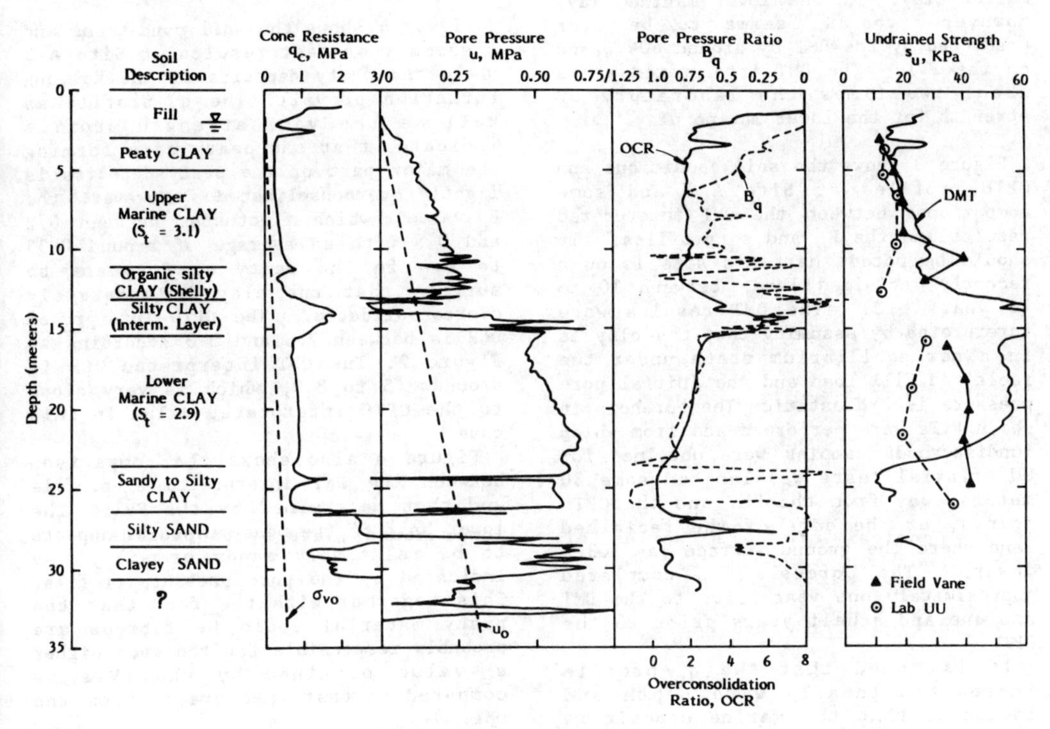

Figure 1 Summary of soil conditions and test results at Site A-1

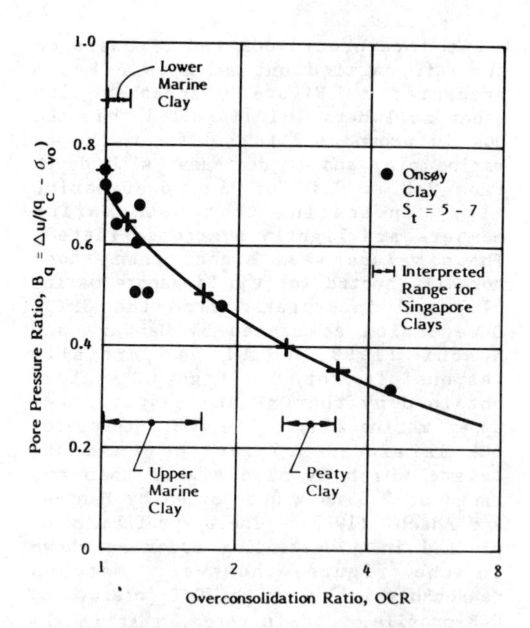

Figure 2 Piezocone pore pressure ratio versus overconsolidation ratio at Onsøy (after Wroth, 1984)

and is slightly higher than the laboratory UU strength, in the upper marine clay. In the lower marine clay, however, the s_u seems to be over predicted by the DMT by around 40% based on the FVT. The DMT interpreted s_u is about two times the laboratory UU strength for the lower marine clay.

Figure 3 shows the soil conditions and CPTU profiles at Site A-2 and some comparisons between the DMT interpreted results and the B_q and s_u profiles. It should be noted that this site is on a recently reclaimed land, between 10 to 13 years old. The DMT results were interpreted by assuming that the clay is in a new equilibrium state under the reclaimed fill load and the initial pore pressure is hydrostatic. The borehole in which FVTs were performed and from which 'undisturbed' samples were obtained for UU triaxial tests was located some 30 meters away from the DMT and the CPTU points, at the edge of the reclaimed land where the ground surface was 1.3 m lower. The borehole was completed approximately one year prior to the DMT and one and a half years prior to the CPTU.

It is noted that the q_c-profile increases linearly with depth and indicates that the marine deposit is practically normally consolidated. The B_q-profile fluctuates drastically between 0.8 and 1.1 in the upper member and assumes more stable values of around 0.8 to 0.9 in the lower member, indicating that the deposit is normally consolidated. Unlike the relatively uniform and steady OCR-profile found at Site A-1, the DMT interpreted OCR-profile at Site A-2 seems to fluctuate around a vertical line representing an OCR of 1.5 in both marine members. This seems be a direct reflection of the nature of the deposit, which is in a non-equilibrium state and is still undergoing consolidation. The OCR-value, which should be less than unity for such a deposit, seems to have been overestimated by the DMT. The general trend of the DMT interpreted OCR-profile, however, appears remarkably similar to the B_q-profile.

The DMT interpreted s_u-profile, as shown in Figure 3, seems to fluctuate and increase roughly linearly with depth. The interpreted s_u-value is consistently higher than the undrained strength measured one year prior to the DMT. The facts that the clay is still undergoing consolidation and that the OCR of the clay is overpredicted by the DMT are mainly responsible for the difference.

Figure 4 shows the soil conditions and a summary of test results at Site A-3 where the peaty deposits of the Kallang Formation prevail. The q_c-profile as well as the vane strength profile indicates that the peaty clay forming the major part of the peaty deposits is lightly overconsolidated. However, the B_q-value, which fluctuates between 0.2 and 0.5 with an average of around 0.35 to 0.4 in the peaty clay, seems to suggest that the clay is moderately overconsolidated. The CPTU interpreted OCR is between 2.6 and 3.3 according to Figure 2. The DMT interpreted OCR is around 2.5 to 3.0, which is very close to the CPTU interpreted value in this case.

Figure 4 also shows the comparison between the DMT interpreted s_u-profile and that determined by the FVT. The lower half of the peaty deposits appears to be relatively sandy or silty, as indicated by the pore pressure profile. This together with the fact that the peaty material could be fibrous are probably responsible for the much higher s_u-value obtained by the FVTs as compared to that interpreted from the DMT.

It is noted that the q_c-value in the

lower marine clay present at Site A-3 is relatively low. The clay also developed a relatively high excess pore pressure during the piezocone test. The corresponding B_q-value is around 1.1, indicating that the material is either meta-stable or extra sensitive. A similar change in the DMT interpreted s_u-profile, which closely resembles the q_c-profile, is also noticeable.

Figure 5 shows the soil conditions at Site A-4 and some comparisons between the DMT interpreted parameters and the corresponding field vane and oedometer values. It is noted that DMT interpreted s_u-profile is very similar to the vane strength profile with the DMT values being slightly higher. The laboratory determined OCR and constrained modulus, M, however, seem to be only half of the corresponding DMT values. Sample disturbance is believed to be partially responsible for the differences, although it is likely that the OCR-value is overestimated by the DMT. For comparison, a M-value corresponding to the in-situ stress state was estimated from the recompression index, C_r, obtained from the oedometer test using a relation: $M = (1 + e_o)C_r/0.435\ \sigma_{vo}'$, where e_o is the initial void ratio and σ_{vo}' is the effective over-burden pressure. The resulting M-profile is also plotted in Figure 5. The calculated M-profile seems to agree fairly well with the DMT interpreted M-profile.

The DMT result also seems to indicate an interesting fact regarding the stress history of Site A-4. Similar to Site A-1, the upper marine clay seems to have been overconsolidated to an extent greater than the lower marine clay. This is reflected in the DMT interpreted s_u-profile as well as the OCR-profile.

Figure 6 shows the variation of the DMT's horizontal stress index, K_D, with the normalized undrained strength, s_u/σ_{vo}', based on the FVTs and the DMTs carried out at Sites A-1, A-3 and A-4. Also shown is the correlation between K_D and s_u/σ_{vo}' proposed by Marchetti (1980). It seems that Marchetti's correlation overpredicts the undrained

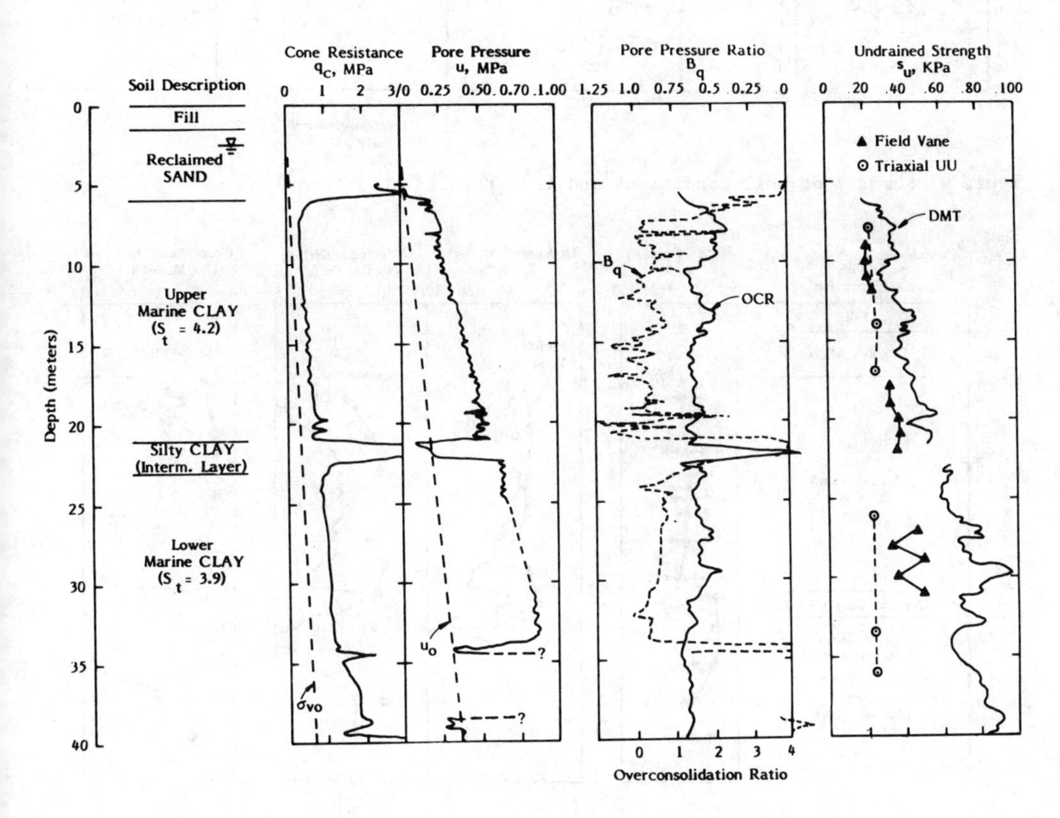

Figure 3 Summary of soil conditions and test results at Site A-2

strength for the marine clay, in general, mainly as a result of the over-estimated OCR. For the peaty deposits, the undrained strength is found to be underestimated by Marchetti's correlation. The measured vane strength could also be too high because the peaty deposits are sandy and probably fibrous.

5. DMT COMPARED WITH CPTU

The CPTU is an in-situ testing method of similar applications and slightly higher level of sophistication to the DMT. It is of great interest to compare the newly developed DMT with this well-established method and to explore the

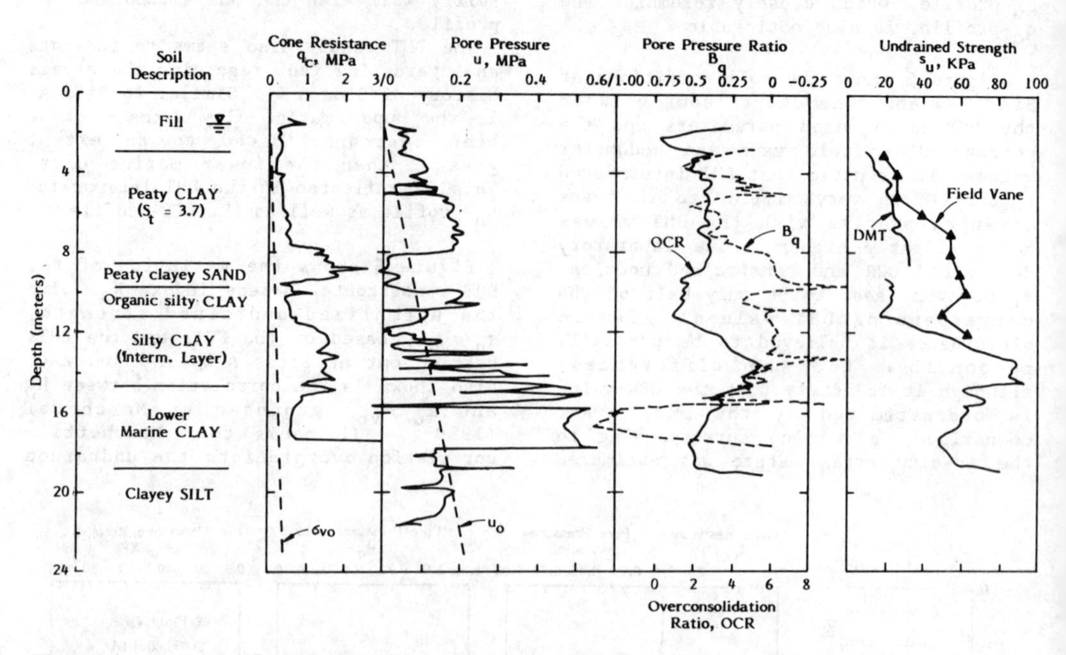

Figure 4 Summary of soil conditions and test results at Site A-3

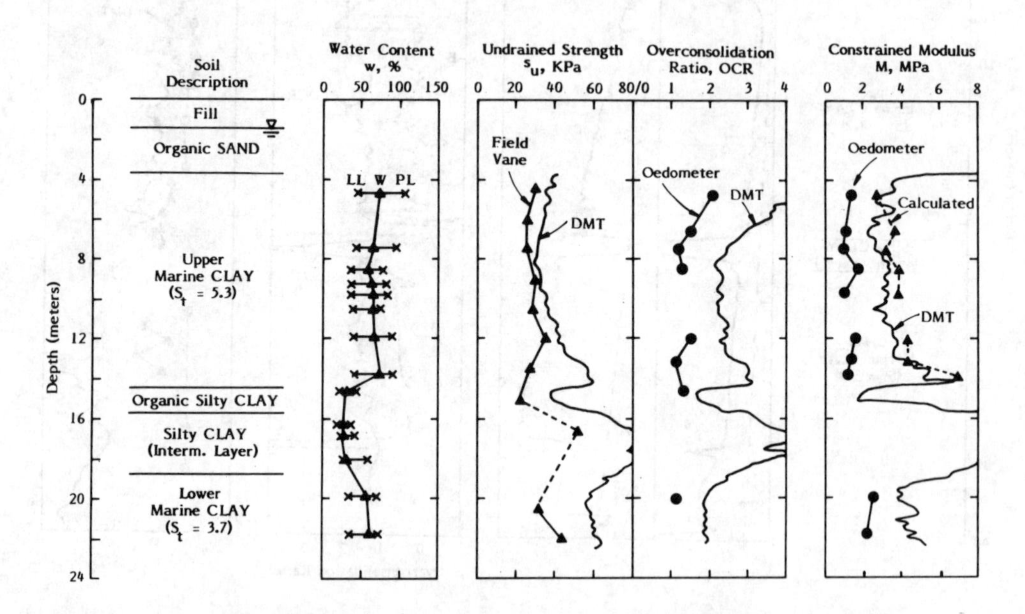

Figure 5 Soil conditions at Test Site A-4 and some interpreted DMT results

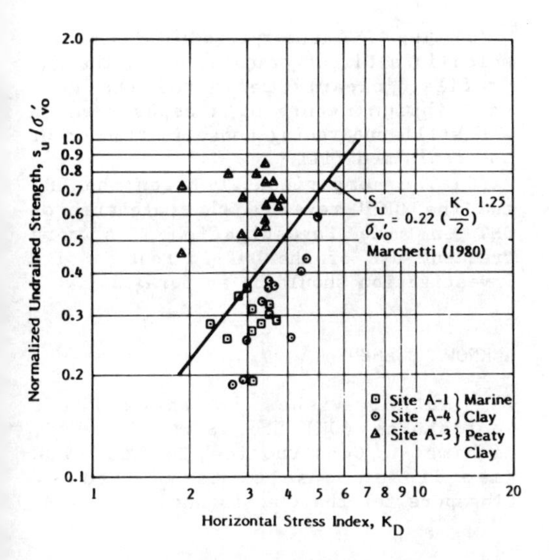

Figure 6 Horizontal stress index versus normalized undrained strength based on FVT for Kallang Formation

merits of each test.

In terms of material characterization, the CPTU, which provides continuous records of q_c, penetration pore pressure, u, and local shaft friction, f_s, with depth, can produce much more accurate soil stratifications than the DMT does. The material index, I_D, on which the DMT's classification of soil type is based, is determined only at every 20 cm depth interval and is semi-continuous with depth.

The CPTU allows a pore pressure dissipation test to be carried out after the penetration is temporarily stopped. This offers a unique way of assessing the coefficient of consolidation of the soil which cannot be as easily accomplished by any other in-situ tests. The dilatometer test, in the current state of its development, provides no information in this regard, although the potential is there (Campanella, et al 1985).

For shear strength evaluation, both the CPTU and the DMT rely on empirical correlations, although the CPTU is a shear strength test and the DMT is basically a deformation test.

As to the evaluation of deformation properties, a handful of empirical correlations have been established based on the CPT. However, there is no unique correlation. As pointed out by Jamiolkowski et al (1985), empirical correlations between q_c and M or Young's modulus cannot be considered either highly reliable or of general validity due to the relatively large deformation involved in the CPT. The DMT, although it is carried out on a soil which has been prestrained by penetration, provides measurement of the stress-strain response of the soil. A more consistent and stronger correlation between the dilatometer modulus, E_D, and the deformation modulus of the in-situ soil is expected. The DMT seems to have more to offer in terms of this vital area of application.

Both the CPTU and the DMT are believed to be capable of detecting the stress history of a deposit. The pore pressure ratio, B_q, produced by the CPTU for a given soil, similar to the pore pressure parameter A_f, is directly related to the OCR of the soil. However, the generated pore pressure is influenced even to a much greater extent by the soil type. It also depends on a number of factors such as the rigidity index, the sensitivity of the soil, and the location of the filter (Campanella et al 1985). The CPTU, at present, can therefore only provide a qualitative measurement of the stress history. On the contrary, the horizontal stress index, K_D, measured in the DMT incorporates the horizontal stress at lift-off after a large strain expansion caused by penetration of the dilatometer blade. It is inherently a closer reflection of the stress history of a deposit. The K_D-profile has been found to be remarkably similar to the OCR-profile for many soil deposits investigated (Marchetti, 1980). For the tests in two Norwegian clays reported by Lacasse and Lunne (1982), the correlation between OCR and K_D as proposed by Marchetti (1980) seems capable of producing a reasonable approximation of the OCR-profile from the K_D-profile. The DMT appears to be a more useful test compared to the CPTU in term of stress history assessment.

The experience gained here in Singapore seems to support the above-mentioned argument. This is clearly illustrated in Figures 1, 3 and 4, which show the comparisons between the B_q profiles and the OCR profiles for the three test sites investigated. It is generally understood that the OCR in a deposit should be relatively independent of the material type and should vary gradually with depth without any abrupt changes. The OCR-profile interpreted by the DMT, such as that shown in Figure 1 for Site A-1, seems to be reasonable and representative of the normal stress history of a sedimentary formation, although the values may be on the high side. The B_q-profile, on the contrary, seems to fluctuate rather drastically, mainly with the variation in soil material with depth, especially in the upper 15 meters, although its general trend seems to agree with the DMT interpreted OCR-profile. The B_q-value is affected so predominately by the variation in material that, in the some cases, the B_q-profile could even give a distorted stress history profile such as that illustrated in Figure 4. The close-to-zero B_q-value observed in the zone between 8 and 14 m depth, if it is not backed-up by material descrip-tion, could be misinterpreted as a sign of heavy overconsolidation. The strata which are relatively sandy, in fact, have much lower OCR-values than the overlaying peaty clay according to the DMT.

The above comparisons seem to suggest that a more frequent use of the DMT in site investigation is justifiable and should be encouraged.

6. CONCLUSIONS

The following tentative conclusions seem relevant based on this study:

(1) The DMT produced OCR profile was qualitatively similar to the pore pressure ratio profile deduced from the CPTU for the recent deposits of Singapore.

(2) The OCR-value interpreted from the DMT based on Marchetti (1980) was found to be higher than those interpreted from the CPTU and determined by the odeometer test for the Singapore marine clay. The OCR-value was probably overestimated by the DMT in these cases.

(3) The undrained shear strength interpreted from the DMT based on Marchetti (1980) was around 5 to 20% and 40% higher than the field vane strength for the upper and lower marine clay, respectively. This was probably because the DMT overestimated the OCR-value of the marine clay. For the peaty clay investigated the DMT's value was around 20 to 40% lower than the field vane strength.

(4) The constrained modulus interpreted from the DMT was found to be higher than the oedometer value but very close to that estimated based on the recompression index for the upper marine clay of Singapore.

(5) The DMT interpreted OCR-profile, which is a direct translation of the K_D-profile, clearly reflected the non-equilibrium nature of a deposit which was still undergoing consolidation under the reclaimed fill.

(6) The comparisons between the DMT and the CPTU revealed the potential of DMT in site investigation. A more frequent use of the DMT in routine site investigation should be encouraged.

ACKNOWLEDGEMENT

The Author wishes to thank Delft Geotechnics (UK) Ltd, SPECS Pte Ltd, Nishimatsu Construction Co Ltd, and Kiso-Jiban Consultants Co Ltd of Singapore for their assistance.

REFERENCES

Campanella, R. G., Robertson, P. K., Gillespie, D. G. and Greig, J. 1985. Recent Developments in In-situ Testing of Soils, Proc. 11th ICSMFE, San Franscisco, Vol. 2, pp. 849-854.

Hanzawa, H. and Adachi, K. 1983. Overconsolidation of Alluvial Clays. Soils and Foundations, Vol. 23, Vol. 4, pp. 106-118.

Jamiolkowski, M., Ladd, C.C., Germaine, J.T., and Lancellotta, R. 1985. New Developments in Field and Laboratory Testing of Soils, Proc 11th ICSMFE, San Francisco, Vol. 1, pp. 57-153.

Lacasse, S. and Lunne T. 1982. Penetration Tests in Two Norwegian Clays, Proc 2nd European Symp. Penetration Testing, Amsterdam, Vol. 2, pp. 661-669.

Marchetti, S. 1980. In-situ Tests by Flat Dilatometer, J. Geotech. Engrg Div., ASCE, Vol. 106, GT3, pp. 299-321.

Senneset, K., Janbu, N. and Svanø, G. 1982. Strength and Deformation Parameters from Cone Penetration Tests, Proc. 2nd European Symp. Penetration Testing. Amsterdam, Vol. 2, pp. 863-870.

Tan S.L. 1983. Geotechnical Properties and Laboratory Testing of Soft Soils in Singapore, Proc 1st NTI Seminar on Construction Problems in Soft Soils, Singapore, pp. TSL1-47.

Wroth, C.P. 1984. The Interpretation of In-situ Tests, 24th Rankine Lecture, Geotechnique. Vol. 34, No. 4, pp. 449-489.

Penetration Testing 1988, ISOPT-1, De Ruiter (ed.)
© 1988 Balkema, Rotterdam, ISBN 90 6191 801 4

The influence of creep on in situ pore pressure dissipation tests

M.Fahey
University of Western Australia, Perth

ABSTRACT: The paper presents the results of pressuremeter holding tests which suggest that creep may have an important effect on the total radial stress during a strain-holding test, and on the cavity strain during a pressure-holding test. These results are difficult to explain using a soil model which does not incorporate strain rate or creep effects additional to consolidation effects. It is postulated that the dissipation tests conducted using a standard piezocone could be even more influenced by creep effects because of the much faster strain rates involved in cone penetration. However, since no measurements of total radial stress are available from a standard cone during a dissipation test, it is unlikely that any indication of the cause of pore pressure dissipation (i.e. whether it is due to drainage or "relaxation") would be obtained. The overall effect of neglecting the effects of creep is likely to be an overestimation of the value of the coefficient of consolidation .

1 INTRODUCTION

The relatively recent development of incorporating pore pressure transducers into electric friction–cone devices has dramatically increased the potential of the cone penetrometer both in identification and classification of soil, and in measurement of "fundamental" engineering properties of soil. The measurements of pore pressure obtained during penetration, coupled with the standard measurements of end stress (q_c) and skin friction (f_s) may be used in a purely empirical fashion to classify the soil. Alternatively, values of engineering parameters such as strength, overconsolidation history, friction angle, relative density etc. may be obtained using either published correlations or local experience. Further information about the consolidation behaviour of the soil may be obtained by halting the penetration at any stage, and monitoring the dissipation of excess pore pressure with time.

This paper is concerned with the use of the piezocone only for determining the coefficient of consolidation of the soil using dissipation tests. In particular, it is postulated that creep effects may have a significant influence on the value obtained for this parameter. The approach which has been adopted is to use the results of pressuremeter holding tests to provide some insight into this problem.

2 INTERPRETING CONE DISSIPATION TESTS

While much of the interpretation of cone data has been performed using an empirical approach, an alternative approach is to base the interpretation of the response on a careful analysis of the deformation of the soil around the advancing cone. Such an approach has been used by Baligh and his co-workers at M.I.T. for a number of years with some considerable success (e.g. Levadoux and Baligh, 1986). Their "strain path method" attempts to follow each soil element through its complete strain history during and subsequent to the cone penetration. A fundamental soil model is then used to predict the stress state (including excess pore pressure) in each element at each stage.

A major problem with any method such as the "strain path method" of Levadoux and Baligh is that the deformation of the soil around an advancing cone is very complex. Given the difficulty of accurately predicting soil response even in a single–element test (e.g. a triaxial test) on laboratory–prepared soil, the task of making accurate predictions where the deformation is complex and the soil is in its natural state is very daunting.

Rather than attempt to model the complex deformations around an advancing cone, a fruitful approach is to concentrate on the simpler problem of the deformation in the soil around expanding cylindrical or spherical cavities. When this problem is fully understood, the more complex problem can then be addressed in light of the findings of the study into the simpler problem. In practice, of course, both approaches can be pursued simultaneously.

3 PRESSUREMETER DISSIPATION TESTS

The deformation of soil around an expanding pressuremeter, especially of the self-boring type, is usually modelled using cylindrical cavity expansion theory. This approach was used by Clarke et al (1979) to analyse the so-called "holding test" in clay using the Cambridge version of the self-boring pressuremeter (a "Camkometer"). In this type of test, the membrane of the instrument is expanded quickly to a set value of cavity strain, ε, where $\varepsilon = \Delta r/r_0$. This results in generation of excess pore pressure which is measured by the pore pressure transducers attached to the expanding membrane. The strain is then held constant using the control features of the system, while the excess pore pressure dissipates.

The "strain-holding" test can be thought of as an idealization of the soil deformation and subsequent pore pressure dissipation in a cone dissipation test. In the cone test, the "cavity strain" is essentially infinite, the deformation may be a combination of end-bearing failure and cylindrical cavity expansion, and, depending on the location of the pore pressure transducer, the drainage direction relevant to the pore pressure being measured may be both radial and vertical. In contrast, the strain in the pressuremeter test is finite, the deformation is likely to be predominantly in a plane strain mode, and the drainage direction is mainly radial. The presssuremeter problem is thus inherently more tractable than the cone penetration problem.

A variation of the holding test has recently been described by Fahey and Carter (1986). In this test, the cavity pressure, rather than the cavity strain, is held constant after initial expansion. The test has thus been designated a "pressure-holding" test as distinct from the "strain-holding" test of Clarke et al.

The method used by Clarke et al for interpreting the results of holding tests is based on an elastic-plastic two phase model to predict the excess pore pressure generated by the expansion, and elastic consolidation theory to model the dissipation stage. The approach adopted by Fahey and Carter (1986) for both strain-holding and pressure-holding tests consisted of using a finite element cavity expansion programme (CAMFE: Carter, 1978) which incorporates both Mohr-Coulomb and Cam Clay soil models.

A feature of elastic consolidation predictions of the dissipation phase of the holding test is that identical behaviour is predicted for strain-holding and pressure-holding tests. That is, a constant cavity strain implies a constant cavity pressure, and vice versa. In fact, however, during a strain-holding test, the cavity pressure must be reduced to maintain a constant strain, while during a pressure-holding test, the cavity strain increases during the dissipation phase. This feature is shown to occur to a limited extent if a simple elastic plastic consolidation model is used for the dissipation phase. If a Cam Clay model is used in a finite element programme such as CAMFE, the correct trends can be more closely modelled.

The greater degree of simplicity of soil deformation associated with the pressuremeter expansion and holding tests, compared to the piezocone penetration and dissipation tests, has already been mentioned. A further important advantage of using the pressuremeter to understand the processes controlling piezocone tests is that the total radial stress, as well as the excess pore pressure, is continually monitored during both the expansion and subsequent dissipation phases. Thus both total and effective radial stress changes are known. The importance of this is further illustrated later.

4 INFLUENCE OF CREEP

In some pressuremeter dissipation tests discussed in the next section, some unusual aspects of the observed behaviour are thought to be due to creep. That is, time–dependant deformations were observed which could not be explained purely on the basis of consolidation. Thus, it is worth evaluating the effects of creep on dissipation tests, even if at this stage this evaluation is purely qualitative.

In the context of creep testing in fields other than geomechanics, the strain-holding and pressure-holding tests described above may be thought of as "relaxation" and "creep" tests, respectively. A cavity expansion test in any material with strain-rate dependant stress-strain behaviour would exhibit a pressure expansion response dependant on the rate of expansion adopted, with a "stiffer" response being observed for faster rates of expansion. Furthermore, strain-holding or pressure-holding tests in such a material would show typical relaxation and creep responses, respectively. Thus, in the strain-holding test, the pressure required to maintain the cavity at a constant size would decrease, while in the pressure-holding test, the cavity size would continue to increase, though at an ever-reducing rate.

In a two-phase, but impermeable, version of this hypothetical material, pore pressure measurements at the cavity wall would show excess pore pressure generated during cavity expansion, and a reduction in excess pore pressure in a strain-holding test as relaxation of total radial stress occurred. The rate of pore pressure reduction might well appear to be very similar to that due to consolidation. Thus, in the absence of measurements of total stress, it would be possible to attribute the observed pore pressure dissipation *to the effects of creep alone*. Of course in practice, the problem is to determine how much of the reduction in pore pressure is due to creep and how much is due to consolidation.

The pore pressure response in a pressure-holding test (or true creep test) in the hypothetical material is more difficult to determine. In such a test, the total radial stress at the cavity wall stays constant, but the hoop stress behaviour would depend on the nature of the stress-strain-time curve for the material. Thus if the peak strength was rate dependant, the hoop stress in the plastic zone would have to increase with time. Thus, in these circumstances, the mean stress, and hence the excess pore pressure could show an increase. If consolidation is also permitted in this test, the exact pore pressure response will be influenced both by the *reduction* in excess pore pressure due to consolidation and by the *increase* due to creep, and it is thus not inconceivable that the pore pressure could remain constant, or even increase, during the early stages of a pressure-holding test in a low-permeability creep-sensitive soil.

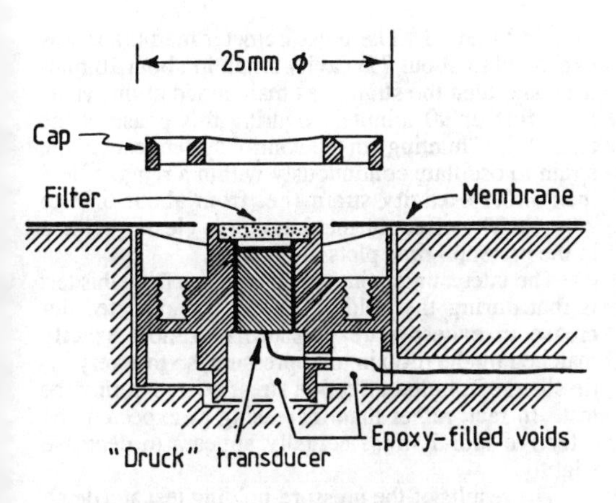

Figure 1 Cross-section through pore-pressure transducer in the Camkometer

5 TEST RESULTS

5.1 Background

The impetus for considering the effects of creep on dissipation tests has come from a recent series of self-boring pressuremeter tests conducted at a soft-clay site in Perth, Western Australia. These tests were complementary to an investigation using a "piezometer probe" device at this and other sites in Perth. This device is described in full by Fahey and Foley (1987). Essentially it consists of a cone penetrometer equipped with two pore pressure transducers located about 4 diameters back on the shaft. The penetrometer is equipped with a 12.7° cone rather than the standard 60° cone. For simplicity, no load measuring function has been incorporated into the cone, since the primary purpose of the device is to study the dissipation phase of the piezocone test, and thus no measurement of end resistance or skin friction were obtained during these tests.

The site for the tests is underlain by soft alluvial clay of high plasticity. The site, at Bayswater in Perth, lies on the bank of the Swan River at a point where river meanders have developed in recent geological time. The consolidation characteristics of the soil at the site are of interest because an embankment has been constructed over the site to form the approach to the Redcliffe Bridge for the Northern Extension of the Tonkin Highway.

5.2 The pressuremeter

The pressuremeter used in the investigation is a modified version of the commercially available Cambridge Self-boring Pressuremeter. The modifications have been described in detail by Fahey et al (1987). One of the main modifications, which were completed before the work described here was undertaken, consisted of incorporating miniature "Druck" pore pressure transducers for measurement of pore pressure. A diagram showing a cross-section through a transducer is included as Figure 1. A system was also devised for de–airing the pore pressure transducers using a portable vacuum pump, and maintaining them in a de-aired state until the instrument is inserted into the water–filled casing of the borehole. Experience with the piezometer probe described earlier has shown that proper de–airing is essential to obtain a satisfactory response from the transducers in this type of testing.

A further modification to the instrument which has recently been made is to incorporate much of the supply voltage regulation and signal amplification into the instrument itself. This has the advantage of practically eliminating errors due to the (changing) resistance of the long signal cable from the surface to the instrument. In addition, variations due to temperature effects, especially in tests of long duration, are much reduced by locating the signal amplifiers in the stable temperature environment of the ground water.

Tests with this instrument are controlled using a microcomputer. Cavity expansion can be carried out at either constant strain rate, or at a constant rate of pressure increase. Strain-holding phases consist of monitoring the output from the strain transducers (at a sampling rate of about 4 Hz), and varying the pressure to maintain a constant strain.

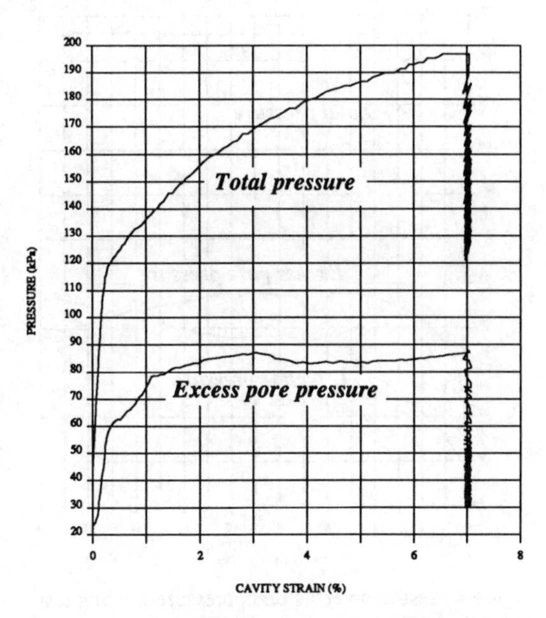

Figure 2 Total pressure and excess pore pressure versus strain, strain-holding test at 5.0 m, Bayswater.

5.3 Results of pressuremeter tests

In this paper, attention is directed mainly to two pressuremeter dissipation tests conducted in adjacent boreholes at similar depths. In Borehole 8, a strain-

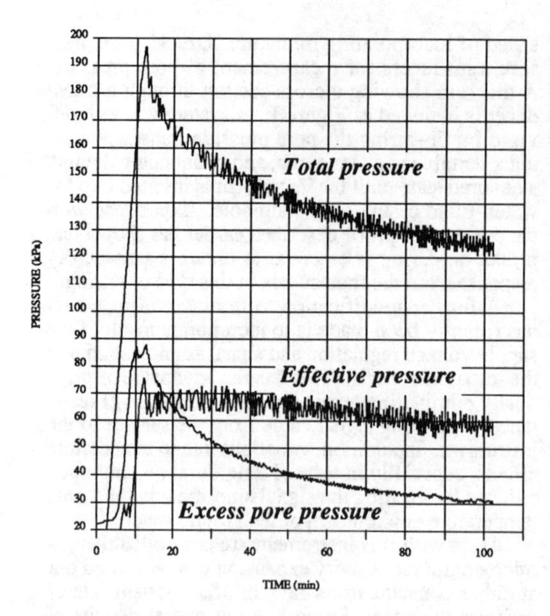

Figure 3 Total pressure, effective pressure and excess pressure versus time (same test as in Figure 2).

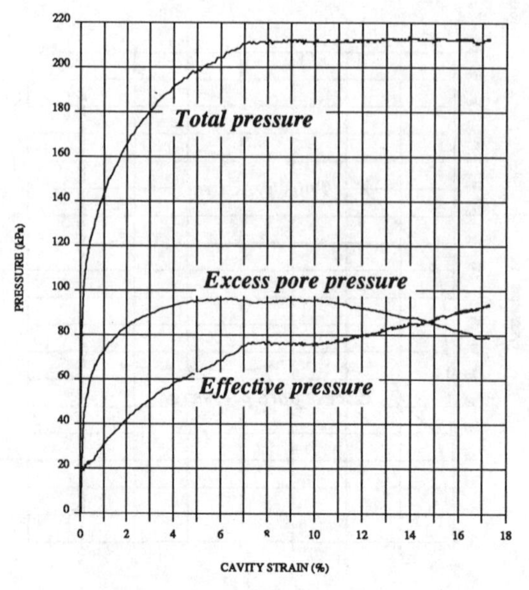

Figure 4 Pressures versus time, pressure-holding test at 5.5 m, Bayswater.

holding test was conducted at a depth of 5.0 m, while in the adjacent Borehole 9, a pressure-holding test was conducted at a depth of 5.5 m.
The results of the strain-holding test in Borehole 8 are presented as total pressure and excess pore pressure versus cavity strain in Figure 2, and as total pressure, excess pore pressure and effective pressure versus time in Figure 3. The pressuremeter membrane was expanded to about 7% cavity strain in about 10 minutes, and then the strain was maintained at this value for a further 90 minutes. During this phase of the test, minor "hunting" in the control system caused the strain to oscillate continuously within a range of less than ± 0.1% cavity strain (i.e. from about 6.93 to 7.07%). The effect of this hunting is clearly evident in the pressure–time plots in Figure 3.

The interesting point which emerges from this test is that during the holding phase, the expected decrease in excess pore pressure is almost exactly matched by the drop in total pressure, so that very little change in radial effective stress occurs during the test. In fact, rather than increasing as expected, the effective radial stress actually appears to decrease slightly.

The results of the pressure-holding test at a depth of 5.5 m in Borehole 9 are shown as total pressure, excess pore pressure and radial effective stress versus cavity strain and versus time in Figures 4 and 5, respectively. The time required to achieve a cavity strain of 7% is effectively about 7 minutes (from the start of expansion), and then a constant cavity pressure is maintained for a further 50 minutes. During this period the cavity continued to expand, and the test was terminated when the strain reached 17%, which is the effective limit of the strain measuring system in the instrument.

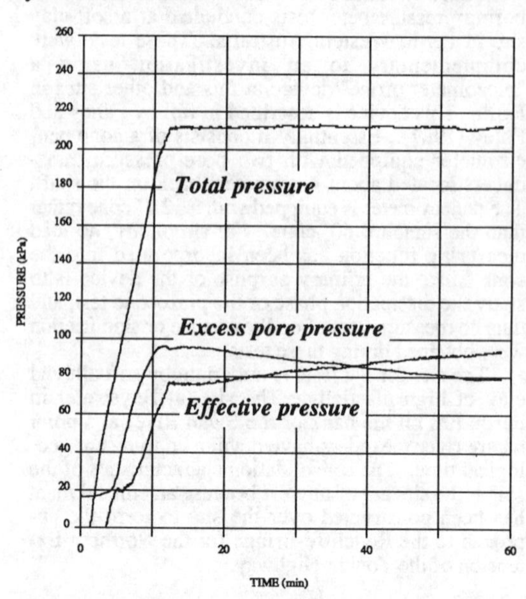

Figure 5 Pressures versus time, pressure-holding test shown in Figure 4.

The cavity strain in the test is shown plotted against time in Figure 6. Though the cavity pressure was stabilized when the strain was 7%, the rate of increase of strain with time only reduced very gradually. In addition, as shown in Figure 4, practically no change in pore pressure (and hence in effective stress) occurred while the strain increased from 7 to 11%.

After this point, some reduction in pore pressure was observed, but this occurred only very slowly. It is interesting to note for this test (and for other pressure-holding tests at the site) that a linear relationship is obtained when cavity strain is plotted against the square root of time (measured from the start of the holding phase).

The CAMFE finite element cavity expansion programme mentioned earlier has been used with some success (Fahey and Carter, 1986) for modelling the contrasting behaviour observed in strain-holding and pressure-holding tests conducted at the Burswood Island site. This site is situated on the banks of the Swan River in Perth, within a few kilometers of the Bayswater site. The CAMFE model was also used in an attempt to reproduce the results obtained from the tests at the Bayswater site. The contrast between strain-holding and pressure-holding tests, as predicted using CAMFE, is illustrated in Figure 7. In this Figure, the CAMFE output from pressure-holding and strain-holding tests are shown by the dashed lines and the solid lines, respectively. The total pressure, pore pressure and effective pressure have been normalized in this plot by the total pressure at the start of the holding phase, and the time is measured from the start of the holding phase. (The output for the initial expansion phases is of course identical for the two tests).

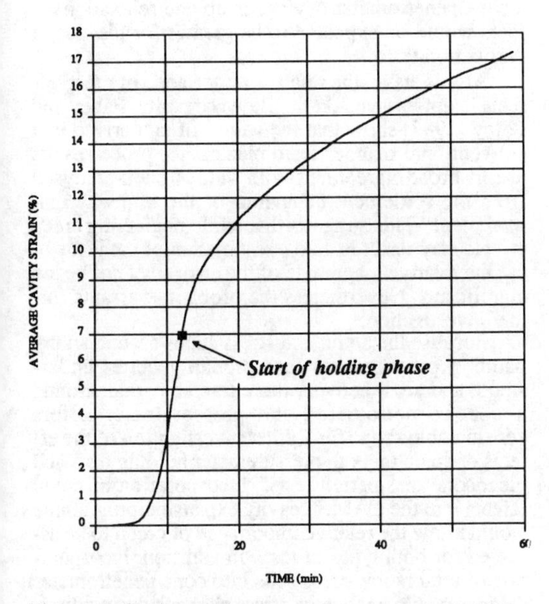

Figure 6 Cavity strain versus time, pressure-holding test shown in Figures 4 and 5.

The model predicts that in a strain-holding test, the total pressure reduces, but the pore pressure reduces more rapidly, so that the effective radial stress is seen to increase at all stages of the holding test. In a pressure-holding test, the reduction in pore pressure is less rapid, but such a reduction occurs right from the start of the holding phase. Note that in this particular plot, the rate of increase in effective pressure is similar for the two tests for the time shown. However, eventually the effective pressure increases at a faster rate for the pressure-holding test, and the final equilibrium value is of course higher than in the strain-holding test. For the pressure-holding test, the model also clearly shows that the cavity does increase in size during the constant-pressure phase of the test, but *only* in response to a reduction in excess pore pressure and consequent increase in effective stress.

While choice of values of the relevant soil parameters does affect the response predicted by CAMFE, the trends shown in Figure 7 are typical of the type of results which can be obtained. These trends are in contrast to the test results shown in Figures 2 to 6. In particular, in the in situ strain-holding test, no increase in effective radial stress was observed during the holding phase, and in the pressure-holding test, a considerable increase in strain occurred without any increase in effective stress. Even when extreme values are input for the relevant soil parameters, it was found to be impossible with the CAMFE model to reproduce these aspects of the observed behaviour. While the programme incorporates all of the features of the Cam Clay model, at present no rate effects, other than those due solely to consolidation, are allowed for. It is believed therefore that the observed behaviour can only be explained if "creep" and "relaxation" effects are also taken into account.

6 IMPLICATIONS FOR INTERPRETATION OF PIEZOCONE TESTS

Dissipation tests using the piezometer probe described above were also carried out at the Bayswater site at

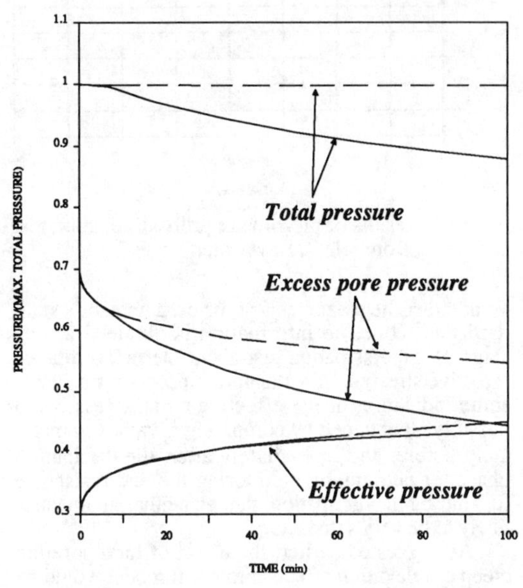

Figure 7 CAMFE output showing dissipation phase of strain-holding (solid lines) and pressure-holding (dashed lines) tests.

the same location as the pressuremeter tests. The results of one such test at a depth of 5.25 m are shown as excess pore pressure versus time in Figure 8. In this test, very close agreement was observed between the outputs from the two pore pressure transducers, though, as discussed by Fahey and Foley (1987), this was not the case for all tests in this series.

The method of interpreting this test used by Fahey and Foley is based on that used by Clarke et al (1979) for interpreting the results of pressuremeter strain-holding tests. For this analysis, the ratio of shear modulus, G, to shear strength, c_u, is required, and this is most conveniently obtained from the results of pressuremeter tests. The time required for the excess pore pressure to decrease to 50% of its maximum value (t_{50}) is obtained from a plot of excess pore pressure versus time such as that shown in Figure 8. In this case, this gives a value of t_{50} of 37 minutes, and a value of coefficient of consolidation (c_h) of 17 m^2/year.

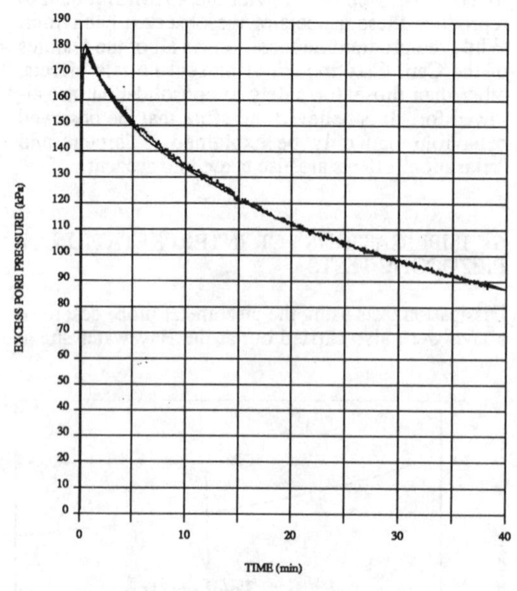

Figure 8 Results of piezometer probe dissipation test at 5.25 m, Borehole 7, Bayswater.

With either the piezometer probe used here, or a standard piezocone, no information is obtained at any stage of the dissipation test about the radial total or effective stresses. If a standard piezocone was used, some indication of the effective radial stress could possibly be obtained by comparing f_s values immediately before, and immediately after, the dissipation phase, or perhaps by monitoring the end resistance (q_c) during the dissipation, though neither approach is likely to be very satisfactory.

As suggested earlier, the effect of incorporating creep or relaxation effects into a soil model would be that the observed rate of pore pressure dissipation would be increased in a pressuremeter strain-holding test. Moreover, if strain rate does have an effect on the stress-strain behaviour, then the very fast rate of cavity expansion during cone penetration, compared to the strain rate in a pressuremeter test, would *increase* the tendency for relaxation in total stress after penetration ceased. Assuming that this relaxation has the effect of reducing the pore pressure independent of radial drainage rate, then any interpretation of the dissipation rate which does not account for creep would give erroneous results.

It is therefore postulated that for a piezocone dissipation test, the overall effect of combining creep or relaxation with consolidation is to produce a faster rate of apparent pore pressure dissipation than would be the case if pore pressure dissipation was caused by consolidation alone. Thus the coefficient of consolidation derived from the results would probably be overestimated if creep effects are not taken into account.

7 CONCLUSIONS

The pressuremeter test results discussed in this paper suggest that for this site at least, creep or relaxation effects may have an important influence on the total stress and hence on the pore pressure observed during pressuremeter holding tests. Because of the higher strain rates involved in creating a cavity using a cone penetrometer device, creep and relaxation effects would be expected to be greater for piezocone dissipation tests.

At this stage, the relative importance of creep effects is not known. At the Bayswater site, Fahey and Foley (1987) show that the values of c_h derived in a conventional manner from piezometer probe results are in broad agreement with values back-analysed from the settlement behaviour of the highway embankment. This suggests that while neglecting creep effects may result in an overestimation of the value of c_h, the relative magnitude of the error may not be too significant. Nevertheless the problem warrants further investigation.

Because the strain and stress states around an expanding pressuremeter are probably better understood, and are inherently more tractable, than around a cone penetrometer, the pressuremeter offers considerable scope for further investigation of the effects of creep on both pressuremeter holding tests and piezocone dissipation tests. Incorporation of creep effects into the CAMFE cavity expansion programme would allow the relative importance of creep to be assessed for both types of test. In addition, incorporation of total radial stress cells into cone penetrometer devices would enable the stress changes during dissipation to be better understood. Such a development is planned for the piezometer probe described in this paper.

8 ACKNOWLEDGEMENTS

The field work described in this paper could not have been performed without the cooperation of the staff of the Materials Engineering Laboratory of the Main

Roads Department of Western Australia, who provided a drilling rig and crew to assist in the field work. In particular the assistance given by Mr Geoff Cocks and Mr Paul Foley is gratefully acknowledged. The continued assistance and advice of Mr Tuarn Brown with development of the electronics, and with performance of the field work, is also gratefully acknowledged.

REFERENCES

Carter, J.P. (1978). *CAMFE, a computer program for the analysis of a cylindrical cavity expansion in soil.* Camb. Uni. Engng. Dept. Internal Report CUED/C–SOILS–TR52.

Clarke, B.G., Carter, J.P. and Wroth, C.P. (1979). In situ determination of the coefficient of consolidation of saturated clays. *Proc. 7th Eur. Conf. on SMFE*, Brighton, **2**, 207-213.

Fahey, M. and Carter, J.P. (1986). Some effects of rate of loading and drainage on pressuremeter tests in clay. *Specialty Geom. Symp: Interpretation of Field Testing for Design Parameters.* Adelaide, 18–19 August. The Institution of Engineers, Australia, 50–55.

Fahey, M. and Foley, P.A. (1987). In situ measurement of the coefficient of consolidation. *Proc. 9th Eur. Conf. SMFE, Dublin*, **1**, 29 - 33.

Fahey, M, Jewell, R.J. and Brown, T.A. (1987). *A Self-Boring Pressuremeter System.* (Submitted to ASCE Geotechnical Testing Journal).

Levadoux, J.N. and Baligh, M.M. (1986). Consolidation after undrained piezocone penetration. I: Prediction. *Jour. of Geotechnical Engineering*, ASCE, **112** (GT7), 707-726.

Penetration Testing 1988, ISOPT-1, De Ruiter (ed.)
© 1988 Balkema, Rotterdam, ISBN 90 6191 801 4

Vertical deformation modulus of sand estimated by pressuremeter test and SPT blow count N

R.Fukagawa
Ehime University, Matsuyama, Japan

H.Ohta
Kanazawa University, Japan

ABSTRACT: Deformation modulus Ev of sandy deposits subjected to vertical loading was estimated in connection with the use of a pressuremeter. The deformation modulus E_{PM} measured by using pressuremeters was related to deformation modulus Ev backcalculated from the in-situ plate loading tests and from the monitored settlement of structures in which SPT blow counts N are known. In order to ensure the Ev-E_{PM} relation, triaxial tests and miniature pressuremeter tests on sand specimens in triaxial chamber were conducted. Plate loading tests and miniature pressuremeter tests were performed on/in artificially prepared sandy deposits. The ratios Ev/E_{PM} obtained from the laboratory tests were found to be related to the SPT blow count N.

Key words: deformation modulus, sandy deposit, pressuremeter test, SPT

1 INTRODUCTION

Pressuremeter tests have been developed as in-situ tests to evaluate not only deformability or initial state variables but also strength characteristics of soils. The self-boring-type pressuremeter developed by Baguelin, Jezequel and Le Mehaute (1973) , Wroth and Hughes(1973) improves the capabilities of the pressuremeter test

This paper aimes at widening the applicability of pressuremeter tests, by proving that the pressuremeter is a reliable tool in estimating the deformability of sandy deposits under vertical loading.

2 ESTIMATION OF IN-SITU DEFORMATION MODULUS UNDER VERTICAL LOADING

In determining the deformation modulus,Ev, of sandy deposits under vertical loading, it is convenient to use the data of SPT tests because SPT is widely used in the world. Ev-values backcalculated from in-situ plate loading tests and those from the monitored settlement of structures are plotted against the SPT blow count N in Fig.1. These two kinds of Ev-values must be of similar nature because of the similarity of loading mechanism. In the backcalculation Ev-values were obtained from Eq.(1) proposed by Janbu, Bjerrum and Kjaernsli(1964). The data collected by Schultze and Sherif(1973), D'Appolonia, D.J., D'Appolonia, E. and Brissete(1970), are also plotted.

$$Ev = \mu_0 \cdot \mu_1 \cdot q \cdot B / S_i \quad (1)$$

in which, S_i : immediate settlement, q: loading pressure per unit area, B : basement width, μ_0 , μ_1 : coefficient determined from penetration depth, basement length, B and depth from bottom of base

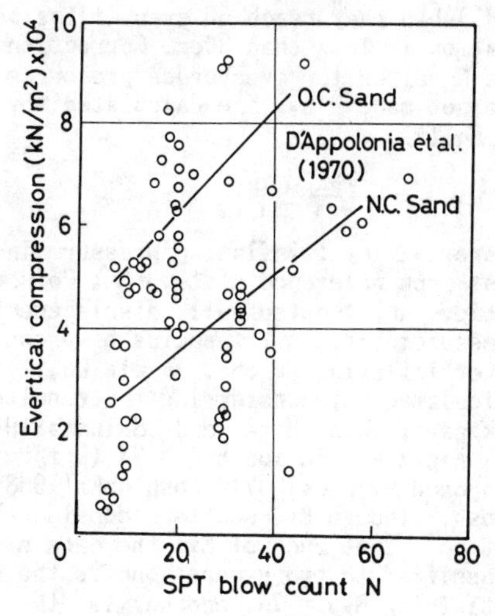

Fig.1 Deformation modulus Ev backcalculated from plate loading tests and field monitoring

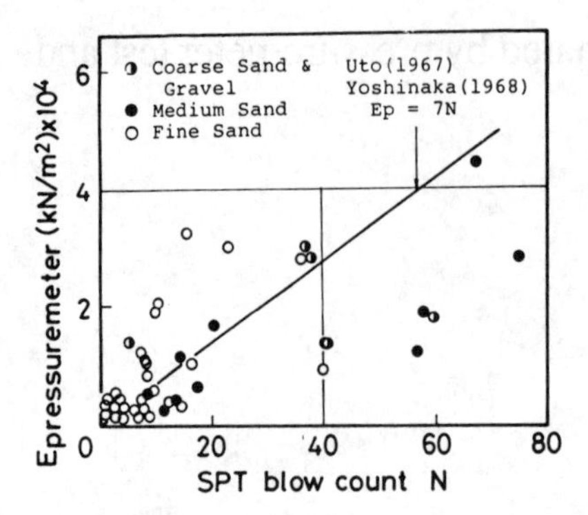

Fig.2 Deformation modulus E_{PM} obtained from pressuremeter tests

ment to stiff soil deposits. In Fig.1 the two straight lines proposed by D'Appolonia et al(1970) are shown. They are ① normally-consolidated sand or sandy gravel deposits --- E=77(N+26) (tf/m²),②overconsolidated or compacted sandy deposits --- E=102(N+41) (tf/m²) respectively. Judging from Fig.1, it is difficult to conclude that the data distribute along these two straight lines. The scatter in Fig.1 is considered to be too large to estimate Ev-values only from SPT blow count N.

E_{PM} obtained from pressuremeter tests are plotted against the N-values in Fig.2. The pressuremeter data in Fig.2 are sampled from 8 sites by use of the pre-boring -type pressuremeter. The N-values greater than 50 in Fig.2 are obtained by extrapolation since N-values are usually stopped when they reach 50 even if the penetration is less than 30cm. Corrections of the N-values for overburden pressures etc. are not made. E_{PM} are calculated by use of Eq.(2).

$$\Delta \sigma_i = \frac{E}{1+\nu} \Delta \frac{u_i}{r_i} \quad (2)$$

Where, $\Delta \sigma_i$: inflating pressure increment from reference state, ν : Poisson's ratio, u_i : outer wall displacement of pressuremeter, r_i : radius of pressuremeter inflating probe. ν in Eq.(2) are calculated by considering the results of Fukagawa, Ohta, Hata and Shikata(1985). In Fig.2 a relation E_{PM} = 7N (kgf/cm²) proposed by Uto(1967), Yoshinaka(1968) is shown. Though E_{PM} scatters considerably, but not that much of Ev, the data may be classified to two groups; one is the data satisfying E_{PM} = 7N, another is the data following E_{PM} = 2-3N. The soils are roughly classified by the average grain size as coarse sand and gravel, medium sand and fine sand in Fig.2. It is difficult to point out a clear correlation between the average diameter of soil particles and the E_{PM}-N relations.

To seek some characteristics of deformation modulus of sand out of the data plotted in Figs.1 and 2, the data are rearranged in a way that the frequency of the ratios Ev/N and E_{PM}/N are investigated. This is demonstarated in Fig.3 in which these ratios are represented by a parameter θ. Both of Ev and E_{PM} clearly have their own two peaks. These peaks may correspond to the two types of sandy deposits suggested by D'Appolonia et al(1970) as seen in Fig.1. The ratio of Ev and E_{PM} at the left peaks of Ev and E_{PM} in Fig.3, is given as

$$\frac{E_v}{E_{PM}} = \frac{\tan 52.5^\circ}{\tan 15^\circ} \doteqdot 4.9$$

while the ratio of E_v/E_{PM} at the right peak is

$$\frac{E_v}{E_{PM}} = \frac{\tan 75^\circ}{\tan 35^\circ} \doteqdot 5.3$$

These two relationships imply that there exists some correlationship Ev = 5E_{PM} regardless of whether the sand layer is normally or overconsolidated if we adopt the expression of D'Appolonia et al. This coefficient is only representative of widely scattered Ev/E_{PM} ratios, and is influenced by many factors, for example, accuracy of N-values and borehole wall disturbance in pressuremeter tests. However, the ratio of 5 seems to be useful in engineering practice.

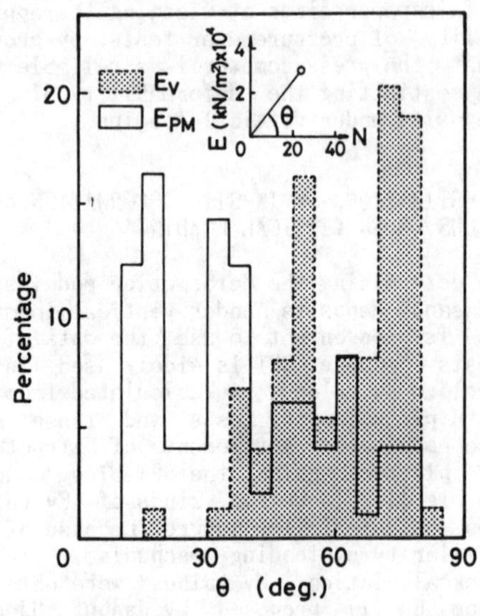

Fig.3 Distribution of deformation modulus /blow count ratio

3 CORRELATION BETWEEN Ev AND E_{PM} IN LABORATORY CONDITIONS

3.1 Sample

Toyoura Sand is used as a sample. Its physical properties are, e_{max} =0.960, e_{min} =0.618, Gs =2.613 and D_{50} =0.23 (mm).

3.2 Experimental apparatus and procedures

a) Triaxial Compression test
A series of isotropically consolidated drained compression tests are conducted. Specimens are 35.5 mm in diameter and about 70 mm in height. The experimental conditions of the triaxial tests are classified into 56, 85, 100 (%) for the relative density Dr and 49, 98, 147(kN/m^2) for the effective consolidation pressure σ_c'.

b) Miniature pressuremeter test in triaxial cell
Miniature pressuremeter test as descrived by Fukagawa, Ohta, Shikata and Hata (1985) are carried on the specimens of 152mm in diameter and about 250mm in height. The inflating probe of pressuremeter is 16.8mm in diameter and 105mm in height (diameter height ratio is 6.25) and the thickness of the rubber membrane covering the inflating probe is 0.2 mm. Relative density Dr and consolidation pressure σ_c' are chosen in a same way in the triaxial compression tests.

c) Miniature pressuremeter test for artificially prepared sandy deposit
The outline of a miniature pressuremeter testing apparatus for artificially prepared sandy deposit is shown in Fig.4. The inflating probe has the same structure and the same diameter with that used in the miniature pressuremeter in triaxial cell, but the length of the probe is 75mm which is 30mm shorter than that of the miniature pressuremeter in triaxial cell. The length/diameter ratio of the inflating probe is 4.12. The experimental procedure is as follows; ① Three pressuremeters are fixed in the test tank. The depth of the center of the inflating probe from the model ground surface of each probe is 75, 125, 200mm respectively. ② Dry sand samples are divided quaterly and are filled into the test tank. The thickness of each layer is about 50, 50, 50, 100mm from top to bottom. Four kinds of uniform ground are prepared. The relative density of each ground is about 40, 60, 75 and 90 %. ③ Pressuremeter tests are carried out in each depth of the ground. Since the

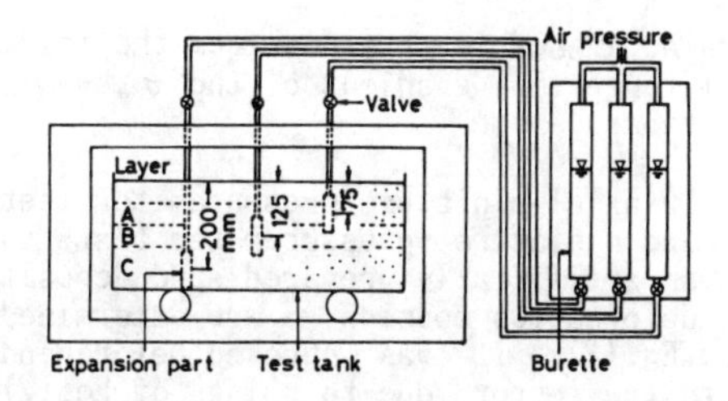

Fig.4 Miniature pressuremeter in artificially prepared sandy deposit

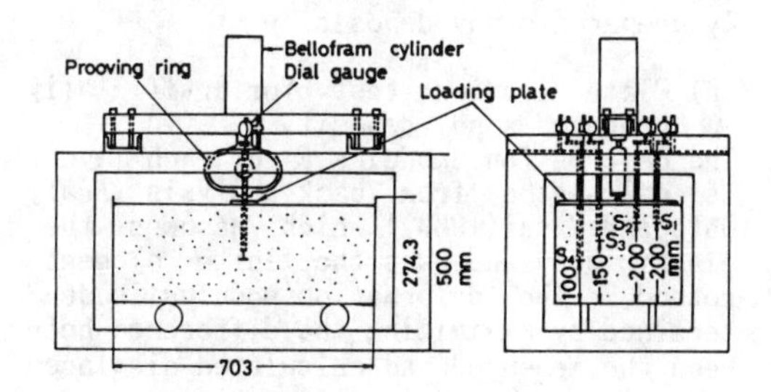

Fig.5 Plate loading tests on artificially prepared sandy deposit

pressure level of the experiments was comparatively low, attention is paid to the measurement of inflating pressure.

d) Plate loading test for artificially prepared sandy deposit
The plate loading tests are carried out on an artificially prepared sandy deposit which is made by the same manner as the case of miniature pressuremeter. The outline of the plate loading tests are shown in Fig.5. The method of analysis of the test results needs plane strain conditions, so the loading plate has almost the same length(325mm in length, 75mm in width and 10mm in thickness) as the width of the test tank(337.3 mm). The measuring points of displacement are the surface of A layer (S1), the surface of B layer (S2) and the surface of C layer (S3). The loading is carried out by the stress control. The increment of loading weight is 49 N, and the displacement after 3 minutes from loading is recorded.

3.3 Determination of deformation modulus

a) Triaxial compression test
The deformation modulus are determined by

$$E_{COMP} = \frac{\Delta \sigma_a}{\Delta \varepsilon_a}$$

in which $\Delta \sigma_a$: the increment of axial stress, $\Delta \varepsilon_a$: the increment of axial strain. E_{COMP} is obtained from the initial tangential gradient of the $\sigma_a - \varepsilon_a$ relations.

b) Triaxial miniature pressuremeter tests and miniature pressuremeter tests in an artificially prepared sand deposit

The deformation modulus E are determined from Eq.(2). It was checked beforehand that the error due to using of Eq.(2) assuming the semi-infinite elastic ground is negligible in the case both of the triaxial miniature pressuremeter and the miniature pressuremeter in the artificially prepared sandy deposit.

c) Plate loading test for artificially prepared sandy deposit

The deformation modulus E of each layer is calculated from back analysis (Arai, Ohta and Yasui(1983)) which assumes the elastic body and uses the finite element method. The deformation modulus is determined by minimizing the difference between the measured and calculated displacement of some measuring points.

3.4 Test results

a) Triaxial compression tests and triaxial miniature pressuremeter tests

E_{COMP} or E_{PM} obtained from the triaxial tests and the triaxial miniature pressuremeter tests are normalized by σ_c' and are plotted against Dr as seen in Fig.6. Both of E_{COMP}/σ_c', E_{PM}/σ_c' increase with the increase in Dr. Then E_{COMP}/E_{PM} are plotted against Dr in Fig.7. The ratio E_{COMP}/E_{PM} shows a tendency to increase gradually with increase in Dr and lies roughly between 1.5 and 2.5. The upper limit values in Fig.7 are obtained from dividing the upper limit values of E_{COMP}

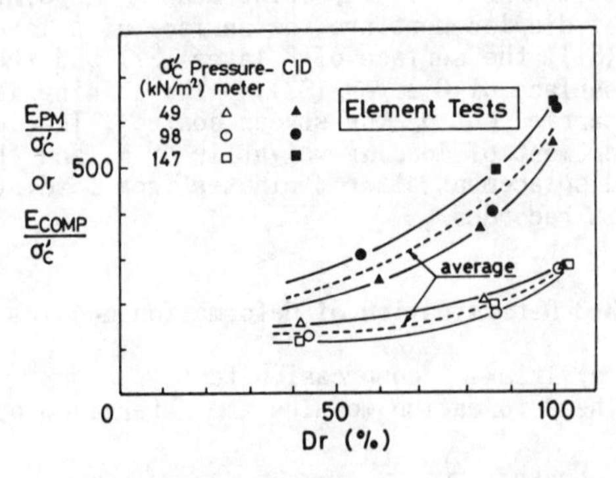

Fig.6 Deformation modulus obtained from triaxial tests and miniature pressuremeter tests in triaxial chamber

by the lower limit values of E_{PM} and the lower limit values are obtained from dividing the lower limit values of E_{COMP} by the upper limit values of E_{PM}. The dotted line in Fig.7 is calculated from the average curves in Fig.6.

b) Plate loading tests and miniature pressuremeter tests for an artificially prepared sandy deposit

The ratio of Ev of each layer backcalculated from the plate loading tests to E_{PM} is obtained against Dr in the same manner as Fig.7 (as shown in Fig.8). The range of Ev/E_{PM} in Fig.8 is a little larger than that in Fig.7 because the scatter of Ev from the back-analysis was relatively large. But the basic trend in which Ev/E_{PM} increases with the increase of Dr is similar to that of E_{COMP}/E_{PM}.

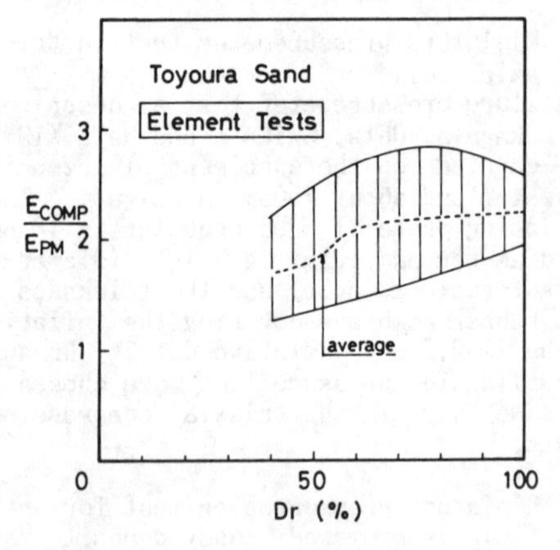

Fig.7 Range of E_{COMP}/E_{PM} plotted against relative density

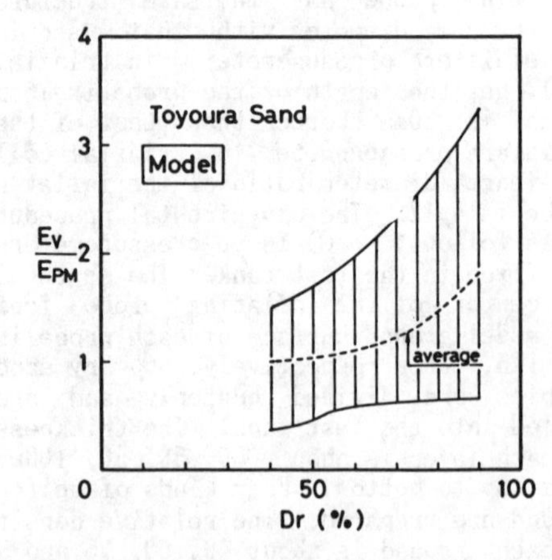

Fig.8 Ev/E_{PM} - Dr relation obtained from artificially prepared sandy deposit

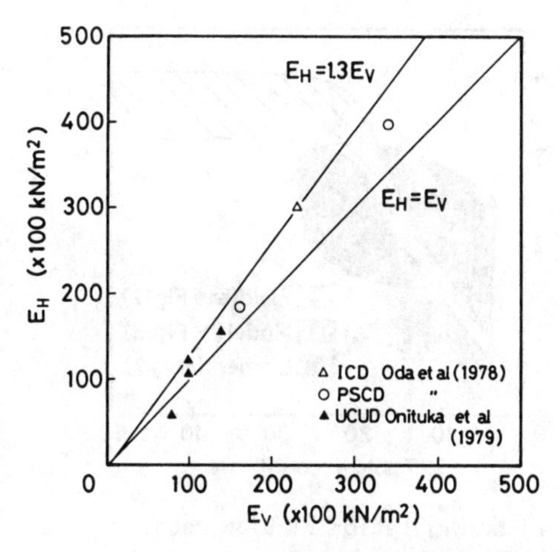

Fig.9 Effect of inherent anisotropy of sand on deformation moduli

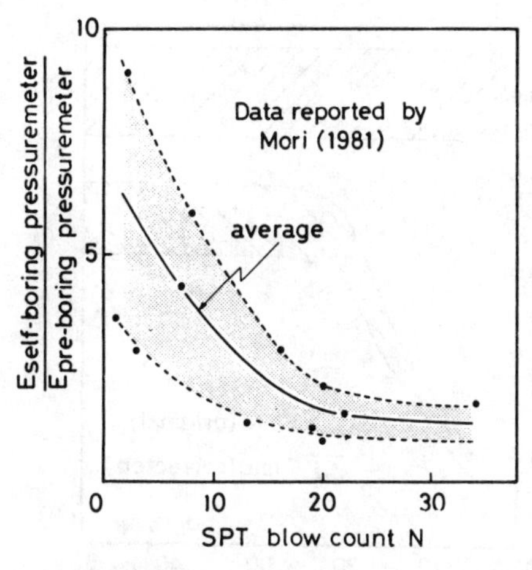

Fig.11 Effect of disturbance of sand in vicinity of bore hole

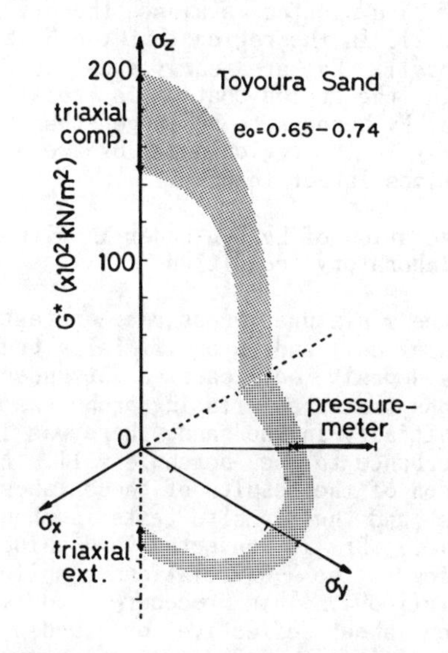

Fig.10 Evaluation of deformation moduli estimated from pressuremeter tests

4 DISCUSSION

4.1 Factors affecting Ev/E_{PM}

Ev/E_{PM} ratio was found to be about 5 from in-situ measurements and was about 2(Figs. 7, 8) in the triaxial tests and laboratory model tests although test results are affected by Dr or magnitude of σ'_c. Such differences between Ev and E_{PM} may be caused by fabric anisotropy, shear mechanism and borehole wall disturbance in the pressuremeter tests etc.. These factors are discussed in the following section.

a) Fabric anisotropy

It is well known that fabric anisotropy gives some effects on the deformation and strength characteristics of sand. Fig.9 shows the effect of the fabric anisotropy on the deformation characteristics (especially in the initial loading stage) of sand. The plots in Fig.9 are rearranged data from the published papers (Oda, Koishikawa and Higuchi(1978), Onitsuka, Hayashi, Yoshitake and Oishi(1979)). E_H, E_V in the figure means the deformation modulus for the H-sample(long axis of soil particles are occupative in horizontal direction) and the V-sample(in vertical direction) respectively.

Judging from these results, Ev of the H-sample for the vertical loading are about 1-1.3 times greater than Ev of the V-sample, so the effect of the fabric anisotropy is not so remarkable.

b) Shear mechanism

The difference of the shear mechanism gives a large effect on Ev and E_{PM}. Fukagawa et al (1986) explained these diferences by the stress path dependency of sand deformation. They determined the deformation moduli corresponding to many stress paths from the rearranged data of true triaxial tests or the plane strain tests for Toyoura Sand published by many researchers, and compared them with E_{PM} as seen in Fig.10. Consequently it is concluded that the relations among the deformation moduli from various tests can be estimated from the stress path dependency

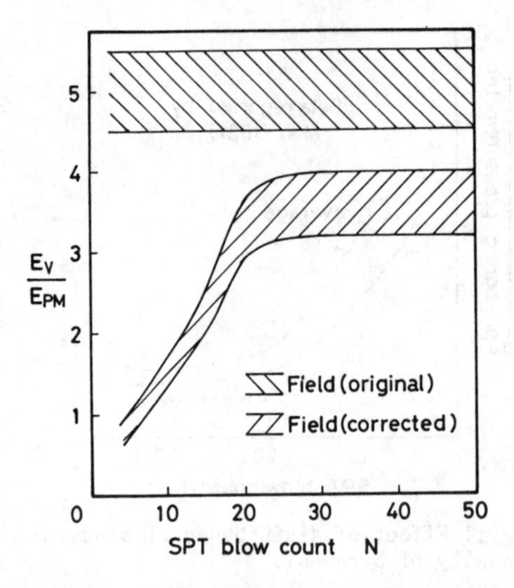

Fig.12 Correction of bore hole disturbance on E_V/E_{PM} ratio

of sand deformation.

Yoshinaka(1968) carried out pressuremeter tests and plate loading tests in the horizontal direction. Both are loading tests to the horizontal direction. The deformation moduli obtained from these two methods are compared to each other. The deformation modulus of the horizontal plate loading test was found to be three times larger than that of the pressuremeter test. This finding also indicates the importance of the shear mechanism in understanding the physical meaning of the deformation modulus obtained from the pressuremeter tests.

c) Effect of borehole wall disturbance on E_{PM}

Mori(1981) carried out pre-boring-type (PBP) and self-boring-type (SBP) pressuremeter tests on loose and dense sandy deposit, and compared the deformation modulus obtained from SBP (E_{SBP}) with the deformation modulus from PBP (E_{PBP}). E_{SBP} / E_{PBP} ratio is plotted against the SPT blow count N (Fig.11). The data in Fig. 11 are rearranged from the data of Mori (1981). Since the borehole wall disturbance is relatively small in the case of SBP, the results of Fig.11 means that the smaller the N-values are, the larger the borehole wall disturbance becomes. Then multiplying E_{SBP} by 5 causes overestimation of the in-situ Ev especially when the N-value is small. The E_V/E_{PM} values corrected for the borehole disturbance can be gained by dividing the average E_{SBP} /E_{PBP} values by E_V/E_{PM} values(assuming the

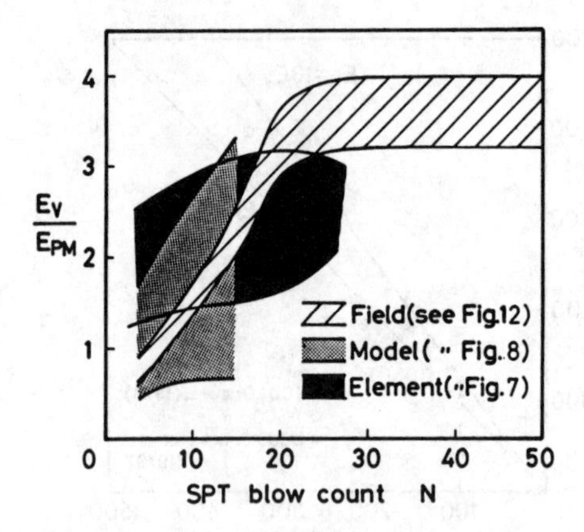

Fig.13 E_V/E_{PM} ratio-SPT blow count relations

band ranges from 4.5 to 5.5). The E_V/E_{PM} values corrected for the borehole disturbance are plotted against the N-values (Fig.12). In the region that the N-values are small, Ev are nearly equal to E_{PM}, and in the region that N is relatively large, Ev becomes to be larger than E_{PM}. Finally in the region N>20, Ev are about 3.5 times larger than E_{PM}.

4.2 Relation of E_V/E_{PM} under in-situ and laboratory conditions

As the miniature pressuremeter tests in triaxial cell and in artificially prepared sandy deposit were carried out under conditions that the inflating probe was buried initially in the sand, there was little disturbance to the borehole wall. A comparison of the results of these laboratory tests and the in-situ tests is shown in Fig.13. Dr is converted to N-value according to the N-Dr relations compiled by Fujita(1980). This procedure needs information about effective overburden pressure, grain size and degree of saturation. The corrected E_V/E_{PM} values obtained from in-situ condition are in a good accordance with the average E_V/E_{PM} curve of laboratory tests. This agreement means that both of in-situ and laboratory experiments were carried out under the similar mechanical conditions.

This agreement in Fig.13 proves the validity of the relation which $E_V/E_{PBP} \fallingdotseq 5$. In other words, the correction of borehole wall disturbance and fabric anisotropy for E_V/E_{PBP} relations will produce the E_V/E_{SBP} relations in the field condition which are ensured by the pressuremeter tests and plate loading tests for artificially prepared sandy deposit.

5 CONCLUSIONS

The deformation modulus Ev backcalculated from the in-situ plate loading tests and from the monitored settlement of structures shows a reasonable correlation with the deformation modulus E_{PM} obtained from the pressuremeters with SPT blow count N as the intermediate parameter, then Ev/ E_{PM} becomes to be about 5. The Ev/ E_{PM} ratio could be explained rationally by conducting some laboratory tests and considering some effects including shear mechanism, borehole wall disturbance and fabric anisotropy.

REFERENCES

Arai,K., Ohta,H. & Yasui,T. 1983. Simple optimization techniques for evaluating deformation moduli from field foundations, S&F, Vol.23, No.1 :107-113.

Baguelin,F., Jezequel,J.F. & Le Mehaute,A. 1973. Etude des pressions in evaluatiielles developpees lors de l'essai pressiometrique, P.8th ICSMFE,1-1 :9-24.

D'Appolonia, D. J., D'Appolonia, E. & Brissete,R.F. 1970. Settlement of spread footings on sand (Discussion), P. ASCE, Vol.96, No.SM2 : 754-762.

Fujita,K. 1980. Interpretation and its application example, Library of soil mechanics and foudation engineering, No.4, :53. (in Japanese)

Fukagawa,R., Ohta,H., Hata,S. & Shikata, H. 1985. Deformation properties of sand in model pressuremeter and tortionmeter tests, S&F, Vol.25, No.3 :113-126.

Fukagawa,R., Ohta,H., Hata,S. & Shikata, H. 1986. Closure to the discussion, S&F, Vol.26, No.4 : 165-166.

Janbu,N., Bjerrum,L. & Kjaernsli,B. 1964. Soil mechanics applied to some engineering problems, NGI Publication number No. 16 :32.

Mori,H. 1981. Studies on the properties of soils in the northern coast of Tokyo bay using a self-boring pressuremeter, S&F, Vol.21, No.3 :83-98.

Oda,M., Koishikawa,I. & Higuchi,T. 1978. Experimental study of anisotropic shear strength of sand by plane strain test, S&F, Vol.18, No.1 :25-38.

Onitsuka,K, Hayashi,S., Yoshitake,S. & Oishi,H. 1979. Compressibility and strength anisotropy of compacted soils, Journal of JSSMFE,Vol.19, No.3 :113-123.

Schultze,E. & Sherif,G. 1973. Prediction of settlement observations for sand, P. 8th ICSMFE, 1-3 :225-230.

Uto,K. 1967. Survey of base ground, Kanto branch ofJSCE :46. (in Japanese)

Wroth,C.P. & Hughes,J.M.O. 1973.An instrument for the in-situ measurement of the properties of soft clays, P.8th ICSMFE, 1-2 :487-494.

Yoshinaka,R. 1968. Coefficient of subgrade reaction for latelal direction, Document for Civil Engineering, 10-1 :32-37. (in Japanese)

Penetration Testing 1988, ISOPT-1, De Ruiter (ed.)
© 1988 Balkema, Rotterdam, ISBN 90 6191 801 4

Analysis of load deflection response of laterally loaded piers using DMT

Mohammed A.Gabr & Roy H.Borden
North Carolina State University, Raleigh, USA

Abstract: The use of the flat dilatometer test to evaluate the coefficient of lateral subgrade reaction, K_{ho}, was investigated. A model is proposed for the evaluation of K_{ho}, in cohesionless soils. Utilizing the K_{ho} values obtained with the proposed model, resulted in reasonable predictions of the lateral load-deflection behavior of drilled piers tested in the field.

1 INTRODUCTION

The Winkler or subgrade reaction approach is one of the most common techniques for the analysis of laterally loaded piles. The Winkler model simulates the soil-pile interaction mechanism by relating the pier deflection at a point to the soil pressure at that point, through a constant of subgrade reaction, referred to as K_{ho}. Because real soil response is non-linear, the springs idealizing the load-deformation response of the soil-pile mechanism are also taken to be non-linear. In general, the shape of a load-deflection curve in cohesionless soil, usually referred to as a p-y curve, is defined by the coefficient of lateral subgrade reaction, K_{ho}, and the ultimate soil resistance, P_u.

The p-y curves used in this study, were formulated using the hyperbolic model proposed by Parker and Reese, (1970). The continuous hyperbolic tangent function is defined as follows, according to Murchison and O'Neill (1984):

$$p = (C_1\ P_u)\tanh \frac{K_{ho}*Z}{C_1\ P_u}\ y \qquad (1)$$

where C_1 = Constant which is a function of pile shape and loading function (static or cyclic), = 3 - 0.8(Z/B) for static loading, Z = Depth below the ground surface, and P_u = Ultimate lateral resistance evaluated using the procedure recommended by Reese et al (1974).

2 COEFFICIENT OF SUBGRADE REACTION

The subgrade reaction coefficient, K_{ho}, has the units of force/length^3. The modulus of subgrade reaction, K_h, is defined as K_{ho} multiplied by the pile width or diameter, B, and it has units of force/length^2. The value assigned to the coefficient of subgrade reaction, as well as its variation along the pile length plays a significant role in the computation of the lateral pile response under a given set of applied loads.

The flat dilatometer test (DMT), developed by Marchetti (1980), has a great potential to give a near continuous profile of K_{ho} values, if indeed they can be inferred from the test data. The soil modulus, E_s, is one of the parameters that could be evaluated using the dilatometer test. To the best knowledge of the authors, only one preliminary attempt has been made so far, Motan and Gabr (1985), to investigate the validity of using the DMT evaluated E_s, to determine the lateral coefficient of subgrade reaction, K_{ho}. The limitations of the laboratory test setup utilized in modeling the field conditions were realized in this study.

2.1 Evaluation of K_{ho}

The existing methods of analysis suggest that K_{ho} can be obtained from the soil Modulus, E_s, as follows:

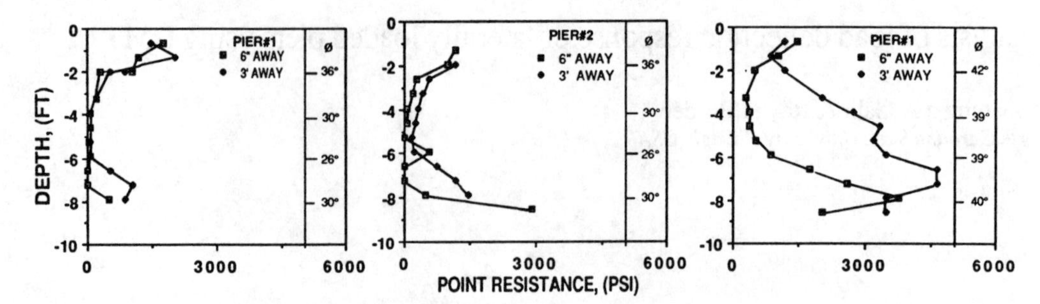

Figure 1. Cone Penetration Test Profiles

a.) $K_{ho} = \dfrac{C\, E_s}{(\text{pile diameter, B})}$ and C= constant

This approach was suggested by Terzaghi (1955), Broms (1964), Baguelin et al (1977,1978), Poulos (1980), and many others who tried to estimate K_{ho} using the soil modulus, E_s. The value of the constant, C, is chosen differently according to the view of the authors and the approach used to obtain the soil modulus, E_s. A similar technique has been used by Menard (1963), Baguelin and Jazequel (1978) and Poulos (1980) to deduce K_{ho}, from the pressuremeter modulus.

b.) $K_h = \dfrac{0.65 E_s}{(1-\nu^2)} \sqrt[12]{\dfrac{E_s B^4}{E\, I}}$

and $K_{ho} = \dfrac{K_h}{B}$

where: ν = Poission's ratio, B = Pile diameter, and EI = Pile stiffness.

This equation was obtained by Vesic (1961), as a result of extending Biot's solution of the problem of an infinite beam resting on a linearly elastic three dimensional subgrade. It is of interest to note that the value of the 12th root radical for many reasonable input pile and soil parameters is close to one. In such cases the relative stiffness between the soil and the pile does not influence the computed K_h values significantly. Accordingly, K_{ho} would mainly be a function of the soil modulus, E_s. The prescribed methods of predicting K_{ho} will be referred to as methods "a" and "b", respectively.

3 IN-SITU PENETRATION TESTS

The use of the techniques described above to evaluate K_{ho} using dilatometer evaluated soil modulus, E_s, has been investigated. Dilatometer field tests were performed adjacent to three drilled piers which were then laterally load tested. Throughout this text, the three piers will be referred to as pier #1, pier #2, and pier #3. A combination of field and laboratory tests indicated that the test site was composed of fine to coarse silty sand to a depth of 15 ft (4.6 m). The friction angles obtained from the dilatometer test, performed adjacent to piers #1 and #2, ranged form 43° for the upper 2 ft (0.61 m) to 34° at depth of 4 ft (1.22 m) to 32° at 7 ft (2.13 m). These friction angle values were determined from the dilatometer test data using the procedure recommended by Schmertmann (1982). This procedure was developed based on Durgunoglu and Mitchell's bearing capacity theory. The CPT profile, given in Figure 1, indicated lower friction angle values. The cone friction angle, for pier #1 and #2, decreased with depth from 36° at the surface to 30° at a depth of 3.5 ft (1.1 m) to 26° at 6 ft (1.83 m), as shown in Figure 1. However, the cone friction angle profile for pier #3 was higher, Figure 1. The values ranged from 42° for the upper 2 ft (.61 m) to 39° at depth of 6 ft (1.83 m). The cone friction angle values were determined according to a procedure recommended by Schmertmann (1975).

The soil modulus profile, obtained from DMT performed approximately 2 ft (.61 m) away from each test pier, is shown in Figure 2. The soil modulus, E_s, was computed as a function of depth using the dilatometer modulus, E_D, and an assumed poission's ratio of 0.3. Table 1 gives the values of K_{ho} obtained from the dilatometer evaluated soil modulus, E_s (DMT), for pier #1. The K_{ho} values were computed using the previously described approaches. Method "a" predicted the higher K_{ho} values, in comparison to method "b", using Vesic's approach. The validity of the different predicted values is examined later.

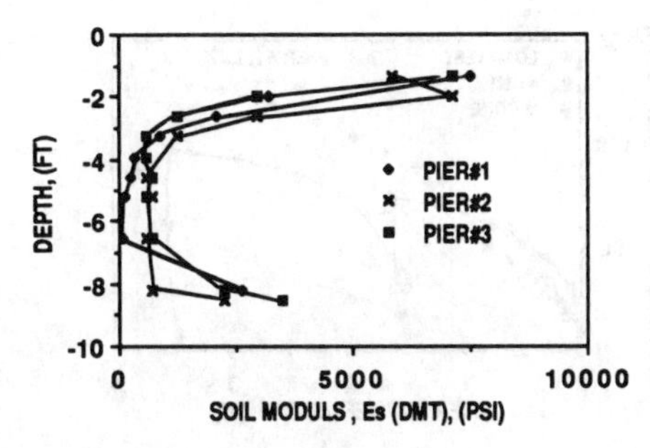

Figure 2. Soil Modulus, E_s (DMT) vs. Depth

Table 1. Coefficient of Subgrade Reaction Evaluated Using Es (DMT)

Depth	Pier #1			Pier #2			Pier #3		
	E_s	Method a	Method b	E_s	Method a	Method b	E_s	Method a	Method b
FT	PSI	PCI	PCI	PSI	PCI	PCI	PSI	PCI	PCI
1.31	7520	250	156				6509	216	134
1.98	3138	106	60	SOIL	EXCAVATED		3974	132	79
2.64	2096	70	40				1122	37	20
3.28	867	29	15	675	22	12	409	13	7
3.94	349	12	6	475	15	8	647	21	11
4.59	259	9	4	475	15	8	633	21	11
5.25	116	4	2	475	15	8	541	18	9
6.56	65	2	1	475	15	8	633	21	11
8.20	2613	871	50	528	17	9	3142	104	61

Table 2. K_o Value Predicted Using Baldi et al Model, "best estimate", Equation (4)

Depth	Pier #1			Pier #2			Pier #3		
FT	q_c/σ_v	K_D	K_o	q_c/σ_v	K_D	K_o	q_c/σ_v	K_D	K_o
1.31	1013	90	4.3				866	65	2.4
1.98	598	31	.56	SOIL	EXCAVATED		290	21	1.1
2.64	132	12	.91	35	4	-.31	131	6	.32
3.28	76	5	.47	34	3.4	.09	108	3	.15
3.94	18	2	.52	23	2.2	.37	120	1.3	-.05
4.59	16	1	.43	22	1.6	.42	105	1.3	.01
5.25	7	1	.43	11	1.8	.32	130	1.1	-.11
6.56	7	1	.44	15	1	.40	357	4.2	-.87

4 FIELD LOAD TESTS

The drilled piers, 30 in (.762 m) in diameter and 7 ft (2.13 m) long, were constructed on a highway embankment with side slopes of 3.5:1. The test piers #1 and #2, were pulled along the direction of the highway (horizontal ground surface). Pier #3 was loaded down the side slope of the embankment. The top 2 ft (.61 m) of soil surrounding pier #2 was excavated in order to obtain a different L/D ratio.

The piers were subjected to an overturning moment accompanied by modest shear. This loading combination was accomplished by applying a lateral load to the top of a 30 ft (9.1 m) column supported by the pier being tested. The load was applied in increments of 250 lb (1.1 kN). Surface instrumentation consisted of four dial gages: one to monitor lateral deflection, one for the out of plane movement, and two to measure normal displacements at the top surface of the pier to monitor rotation. The measured load-deflection response for the three piers is shown in Figures 3, 4 and 5.

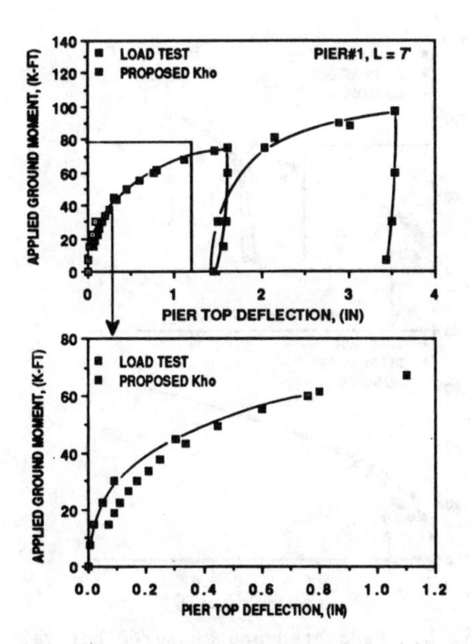

Figure 3. Predicted and Measured Lateral Response for Pier #1

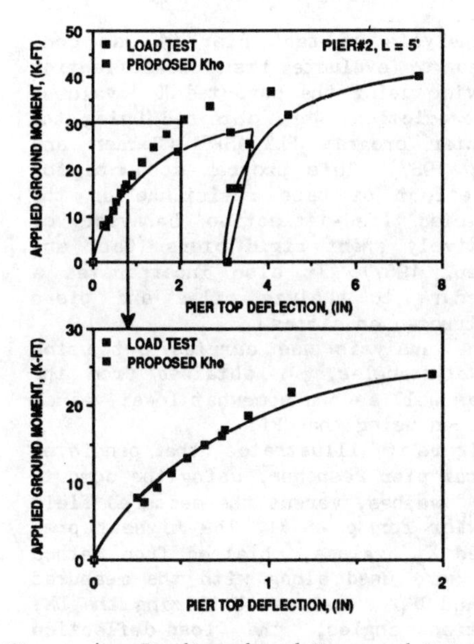

Figure 4. Predicted and Measured Lateral Response for Pier #2

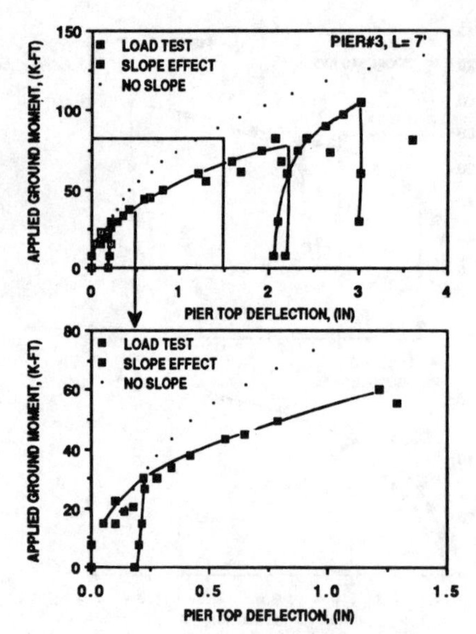

Figure 5. Predicted and Measured Lateral Response for Pier #3

5 VALIDITY OF THE EVALUATED K_{ho}

An analysis of test pier #1 was conducted to evaluate its load-deflection behavior using the computed K_{ho} values. The predictions were obtained using the computer program "LTBASE" (Borden and Gabr, 1987). This program accounts for the effect of base resistance on the predicted load-deflection behavior of relatively short rigid piers (Gabr and Borden, 1987). It also incorporates a procedure to analyze piles and piers constructed on slopes.

The analysis was carried out using friction angles, ϕ, obtained from the DMT as well as the somewhat lower values obtained using the CPT.

Figure 6 illustrates the predicted lateral pier response, using the computed K_{ho} values, versus the measured field behavior for pier #1. The highest predicted K_{ho} values, obtained from method "a", were used along with the measured CPT and DMT values. Utilizing the DMT friction angles, the load-deflection behavior was underestimated at the early stage of the curve and overestimated toward the ultimate load. This could be interpreted as K_{ho} values being low and friction angle values being high.

Using the same K_{ho} values with the CPT friction angles underestimates the measured load-deflection behavior at the early stage of the curve. However, good agreement between the measured and predicted ultimate behavior is obtained.

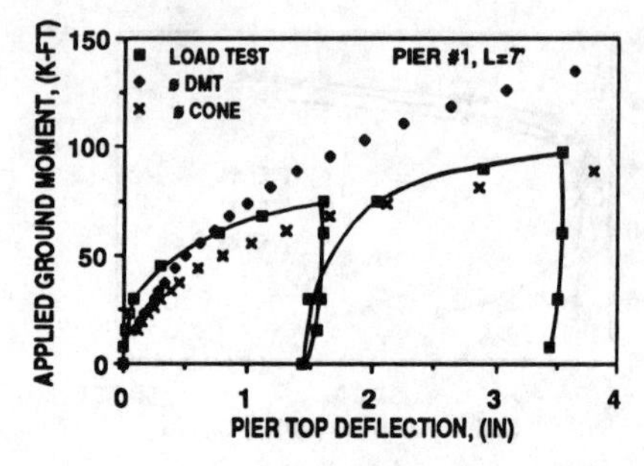

Figure 6. Predictions Based on Utilizing E_s (DMT) vs Measured Response

This suggests that while the evaluated K_{ho} values are low, the CPT friction angles are reasonable.

The low values of K_{ho} may be attributed to the dilatometer modulus. Being a tangent modulus, it is evaluated at a strain level corresponding to .27 in (6.85mm) separation of the soil (half the blade thickness). A more representative K_{ho} is determined based on the initial secant modulus of the soil. A new procedure is proposed below to determine K_{ho} using the dilatometer data.

6 PROPOSED SUBGRADE REACTION MODEL

Using the dilatometer "P_o" reading, the subgrade reaction coefficient, K_{ho}, is defined as follows:

$$K_{ho} = \frac{P_o - \sigma_h}{\text{half blade thickness}} \quad (2)$$

where P_o = Dilatometer reading, A, corrected, σ_h = In-situ lateral at rest pressure, and half the blade thickness is equal to .27 in. if P_o and σ_h are in psi, and .00685 m if P_o and σ_h are in kPa.

This model is an approximation of the field situation assuming that the lateral in-situ stresses after the pile installation are comparable to those existing at the time of DMT testing. In this study the test piers were existing at the time of DMT testing, and therefore no installation effects were investigated.

6.1 In-situ lateral stresses

An essential component of the proposed model is determination of σ_h. The evaluation of σ_h can be achieved providing that the coefficient of at-rest earth pressure, K_o, is known. A model to evaluate K_o using the dilatometer test was presented by Baldi et al (1986). The model was obtained from a statistical analysis of calibration chamber test data. The tests were performed at ENEL CRIS (Milano, Italy) and ISMES (Bergamo). The original equation developed to fit the test data was as follows:

$$K_o = .376 + .095\, K_D - .00172 \frac{q_c}{\sigma_z} \quad (3)$$

where K_o = Coefficient of earth pressure at-rest, K_D = Dilatometer lateral stress index, q_c = Cone point resistance, and σ_z = In-situ effective overburden stress.

However, modifications were necessary in order to correctly predict the K_o values for the Po river site from which the sand samples, used in the calibration chamber, had been taken. This was achieved by searching , using trial and error, for a multiplier for the constant ".00172" used in the above equation. The "best estimate" equation obtained was :

$$K_o = .376 + .095\, K_D - .00461 \frac{q_c}{\sigma_z} \quad (4)$$

Another model to obtain the in-situ lateral stresses prior to the insertion of the dilatometer blade was developed using NCSU calibration chamber test results. This model is different from the above in the context of directly relating the dilatometer "P_o" reading to the octahedral stress state, σ_{oct}. The lateral stress, σ_h, was inferred using the vertical overburden stress, σ_z.

6.2 NCSU Calibration Chamber

The North Carolina State University (NCSU) calibration chamber is capable of accommodating insertion of the dilatometer blade, and has the facility for the independent application of horizontal and vertical pressures. The chamber, Figure 7, is about 4 ft (1.22 m) in diameter and 5 ft (1.52 m) high and houses a sample approximately 3 ft (.91 m) in diameter and 3 ft (.91 m) high. The pressures are applied and monitored using a data acquisition system which consists of Analog/Digital, Digital/ Analog control board, Voltage/Pressure regulators, Pressure/Voltage transducers, inductance coils for displacement measurements and a Zeinth computer. Control software, developed by the first author, permits the monitoring of applied pressures such that a specified stress path can be followed. The lateral and vertical displacements are also monitored using the developed software.

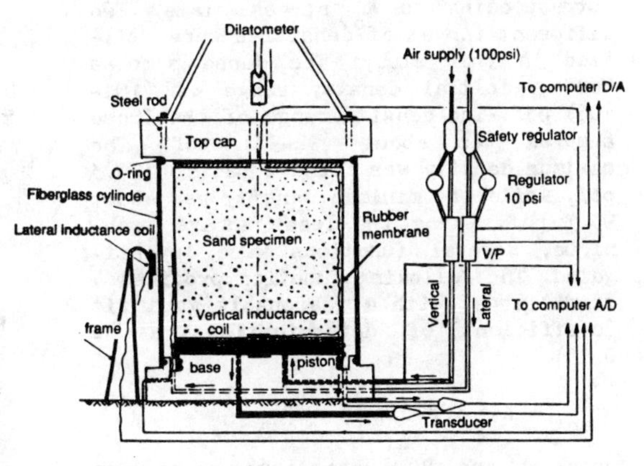

Figure 7. NCSU Calibration Chamber

The sand rain device used to prepare the samples consists of a plastic cylinder 3 ft (.94 m) in diameter and 3.75 ft (1.14 m) high. The bottom of the cylinder is composed of two perforated plates, one fixed and the other movable with the same hole patterns. Uniform samples are formed by pluvial deposition using a movable diffuser which maintains a constant drop height from the diffuser to the sand surface. The diffuser, composed of three metal meshes which are 4in(.1 m) apart and rotated 45° with respect to each other is used to disperse the sand before it reaches the bottom of the calibration chamber. Different relative densities are achieved by changing the size of the holes on the bottom plates of the sand rain device.

6.3 NCSU lateral stress model

Calibration chamber tests on normally consolidated Cape Fear River sand samples were performed. The tests proceed by incrementing vertical and horizontal pressures to achieve K_o consolidation of the sample, comparable to particular insitu stress conditions. The dilatometer blade is then inserted into the sample at a controlled rate typical of field testing, and the dilatometer test is performed. The confining pressure in the chamber is then increased in increments while the vertical pressure is constantly maintained. After each of these subsequent increases in lateral stress, the dilatometer is pushed to the next depth and another DMT is performed. Figure 8(a and b) show the relationship between the dilatometer reading, P_o, and the applied octahedral stress, σ_{oct}, corresponding to K_o stress state. Two different ranges of densities were utilized in this study. The dense samples had an initial density range of (101-106) pcf. The density range of the loose samples was about (92-97) pcf. The maximum density was found equal to 109.3 pcf, while the minimum density was about 93.1 pcf. Using the least square technique, several functions were investigated. The following function provided a simple model with a reasonably good fit (coefficient of determination, R^2 = 0.85)

$$\sigma_{oct} = A * (Po-uo)^B \qquad (5)$$

where A and B = constants which are functions of the relative density and sand characteristics. For this sand:
A= 0.97, B= 0.57 => dense sand
A= 1.54, B= 0.54 => loose sand

These numerical values of the constants, A and B, are only valid when Po, uo, and σ_{oct}, are expressed in psi. For different units, new constants could be found by introducing the appropriate conversion factors, from psi to the desired units. The octahedral stresses, σ_{oct}, can be expressed in terms of σ_1 and σ_3:

$$\sigma_{oct} = \frac{\sigma_1 + 2\,\sigma_3}{3} \qquad (6)$$

Substituting equation (6) in equation (5), and setting σ_h equal to σ_3, and σ_z equal to σ_1, P_o is related to the in-situ lateral stress as follows:

$$\sigma_h = \frac{3A* (Po-uo)^B - \sigma_z}{2} \qquad (7)$$

where σ_h = In-situ effective lateral stresses prior to the insertion of the dilatometer blade (psi), σ_z = Vertical effective overburden pressure (psi), and P_o = dilatometer reading (psi).

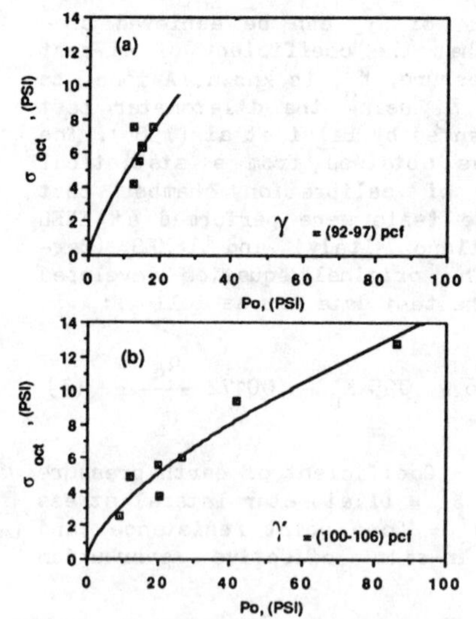

Figure 8. Relationship Between Po and σ_{oct} in CCT

Table 3. K_o Value Predicted Using Baldi et al Model, Equation (3)

Depth	Pier #1			Pier #2			Pier #3		
FT	q_c/σ_v	K_D	K_o	q_c/σ_v	K_D	K_o	q_c/σ_v	K_D	K_o
1.31	1013	90	7.2				866	65	5
1.98	598	31	2.3	SOIL	EXCAVATED		290	21	1.9
2.64	132	12	1.3	35	4	.35	131	6	.70
3.28	76	5	.68	34	3.4	.47	108	3	.46
3.94	18	2	.57	23	2.2	.50	120	1.3	.29
4.59	16	1	.47	22	1.6	.49	105	1.3	.32
5.25	7	1	.45	11	1.8	.46	130	1.1	.26
6.56	7	1	.47	15	1.1	.45	357	4.2	.16

Table 4. Coefficient of Subgrade Reaction Based on the Above Evaluated σ_h

Depth	Pier #1			Pier #2			Pier #3		
	P_o	σ_h	K_{ho}	P_o	σ_h	K_{ho}	P_o	σ_h	K_{ho}
FT	PSI	PSI	PCI	PSI	PCI	PCI	PSI	PCI	PCI
1.31	113	8	380				84	6	278
1.98	56	4	189	SOIL	EXCAVATED		40	3	132
2.64	29	3	96				14	2	45
3.28	14	2	42	11	1	34	8	1	25
3.94	8	2	22	8	2	23	5	1	14
4.59	5	2	12	7	2	22	2	1	16
5.25	4	1	6	9	2	12	6	1	16
6.56	5	1	9	5	2	15	24	1	83

Table 5. Coefficient of Subgrade Reaction Based on NCSU Lateral Stress Model

Depth	Pier #1			Pier #2			Pier #3		
	P_o	σ_h	K_{ho}	P_o	σ_h	K_{ho}	P_o	σ_h	K_{ho}
FT	PSI	PSI	PCI	PSI	PSI	PCI	PSI	PSI	PCI
1.31	113	22	340				84	18	245
1.98	56	14	160	SOIL	EXCAVATED		40	11	108
2.64	29	9	75				14	5	33
3.28	14	5	33	11	4	24	8	3	19
3.94	8	3	19	8	3	18	5	2	12
4.59	5	2	12	7	2	15	6	2	14
5.25	4	1	12	9	3	22	6	1	15
6.56	5	1	16	5	1	15	24	6	67
8.20	30	6	88	23	5	65	28	6	82

6.4 Predictions using Baldi et al model

The predictions of K_o for the test site were performed using the DMT and CPT data adjacent to each of the three piers. The DMT data were obtained approximately 2 ft (0.61m) away from each pier. The average of two CPT profiles, obtained 3in (.076 m) , and 3 ft (.91 m) away from each pier was used in conjunction with the DMT data to predict Ko. Using the recommended model, equation (4), some negative values for K_o along the pier length for piers #2 and #3, were obtained as shown in Table 2. This was due to the high values of the cone point resistance, q_c at these locations. However, using the original model, equation (3) resulted in a logical estimates of K_o. These values were used along with σ_z to determine the lateral at rest pressure, σ_h, given in Table 3.

The predicted σ_h values were used to deduce the coefficient of subgrade reaction, K_{ho}, using the proposed model, equation (2). The computed K_{ho} values and their variation as a function of depth are given in Table 4. It is of interest to note that, for this particular site, the predicted values of σ_h are approximately one order of magnitude smaller than the dilatometer P_o reading. Although, for practical purposes σ_h may be neglected, conceptually the proposed model is based on including the σ_h value. Furthermore, it can not be assumed that P_o will always be this order of magnitude higher than σ_h, thus making the estimate of σ_h of minor significance.

6.5 Stress predictions using NCSU model

Utilizing the predicted σ_h value along with the DMT, P_o, the coefficient of subgrade reaction, K_{ho}, is computed using Equation (2). The values of P_o, σ_h and K_{ho} as a function of depth are given in Table 5, for the three piers tested. As was the case for Baldi et al model, the new estimated K_{ho} values are also higher than those estimated previously using methods "a" and "b". It is of interest to notice that adjacent to pier #1 the K_{ho} values are about 1.5 times higher than those determined for pier #3. Although the soil is more dense around pier #3, it is believed that the top 2 ft (.61 m) of the soil surrounding pier #1 was compacted, see CPT profile, Figure 1. This resulted in higher residual lateral stresses in the soil. Consequently, higher K_{ho} values were determined. The K_{ho} values based on σ_h predicted using NCSU model are reasonably close to those obtained based on Baldi et al model. This is believed to be partly due to the order of magnitude difference between P_o and σ_h. In this particular case, an educated guess of K_o values using the angle of internal friction, ϕ , would have been sufficient. However, it has not yet been shown that this will hold true for every situation.

The K_{ho} values, computed using the proposed model and utilizing σ_h computed using NCSU model, were utilized to predict the lateral response of the test piers. The angles of internal friction from the CPT were used in the analysis.

7 RESULTS AND DISCUSSION

The predicted and measured responses for pier #1 are shown in Figure 3. In general, the predicted response of pier #1 agreed exceptionally well with the field measured behavior. The lateral response of pier #2 was somewhat over-estimated at the ultimate, as shown Figure 4. Bearing in mind that the DMT and the CPT profiles were obtained prior to the excavation of the top 2 ft (.61 m), it is believed that the excavation process might have disturbed the soil below. The measured load-deflection response of pier #3 was overpredicted when ignoring the presence of the sloping ground conditions. However, good agreement between the measured and predicted behavior was obtained when the slope effect was included, as shown in Figure 5.

Assuming the point of rotation lies in the middle third of the tested piers, the majority of resistance will be mobilized in the upper 4 ft (1.26 m), for the 7 ft (2.13 m) long piers. The K_{ho} values computed for the upper 4 ft (1.26 m) play a significant role in predicting the load deflection response. This fact emphasizes the importance of obtaining the distribution of K_{ho} with depth. The good agreement between the measured and predicted response endorses such necessity.

8 SUMMARY AND CONCLUSION

A model is proposed for the evaluation of the coefficient of subgrade reaction, K_{ho}, in cohesionless soils, using the flat dilatometer test. Since this model requires the knowledge of the in-situ lateral stresses prior to the insertion of the DMT blade, a lateral stress model is also constructed, based on calibration chamber test data. Based on the results of predictions performed using the developed models, the following conclusions can be advanced:

1. The magnitude of the in-situ lateral stress determined using the developed model and the model proposed by Baldi et al was an order of magnitude smaller than the DMT, "P_o" reading. This indicated that, in this particular case, an educated guess for the estimation of the in-situ lateral stresses, σ_h, would have been sufficient.

2. Utilizing the K_{ho} values obtained using the proposed model resulted in good predictions of the load deflection behavior of piers tested in the field.

3. The DMT angle of internal friction, determined based on Schmertmann procedure, seems to be high. It is recommended to use cone evaluated friction angles in conjunction with this model.

4. While the developed model is expected to produce good results for drilled piers and bored piles, it remains to be seen if it will hold true for driven piles.

5. An enlarged data base is required to further assess model applicability.

ACKNOWLEDGMENTS

This research was sponsored by the North Carolina DOT in cooperation with the U.S. Department of Transportation, Federal Highway Administration (FHWA). A technical advisory committee chaired by Mr. W.G. Marley, and including Messrs. D. Bingham, J. Ledbetter, R. Martin, P. Strong, J. Wilder of the NCDOT, and Mr. J. Wadsworth of FHWA aided in the coordination of the research, and are gratefully acknowledged. We also wish to thank Mr. D. Milton and the NCDOT personnel who supported the geotechnical investigation and load testing. Thanks are due to Dr. Peter K. Robertson for reviewing this paper and providing several helpful comments.

REFERENCES

Baguelin, F., Frank R., and Said, Y. H., 1977, " Theoretical Study of Lateral Reaction Mechanism of Piles," Geotechnique, Vol. 27 , pp. 405-434.

Baguelin, F., Jezequel, J. F., and Shields, D., 1978, The Pressuremeter and Foundation Eng., Aedermannsdorf, Switzerland, Trans Tech Publ.

Baldi, G. , R. Bellotti, V. Ghionna, M. Jamiolkowski, S. Marchetti and E. Pasqualini, 1986" Flat Dilatometer Tests in Calibration Chambers" Proc., In Situ 86, Special Publication No. 6, ASCE, pp. 431-446.

Borden, R. H., and Gabr, M. A., 1987," LTBASE: Lateral Pier Analysis Including Base and Slope Effect," Research Report No. HRP 86-5, Cent. for Transp. Eng. Studies, NCSU, Raleigh.

Broms, B. , 1964 " Lateral Resistance of Piles in Cohesive Soils" J. of Soil Mech. and Foun, ASCE, V 90, pp 27-63.

Gabr, M. A., and Borden, R. H., 1987 " Influence of the Base Resistance on the lateral Load Deflection Behavior of Rigid Piers in Sand,"Int. Conf., Soil-Struc. Interact., Paris.

Marchetti, S., 1980, "In Situ Tests by Flat Dilatometer, "J of the Geo. Eng. Div ASCE, V 106, NO. GT3, pp 299-321.

Menard, L., 1963, "Comportement d'une Foundation Profonde Soumise a' des Efforts de Renversement,"Sols-Soils.

Motan, E. S., and Gabr, M. A., 1985, " Flat Dilatometer and Lateral Soil Modulus," Tran Res Rec No 1022, Wash.

Murchison, J.M., O'Neill, M.W., 1984, "Evaluation of P-Y Relationships in Cohesionless Soils," Analy. and Des. of Pile Foun. ,ASCE, San Frans.

Parker, F., Jr., and Reese, L.C., 1970, " Experimental and Analytical studies of Behavior of Single Piles in Sand under Lateral and axial Loading," Res Report 117-2, Cent. for Highway Res., Univ. of Texas at Austin.

Poulos, H. G., and Davis, E. H., 1980, Pile Foundation Analysis and Design, John Wiley and Sons, N.Y.

Reese, L. C., Cox, W.R., Koop, F. D., 1974, "Analysis of Laterally Loaded Piles in Sand," 6th Annual Offshore Technology Conf., V 2, Houst., TX.

Schmertmann, J.,"The Measurement of In-Situ Shear Strength," State-Of-The-Art, ASCE Spec. Conf. on In-Situ Meas, of Soil Prop., NCSU, Apr 1975.

Schmertmann, J. H., 1982, "A Method for Determining the Friction Angle in Sands from Marchetti Dilatometer Test," Proc. of ESOPT II, Amestrdam .

Terzaghi, K., 1955, " Evaluation of Coefficient of Subgrade Reaction," Geotechnique, V 5, pp. 297-326.

Vesic, A. B., 1961, "Bending of Beams Resting on Isotropic Elastic Solid," J. of Eng Mech, ASCE, V 87, No. EM-2.

Penetration Testing 1988, ISOPT-1, De Ruiter (ed.)
© 1988 Balkema, Rotterdam, ISBN 90 6191 801 4

CPT and DMT in evaluation of blast-densification of sand

Roman D.Hryciw
University of Michigan, Ann Arbor, USA
Charles H.Dowding
Northwestern University, Evanston, Ill., USA

ABSTRACT: Electric cone penetration and Marchetti dilatometer tests were performed to evaluate changes in soil properties after detonation of explosives in sand. Despite apparent densification, cone tip resistance, dilatometer modulus and the dilatometer horizontal stress index decreased after blasting. Cone penetration tests were performed at five distances from each shot to evaluate the zone of influence from a given charge.

1 INTRODUCTION

Although the use of explosives to densify loose saturated cohesionless soils (blast-densification) is a viable and cost-efficient treatment technique (Lyman, 1941; Kummeneje and Eide, 1961; Hall, 1962; Prugh, 1963; Ivanov, 1967; Dembicki and Kisielowa, 1983; Barendsen and Kok, 1983; Solymar, 1984; Carpenter et al, 1985; Hryciw, 1986), it has enjoyed only marginal acceptance by the geotechnical community. The reasons for the apparent lack of confidence in blast-densification stem from two commonly raised questions: "is it safe?" and "will it work?". Without satisfactory answers, engineers will generally opt for methods which are more expensive but perceived to be safer such as vibroflotation or deep dynamic compaction.

Engineers fear using explosives for two reasons: fear of accidental detonation and fear of blast generated vibrations to nearby structures and facilities. In the past, explosives such as dynamites were truly dangerous materials to handle and transport. These fears should be alleviated somewhat today by the availability of safe, packaged "two component" explosives. They consist of a solid oxidizer such as ammonium nitrate and a combustible liquid. Neither component is an explosive by itself and therefore each can be handled and shipped separately without special precautions. Even after "arming", the explosive can only be detonated by a high detonation pressure blasting cap or primer. Seismic prospecting crews routinely use these compounds and report excellent safety records.

The second fear, that of blast vibrations can be relieved by referring to previously documented attenuation charts for blast vibrations and designing the shots to maintain vibration below allowable limits. Such attenuation charts are commonly available from rock blasting operations (Langefors and Kihlstrom, 1978; Duvall et al 1963; and Dowding, 1985) and are now also available for blast densification of sands (Hryciw, 1986).

The answer to the second question, "will it work?" has been somewhat debatable until recently. The debate centered around numerous reports of decreased cone penetration resistance and decreased standard penetration N-values after blasting. These apparent decreases in stiffness and strength were observed despite obvious densification as attested to by surface settlements. Mitchell and Solymar (1984) and Dowding and Hryciw (1986) have now shown that the decreased penetration resistances are only temporary and that significant increases in strength occur with time after blasting. The present paper presents new CPT and DMT results from a small scale blast-densification program in which decreases in cone penetration resistance, q_c, dilatometer modulus, E_D and horizontal stress index were observed despite surface settlement.

2 THE BLASTING PROGRAM

The test program was conducted on the Atlantic Coast at Harriets Bluff, Georgia. The purpose of the test program was to evaluate the effectiveness of explosives in densifying a loose sand layer at depth.

The soil was a loose to medium fine sand with silty sand lenses to a depth of 6 m. A layer of denser sand or clay was sometimes encountered at a depth of 2 to 3 m. Below 6 m, medium to dense, rigid sand and silty sand were sometimes observed. Groundwater was encountered at a depth of 0.6 m. The grain size distribution is shown in Figure 1. D_{50} ranged between 0.10 and 0.15 mm. Typical pre-blast CPT's are given in Figure 2. According to Robertson and Campanella's (1984) CPT-liquefaction criteria, the soil was loose and liquefiable.

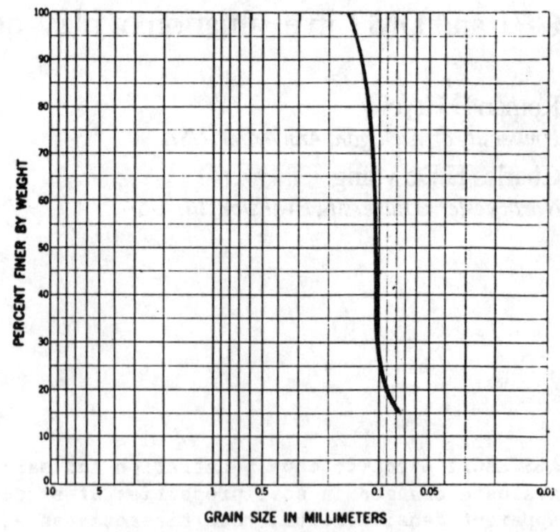

Fig. 1 Grain size distribution of Harriet's Bluff sand

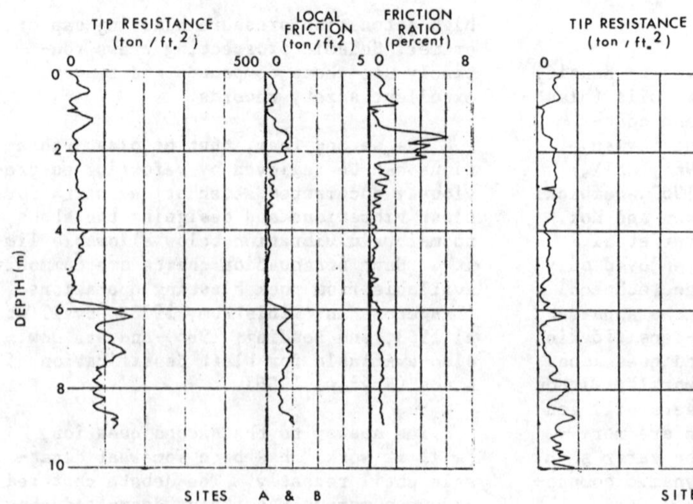

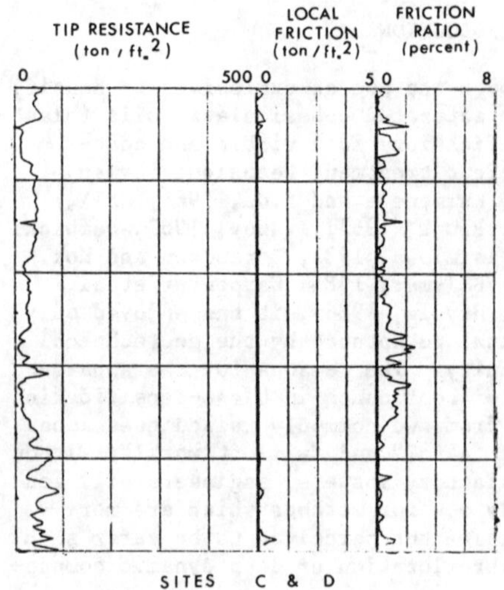

Fig. 2 Typical pre-blast cone penetration tests.

In all, five test shots were fired. Boreholes could not be kept open in the loose saturated sand, therefore, 6.35 cm plastic casing was jetted into the soil for the charges. A 60% straight dynamite was used as the explosive. One to three 1.75 kg sticks of explosive were used in each test shot. Table 1 lists the depths and time delays for the charges at each test location.

In a full scale blast-densification program charges would be placed at a 5 to 10 m spacing and detonated in a predetermined timing pattern. However, the object of the present test program was to to observe the radius of influence and the changes in soil properties from a single charged borehole. As such, the holes were spaced at least 50 m apart and only one hole was detonated at a time. Charges were detonated in various delays as listed in Table 1. An attempt was made to generate increases in residual pore water pressure and therefore increase the radius of liquefaction-densification by repetitive impulse loads.

Table 1 Charge depths and delays

TEST SITE	CHARGE DEPTHS (m)	DELAY (msec)	CHARGE PER DELAY (kg)	TOTAL CHARGE FIRED (kg)
S	7.3	0	1.75	1.75
A	4.0	0	3.50	3.50
	6.4	0		
B	4.0	0	1.75	3.50
	6.4	25		
C	3.7	0	5.25	5.25
	5.5	0		
	7.3	0		
D	3.7	0	1.75	5.25
	5.5	25		
	7.3	50		

3 TEST RESULTS

Cone penetration tests were conducted at radial distances of 0.9, 3.0, 5.5, 7.6 and 12.2 m prior to blasting, 1 to 2 days after blasting and once again 30 days after blasting to observe changes in cone penetration resistance with distance and time from the blast. A typical cross section showing the location of charges, cone penetration tests and dynamic sensors is presented in Figure 3.

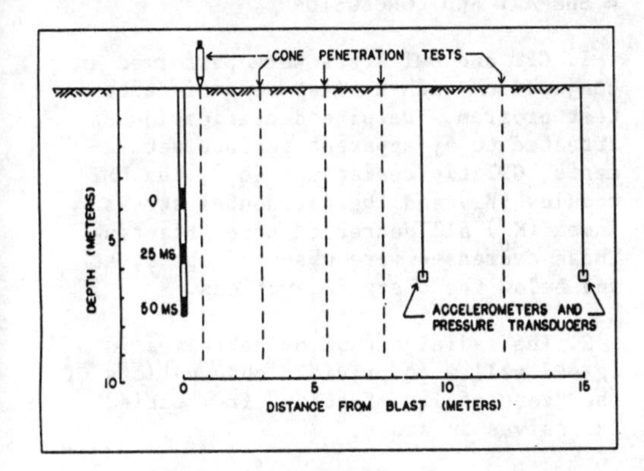

Fig. 3 Typical cross section showing the location of charges, CPT's and dynamic sensors

Although surface settlements could not be measured because of a highly irregular surface and vegetation, the area around the blast hole appeared to subside in localized spots to a radius of 3 to 4.5 m. Nevertheless, the 1 to 2 day post-blast cone penetration tests exhibited decreases in tip resistance over the pre-blast values. This observation was consistent with the findings of Mitchell and Solymar (1984) and others. The cone penetration tests performed at 30 days still showed no appreciable change over the 1 to 2 day readings. No further tests were performed.

Although the CPTs did not verify soil improvement, they furnished valuable information about the radius of influence from each charge. The percent change in tip resistance from the pre-blast values is presented in Figure 4 for each of the test sites. In the case of the 0-0-0 shot (Test C), the hole was not adequately stemmed with inert material and a partial blowout resulted. The low changes in q_c at 3 m in this shot confirmed that not all of the explosive energy was contained underground. The largest decreases in q_c were observed immediately adjacent to each charge. Increases in q_c were observed only in localized patches beyond 3 m from each shot.

From blasting experience in the Netherlands, Studer and Kok (1980) reported that on the average, the liquefaction coefficient, defined as the increase in pore residual pore water pressure, Δu, divided by the effective octahedral stress, σ'_o, was given by:

$$\frac{\Delta u}{\sigma'_o} = 1.65 + 0.64 \ln \frac{W^{0.33}}{R}$$

where W is the charge weight in kilograms and R is the distance from the blast in meters. Liquefaction would occur when the liquefaction coefficient was approximately equal to 1.0. It can be shown that for 1.75 kg charges, the liquefaction coefficient would be 1.0 at a distance of 3.4 m which is in excellent agreement with the observed extent of change in q_c.

Most perplexing are the decreased resistances below the charges. Increases in void ratio in this area would not be expected unless accompanied by decreases elsewhere. Since increases in q_c were not observed, a loosening of the soil could not be responsible for the decreased q_c in

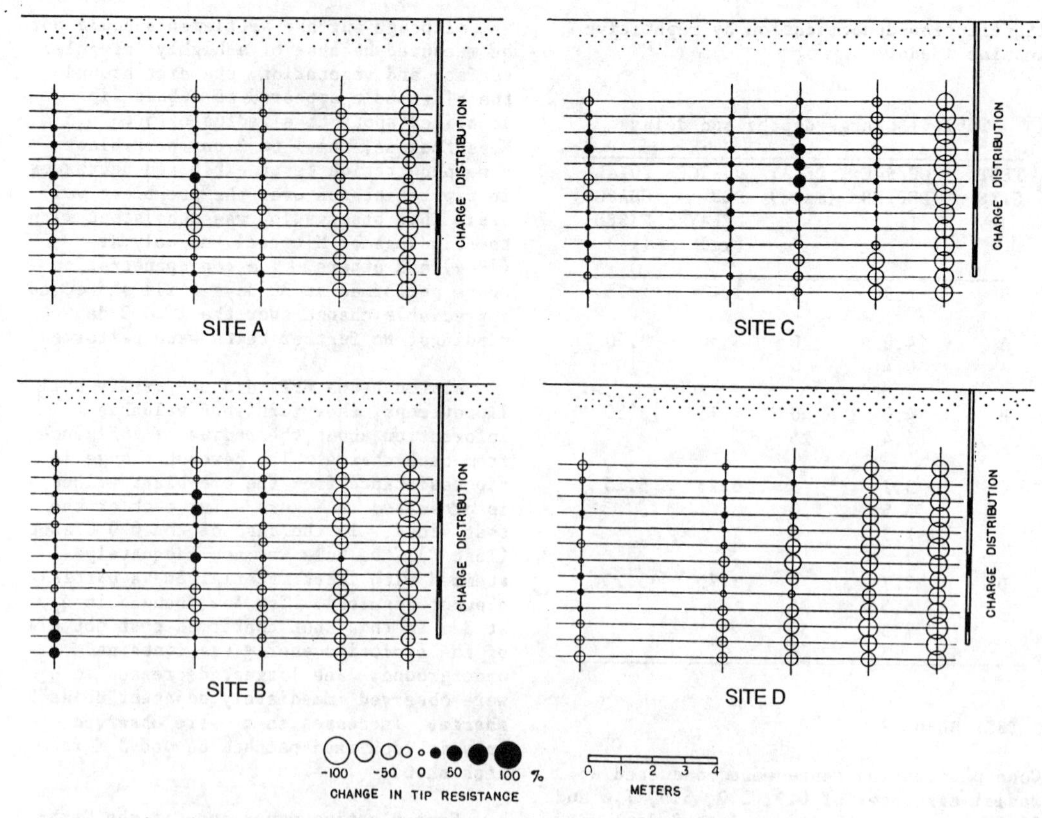

Fig. 4 Percent change in CPT tip resistance from pre-blast values

this zone. Therefore, it is apparent that changes in lateral earth pressures as suggested by Schmertmann (1987), disruption of cementatious intergranular bonds as suggested by Mitchell and Solymar (1984), or both, resulted in the lower q_c's.

Flat-plate dilatometer tests (DMTs) were performed on site A at 3.4 m. The before and after dilatometer modulus, E_D and horizontal stress index are shown in Figures 5 and 6. They clearly agree with the cone penetration results and suggest increased soil compressibility and decreased lateral earth pressures. Two DMT tests were conducted at 3.4 m. One of these was in the radial direction (diaphragm expanded toward the blast location) and the other in the tangential direction. The in situ stresses in the radial direction were slightly higher than those in the tangential direction, as shown in Figure 7, suggesting that a strong radial blast loading may produce a residual anisotropic stress field in the horizontal plane. This observation however is based on limited data and requires further study.

4 SUMMARY AND CONCLUSIONS

1. CPT and DMT tests were performed in conjunction with a blast-densification test program. Despite densification as attested to by apparent surface settlements, CPT tip resistance (q_c), the DMT modulus (E_D) and the horizontal stress index (K_D) all decreased after blasting. These decreases were observed above, at and below the charge elevations.

2. The radial extent of decreases in q_c agreed well with previous observations of the "zone of liquefaction" from buried explosives in sands.

3. DMT tests performed after blasting in both the radial and tangential directions indicated that a residual stress anisotropy may be exist after blasting.

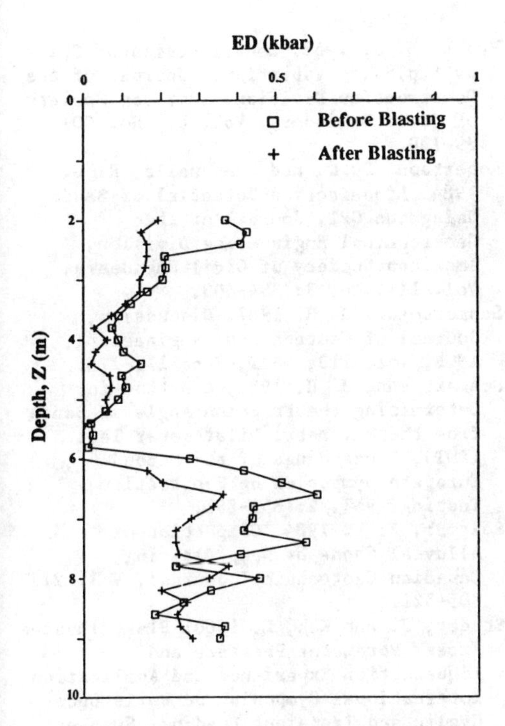

Fig. 5 Dilatometer modulus for site A at 3.4 m before and after blasting

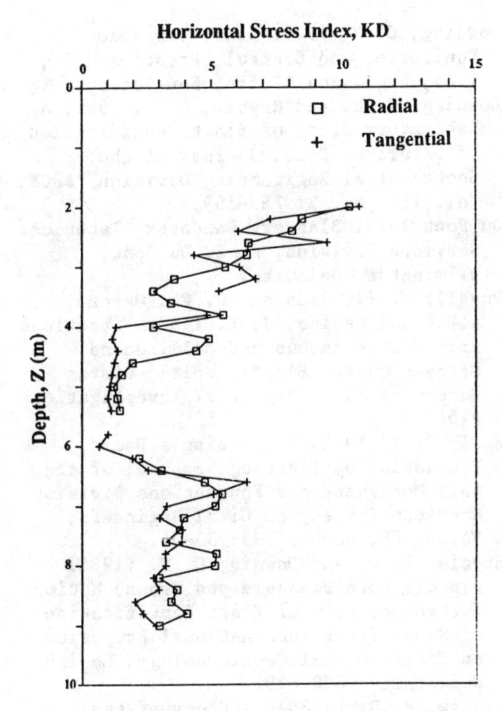

Fig. 7 Horizontal stress index for site A at 3.4 m after blasting in radial and tangential directions

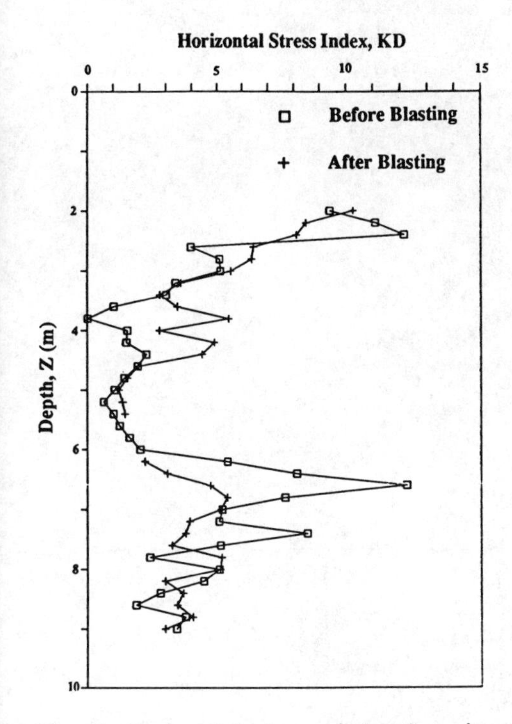

Fig. 6. Horizontal stress index for site A at 3.4 m before and after blasting

ACKNOWLEDGEMENTS

The CPT and DMT field tests were performed by the GKN-Hayward Baker Company. Partial Funding for the program was also obtained from NSF Grant CEE-8414471.

REFERENCES

Barendsen, D. A. and Kok, L. 1983. Prevention and Repair of Flow-Slides by Explosion Densification, Proceedings of the 8th European Conference on Soil Mechanics and Foundation Engineering, Helsinki: 205-208.

Carpenter, R., De Wolf, P., Van Damme, L., De Rouck, J. and Bernard, A. 1985. Compaction by Blasting in Offshore Harbour Construction, Proceedings of the XI ICSMFE, Vol. 3, San Fransisco: 1687-1692.

Dembicki, E. and Kisielowa, N. 1983. Technology of Soil Compaction by Means of Explosion", Proceedings of the 8th European Conference on Soil Mechanics and Foundation Engineering, Helsinki: 229-230.

Dowding, C. H. 1985. Blast Vibration Monitoring and Control, Prentice-Hall, Inc., Englewood Cliffs, New Jersey.

Dowding, C. H. and Hryciw, R. D. 1986. A Laboratory Study of Blast-Densification of Saturated Sand, Journal of the Geotechnical Engineering Division, ASCE, Vol. 112, No. 2: 187-199.

Du Pont 1977. Blasters Handbook, Technical Services Division, E. I. Du Pont, Wilmington, Delaware.

Duvall, W. I., Johnson, C. F., Meyer, A.V.C and Devine, J. F. 1963. Vibrations from Instantaneous and Millisecond Delayed Quarry Blasts, United States Bureau of Mines Report of Investigations 6151.

Hall, C. E. 1962. Compacting a Dam Foundation by Blasting, Jpurnal of the Soil Mechanics and Foundations Division, American Society of Civil Engineers, Volume 88, Number SM3: 33-51.

Hryciw, R. D. and Dowding C. H. (1986) Dynamic Pore Pressure and Ground Motion Instrumentation of Blast Densification of Sand, First International Symposium on Environmental Geotechnology, Lehigh University: 620-629.

Hryciw, R. D. (1986), A Study of the Physical and Chemical Aspects of Blast Densification of Sand, submitted to Northwestern University in Partial Fulfillment of the requirements for a Ph.D. Degree in Civil Engineering, Evanston, Illinois.

Ivanov, P. L. 1967. Compaction of Noncohesive Soils by Explosions, Translated from Russian, Indian National Scientific Documentation Center, Available from the United States Water and Power Resources Service, Denver, Colorado.

Kummeneje, O. and Eide, O. 1961. Investigation of Loose Sand Deposits by Blasting, Proceedings of the 5th International Conference of Soil Mechanics and Foundation Engineering, Volume 1: 491-497.

Langefors, U. and Kihlstrom, B. 1978. The Modern Technique of Rock Blasting, John Wiley and Sons, New York.

Lyman, A. K. B. 1941. Compaction of Cohesionless Foundation Soils by Explosives, Proceedings of the American Society of Civil Engineers, Volume 67, May: 769-780.

Mitchell, J. K. and Solymar, Z. V. 1984. Time-Dependent Strength Gain in Freshly Deposited or Densified Sand, Journal of Geotechnical Engineering, American Society of Civil Engineers, Vol. 110, No. 11: 1559-1576.

Prugh, B. J. 1963. Densification of Soils by Explosive Vibrations, Journal of the Construction Division, American Society of Civil Engineers, Vol. 89, No. CO1: 79-100.

Robertson, P. K. and Campanella, R. G. 1984. Liquefaction Potential of Sands Using the CPT, Journal of the Geotechnical Engineering Division, American Society of Civil Engineers, Vol. 111, No. 3: 384-403.

Schmertmann, J. H. 1987. Discussion, Journal of Geotechnical Engineering, ASCE, Vol. 113, No.2, Feb: 173-175.

Schmertmann, J. H. 1982, A Method for Determining the Friction Angle in Sands from the Marchetti Dilatometer Test (DMT), Proceedings of the Second European Symposium on Penetration Testing, Vol. 2: 853-861.

Solymar, Z. V. 1984. Compaction of Alluvial Sands by Deep Blasting, Canadian Geotechnical Journal, Vol. 21: 305-321.

Studer, J. and Kok, L. 1980. Blast Induced Excess Porewater Pressure and Liquefaction Experience and Application, International Symposium on Soils Under Cyclic and Transient Loading, Swansea: 581-593.

Penetration Testing 1988, ISOPT-1, De Ruiter (ed.)
© 1988 Balkema, Rotterdam, ISBN 90 6191 801 4

DMT-cross hole shear correlations

Roman D.Hryciw & Richard D.Woods
University of Michigan, Ann Arbor, USA

ABSTRACT: Dilatometer tests (DMT) were performed in conjunction with cross-hole shear (V_s-CH) testing to compare the dilatometer modulus, E_D, with the maximum shear modulus, G_o, as computed from elastic wave propagation velocity measurements. Excellent agreement was observed between G_o and confining stresses as determined from dilatometer tests. The ratio of G_o to E_D, defined as R_G, was shown to be inversely related to the coefficient of lateral earth pressure, K_o.

1 INTRODUCTION

Since its development in the seventies, the Marchetti dilatometer (DMT) has been the focus of much research to develop empirical and semi-empirical correlations between the DMT indices and various soil properties. Correlations are now routinely used to determine soil type, coefficient of lateral earth pressure, preconsolidation pressure, overconsolidation ratio, phi-angle for sands, undrained shear strength in clays and constrained modulus.

One of the remaining tasks is development of correlations between the dilatometer modulus and the small strain shear modulus. The small strain shear modulus plays an important role in the design of foundations for dynamic loads. Existing methods for determining shear modulus in situ are well developed and generally based on seismic wave propagation techniques. A particularly useful test in this regard is the seismic cross-hole test. The present paper presents a comparison of the dilatometer modulus with the small strain cross-hole shear modulus.

1.1 The Dilatometer (DMT) Test

The DMT test has been described in detail by Marchetti (1980); Marchetti and Crapps (1981); Baldi et al (1986) and others. In brief, a steel blade with a pneumatically expandable steel diaphragm is pushed or driven into the soil. Two readings "A" and "B" are taken at each test depth. The A reading is the pressure to cause "lift-off" of the dilatometer diaphragm while the B reading is the pressure required to expand the center of the diaphragm by 1 mm. After appropriate corrections, the A reading yields P_o, the "first corrected reading" and B yields the "second corrected reading".

The closing pressure required to return the diaphragm to the original A position has been shown to closely match the in-situ pore water pressure, u_o in sands (Campanella et al., 1985). In clays, the closing pressure is the in-situ hydrostatic pressure plus the excess pressure caused by blade insertion. Schmertmann (1986) has labeled the third reading as "C" and suggests a method for determining and correcting "C" to obtain u_o.

The following three indices are determined from the corrected readings:

$$I_D \text{ (Material Index)} = \frac{P_1 - P_o}{P_1 - u_o}$$

$$K_D \text{ (Horizontal Stress Index)} = \frac{P_o - u_o}{\sigma'_{vo}}$$

$$E_D \text{ (Dilatometer Modulus)} = 34.7(P_1 - P_o)$$

where σ'_{vo} is the effective vertical stress at the test depth. The three indices form the basis for empirical or semi-empirical correlation to various geotechnical parameters including soil type, K_o, ϕ, OCR, C_u, and M, the constrained modulus. The recommended current correlations are documented by Baldi et al (1986). The focus of the present paper will be on E_D and its correlation to the small strain shear modulus, G_o.

1.2 The Cross-Hole Shear Test

The seismic cross-hole test is a relatively simple technique for determining the shear wave velocity of a soil. The test has been described by Stokoe and Woods (1972), Woods (1978), Stokoe and Hoar (1978), Woods and Stokoe (1985) and Woods (1986). In this technique, two or three holes are drilled in line a known distance apart. One borehole contains a source, typically an impulse rod or hammer, while the remaining hole or holes are equipped with geophones to detect the arrival of the seismic wave. Most commonly, the source and receivers are located at the same elevation. A trigger from the hammer is used to record the impulse time and therefore facilitate computation of the transit time. After appropriate correction for possible raypath curvature or wave refraction by higher velocity layers, the shear modulus is computed from:

$$G_o = V_s^2 \rho, \quad (1)$$

where V_s is the computed propagation velocity and ρ is the mass density of the soil. The dilatometer indices I_D and E_D may be utilized to determine ρ as suggested by Marchetti (1980) and Marchetti and Crapps (1981).

2 CROSS-HOLE SHEAR AND DILATOMETER MODULI

Since the objective of this study is to compare cross-hole shear and dilatometer moduli, it will prove useful to define the ratio of shear modulus to dilatometer modulus by:

$$R_G = \frac{G}{E_D}. \quad (2)$$

Marchetti (1980) has shown that E_D can be related to Young's modulus through the theory of elasticity:

$$E_D = \frac{E}{(1-\nu^2)}, \quad (3)$$

where E is Young's modulus and ν is Poison's ratio. Since the shear modulus for linearly elastic, homogeneous, isotropic medium is related to E and ν by:

$$G = \frac{E}{2(1+\nu)}, \quad (4)$$

it can be shown that

$$R_G = \frac{(1-\nu)}{2}. \quad (5)$$

Jamiolkowski et al (1985) suggested defining R_G somewhat differently as:

$$R_G = \frac{G}{G_D}, \quad (6)$$

where

$$G_D = \frac{E_D}{2(1+\nu)}. \quad (7)$$

Thus, by their definition,

$$R_G = \frac{2G(1-\nu^3-\nu^2+\nu)}{E}. \quad (8)$$

Since $2(1-\nu^3-\nu^2+\nu)$ will only range between 2.25 and 2.37 for the range of Poisson's ratios encountered in soils and because neither E nor ν are actually determined in either the DMT or the VS-CH test, the authors have chosen to avoid the somewhat extraneous G_D and prefer to define R_G by equation (2).

Two factors will invalidate the simple predictions for R_G given by equation (5). First, whereas the shear modulus obtained by seismic cross-hole testing is a small strain modulus, the dilatometer modulus is defined arbitrarily by two points on a non-linear diaphragm expansion curve. As indicated earlier, the P_1 value corresponds to a soil displacement of 1 mm. This displacement is associated with appreciable strain softening and is much greater than the displacements associated with elastic wave propagation. Also, whereas the cross-hole shear test can genuinely be classified as non-destructive, the 15 mm thick DMT blade must disturb the soil during insertion. As such, equation (5) should not be expected to adequately predict R_G and must be modified by a correction for strain softening and soil disturbance.

A second equally important factor stems from soil anisotropy and the fundamental parameters controlling G_o and E_D. Whereas the dilatometer modulus reflects soil stiffness only in the direction of diaphragm expansion, the seismic shear modulus reflects the in-situ stiffness in both the direction of wave propagation and the direction of particle motion. For cross-hole SV tests, these two directions are parallel and perpendicular to the ground surface. It is also known that the shear modulus is approximately proportional to the square root of the effective confining pressure (Hardin and Richart, 1963). As such,

$$G_o \propto \sqrt{\frac{\sigma_v'(1+K_o)}{2}}. \quad (9)$$

Therefore, R_G should also be expected to be a function of K_o. Since G_o depends on both σ_h' and σ_v' while E_D is only controlled by σ_v', high K_o conditions should yield lower R_G values than low K_o conditions. A benchmark value of R_G at K_o=1.0 would eliminate the anisotropy factor and focus on strain softening and soil disturbance.

The value of K_o can be empirically obtained from K_D (Marchetti, 1985; Baldi et al, 1986) by:

$$K_o = 0.376 + 0.095\,K_D - 0.00461\,q_c/\sigma_v', \quad (10)$$

where q_c is the static penetration resistance or according to Schmertamann (1983) by:

$$K_o = \frac{40 + 23K_D - 86K_D(1-\sin\phi) + 152(1-\sin\phi) - 717(1-\sin\phi)^2}{192 - 717(1-\sin\phi)}, \quad (11)$$

where ϕ is the drained axisymmetric (triaxial) friction angle.

3 PREVIOUS MODULUS CORRELATIONS

In comparisons of dilatometer, triaxial and resonant column tests, Baldi et al (1986) found:

$R_G = G_o(RCT)/E_D = 2.72$ +/- 0.59 for NC Ticino Sand,

$R_E = E_{25}/E_D = 0.88$ +/- 0.27 for NC Ticino Sand,

$R_E = E_{25}/E_D = 4.29$ +/- 0.62 for OC Ticino Sand and

$R_E = E_{25}/E_D = 2.49$ +/- 0.74 for OC Hokksund Sand

where $G_o(RCT)$ is the small strain shear modulus obtained from resonant column testing and E_{25} is the drained modulus from triaxial CK_oD tests evaluated at one fourth of the deviator stress at failure. Field results presented by Bellotti et al (1986) and Jamiolkowski et al (1985) revealed similar ratios for Po River Sand:

$R_G = G_o(V_s\text{-CH})/E_D = 2.2$ +/- 0.7

$R_G = G(SBPT)/E_D = 2.3$ +/- 0.5

where $G_o(V_s\text{-CH})$ is the small strain shear modulus determined from cross-hole shear tests and $G_o(SBPT)$ is the unload-reload shear modulus determined from self-boring pressuremeter tests. Bellotti et al (1986) also reported for PO River Sand:

$G_o(V_s\text{-CH})/G(SBPT) = 3.6$ to 4.8 (Camkometer test)

$G_o(V_s\text{-CH})/G(SBPT) = 4.6$ to 5.9 (Pafsor test)

Baldi et al (1986) noted that the dilatometer modulus is not as sensitive to prestraining as moduli obtained from triaxial tests. The higher values of R_E for OC Ticino sand than for the NC Ticino sand confirmed this hypothesis. The apparent dependency of R_E on K_D and OCR led Baldi et al (1986) to speculate that a similar relationship between R_G and K_D may exist as well.

4 TEST PROGRAM AND RESULTS

The DMT and V_s-CH tests were performed at the Beal Street Test Site on the University of Michigan campus. The soil at the site is a glacially overconsolidated silty sand to a depth of at least 8 m. Cross-hole tests were performed using a three-hole arrangement as shown in Figure 1. The source to receiver distances were 1.79 m and 4.54 m. An in-hole Bison hammer was employed to generate the SV waves (Woods

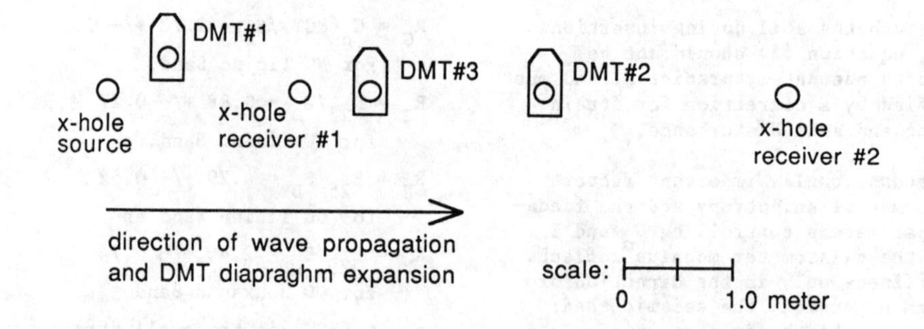

Fig. 1 Plan showing location of DMT soundings and cross-hole shear boreholes

and Stokoe, 1985). Tests were performed at 0.5 m intervals. The propagation velocities and density values (obtained from dilatometer E_D-I_D correlations) were then used to compute G_o by equation 1.

The DMT sounding were performed at three locations as shown on Figure 1. The blade was pushed with a CPT rig and tests were performed at 10 cm intervals. The diaphragm was expanded in the direction of wave propagation. Beyond a depth of approximately 2.5 m the B reading exceeded 40 bars and soundings #1 and #2 were suspended. An 80 bar dilatometer control unit was therefore procured and employed for sounding #3 which permitted testing to a depth of 4.2 m.

Table 1. DMT and Cross-Hole Shear data

Depth (m)	V_S (m/s)	Unit Wt. (T/m^3)	K_o	G (bar)	ED (bar)	R_G
0.50	166	1.90	4.34	521	326	1.61
1.00	177	1.80	1.59	563	160	3.89
1.50	194	1.80	0.89	680	147	4.80
2.00	236	1.80	1.12	1001	185	6.08
2.50	245	2.00	2.42	1197	630	1.93
3.00	280	2.15	3.30	1693	1023	1.65
3.50	299	2.15	3.37	1925	1260	1.53
4.00	317	2.15	3.25	2159	1351	1.60

The variations of E_D and $G_o(V_S$-CH) with depth are shown in Figure 2. Table 1 summarizes the findings at each cross-hole shear test depth and indicates that R_G ranged from 1.53 to 6.08. Upon initial inspection of Figure 2 it would appear that the cross-hole shear tests failed to disclose a relatively low stiffness layer between 1.0 and 2.0 m which the dilatometer test did reveal. Accordingly, higher R_G values were computed at 1.0, 1.5 and 2.0 m. However, quite the contrary, the cross-hole shear results were in excellent agreement with the dilatometer information as shown in Figure 3. Here the small

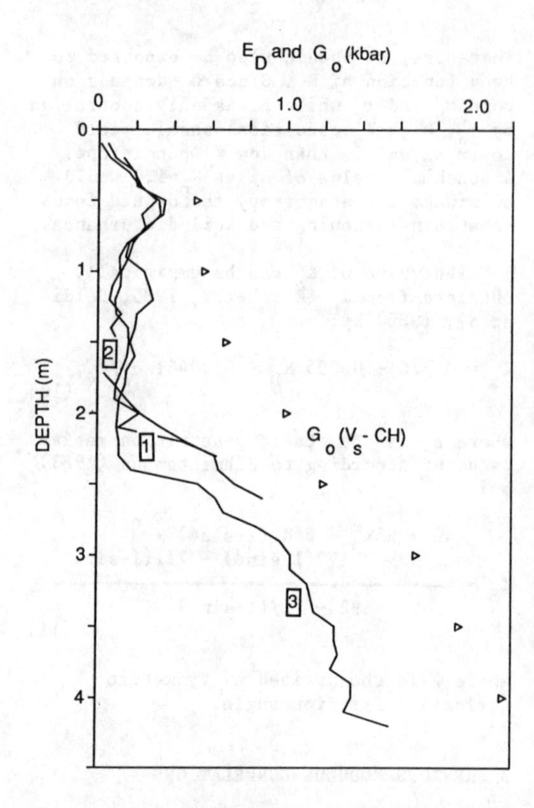

Fig. 2 Dilatometer modulus and cross-hole shear modulus versus depth.

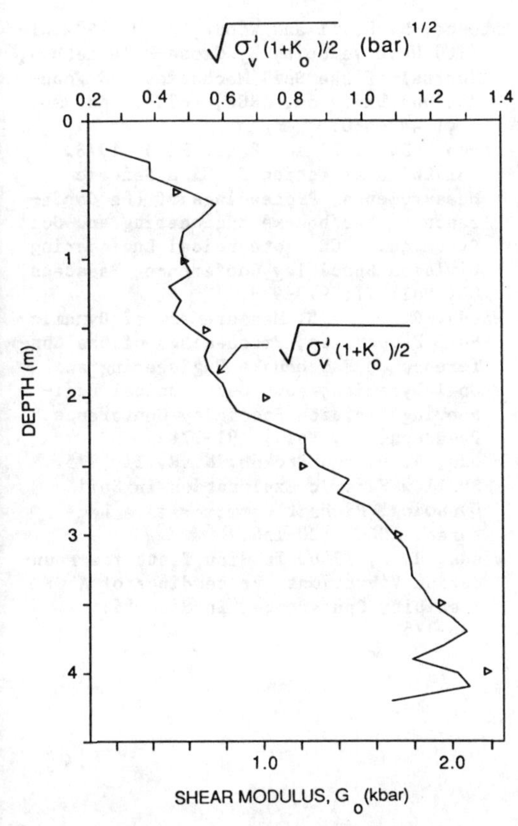

Fig. 3 Confining pressure and cross-hole shear modulus versus depth.

strain shear modulus is once again shown versus depth along with a plot of $\sqrt{\sigma_v'(1+K_o)/2}$. The vertical stresses were obtained by integrating the empirically determined soil densities and K_o was determined by equation (11). Clearly, the information is in excellent agreement. The deviation of G_o from the curve at a depth of 2 m may be due to wavepath curvature through the stiffer soil immediately below.

The excellent agreement between the cross-hole shear modulus and the square root of the confining pressures confirms that R_G is a function of K_o. Figure 4 reveals this inverse relationship. These findings, although based on a small data base, offer hope for developing semi-empirical relationships for small strain shear modulus as a function of the DMT indices.

5 CONCLUSIONS

1. Field DMT and cross-hole shear tests were performed in a dry, glacially overconsolidated silty sand deposit to study the relationship between the small strain shear modulus and the dilatometer modulus. The ratio of the former to the latter was defined as R_G. In the tests performed, R_G ranged from approximately 1.5 to 6.0.

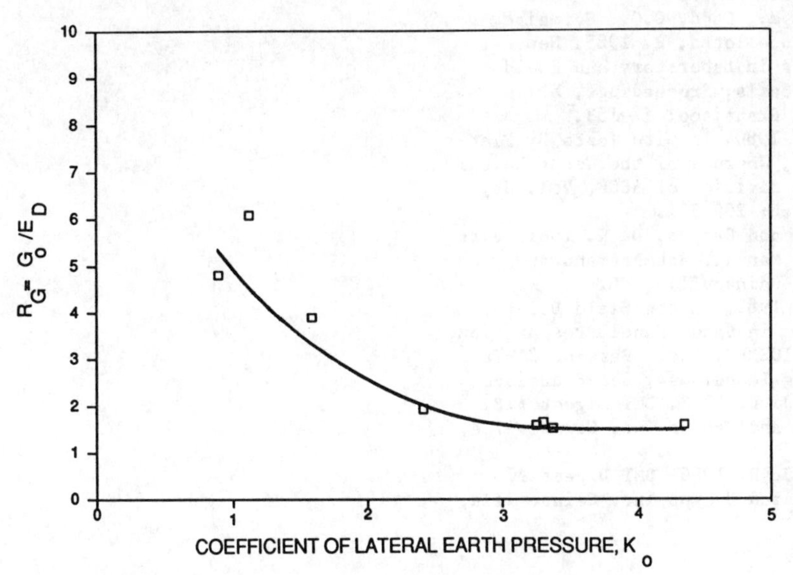

Fig. 4 R_G versus coefficient of lateral earth pressure.

2. Excellent agreement was observed between the cross-hole shear modulus and $\sqrt{\sigma'_v(1+K_o)/2}$ where K_o and σ'_v were obtained through existing empirical relationships based on DMT indices.

3. R_G was shown to be inversely related to K_o. This suggests that the dilatometer indices K_D and E_D may eventually be used to predict the small strain shear modulus, G_o.

REFERENCES

Baldi, G., Bellotti, R., Ghionna, V., Jamiolkowski, M., Marchetti, S. and Pasqualinini, E. 1986. Flat Dilatometer Tests in Calibration Chambers, Proceedings of ASCE Specialty Conference In Situ 86: 431-446.

Bellotti, R., Ghionna, V., Jamiolkowski, M., Lancellotta, R. and Manfredini, G. 1986. Deformation Characteristics of Cohesionless Soils From In Situ Tests, Proceedings of ASCE Specialty Conference, In Situ 86: 47-73.

Campanella, R. G., Robertson, P. K., Gillespie, D. G. and Grieg, J. 1985. Recent Developments in In Situ Testing of Soils, Proceedings, XI ICSMFE, San Fransisco: 849-854.

Hardin, B. O. and Richart, F. E.,Jr. 1963. Elastic Wave Velocities in Granular Soils, Journal of the Soil Mechanics and Foundations Division, ASCE, Vol.89, No.SM 1, Feb.: 33-65.

Jamiolkowski, M., Ladd, C.C., Germaine, J.T. and Lancellotta, R. 1985. New Developments in Laboratory and Field Testing of Soils, Proceedings, XI ICSMFE, San Fransisco: 57-153.

Marchetti, S. 1980. In Situ Tests By Flat Dilatometer, Journal of the Geotechnical Engineering Division of ASCE, Vol.106, No.GT3, March: 299-321.

Marchetti, S. and Crapps, D. K. 1981. Flat Dilatometer Manual. Schmertmann and Crapps Inc. Gainseville, FL.

Marchetti, S. 1985. On the Field Determination of K_o in Sand, Panel Presentation at the XII ICSMFE, Disc. Session 2A-In Situ Testing Techniques, San Fransisco.

Schmertmann, J. H. 1983. DMT Digest #1B, Schmertmann and Crapps Inc. Gainseville, FL.

Schmertmann, J. H. 1986. DMT Digest #7, Schmertmann and Crapps Inc. Gainseville, FL.

Stokoe, K. H. II and Woods, R. D. 1972. In Situ Wave Velocity by Cross-Hole Method, Journal of the Soil Mechanics and Foundations Division, ASCE, Vol.98, No.SM5, May: 443-460.

Stokoe, K. H. II and Hoar, R. J. 1978. Variables Affecting In Situ Seismic Measurements, Proceedings of the Conference on Earthquake Engineering and Soil Dynamics, ASCE Geotechnical Engineering Division Specialty Conference, Pasadena, CA, Vol. II: 919-939.

Woods, R. D. 1978. Measurement of Dynamic Soil Properties, Proceedings of the Conference on Earthquake Engineering and Soil Dynamics, ASCE Geotechnical Engineering Division Specialty Conference, Pasadena, CA, Vol.I: 91-178.

Woods, R. D. and Stokoe, K. H. II 1975. Shallow Seismic Exploration in Soil Dynamics, Richart Commemorative Lectures, ASCE: 120-156.

Woods, R. D. 1986. In Situ Tests for Foundation Vibrations, Proceedings of ASCE Specialty Conference, In Situ 86: 336-375.

Penetration Testing 1988, ISOPT-1, De Ruiter (ed.)
© 1988 Balkema, Rotterdam, ISBN 90 6191 801 4

A push-in pressuremeter / sampler

A.-B.Huang & K.C.Haefele
Clarkson University, Potsdam, N.Y., USA

ABSTRACT: A push-in pressuremeter/sampler has been developed for on land testing and taking representative samples in soft soil deposits. The device is made of a 75.2 mm O.D., 62.5 mm I.D. stainless steel tube with a 7.7° inward angle of taper at the tip. The pressuremeter was designed to be pushed into the soil from the bottom of the borehole prior to its expansion. Preliminary push-in pressuremeter tests were performed in a marine clay deposit. This paper describes the new pressuremeter design and presents available data. Comparisons are also made with pre-bored, self-bored and full displacement pressuremeter test data obtained at the same site.

1 INTRODUCTION

Since the development of Menard pressuremeter in 1950's, many modifications have been introduced to the Geotechnical industry. Currently available pressuremeters can be divided into four groups according to the method of probe insertion as shown in Table 1. Displacement pressuremeters (groups 3 and 4 in Table 1) are more recent to this class of in situ testing devices. They are relatively simple and efficient to use, but at the cost of more soil disturbance (as opposed to self-boring pressuremeter).

As part of a research program at Clarkson University to study the use of displacement pressuremeters in soft clays, an open-ended push-in pressuremeter (PIPM) was developed and preliminary tests were performed in a marine clay deposit at a test site in Massena, New York. The main purpose of these tests was to evaluate the feasibility of the new PIPM equipment and characteristics of the test results. A series of Menard pre-bored, self-boring (SBPM) and full displacement pressuremeter (FDPM) tests were also conducted at the same site for comparison. This paper describes the design of the new

Table 1. Classification of pressuremeters by method of insertion.

Group No.	Insertion method	References
1	Pre-bored	Menard pressuremeter (Baguelin et al., 1978); OYO LLT (Suyama et al., 1982)
2	Self-bored	Camkometer (Wroth and Hughes, 1973); PAFSOR (Baguelin et al., 1978)
3	Full displacement	FDPM (Hughes and Robertson, 1985)
4	Push in	Stressprobe (Reid and Fyffe, 1982)

push-in pressuremeter and presents available test results. Evaluations of this device are made by comparing data with those of other types of pressuremeters mentioned above and laboratory experiments of the same soil deposit.

2 DESIGN OF THE PIPM

The pressuremeter described herein is similar to that developed by Reid and Fyffe (1982) with the exception that it is mainly for on land testing. The design was made with an intention to provide a low cost and efficient pressuremeter so that it can be used for routine testing. Figure 1 shows the engineering drawing of this single cell and water inflated pressuremeter. The probe consists of only eight parts. The main body of the probe was machined from a 76.2 mm OD, 6.35 mm wall stainless steel tube. The surface of the core (see Figure 1) was grooved to prevent membrane slippage during probe insertion. The same adiprene membrane as used in Camkometers is applied in the push-in pressuremeter. When deflated, it has a length to diameter ratio of 5 with a 76.2 mm OD. The probe is essentially like a thick wall shelby tube with an area ratio of 44%. A piece of 76.2 mm shelby tube was attached to the upper end of the pressuremeter to provide additional room for the excess soil recovery. It also allows the pressuremeter be attached to a regular shelby tube adapter which are readily available with most drill rigs. The cutting shoe (Figure 1) has an inward 7.7^{o} angle of taper. This is less than that suggested by Hvorslev (1949) for thick-wall samplers. A catcher such as the one used in a split spoon sampler can be added to hold the sample in extremely soft or loose soils. The soil sample was extruded from the pressuremeter probe using the hydraulic piston on the drill rig.

Two tubings are used to connect the probe to the control console (Figure 1). One of the tubings is connected to the standpipe for water injection and the other to a pressure transducer. Preliminary tests indicated that there was a significant pressure difference between the standpipe and pressuremeter probe during expansion. This was mainly due to the movement of water in the tubing and head losses from the movement. The second tub-

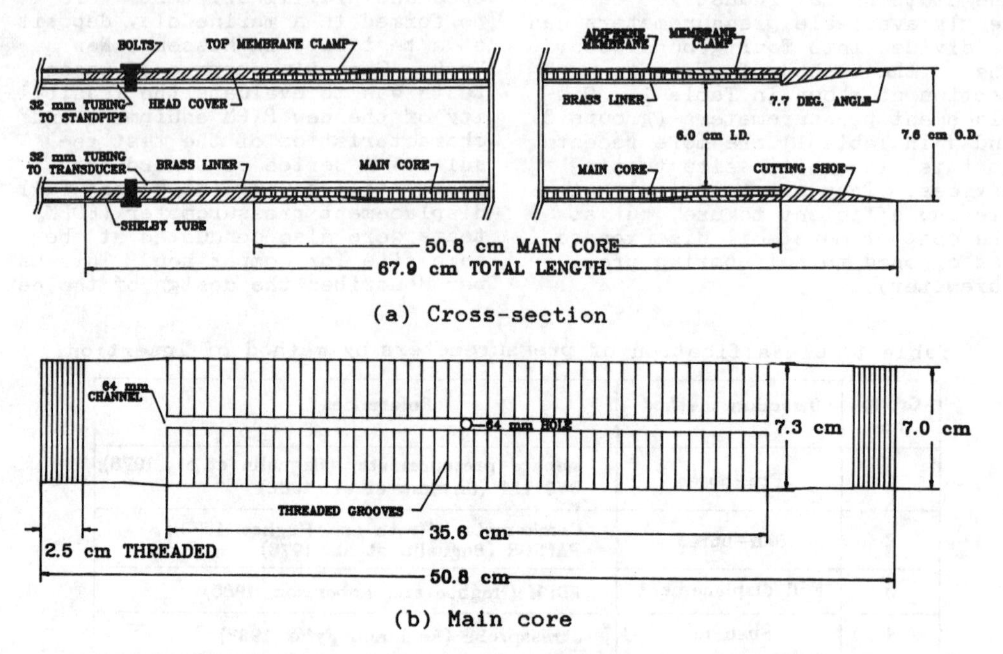

Fig. 1. Engineering Drawing of the push-in pressuremeter.

ing insures that water remains stationary between the probe and pressure transducer and hence, zero head loss.

The probe expansion is currently configured to be stress controlled. A differential pressure transducer is installed at the standpipe to sense the water level (and therefore, the pressuremeter radial strain) variations. All instruments are monitored by a personal computer based data logging system.

3 GEOTECHNICAL CHARACTERISTICS OF THE TEST SITE

The test site is located east of the town of Massena, New York across Route 37 from the Massena High School. Previous studies performed at the test site (Lutenegger and Timian, 1986) indicated that there was approximately 12 m of a marine clay deposit underlain by glacial till. The marine clay was soft and lightly overconsolidated (OCR = 1.5 to 3) below a crust of approximately 1.5 m. Figure 2 shows a profile of this marine clay deposit at the test site. The moisture contents and undrained shear strengths, su, were fairly uniform in the marine clay deposit at depths of between 4 m and 6.5 m. Most of the pressuremeter tests presented herein were performed within this layer.

4 PUSH-IN PRESSUREMETER TESTS

4.1 Test Procedures

The borehole was advanced using a 101.6 mm diameter solid stem auger. In all the tests, the PIPM probe was pushed at a rate of 250 mm/minute. The pushing stopped as the pressuremeter tip reached approximately 700 mm from the bottom of the borehole. Vacuum was applied to hold the membrane during pushing. The probe expansion was stress controlled with each stress increment conducted for one minute. The probe pressure and radial strain readings were recorded in two second intervals using the data logging system. Upon completion of the pressuremeter test, the probe was deflated and vacuum applied during removal of the device.

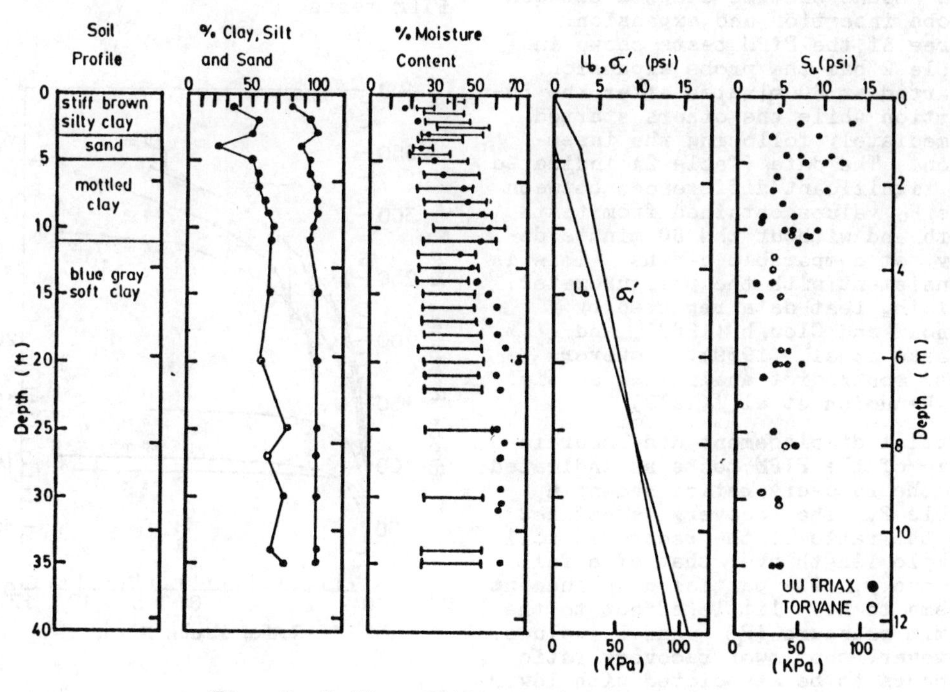

Fig. 2. Soil profile at the test site.

4.2 Test Results and Discussion

Figure 3 shows the probe pressure versus radial strains of three PIPM tests. SBPM tests performed at comparable depths are shown in Figure 4. Interpretations of all the pressuremeter data were performed using a simplex curve fitting technique (Huang et al., 1986). The derived soil stress-strain relationships are also included in Figures 3 and 4. Key parameters obtained from these and other PIPM tests are shown in Table 2. The tests were all performed in the bluish gray, soft marine clay (Figure 2) according to the samples recovered.

It is likely that the insertion of PIPM prestresses the soil and results in a pore pressure increase. A situation similar to a pressuremeter holding test (Clarke et al., 1979) is then created by allowing pore pressure dissipation and not expanding the probe. If the total stress at the probe boundary varies during this pore pressure dissipation, the lift-off pressure (P_o) may change by varying the amount of time elapsed between probe insertion and expansion. Three of the PIPM tests shown in Table 2 had the probe expansion started at 60 minutes after the insertion while the others started immediately following the insertion. The data (Table 2) indicated no significant differences between the P_o values obtained from tests with and without the 60 minute delay, at comparable depths. This is consistent with the pressuremeter holding test data reported by Benoit and Clough (1986); and Clarke et al. (1979). However, they contradict analytical studies by Randolph et al. (1979).

Partial displacement did occur in some of the PIPM tests as indicated by the recovery ratios shown in Table 2. The recovery is defined as the ratio of the recovered soil sample length over that of a full recovery. The partial displacement seems to have little effect to the limit pressure (P_l) and P_o values. However, the lower recovery ratio appears to be associated with lower shear moduli (Gr) obtained in unload-reload loops.

Key parameters derived from Menard pressuremeter, FDPM and SBPM tests at the same site are shown in Table 3. These data indicate that the

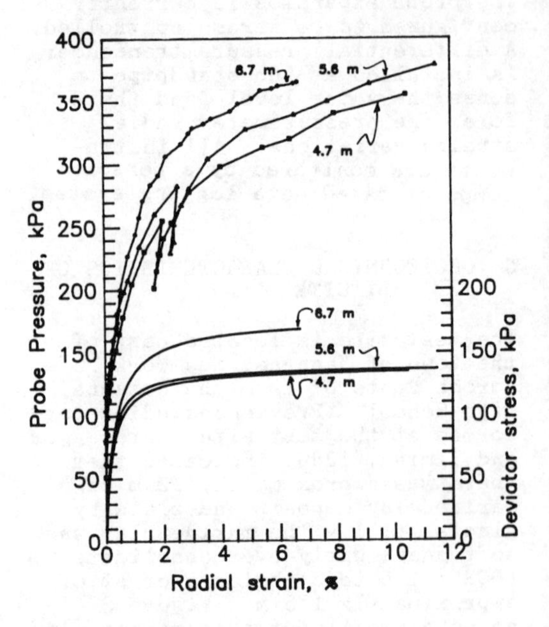

Fig. 3. Results and interpretation of PIPM tests.

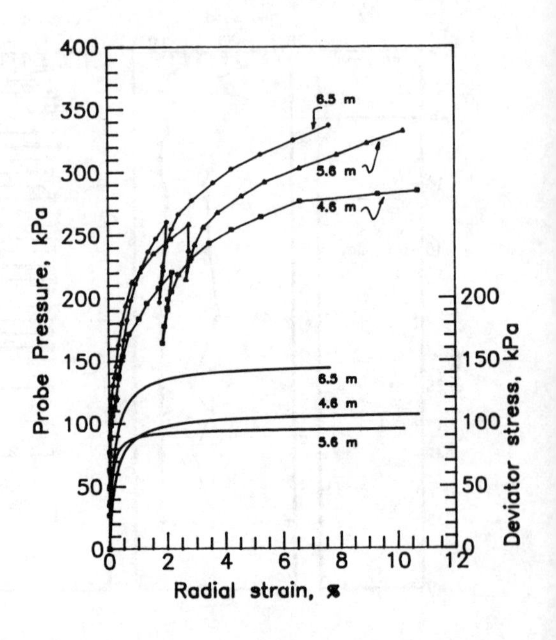

Fig. 4. Results and interpretation of SBPM tests.

insertion method may affect the value of P_0 by as much as one order of magnitude among the four pressuremeters. The P_0 from PIPM represents approximately an average of those from Menard and FDPM tests and are about 50% higher than those from SBPM. The initial shear moduli (Gi) in Tables 2 and 3 are taken as the derivative of soil stress-strain function when radial strain approaches zero. The Gi and Gr values from Menard and FDPM tests are less than 50% of those from PIPM tests. This is obviously due to the different levels of disturbance involved in insertion methods. The shear moduli from PIPM are fairly comparable to those from SBPM, although more studies are needed to substantiate this conclusion.

Based on the undrained shear strengths shown in Figure 2, the (P_1-P_0)/su consistently ranged from 4 to 6 for PIPM and SBPM tests which are comparable to those reported by others (Baguelin et al., 1978). The (P_1-P_0)/su ratios are significantly higher for Menard pressuremeter tests and rather erratic for the available FDPM tests.

5 CONCLUSIONS

A simple push-in monocell pressuremeter/sampler has been described. Because of its simplicity, a PIPM can be made at a fraction of the cost of an SBPM such as the Camkometer. A series of field experiments using the new push-in pressuremeter and reference tests using Menard pre-bored, self-boring and full displacement pressuremeters have been performed. Results show that successful tests can be performed with the new PIPM and the data seem to be more consistent than those of Menard pressuremeter and FDPM. The reduced soil disturbance in PIPM tests resulted in significant differences in P_0 and

Table 2. PIPM test data.

Depth (m)	Recovery (%)	M.C. (%)	P_0 (kPa)	P_1 (kPa)	G_i (MPa)	G_r (MPa)	Su (kPa)
*4.3	45.8	52.5	149	375	27.3	10.5	43.2
4.7	87.5	58.3	126	375	24.0	14.4	54.2
*4.7	100	63.9	112	395	25.8	16.4	69.3
*5.2	100	60.4	130	445	32.3	16.8	67.8
5.6	100	60.3	124	405	35.3	17.8	68.5
6.7	100	61.6	133	450	24.3	–	88.2

* - 60 minute delay before probe expansion

Table 3. Other pressuremeter test data.

Test	Depth (m)	P_0 (kPa)	P_1 (kPa)	G_i (MPa)	G_r (MPa)	Su (kPa)
Menard	4.6	42	290	6.5	–	85.2
Menard	6.1	66	460	6.1	–	97.1
FDPM	4.6	208	260	–	6.3	38.3
FDPM	7.7	378	475	–	6.1	103.3
SBPM	4.6	82	310	28.4	9.6	51.7
SBPM	5.6	87	360	55.1	11.8	47.9
SBPM	6.5	75	365	35.1	13.3	73.8

shear moduli as compared with Menard and FDPM tests. The P_o values from PIPM are approximately 50% higher than those from SBPM. However, the moduli from PIPM are comparable to those of SBPM.

It can be argued that the PIPM is not suitable for continuous testing. It does, however, provide representative soil samples which may be advantageous in certain cases.

6 ACKNOWLEDGEMENTS

The soil profile and FDPM data are based on tests performed by Prof. A.J. Lutenegger of Clarkson University. The Menard pressuremeter tests were conducted by Mr. S.W. Perkins, a former graduate student of Clarkson University. The project was funded by a Clarkson University research initiation grant.

REFERENCES

Baguelin, F., Jezequel, J.F., and Shield, D.H., (1978), Pressuremeter and Foundation Engineering, Trans Tech Publications, 617p.

Benoit, J., and Clough, W., (1986), "Self-Boring Pressuremeter Tests in Soft Clays," Journal of Geotechnical Engineering Division, ASCE, Vol.112, pp. 60-78.

Clarke, B.G., Carter, J.P., and Wroth, C.P., (1979), "In Situ Determination of the Consolidation Characteristics of Saturated Clays," Proceedings, 7th European Conference on Soil Mechanics, Vol.2, pp.207-213.

Huang, A.B., Chameau, J.L., and Holtz, R.D., (1986), "Interpretation of Pressuremeter Data in Cohesive Soils by Simplex Algorithm," Geotechnique, Vol. 36, pp.599-604.

Hughes, J.M., and Robertson, P.K., (1985), "Full-Displacement Pressuremeter Testing in Sands," Canadian Geotechnical Journal, Vol. 22, pp.298-307.

Hvorslev, M.J., (1949), Subsurface Exploration and Sampling of Soils for Civil Engineering Purposes, Waterways Experimentation Station, Vicksburg, Mississippi, 521p.

Lutenegger, A.J., and Timian, D.A., (1986), "In Situ Tests with Ko Stepped Blade," Proceedings, In Situ ´86, ASCE, Blacksburg, Virginia, pp.730-751.

Randolph, M.F., Carter, J.P., and Wroth, C.P., (1979), "Driven Piles in Clay-The Effects of Installation and Subsequent Consolidation," Geotechnique, Vol.29., pp.361-393.

Reid., W.M., and Fyffe, S., (1982), "The Push-in Pressuremeter," Proceedings, Symposium on the Pressuremeter and Its Marine Applications, Paris, pp.247-262.

Suyama, K., Imai, T., and Ohya, S. (1982), "Development of LLT and Its Application in Prediction of Pile Behavior Under Horizontal Load," Proceedings, Symposium on the Pressuremeter and Its Marine Applications, Paris, pp.61-67.

Wroth, C.P., and Hughes, J.M., (1973), "An Instrument for the In Situ Measurement of the Properties of Soft Clays," Proceedings, 8th ICSMFE, Moscow, Vol.1.2, pp.487-494.

Penetration Testing 1988, ISOPT-1, De Ruiter (ed.)
© 1988 Balkema, Rotterdam, ISBN 90 6191 801 4

Calibration of dilatometer correlations

Suzanne Lacasse & Tom Lunne
Norwegian Geotechnical Institute, Oslo

ABSTRACT: The results of the dilatometer test combined with empirical correlations are used to estimate unit weight, coefficient of lateral earth pressure, overconsolidation ratio, constrained modulus, undrained shear strength for clays, and effective friction angle for sands. Recent work also suggests obtaining relative density, Young's modulus, liquefaction potential for sands, limit skin friction for piles in clay and constrained modulus for settlement analysis in sand.

The paper evaluates the dilatometer correlations for clays and sands in light of the results obtained in a four-year research program at NGI. The dilatometer is one of the only methods to provide estimates of unit weight, coefficient of lateral earth pressure and overconsolidation ratio simultaneously and very cost-effectively. Important uncertainties still exist, but the evaluation of well-documented sites and the use of consistent reference parameters should reduce the scatter observed in the data underlying the interpretation charts. The correlations are fairly well documented for clay deposits, but need considerable validation for silt and sand deposits.

Keywords: In situ testing, dilatometer, coefficient of lateral earth pressure, overconsolidation ratio, constrained modulus, undrained shear strength, soil classification.

INTRODUCTION

Use of the Marchetti (1980) dilatometer test has spread widely since the apparition of the device in the late seventies. A modified version of Marchetti's equipment is now in use offshore (Lunne et al., 1987). Reasons for the popularity of the test are simplicity of operation, cost-effectiveness and its potential for derivation of several soil parameters.

The results of the dilatometer test are used to estimate unit weight, coefficient of lateral earth pressure, overconsolidation ratio, constrained modulus, undrained shear strength for clays, and effective friction angle for sands. Recent work investigated the possibility of also relating the measured parameters to relative density, Young's modulus (Jamiolkowski et al., 1985; Bellotti et al., 1986), liquefaction potential for sands (Robertson and Campanella, 1986), limit skin friction for piles in clay (Marchetti et al., 1986) and constrained modulus for settlement analysis in sand (Schmertmann, 1986; Lacasse and Lunne, 1986). These relationships are not evaluated in this paper.

Since its introduction, the Marchetti dilatometer has expanded its data base. The interpretation of the test is in general made on an empirical basis and the correlations used by the profession are essentially still those developed by Marchetti (1980) with data from predominantly Italian clayey deposits. Schmertmann (1982, 1983) provided an alternative for interpreting dilatometer test results in sands, which is an improvement to the correlations for sands proposed in 1980. Marchetti (1985) extended the Schmertmann method and provided interpretation charts which are easier to use than Schmertmann's equations.

The present paper evaluates the existing correlations for clays and sands in light of the results of a research program carried out between 1982 and 1986 at the Norwegian Geotechnical Institute (NGI).

RESEARCH PROGRAM

The testing program used the conventional Marchetti dilatometer blade, 14 mm thick, 95 mm wide and 220 mm long, with a 60 mm stainless steel membrane on its face. Tests were carried out at several sites where extensive in situ and laboratory testing programs were carried out in parallel, to establish the reference soil parameters needed for the evaluation of the dilatometer test results.

To evaluate the performance of the dilatometer tests for a wide range of materials, the results of dilatometer tests run by NGI at several other sites were also evaluated. Table 1 lists the sites investigated and describes typical soil properties.

RESULTS OF RESEARCH

Recommended testing and logging procedures

The experience obtained from testing at the different sites suggests that the following testing procedures give the best results:

1. One should exercise in air a new dilatometer membrane sufficiently to ensure constant membrane resistance throughout a profiling. As many as 100 expansion cycles can be required.

2. Calibrations of membrane stiffness should be done in air before and after test.

3. The dilatometer penetration rate should be approximately constant, and about 2 cm/s.

Table 1. Description of soils tested.

Test site / Testing org.[1]	No. of profiles	Soil type	w_n %	I_p %	S_t	OCR[2]
Haga/NGI	5	Lean clay and plastic clay	34-40 55-60	12-16 35-40	4-6	2-30
Onsøy/NGI		Plastic sensitive aged clay	60-65	30-55	6-11	≃1.2
Drammen/NGI	5	Lean clay and plastic aged clay	≃30 40-55	≃10 27	4-7 6-8	1.2-1.5
Holmen/NGI	12	Medium to coarse loose to medium dense sand	≃20	-	-	(1.0)
Verdal/NGI-OK	12	Fine to coarse sand	≃30	-	-	-
Eberg/NTH[3]	2	Medium plastic sensitive clay	55±5	10-22	8-14	≃2.0
Risvollan/NTH[3]	2	Lean sensitive clay	42±4	8-14	6-13	2-3.0
Barnehagen/NTH[3]	3	Med. plastic clay, silty	40-25	8-12	3-20	2-7
Glava-Stjørdal/NTH	1	Lean layered clay, silty	32±5	5-12	5-9	2-10
Brent Cross/ BRE-NGI[4]	3	Weathered fissured stiff to hard clay	≃30	43-55		OC very high
Cowden/BRE-NGI[4]	5	Weathered and unweathered clay till	≃15	20±4	-	≃2 (?)
Madingley/BRE-NGI[4]	2	-	30-33	45-55	-	High
Rio/COPPE-NGI[5]	4	Highly plastic clay, 5% organics	170-130	30-70	2-4	1.5-4

[1] OK = Ottar Kummeneje, Trondheim
NTH = Norwegian Institute of Technology, University of Trondheim
BRE = Building Research Establishment, Watford, England
COPPE = Federal University of Rio de Janeiro, Brazil

[2] From good quality oedometer tests

w_n = natural water content
I_p = plasticity index
S_t = sensitivity

[3] Mokkelbost (1984)
[4] Powell and Uglow (1986)
[5] Soarès et al. (1986)

4. The delay between stop of penetration and start of membrane expansion should be kept to a minimum for all soils.

5. The expansion rate both for reaching contact pressure and for deforming the soil by 1 mm, should be about 1 mm/15 s. (In the case of special testing for estimating limit skin friction for piles, the excess pore pressure is allowed to dissipate between the measurements of contact pressure and 1-mm expansion pressure.)

6. Pressure gauges with different sensitivities should be selected in function of the expected maximum pressure.

7. Regular checks of the linearity of the blade should be made.

8. The verticality of the dilatometer during penetration should be checked.

Among the 30 profilings done at 4 sites, only one test in clay was unusable, and two tests in sand yielded suspect results at certain depths. In dense sand deposits or gravel, puncturing problems were encountered if one used "standard" 0.2 mm thick membranes. This problem was solved with the use of stronger membranes in coarse and dense sand deposits (Lacasse and Lunne, 1986).

Interpretation of test results

Marchetti (1980) and Marchetti and Crapps (1981) proposed correlations to reduce and interpret the dilatometer test data. A review of Marchetti's data base indicates that:

1. Nine Italian clays and two sands, one Italian, one from Saudi Arabia, form the basis for the correlations. The clays have plasticity index between 25 and 50%, and sensitivity between 1 and 4. Four clays are silty, one is sandy. The two sands are medium loose, normally consolidated sands.

2. The reference K_o value used to obtain the dilatometer correlations for coefficient of lateral earth pressure, is essentially based on the Brooker and Ireland (1965) relationship among K_o, overconsolidation ratio and plasticity index, and not on measured values of in situ lateral earth pressure.

3. Overconsolidation ratio in the majority of cases was not well defined.

4. One sand and three clays were used to establish the correlation for constrained modulus.

5. The reference undrained shear strengh for clays was either the field vane, the unconfined compression or the unconsolidated undrained triaxial compression strength.

In the following, the dilatometer results at the NGI research sites are combined with data from other sites to illustrate the performance of the device in different soil types. The results are put into plots correlating measured variables with reference soil parameters. Where possible, algebraic relations are established and the uncertainty (scatter of data) about a best estimate is quantified. The dilatometer test results are interpreted via the material index, I_D, horizontal stress index, K_D, and dilatometer modulus, E_D, defined by Marchetti (1980):

Material index $I_D = (P_1-P_0/P_0-u_o)$

Horiz. stress index $K_D = (P_0-u_o)/\sigma'_{vo}$

Dilatometer modulus $E_D = 34.6(P_1-P_0)$

where P_0 = contact pressure

P_1 = 1-mm expansion pressure

u_o = in situ pore pressure prior to dilatometer insertion

σ'_{vo} = effective overburden stress

The material index is used to determine soil type. Variations within one profiling and from one test to another are indicative of soil layering and spatial variability.

The horizontal stress index is related to the in situ coefficient of lateral earth pressure. The index is also used to obtain the overconsolidation ratio and the effective friction angle.

The dilatometer modulus characterizes the stress-displacement curve during the 1-mm expansion, and is related to the stiffness of the soil.

In clays, the one mm expansion most certainly takes place under undrained conditions. In clean sands, the expansion process is probably slow enough to allow for full or nearly full drainage. In silts, the test is believed to occur under undrained conditions, although little experience exists for dilatometer tests in silty soils.

Test repeatability

The dilatometer test, shows remarkable repeatability, in both sands, clays and silts, as illustrated by Fig. 1. The material index and horizontal stress index tend to be identical test after test.

Soil variability

Originally, soil type was defined as a function of material index only, with

$I_D < 0.6$: clay

$0.6 \leq I_D \leq 1.8$: silt

$I_D > 1.8$: sand

Based on recent experience, it is recommended to combine the knowledge from P_0

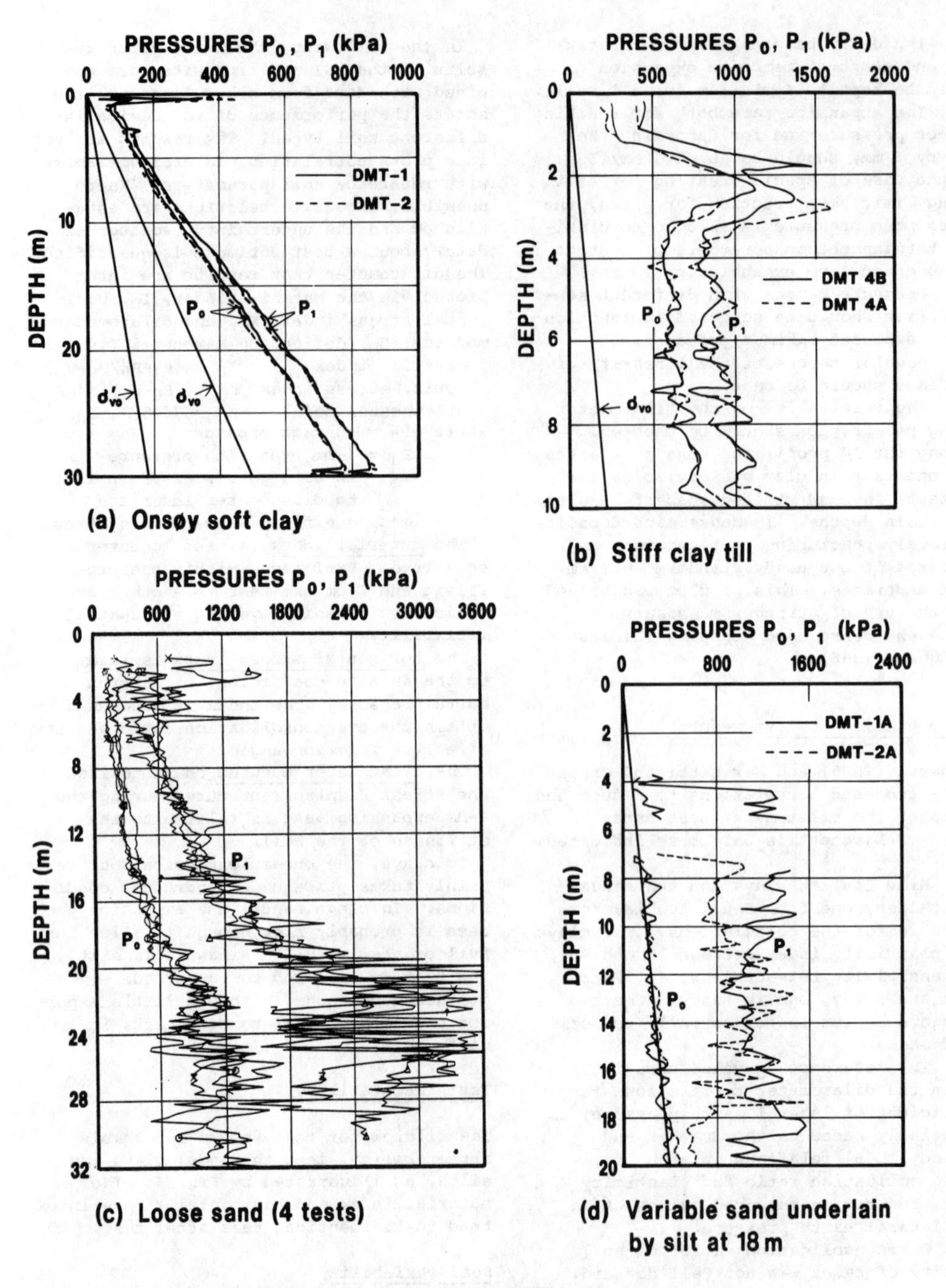

Fig. 1 Examples of dilatometer test repeatability.

and P_1 (Fig. 1), material index and dilatometer modulus to obtain soil classification on the basis of the in situ dilatometer profiling.

Figure 2 summarizes the positions of the soils tested by NGI on the dilatometer soil classification chart proposed by Marchetti and Crapps (1981). The newer information enables one to illustrate qualitatively the effects of overburden, overconsolidation ratio and density on the dilatometer modulus. For Norwegian soils, material indices between 0.05 and 0.1 have been obtained. The original chart was therefore extended in this direction.

Comparisons of the unit weight predicted by the Marchetti and Crapps (1981) dilatometer soil classification chart (Fig. 2) and reference unit weights measured in the laboratory, are made in Fig. 3. The existing chart tends to underpredict the unit weight in soft clays.

Coefficient of lateral earth pressure

Figure 4 establishes the relationship between the horizontal stress index from the dilatometer test and reference values of coefficient of earth pressure at rest, K_o, at the NGI test sites. In situ tests such as total stress cells, hydraulic fracturing, self-boring pressuremeter and field vane tests (Aas et al., 1986) were used to assess the lateral stress. In a few cases, the results of instrumented K_o-oedometer and K_o-triaxial tests run in the laboratory were used. For Holmen sand, only a few self-boring-pressuremeter tests and the well-known Jaky relationship, $K_o = 1-\sin\phi'$, were used.

Not all reference values are equally reliable, and the true in situ value of lateral stress is unknown. For the sites investigated, it is believed that:

1. K-values from K_o-oedometer tests and in situ stress cells are reliable.
2. K-values from hydraulic fracturing in Onsøy clay are very reliable.
3. K-values from field vane tests are also very reliable (Aas et al., 1986).
4. K-values from self-boring pressuremeter are very realistic in soft clays, i.e. OCR $\leq$ 2.0, (Lacasse and Lunne, 1982; Jamiolkowski et al., 1985), but probably too high in highly overconsolidated clays (Ghionna et al., 1983).
5. Test data in the desiccated crust can often be questionable.

Based on the data shown, the Marchetti relationship shown in Fig. 4 tends to overestimate K_o, at least for K_D between 1.5 and 4. It is recommended to estimate K_o in clays from dilatometer tests in the following manner:

1. Clays: use the revised relationship in Fig. 4, $K_o = 0.34\ K_D^m$ for $K_D<4$ with m between 0.44 and 0.64. A value of m = 0.44 may be associated with highly plastic clays and a value of m = 0.64 corresponds to low plasticity clays. In clays with

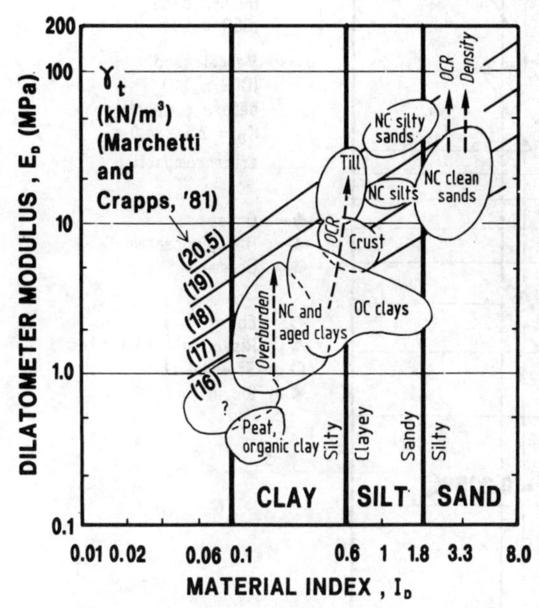

Fig. 2 Classification chart for soils tested. Effects of overburden, overconsolidation ratio and density.

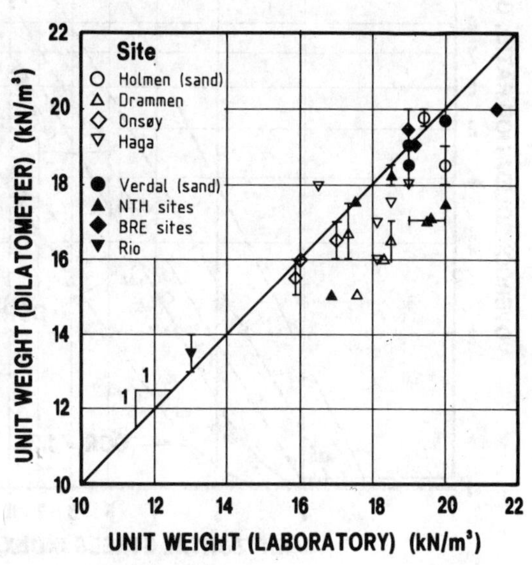

Fig. 3 Prediction of in situ unit weight from dilatometer parameters.

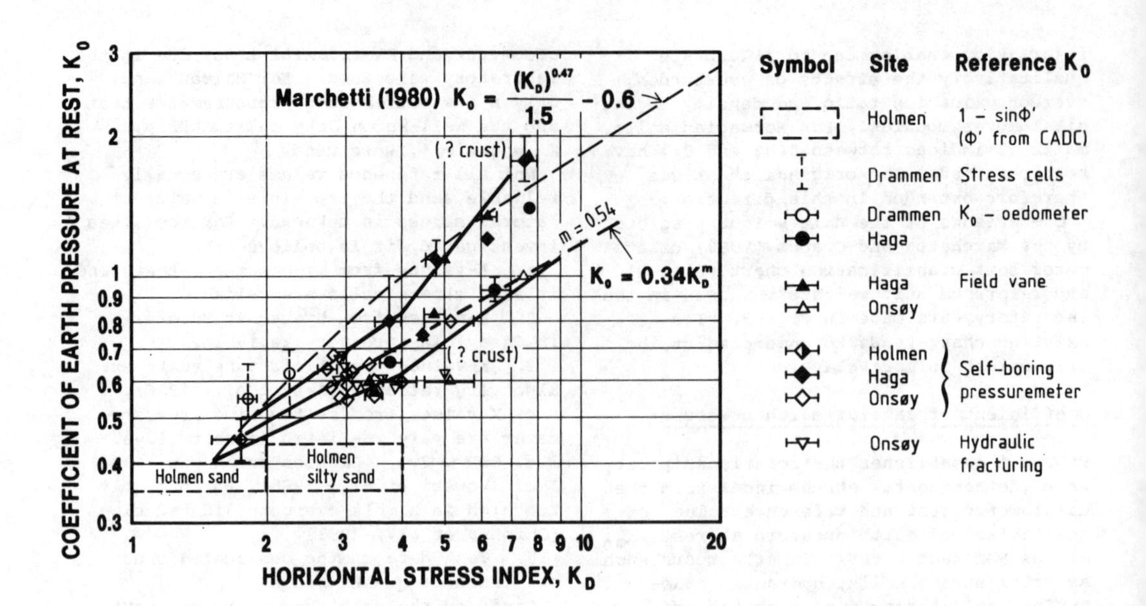

Fig. 4 Relationship between horizontal stress index, K_D, and coefficient of earth pressure at rest at NGI sites.

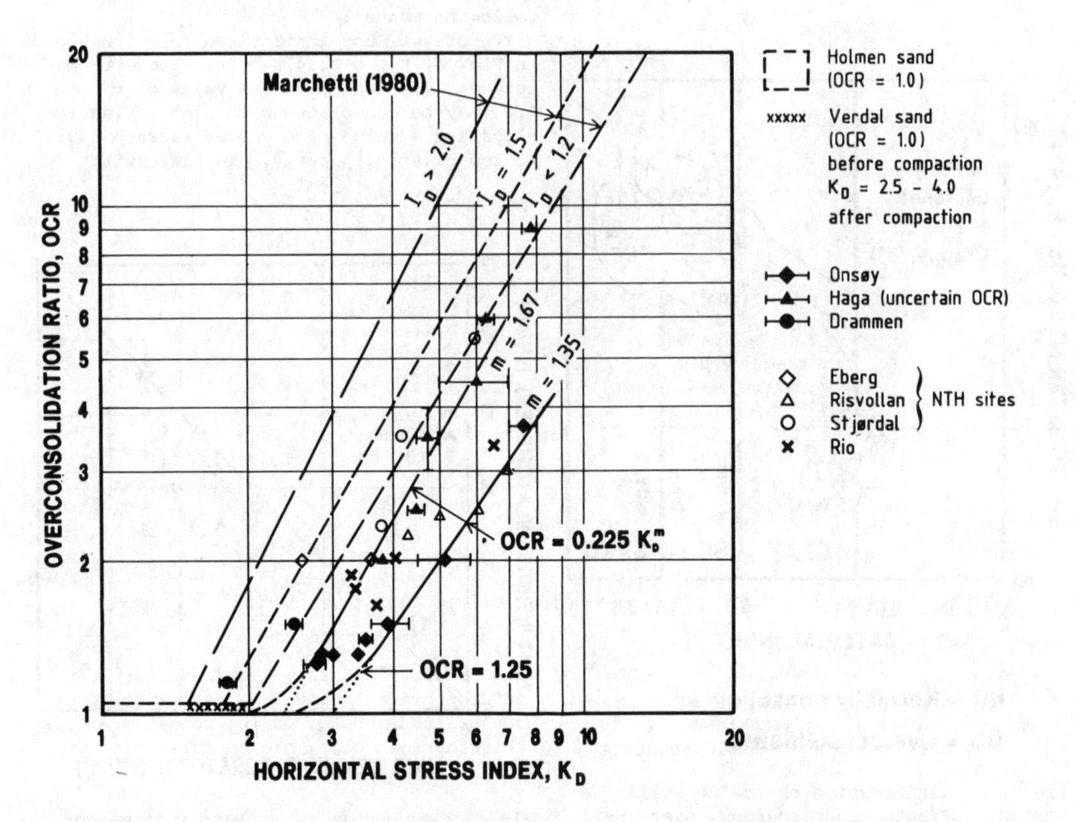

Fig. 5 Relationship between horizontal stress index, K_D, and overconsolidation ratio.

K_D>4, more evaluated experience is needed, but one may want to use both Marchetti's (1980) and the revised correlations shown in Fig. 4 to estimate a range of K-values.

2. Sands: use the Marchetti (1985) chart with the Po River sand modifications to estimate lateral stress. The Marchetti chart with cone penetration test results is preferred to the more complex algebraic functions proposed by Schmertmann (1982, 1983). The data base on sand is very limited and the reference K-values are generally uncertain.

3. Silts: use the relationship for either clay or sand, depending on whether one expects the material to behave like a clay or a sand.

The need for empirical relationships to derive the coefficient of lateral stress is a serious drawback, but this drawback is shared by the majority of the in situ testing devices.

Overconsolidation ratio

In the same manner as for coefficient of lateral earth pressure, Fig. 5 establishes a correlation between horizontal stress index, K_D, and overconsolidation ratio, OCR. Some uncertainty underlies the reference OCR values, since they are based on oedometer tests which may be affected by sample disturbance. Among the different test sites, the Onsøy data are believed to provide the most reliable reference overconsolidation ratios.

For soils with material index less than 1.2, it is recommended to use the modified correlation in Fig. 5, $OCR = 0.225\ K_D^m$, with m between 1.35 and 1.67. The limited data available suggest that plastic clays tend towards the lower m-values, whereas low plasticity clays tend towards the higher m-values. The functions given apply to OCR greater than 1.25. The functions on log-log scales are not linear below this OCR, mainly to account for high quality data presented by Marchetti (1980) for Porto Tolle clay ($OCR \approx 1.0$, $K_D = 2.0$).

For sands ($I_D > 1.8$), the Schmertmann approach is believed more reliable than the Marchetti correlation, but the NGI data validate the theory only for loose sands (OCR = 1).

Soil stiffness

Marchetti (1980) correlated the dilatometer modulus to the constrained modulus obtained from oedometer tests, whether the expansion occurred under drained or undrained conditions. Even though the dilatometer test can be undrained, empirical correlations to the constrained modulus M from a drained oedometer test have been preferred. This is because the laboratory M-values are easy to obtain, generally more consistent than other moduli, and easier to judge whether reliable or not if one uses reference data from the literature. The actual dilatometer loading, however, has very little in common with the manner in which the reference modulus is obtained in the laboratory.

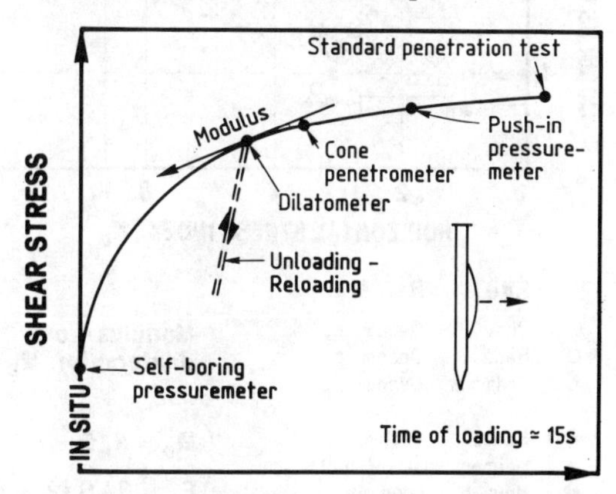

Fig. 6 Qualitative estimate of insertion of in situ testing devices.

Figure 6 presents graphically the probable in situ stress-strain curve of the soil tested. The strain levels at which moduli are estimated by other in situ devices are also indicated. In the vicinity of the probe, the flat dilatometer blade is expected to produce smaller strains than cylindrical probes. At some distance of it, the dilatometer blade is expected to cause larger strains than for example the piezocone or the push-in pressuremeter.

The dilatometer membrane expansion takes place after large strains have already occurred, which probably obliterates the effects of stress and strain history, and makes it impossible to actually "measure" an undisturbed modulus. Such discrepancy requires that the test be interpreted empirically. However, as evidenced by recent published results (Lacasse and Lunne, 1982; Lacasse, 1985), the dilatometer provides reasonable estimates of the constrained modulus during reloading, and this for soft clay, overconsolidated clay and loose sands.

Figure 7 summarizes the results obtained by NGI and presents Marchetti's correlation. It is recommended to continue using

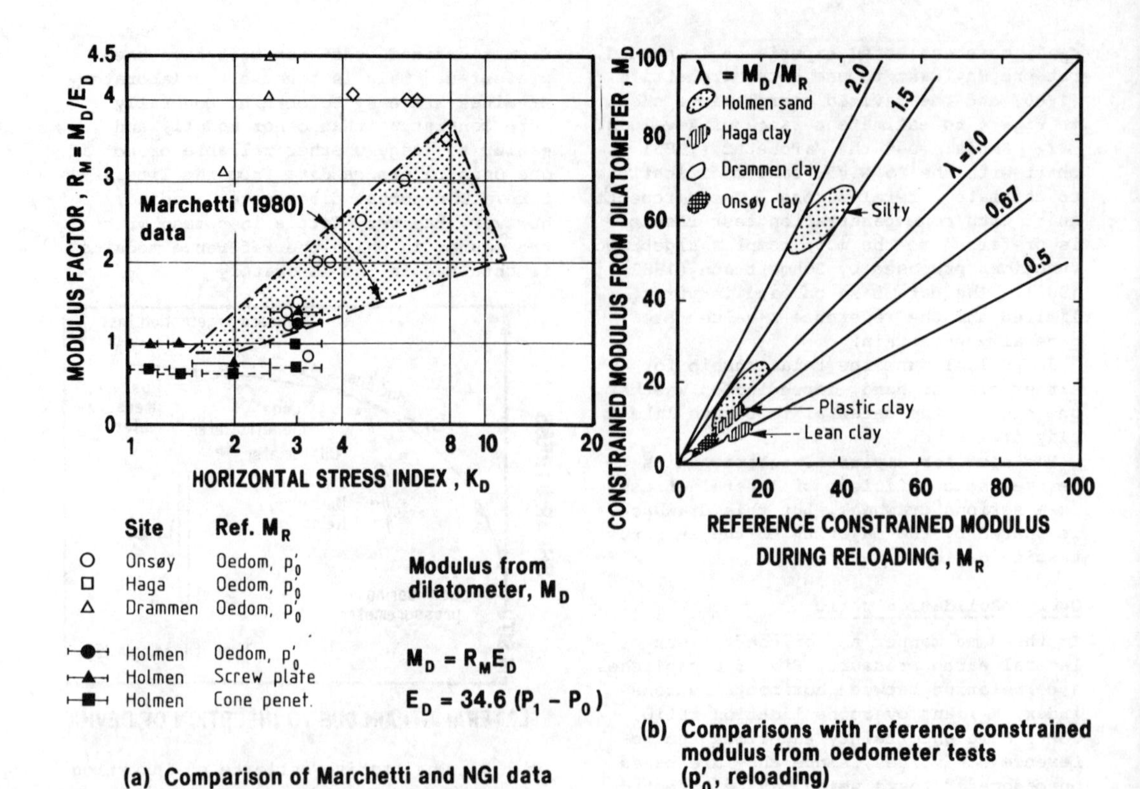

Fig. 7 Prediction of constrained modulus from horizontal stress index, K_D.

the Marchetti (1980) correlation for determination of constrained modulus from the dilatometer test. Because of the simplicity, repeatability and cost-effectiveness of the dilatometer test, it is recommended that further work be directed towards the evaluation of Young's and shear modulus. However one should also consider that the reference laboratory moduli can themselves be considerably affected by sample disturbance.

Undrained shear strength

Figure 8 correlates horizontal stress index to the undrained shear strength of the clays tested. Field vane, anisotropically or isotropically consolidated triaxial compression and direct simple shear strength values were used as reference strength. Recommended correlations between K_D and undrained shear strength are given on the figures. In practice the user should select the diagram corresponding to the strength needed for design (i.e. triaxial compression, extension and simple shear values).

Effective friction angle

The NGI data on sands indicate that the effective friction angle plot proposed by Marchetti (1985) results in friction angles which compare well to the results of drained triaxial compression tests. More data, especially in denser sand, are needed to further evaluate the interpretation method. Meanwhile it is recommended to use the Marchetti (1985) chart to obtain the friction angle of sands. To interpret friction angle in sand, the cone penetration resistance beside a dilatometer test rather than the thrust at the top of the rods should be used.

CONCLUSIONS

There are few sources of error in running the dilatometer test, other than calibration or pressure gauge error. In soft clays, the measured contact and 1-mm expansion pressures and the difference P_1-P_O can be very small, and small errors in the in situ vertical stress and equilibrium

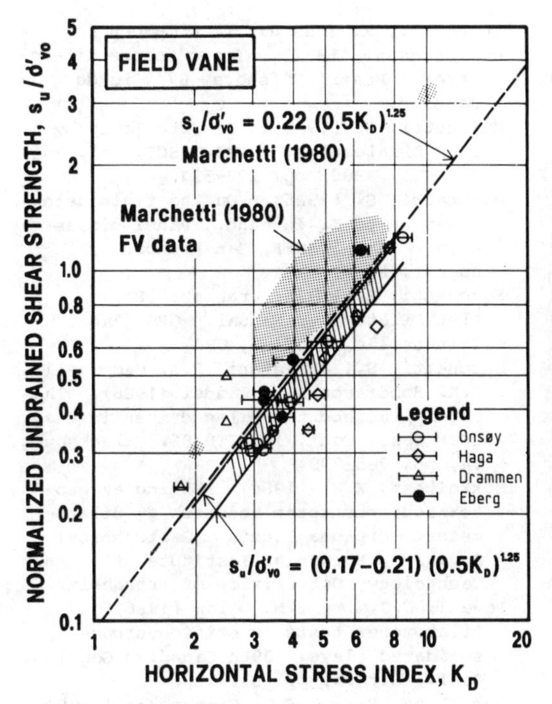

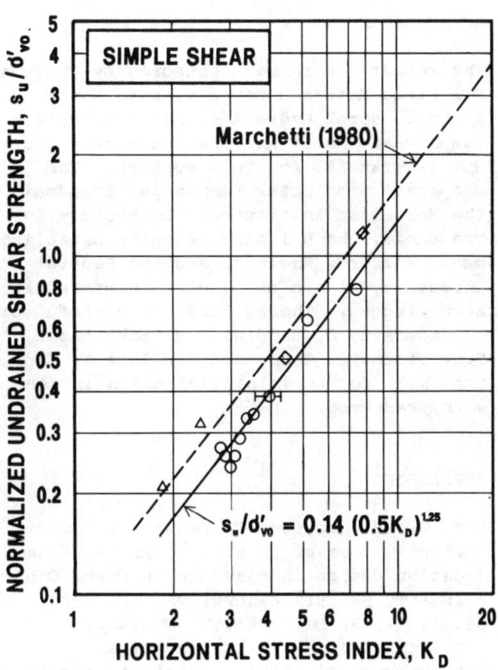

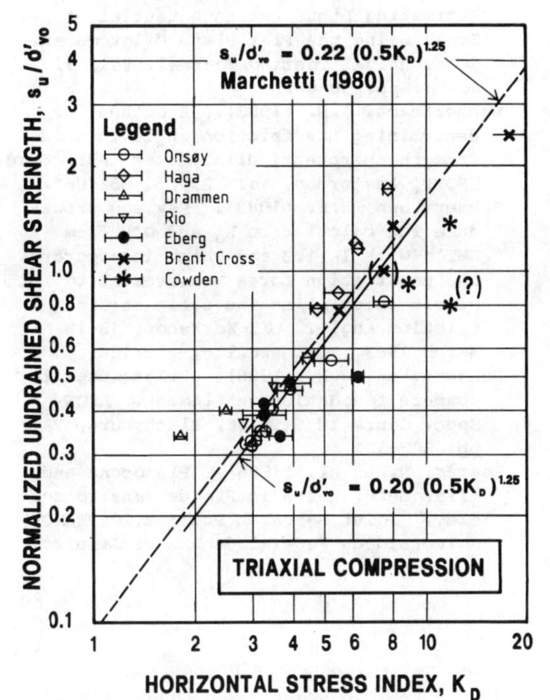

Fig. 8 Relationship between horizontal stress index, K_D, and undrained shear strength from field vane, triaxial compression and direct simple shear tests.

pore pressure can be significant. In sands, the measured contact and 1-mm expansion pressures and the difference P_1-P_0 are large and errors in calibration constants or in situ stresses have little or no effect on the results.

The major drawback of the dilatometer test is its empirical character, and the lack of understanding, until now, of the reasons why the test has provided reasonable estimates for a number of engineering parameters.

The device can provide a classification of the soil tested, and is one of the only methods to provide estimates of unit weight, coefficient of lateral earth pressure and overconsolidation ratio simultaneously and very cost-effectively. Important uncertainties still exist, but the evaluation of well-documented sites and the use of consistent reference parameters should reduce the scatter observed in the data underlying the interpretation charts. The correlations are fairly well documented for clay deposits, but need considerable validation for silt and sand deposits. Calibration tests with dilatometer tests in sand (Baldi et al., 1986) should help bridge the gap in evaluated data. Because of this lack of evaluated experience, it is recommended to consider the cone/piezocone penetration (or field vane) test and the dilatometer test as complementary rather than as alternatives.

ACKNOWLEDGEMENT

The research work was sponsored by Chevron Oil Field Research Company, Esso Norge a.s, Statoil, Norsk Hydro A/S, Sohio Petroleum Company and Mobil Exploration Norway Inc. NGI is grateful for this support. The cooperation of Ottar Kummeneje, Trondheim, The Norwegian Institute of Technology in Trondheim, the Building Research Establishment, Watford, England, and the Federal University of Rio de Janeiro is gratefully acknowledged. The authors are grateful to C. Andersen, C. Bandis, K.H. Mokkelbost, T.S. Olsen, B. Thune and the late T. Våge, from NGI, for their participation in the work presented.

REFERENCES

Aas, G., S. Lacasse, T. Lunne & K. Høeg (1986). Use of in situ tests for foundation design in clay. ASCE Spec. Conf. IN SITU 86. Blacksburg, VA, pp. 1-30.

Baldi, G., R. Bellotti, V. Ghionna, M. Jamiolkowski, S. Marchetti & E. Pasqualini (1986). Flat dilatometer tests in calibration chambers. ASCE Spec. Conf. IN SITU 86. Blacksburg, VA, U.S.A, pp. 431-446.

Bellotti, R., V. Ghionna, M. Jamiolkowski, R. Lancelotta & G. Manfredini (1986). Deformation characteristic of cohesionless soils from in situ tests. ASCE Spec. Conf. IN SITU 86, Blacksburg, VA, pp. 47-73.

Brooker, E.W. & H.O. Ireland (1965). Earth pressures at rest related to stress history. Canadian Geotechnical Journal, Vol. 2, No. 1, pp. 1-15.

Ghionna, V., M. Jamiolkowski, S. Lacasse, C.C. Ladd, R. Lancellotta & T. Lunne (1983). Evaluation of self-boring pressuremeter. International Symposium on Soil and Rock Investigations by In Situ Testing. Paris, Vol. 2, pp. 239-301.

Jamiolkowski, M., C.C. Ladd, J.T. Germaine & R. Lancelotta (1985). New developments in field and laboratory testing of soils. 11th ICSMFE San Francisco, Vol. 1, pp. 57-154.

Lacasse, S. (1985). In situ determination of deformation parameters. Panel disc. Session 2A, 11th ICSMFE, San Francisco.

Lacasse, S. & T. Lunne (1982a). Dilatometer tests in two soft marine clays. Proc. Conf. Updating subsurface sampling of soils and rocks and their in situ testing, Santa Barbara, CA, pp. 286-292.

Lacasse, S. & T. Lunne (1986). Dilatometer tests in sands. ASCE Spec. Conf. IN SITU 86, Blacksburg, VA, pp. 689-699.

Lunne, T., R. Jonsrud, T. Eidsmoen & S. Lacasse (1987). The offshore dilatometer. Brazil Offshore -87, Rio de Janeiro.

Marchetti. S. (1980). In Situ Tests by Flat Dilatometer. JGED, ASCE, Vol. 106, No. GT3, 1980, pp. 299-321.

Marchetti, S. (1985). On the field determination of K_o in sand. Panel discussion. 11th ICSMFE, San Francisco, August, Session 2A.

Marchetti, S. & D.K. Crapps (1981). Flat Dilatometer Manual. GPE, Inc., Gainesville, Florida, USA.

Marchetti, S., G. Totani, R.G. Campanella, P.K. Robertson & B. Taddei (1986). The DMT-σ_{hc} method for piles driven in clay. ASCE Spec. Conf. IN SITU 86. Blacksburg, VA, pp. 765-779.

Mokkelbost, K.H. (1984). Måling av geotekniske parametre ved bruk av dilatometer. Diploma Thesis. Geotechnical Division, Norwegian Institute of Technology, Univerisity of Trondheim.

Powell, J.J.M. & I.M. Uglow (1986). Dilatometer tests in stiff overconsolidated clays. 39th Canadian Geotech. Conf., Ottawa, pp. 317-326.

Robertson, P.K. & R.G. Campanella (1986). Estimating Liquefaction Potential of Sands Using the Flat Plate Dilatometer. Geotechnical Testing Journal, Vol. 9, No. 1, pp. 38-40.

Schmertmann, J.H. (1982). A method for determining the friction angle in sands from the Marchetti dilatometer test. 2nd ESOPT, Amsterdam, Vol. 2, pp. 852-861.

Schmertmann, J.H. (1983). Revised procedure for calculation K_o and OCR from DMT's with I_D>1.2 and which incorporates the penetration force measurement to permit calculating the plain strain friction angle. DMT-Workshop, 16-18 March 1983. Gainesville, Florida.

Schmertmann, J.H. (1986). Dilatometer to compute foundation settlement. ASCE Spec. Conf. IN SITU 86, Blacksburg, VA, pp. 303-321.

Soarès, M. et al. (1986). Piezocone and dilatometer tests in Rio de Janeiro soft clay. Joint Research Project COPPE/NGI, Universidade Federal do Rio de Janeiro.

Penetration Testing 1988, ISOPT-1, De Ruiter (ed.)
© 1988 Balkema, Rotterdam, ISBN 90 6191 801 4

Dilatometer C-reading to help determine stratigraphy

A.J.Lutenegger & M.G.Kabir
Clarkson University, Potsdam, N.Y., USA

ABSTRACT: The use of the Dilatometer membrane closure pressure following a conventional inflation test is described. This pressure reading, designated as the C-Reading, is corrected for membrane resistance and provides an additional test pressure, p_2. The Dilatometer pore pressure parameter, defined as $U_D = (p_2-u_o)/(p_o-u_o)$, is shown to be useful for defining site stratigraphy and other engineering parameters. Comparisons are presented between DMT C-reading dissipation tests and dissipation tests obtained from Piezocone and Piezoblade.

Keywords: Dilatometer, pore pressure, stratigraphy, stress history, dissipation tests

1 INTRODUCTION

The Flat Dilatometer (DMT) is fast becoming one of the most popular in situ tests available to geotechnical engineers. The test is simple in concept, easy and fast to perform, and provides reliable estimates of several important soil parameters. Therefore, it may soon become an economical and integral part of many geotechnical investigations. In its original form (Marchetti, 1980) and in the form proposed for standardization in the U.S. by ASTM (Schmertmann, 1986) the DMT is a total stress instrument and a pore pressure sensor is not a part of the instrument. Specially designed DMT blades (Davidson and Boghrat, 1983; Campanella et al., 1985) have incorporated a porous-stone electrical transducer pore pressure element at the position of the DMT diaphragm in order to provide both a measure of penetration pore pressure (Davidson and Boghrat, 1983) and changes in pore pressure during inflation of the diaphragm from the p_o to p_1 position (Campanella et al., 1985).

These investigations clearly demonstrated that in sands and other fast draining materials, the DMT pressure readings represent essentially drained conditions. This has led to the suggestion that a controlled deflation of the membrane could be used to obtain in situ pore pressures in sands or sand layers (GPE, 1985). In contrast, in clays large excess pore water pressures may be generated during penetration, much like during cone testing. Therefore, in clays where the test is largely undrained, the DMT expansion pressures may be dominated in large part by water pressures.

The measurement of excess pore water pressures in the DMT and most in situ tests is important not only for understanding the mechanics of the test but also for helping interpret data. However, the use of a transducer pore pressure measuring system with the DMT unnecessarily complicates the test and presents several disadvantages:

1. Equipment modifications are required.
2. Additional cost in obtaining the transducer and readout devices and the potential inconvenience and costs involved in maintaining their reliability.
3. Deairing of the porous-stone system.
4. Cavitation of the system during penetration.

Many of these problems are also inherent in Piezocone testing and special procedures and care are necessary to obtain high quality results. Nonetheless, it may be especially desirable to have an estimate of pore pressures during the DMT test.

The purpose of this paper is to present penetration pore presure data as measured from Piezoblade soundings vs. those estimates from DMT diaphragm closing pressure "C-Reading" at several clay sites. For a number of sites, these data are compared with penetration pore pressures obtained from adjacent Piezocone soundings where pore pressures were measured behind the base of the cone. The use of the C-Reading for estimating site stratigraphy and as an aid in interpreting soil parameters is also described. Dissipation tests from the DMT, Piezoblade and Piezocone are also presented and discussed.

2 TEST SITES AND PROCEDURES

A number of test sites have been investigated by the writers to study the usefulness of the C-Reading pressure. Most of the sites represent cohesive soils of varying stress history and strength. For the current discussion only a limited number of sites which best summarize the writers' work will be described.

Tests were conducted at three sites in northern New York along the St. Lawrence lowlands and at a fourth site in the Missouri River floodplain, near Omaha, Nebraska. Soil deposits at the three sites in New York represent late Pleistocene/Holocene marine clays (Leda Clay) as part of the Champlain Sea. At all four sites a thin veneer (<2m) of sandy/silty material overlies the marine clay. Generally, a weathered crust grades into soft to very soft sensitive clay with increasing depth. At the Nebraska site, the deposits consist of fine-grained alluvial materials from the Missouri River. A desiccated-overconsolidated crust overlies soft clay.

2.1 Dilatometer C-Reading

Following the normal pressure expansion sequence to obtain the conventional DMT A and B readings, an additional pressure reading may be taken to obtain the pressure at which diaphragm recontact takes place during controlled deflation. This may be done by installing a simple flow control valve and sensitive pressure gage in the pressure vent line. C-Readings were generally obtained in less than 15 sec. following the DMT B-reading. Schmertmann (1986) recommends that DMT A and B readings be obtained within about 1 min. after penetration. In general, all three pressure readings A, B and C were obtained in a time span of about 1 min.

An alternative procedure to obtain some estimate of pore pressures might be to reinflate the diaphragm following venting after the normal A and B readings to obtain a second A-reading. In this case, as before, all three readings, A, B and reinflation A, were obtained within about 1 min.

At several sites dissipation tests were conducted with the DMT by performing a full sequence of A, B and C-Readings at various times following penetration.

2.2 Piezoblade

Davidson and Boghrat (1983) and Boghrat (1987) described the use of a Piezoblade which was fabricated from a DMT blade. Modifications included replacing the flexible DMT diaphragm with a flush-mounted porous stone/transducer system in the same position.

For each site investigated in the current study a Piezoblade profile was conducted adjacent to the DMT sounding with penetration stopped at each elevation where a DMT test had been performed. Penetration pore pressure readings were taken for each test as well as the pressure at 15 sec., 30 sec. and 1 min. after penetration. Dissipation tests in which pore pressure was monitored for several hours were conducted at selected depths at each site for comparison with DMT and Piezocone dissipation tests.

2.3 Piezocone

Adjacent Piezocone profiles were also conducted at several of the sites using a standard 35.6 mm diameter 60° apex cone with a 5 mm thick porous element located 5 mm behind the base of the cone. It has been well established that the location of the porous element significantly influences the magnitude of pore pressures measured. However, this effect is less pronounced in soft normally consolidated clays. Ideally, a porous element located on the cone shaft at the same aspect from the penetration wedge as the DMT diaphragm and Piezoblade transducer should be used. Figure 1 presents the geometry and dimensions of all three instruments used.

3 PENETRATION PORE PRESSURE

The deflation C-Reading provides a measure of the external pressure forcing the diaphragm to recontact the DMT blade. This pressure reading, corrected for membrane stiffness, gives a pressure referred to as p_2 such that:

$$p_2 = C - A \text{ correction} \qquad [1]$$

It is of interest for this discussion to determine how accurately the value of p_2 estimates penetration pore pressure generated by advancing the DMT.

Profiles of penetration pore pressure measured with the Piezocone and Piezoblade are compared with DMT p_2 values in Figure 2 for each test site. Piezocone pressures represent instantaneous pore pressure. The trend noted for all sites indicates that Piezocone pore pressures are larger than Piezoblade pore pressures, however, the value of p_2 closely matches the Piezoblade penetration pore pressures. This is especially true in the soft clays. Since p_2 is obtained about 1 min. after penetration, it is reasonable to expect the pressure to be lower than the Piezoblade, as some dissipation may have taken place.

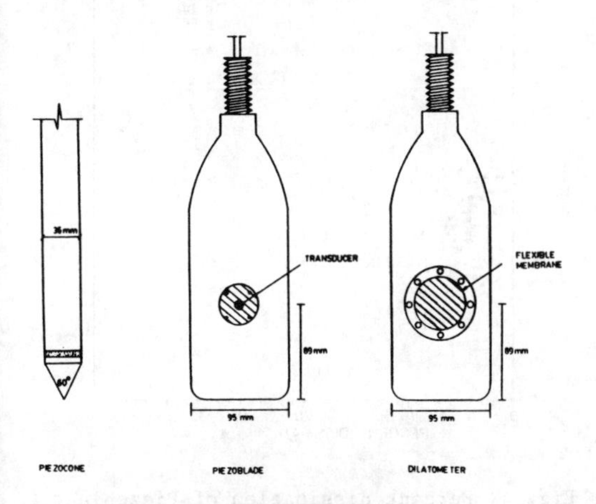

Fig. 1 Comparative geometry of Piezocone, Piezoblade and DMT

A comparison between Piezoblade pore pressure and DMT p_2 for all sites investigated by the writers is shown in Figure 3. Both penetration pore pressures and pore pressures obtained 1 min. after penetration from the Piezoblade are shown. It can be seen that nearly all data points fall within this bound. In some cases p_2 is greater than u_o from the Piezoblade. These values may be due to slight soil variability or a component of soil pressure pushing on the diaphragm. It should be expected that soils with higher permeability would give p_2 values closer to the 1 min. Piezoblade pressure while those with lower permeability give p_2 values closer to Piezoblade penetration pore pressure. Thus, it appears important to identify soil type in assessing the usefulness of p_2 to indicate penetration pore pressure.

It has previously been shown by Davidson and Boghrat (1983) that the percent dissipation of penetration pore pressure obtained by the Piezoblade after 1 minute was related to the DMT I_D value. The writers prepared a similar comparison which is shown in Figure 4. As can be seen, for I_D values between 0.1 and 0.6 there is a relatively small amount of pore pressure dissipation within the first minute.

The magnitude of pressures indicated in Figure 2 are in excess of assumed hydrostatic pressure by a factor of about 2 to 5, and are a function of the OCR, s_u, and G. The depth to the static groundwater table, determined from long-term observa-

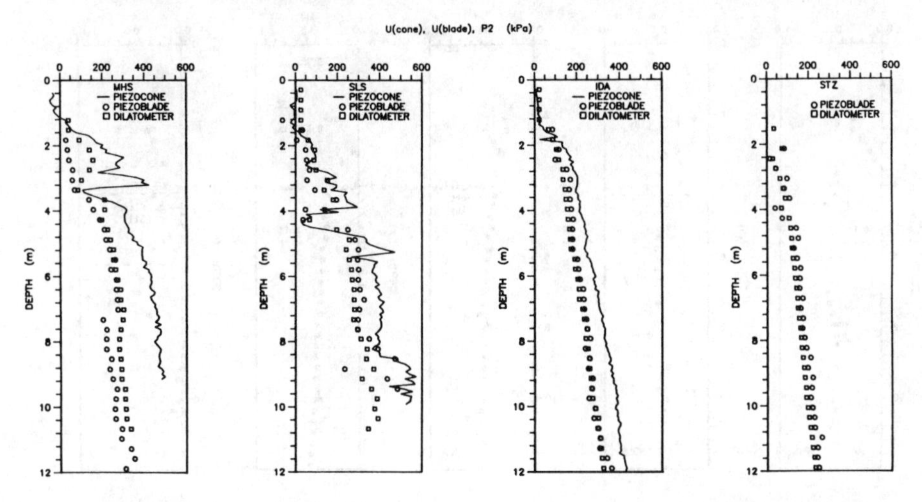

Fig. 2 Profiles of Piezocone and Piezoblade penetration pore pressure and DMT p_2

tions of open boreholes is easily identified by the first positive value of p_2 as noted in the profiles.

4 U_D AND SITE STRATIGRAPHY

The sensitivity of p_2 to changes in material was made evident in the previous section. Since the magnitude of pore pressures existing at the time p_2 is obtained is a function of both generated pore pressure and drainage characteristics, direct material identification is somewhat complicated. As an initial indication of stratigraphy, one can make a comparison of p_2 in relation to the DMT p_o reading. This value is hereafter referred to as the DMT pore pressure index, U_D, and is defined as:

$$U_D = \frac{p_2 - u_o}{p_o - u_o} \quad [2]$$

where u_o = in situ pore pressure. Thus, a site profile of U_D may be used to give an indication of stratigraphy. Figure 5 presents profiles from the sites previously

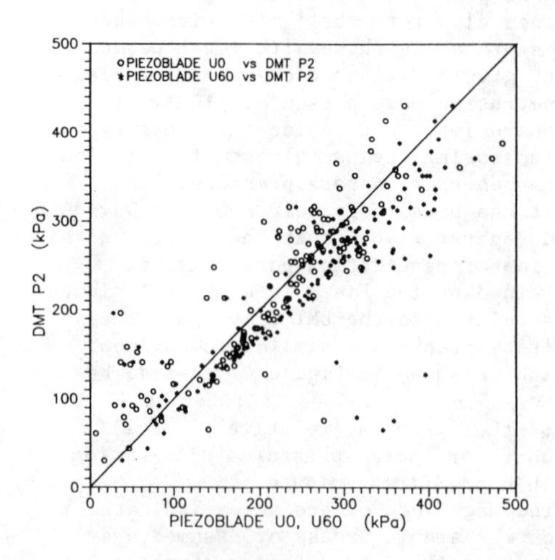

Fig. 3 Comparison between Piezoblade U_{excess} and DMT p_2

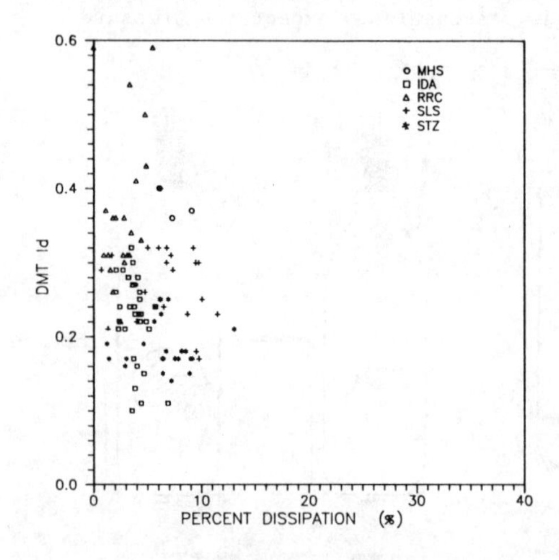

Fig. 4 Percent dissipation of Piezoblade pore pressure vs. DMT I_D.

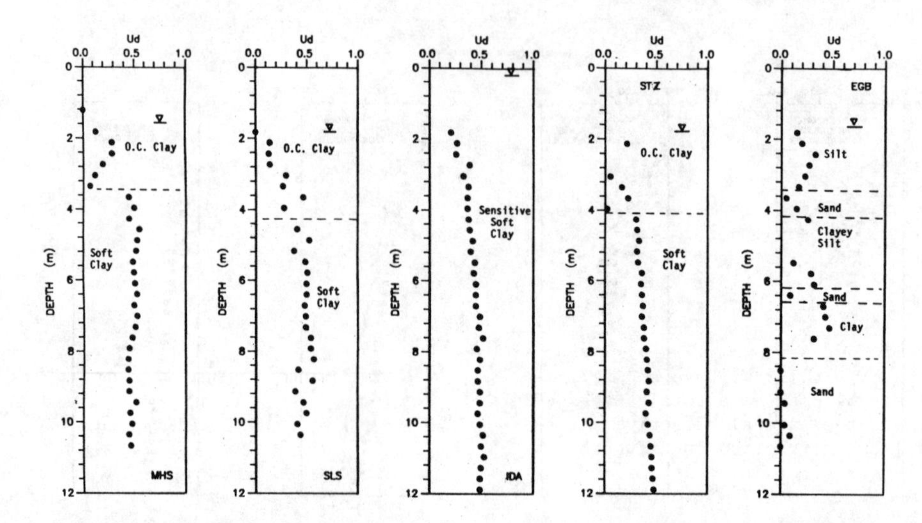

Fig. 5 Profiles of U_D to help determine site stratigraphy

shown in Figure 2 and an additional site from an alluvial flood-plain which contained a complex sequence of sand, silty and clay layers. Clearly, the U_D ratio can be useful for identifying changes in stratigraphy.

Variations in U_D for a given soil also reflect the tendency for generating positive pore pressures, which one can expect to vary with the stress history of the soil. Therefore, one should logically expect a relationship between U_D and OCR. Available data from oedometer tests on undisturbed Shelby Tube and piston samples (76 mm) for sites previously described are shown in Figure 6. If both U_D and I_D are

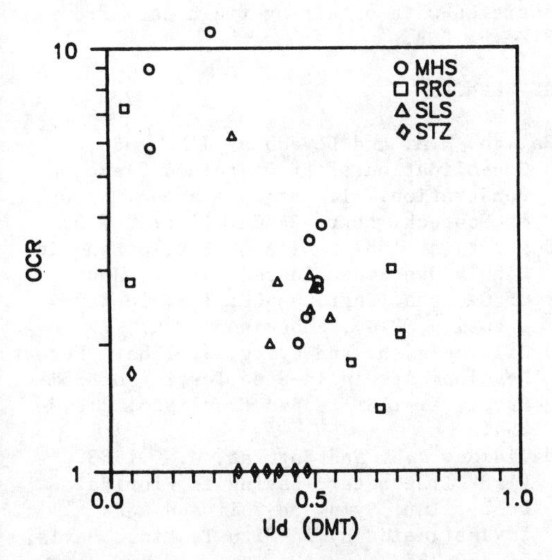

Fig. 6 Relationship between OCR and U_D

an indication of soil type one should expect a relationship to exist between these two parameters. Combined data from the test sites are shown in Figure 7.

5 DISSIPATION TESTS

Observations of the dissipation of excess pore pressure after interruption of a Piezocone profile have been used to estimate the in situ coefficient of consolidation using several theoretical approaches (Torstenson, 1977; Baligh and Levadeaux, 1985; Gupta and Davidson, 1986). Recently Marchetti et al. (1986) have shown that the DMT may be used as a total stress cell to estimate the lateral effective stress for local shaft friction on driven piles. This is done by obtaining a timed series of DMT p_o measurements without conducting a complete 1 mm expansion. The results

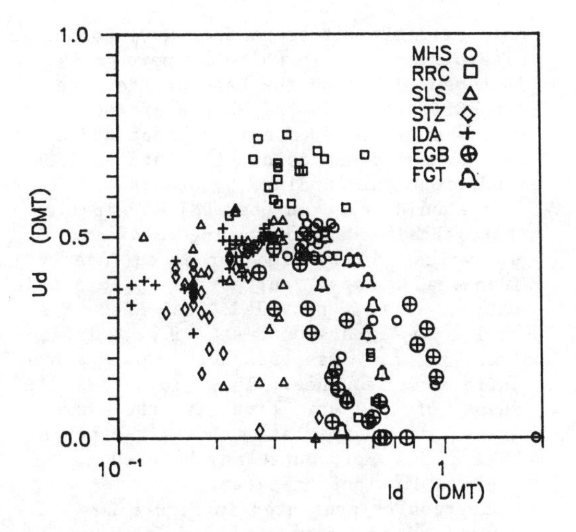

Fig. 7 Relationship between U_D and I_D

would then be similar to those obtained from other push-in spade cells (Massarsch, 1975; Tedd and Charles, 1981). In light of the current discussion regarding p_2 the writers made a comparison between the pore pressure dissipation results obtained from the Piezocone and Piezoblade and the change in p_2 with time.

As shown in Figure 8, typical results indicate that the shape of pore pressure dissipation test from the Piezoblade closley matches that from the p_2 dissipation test. Additionally, these curves are

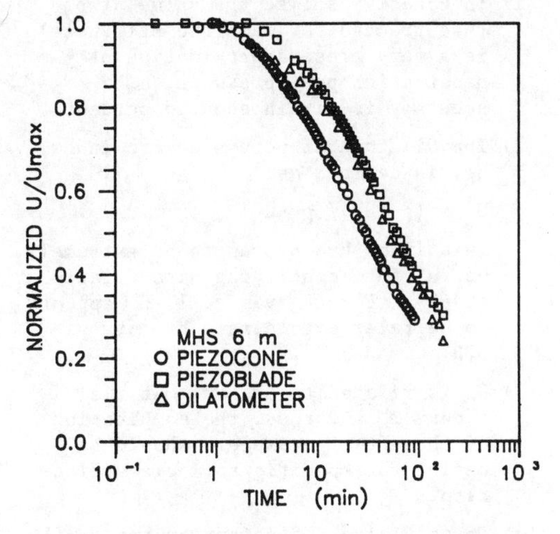

Fig. 8 Comparison of Piezocone and Piezoblade dissipation tests with p_2 dissipation

only slightly different from a Piezocone dissipation test in which the pore pressure is measured behind the base of the cone. The differences in the shapes of the curves obtained from a flat probe and a cylindrical probe no doubt relate to flow conditions surrounding the probes.

It should be noted that DMT dissipation tests obtained by conducting reinflation p_o values with time produced nearly identical shaped curves as p_2 dissipation tests, provided that the full 1 mm expansion test, i.e., inflation to p_1, was conducted on each inflation sequence. That is, the shape of p_o vs. time is the same as p_2 vs. time and thus it appears that either procedure may be used to perform dissipation tests.

The results presented in Figure 8 suggest that it may now be possible to obtain an estimate of the horizontal rate of consolidation from a suitable DMT dissipation test, provided that an appropriate analytical solution may be derived for the proper pore pressure distribution and drainage conditions.

6 SUMMARY AND CONCLUSIONS

The Dilatometer recontact pressure measurement (C-reading) obtained by controlled membrane deflation following the normal A and B pressure readings has already proven to be useful for obtaining in situ water pressures in sand layers. The writers herein have found new information about the C-reading and a new use.

(1) In cohesive soils, the value of p_2 (C corrected for membrane stiffness) is a very close approximation of penetration pore pressure. The accuracy is within about 5 percent.

(2) The Dilatometer pore pressure index, U_D, is defined as:

$$U_D = (p_2-u_o)/(p_o-u_o)$$

This index has proven to be extremely useful for identifying site stratigraphy. For a given soil, U_D appears to be related to stress history via OCR.

(3) U_D is related to the DMT material index, I_D and thus, the combination of these two parameters should be useful for specific soil classification.

(4) In soft clays, dissipation tests of p_2 closely match Piezoblade dissipation tests. The shape of these curves is only slightly different than that obtained from a Piezocone, and therefore one should be able to obtain an estimate of the horizontal coefficient of consolidation.

Based on the work reported herein, and previous acknowledged use of the DMT closure pressure, the writers are of the opinion that the C-reading should become a routine part of the DMT procedure.

7 ACKNOWLEDGEMENTS

The authors wish to thank Dr. J. Davidson of the University of Florida for load of the Piezoblade and Dr. John Schmertmann for first suggesting the writers consider calling their pore pressure parameter U_D. Mr. S. Saye, Woodward-Clyde Consultants, assisted with obtaining field data from STZ and EGB.

REFERENCES

Baligh, M.M. and Levadous, J.N. 1986. Consolidation after Undrained Piezocone Penetration. II: Interpretation. Jour. of Geotech. Engr., ASCE: 112:727-745.

Boghrat, A. 1987. Dilatometer Testing in Highly Overconsolidated Soils. Jour. of Geotech. Engr., ASCE: 113:516-519.

Campanella, R.G., Robertson, P.K., Gillespie, D. and Grieg, J. 1985. Recent Developments in In-Situ Testing of Soils. Proc., XI ICSMFE, San Francisco, 2:849-854.

Davidson, J.L. and Boghrat, A.G. 1983. Flat Dilatometer Testing in Florida. Proc., Int. Symp. on Soil and Rock Investigation by In-Situ Testing, Paris, II:251-255.

Gupta, R.C. and Davidson, J.L. 1986. Piezoprobe Determined Coefficient of Consolidation. Soils and Foundations, 26:12-22.

Marchetti, S. 1980. In-Situ Tests by Flat Dilatometer. Jour. of Geotech. Engr., ASCE: 106:229-321.

Torstensson, B.A. 1977. The Pore Pressure Probe. Nordiske Geotekniske Mote, Oslo, Paper No. 34.1-34.15.

Penetration Testing 1988, ISOPT-1, De Ruiter (ed.)
© 1988 Balkema, Rotterdam, ISBN 90 6191 801 4

Marchetti dilatometer testing in UK soils

J.J.M.Powell & I.M.Uglow
Building Research Station, Garston, Herts, UK

ABSTRACT: The use of the Marchetti flat-blade dilatometer (DMT) has been investigated on a number of test-bed sites, ranging from soft normally consolidated clays to heavily over-consolidated stiff clays and one sand site. The device has been found to be simple and robust in use, and to give repeatable results. However, the use of 'standard' correlations in order to assess soil properties is questioned, gross errors occurring in some soils. Site specific correlations are shown to exist.

The usefulness of the 'closing pressure' from tests with controlled unloading has been investigated. It is shown that the closing pressure is a function of variations in test procedure, the maximum pressure attained during the test and soil type; it is not unique for a given soil type. The use of dissipation type testing is discussed.

KEY WORDS: Dilatometer testing, Clays, Sands, Field tests, Pore pressures

1 INTRODUCTION

The flat dilatometer was developed in Italy in the late 1970's by Marchetti (Marchetti, 1980). It is a simple and robust tool used to assess the properties of soils in-situ. More recently it has found use in other parts of the world (Lacasse and Lunne, 1982, 1986; Schmertmann, 1982; Campanella and Robertson, 1983; FICFPD, 1983; Aas et al. 1984; Campanella et al, 1985; Lutenegger and Timian, 1986). It has now started to be used in Britain and, as a result, the Building Research Establishment (BRE) is undertaking a programme of work to assess the usefulness of the dilatometer in typical UK deposits (Powell and Uglow, 1986).

Marchetti (1980) and other workers (see FICFPD, 1983) have defined three basic index parameters based on dilatometer readings; these have been correlated with various soil properties. Much of the published work to date has been on sands and normally or lightly overconsolidated clays, with little work on stiff overconsolidated soils. The first phase of the BRE investigation was concerned with filling this gap, work being undertaken on three well documented test bed sites in stiff clays in the UK (Powell and Uglow, 1986). This has now been extended to other test bed sites (3 soft clay, 2 more stiff clays and 1 sand). This present paper summarises the results to date, compares the deduced with the known soil properties and discusses the findings. In addition, work is presented from an investigation, on the above sites, into the significance of the "closing pressure" in the dilatometer test (Campanella et al, 1985).

2 DILATOMETER TESTING

Table I lists the sites for which data are currently available, they cover stiff overconsolidated clays, normally and lightly over consolidated soft clays and one sand site. Space only allows for a summary of the findings to be given here (the Cowden, Brent Cross and Madingley sites are discussed in detail by Powell and Uglow, 1986). The results will be discussed by reference to figures of the same type as those used by Marchetti in establishing his correlations between dilatometer (DMT) parameters and soil properties. The "known" soil properties for Cowden, Brent Cross and Madingley are given by Powell and Uglow 1986 based on field and laboratory testing; for the other sites shear strengths are in general based on laboratory triaxial and field vane tests, whilst the

Table I: Soil properties of the sites

Location	ρ t/m³	W %	Ip %	Clay %	OCR	Description
Cowden, Glacial Till						
0 - 15 m	2.15	16-20	17-20	30-40	>3	Brown and blue grey glacial tills containing lumps of chalk. A few horizontal bedding features. Very few macrofabric features below 5m.
Brent Cross, London Clay						
0 - 9 m	2.0	26-30	53	52-62	>40	Brown weathered clay; intact lumps 6-25 mm in
9 - 15 m	2.0	28	48	59	>30	Grey blue unweathered highly fissured clay lumps
15 - 25 m	2.0	26	42	55-60	>20	Grey blue unweathered fissured clay lumps 25-325mm bedding planes containing silt and sand.
Canons Park, London Clay						
0 - 2 m	2.0	20-28	28-48	35	>10	Gravely brown clay.
2 - 5 m	2.0	20-28	28-48	45	>10	Re-worked remoulded brown London clay.
5 - 8 m	1.8-2.2	22-28	40-44	44	>10	Stiff brown silty fissured clay.
8 - 20 m	2.0	28	44		>10	Blue London clay
Madingley, Gault Clay						
0 - 4 m	1.9	30-32	46-51	60	>50	Firm intact grey/green mottled weathered clay.
4 - 7 m	1.9	30-31	40-48	60	30-50	Stiff fissured weathered silty clay with pockets of glauconitic sand.
7 - 25 m	1.94-2.0	28-30	40-48	60	10-30	Very stiff dark grey closely fissured silty clay.
BRE/Herts, Glacial Till						
0 - 6 m	2.15	17-20	28	41		Highly fissured reddish brown chalky till.
6 - 12 m	2.25	22-24	26	36		Very highly fissured brown chalky till.
Grangemouth, alluvial clay						
0 - 1 m	1.74	38-50	26-40	40		Firm brown silty clay with roots.
1 - 6 m	1.6-1.7	60-76	32-44	30-35	1-2	Soft black silty clay.
6 - 16 m	1.6-1.7	60-80	40-50	25-30	1-2	Soft dark grey micaceous clay with thin silt laminations, more silty with depth.
Dartford, alluvium						
0 - 4.5 m	1.5	85-120	60-70	40-44	1-3	Soft blue grey silty organic clay: numerous partially decayed root fibres and reeds.
4.5 - 5 m	1.2	260			1	Peat.
5 - 9 m	1.3-1.4	95-140	75-120	40-52	1	Soft silty clay: numerous partially decayed root fibres and reeds.
9 - 12 m						Silty sandy gravel.
Gorpley, puddle clay						
0 - 10	1.6-1.9	30-40	19-22	32-40	1 ?	Very soft, highly remoulded, homogeneous brown puddle clay.
Manchester, sand						
0 - 5 m	1.7			sand		Loose to medium dense medium sand.

overconsolidation ratio, OCR, and the insitu coefficient of earth pressure at rest, K_o, have been derived from oedometer, pressuremeter and field push-in earth pressure cells. In this part of the paper all results are based on routine DMT tests recording 'lift off' pressure, p_0, and the pressure at 1mm displacement, p_1,(see Powell and Uglow, 1986 for description of equipment and test procedures used). Reference will be made to the DMT parameters, the horizontal stress index (K_D), the Material index (I_D), and the Dilatometer modulus (E_D).

For all sites the test results were found to be reproducible and produced a profile, in terms of both p_0 and p_1, very similar to that obtained from a cone penetrometer. In figure 1 the undrained shear strength, c_u, divided by the effective vertical overburden pressure, σ_{vo}, is plotted against K_D. The data for each site are seen to plot in well defined groups, with little of the data fitting the Marchetti correlation line. The use of this line could have resulted in gross errors in estimations of c_u in many deposits (e.g. up to 100% overestimation at Brent Cross, and up to 50% underestimation at Cowden and Gorpley). The presentation of data on a log log plot, whilst convenient for correlation purposes, can mask the true size of discrepancies. However it should be said that Marchetti's original data tended to

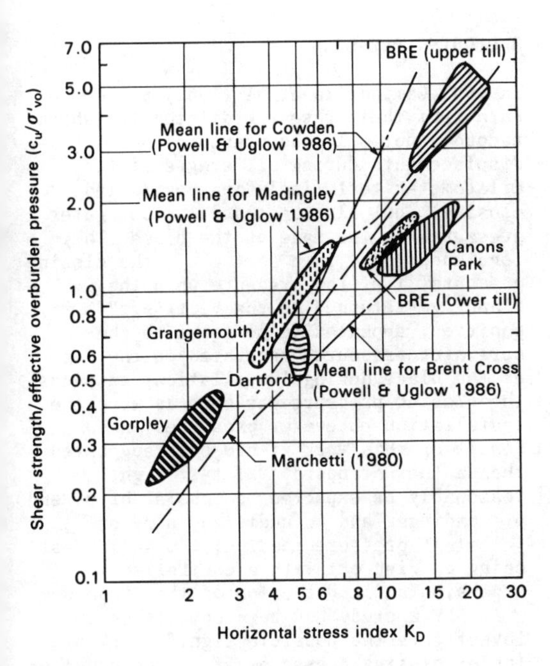

Figure 1 Shear strength/effective overburden pressure verses horizontal stress index K_D.

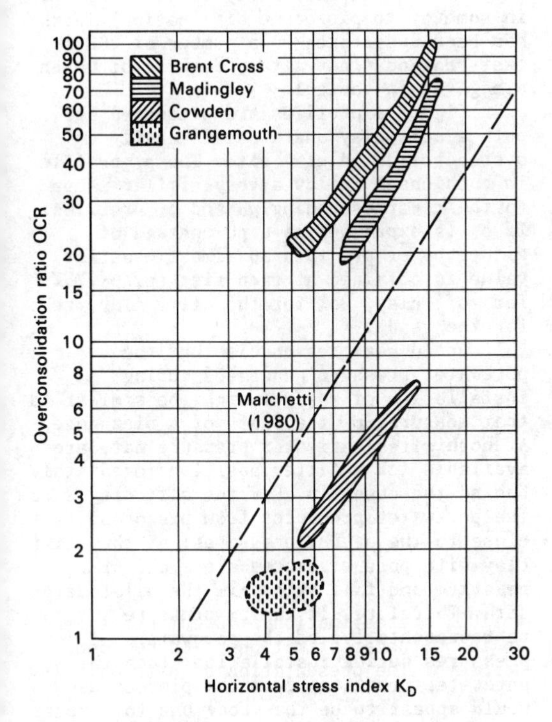

Figure 2 Overconsolidation ratio verses horizontal stress index K_D.

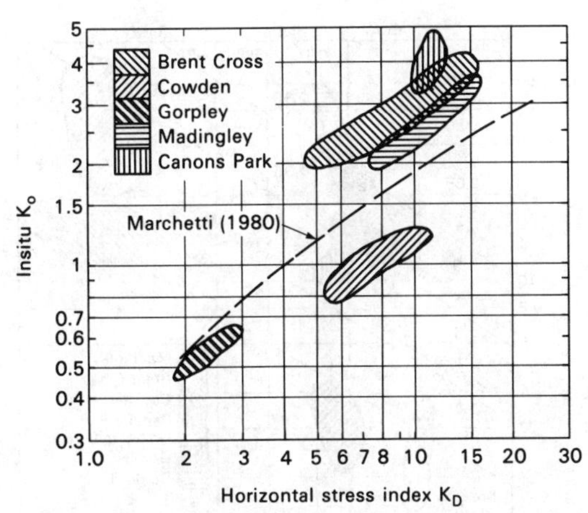

Figure 3 Insitu K_0 verses horizontal stress index K_D.

fall above the correlation line, the line itself intended to be on the safe side, i.e. underestimating c_u.

The plot of overconsolidation ratio, OCR, against K_D in figure 2 shows a very similar behaviour to figure 1 above, as does the plot of the insitu coefficient of earth pressure at rest, K_0, against K_D in figure 3. Significant differences can be seen between the correlation lines and the known data, both under and over estimates can occur. Even when deposits similar to those studied by Marchetti are considered disagreements are seen (e.g. Grangemouth and Gorpley). Some of the disagreements could of course be due to the different methods that can be used for measuring the 'known' soil properties. The masking effect of log log plotting mentioned above, still being a problem.

In a location at Cowden site DMT tests have been performed close to and inland from a cliff edge, the DMT failed to detect the reductions in horizontal stress due to the cliff edge which were detected by other devices. It is suggested that this may well be due to the strength and stiffness (shear modulus G) of the soil mass controlling the short term behaviour of the DMT and the insitu stresses.

The soil classification chart, figure 4, based on E_D and I_D also seems to have mixed success in identifying soil type, density and consistency. All the soils studied (except the sand), plot above the Casagrande 'A-line' and would therefore be described as clays (all have >65% fines). The soft clays at Grangemouth, Dartford and Gorpley seem to identify correctly. The Cowden and Gorpley soils have almost equal

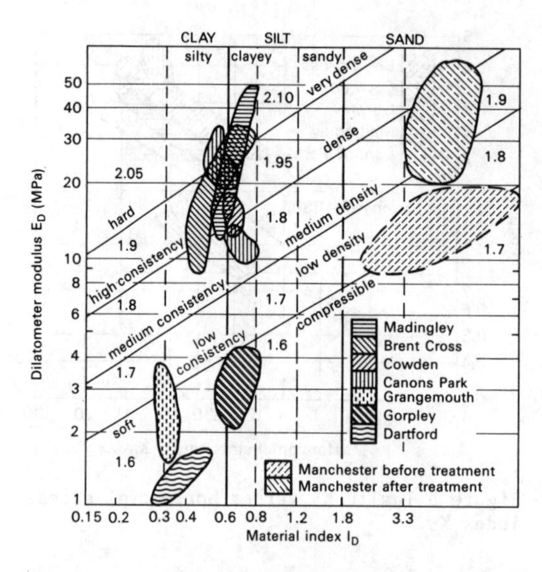

Figure 4 Soil classification: E_D verses I_D.

percentages of clay and silt, and so it is not surprising that they fall close to the clay/silt border in figure 4 (total fines content of 70-75% for both). The other clay soils all have fines in excess of 90%, with clay contents >50% except for Grangemouth which has 30-40% clay. It is therefore somewhat surprising that both the Dartford and Grangemouth soils plot as clays whilst the rest plot on the clay/silt border. It is suggested that the high degree of overconsolidation of these other clays might well be affecting the I_D values and could explain these anomalies. The ability of the chart to estimate the densities of the stiffer clays also seems to be in doubt; in general they are over estimated.

Data from the sand site are so far limited, however, as seen in figure 4 the deposit is correctly identified as sand. The site has, since the original investigations, been subjected to ground improvement, the DMT successfully recorded this as shown in figure 4.

Marchetti suggested that the constrained modulus of the soil could be estimated from E_D; when this has been tried for the stiffer clays, the constrained modulus appears to be significantly overestimated by Marchetti's correlations although site specific correlations do appear to exist.

The site specific correlations, identified by Powell and Uglow (1986), between the p_1 pressure and the pressuremeter limit pressure were again found for other sites included in this paper.

3 CLOSING PRESSURE

In 1985 Campanella et al (1985) presented data from their research dilatometer which recorded not only the pressures and displacements during all stages of the dilatometer test (including controlled closing), but also recorded the porewater pressures on the face of the blade. They concluded that, in a soft clay, the closing pressure, i.e. the pressure when the diaphragm returned to the "lift-off" position, appeared to be equal to the porewater pressures generated on the face of the blade during installation; in sands the closing pressure was zero as were the installation porewater pressures. As a result of this work it had been suggested that a 'new standard' DMT test might reasonably be expected to consist of p_0 and p_1 readings, and in addition, a p_2 or 'closing' pressure reading; the whole test being carried out with a controlled expansion and contraction of the diaphragm.

At BRE a study has been undertaken to investigate the possible significance of the p_2 closing pressure. If p_2 was equal or related to the generated porewater pressures, then it should be possible to study the decay or dissipation of p_2 with time (with the possibility of relating this in someway to piezocone dissipation). Work has been undertaken on a number of BRE's test-bed and other sites, details of which are given in table 1.

In figure 5 profiles are presented for both a soft clay and a stiff heavily overconsolidated clay site. The p_2 profiles in both cases follow a very similar shape to their corresponding p_0 and p_1 profiles. If p_2 is expressed as a percentage of either p_0 or p_1, then an almost constant value is obtained at each site (p_2/p_0=75% for soft clay, 35% for the stiff clay, zero for the sand).

It would seem reasonable that the porewater pressure generated during installation of the DMT might be similar to that measured on the shaft of a piezocone. At both sites porewater pressure data are available for a filter position immediately behind the cone tip. For the soft clay site the porewater pressures from piezocones are close to the p_0 DMT pressures; at the stiff clay site pore pressures are small or negative and fall below all the dilatometer data. Therefore, if the p_2 pressure was to be representative of the porewater pressures during installation, then the porewater pressures from the piezocones would appear to be the wrong one to compare with, unless the pore pressures further along the shaft significantly reduce in

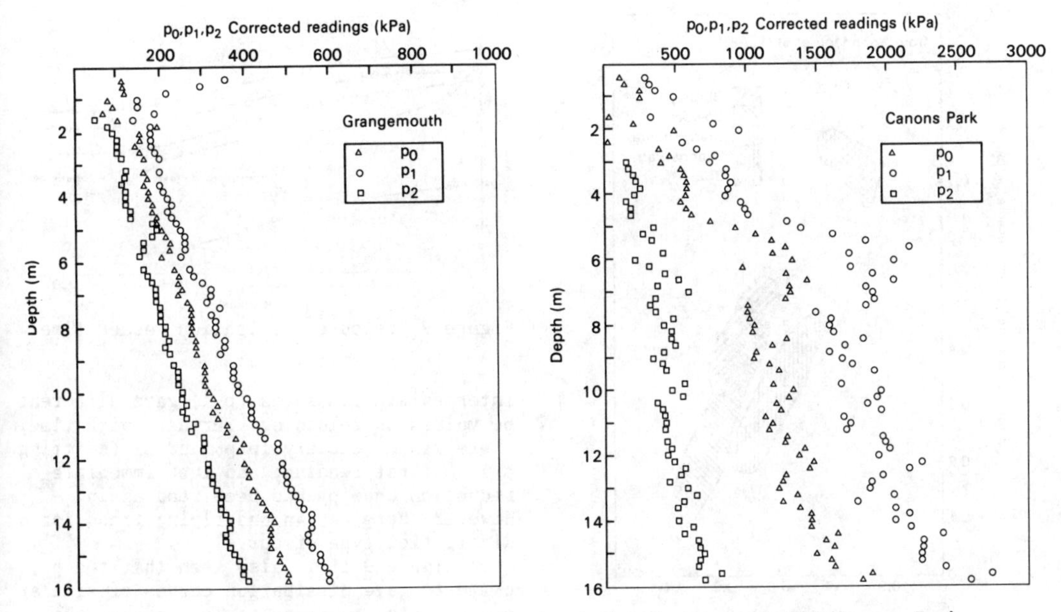

Figure 5 Corrected pressure readings verses depth (a) Grangemouth (b) Canons Park

soft clays and increase in stiff clays (Coope 1987, shows that in stiff clays the pore pressures remain small or negative for some distance behind the tip.

From the data above and that obtained from the other sites investigated (table 1), the closing pressure p_2 was found to be reasonably reproducible for tests at any one site and depth. Further, when expressed as the ratio p_2/p_0, it was found to be reasonably constant for any particular soil type. However, if the closing pressure is a unique value, then it should be independent of slight variations in the test procedure. To investigate this, a series of tests was performed where the procedure was to measure p_0, p_1^* and p_2^*, where p_1^* was a pressure equal to or in excess of p_0, and p_2^* was the closing pressure resulting after closing from p_1^*. In all cases a controlled application and bleed-off of pressure was carried out.

In figure 6 the results of these tests for the stiff clays at Canons Park are shown as (p_2^*/p_0) against $[(p_1^*-p_0)/(p_1-p_0)]$, where p_1 is the standard p_1 value relevant to the particular test depth. The effect on p_2 of the intermediate pressure p_1^* is immediately evident; the closing pressure is controlled by the maximum pressure reached. An overshoot of p_1 in a standard test results in a reduction in p_2 (i.e. when $(p_1^*-p_0)/(p_1-p_0)$ is greater than 1, p_2^* is less than p_2). Data for the other

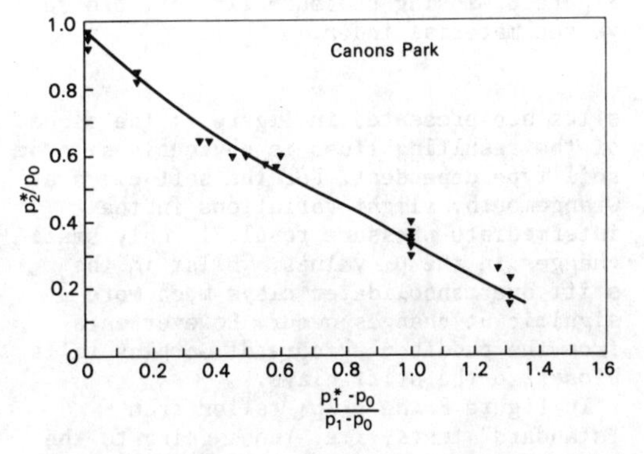

Figure 6 Variation of closing pressure with intermediate excess pressure: Canons Park.

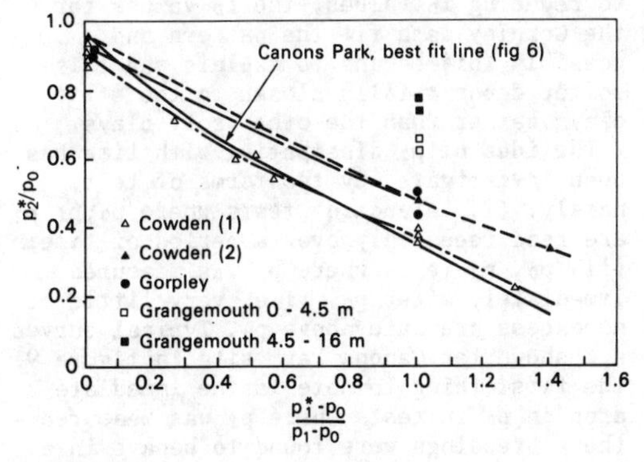

Figure 7 Variation of closing pressure with intermediate excess pressure: all sites.

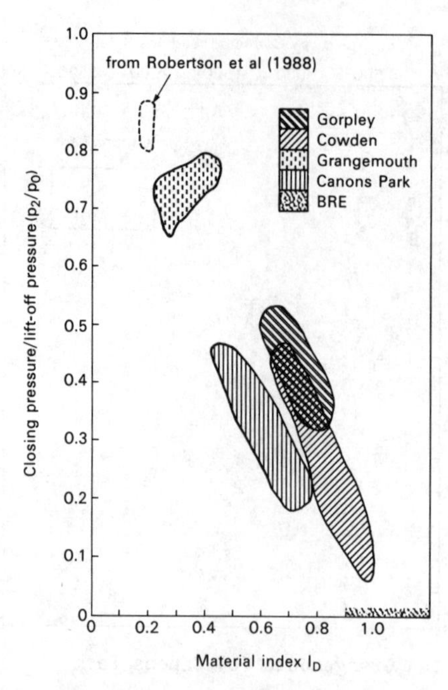

Figure 8 Closing pressure/lift-off pressure verses material index.

sites are presented in figure 7; the slope of the resulting lines is obviously site or soil type dependent. For the soft clays at Grangemouth, slight variations in the intermediate pressure result in only small changes in the p_2 values, whilst in the stiff overconsolidated clays much more significant changes occur. However data from the puddle clay core at Gorpley falls closer to the stiff clays.

In figure 8 the p_2/p_0 ratios from "standard" tests, i.e. those going to the normal p_1 state, are plotted against the DMT parameter I_D. A clear trend exists, with increasing p_2/p_0 values corresponding to reducing I_D values; the I_D values for the Gorpley data fit the pattern and possibly this begins to explain why this softer deposit falls closer to the stiff clays rather than the other soft clays.

The idea of p_2 dissipating with time has been investigated by two forms of test, namely: (i) "standard" tests where p_0, p_1, p_2 are read repeatedly over a period of time; (ii) p_0, p_2 tests where p_2 was measured immediately after p_0, i.e. very little or no excess pressure above p_0. Typical curves are shown for Canons Park site in figure 9. The first thing to note is the immediate drop in p_0 in tests where p_1 was measured - the p_0 readings were found to behave in a similar way to p_2 readings i.e. different

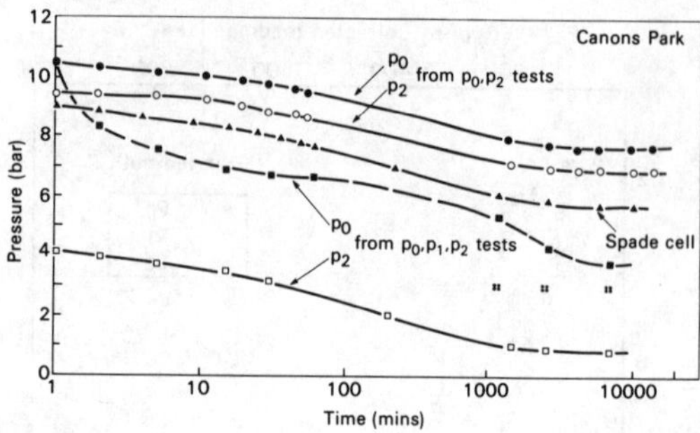

Figure 9 Pressure dissipation verses time.

intermediate pressures (p_1^*) gave different p_0 values on reloading. Further, with time, there was a recovery in p_0 and p_2 (omitting p_1) on first reading, and then immediate reduction once p_1 had been read again. However there was an underlying trend for a dissipation type of curve.

In figure 9 it is also seen that the p_0, p_2 tests gave dissipation curves of similar shape to the p_0, p_1, p_2 tests, however the long term or equilibrium values of p_0 and p_2 are markedly different and therefore depend on test procedure. If the data for p_2 from the two types of test are replotted as the percentage dissipation (i.e. reduction in p_2 from the initial value as a percentage of the difference between initial and final p_2 readings) with time, very similar curves are obtained, figure 10. It is suggested that p_2 is still a function of test procedure but that its dissipation is controlled by soil behaviour. The curves are very similar in shape to that obtained from push-in total stress spade cells (described by Tedd and Charles, 1983) at the same depth (see figure 10).

In some of the softer soils where there is less of a difference in p_2 readings with and without p_1, then these marked differences in actual values are less pronounced.

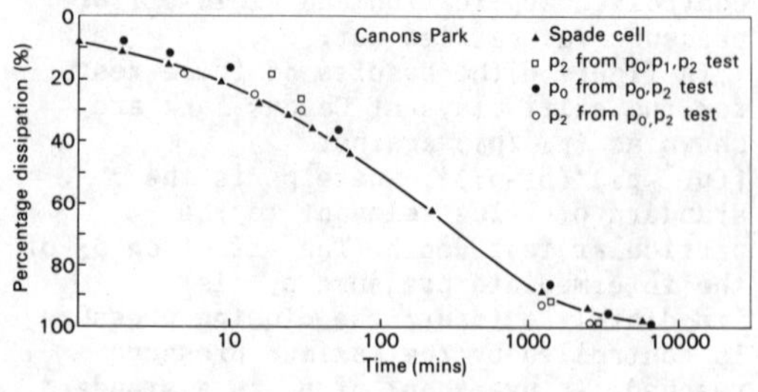

Figure 10 Percentage dissipation verses time.

This behaviour in p_2 would imply that there is no basis for assuming that the value after dissipation should equal the insitu porewater pressure. Further, if after reaching equilibrium during a p_0, p_2 dissipation test, the test is continued as a p_0, p_1^*, p_2 dissipation test then a new set of equilibrium pressures are reached, and so on if repeated again with a higher value of p_1^*.

4 CONCLUSIONS

The Marchetti dilatometer has been found to be a simple, easy to operate and robust tool. It gave reproducible test results in all the deposits investigated. However, the Marchetti correlations for obtaining soil properties seem to be in error for most of the soils investigated, resulting in errors of up to 100%. Site specific correlations do, however, appear to exist for all the soils investigated, but will themselves depend on the test type used to establish the 'known' soil properties. It may be that alternative 'dilatometer' parameters could give more unique correlations. The device is able to detect changes in sands due to ground improvement. It fails, however, to detect differences in insitu horizontal stress due to stress relief at a cliff edge. It is suggested that this may be due to the strength and stiffness of a deposit controlling the measured pressures. Long term readings of p_0 or p_2 may be better indicators of the stress differences.

The instrument has shown considerable potential as a profiling tool. Consideration needs to be given as to when Marchetti's correlations cease to work and to what extent it is a function of stress history, soil type or scale effects that cause the breakdown. It is suggested that care should be taken when transferring experience built up on certain soil types in one part of the world and expecting them to work elsewhere. Be aware of the limitations in the original data and any riders given to that data.

The closing pressure, p_2, has been shown to be a function of soil type and probably stress history, but more importantly a function of the excess or maximum pressure, attained during the test. It is a reproducible value for a given soil type, test procedure and test depth. Whilst the porewater pressures generated during installation of the DMT will also be a function of soil type it is thought that to consider the closing pressure to be equal to the porewater pressure is somewhat optimistic at this stage (agreement may be fortuitous when it does occur). However, it is not surprising that a relationship between porewater pressure and closing pressure can be obtained as both are likely to be controlled by the same parameters, i.e. soil type, stress history, and G/c_u, but the relationships will change with any variations in the DMT test method.

These problems are more acute in the stiff clays (or those having higher I_D values) where variations in the excess pressures will cause significant variations in p_2.

The possible use of the p_2 dissipation to establish some sort of curve to assess the coefficient of horizontal consolidation, c_h, seems to have potential. The dissipation curves are repeatable for a given test procedure. However, in the stiff soils it is suggested that p_0, p_2 tests be used to avoid inaccuracies in p_0, p_1, p_2 dissipation test results due to varying amounts of recovery with the time interval between readings. The p_2/p_0 ratio has the potential to aid soil identification.

5 ACKNOWLEDGEMENTS

The work described in this paper forms part of the research programme of the Building Research Establishment and is published by permission of the Director of BRE. The authors wish to thank their colleagues who have helped in this work.

6 REFERENCES

Aas, G., Lacasse, S., Lunne, T. and Madshus, C.(1984):"In-situ testing: New developments". Nordisk, Geotekniker note, Linhoping, Vol II pp 705-716.

Campanella, R.G. and Robertson, P.K.(1983): "Flat plate dilatometer testing; research and development". Proc FICFPD, Edmonton, pp 69-112.

Campanella, R.G., Robertson, P.K., Gillespie, D.G. and Greig, J.(1985): "Recent developments in in-situ testing of soils". XIth ICSMFE, San Francisco, pp 849-854.

Coope, M.R. (1987):"The axial capacity of driven piles in clay". Submitted for D. Phil. degree, Univ. of Oxford, England.

FICFPD (1983):"First International conference on the flat plate dilatometer". Edmonton, Alberta, Canada, February 4th 1983.

Lacasse, S. and Lunne, T.(1982): "Dilatometer tests in soft marine clays" Updating surface sampling of soils and

rocks and their in-situ testing". Santa Barbara, pp 369-379.
Lacasse, S. and Lunne, T.(1986): "Dilatometer tests in sand". Proc ASCE Spec. Conf. "IN-SITU 86" Blacksburg, pp 686-699.
Lutenegger, A.J. and Timian, D.A.(1986): "Flat-plate penetrometer Tests in marine clays. Proc. 39th. Can. Geot. Conf. on "Insitu testing and field behaviour". Ottawa, Canada, pp 301-309.
Marchetti, S.(1980):"In-situ tests by flat dilatometer". ASCE, J. GED, Vol 106, No G.T.3 pp 299-321.
Marchetti, S. and Crapps, D.K.(1981): "Flat dilatometer manual". Schmertmann and Crapps INC., Gainsville, Florida.
Powell, J.J.M. and Uglow, I.M.(1986): "Dilatometer testing in stiff overconsolidated clays". Proc. 39th. Can. Geot. Conf. on "Insitu testing and field behaviour". Carleton Univ., Ottawa, Canada. pp 317-326.
Robertson, P.K., Campanella, R.G., Gillespie, D. and By,T.(1988): "Excess pore pressures and the DMT". Proc. ISOPT.
Tedd, P. and Charles, J.A.(1983): "Evaluation of push-in pressure cell results in stiff clay". Proc. Int. Symp. on Soil and Rock Investigations by in-situ testing. Paris, Vol. 2, pp 579-584.

Penetration Testing 1988, ISOPT-1, De Ruiter (ed.)

Acoustic emission in a quarry

M.C.Reymond
ER Mécanique des Surfaces CNRS-LCPC, Paris, France

ABSTRACT: This paper describes measurements for one year of the deformation behaviour and acoustic emission in an underground quarry loaded at the ground surface. The pillar deformation is accompanied by A.E. as a result of the formation and growth of cracks. The results of acoustic emission and deformation measurements both indicate two mecanisms of failure.

1 INTRODUCTION

The changes in rocks in quarries can be established by acoustic emission (Reymond I983) . One use of this method is to investigate the state of .underground quarries with an increase in rock pressure resulting from an loading by ambankments on the surface (Figure 1)
These experiments were carried out in a limestone quarry with the co-operation of the Laboratoire des Ponts et Chaussées (Morat, Rochet,Thorin I985).This paper describes the instrumentation techniques and microseismic results current.

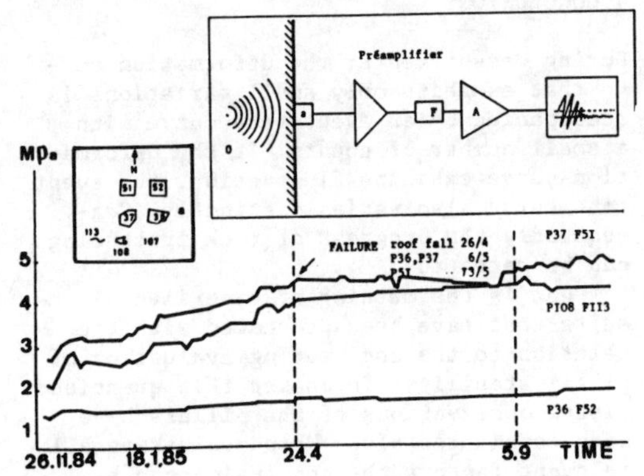

Fig.I Variations of the stress in pillars P_{36}, P_{I08}, P_{37} during I985.
a) first period of increasing load
b) period of constant load
c) period of second increasing loading

2 INSTRUMENTATION

The system (Sté Sofratest) was used for monitoring and recording the acoustic emissions. Acoustic emissions signals were de tected by accelerometers attached to the one of the pillars in the quarry and data were recorded after suitable filtering and amplification. The accelerometers used in the study had the following specifications: sensitivity, 60 mV/g; and frequency response, 2 to I6 kHz. Most acoustic emission events observed during the investigation had acceleration levels of the order of 0.01 g. The AE rate was monitored under conditions of increasing load and with a constant loading. Jacks and extensometers fixed to the pillars provided strain and deformation data. The monitoring station is about I00 m away from the pillars.

3 RESULTS

The AE pattern was obtained from two loading phases:

1 with increasing rock pressure the first phase of AE was characterized by a high rate of AE corresponding to fracturing of the roof rock (Figure 2)

2 The second phase recorded during constant load was characterized by a mean activity for a long time, followed by a peak and a decrease in the activity, corresponding to visual observation of progressive failure in the pilars (Figure 3)
There is a good correlation between the stress strain characteristics and the rate

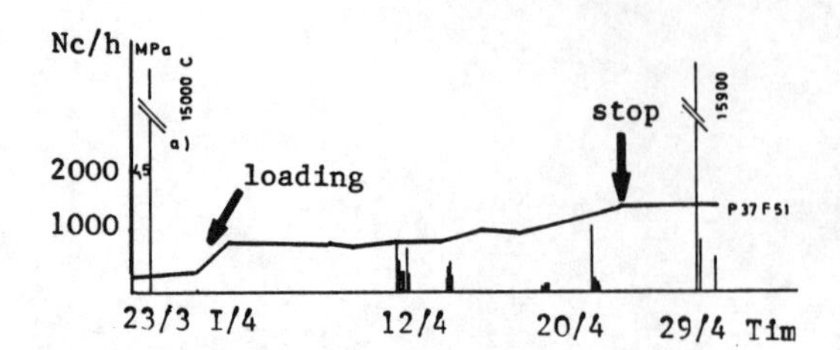

Fig.2 Number of counts per hour registered during loading april I985.

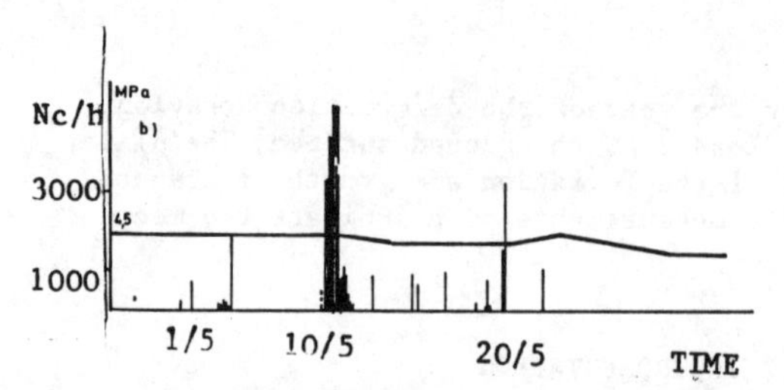

Fig.3 Number of counts per hour with constant loading during May I985.

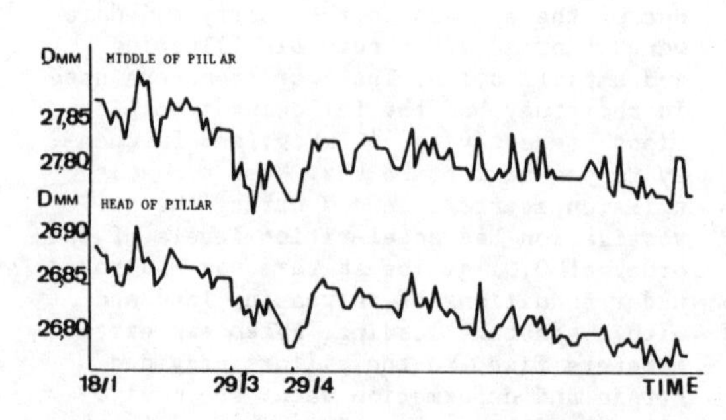

Fig.4 Stress-Strain data from Pillar 36 on which is attached the accelerometer.29March 29 April correspond to a high rate of A.E.

of AE during the loading history of the calcareous quarry.(Figure 4)
When considering the relationship between Pillars deformation (compression) and time,(Figure 4), the deformations measurements decrease significantly over a long time (29/3 - 29/4).

Pillars deformation occurs by a creep process fracturation of the roof rock. During the same time period, the acoustic emission event rate is characterized by a very important unmber of counts

29 March, 15.000 counts
29 April, 15.900 counts

In contrast to the event rate during May I985 (Figure 3) the number of counts is not very high (I0 May, 5200 counts) corresponding to a progressive failure in the pillar.

The deformation shows a long term linear time dependence. In addition measuring point was chosen to provide an example of the compression process.

A correlation coefficient calculation for P36, P37, P I08 during a long period resulted in Table I

Table I Correlation coefficients of deformation curves.

PILLARS	PHASES OF LOAD	r
P36 Face I07	Increasing load	- .89
P37 Face 5I	" "	- .6I
P37 Face I08	" "	- .70
PI08 Face II3	" "	- .77
PI08 Face I07	" "	- .86
PI08 Face I07	Constant load	- .54
PI08 Face I07	2nd increasing load	- .64

4 CONCLUSIONS

During measurements, the deformation curve that exhibits only small variations is accompanied by an event rate curve with a small number of counts. If the deformation curve exhibits fluctuation, the event rate curve also varies distinctly. Consequently the progress of rock fracturing can be detected.

What is the meaning of experimental results that have been presented with the relation to the engineering evaluation of pillar stability? To answer this question, visual observations of the pillars have been considered also. When evaluating all relevant factors the conclusion can be drawn for the present that the underground quarry is stable.

Determination of the state of the pillars and conclusions concerning its stability have a basis in the results of pillars deformation measurements and in the monitoring of its acoustic emissions. The combination of both methods provided logical approach to quarry stability surveillance.

REFERENCES

Morat, P., Rochet,L., Thorin, R.(1985) Comportement d'une carrière souterraine (abandonnée) de calcaire grossier soumise à une surcharge.Site de Villiers Adam. Journées de Versailles, Inspection Générale des carrières.

Reymond, M.C. (1983) Comportement des carrières après tirs de mines.Bull.de l'Ass. Intern. de Géologie de L'ingénieur, n° 26,27.

bility have a basis in the [illegible] of il-
[illegible] deformation measurements [illegible]
[illegible]
combination of both methods provided [illegible]
[illegible]
tions.

REFERENCES

[illegible] (1985) [illegible]

[illegible] (1978) [illegible]

Penetration Testing 1988, ISOPT-1, De Ruiter (ed.)
© 1988 Balkema, Rotterdam, ISBN 90 6191 801 4

Excess pore pressures and the flat dilatometer test

P.K.Robertson, R.G.Campanella & D.Gillespie
University of British Columbia, Civil Engineering Department, Vancouver, Canada
T.By
Norwegian Geotechnical Institute, Oslo

ABSTRACT: Data is presented comparing pore pressures measured during penetration with UBC's and NGI's specially designed flat dilatometers and a piezocone. Representative data is presented for sands and soft clays and shows that the pore pressures in a DMT during penetration are similar to those measured behind the friction sleeve in a CPTU. Data is also presented to show that the closing pressure using a standard Marchetti DMT is very similar to the penetration pore pressures in sands and soft clays. Dissipation data during a pause in penetration with the DMT and CPTU is presented and discussed. A tentative procedure to estimate c_h using a standard DMT is proposed.

KEY WORDS: Flat dilatometer, pore pressures, CPTU, dissipation.

INTRODUCTION

The flat dilatometer test (DMT) was first introduced in 1980 (Marchetti, 1980) and has become a simple, cost-effective in situ test. The dilatometer is a flat plate 14 mm thick, 95 mm wide by 220 mm in length. A flexible stainless steel membrane 60 mm in diameter is located on one face of the blade. Beneath the membrane is a measuring device which turns a buzzer off in the control box at the surface when the membrane starts to lift off the sensing disc and turns a buzzer on again after a deflection of 1 mm at the centre of the membrane. Readings are made every 20 cm in depth. The membrane is inflated using high pressure nitrogen gas supplied by a tube pre-threaded through the rods. As the membrane is inflated, the pressures required to just lift the membrane off the sensing disc (reading A), and to cause 1 mm deflection at the centre of the membrane (reading B), are recorded. Readings are made from a pressure gauge in the control box and entered on a standard data form. However, one disadvantage with the test was the fact that no pore pressure measurements were made. Several specially designed flat dilatometers have been developed that incorporate pore pressure elements (Davidson and Boghrat, 1982; Campanella et al., 1985; Lunne et al., 1987). However, these devices detract somewhat from the primary advantage of the standard Marchetti device, that is, simplicity.

In 1985 Campanella et al., (1985) presented data from a sophisticated research dilatometer that showed the following main points:

- In soft clays the basic DMT data (P_o, P_1) is dominated by large penetration pore pressures.
- The pressure when the membrane returns to the closed position is the same as the penetration pore pressure.
- In clean sands no excess pore pressures are generated during penetration. Therefore, the closing pressure is approximately equal to the static equilibrium pore pressure (u_o).

Examples of the results are shown in Figure 1.

This paper will present data comparing penetration pore pressures using the University of British Columbia (UBC) research DMT, the Norwegian Geotechnical Institute (NGI) offshore DMT and the piezometer cone penetration test (CPTU). DMT penetration pore pressures will also be compared to the closing pressures using the standard Marchetti DMT.

EQUIPMENT AND FIELD TESTING

Research at UBC and NGI involving the DMT has included the development and use of research flat dilatometers. The UBC research dilatometer is identical in size, shape and operation as the Marchetti design except for the following passive measurements:

1. pore water pressure at the center of the moving membrane,
2. deflection at the center of the membrane,
3. gas pressure activating the membrane,
4. verticality of the dilatometer blade during penetration, and,
5. the penetration force for the blade of the dilatometer.

Typical results from the UBC research dilatometer for tests in a clean sand and soft clay deposit are shown in Figure 1.

In order to fit inside the standard offshore drill string NGI's offshore dilatometer has a slightly different size (width = 77 mm, thickness = 16 mm) and shape compared with the Marchetti design (Lunne et al., 1987). As with the UBC dilatometer, the pore pressure is measured, but the filter is located on the reverse side of the blade, directly opposite the center of the membrane. The operation pressure (oil) is generated by means of a piston pump and the movement of the piston is achieved by a D.C. electric motor. The pressure is recorded continuously in addition to P_o and P_1. Calibration tests at NGI's research sites in Norway and at UBC's research sites in Canada have shown that the offshore dilatometer gives results very similar to the standard Marchetti device.

The data presented in this paper was obtained at two sites near Vancouver, B.C.. One site is the McDonald Farm site which is located in the Fraser River delta. The site consists of relatively clean sand from a depth of about 2 m to 15 m underlain by an approximately normally consolidated clayey silt deposit.

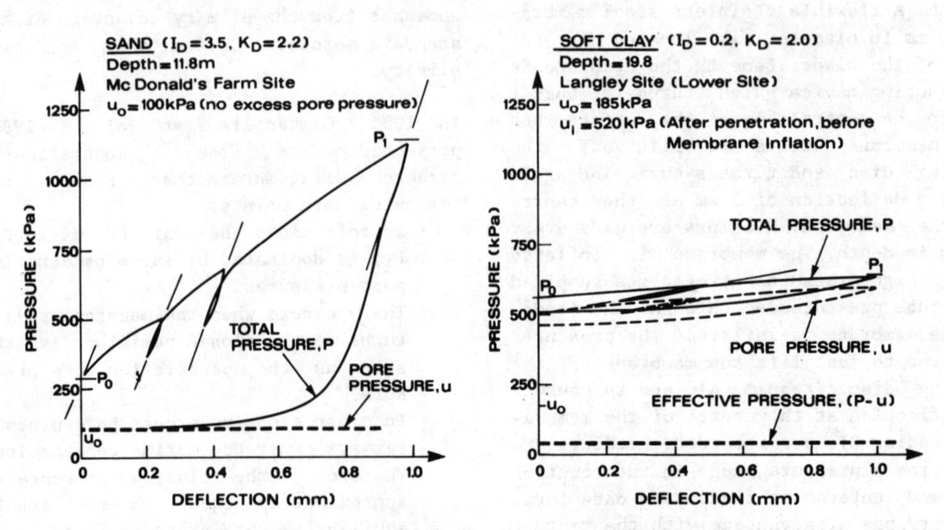

Fig.1 Typical results from UBC research DMT (After Campanella et al., 1985)

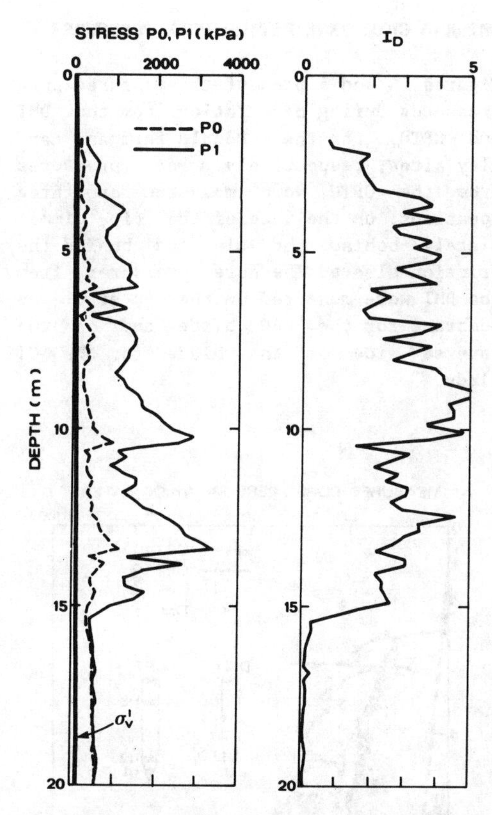

Fig.2 Basic DMT data for McDonald Farm Site obtained with NGI offshore DMT

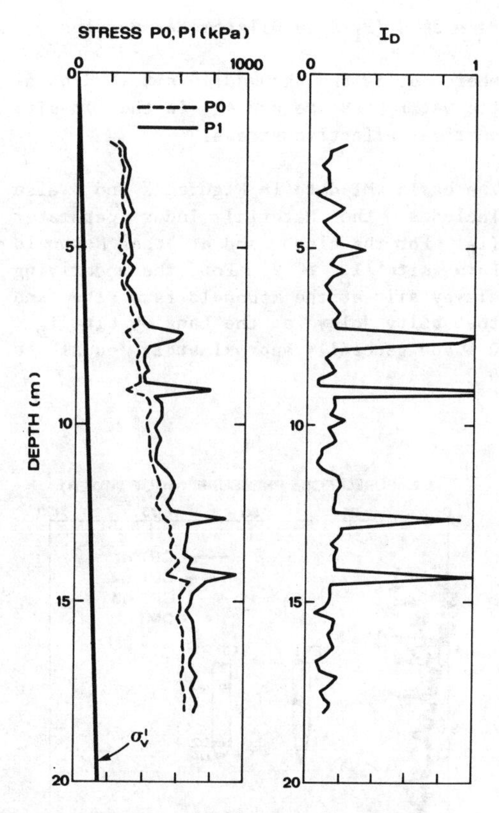

Fig.3 Basic DMT data for Langley Site obtained with NGI offshore DMT

Table 1. Summary of Soil Index Properties for Fine Grained Soils

Site	w %	w_L %	PI %	s_t	OCR
McDonald Farm	23-40	25-42	3-20	2-7	1-2
Langley	27-53	32-59	16-34	7-10	1-3

w - Water Content
w_L - Limit Liquid
PI - Plasticity Index
s_t - Sensitivity
OCR - Overconsolidation Ratio

Basic DMT data for the McDonald Farm site is shown in Figure 2. Full details of the site are given by Campanella et al., (1983).
The second site is near Langley, B.C. and consists of lightly overconsolidated glaciomarine silty clay interbedded with thin lenses of silt and sand. Basic DMT data for the Langley site is shown in Figure 3. A summary of the index properties of the fine grained soils is given in Table 1.
The dilatometers were pushed into the ground at a constant rate of penetration of 2 cm/sec. Before and after each sounding the dilatometers were calibrated for membrane stiffness.
The basic dilatometer data (readings A and B) are corrected for offset in the measuring gauge and for membrane stiffness (P_o, P_1).

Using P_o and P_1 the following three index parameters were proposed by Marchetti:

$$I_d = \frac{(P_1 - P_o)}{P_o - u_o} \quad \text{- Material Index}$$

$$K_d = \frac{P_o - u_o}{\sigma'_{vo}} \quad \text{- Horizontal Stress Index}$$

$E_d = 34.6 (P_1 - P_o)$ = Dilatometer Modulus

where u_o is the assumed in-situ hydrostatic water pressure and σ'_{vo} is the in-situ vertical effective stress.

The basic DMT data in Figures 2 and 3 also includes the Material Index parameter (I_D). For the clean sand at the McDonald Farm site $I_D > 2$. For the underlying clayey silt at the McDonald farm site and the silty clay at the Langley site $I_D < 0.6$ and generally approximately equal to 0.2.

DMT AND CPTU PENETRATION PORE PRESSURES

Figures 4 and 5 present the measured pore pressures during penetration from the DMT and CPTU, for the McDonald Farm and Langley sites, respectively. Pore pressures from the CPTU were measured at three locations; on the face of the tip, immediately behind the tip and behind the friction sleeve. The pore pressures from the DMT were measured on the center of the membrane for the UBC blade and on the reverse side of the blade for the NGI blade.

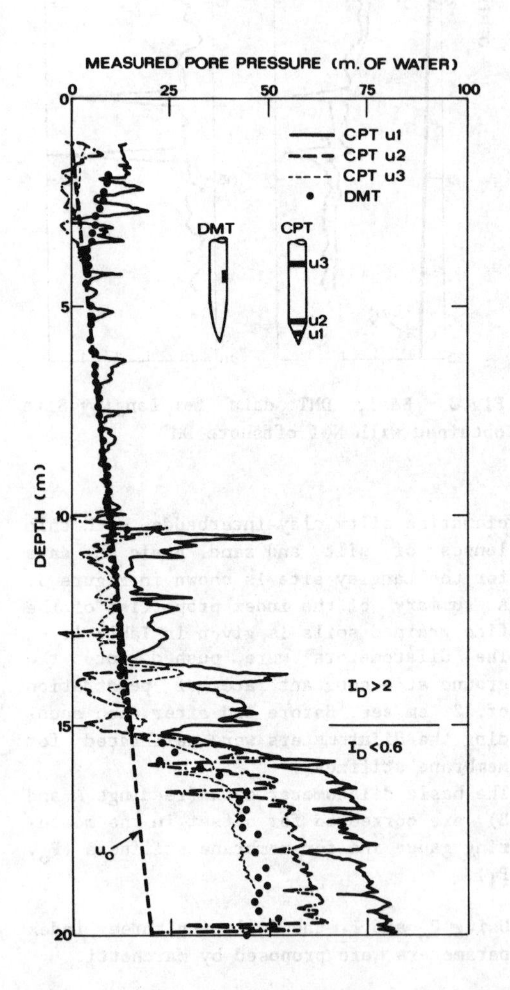

Fig.4 Penetration pore pressures from NGI offshore DMT and CPTU at McDonald Farm Site

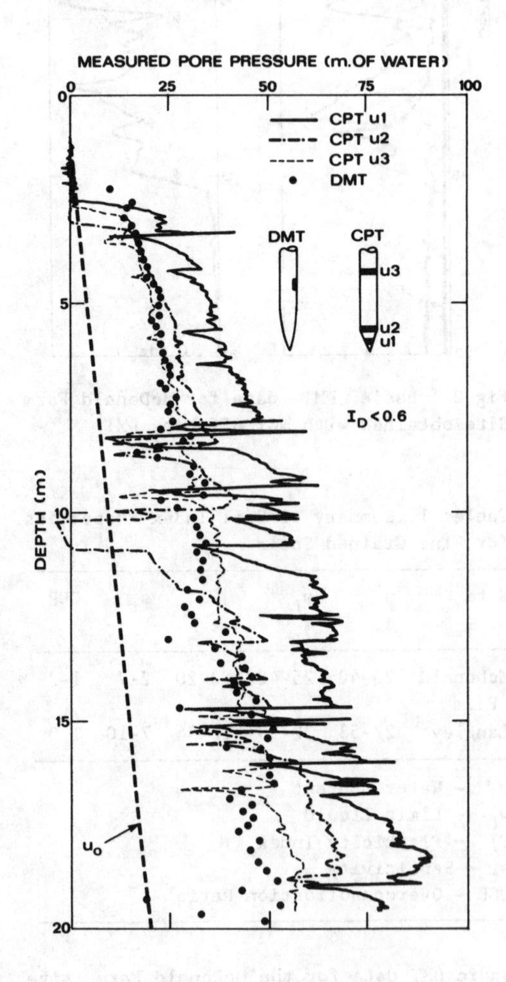

Fig.5 Penetration pore pressures from UBC research DMT and CPTU at Langley site

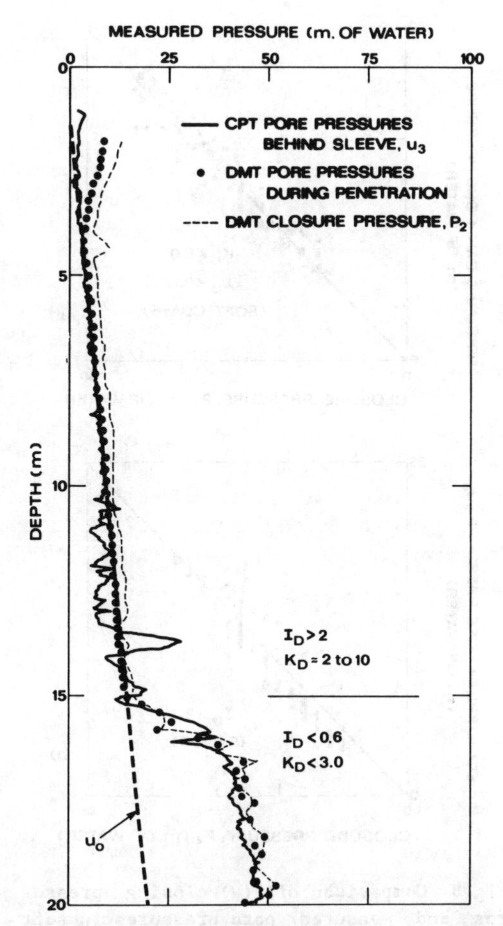

Fig.6 Comparison between penetration pore pressures from DMT and CPTU and closing pressures P_2 from NGI offshore DMT at McDonald Farm Site

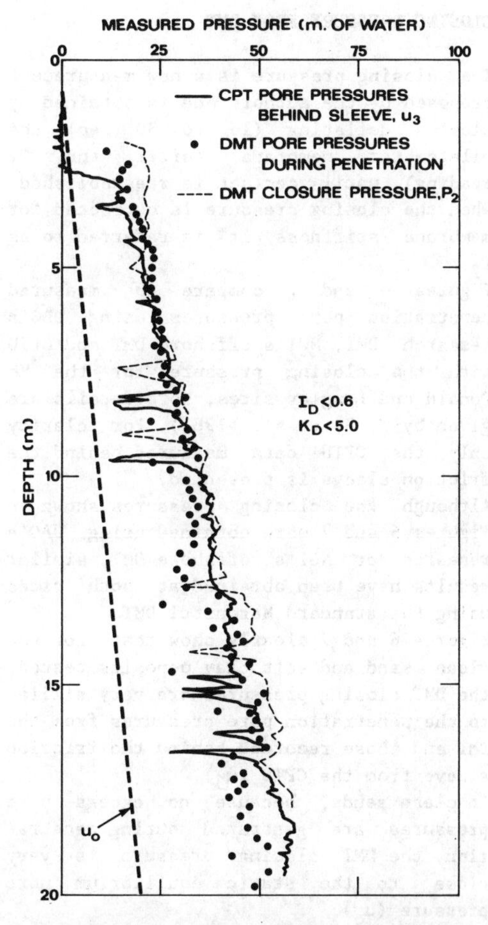

Fig.7 Comparison between penetration pore pressures from DMT and CPTU and closing pressures P_2 from UBC research and NGI offshore DMT at Langley site

Figures 4 and 5 clearly show that the DMT penetration pore pressures are very similar to the pore pressures measured behind the friction sleeve in the CPTU.
In the clean sand deposit (2 m to 15 m) at the McDonald Farm site no excess pore pressures were generated during the DMT penetration. In the clayey silt and silty clay deposits large excess pore pressures were generated during penetration in both the DMT and CPTU. The difference in pore pressures measured at different locations on the cone is consistent with results recorded elsewhere (Campanella et al., 1985; Jamiolkowski et al., 1985; Lunne et al., 1986). The results from the Langley site are complicated by the interlayering of the silt and sand lenses.
These results confirm the previous findings (Campanella et al., 1985) that in soft clays ($I_D \leq 0.6$; $K_D \leq 5.0$), the basic DMT data (P_o, P_1) is dominated by large penetration pore pressures. These pore pressures are influenced by the soils undrained shear strength, stress history and stiffness and macrofabric. Therefore, it is not surprising that the basic DMT data can be related to undrained shear strength and stress history for some soils. But that any correlation relating them will not be unique for all soils.

CLOSING PRESSURE FROM DMT

The closing pressure is a new measurement proposed by the authors and is obtained by slowly deflating (15 to 30 sec) the dilatometer membrane (after the P_1 reading) until contact is reestablished. When the closing pressure is corrected for membrane stiffness it is referred to as P_2.

Figures 6 and 7 compare the measured penetration pore pressures using UBC's research DMT, NGI's offshore DMT and CPTU with the closing pressure for the Mc Donald and Langley sites. More details are given by T. By et al. (1987). For clarity only the CPTU data measured behind the friction sleeve is presented.

Although the closing pressures shown in Figures 6 and 7 were obtained using UBC's research or NGI's offshore DMT, similar results have been obtained at both sites using the standard Marchetti DMT.

Figures 6 and 7 clearly show that, for the clean sand and soft clay deposits tested, the DMT closing pressures are very similar to the penetration pore pressures from the DMT and those recorded behind the friction sleeve from the CPTU (u_3).

In clean sands, because no excess pore pressures are generated during penetration, the DMT closing pressure is very close to the static equilibrium pore pressure (u_o).

It is interesting to note from Figure 7 that the DMT closing pressure also records low values within the silt or sand lenses. There is clearly the potential to reinforce or possibly improve soil classification with the standard DMT if the additional closing pressures are also recorded.

Figure 8a presents a summary of DMT penetration pore pressures and closing pressures for soils where $I_D \leq 0.6$.

Since the soils tested where generally soft, the relationship shown in Figure 8a may apply only to soils where the DMT horizontal stress index $K_D \leq 5.0$.

Figure 8b presents a summary of the hydrostatic pore pressures and DMT closing pressure for soils where $I_D > 2$, (i.e. sands).

The relationships shown in Figure 8 are remarkably consistent and indicate the potential usefulness of recording the closing pressure (P_2) in standard DMT.

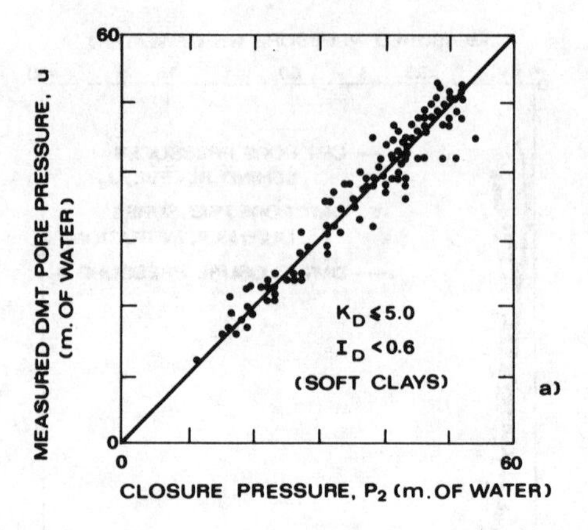

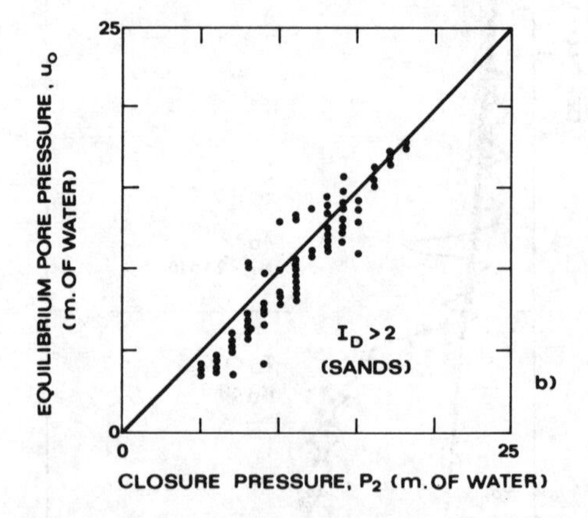

Fig.8 Comparison of: (a) closing pressures and measured pore pressures in soft clay, and, (b) closing pressures and equilibrium static pore pressures in sand

DISSIPATION RATES

Figure 9 presents a typical example of the dissipation of excess pore pressures during a stop in penetration for the research DMT and CPTU. The example shown is from the McDonald Farm site at a depth of 20 m. The CPTU dissipation data is for the pore pressure element located immediately behind the tip.

Figure 9 shows that, when the dissipation data is normalized with respect to the equilibrium static pore pressure (u_o), the rate of dissipation for pore pressures around the DMT is slower than that around a 10 cm^2 cone.

At the McDonald Farm site at a depth of 20 m the times for 50% dissipation are:

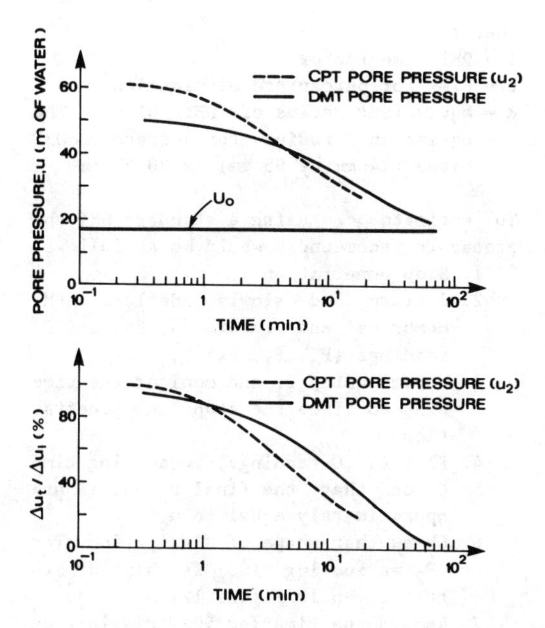

Fig.9 Comparison of pore pressure dissipations using UBC research DMT and 10 cm^2 CPTU at McDonald Farm site, Depth=20 m

DMT t_{50} = 9.5 mins
CPTU (10 cm^2) t_{50} = 4.5 mins

The time for 50% dissipation is therefore approximately twice as fast for the CPTU (10 cm^2) as it is for the DMT.

The slower rate of dissipation is probably related to the approximate 2-D shape of the flat dilatometer blade.

Marchetti et al. (1986) suggest the possibility of recording the DMT A-reading with time during a stop in penetration, and plotting the A-reading versus log time similar to a dissipation test. This type of test was performed using a standard DMT at a depth of about 20 m at the McDonald Farm Site, and the results are shown in Figure 10.

The contact pressure, A-reading, when corrected for membrane stiffness is the lift-off stress, P_o. The difference between the P_o and the penetration pore pressure (u) is the effective stress acting on the membrane, as shown in Figure 10. In soft clays the effective stress immediately after penetration ($\bar{\sigma}_{hi}$) is very low. As the excess pore pressures dissipate the effective stresses increase. Therefore, immediately after the stop in penetration the A-reading is slightly higher than the pore pressure (DMT-u). At the end of dissipation when all the excess pore pressures have dissipated the effective stress has increased ($\bar{\sigma}_{hc}$) and the difference between the A-reading and u_o is larger than at the beginning of dissipation.

Marchetti et al. (1986) suggested using this technique to estimate the effective stress acting on a driven displacement pile after consolidation ($\bar{\sigma}_{hc}$).

Since the closing pressure (P_2) closely represents the pore pressure on the DMT membrane (see Figures 6 and 7) it should be possible to record the C-reading with time and obtain a dissipation curve from a standard Marchetti DMT.

Figure 11 shows the results of repeated A, B and C-readings (P_o, P_1, P_2) obtained using a standard DMT at a depth of 22 m at the McDonald Farm site. For comparison the pore pressures recorded using the research DMT at a depth of 20 m are also shown in Figure 11. For the research DMT pore pressure dissipation no membrane expansion was performed. Figure 11 shows that the P_2 readings follow a very similar dissipation curve to that of the actual DMT pore pressures. Also interesting is the fact that the P_o value (A-reading) decreases in a different manner if the membrane is

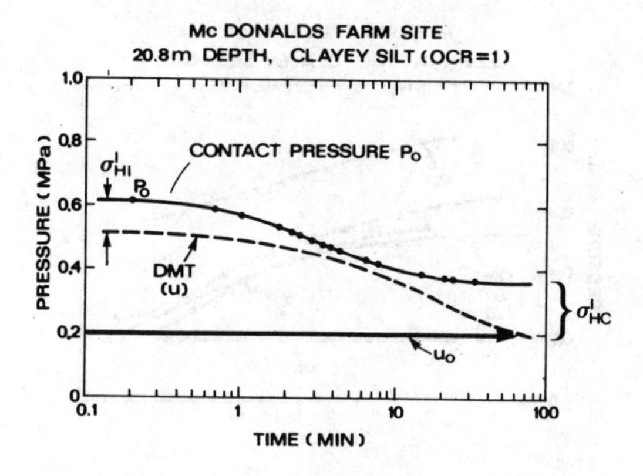

Fig.10 Repeated A-readings (corrected for membrane stiffness, P_o) using standard DMT compared with DMT pore pressure dissipation (McDonald Farm Site)

expanded to the full 1 mm deflection (i.e. B-reading). Whereas, in Figure 10, the P_o reading was performed with no expansion. This difference in response is probably related to the changes in effective and total stress around the membrane during dissipation. The increases in effective stresses are reflected in the progressive increase between P_o and P_1 as dissipation occurs (i.e. an apparent increase in I_D).
Because of the similarity between the dissipation curves of the DMT pore pressures and the C-readings, it should be possible to estimate the coefficient of consolidation, c_h, using a standard DMT and repeating the A, B and C-readings (P_o, P_1, P_2) with time. The resulting plot of P_2 versus log time should be similar to the dissipation of excess pore pressures around the dilatometer membrane. The final P_2 value after complete dissipation also represents a measure of the equilibrium pore pressure, u_o.
By comparing the shape of the DMT pore pressure dissipation curve, shown in Figure 9, with the theoretical curves for CPTU (Torstensson, 1977, cylindrical solution) it is possible to produce empirical dissipation curves for DMT data, as shown on Figure 12. The time factor, T, in Figure 12 is taken to be:

$$T = \frac{c_h \cdot t}{R^2} \qquad \ldots (1)$$

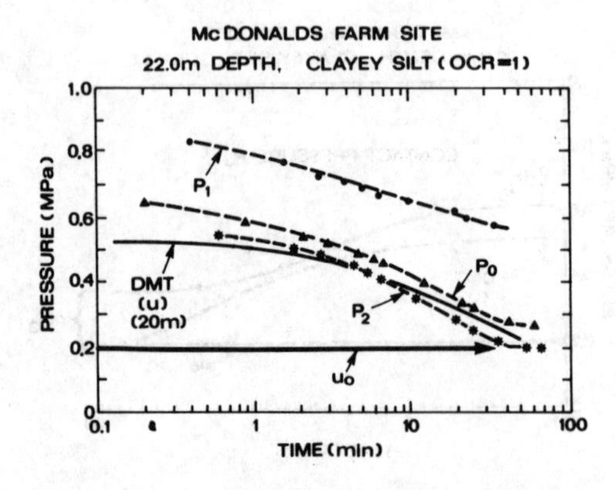

Fig.11 Repeated A, B and C-readings (corrected for membrane stiffness P_o, P_1, P_2) using standard DMT compared with DMT pore pressure dissipation (McDonald Farm site)

where:
T - DMT time factor
t - time for percentage dissipation
R - equivalent radius of DMT blade. The equivalent radius for a standard DMT blade (14 mm by 95 mm) is 20.57 mm.

To estimate c_h using a standard DMT the procedure recommended would be as follows;

1. Stop penetration
2. Inflate and slowly deflate DMT membrane and record A, B,, and C-readings (P_o, P_1, P_2)
3. Repeat step 2. and monitor the time elapsed since the stop in penetration.
4. Plot P_2 (C-readings) versus log time
5. Check that the final P_2 values are approximately equal to u_o
6. Check that shape of dissipation plot (P_2 versus log time) is similar to that given in Figure 12.
7. Record the time for 50% dissipation (t_{50}).
8. Use equation (1) above to estimate c_h, use $T_{50} \approx 4$, and R - 20.57 mm.

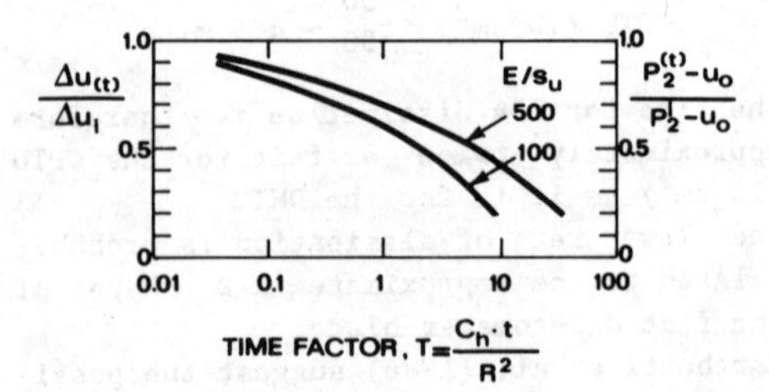

u_o = EQUILIBRIUM PORE PRESSURE
$\Delta u_{(t)}$ = EXCESS PORE PRESSURE AT TIME t
Δu_i = INITIAL EXCESS PORE PRESSURE
R = EQUIVALENT RADIUS (20.57mm)
C_h = COEFFICIENT OF CONSOLIDATION
E = YOUNG'S MODULUS
s_u = UNDRAINED SHEAR STRENGTH
$P_2^{(t)}$ = P_2 VALUE AT TIME t
$P_2^{(i)}$ = INITIAL P_2 VALUE (t=0)

Fig.12 Tentative empirical Time Factors for Dissipation test using DMT

The proposed method given above is preliminary and requires further field verification. It is probable that the above approach is only valid for soft normally to lightly overconsolidated clays (i.e. $I_D \leq 0.6$ and $K_D \leq 5.0$).

The major advantages in using the DMT C-reading (P_2) approach to obtain dissipation data are:

- no problems related to saturation of porous elements, as in CPTU,
- P_2 values closely resemble pore pressure values, when $I_D \leq 0.6$ and $K_D \leq 5.0$.,
- P_2 values tend towards equilibrium pore pressure (u_o), therefore, providing important additional hydro-geologic data.

The observation that P_2 values tend towards the equilibrium pore pressure (u_o) as dissipation proceeds can be useful for dissipation tests in low permeability clays. Dissipation can be stopped when the P_2 values have decreased by 50% based on an estimate of u_o.

This can significantly improve the time required for DMT dissipation tests.

SUMMARY

Data has been presented comparing penetration pore pressures using the UBC research DMT, the NGI offshore DMT and CPTU.

Results show that in clean sands and soft clays the DMT penetration pore pressure is very similar to the pore pressures measured behind the friction sleeve of the CPTU.

The closing pressure (P_2) from a standard Marchetti DMT is very similar to the DMT penetration pore pressure for clean sands and soft clays. In clean sands no excess pore pressures are generated and the closing pressure represents a good approximation of the equilibrium piezometric pressure (u_o).

The dissipation of excess pore pressures during a stop in penetration is slower for the DMT than for the CPTU. Time for 50% dissipation for the DMT is approximately twice that of a 10 cm^2 cone.

If the DMT inflation is repeated at one depth in a soft clay the plot of P_2 versus log time is almost the same as the dissipation of the DMT pore pressures. A tentative procedure to estimate c_h from the repeated C-readings using a standard Marchetti DMT is proposed. However, this procedure is tentative and requires further field verification. The procedure, at present, is only recommended for tests in soft clays, where $I_D \leq 0.6$ and $K_D \leq 5.0$.

ACKNOWLEDGEMENTS

The assistance of the Natural Sciences and Engineering Resarch Council of Canada and the technical staff of the Civil Engineering Department, University of British Columbia is much appreciated. The valuable work of C. Tsang a graduate student at UBC is also appreciated. We also appreciate the grant from the Norwegian Geotechnical Institute which made it possible for T. By to spend 4 weeks at UBC.

REFERENCES

By, T., D. Gillespie, T. Lunne and R.G. Campanella (1987). "Comparison of UBC's Research Dilatometer, NGI's Offshore Dilatometer and Marchetti's Standard Dilatometer". The University of British Columbia and Norwegian Geotechnical Institute NGI, Report 52157.

Campanella, R.G., Robertson P.K., Gillespie D. and Greig J. (1985). "Recent Developments In-Situ Testing in Soils", Proceedings of XI ICSMFE San Francisco.

Campanella, R.G., Robertson P.K. and Gillespie D. (1983). "Cone Penetration Testing in Deltaic Soils", Canadian Geotechnical Journal, Vol.20, No.1.

Davidson, J.H. and Boghrat A.G. (1983). "Flat Dilatometer Testing in Florida", Proceedings of the International Symposium on Soil and Rock Investigation by In-Situ Testing, Paris, Vol.II.

Jamiolkowski, M., Ladd C.C, Germaine J.T. and Lancellotta R. (1985). "New Developments in Field and Laboratory Testing of Soils", State-of-the-Art Paper at 11th International Conference SMFE (ICSMFE), San Francisco.

Lunne, T., Eidsmoen T., Gillespie D. and Howland J.D. (1986). "Laboratory and Field Evaluation of Cone Penetrometers", Proceedings of In-Situ 86, Specialty Conference, ASCE, Blacksburg, Virginia.

Lunne, T., R. Johnsond, T. Eidsmoen and S. Lacasse (1987). "The Offshore Dilatometer". Paper submitted to the International Symposium on Offshore Engineering, Brazil Offshore 87. 24-28 August 1987, Rio de Janeiro.

Marchetti, S. (1980). "In-Situ Tests by Flat Dilatometer", ASCE Journal of Geotechnical Eng. Div., No.106, GT3.

Marchetti, S., Totani G., Campanella R.G., Robertson P.K. and Taddei B. (1986). "The DMT-σ_{hc} Method for Piles Driven in Clay", Proceedings on In-Situ 86, Specialty Conference, ASCE, Blacksburg, Virginia.

Torstensson, B.A., (1977). "The Pore Pressure Probe", Nordiske Geotekniske Møte, Oslo, Paper No.34.1-34.15.

Penetration Testing 1988, ISOPT-1, De Ruiter (ed.)
© 1988 Balkema, Rotterdam, ISBN 90 6191 801 4

Basic interpretation procedures of flat dilatometer tests

Reynaldo Roque*, Nilmar Janbu & Kaare Senneset
Norwegian Institute of Technology, Trondheim
**NTNF Postdoctoral Fellow*

ABSTRACT: Dilatometer tests were performed in five natural soil deposits of known characteristics to evaluate the potential of the dilatometer and existing correlations for use in Norwegian soils. The soils tested ranged from soft to medium-stiff clay and clayey silt with sensitivities ranging from low to medium high (St = 7 to 20). Non-conventional dilatometer tests were also carried out at selected depths to evaluate the potential of the dilatometer for determining basic stress-strain-time response parameters of in situ soils. Theoretical analysis procedures were developed and/or identified for this purpose. The measurements and analyses were also used to evaluate the effects of dilatometer insertion and testing on the state and the response of the soils tested. Pore pressure dissipation tests performed with the piezocone also aided in this analysis.

This paper briefly describes the test and theoretical calculation procedures applied. Because of space limitations, detailed test results are presented for one site only, and the theoretical analysis procedures are presented with little background or discussion. A comparison of soil parameters from different interpretation methods with laboratory test results is presented for this one site. Partial test results are presented for a second site to illustrate the effect of penetration on different soil types. Derivations and discussions of the theoretical analysis procedures as well as complete results for the other four sites will be found in the internal report for this project (Roque, 1987, Mokkelbost, 1984).

1. INTRODUCTION

Insertion of the dilatometer blade results in a soil deformation immediately adjacent to the blade equal to one-half the blade thickness (i.e. approximately 7 mm). In saturated soils, this horizontal deformation results in excess pore pressures and effective stress changes. Depending on soil type and depth of testing, this 7 mm deformation and corresponding stress changes may result in anything from near elastic response to bearing capacity failure. Shear stresses and rotation of principal planes during insertion and possible liquefaction of collapsible of highly sensitive soils may cause additional disturbance. Thus, measurements during membrane expansion may be obtained in a soil state which may either be very similar or have very little resemblance to the original in situ soil condition.

Since there is presently no way to evaluate or account for these effects using conventional dilatometer test measurements, empirical correlations have been developed between membrane expansion pressures and various soil parameters. These correlations have been shown to give satisfactory results in a variety of soil types within given geological regions.

However, there are always some uncertainties and even possible dangers involved with using empirical relationships, and extrapolation to different soil types and/or geological regions may be difficult, if not impossible. Also, since classical concepts are not used to obtain the parameters, it may be difficult for practicing engineers to get a feel for how soil parameters were determined, and thus make an assessment of their significance for different problems.

This investigation was carried out in an attempt to determine the following:

1. To evaluate the usefulness of the dilatometer and the applicability of exisiting correlations in Norwegian soils.
2. To evaluate the potential of the dilatometer for determining fundamental stress and stress-time response of soils in

TABLE 1: BASIC PROPERTIES AND DESCRIPTION OF PREDOMINANT SOIL AT SITES TESTED BY DMT

LOCATION	SOIL DESCRIPTION	w (%)	γ (kN/m³)	s_u (kPa)	S_t	$a=\frac{c}{\tan\phi'}$ (kPa)	$\tan\phi'$	K_o'	M (MPa)	m	c_v (m²/yr)	σ_c' (kPa)	% CLAY ≤2μ
GLAVA (STJØRDAL)	COARSE OC CLAY LOW-MIDDLE SENSITIVE HOMOGENEOUS	32	19.4	25 to 50	7	15	0.58	0.60	5-10	16-20	13-34	220-500	40
HALSEN (STJØRDAL)	OC CLAYEY-SANDY SILT SENSITIV LAYERED	30	19.0	20 to 70	20	6	0.75	0.45	4-10	20-30	50-3300	150-250	0-20

situ, using strictly theoretical analysis procedures.
3. To develop and/or identify theoretical calculation procedures based on classical geotechnical engineering concepts for determining soil strength, deformation and consolidation parameters from dilatometer measurements.
4. To identify testing procedures and data interpretation methods from which the values necessary to carry out such analyses can be obtained.
5. To evaluate the effect of dilatometer insertion and testing on soil response, especially as it relates to the true significance of the measurements in terms of their validity for determining strength and deformation parameters.

To accomplish these goals, conventional and non-conventional dilatometer tests were carried out in five soil deposits of known characteristics in the area of Trondheim, Norway. The soils ranged from soft to medium stiff clays and clayey silt.

2. DESCRIPTION OF TEST SITES

A brief description and summary of the basic soil properties for the two test sites discussed in this paper is given in Table 1. The table is presented for easy reference and includes only the soil properties which are characteristic of the predominant soil in each deposit. Detailed soil profiles, including routine data, and results of oedometer and triaxial tests performed on undisturbed samples are available for each site and presented in the internal report for this project (Roque, 1987).

3. TEST PROCEDURES

Conventional dilatometer tests were performed at two locations at each of the sites tested. The Norwegian Institute of Technology's tractor-mounted boring rig was used to insert the dilatometer at a constant rate of 2 cm/sec. Dilatometer tests were performed every 20 cm following the procedures suggested in the User's Manual as closely as possible.

For several depths at each of the sites tested, the dilatometer blade was not advanced immediately after the conventional test was performed. Instead, dilatometer tests were performed at specified time intervals after insertion, so that a record of p_0 and p_1 was obtained as a function of time, see Fig.1. Long-term (>16 hrs.) measurements were obtained at one depth for the Glava and Halsen sites. A limited number of tests were also performed to evaluate the effect of repeated expansion on the measurements, but these are not presented herein. These tests were first performed in 1985 as part of a diploma thesis, (Stordahl, 1985).

Pore pressure dissipation tests were performed with the piezocone at both Glava and Halsen. The filter for the piezocone used was located immediately behind the cone. The cone was driven at the same constant rate as the dilatometer (2 cm/sec).

4. CONVENTIONAL INTERPRETATION

Soil parameters were determined according to existing empirical correlations. No further explanation of these are given here, since these are well documented elsewhere (see Marchetti and Crapps, 1981 and Bullock, 1983). The results shown in Fig.2 for Glava clay are:

1. Contact pressure (p_0) and 1-mm expansion pressure and
2. Dilatometer parameters I_d, K_d, and E_d, as determined from the equations proposed by Marchetti (1980).

Each profile shows two sets of values, corresponding to two parallel holes. Similar profiles were obtained for the other test sites.

Very high values of p_0 and p_1 were measured in the weathered zone (one to three meters depth) of all the test sites. In general, the repeatability was good for all

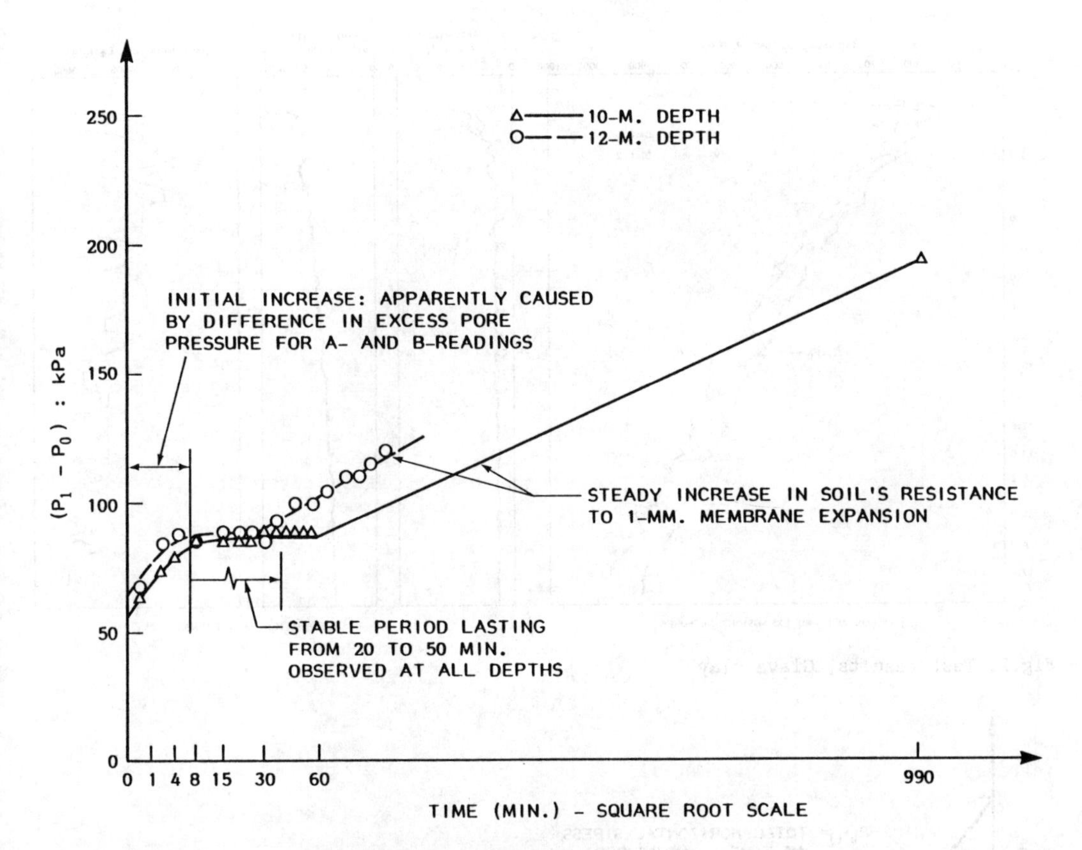

Fig.1. (p_1-p_0) as a function of time: Glava clay

sites tested. The material index (I_d) of all the soils tested was relatively low (sometimes below 0.2), compared to previous values reported by Marchetti (1980). This observation was also made by Lacasse and Lunne (1983) and by Marchetti (1980) for dilatometer tests in the Onsøy and Drammen clay deposits near Oslo, Norway.

5. NEW INTERPRETATION

The basic ideas behind the theoretical procedures developed for use with dilatometer are presented in this section. However, because of space limitations, these procedures are presented with little background or discussion. Detailed derivations and further discussions on these may be found in the internal report for this project (Roque, 1987). This will also be the subject of a future article.

5.1 Pore Pressure and Effective Stresses From New Test Procedures

Two methods are proposed for determining the excess pore pressure using the conventional dilatometer blade (i.e. without a pore pressure transducer.)

Method I: The basic assumption in this method is that the measured reduction in p_0 is only caused by a reduction in excess pore water pressure (i.e. the horizontal effective stress in the soil remains constant). It will later be shown that this appeared to be true when the soil did not liquefy during insertion. For this condition, the stresses acting on the face of the dilatometer at any time, are as shown in Fig.3, and the excess pore pressure at time t can be calculated as follows:

$$\Delta u_{(t)} = p_{o(t)} - p_{of} \quad (1)$$

Furthermore, if the in situ horizontal effective stress can be estimated, the effective stress change resulting from dilatometer insertion can be determined as follows:

$$\Delta\sigma_h' = p_{of} - \sigma_{ho} \quad (2)$$

The symbols used in these equations are defined in Fig.3.

Method II: Schmertmann (1986) and others have proposed a closure reading (U-read-

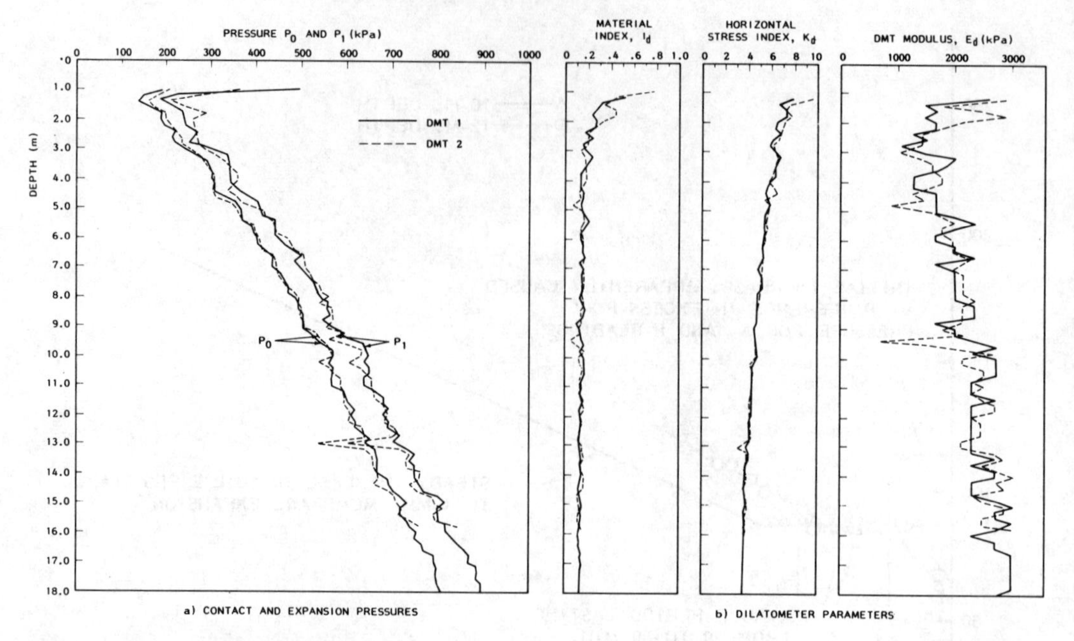

Fig.2. Test results, Glava clay

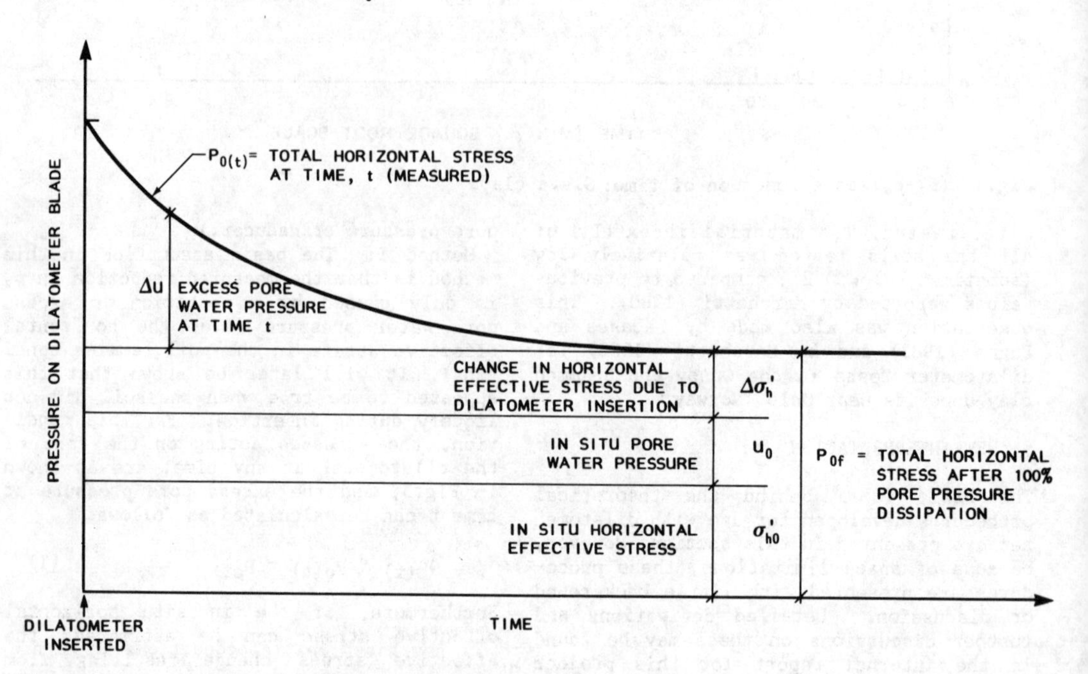

Fig.3. Stresses acting on dilatometer face

ing), which gives the pressure (C-pressure) measured after the standard dilatometer test is performed, and the membrane returns to its lift-off position of 0.01 mm. Several researchers have suggested that this pressure is very close to the ambient pore pressure for sands, and the ambient plus excess pore pressure for clay (see Schmertmann, 1986). Assuming that this is true, and that exact times to the U-readings can be recorded, a plot of C- pressures with time can be used to obtain pore pressure and effective stress. The U-reading had not been proposed when these measurements were obtained so this method was not evaluated.

5.2 Undrained Shear Strength

Overall stress equilibrium must be met regardless of the state of the soil immediately adjacent to the blade. Thus, if dilatometer insertion is considered as a footing loaded horizontally to failure, classical bearing capacity formulas may be used to calculate the undrained shear strength of the soil. Although failure may not be reached by insertion, the authors feel that a strength determined in this way should relate more directly to the actual undrained strength of the soil.

For a strip load, (p_1) on a weightless soil with a surcharge equal to the total in situ horizontal stress σ_{ho}, the following equation may be used to determine undrained shear strength s_u

$$s_u = \frac{p_1 - \sigma_{ho}}{N_c} \qquad (3)$$

where p_1 = initial (maximum) DMT expansion pressure and σ_{ho} = total horizontal stress in situ, estimated by the formula

$$\sigma_{ho} = K_o' \sigma_{vo}' + u_o \qquad (4)$$

when σ_{vo}' = in situ vertical, effective overburden $(\gamma z - u_o)$, u_o = in situ pore pressure and K_o' = coefficient of lateral earth pressure at rest in terms if effective stress (assuming a=0).

The basic assumption is made that the bearing capacity factor N_c varies from about 5 to 9 in this equation for almost all cases in geotechnical engineering. Thus, if it is possible to classify the soil as brittle, medium, or plastic, the following table is proposed to obtain N_c for use in equation (3):

Soil Type	N_c
Brittle clay and silt	5
Medium clay	7
Nonsensitive plastic clay	9

5.3 Coefficient of Lateral Earth Pressure at Rest K_o'

A purely theoretical determination of K_o' cannot be made from the dilatometer measurements. However, more direct relationships may be obtained by using the total horizontal effective stress after dilatometer insertion ($\sigma_{ho}'+\Delta\sigma_h'$ in Fig.3) to obtain the following K-parameter:

$$K_N = \frac{(\sigma_{ho}' + \Delta\sigma_h') + a}{\sigma_{vo}' + a}$$

where σ_{vo}' = in situ vertical, effective overburden $(\gamma_z - u_o)$, u_o = in situ pore pressure, and a = soil attraction. The inherent assumption is that the change in horizontal effective stress resulting from dilatometer insertion is related to the in situ horizontal effective stress. This may or may not be true. This approach is similar to Marchetti's original correlations using K_d, which was defined as follows:

$$K_d = \frac{p_o - u_o}{\sigma_{vo}'}$$

where p_o = lift of pressure from DMT.

5.4 Modulus From Plasticity Approach

Theoretical solutions were derived for determining in situ soil modulus from effective stress changes resulting from dilatometer insertion. The solutions were obtained for the settlement of a vertical rectangular plate loaded horizontally. The concept of tangent modulus and soil adopted stress distribution (Janbu, 1967) was used for the derivation of the settlement formula, containing the modulus number. The modulus is defined in the idealized stress formulation of one-directional moduli (Janbu, 1963):

$$M = m\, p_a \left(\frac{p'}{p_a}\right)^{1-a} \qquad (5)$$

where
M = soil modulus at effective stress p'
p_a = reference stress = 100 kPa
m = modulus number
a = stress exponent

The required integrations were solved for three cases (soil types). The solutions are presented below.

Case I: For overconsolidated clays and for undrained initial conditions in saturated clays, where a constant modulus is a fair approximation (i.e. M= m p_a= constant) one gets:

$$M = \tfrac{1}{2}\, m_r\, \sigma_{ho}'\, s^2 \qquad (6)$$

Case II: For normally consolidated, saturated clay and very fine silt, where modulus that increases linearly with effective stress is a good approximation (i.e. M = m σ') one finds:

$$m = m_r[(1+2s)\ln(1+2s)-(1+s)\ln(1+s)-s] \quad (7)$$

Case III: For sandy and silty sediments where a modulus that increases with the square root of stress is a good approximation (i.e. $M = m\sqrt{\sigma' p_a}$), one obtains:

$$m = \frac{2}{3} m_r \sqrt{\frac{\sigma_{ho}'}{p_a}} \, [2(1+s)^{3/2}-(3s+2)] \quad (8)$$

In Eqs.(6 to 8) the following abbreviations are used:

$s = \Delta\sigma_h'/\sigma_{ho}'$ = stress ratio, see Fig.3.

$$m_r = \frac{B\,\sigma_{ho}'}{\delta\,\tau_m(1+B/L)} = \text{DMT-parameter}$$

τ_m = the maximum shear stress in the soil

The dilatometer blade dimensions are: width (B = 94 mm), length (L ≅ 235 mm), and one-half thickness (δ = 6.85 mm). The maximum shear stress is obtained by using earth pressure theory, and a method for determining $\Delta\sigma_h'$ due to dilatometer insertion was described earlier.

5.5 Modulus From Theory of Elasticity

An elastic modulus can also be obtained from the effective stress changes caused by dilatometer insertion by using the closed form solution for the deflection of a smooth rigid rectangle on a semi-infinite mass (Poulus and Davis, 1974). For elastic conditions, the dilatometer pressures should be adjusted to account for the non-uniform stress distribution on the face of the blade. The measured effective stress change ($\Delta\sigma_h'$) should be increased by a factor of 1.4 if plane strain conditions are assumed and by a factor of 1.7 for L/B = 2.5, as herein.

A second elastic modulus may be obtained from the stress changes caused by membrane expansion. The closed form solution given in the User's Manual for obtaining the dilatometer modulus may be used for this purpose.

5.6 Coefficient of Consolidation

Cylindrical cavity expansion theory was used to calculate a coefficient of consolidation from plots of p_0 vs. time. The dilatometer blade was modelled as a vertical cylinder with an equivalent radius calculated on the basis of the cross-sectional area of the blade. Radial drainage on horizontal planes was assumed. The models used were based on the differential equations for the radial dissipation of excess pore pressure induced by an expanding cylindrical cavity in an ideal elasto-plastic material (Svanø, 1982). Two models were developed: one based on the degree of dissipation, and the second based on the dissipation rate.

Model I: Degree of Dissipation.

The coefficient of consolidation was determined using the following equation:

$$c_h = T\,\frac{r_0^2}{t} \quad (9)$$

where T = time factor, r_0 = radius of cylindrical cavity, and t = time for specified degree of dissipation (determined from p_0-t curves).

Curves for determining the time factor which is a function of degree of dissipation and soil stiffness were presented by Svanø (1982).

Model II: Dissipation Rate.

The coefficient of consolidation was determined using the following equation:

$$c_h = \lambda_c\, r_0^2\, \frac{\Delta\dot{u}_t}{\Delta u_c} \quad (10)$$

where λ_c = rate factor, r_0 = radius of cavity, $\Delta\dot{u}_t$ = measured pore pressure dissipation rate, Δu_c = measured excess pore pressure at time t.

The last three variables are determined from p_0-t curves. Svanø (1982) presented curves for obtaining the rate factor, which is a function of degree of dissipation and soil stiffness. Svanø's diagrams may also be found in Senneset et al. (1986).

6. EVALUATION OF DILATOMETER TEST RESULTS

The results of conventional interpretations of DMT parameters are shown in Fig.2 for the Glava clay. These parameters were used to obtain soil properties for comparison with the non-conventional interpretations described above.

6.1 Undrained Shear Strength

Fig.4 shows that undrained shear strengths determined from a bearing capacity analysis of dilatometer insertion, were in good correspondence with strength values from triaxial compression tests. Apparently, insertion of the dilatometer blade resulted in a bearing capacity failure in the soil adjacent to the blade. This implies that dilatometer tests were performed within a zone of failure. This agrees with Campanella et al. (1985) who found that some limit pressure may be reached during insertion in soft clay. Apparently, this limit pressure may be the pressure required to induce bearing capacity failure.

Undrained strengths determined from existing correlations were high compared to unconfined compressive strengths, to which these values should correspond (see Fig.4). Both the magnitude and trend with depth of these values, were in better agreement with triaxial test results.

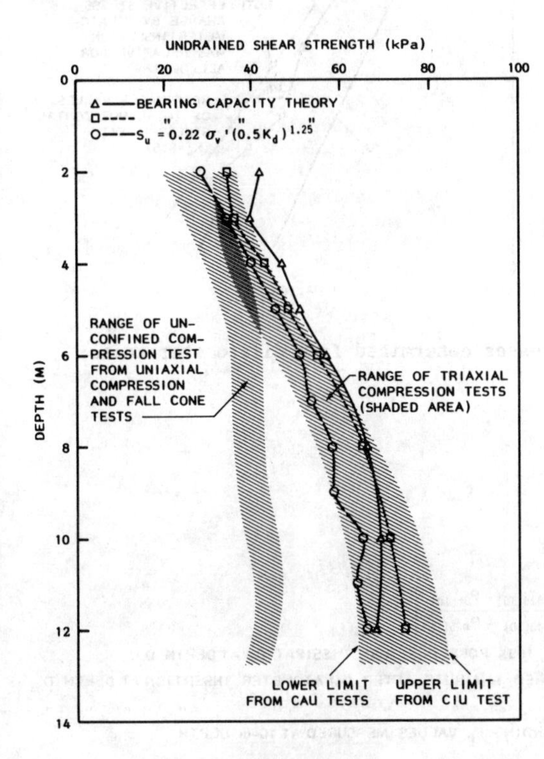

Fig.4. Undrained shear strength from dilatometer and laboratory tests: Glava

6.2 Coefficient of Lateral Earth Pressure at Rest

Fig.5 shows that existing correlations overpredicted K_0' for this clay. Values calculated on the basis of the horizontal effective stress after dilatometer insertion were also high. However, this was expected, since it seems reasonable to assume that the horizontal effective stress increases after dilatometer insertion. A comparison of the residual horizontal total stresses determined from dilatometer measurements (p_{of} in Fig.3) with an estimated reasonable range of in situ horizontal stresses for Glava, indicated that this assumption was indeed reasonable. This comparison is shown in Fig.6a, which illustrates that the residual effective (and total) stresses were significantly higher than the original in situ stresses. Therefore, it can be said that the in situ K_0' must be less than values calculated from the residual stresses determined from dilatometer measurements in this soil.

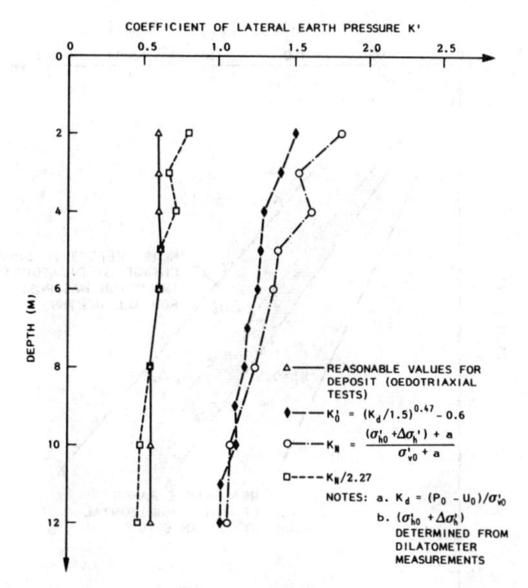

Fig.5. K-values from dilatometer and laboratory tests: Glava

However, this assumption may not be correct for all soils. Fig.6b shows that the residual stresses measured in the Halsen (sensitive clayey silt) deposit were lower than reasonable estimates of in situ stresses. Thus it appears that a break-down in the soil's structure resulted from dilatometer insertion, probably causing liquefaction. It should be noted that the total pressure p_0 was about equal to the maximum pore water pressure measured by piezocone in Halsen, while in Glava p_0 was significantly greater than the maximum pore pressure measured by piezocone.

Fig.5 also shows that when the K-parameters determined from residual stresses were divided by an arbitrary constant, the resulting values agreed well with laboratory values for this deposit. Thus, it appears that direct correlations between these values and in situ K_0' may be possible. The real in situ K_0-value may be closer to 1.0.

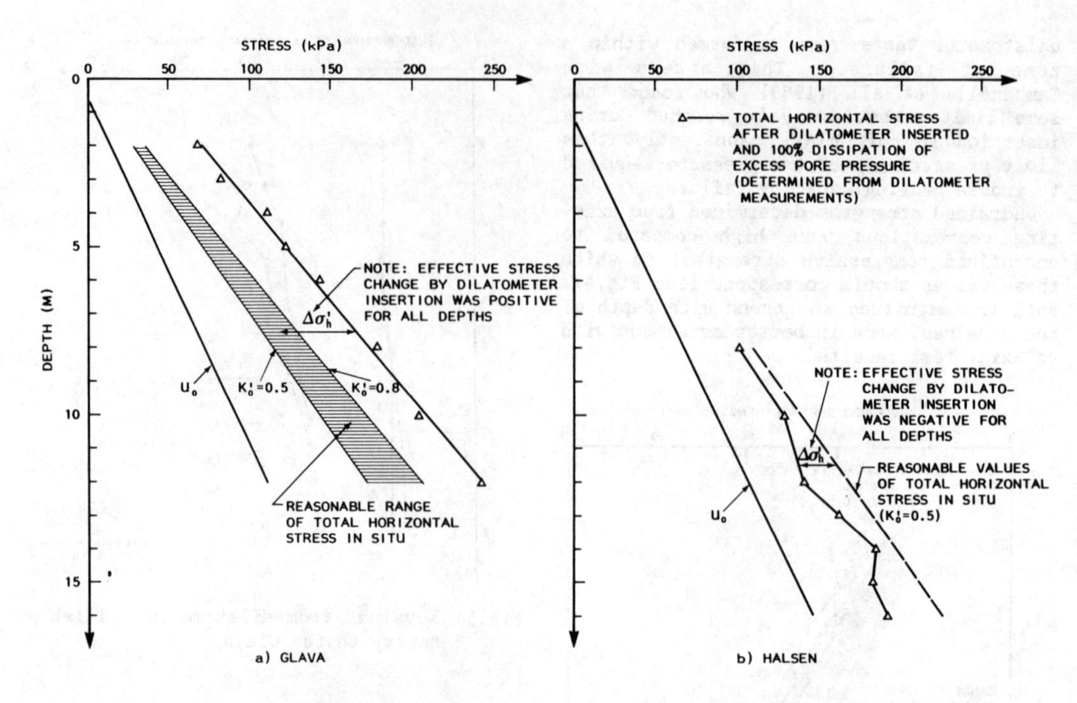

Fig.6. Estimated vs. residual horizontal stresses determined from dilatometer tests: Glava and Halsen

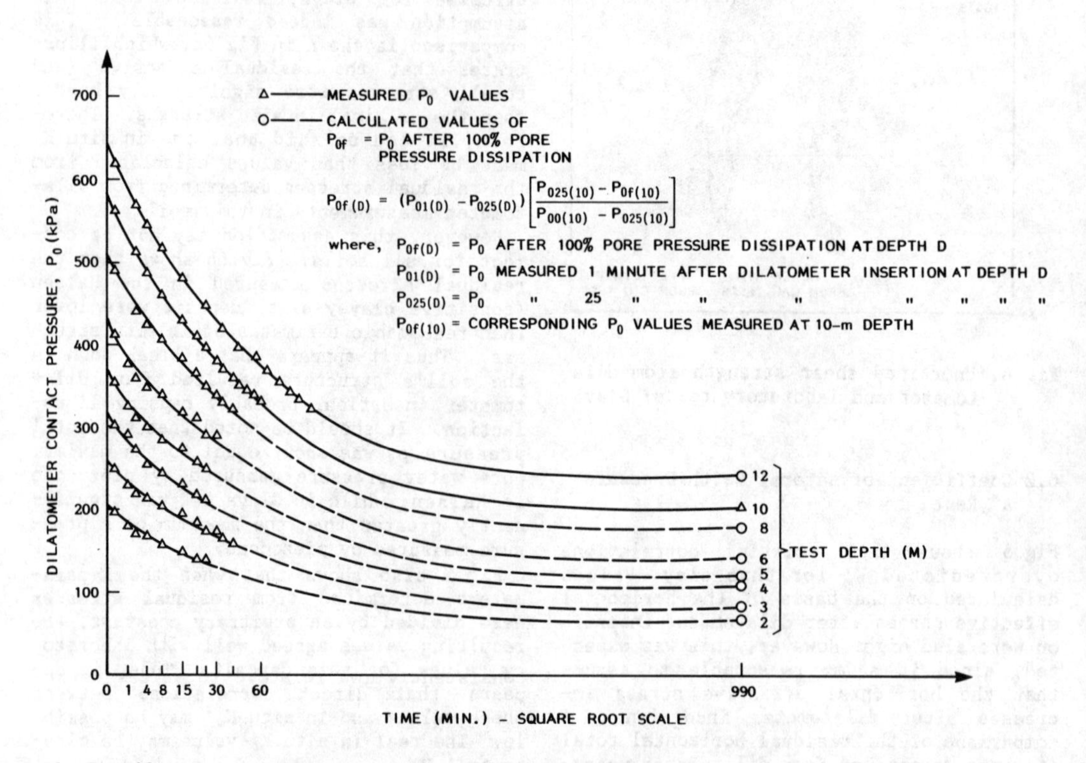

Fig.7. Dilatometer contact pressures with time: Glava

6.3 Coefficient of Consolidation

Coefficients of consolidation determined from p_0-t curves (Fig.7) and similar pore pressure dissipation curves from piezocone were almost identical. The comparisons are presented in Fig.8. The values were determined from expanding cavity theory (Svanø's models I and II) at 50% degree of dissipation. The consistency between these values was remarkable, considering that the dilatometer and the cone have different shapes, volumes, and disturbance characteristics. These results appear to indicate two things: 1) that the theory used accounted properly for the difference in volume between the dilatometer blade and the cone; and 2) that the reduction in p_0 is a direct measure of the pore pressure change with time.

This second finding is very important, because it implies that the horizontal effective stress in the soil did not change during pore pressure dissipation (i.e. the horizontal effective stress in the soil was the same after all pore pressures had dissipated as they were immediately after the dilatometer blade was inserted). Thus, the effective stress change caused by dilatometer insertion may be calculated by using equation (2) and this value may then be used to obtain an effective stress modulus for dilatometer insertion.

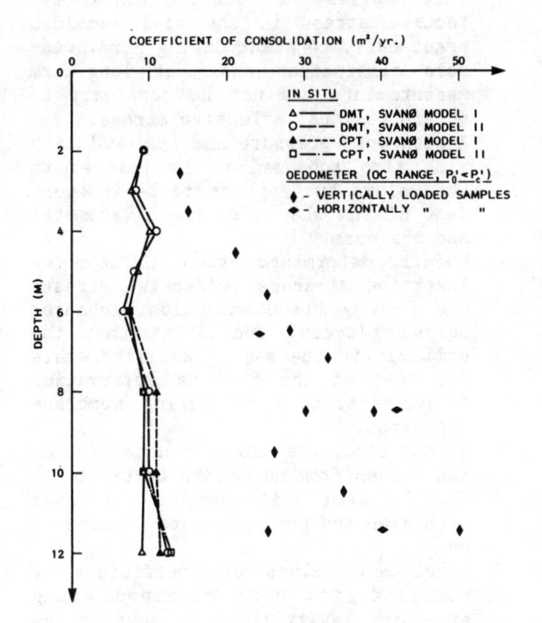

Fig.8. Coefficient of consolidation from dilatometer and laboratory tests: Glava

These results seem to agree with the inherent findings of Baligh (see Jamiolkowski et al., 1985) that total pressures and their evolution with time are mainly dependent on the volume and not the shape of the protruding instrument. It appears that this statement applies to pore pressures as well.

Fig.8 also shows that the actual magnitude of the values determined grossly underestimated values measured within the overconsolidated range in the oedometer. There are several possible reasons for this. First, expanding cavity theory is not strictly correct for the dilatometer or the cone, since constant horizontal stresses are assumed in the theory. As shown in Fig.7, this was clearly not the case for the dilatometer. Second, if a bearing capacity failure resulted from dilatometer insertion, then the preconsolidation pressures were probably exceeded, so that these two values should not correspond. In addition, a large part of the deformations during insertion may have occured at constant stress within the plastified failure zone, which would result in an underestimate of the coefficient of consolidation.

However, the findings show that dilatometer measurements with time may be used to obtain theoretically based coefficients of consolidation which appear to be at least as good as those determined from piezocone dissipation tests. Furthermore, the resulting values were consistent and may possibly be related semi-empirically to more reasonable values for this soil.

6.4 Modulus

Fig.9 shows that moduli determined theoretically from dilatometer insertion and 1-mm expansion pressures grossly underpredicted values determined within the overconsolidated range in the oedometer.

If as indicated earlier, a bearing capacity failure resulted from dilatometer insertion, then these results are not surprising, since very high deformations may develop at constant stress within a plastified zone. This is totally consistent with the low values of coefficent of consolidation presented earlier. It is significant that these two parameters were determined from two totally different and independent sets of measurements (i.e. one from effective stress changes and deformations and the other from pressure dissipation curves) and that the values obtained were consistently low relative to values determined in the laboratory (compare Figs.8 and 9).

The close agreement between the insertion and expansion moduli shown in Fig.8, is

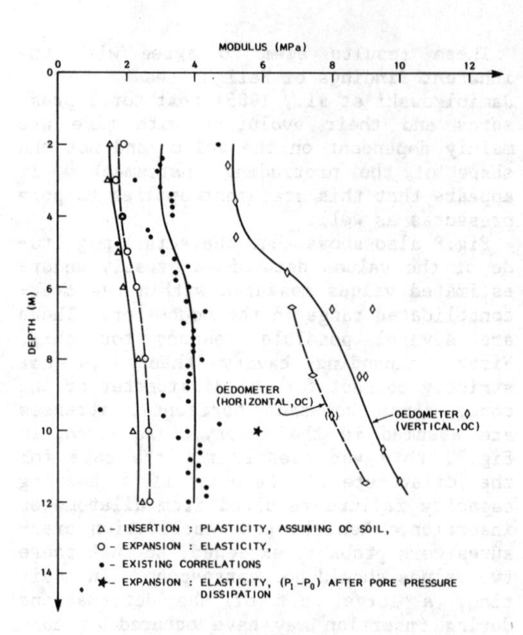

Fig.9. Modulus from dilatometer and laboratory tests: Glava

further evidence that failure was reached during insertion and implies that the theoretical moduli obtained were reasonable. Apparently, the soil was in the same plastified state for most of the 6.85 mm deformation during insertion as during the 1-mm membrane expansion.

A plot of (p_1-p_0) indicated that the soil's resistance to the 1-mm membrane expansion increased with time. After 16.5 hrs. the modulus as determined from membrane expansion had increased by almost 300% at the 10-m depth. This value is plotted in Fig.9, and although it is not as high as the oedometer values, it is much closer to the results expected for this soil. It is not known whether the soil's resistance would continue to increase with time and approach values in the range of those determined in the oedometer. However, lower dilatometer values should probably be expected because of the higher confinement stresses in the oedometer. Additional differences may be expected due to soil anisotropy.

Fig.9 also shows that existing correlations underpredicted the oedometer values. This was also found to be the case for dilatometer tests performed by NGI in the Drammen clay deposit near Oslo, Norway (Lacasse and Lunne, 1983). However, both the theoretical values and values from existing correlations seemed to have the same trend with depth as the oedometer values. Therefore, it appears that correlations between in situ and laboratory values are possible for this soil.

7. SUMMARY

It appears that existing correlations for determining soil parameters were not applicable for the Norwegian soils tested. This was not surprising, since these soils were probably outside the limitations specified for these correlations. The new test and theoretical interpretation procedures presented in this paper appeared to give reasonable and consistent results. The procedures led to the following findings in a medium-stiff clay of medium sensitivity:

1. Dilatometer insertion probably resulted in a bearing capacity failure in the soil immediately adjacent to the blade.
2. Strength values determined from a bearing capacity analysis of dilatometer insertion agreed closely with undrained strengths determined in triaxial tests.
3. A new horizontal stress indicator, calculated from the horizontal effective stress after dilatometer insertion, appears to provide an upper limit for in situ K_o'.
4. It appears that measurements of p_0 with time were a direct reflection of the excess pore pressure dissipation. This implies that the horizontal effective stress in the soil remained practically constant during pore pressure dissipation and that long-term measurements may not be necessary to obtain residual effective stress.
5. Excess pore pressure and its evolution with time appeared to be related to the volume but appears to be independent of the shapes of the dilatometer and the cone.
6. Moduli determined from dilatometer insertion stresses (effective stress) and 1-mm membrane expansion pressures agreed closely, indicating that the soil was in the same plastified state for most of the 6.85 mm deformation during insertion as during membrane expansion.
7. As expected, these moduli underpredicted values from oedometer tests.
8. The in situ soil modulus increased with time and pore pressure dissipation.
9. Consistent values of coefficient of consolidation were obtained using expanding cavity theory. However, as was the case for the moduli, the values underpredicted values determined in the oedometer.

In the research to follow we will investigate the possibility of obtaining effective shear strength parameters and improved values of in situ moduli by expanded theory and interpretations.

REFERENCES

Bullock,P.J. (1983). The dilatometer: current test procedure and data interpretation. Master's Thesis , University of Florida, Gainsville. Fla. USA.

Campanella, R.G., Robertson,P.K., Gillespie,D.G., and Grieg,J. (1985). Recent developments in in situ testing of soils. XI ICSMFE, San Francisco, USA.

Jamiolokowski,M, Ladd,C.C., Germaine,J.T., and Lancelotta,R. (1985). New developments in field and laboratory testing of soils. XI ICSMFE, San Francisco, USA.

Janbu,N.(1963). Soil compressibility as determined by oedometer and triaxial tests. Proceedings, Third European Conference, SMFE Wiesbaden, West Germany.

Janbu,N.(1967). Settlement calculations based on tangent modulus concept. Bulletin No.2, Geotechnical Division, Norwegian Institute of Technology, Trondheim, Norway.

Lacasse,S., and Lunne,T.(1983). Dilatometer tests in two soft marine clays. NGI Publication No.146, Oslo, Norway.

Marchetti,S.(1980). In situ tests by flat dilatometer. Journal of Geotechnical Engineering Division, ASCE.

Marchetti,S and Crapps,D.K. (1981). Flat dilatometer manual. Internal Report, Schmertmann and Crapps, Inc., Gainesville, Fla. USA.

Mokkelbost, K.H. (1984). Measurement of geotechnical parameters by use of dilatometer. Diploma thesis. The Norwegian Institute of Technology, Geotechnical Division, Trondheim, Norway. (In Norwegian).

Poulus, H.G., and Davis,E.H. (1974). Elastic solutions for soil and rock mechanics, John Wiley and Sons, New York.

Roque, R. (1987). Evaluation of flat dilatometer for use in Norwegian soils. NTH Report, No.F.87.02, NTH, Trondheim, Norway.

Schmertmann, J.H. (1986). Some 1985-86 Developments in dilatometer testing and analysis. Proceedings, Innovations In Geotechnical Engineering, Harrisburgh, Pa. USA.

Senneset et al. (1986). Interpretation of piezocone tests in silt. Recommended methods of interpretation. Contract Report to Statoil and Norsk Hydro (Report No.84244-3, NGI and SINTEF), Trondheim Norway.

Stordahl, P. (1985). Evaluation of geotechnical parameters by use of dilatometer. Diploma thesis. The Norwegian Institute of Technology, Geotechnical Division, Trondheim, Norway. (In Norwegian).

Svanø,G. (1982). Interpretation of c_h from piezocone dissipation tests. Internal Report No.O.81.19, Geotechnical Division, Norwegian Institute of Technology, Trondheim, Norway. (In Norwegian).

Penetration Testing 1988, ISOPT-1, De Ruiter (ed.)
© 1988 Balkema, Rotterdam, ISBN 90 6191 801 4

Site assessment and settlement evaluation of firm alluvial silts and clays with the Marchetti flat dilatometer

Steven R.Saye
Woodward-Clyde Consultants, Omaha, Nebr., USA
Alan J.Lutenegger
Clarkson University, Potsdam, N.Y., USA

ABSTRACT: Marchetti Flat Dilatometer (DMT) tests conducted in conjunction with settlement observations of fills for a site development in Omaha, Nebraska are compared with laboratory oedometer data, field vane tests and boring records to assess site conditions and compressibility of the foundation soils. Supplemental DMT tests were conducted through the completed fills following primary settlement to compare the DMT stress history estimates with known stress conditions.

The authors conclude that the DMT provides a sensitive evaluation of the site conditions and material changes in stratigraphy with depth. The DMT measurements of the maximum past vertical preconsolidation pressure show reasonable but slightly high values when compared to laboratory data and known variable, estimates of soil compressibility suitable for preliminary evaluation of settlements. The adjustment of the DMT modulus values for low OCR soils recommended by Schmertmann (4) was necessary to match the field observations at this site.

1 INTRODUCTION

Instrumentation to monitor settlement of general fill and building surcharges for a site development in Omaha, Nebraska underlain by about 12 meters of firm and stiff alluvium provides the basis for a comparison of site assessment and settlement evaluations from the Marchetti Flat Dilatometer (DMT) with laboratory data, boring records and actual settlements. Site development included placement of about 2 meters of general site fill and up to 3 additional meters of surcharge in building locations. DMT data and settlement observations are presented in detail for two locations on the site with summarized data from 3 additional locations.

2 SITE AND SUBSURFACE CONDITIONS

The site is located within a one-half mile wide alluvial valley in the loess hills of eatern Nebraska. Original site grades are relatively uniform with a hydrostatic water table between 0.6 to 3 meters in depth. The foundation soils consist of thick topsoils formed in low plastic silty clay alluvium about 6 meters thick. The surface of the alluvium exhibits the effects of desiccation with a stiff consistency at the surface becoming firm with depth. Near depths of 7 to 8 meters a stiff medium plastic silty clay is generally encountered representing the desiccated surface of an older stream level. The stiff clay is variably eroded and about 1 meter thick at locations 101 and 108. Low plastic clayey silt forms the lower 3 to 6 meters of alluvium at this site. Bedrock was encountered near a depth of 12 meters. Figures 1 and 2 summarize the general soil properties at two representative locations.

3 SITE STRATIGRAPHY

DMT evaluations of the site stratigraphy were compared with geologist records, field vane tests and laboratory data to identify site changes due to changes in materials and stress history. Figures 1 and 2 illustrate the general variations in site conditions observed with depth and representative DMT profiles for locations 101 and 108. The field vane data, laboratory consolidation tests and DMT profiles of preconsolidation stress and modulus identify a surface desiccated crust extending to a depth of 4 to 5 meters underlain by lightly overconsolidated soils. The DMT profiles appear

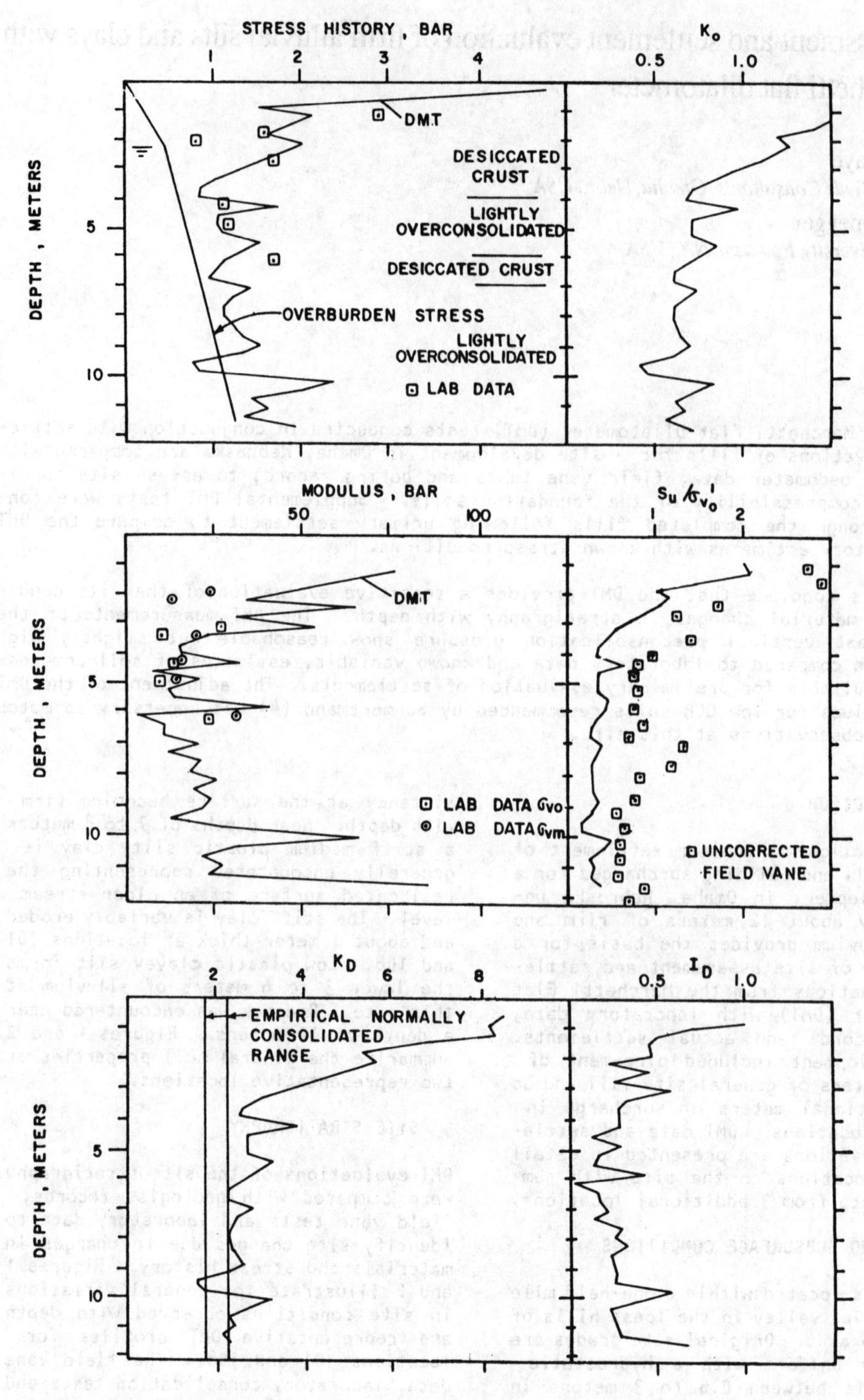

Fig. 1 DILATOMETER DATA BEFORE CONSTRUCTION LOCATION 101

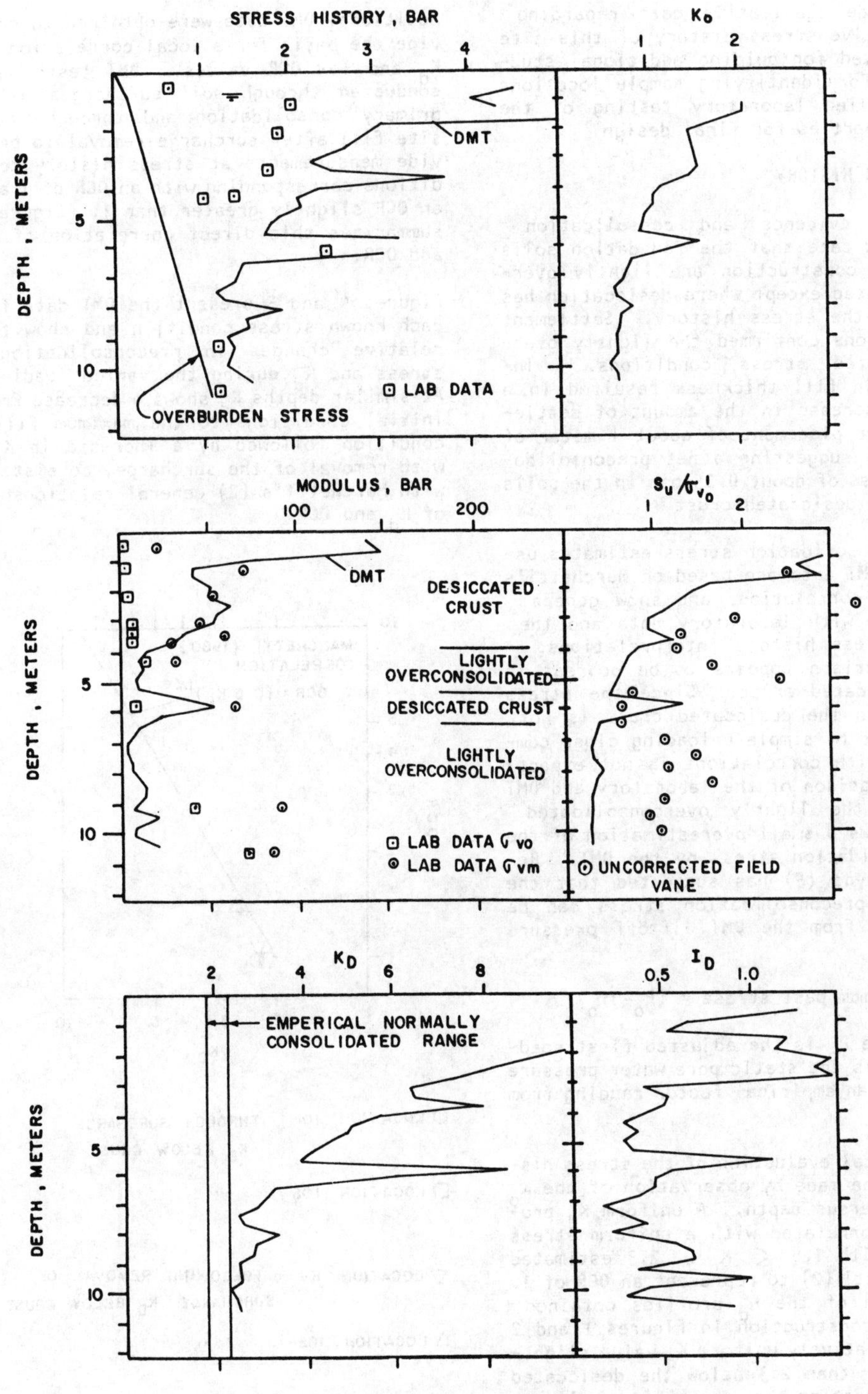

Fig. 2 DILATOMETER DATA BEFORE CONSTRUCTION LOCATION 108

to provide qualitative data regarding the relative stress history of this site well suited for guiding additional studies and for identifying sample locations for detailed laboratory testing of the soil properties for final design.

4 STRESS HISTORY

Geologic evidence and consolidation tests indicate that the foundation soils prior to construction are lightly overconsolidated except where desiccation has improved the stress history. Settlement observations confirmed the lightly overconsolidated stress conditions. Increases in fill thickness resulted in a marked increase in the amount of settlement after placement of about 1 meter of site fill suggesting a net preconsolidation stress of about 0.25 bar in the soils below the desiccated crust.

The preconsolidation stress estimates using the DMT data are based on Marchetti's (2) 1980 correlations and show general agreement with laboratory data and the above stress history interpretations. The comparison appears to be poorest in the desiccated crust. Since the stress history in the desiccated crust is not determined by simple unloading close comparison with correlations is not expected. Comparison of the laboratory and DMT data in the lightly overconsolidated soils shows a small overestimation of the preconsolidation stress by the DMT. Recently Mayne (5) has suggested that the vertical preconsolidation stress can be predicted from the DMT liftoff pressure as:

$$\text{Maximum past stress} = (P_o - U_o)/\Delta$$

Where P_o is the adjusted first reading U_o is the static pore water pressure and Δ is an empirical factor ranging from 1 to 3.

An emperical evaluation of the stress history can be made by observation of the K_o profile versus depth. A uniform K_d profile is correlated with a uniform stress history with $1.8 < K_d < 2.3$ estimated by Marchetti (2) to represent an OCR of 1. Comparison of the K_d profiles obtained prior to construction in Figures 1 and 2 show a relatively uniform K_d value slightly greater than 2.3 below the desiccated crust suggesting that the sites are lightly overconsolidated.

Additional DMT data were obtained to provide the basis for a local correlation of K_d and low OCR values. DMT tests were conducted through soil surcharges after primary consolidation and through the site fill after surcharge removal to provide measurements at stress history conditions corresponding with an OCR of 1 and an OCR slightly greater than 1. Figure 3 summarizes this direct correlation of K_d and OCR.

Figures 4 and 5 present the DMT data for each known stress condition and show the relative changes in preconsolidation stress and K_d during the varied loading. At similar depths K_d shows a decrease from initial conditions to the maximum fill condition followed by a increase in K_d with removal of the surcharge, consistent with Marchetti's (2) general relationship of K_d and OCR.

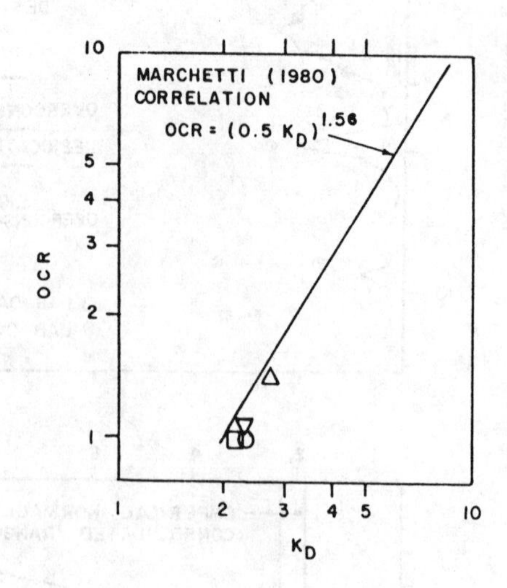

○ LOCATION 101 THROUGH SURCHARGE K_D BELOW CRUST

□ LOCATION 108 " "

▽ LOCATION 101 FOLLOWING REMOVAL OF SURCHARGE K_D BELOW CRUST

△ LOCATION 108 " " "

Fig. 3 SUMMARY OF KD VS. OCR OBSERVATIONS AT KNOWN STRESS CONDITIONS

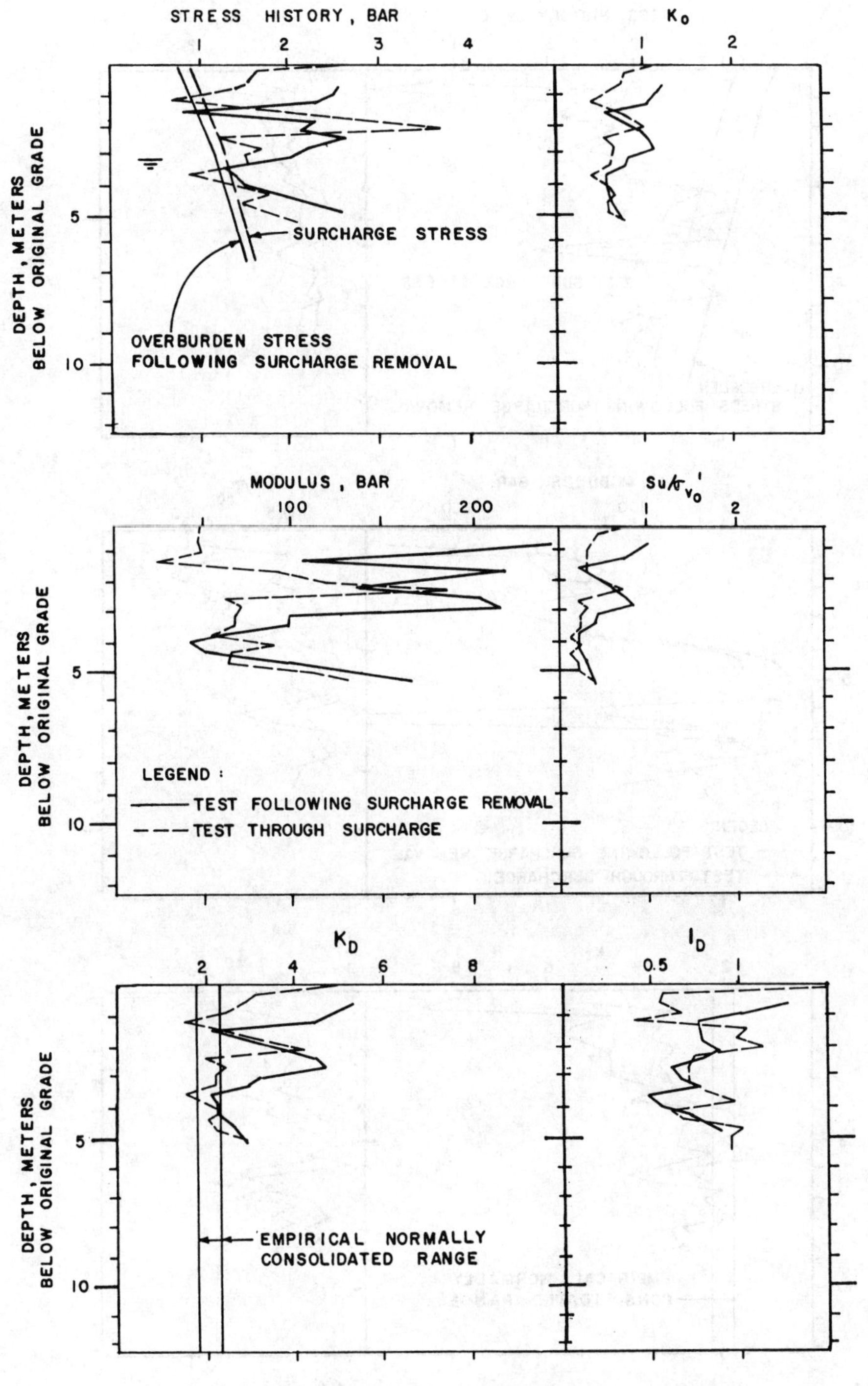

Fig. 4 DILATOMETER DATA FOLLOWING SURCHARGE REMOVAL LOCATION 101

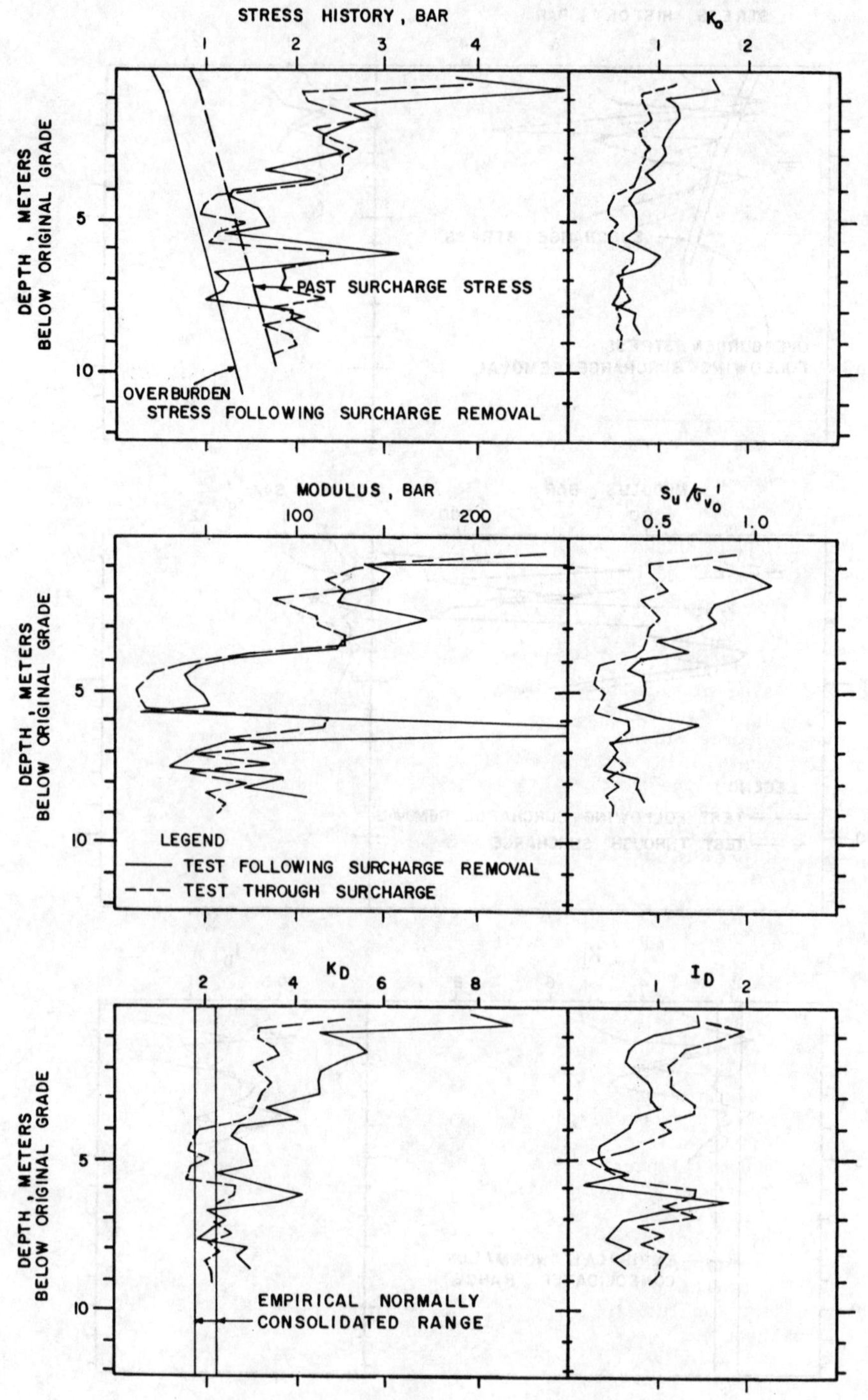

Fig. 5 DILATOMETER DATA FOLLOWING SURCHARGE REMOVAL LOCATION 108

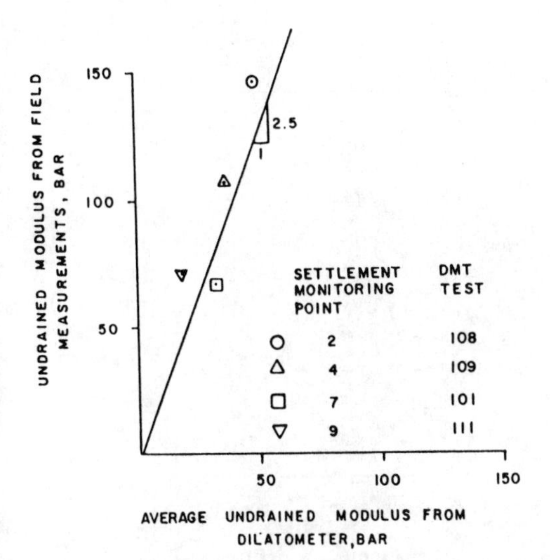

Fig. 6 COMPARISON OF UNDRAINED MODULUS ESTIMATES

Figure 3 suggests that Marchetti's (2) 1980 correlation slightly overestimates the actual OCR value at this site. Some variation of the K_d measurement is apparent suggesting a degree of uncertainty in both the measurement of K_d and the correlation with OCR for these soils. The cause of the variations is under study. The measurements of stress history appear reasonable for a preliminary evaluation of site behavior.

5 COMPRESSIBILITY

The soils at this site were relatively compressible after the preconsolidation stress was exceeded. Fill loads below the interpreted preconsolidation stress generally resulted in settlements less than 30 millimeters. Maximum applied loads were about 0.9 bar and exceeded the preconsolidation stress resulting in settlements of 300 to 400 millimeters.

The DMT provides an estimate of the soil compressibility by direct measurement of the undrained modulus and an empirical correlation (R_m) to the one dimensional constrained tangent modulus (M). The undrained modulus has been back calculated from the initial loading portion of fill thickness - settlement curves using the procedures for immediate settlement in NAVFAC DM 7.1 and Poissions ratio = 0.5. Figure 6 summarizes the undrained modulus comparisons. Significantly higher undrained modulus values are seen for the field loading than the DMT data. The DMT

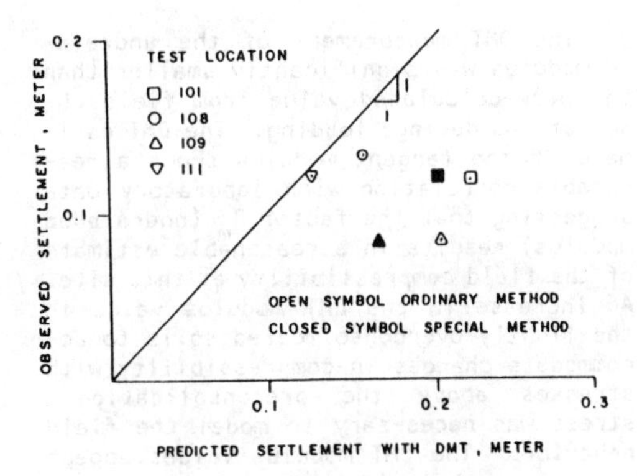

Fig. 7 OBSERVED SETTLEMENT VS. DTM PREDICTION

measurement of undrained modulus may be affected by remolding of the clays during penetration of the blade or high strain levels.

The soil compressibility can be evaluated on the basis of Janbu's tangent modulus concept (1). In clays the modulus equals the stress times a modulus number, m. Figures 1 and 2 summarize tangent modulus measurements with the DMT and laboratory data based on a reference stress of the maximum past stress and the overburden stress. A reasonable correlation is seen in the lightly overconsolidated clays.

The settlement of the fills were estimated using the procedure suggested by Schmertmann (4) for DMT data. Figure 7 shows relatively good correlation of observed and estimated settlements using the DMT data following an adjustment of the DMT modulus in the lightly overconsolidated soils to accommodate changed stress levels associated with virgin compression.

6 CONCLUSIONS

1. The DMT provides a reasonable evaluation of soil variations with depth at this site identifying overconsolidated and lightly overconsolidated zones within the foundation clays.
2. DMT testing at known stress conditions and comparison with laboratory data shows a reasonable estimate of the stress history of the foundation clays suitable for a preliminary investigation. Some variability of the DMT estimate of stress history is apparent and local correlations of K_d and OCR appear to be appropriate. The DMT appears to make a variable estimate the preconsolidation stress in the desiccated crust.

3. The DMT measurement of the undrained modulus was significantly smaller than the back-calculated value from field observations duringl loading. The DMT estimate of the tangent modulus shows a reasonable correlation with laboratory data suggesting that the factor R_m (undrainded modulus) results in a reasonable estimate of the field compressibility at this site. An increase in the DMT modulus value in the lightly overconsolidated soils to accommodate changes in compressibility with stresses above the preconsolidation stress was neces-sary to model the field behavior. The DMT modulus values appear suitable for preliminary estimates of settlements.

REFERENCES

1. Janbu, N 'Settlement Calculations Based on the Tangent Modulus Concept' Three Guest Lectures at Moscow State University, Bulletin No. 2, Soil Mechanics, Norwegian Institute of Technology, 1967, pp. 1-57

2. Marchetti, S., 'In Situ Tests by Flat Dilatometer', Journal of the Geotechnical Division, ASCE, Vol. 106, No. GT6, June 1980, pp. 299-321.

3. NAVFAC, 'Soil Mechanics Design Manual 7.1', Department of the Navy, May, 1982

4. Schmertmann, J.H., 'Dilatometer to Compute Foundation Settlement,' Proceedings, Use of In Situ Tests in Geotechnical Engineering, GSP No. 6, ASCE, 1986, pp. 303-321

5. Mayne, P.W., 'Determining Preconsolidation Stress and Penetration Pore Pressures from DMT Contact Pressures' Geotechnical Testing Journal, ASTM, Vol. 10, No. 3, 1987, pp. 146-150.